BEAM INSTRUMENTATION WORKSHOP

BEAM INSTRUMENTATION WORKSHOP

Stanford, CA May 1998

EDITORS
Robert O. Hettel
Stanford Synchrotron Radiation Laboratory

Stephen R. Smith
Jennifer D. Masek
Stanford Linear Accelerator Center

AIP CONFERENCE PROCEEDINGS 451

American Institute of Physics
Woodbury, New York

Editors:

Robert O. Hettel
Stanford Synchrotron Radiation Laboratory
Stanford Linear Accelerator Center
MS 69 P.O. Box 4349
Stanford, CA 94309

E-mail: hettel@ssrl.slac.stanford.edu

Stephen R. Smith
Stanford Linear Accelerator Center
MS 50 P.O. Box 4349
Stanford, CA 94309

E-mail: ssmith@slac.stanford.edu

Jennifer D. Masek
Technical Publications
Stanford Linear Accelerator Center
MS 68 P.O. Box 4349
Stanford, CA 94309

E-mail: jmasek@slac.stanford.edu

L.C. Catalog Card No. 98-88669
ISBN 1-56396-794-4
ISSN 0094-243X
DOE CONF- 980573

Printed in the United States of America

CONTENTS

CONTRIBUTED PAPERS—POSTERS

DISCUSSION GROUP AND CLOSEOUT SUMMARIES

APPENDICES

PREFACE

The eighth Beam Instrumentation Workshop (BIW 98), hosted by the Stanford Synchrotron Radiation Laboratory (SSRL) and the Stanford Linear Accelerator Center (SLAC), took place at SLAC from May 4–7, 1998. The Workshop had 149 registered participants, of which about 25% were from Europe, Great Britain and Asia. Nine vendor companies were represented. The Workshop welcome was given by Ewan Paterson, head of the SLAC Technical Division, on behalf of Burton Richter, the Director of SLAC. The Workshop included 4 tutorials, 5 invited talks (including the Faraday Cup Award speaker), 14 contributed talks, and 44 poster displays. Seven discussion sessions were held on topics selected by the registrants from a larger list. The tutorials and talks were given in the SLAC Auditorium, discussion groups were held in various smaller meeting rooms, and the posters were displayed in a tent on the lawn outside the auditorium. A tour of SLAC facilities, including the Linac Klystron Gallery, the PEP-II tunnel, the Babar detector, the SLC Detector Hall, the NLC Test Accelerator, and SSRL, was given at the end of the Workshop.

A reception was held on Sunday evening before the Workshop in a moorish-style courtyard, complete with palm trees, at the Schwab Residential Center on the Stanford Campus. Participants were welcomed to sunny California as clouds gathered and the first drops of what was to be a week of on and off rain fell from the sky. Luck prevailed for most of the Workshop, however, since the site power outage during one of the invited talks lasted less than a half hour, the flooding bathroom was repaired in less than a day, the main poster sessions in the outdoor tents were completed before the wind and rain destroyed most of the posters, and the first Bay Area tornadoes in almost 50 years touched down more than ten miles away. Some of the high-quality poster paper, washed clean by the rain, was salvageable and could be used for future posters. Spirits remained high, especially after participants were treated to a little peek at the sun as it set beyond the Golden Gate during the banquet at the Golden Gate Yacht Club overlooking the San Francisco Bay. In fact, spirits were so aroused that banquet tables were seen to levitate towards the ceiling as the evening progressed. The end of the banquet was just the beginning of a night of music club- and pub-hopping through San Francisco for some participants; a couple of these hearty revelers were actually seen at the first talk the next morning.

The eighth Workshop was organized by a committee comprised of fourteen members from US accelerator laboratories and two members from CERN. This year, one of the two original organizers of the first Workshop, Dick Witkover, announced his retirement from BNL and from the Organizing Committee. Another original Workshop contributor and long-time Organizing Committee member, Bob Shafer from LANL, also announced his retirement. These two gentlemen have been instrumental in developing the instructional format and engineering-oriented philosophy of the Workshops over the years. Two more committee members, Jim Hinkson from LBL and Alex Lumpkin from the APS, announced their resignations from the committee, but they still have many more years of laboratory duties and BIW participation ahead of them. On behalf of the Organizing Committee, we would like to thank these former members for their years of service on the committee and their valued contributions to the Workshop.

In addition to the members of the Workshop Organizing Committee, many people at SSRL and SLAC contributed to the success of BIW 98. Suzanne Barrett, the Workshop Secretary, together with Todd Slater, Michelle Steger, and Diana Viera worked tirelessly to handle the mailings, hand outs and registration, and to arrange food, accommodations, reception, banquet, poster boards and tents, and numerous other Workshop details. Vern Smith located and procured the Workshop document bags. Nina Adelman-Stolar was instrumental in arranging for auditorium audio/visual support, meeting rooms, and for tour plans. Heinz-Dieter Nuhn created and managed the BIW 98 Web site, and Clemens Wermelskirchen implemented the Web registration database. Raymond Muller from the Hamamatsu Corporation contributed the wine for the banquet. Kathryn Henniss, head of the SLAC Publication Department, together with Jennifer Masek, Vibha Akkaraju, Ruth McDunn, and Bridgitt Ahern were responsible for producing the camera-ready Workshop Proceedings submitted for publication. Terry Anderson produced the graphical designs and the Workshop poster. El Niño provided the weather. We are deeply indebted to all these contributors for their efforts and for making BIW 98 an enjoyable and rewarding experience.

We wish to acknowledge the staffing and funding support contributed by the SSRL and SLAC Associate Directors, including Keith Hodgson from SSRL, his predecessor, Art Bienentstock, and Ewan Paterson from the SLAC Technical Division. We are grateful to Dave Sutter from the High-Energy Physics Division of the Department of Energy for providing funding assistance, and to Burton Richter, Director of SLAC, for his overall support of the Workshop at SLAC.

Finally, we wish to thank the participants who contributed such a wealth of fine technical information to the Workshop and its proceedings and who made the Workshop an interesting, engaging and educational experience. We commend them for their accomplishments and for continuing to advance the field of beam instrumentation to higher levels of performance and sophistication.

Robert Hettel
Stephen Smith
BIW 98 co-Chairmen

BIW 98 Organizing Committee

Robert Hettel, SSRL/SLAC, co-chair
Steve Smith, SLAC, co-chair
Robert Averill, MIT Bates
Claude Bovet, CERN
Jean-Claude Denard, TJNAF
Jim Hinkson, LBNL
Heribert Koziol, CERN
Alex Lumpkin, APS
Ralph Pasquinelli, FNAL
Mike Plum, LANL
Robert Shafer, LANL
Gary Smith, BNL
Gregory Stover, LBNL
Robert Webber, FNAL
Richard Witkover, BNL
Jim Zagel, FNAL

Local Arrangements Committee

Nina Adelman-Stolar, SLAC
Terry Anderson, SLAC
Suzanne Barrett, SSRL
Kathryn Henniss, SLAC
Robert Hettel, SSRL
Heinz-Dieter Nuhn, SSRL
Todd Slater, SSRL
Steve Smith, SLAC
Michelle Steger, SSRL
Diana Viera, SSRL
Clemens Wermelskirchen, SSRL

8th BEAM INSTRUMENTATION WORKSHOP

Hosted by the Stanford Linear Accelerator Center (SLAC)

	May 3 Sunday	May 4 Monday	May 5 Tuesday	May 6 Wednesday	May 7 Thursday
8:00		Registration at SLAC			
8:30		Welcome/ Opening	Tutorial #2:	Tutorial #3:	Tutorial #4:
9:00		Invited Talk #1: PEP-II Instrumentation A. Fisher/SLAC	Polarimeters and their Applications C. Sinclair/TJNAF Y. Makdisi/BNL	Cavity BPMs R. Lorenz/ Tech. U. of Berlin	Camera Technology & Image Processing R. Jung/CERN
9:30					
10:00		Break	Break	Break	Break
10:30		Tutorial #1: Beam Diagnostics and Applications A. Hofmann	Invited Talk #2: High Power Proton Beam Diagnostics D. Gilpatrick/LANL	Invited Talk #4: RHIC Instrumentation T. Shea/BNL R. Witkover/BNL	Contrib. Talk: H. Schmickler
11:00					Faraday Cup Speaker (A. Peters)
11:30			Invited Talk #3: Real-Time Orbit Feedback at APS J. Carwardine/APS	Contributed Talks: G. Decker C. Adolphsen G. Vismara	
12:00		Contributed Talks: A. Ghigo R. Shafer			Closeout
12:30		Lunch (12:40 pm)	Lunch	Lunch	Lunch
1:00					
1:30					SLAC Tour (2 hours)
2:00		Poster and Vendor Display	Contributed Talks A. Lumpkin W. Graves A. Lumpkin D. Teytelman B. Yang	Contributed Talks: T. Powers P. Tenenbaum P. Rojsel	
2:30					
3:00				Break (3:00 pm) Discussion Group 7 (3:20 pm)	
3:30		Break (3:40 pm)	Break (3:40 pm)		
4:00		Discussion Group 1	Discussion Group 4		
4:30					
5:00		Discuss. Group 2 Discuss. Group 3	Discuss. Group 5 Discuss. Group 6	Buses to Banquet (5:00 pm)	
5:30					
6:00	Reception/ Registration: Schwab Residential Center at Stanford Uni.				
6:30				Banquet: Golden Gate Yacht Club in San Francisco	

Andreas Peters (third from left) received the Faraday Cup Award from Julien Bergoz (left), the sponsor of the award, Steve Smith (second from left), BIW co-chairman, and Robert Hettel, BIW co-chairman.

Faraday Cup Award

The Faraday Cup Award, donated by Bergoz Inc., Crozet, France, is intended to recognize and encourage innovative achievements in the field of accelerator beam instrumentation. It is presented to those who have made an outstanding contribution to the development of an innovative beam diagnostic instrument of proven workability. The prize is only awarded for demonstrated device performance and published contribution. The award consists of a US $5000 prize and a certificate that is presented at the US Beam Instrumentation Workshop (BIW). Winners participating in the BIW share a $1000 travel allowance. The selection of recipients is the responsibility of the BIW Program Committee.

The 1998 Faraday Cup Award was presented to Andreas Peters, from GSI, Darmstadt, for his development of the Cryogenic Current Comparator, which is capable of measuring nA DC beam currents.

TUTORIALS

Beam Diagnostics and Applications

A. Hofmann

Chemin de l'Erse 20, CH-1218 Grand Saconnex, Switzerland

Abstract. Particle beams in accelerators are detected through the electromagnetic fields they create. Position and intensity monitors are based on the near field which stays attached to the charges. A large variety of measurements can be carried out with these devices. The closed orbit is obtained by reading out the position averaged over many turns. Change in the orbit resulting from a controlled deflection reveals the lattice functions. With a fast position monitor the betatron frequency can be measured. Its dependence on energy deviation, current, and quadrupole strength gives information on chromaticity, impedance, and local beta function. Turn-by-turn reading in all monitors allows one to check the optics and to measure the beta function and phase advance around the machine. Diagnostics based on the far field is done with synchrotron radiation. It is used to form an image of the beam cross section and to get its dimensions. Due to the small natural opening angle of the radiation, diffraction effects are important and limit the resolution. The angular spread of the particles in the beam can be measured by a direct observation of the emitted radiation.

ELECTROMAGNETIC FIELDS USED FOR BEAM DIAGNOSTICS

The beam diagnostics considered here are based on the electromagnetic fields created by the charged particles. We distinguish between the 'near field' and the 'far' or 'radiation field'. The near field is the Lorentz-transformed Coulomb field which now also contains a magnetic field. The electric field of a point charge on axis of a circular conducting chamber of radius a induces on the wall a charge distribution $q_w(s)$ having a rms width of $\sigma = a/\sqrt{2}\gamma$. Since this wall current pulse is very short for a relativistic particle, a bunch with longitudinal current distribution $I(t)$ induces on the wall a current $I_W(t)$ having practically the same form (see Figure 1). However, the wall current is of the opposite sign and does not contain the average beam current. The latter induces just a static charge which does not represent a current

$$I_w(t) \approx -(I(t) - \langle I \rangle).$$

Most beam position or intensity monitors are based on a measurement of the wall current. The loop or strip-line monitor shown in Figure 2 has a mixture of

CP451, *Beam Instrumentation Workshop*
edited by R. O. Hettel, S. R. Smith, and J. D. Masek

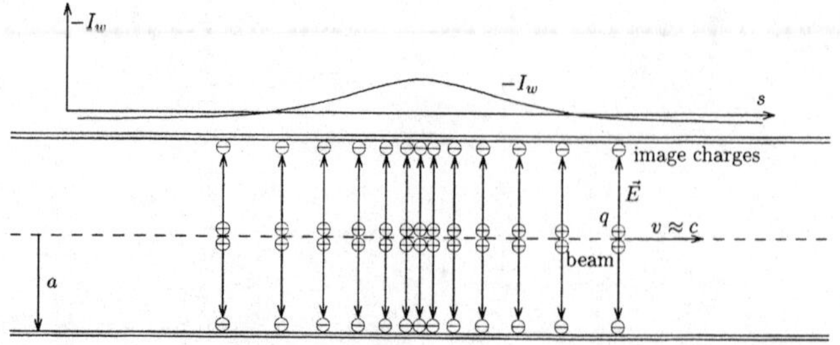

FIGURE 1. Wall current induced by a relativistic bunch.

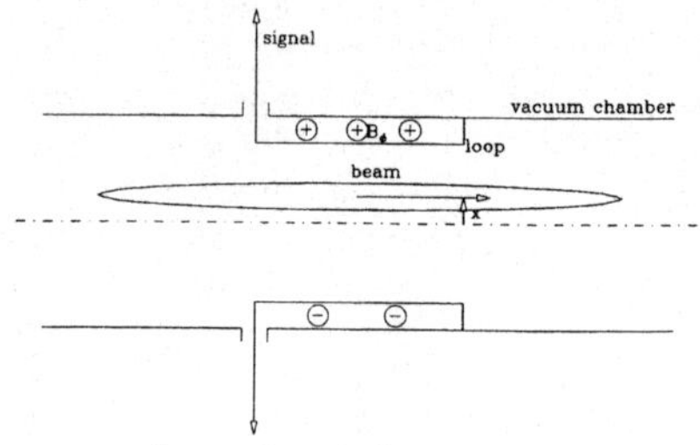

FIGURE 2. Loop monitor using, inductive and capacitive coupling.

inductive and capacitive coupling to the beam which depends on the width of the band forming the loop. The original wall current signal is generally distorted by the type of coupling used and the front end of the read-out system. Although the wall current induced by the beam does not contain the DC part, the latter can be estimated from the monitor reading between widely spaced bunches (see Figure 3) and determined directly from the average magnetic field produced by the beam outside the chamber.

Synchrotron light is the most common radiation field applied to beam diagnostics. It propagates through space and is not attached to the charges. In the majority of cases, it is used to form an image of the beam cross section to measure its dimensions. By observing the radiation directly we can determine its opening angle and measure the angular spread of the particles in the beam. Finally, from the time structure of the radiation we can determine the bunch length using a fast photon detector.

A position monitor uses the near-field which is attached to the beam and deter-

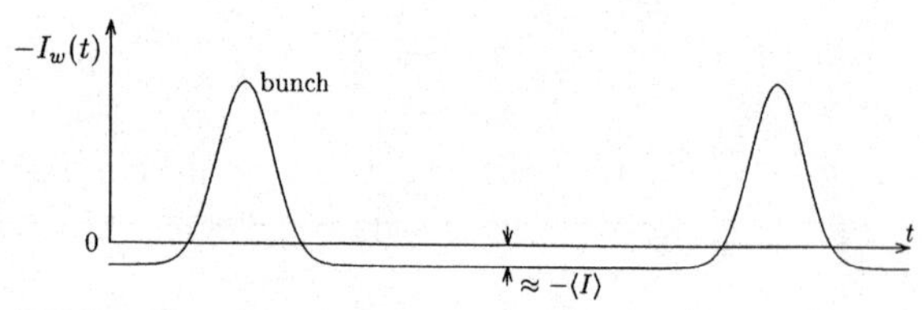

FIGURE 3. Monitor reading for widely spaced bunches.

mines the beam properties at the location of this monitor. However, monitors can also be sensitive to propagating radiation fields created far upstream in a bending magnet or in an aperture change close the monitor in form of diffraction radiation. Most monitors have a limited bandwidth lying below the cut-off frequency of the chamber and are therefore not sensitive to the propagating radiation fields.

MEASUREMENTS WITH BEAM POSITION MONITORS

We concentrate now on beam measurements which can be carried out with beam position monitors combined with deflectors. The accuracy and stability of position monitors have been greatly improved over the past years by precise and stable mechanical construction as well as by the use of state-of-the-art electronics. Furthermore alignment of the monitors with respect to the axis of the adjacent quadrupoles is now more precise. It is, in many machines, checked by moving the beam slowly in a quadrupole while observing the response to a strength modulation of this quadrupole. The observed position change due to the modulation is directly proportional to the distance of the beam from the quadrupole axis. The spatial difference between the electric center of the monitor and the magnetic axis of the quadrupole can be determined by this method and entered in the data analysis. This method, called often 'K-modulation', has been used in different laboratories [1]. With these improvements beam position measurements can reach an absolute accuracy of less than 0.1 mm and a resolution of a few microns. Many beam parameters can be determined by difference measurements where the beam position is obtained for two machine settings within a short time interval. The accuracy is in this case very high, since long-term drifts are avoided.

Position monitors can be used to give a turn-by-turn read out, allowing one to directly observe betatron oscillations and measure the tunes and their dependence on different parameters. At least one such monitor should be available in a storage ring. The demands in accuracy can be more relaxed then for the case of an orbit measurement. A bandwidth as large as the bunch frequency is desirable to check the filling pattern of the ring and to analyze coupled-bunch mode instabilities. Such a wide-band monitor can also observe phase oscillations directly without the need of dispersion. In the past, most machines had such monitors with a simple analog output displayed on a spectrum analyzer or a scope. In some proton machines a monitor of even larger bandwidth can be useful to measure the bunch length and to observe head-tail modes. For electron machines it is difficult to resolve the very short bunches. The corresponding measurements are usually carried out by detecting the synchrotron light with a fast device like a photodiode or a streak camera.

Some larger machines now have a possibility of turn-by-turn measurement in all position monitors to observe betatron oscillations and measure betatron phase advance and beta functions around the machine for optics checks.

The most important application of beam position monitors is the determination of the closed orbit around the ring followed by its correction with dedicated dipole magnets using special codes. We will not describe orbit corrections here but discuss a series of beam dynamics measurements which can be carried out with the help of the improved position monitors. Such experiments have been done in many machines. Here we select examples of a few rings for which the conditions and results were easily available.

AVERAGED READOUT TO MEASURE ORBIT

Measuring Optics from the Orbit Response to Local Deflections

We start with a system of monitors which measure the beam position averaged over many turns. This determines the closed orbit for stationary conditions by giving the horizontal and vertical coordinates

$$x_i = x_i(s_i) \quad , \quad y_i = y_i(s_i)$$

of all monitors i located at the longitudinal position s_i around the ring. As mentioned above, the most useful application of this information is the measurement of the closed orbit in a ring. From this, bending errors might be identified and corrected or the settings of the corrector magnets may be calculated, bringing the orbit as close as possible to the ideal.

We like to measure the effect of an applied deflection by a angle θ with a corrector magnet located at $s = 0$. Outside this deflector the orbit has the form of a betatron oscillation

$$x(s) = x(0)\sqrt{\frac{\beta(s)}{\beta(0)}}\cos(\phi(s) - \phi_0).$$

A local deflection by the angle θ imposes the condition

$$x(0) = x(2\pi R)\ ,\ x'(0) - x'(2\pi R) = \theta.$$

Using the relation

$$\frac{d\phi}{ds} = \frac{1}{\beta}$$

we get for the orbit distortion created by a single deflection

$$x(s) = \frac{\theta}{2}\sqrt{\beta(s)\beta(0)}\frac{\cos(\pi Q - \phi(s))}{\sin(\pi Q)}.$$

In presenting this orbit distortion it is helpful to normalize the position monitor reading with $\sqrt{\beta}$ and display it against the betatron phase $\phi(s)$, as shown in Figure 4. The orbit has the form of a cosine with a cusp at the deflection error. This is, of course, only exact if the actual beta function agrees with the calculated one.

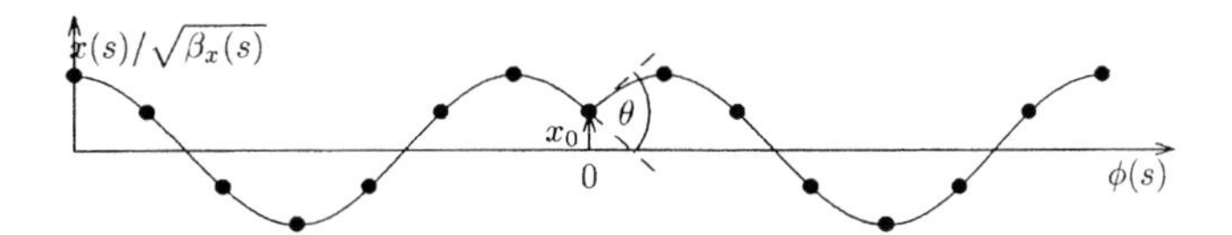

FIGURE 4. Orbit distortion caused by a single deflection.

A measurement of an orbit difference due to a known deflection θ determines the beta function. With a monitor at the location of the deflection magnet we measure a displacement $x_0 = x(0)$ given by

$$\beta(0) = \frac{2x_0 \tan(\pi Q)}{\theta}.$$

From this we obtain $\beta(0)$ if the tune Q is known. By measuring the position change at all monitors for sequential changes of all corrector magnets it is possible to determine the lattice functions around the ring. Since this problem is overdetermined also some checks of the monitor and corrector calibrations can be made. Such experiments have been successful to obtain the lattice function in storage rings with good accuracy [2].

Measuring Impedance from the Current Dependence of the Orbit

A circulating bunch induces a current in the chamber wall. This can lead to an energy loss if the beam surrounding has a resistive impedance. The loss per particle is proportional to the bunch current and is replaced by the rf cavities. For this reason a bunch has excessive energy when leaving the cavity and lacks energy entering the cavity after one turn. Beam position monitors located at finite dispersion D_x can measure this energy deviation from the relation $\Delta x = D_x \Delta p/p \approx D_x \Delta E/E$. By observing the orbit difference for two bunch currents, the resistive impedance distribution around the ring can be determined.

Such an experiment has been carried out in LEP [3] having two groups of cavities as shown left of Figure 5. The group in point 6 is powered while the other in point 2 is passive. The impedance of the latter and the one of the long arcs containing bellows, aperture changes, resistive walls, etc. can be determined separately. With a monitor in a dispersion-free region, it can be checked that there is no systematic current dependence of the monitor reading. The measured orbit difference for a change in bunch current is shown on the right of Figure 5. To increase accuracy, the readings of all monitors situated at the same dispersion value in one arc have been averaged. Since all arcs have the same properties, the obtained points are fitted with lines of the same slope. At point 2 there is a drop in orbit due to the

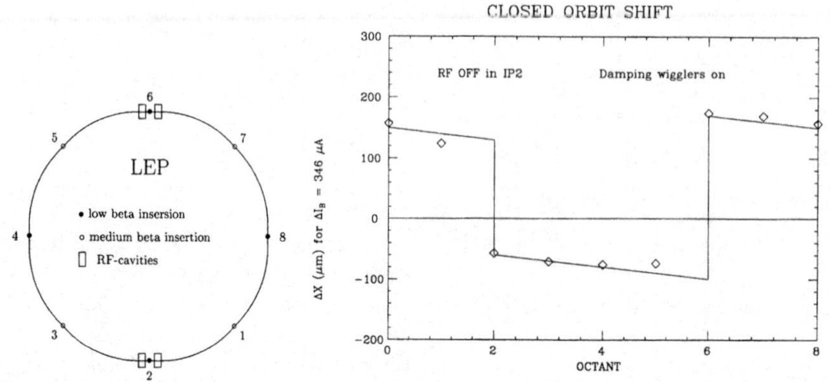

FIGURE 5. Difference orbit for two bunch currents in LEP having two rf sections. The cavities at point 6 are active, the ones at point 2 passive.

impedance of the cavities located there. The lost energy is replaced by the active cavities at point 6 and the orbit moves to the outside.

TURN-BY-TURN MEASUREMENTS IN ONE MONITOR

Observing Betatron Oscillations

A storage ring needs at least one monitor that can read the beam position turn by turn to observe betatron oscillations. This can also be achieved with an analog output of a bandwidth larger than the revolution frequency.

At the top of Figure 6 the readout of a position monitor at each turn k is shown for the observation of a betatron oscillation. Different harmonic fits can be made through the points. Since only one sample is taken at each revolution of the bunch, there is an ambiguity in the observed frequency ω_β which is related to the betatron tune Q by

$$\omega_\beta = \omega_0(n \pm Q).$$

The corresponding spectrum is shown at the bottom of the figure in units of the revolution frequency ω_0.

The most important application of this fast monitor is the measurement of the betatron tunes. To do this, we need a fast magnet which can produce a horizontal or vertical deflection within one revolution to excite betatron oscillations. This can be done either by giving the beam a relatively large kick and observing the free betatron oscillation or by a continuous harmonic excitation with a swept frequency. Noise-exciting and FFT spectrum analysis can also be used. The tune measurement

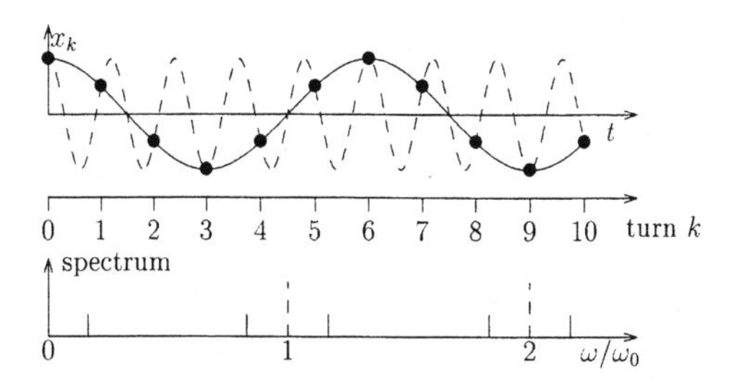

FIGURE 6. Betatron oscillation observed with a monitor having turn by turn read out.

is, in most machines, automated and can be used to study its dependence on different parameters:

- Measurement of the tune as a function of momentum $Q = Q(\Delta p/p)$ gives the chromaticity $Q' = dQ/(dp/p)$. The momentum, or energy, can be changed by a small variation of the rf frequency. This relation is given by the momentum compaction factor α_c and the Lorentz factor $\Delta p/p = -\eta \Delta f_{RF}/f_{RF}$ with $\eta = \alpha_c - 1/\gamma^2$.

- One of the fastest ways to obtain information about the impedance is a measurement of the tune as a function of bunch current $Q = Q(I_b)$. The field induced in the reactive transverse wall impedance by the bunch affects the focusing and therefore the tune.

- By measuring the tune as a function of oscillation amplitude $Q = Q(\hat{x})$ we obtain information about the non-linearity of the focusing. Increasing this amplitude until some beam loss occurs gives the acceptance of a ring, also called dynamic aperture.

- The beta function at a quadrupole can be measured by making a small variation ΔK of its strength parameter and observing the resulting change of the tune ΔQ. The short lens relation

$$\Delta Q = \frac{\beta \Delta K L}{4\pi}$$

directly gives the beta function. This is a very quick and important experiment to check the optics at some important locations, such as the low-beta insertion or the source point of synchrotron radiation. It is necessary that the quadrupoles involved have some independent powering. Some correction has to be made in the above relation for the finite length of the quadrupoles. In practice, the expected tune change for a given quadrupole variation is computed with an optics code and checked by the experiment. Hysteresis effects are important for small quadrupole strength variations. They can be checked with an integrator connected to an induction loop in the quadrupole which measures the actual field change.

- Coupling can be measured by kicking the beam in one plane and observing the oscillation in the other plane.
- If the bandwidth of the monitor is larger than the bunch frequency the filling pattern of the bunches can be checked. Furthermore, coupled bunch instabilities can be observed and the phase difference between the oscillation of individual bunches (coupled mode number), as well as the growth rate, can be determined. This information is often helpful to find the element driving the instability.

Beam Response — Transfer Function

The wide-band monitor combined with a deflection magnet can be used to measure the beam response to a pulse or a harmonic excitation. In addition to the observation of just the tune or the amplitude, we can measure the detailed response of the beam. In case of pulse excitation, this measurement contains a reading of the beam position at each turn after the kick. In the case of harmonic excitation we measure the amplitude and phase with respect to the driver, also called the beam transfer function. The time domain measurement is more popular with bunches while the transfer function is more suitable for continuous beams.

The beam transfer function measurement can be carried out with a network analyzer as indicated in Figure 7. For coasting (unbunched) beams, the theory and the measurement of the beam transfer function is particularly simple and therefore quite popular. In the transverse case the real (in-phase) part of the response gives directly the particle distribution in incoherent betatron frequency. An example of such a measurement in the ISR [4] is given in Figure 7. It shows that the phase changes by about π while sweeping through the betatron frequency distribution similar to the response of damped oscillator. In the longitudinal case one obtains the derivative of the energy distribution. Measuring the transfer function for different currents or distribution widths can give the complex impedance of the surroundings. A feedback system can be checked in phase and gain by taking the transfer function with and without it.

Experimental Tracking

The non-linear optics of a storage ring is often studied theoretically by particle tracking. A particle is launched with some initial offset and angle (x_0, x'_0) and its trajectory calculated. After each revolution the coordinate pair (x_k, x'_k) is plotted. In a linear machine these points are on an ellipse, however, non-linearities distort this curve. For some initial conditions particles get lost after some turns. By varying (x_0, x'_0), the dynamic acceptance can be determined.

We can study an existing machine experimentally with the same method. An angular deflection is given at some time and the beam position is observed turn by

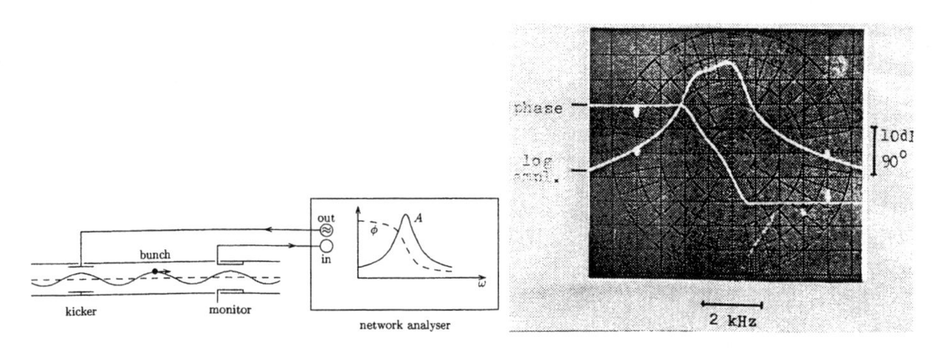

FIGURE 7. Transverse transfer function of a coasting beam.

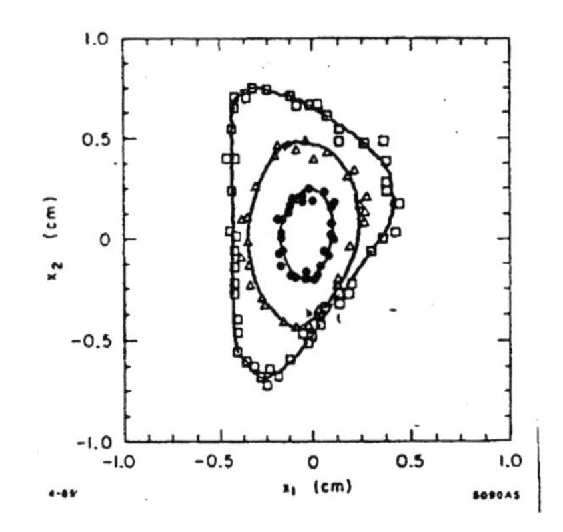

FIGURE 8. Experimental tracking in SPEAR.

turn, using two position monitors separated by a betatron phase of 90^0. These two monitors are equivalent to a measurement of position and angle at one location. In the case of electron beams, one probes the optics with different amplitudes due to the radiation damping. For protons the coherent signal disappears due to phase mixing but the amplitude of the individual particle stays constant. This makes the interpretation difficult. Usually one observes the beam for a limited number of turns and changes the initial kick to vary the amplitude. An example of such a measurement [5] with an electron beam in SPEAR is shown in Figure 8. An x, y-plot of the signals from the two monitors is made for a few turns at different amplitudes. The oscillation is quite linear at small amplitudes resulting in an ellipse on the plot. For larger amplitudes this figure has a triangular distortion due to the non-linearity and the proximity of the third-order resonance.

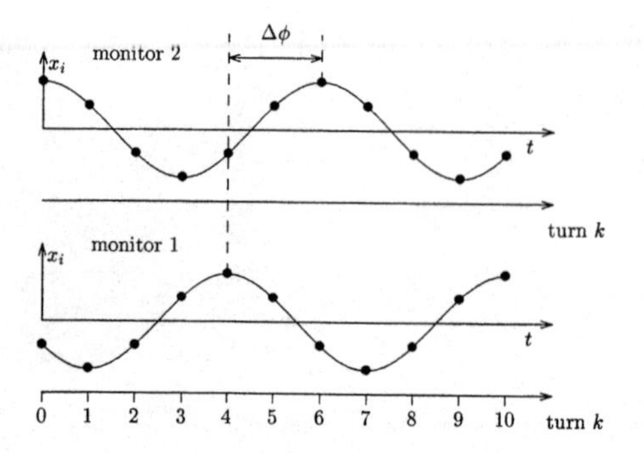

FIGURE 9. Betatron oscillation phase at two position monitors.

TURN-BY-TURN MEASUREMENTS IN ALL MONITORS

Measuring Phase Advance and Optics Checks

Turn-by-turn readout in all position monitors allows one to follow a betatron oscillation around the ring and measure the lattice functions. A betatron oscillation is excited and measured with many position monitors i around the ring every revolution k giving the readings

$$x_{ik} = \frac{\hat{x}}{\sqrt{\beta_{x0}}}\sqrt{\beta_{xi}}\cos(2\pi Q_x k + \mu_{xi}).$$

Comparing the phase of the oscillation observed in two monitors gives the betatron phase advance between them

$$\Delta\phi = \mu_{i+1} - \mu_i$$

as illustrated in Figure 9. The ratio of the beta functions is obtained from the square of the amplitude ratio observed between the two monitors.

$$\frac{\beta_{x,i+1}}{\beta_{x,i}} = \left(\frac{\hat{x}_{i+1,k}}{\hat{x}_{i,k}}\right)^2.$$

The measurement of the phase advance has the advantage of having very small systematic errors. Once the signal is sufficiently clean to determine its phase it is unlikely to introduce errors in the form of delays. The measurement of the beta function ratio however depends on the monitor calibration. The relation between the optics functions

$$\Delta\phi = \mu_{i+1} - \mu_i = \int_{s_i}^{s_{i+1}} \frac{ds}{\beta(s)}$$

can be used to relate the two measurements.

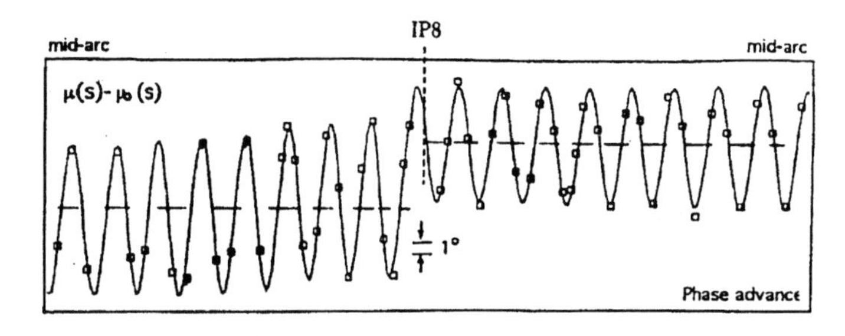

FIGURE 10. Beta beating caused by a focusing error.

By measuring the betatron phase around the ring and comparing it with the calculated one, we can check the beam optics and find errors. A local focusing deviation creates a beating of the beta function and the betatron phase advance around the ring. This deviation advances at twice the betatron wave number around the machine, that is, points being separated by $n\pi$ will keep this separation unless the error is located between them. An example of such a measurement in LEP [6] is shown in Figure 10. The difference between the measured and calculated betatron phase advance is plotted against the latter for about one eighth of LEP including a low-beta insertion at point 8. In the ideal case this plot should result in a straight line, however, due to a focusing error we observe a modulation of the line at twice the betatron phase advance with a phase jump at the interaction point IP8. Clearly an error is introduced by one or both strong quadrupoles of the low-beta insertion located around this point.

Measuring the Local Chromaticity and its Correction

The chromaticity gives the change of tune with momentum $Q' = dQ/(dp/p)$. In the absence of sextupole magnets, the chromaticity is negative since the focusing strength of the quadrupoles is smaller for particles with excess energy. With sextupole magnets located at finite dispersion this natural chromaticity can be corrected. To measure the chromaticity we make a small momentum change by varying the rf frequency and registering the difference in tune. Most rings operate with slightly positive chromaticity to stabilize head-tail modes. Strongly focusing sections, like low-beta insertions, produce negative chromaticity and are often at places of vanishing dispersion. The chromaticity cannot therefore be corrected locally but only at some distance. It might be interesting to study the local chromatic effects by measuring the phase advance as a function of momentum. Such an experiment was carried out in LEP [7] which has eight arcs with dispersion where the sextupoles are located and eight dispersion-free straight sections with low or medium beta insertions as shown on the left of Figure 11. The calculations and measurements of vertical chromatic phase advance $d\mu_y/(2\pi)/(dp/p)$ are shown on

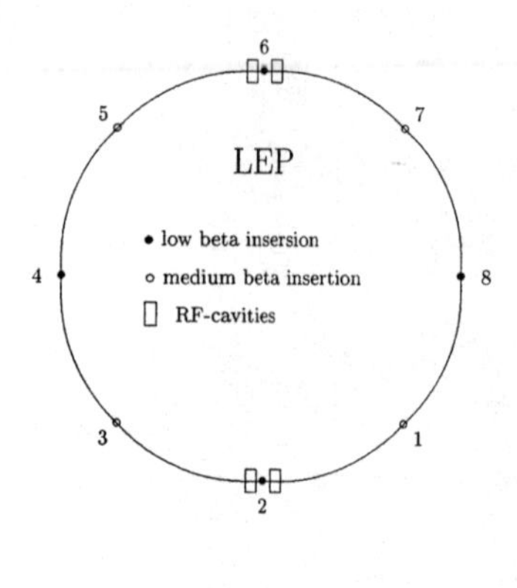

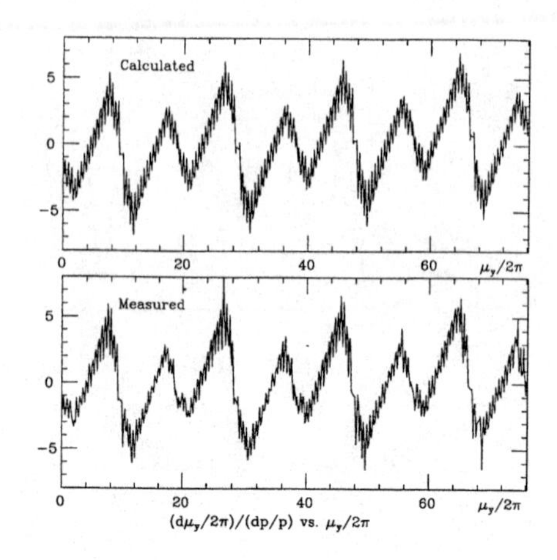

FIGURE 11. Calculation and measurement of the chromatic phase advance in LEP.

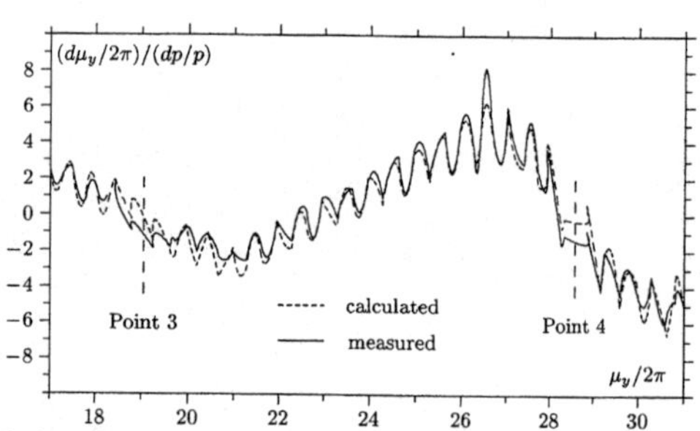

FIGURE 12. Detail of the chromatic phase advance showing a beta beating due to an optical mismatch of the off energy beam.

the right in Figure 11. Clearly visible are the negative slopes of this quantity in the four straight sections having strong focusing and the other four having medium focusing. In the arcs containing the sextupoles, this function has a compensating positive slope leading to a slightly positive chromatic phase advance around the whole ring. The agreement between calculation and measurement is very good and gives no indication of any error in the sextupole arrangement. There is a small modulation of this chromatic phase advance at twice the betatron wave number for both the calculated and the measured data shown in Figure 12 in more detail. This is due to the fact the chromatic effects are not corrected exactly where they occur. Particles with an energy deviation have therefore local focusing errors resulting in a modulation of the beta function which can only be corrected for some strategic places, like the interaction points.

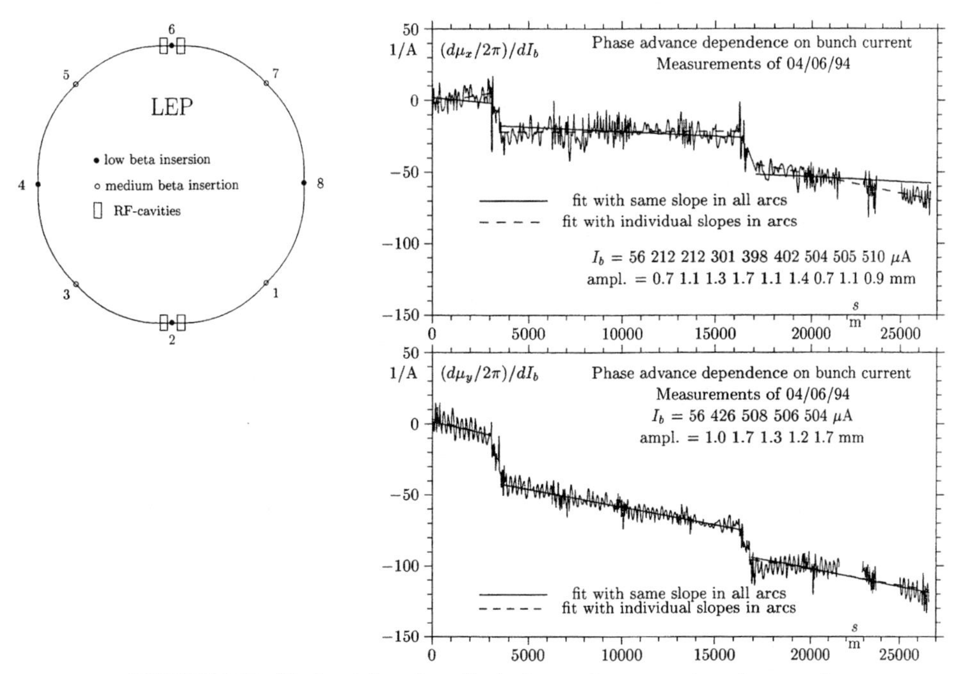

FIGURE 13. Horizontal and vertical phase advance vs. bunch current.

Phase Change with Current Due to the Transverse Reactive Impedance

The wall current induced by a circulating bunch, executing a betatron oscillation, produces fields through the transverse reactive impedance, which modify the focusing. This effect leads to a tune dependent on current and is often measured to estimate of the transverse impedance. By observing the betatron phase advance as a function of bunch current, we find the distribution of this impedance around the ring. In such an experiment [8] the betatron phase advance was measured for different currents in LEP which had at the time two groups of rf cavities as indicated on the left of Figure 13. The resulting dependence $d\mu/dI_b$ is plotted in the right part of this figure for both planes. It clearly shows a strong decrease of this quantity at the rf cavities and a more smooth change in the arcs. The cavities have a circular cross section and are expected to have the same transverse impedance in the horizontal and vertical plane. The arcs, however, have an elliptical chamber leading to an impedance being predominantly vertical. This expectation is clearly reproduced by the measurements which show no difference between the two planes at the cavities but have clearly a stronger slope in the arcs for the vertical plane.

DIAGNOSTICS WITH SYNCHROTRON RADIATION

Types of Measurements

Three main types of measurements are done with synchrotron radiation: imaging to measure the beam cross section, direct observation to measure the angular spread of the particles, and observing the longitudinal structure of the radiation to obtain the bunch length.

For the most common measurement, the radiation emitted tangentially in the bending magnet is extracted from the vacuum chamber through a window. A lens is then used to form an image of the source point on a screen, as illustrated in Figure 14.

It is also possible to observe the synchrotron radiation directly without using focusing elements, as shown in Figure 15. In this case one measures the angular distribution of the particles in the beam. If a long horizontal bending magnet serves as the source of the radiation, only the vertical angles can be measured. The resolution is limited by the natural opening angle of the radiation itself.

The bunch length can be measured by observing the time structure of the emitted radiation. Since the light pulse emitted by each individual particle is very short, the time distribution of the radiation directly reflects the longitudinal bunch shape. Such measurements depend on a fast photon detector and will not be discussed further here.

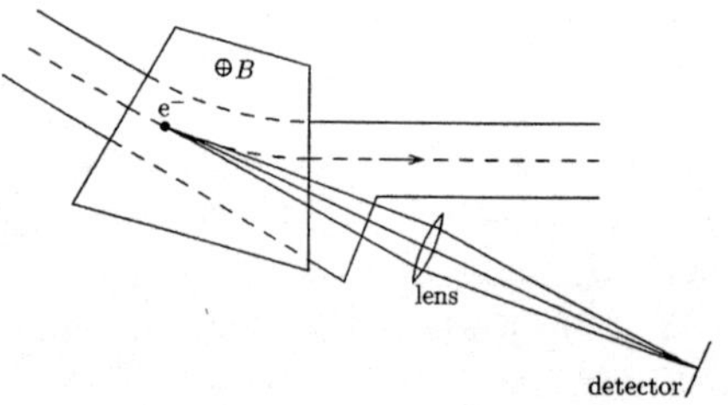

FIGURE 14. Imaging of the beam cross section with synchrotron radiation.

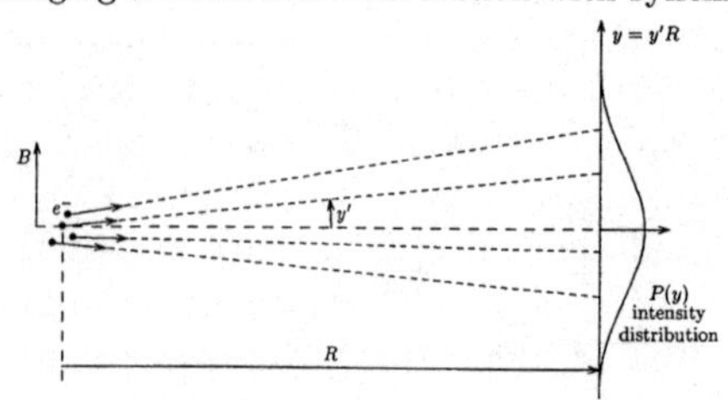

FIGURE 15. Direct observation of synchrotron radiation.

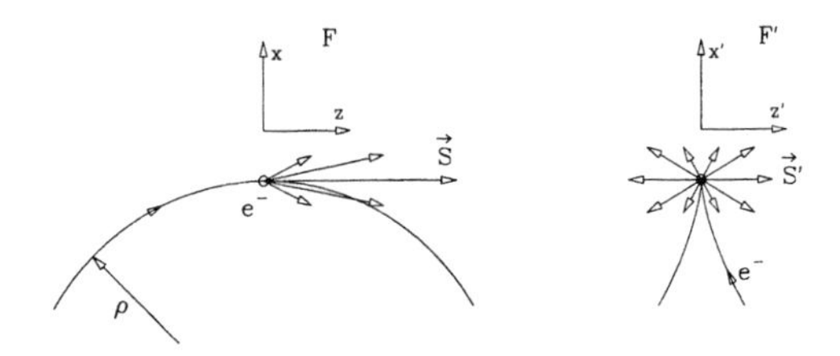

FIGURE 16. Opening angle of synchrotron radiation caused by the moving source.

Properties of Synchrotron Radiation

Many properties of synchrotron radiation can be understood from a qualitative treatment.

To estimate the opening angle we consider an electron moving in the laboratory frame F on a circular orbit and emitting synchrotron radiation (Figure 16). In an inertial frame F', which moves at one instant with the same velocity $\mathbf{v} = \boldsymbol{\beta} c$ as the electron, the particle trajectory has the form of a cycloid with a cusp where the electron undergoes acceleration in the $-x'$ direction. It will emit radiation which in this frame F' is approximately uniformly distributed. Going back to the laboratory frame F, by applying a Lorentz transformation, this radiation will be peaked forward. A photon emitted along the x'-axis in the moving frame F' will appear at an angle $1/\gamma$ in the laboratory frame F. The typical opening angle of synchrotron radiation is therefore of order $1/\gamma$ which is very small for ultra-relativistic particles where $\gamma \gg 1$.

The spectrum of synchrotron radiation depends on the type of magnet from which it originates. We consider first the usual case of a long magnet and estimate the typical frequency emitted by an electron and received by an observer P (Figure 17). The received radiation pulse is very short due to the small opening angle. The radiation seen first was emitted at point A, where the electron trajectory has an angle of $1/\gamma$ with respect to the direction towards the observer, and last at point A', where this angle is $-1/\gamma$. The length of the observed radiation pulse is the difference in travel time between electron and photon in going from A to A':

$$\Delta t = t_e - t_\gamma = \frac{2\rho}{\beta\gamma c} - \frac{2\rho\sin(1/\gamma)}{c} \approx \frac{2\rho}{\beta\gamma c}\left(1 - \beta + \frac{\beta}{6\gamma^2}\right) \approx \frac{4}{3}\frac{\rho}{c\gamma^3},$$

where we assumed the ultra-relativistic case $\gamma \gg 1$ and expanded the trigonometric function for small angles. The typical frequency is approximately

$$\omega_{typ} \sim \frac{2\pi}{\Delta t} \sim \frac{3\pi c\gamma^3}{2\rho}.$$

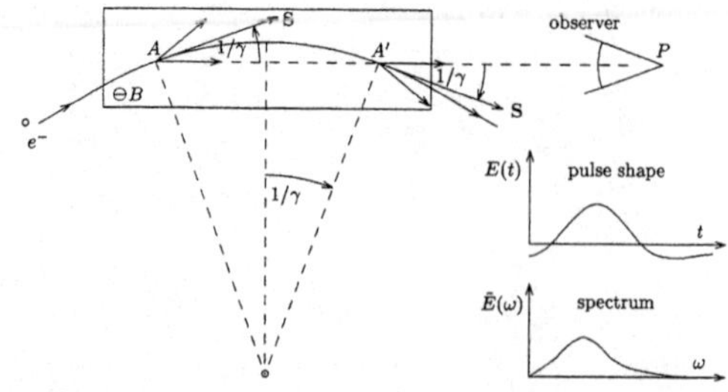

FIGURE 17. Spectrum of synchrotron radiation emitted in a long magnet.

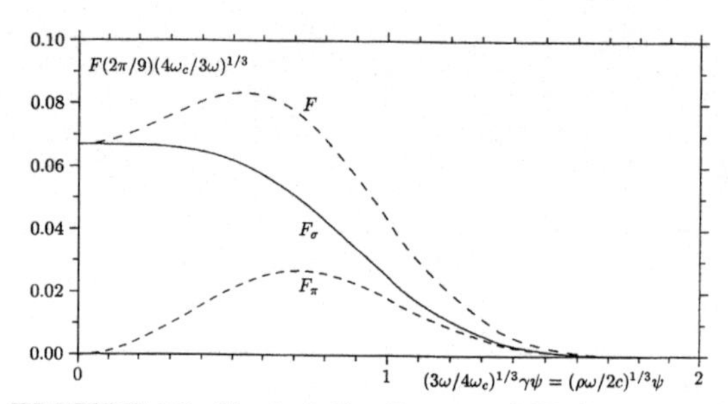

FIGURE 18. Vertical distribution of SR for $\omega \ll \omega_c$.

A quantitative treatment results in a critical frequency of similar magnitude

$$\omega_c = \frac{3c\gamma^3}{2\rho}$$

which divides the spectrum into two parts of equal power.

For diagnostics we like to use visible (or close to visible) light. This part of the spectrum is usually well below ω_c and approximations can be made. The calculated angular distribution for $\omega \ll \omega_c$ is shown in Figure 18 for the horizontal (σ-mode) and vertical (π-mode) polarization and for the total radiation. Here, it is independent of γ. The vertical rms opening angle for the σ-mode is

$$\sqrt{\langle \psi_\sigma^2 \rangle} \approx 0.41\,(\lambda/\rho)^{1/3}.$$

An undulator is an interesting source of synchrotron radiation. It consists of a spatially periodic magnetic field with period length λ_u in which the particle moves on a sinusoidal orbit (Figure 19). Each of the periods represents a source of radiation. These contributions emitted towards an observer at an angle θ will interfere with each other. We get maximum intensity at a wavelength λ for which the contributions from different undulator periods are in phase. The time difference ΔT between the arrival of adjacent contributions is in ultra-relativistic approximation

$$\Delta T = \frac{\lambda_u}{\beta c} - \frac{\lambda_u \cos\theta}{c} = \frac{\lambda_u(1-\beta\cos\theta)}{\beta c} \approx \frac{\lambda_u}{\beta c}\left(1-\beta+\frac{\theta^2}{2}\right) \approx \frac{\lambda_u}{2c\gamma^2}(1+\gamma^2\theta^2).$$

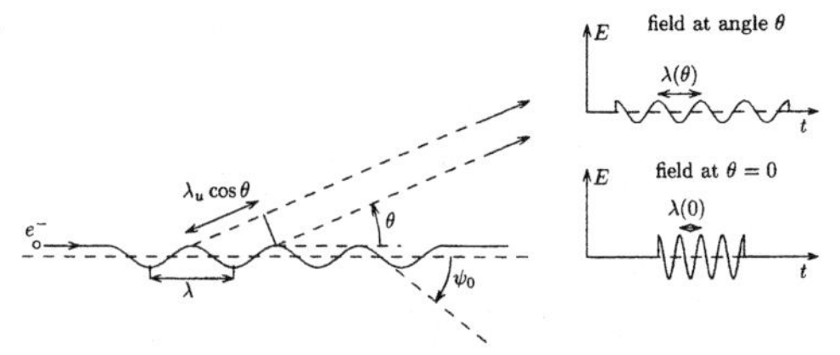

FIGURE 19. Spectrum of synchrotron radiation emitted in an undulator.

The frequency for which we get constructive interference is just $\omega = 2\pi/\Delta T$:

$$\omega = \frac{4\pi c\gamma^2}{\lambda_u(1+\gamma^2\theta^2)}.$$

Harmonics of this frequency might also be emitted.

Since for undulator radiation both the horizontal and vertical opening angles are small, direct observation gives the angular spread of the particles in both planes.

Imaging with Synchrotron Radiation

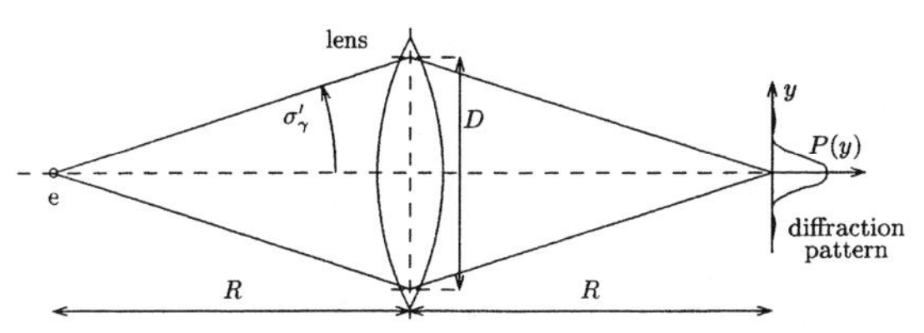

FIGURE 20. Imaging the beam cross section with synchrotron radiation.

We use synchrotron radiation to form a 1:1 image of the beam cross section with a single lens (Figure 20). Due to the small vertical opening angle of $\psi \approx 1/\gamma$ only the central part of the lens is illuminated. Similar to optical imaging with limited lens aperture D, this leads to a diffraction limit in the resolution d (half image size):

$$d \approx \frac{\lambda}{2D/R}.$$

For horizontally polarized synchrotron radiation from long magnets we found an rms opening angle $\sigma'_\gamma = \psi_{\sigma-rms} \approx 0.41(\lambda/\rho)^{1/3}$. Taking $D \approx 4\sigma_\gamma R$ we get $d \approx 0.3(\lambda^2\rho)^{1/3}$. A more quantitative treatment using Fraunhofer diffraction gives an rms image size

$$\sigma_y = 0.21(\lambda^2\rho)^{1/3}.$$

The resolution improves with small λ and ρ and is limited in large machines.

Measurement Examples

Imaging with Synchrotron Radiation in LEP

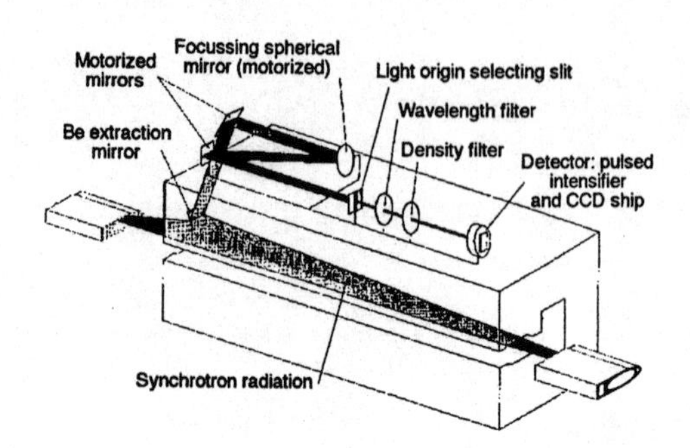

FIGURE 21. Telescope for imaging in LEP.

In LEP ($\rho = 3096$ m) a telescope ($\lambda \geq 200$ nm) is used (Figure 21) to image the beam [9]. Fraunhofer diffraction gives a vertical resolution of $\sigma_y \approx 0.1$ mm which is comparable to the beam size at E=45 GeV. Using a CCD camera and image processing, 3D plots, projections, and turn-by-turn reading are obtained.

Imaging with X-Ray Pin-Hole Camera

Since the resolution is $\propto (\lambda^2 \rho)^{1/3}$ the use of x-rays gives better results but optics elements are more difficult to implement. A simple pin-hole was used at CEA, [10]. The radiation originates in magnet with $\rho = 26.2$ m, reaches a pin-hole of 0.07 mm diameter after 8 m, and is detected 16 m further on a film (Figure 22). Absorption in 1.3 mm of Al and 6 m of air and film sensitivity gives $\lambda \approx 0.05$ nm as the dominant wave length. Diffraction effects are small but pin-hole size and film give a resolution of about 0.1 mm. It was used to minimize coupling by powering some quadrupoles to separate the tunes.

Direct Observation of Undulator Radiation in PEP

The radiation from an undulator in PEP was observed directly to get the horizontal and vertical angular spread of the electrons [11]. The emitted radiation passes through a monochromator to a screen (Figure 23). The undulator spectrum is measured and the beam picture on the screen observed. The one taken close to the undulator peak is scanned and corrected for the known natural distribution of the undulator radiation, the instrument resolution, and the beam size due to energy spread and dispersion at the source. With the known value of the electron beam lattice function at the source, the emittance of the electron beam is obtained.

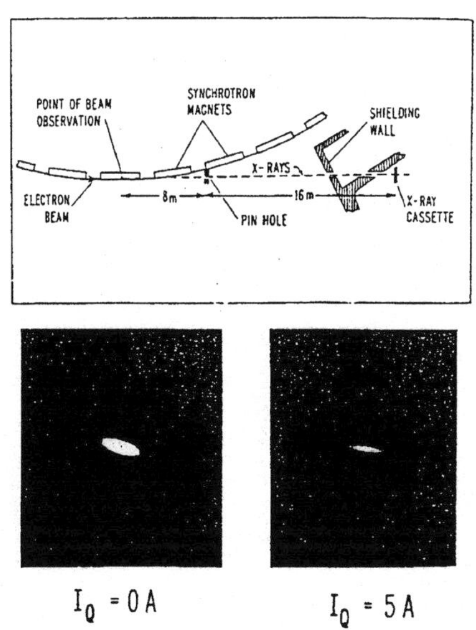

FIGURE 22. Imaging with an x-ray pin-hole camera.

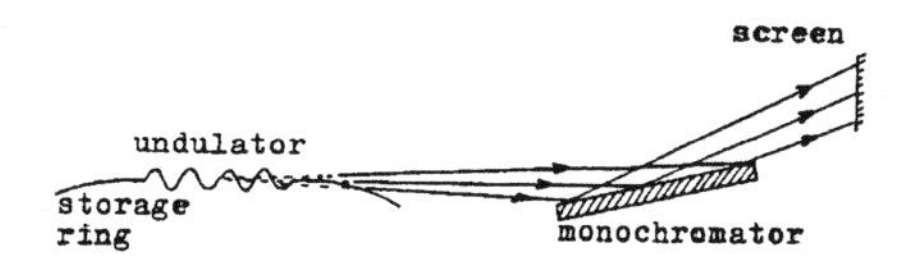

FIGURE 23. Direct observation of monochromatized x-rays in PEP.

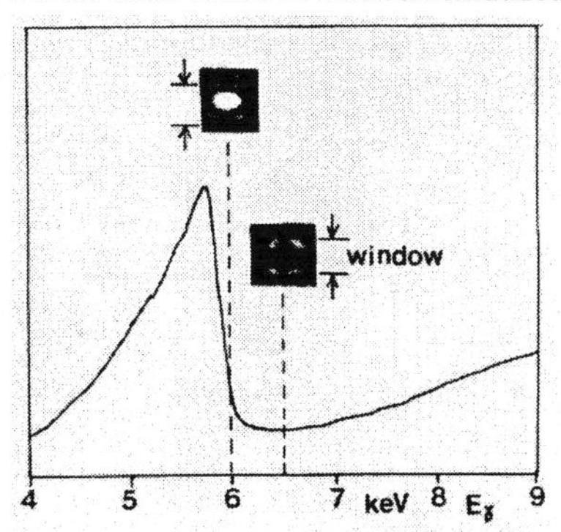

FIGURE 24. Measured undulator spectrum and photon beam picture.

REFERENCES

1. Tecker, F., B. Dehning, P. Galbraith, K. Hendrichsen, M. Placidi, and R. Schmidt, "Beam Position Monitor Offset Determination at LEP," Proceedings of the 1997 Particle Accelerator Conference, Vancouver (1997).
2. Robin, D., G. Portmann, H. Nishimura, and J. Safranek; "Model Calibration and Symmetry Restoration of the Advanced Light Source," in *Proceedings of EPAC96* (conference held in Sitges (Barcelona)), pp. 971–973 (1996).
3. Brandt, D., P. Castro, K. Cornelis, A. Hofmann, G. Morpurgo, G. L. Sabbi, J. Wenninger, and B. Zotter; "Measurement of Impedance Distributions and Instability Thresholds in LEP," in *Proceedings of the 1995 Particle Accelerator Conference* (conference held in Dallas), pp. 570–572 (1995).
4. Hofmann, A. and B. Zotter, "Measurement of Beam Stability and Coupling Impedance by RF Excitation," in *Proc. of the 1977 Particle Accelerator Conference*, IEEE Trans. on Nucl. Sci. NS 24-3, pp. 1487–1489 (1977).
5. Morton, P. L., J. L. Pellegrin, T. Raubenheimer, L. Rivkin, M. Ross, R. D. Ruth, W. L. Spence; "A Diagnostic for Dynamic Aperture," in *Proceedings of the 1995 Particle Accelerator Conference*, IEEE Trans. on Nucl. Sci., NS-32-5, pp. 2291–2293 (1985).
6. Borer, J., C. Bovet, A. Burns, and G. Morpurgo; "Harmonic Analysis of Coherent Bunch Oscillations in LEP," in *Proceedings of EPAC92* (conference held in Berlin), pp. 1082–1084 (1992).
7. Brandt, D., P. Castro, K. Cornelis, A. Hofmann, G. Morpurgo, G. L. Sabbi, and A. Verdier; "Measurement of Chromatic Effects in LEP," in *Proceedings of the 1995 Particle Accelerator Conference* (conference held in Dallas, 1995).
8. Brandt, D., P. Castro, K. Cornelis, A. Hofmann, G. Morpurgo, G. L. Sabbi, J. Wenninger, B. Zotter; "Measurement of Impedance Distributions and Instability Thresholds in LEP," in *Proceedings of the 1995 Particle Accelerator Conference*, Dallas (1995).
9. Bovet, C., G. Burtin, R. J. Colchester, B. Halvarsson, R. Jung, S. Levitt, and J. M. Vouillot, "The LEP Synchrotron Light Monitors," CERN SL/91-25 and *Proceedings of the 1991 IEEE Particle Accelerator Conference*, p. 1160.
10. Hofmann, A., and K. W. Robinson, "Measurement of the Cross Section of a High-Energy Electron Beam by means Portion of the Synchrotron Radiation", in *Proceedings 1971 Particle Accelerator Conference, IEEE Trans. Nucl. Sci.* NS 18-3, p. 973 (1971)
11. Brendt, M., G. Brown, R. Brown, J. Cerino, J. Christensen, M. Donald, B. Graham, R. Gray, El Guerra, C. Harris, A. Hofmann, C. Hollosi, T. Jones, J. Jowett, R. Liu, P. Morton, J. M. Paterson, R. Pennacchi, L. Rivkin, T. Taylor, T. Troxel, F. Turner, J. Turner, P. Wang, H. Wiedemann, and H. Winick, "Operation of PEP in a Low Emittance Mode," in *Proceedings of the 1987 IEEE Particle Accelerator Conference*, p. 461 (1987).

Electron Beam Polarimetry*

Charles K. Sinclair

Jefferson Laboratory 12000 Jefferson Avenue,
Newport News, Virginia, 23606

Abstract. Along with its well known charge and mass, the electron also carries an intrinsic angular momentum, or *spin*. The rules of quantum mechanics allow us to measure only the probability that the electron spin is in one of two allowed spin states. When a beam carries a net excess of electrons in one of these two allowed spin states, the beam is said to be *polarized*. The beam polarization may be measured by observing a sufficient number of electrons scattered by a spin-dependent interaction. For electrons, the useful scattering processes involve Coulomb scattering by heavy nuclei, or scattering from either polarized photons or other polarized electrons (known as Mott, Compton, and Møller scattering, respectively). In this tutorial, we will briefly review how beam polarization is measured through a general scattering process, followed by a discussion of how the three scattering processes above are used to measure electron beam polarization. Descriptions of electron polarimeters based on the three scattering processes will be given.

INTRODUCTION

Along with its well-known charge and mass, the electron also carries an intrinsic angular momentum, or *spin*. The magnitude of the angular momentum carried by each electron is an exact number $-3h/8\pi$, where h is Planck's constant. Sensibly enough, the electron also has a magnetic moment directly proportional to the spin. However, the electron spin is a quantum mechanical quantity—there is no classical analog for electron spin.

Just as in classical mechanics, there is a direction as well as a magnitude associated with the angular momentum. In classical mechanics the angular momentum may be oriented in any direction in space, while in quantum mechanics, only certain discrete possibilities are allowed for this orientation. For electrons there are only two allowed orientations for the spin. The projection of the spin along a quantization axis may be only + or $-h/4\pi$. The quantization axis is defined by the physical situation at hand, as will become clear later. The two possible spin orientations are often referred to as "parallel" or "up," and "antiparallel" or "down."

In a beam of electrons from a conventional electron source (e.g., a thermionic emission cathode), the numbers of electrons with positive and negative spin

* Work supported by Department of Energy contract DE-AC05-84ER40150.

CP451, *Beam Instrumentation Workshop*
edited by R. O. Hettel, S. R. Smith, and J. D. Masek
 1-56396-794-4/98/$15.00

projections along *any* axis are equal, with the result that the beam electrons carry no net angular momentum along any axis. Such a beam is said to be *unpolarized.* If, by some means, an electron beam is created with a net difference in the numbers of positive and negative spin projections along some axis, the beam is said to be *polarized* along that axis. The polarization of an electron beam along an axis is measured by counting the difference between the numbers of electrons with positive and negative spin projections along that axis, divided by the sum, i.e.:

$$P = \frac{n_+ - n_-}{n_+ + n_-} \tag{1}$$

in an obvious notation.

The rules of measurement in quantum mechanics tell us that it is *fundamentally impossible* to measure the orientation of the spin of an individual electron. Rather, one can only measure the *probability* that an electron is in one or the other of the two allowed spin orientations. Thus, the measurement of beam polarization implies that we must measure, or sample, the spin projection probabilites of a sufficiently large number of the beam electrons.

The physics programs at almost all electron accelerators dedicated to basic research in nuclear and high-energy physics demand polarized beams. In general, they require longitudinal beam polarization—i.e., an electron spin orientation either parallel or antiparallel to the beam momentum. The methods employed to produce polarized electrons do not provide a precisely known beam polarization. Furthermore, the orientation of the electron beam polarization does not stay fixed with respect to the beam momentum as the beam moves through the electromagnetic fields of an accelerator and its transport lines. Measurement of both the magnitude and the orientation of the beam polarization is thus essential. To date, essentially all electron beam polarimeters have been developed by the research groups using the polarized beams. This is true in part because the techniques involved in measuring electron polarization are very similar to those employed in the physics experiments themselves —i.e., clean identification of scattered electrons or photons and the rejection of scattered particle backgrounds from unwanted sources.

Rather than provide references for statements made throughout the text of this tutorial article, an annotated bibliography is provided at the end. The references in this bibliography cover in some detail all material presented in this article.

Electron Polarization Measurement by Scattering

All techniques devised to date for the measurement of electron beam polarization at accelerator energies involve measuring a difference in the scattering rate of electrons in the two possible polarization states. Three different scattering targets have been used—heavy nuclei, magnetized materials, and optical photons from a laser—and the three scattering processes are known as Mott, Møller, and Compton scattering, respectively. Before describing polarimeters based on these scattering processes, it is useful to work through the algebra underlying polarization measurement by scattering.

Consider scattering an electron into a detector by a process which has a spin dependent scattering probability; that is to say a scattering probability which depends on the spin orientation of the incident electron. In general, only a fraction of the total scattering probability depends on the spin orientation, so we split the total scattering probability into two pieces, one spin independent, S_0, and the other spin dependent, AS_0. Thus the probability of scattering an electron with a positive or a negative spin projection into the detector is $S_+ = S_0(1+A)$ and $S_- = S_0(1-A)$, respectively.

Now consider scattering of a beam of polarization P. We assume that we are able to reverse, or "flip" the polarization of the beam in some way, and that on reversal, the number of positive and negative electrons are simply exchanged, i.e., the polarization P is simply changed in sign. It is easy to show that the number of positive and negative electrons in the beam are given by:

$$n_+ = \frac{n_0}{2}(1+P) \text{ and } n_- = \frac{n_0}{2}(1-P), \text{ where } n_0 = n_+ + n_- \tag{2}$$

When the beam polarization is $+|P|$, we detect scattered electrons in the detector, and when the polarization is reversed to $-|P|$, we detect scattered electrons, where:

$$R_+ = S_0(1+A)(1+P)\frac{n_0}{2} + S_0(1-A)(1-P)\frac{n_0}{2}$$

$$R_- = S_0(1+A)(1-P)\frac{n_0}{2} + S_0(1-A)(1+P)\frac{n_0}{2}. \tag{3}$$

A little algebra then shows that:

$$\frac{R_+ - R_-}{R_+ + R_-} = AP. \tag{4}$$

Thus, by measuring the difference in counting rates in a single detector as the polarization is reversed in sign, we can measure the magnitude of the polarization. The counting rate difference is often called the asymmetry, and the quantity A is known as the "analyzing power" of the particular scattering process. Clearly a larger A, which gives a greater difference in the two counting rates, is desirable. It is also worth noting that we could have obtained the same result if we were able to reverse the sign of A, instead of reversing the beam polarization.

In assessing the precision with which the polarization is measured, one needs to consider both statistical and systematic uncertainties. An obvious statistical uncertainty is the counting statistics associated with measuring and . A quick estimate of the number of counts required to obtain a particular statistical error in P can be made by assuming that $AP = 0$. This makes and equal. If we accumulate a total number of counts N, equally divided between the two cases, it is easy to show that the statistical uncertainty in P is P is $\delta P = (1/A)N^{-1/2}$. Thus, for example, if A were 0.10, and one wanted a measurement of polarization with a statistical uncertainty of 0.01, 10^5 counts would be required. More often, it is necessary to measure P to a certain fraction of itself. To measure a P of 0.1 with a statistical precision 3% of itself, again with an A of 0.1, would require over 10^7 counts. Measuring polarization with good statistical precision can require large numbers of counts in practice, and, in some cases, the uncertainty in a polarization measurement is dominated by counting statistics.

In measuring polarization by scattering, it is important that the detector count only electrons scattered by the desired process. Electrons scattered by other processes are "backgrounds" which, in general, have a different, or even no, analyzing power. Good polarimeter designs allow the user to measure how well contributions from background are eliminated, thereby reducing the systematic uncertainties arising from background contributions.

Mott Scattering Polarimeters

In Mott polarimeters, electrons are scattered by the Coulomb field of a heavy nucleus. The scattered electrons have an orbital angular momentum about the scattering nucleus. The scattering probability depends upon whether the electron spin is parallel or antiparallel to this orbital angular momentum. Since the orbital angular momentum is perpendicular to the scattering plane, Mott scattering analyzes only the component of the spin which is also perpendicular to the scattering plane, and thus transverse to the electron momentum. One measures the difference in scattering rate for electrons scattered to the left and to the right. This difference is largest for electrons scattered at large angles from high-charge nuclei.

The calculated analyzing power for Mott scattering from single free-atoms is known as the Sherman function. This is shown in Figure 1 as a function of laboratory scattering angle and electron energy. At the present time, Mott scattering is the only practical way to measure electron beam polarization at the beam energies typical of electron guns (~50 to 100 keV) and electron injectors (a few MeV). Above beam energies of about 10 MeV, the Mott scattering probability is very small, and the scattering angle for maximum analysing power becomes impractially close to 180 degrees.

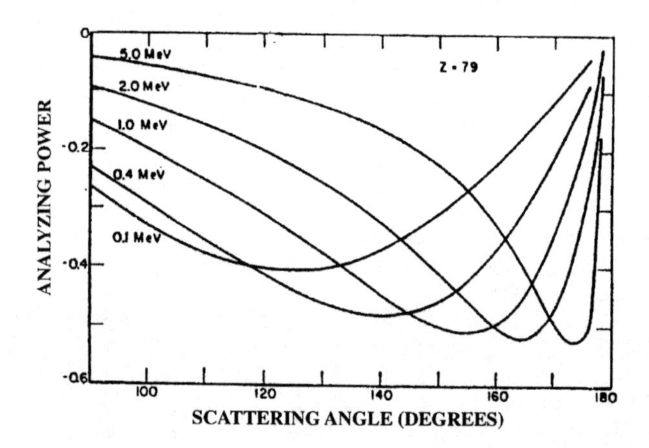

FIGURE 1. The Mott scattering analyzing power for gold as a function of scattering angle and electron energy, from J. Kessler, *Rev. Mod. Phys.* **41**, p. 3 (1969).

Mott scattering probabilities, particularly at lower beam energies, are very large, leading to the use of exceptionally thin scattering targets, and very low-average electron beam currents. Gold is the common scattering target, as it offers a high nuclear charge and is easy to fabricate into very thin targets. Target thicknesses from a few hundred to about 1000 angstroms are typical. Even with such thin targets, multiple and plural scattering is common, leading to substantial uncertainties in the analyzing power of the real target. It is normal to measure Mott scattering asymmetries for a range of target foil thicknesses, and use this information to extrapolate to zero target thickness. It was long believed that the theoretically calculated single atom analyzing power was the correct number to use for the zero target thickness extrapolation. More recently, it has been demonstrated that it is necessary to also verify that the electron has lost no energy in the scattering process for this theoretical analyzing power to be correct.

Small changes in beam steering on the Mott scattering target can cause large changes in counting rates. In low-energy Mott polarimeters, the usable beam current is too small to be monitored, either in position or intensity. To reduce the systematic effects associated with small changes in intensity or beam steering, one normally uses two nominally identical scattered electron detectors, located in the scattering plane at equal scattering angles, left and right, to the incident beam. Defining L_+ as the counting rate into the left detector with $|P|$ positive, with an obvious extension to the other three rates L_-, R_+, and R_-, one can show that:

$$\frac{X_+ - X_-}{X_+ + X_-} = AP \tag{5}$$

where $X_+ = \sqrt{L_+ R_-}$, and $X_- = \sqrt{L_- R_+}$, and A is the analyzing power. The advantage of using two nominally identical detection channels is that the polarization calculated from the above relation is insensitive, in first order, to systematic effects arising from beam steering and intensity fluctuations and target foil inhomogeneities.

Low-energy electrons scattered from heavy nuclei by Mott scattering have lost essentially no energy. The use of scattered electron detectors which give a signal proportional to the detected electron energy thus makes good sense. Such detectors provide one way to discriminate against background electrons which have lost energy, and background photons. At beam energies of a few tens of keV, silicon detectors are a good choice, while at MeV energies, a plastic scintillator makes a good total energy detector. The energy resolution of these detectors is not good enough to assure that the analyzing power of low-energy Mott polarimeters is undegraded by small energy losses.

Mott scattering polarimeters require only a modest vacuum—10^{-6} torr or so. However, vacuum venting and pumpdown must be done with great care to prevent destruction of the exceptionally thin target foils. It is useful to make the vacuum chamber walls and internal components from low-*Z* materials as much as possible, to minimize backscattering from these surfaces. Be, C, Al, and CH_2 are all useful. Collimators internal to the scattering chamber are commonly employed to define the detector acceptance for scattered electrons, and must be designed with care. The old maxim (attributed to Alvin Tollestrup) that, "You can't collimate electrons; you can only make them angry," must be understood and respected. A viewscreen which can be placed in the plane of the target foils is important for both steering and focusing the beam at the target. This is particularly important for low-energy Mott polarimeters, where the beam cannot be otherwise observed. Similarly, a no-target position, followed by a Faraday cup, is very useful in setup.

Until very recently, Mott polarimeters were routinely used only with beam energies no greater than 100 to 120 keV. Mott scattering at significantly higher energies—a few MeV—offers a number of advantages. The total scattering probability is much smaller, which greatly reduces plural scattering, making the results of the foil thickness extrapolation much less uncertain. The basic analyzing power is quite large, ~52%. The small scattering probability allows the use of much higher beam currents, which are easier to monitor. At MeV beam energies, there is typically rf micro-structure on the beam, permitting excellent monitoring of both beam position and current. At a few MeV, optical transition radiation produces a visible beam spot on the target foil. This spot may be measured with a CCD camera, and very small changes in spot size, shape, and position associated with polarization reversal can thereby be detected. It is even practical to consider measuring the beam polarization with foils of differing Z (e.g., Cu, Ag, and Au) to obtain an absolute calibration of the polarimeter.

At Jefferson Laboratory, we have constructed a 5 MeV Mott polarimeter. This is now in routine use for measuring beam polarization at the exit of the low-energy part of the injector. Its design is shown in Figure 2. The scattering angle for maximum analyzing power is ~172.5°, making it easy to incorporate four detectors to measure both transverse components of the polarization. Plastic scintillators coupled to photomultipliers are used as total-energy counters. Internal collimators are installed to assure that each scintillator detects electrons from only the central part of the target foil. The largest difficulty with a polarimeter like this is reduction of backgrounds. The Mott scattering probability is so low that a very large number of beam electrons must transit the target foil to produce a single useful scatter. Dumping the beam electrons may result in high background rates in the detectors. Since these background events must originate from the walls of the vacuum chamber, rather than the target foil itself, they arrive out of time with the good events. This allows time-of-flight analysis for separating signal events from the background. In tests with the Jefferson Lab 5 MeV polarimeter, this time-of-flight rejection has proven quite effective.

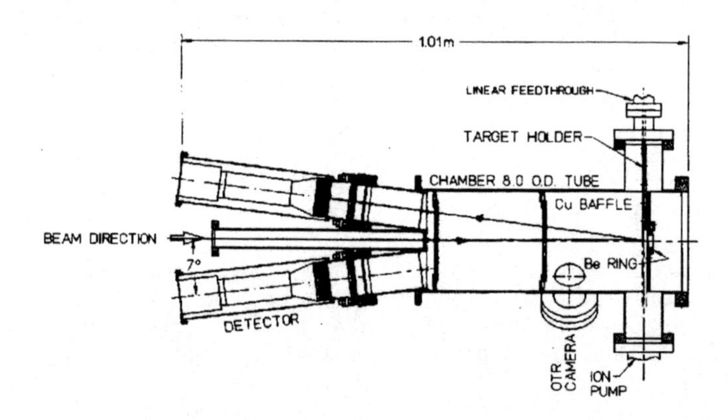

FIGURE 2. A schematic view of the Jefferson Lab 5 MeV Mott polarimeter.

MØLLER SCATTERING POLARIMETERS

Electron polarimeters based on Møller scattering are the "work horse" polarimeters for fixed target experiments at full accelerator beam energies. They have been used for beam energies between ~100 MeV to ~50 GeV. In these polarimeters, the polarized beam electrons are scattered from other polarized electrons in a target. To date, all Møller targets have employed magnetized foils. In such foils, only a small fraction of all the target electrons are polarized, leading directly to a small analyzing power.

The analyzing power of the Møller scattering process is exactly calculable in quantum electrodynamics. At high beam energies, both the analyzing power and the scattering probability in the center-of-mass system become constant, independent of beam energy. The maximum analyzing power for scattering longitudinally polarized electrons on longitudinally polarized electrons is 7/9, for scattering at 90° in the center-of-mass system. Similarly, Møller scattering of transversely polarized electrons can be used to analyze transverse beam polarization; although in this case the maximum analyzing power is only 1/9. These maximum analyzing powers are diluted by the fraction of the electrons in the target which are polarized, so the measurement of transverse beam polarization by Møller scattering is problematic in practice. The analyzing power as a function of center-of-mass scattering angle is shown in Figure 3.

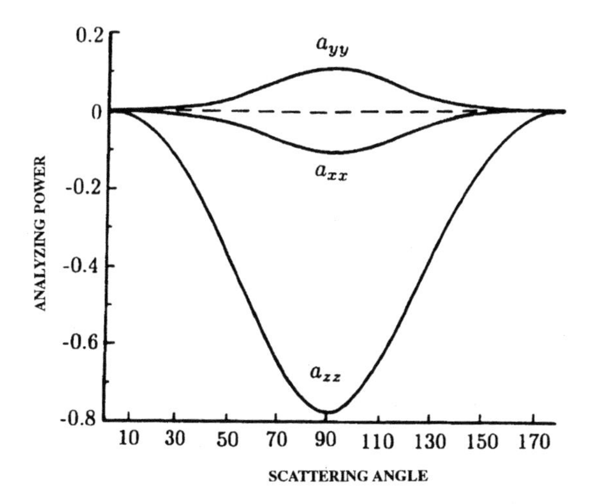

FIGURE 3. The Møller scattering analyzing power for transverse and longitudinal polarization from a single electron, as a function of center-of-mass scattering angle, from B. Wagner et al., *Nucl. Instr. Meth. A*, **294**, 541 (1990).

At the maximum analyzing power, the beam and target electrons are each scattered through 90° in the center-of-mass system. When we do the Lorentz transformation from the center-of-mass system to the laboratory system, the result is two electrons with equal energies (each having half of the incident beam energy), moving at equal and opposite small angles to the incident beam direction in the scattering plane. These facts lead very naturally to the use of magnetic fields to separate the scattered electrons from the beam electrons.

Two different magnet arrangements have been used to separate the scattered and beam electrons. In the first, one or more quadrupoles deflect the Møller scattered electrons to larger angles, while allowing the primary beam to pass through undeflected on the quadrupole axis. In the second, a dipole is constructed with a central magnetic shunt plate. A hole through the shunt plate allows the primary beam to pass through undeflected, while the scattered electrons are deflected by the full dipole field. Hybrid magnet arrangements, using both quadrupoles and dipoles, may also be used. All of these schemes clearly require locating the magnets sufficiently far downstream of the Møller target so that the small scattering angle has separated the scattered electrons adequately from the primary beam. Collimators are often utilized in front of and/or between the magnetic elements and the electron detectors. These serve to restrict the acceptance of the electron detectors to the center-of-mass scattering region with the highest analyzing power.

Once the Møller scattered electrons have been physically separated from the primary beam electrons, they are detected in counters. Counters which give unique signals for electrons, such as lead glass total absorption Cherenkov counters, are commonly used. The two-body kinematics of Møller scattering provides a strong correlation between the electron scattering angle and the electron energy. This correlation can be exploited by using position-sensitive detectors, such as scintillation counter hodoscopes or drift chambers, as part of the scattered electron detection package.

At low-duty-factor accelerators, it is common to detect just one of the two scattered electrons, while at high-duty-factor machines, coincidence detection of both electrons is essentially always employed. Coincidence detection has been employed at low-duty-factor machines, but the necessary low event rate required to obtain low accidental coincidence rates coupled with the relatively low analyzing power of Møller polarimeters means that it can be time-consuming to obtain good statistical precision. With coincidence detection at high-duty-factor accelerators, factors other than counting statistics usually limit the precision of the polarization measurement.

The primary background to Møller scattering is radiative Mott scattering from the Møller target nuclei. Though Mott-scattered electrons have energies very close to the beam energy, radiation allows them to lose enough energy to contribute to the counting rate in single-arm Møller polarimeters. It is possible to substantially reduce the radiative Mott background in coincidence Møller polarimeters through a combination of good coincidence timing, energy selectivity, and careful collimation.

All Møller polarimeter targets used to date have employed magnetized foils of either pure iron, or vanadium permendur. The analyzing power of these targets is not large simply because so few target electrons have their spins oriented. In iron, for example, only two of the 26 electrons per iron atom are spin-oriented in the magnetized material. This fact, along with the maximum single electron analyzing power of 7/9, leads to a net longitudinal analyzing power of about 0.06, and a transverse analyzing power below 0.01. In any of these targets, it is essential to know the relationship between the electron spin polarization in the target and the magnetic field applied to the target.

Two target configurations have been used. In the so-called "easy" magnetization method, the target foil is magnetized in its plane by a relatively low field. These fields, typically ~100 Gauss, are easily provided by Helmholz coils in air. In this arrangement, it is easy to reverse the magnetizing field, and thus the target polarization. The net magnetization in the foil is measured by placing pickup coils around the foil, and measuring the induced flux change with an integrating voltmeter as the magnetizing field is reversed. The foils are oriented at a small angle, typically about 20 degrees, to the beam direction, reducing the effective analyzing power by the cosine of this angle. The use of small magnetizing fields requires the use of magnetic materials which are easily saturated at low fields, such as vanadium permendur. Unfortunately, the relationship between the net electron-spin polarization and the foil magnetization is not well known in these materials, leading to a systematic uncertainty in the analyzing power of these targets. Corrections due to foil end effects and thickness inhomogeneities also lead to systematic uncertainties in the analyzing power. Finally, demagnetization from the beam heating the foil essentially goes undetected, since the area heated by the beam is very small compared to the total area of the foil.

In the "hard" magnetization scheme, the target foil is magnetized by the "brute force" application of a strong magnetic field perpendicular to the plane of the foil. The fields required are very large—several tesla—requiring the use of superconducting magnets. This magnetization scheme allows the use of pure iron foils, which, when magnetically saturated, have a well known net spin polarization. This is a considerable advantage, as it reduces a major source of systematic uncertainty. Furthermore, since the entire foil is fully magnetically saturated, it is not necessary to measure the magnetization in situ and variations in foil thickness are unimportant. Clearly the target magnetization is not easily reversible in this scheme. The depolarization caused by beam heating can, in principle, be measured by the Kerr magneto-optic effect. One observes the rotation of the plane of polarization of laser light reflected from the spot where the beam hits. This technique has yet to be implemented in an operating Møller polarimeter.

Only one hard-magnetization Møller polarimeter has been built to date, by a University of Basel group for use at Jefferson Lab. This is a coincidence polarimeter, employing two quadrupoles and a system of collimators to separate the Møller scattered electrons from both the primary beam and the radiative Mott background. With the systematic uncertainties associated with a fully saturated pure iron foil, and the high counting rate provided by the 100% duty factor CEBAF accelerator, this polarimeter should ultimately be capable of a combined statistical and systematic uncertainty in the measured beam polarization below 1%, and thus is the highest precision Møller polarimeter yet developed. It has an analyzing power of 0.06426 ± 0.00082, with all uncertainties included. The uncertainty in the analyzing power may be further reduced by use of the Kerr effect to measure the target polarization at the point of beam incidence. A schematic view of this polarimeter is given in Figure 4. The power of collimation and coincidence timing to separate Møller scattering events from backgrounds is illustrated in Figures 5 and 6, which clearly show the clean separation of signal and background.

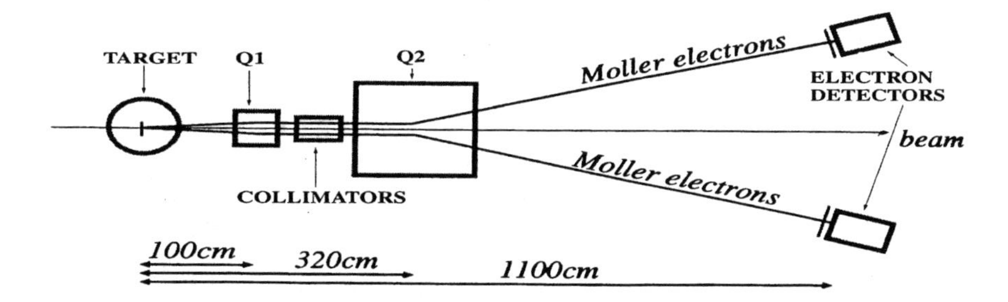

FIGURE 4. A schematic view of the hard-magnetization Møller polarimeter built by the University of Basel group for Jefferson Lab. Two quadrupoles are used to separate the Møller scattered and beam electrons, with a set of collimators between them to reduce backgrounds. The two Møller-scattered electrons are detected in time coincidence.

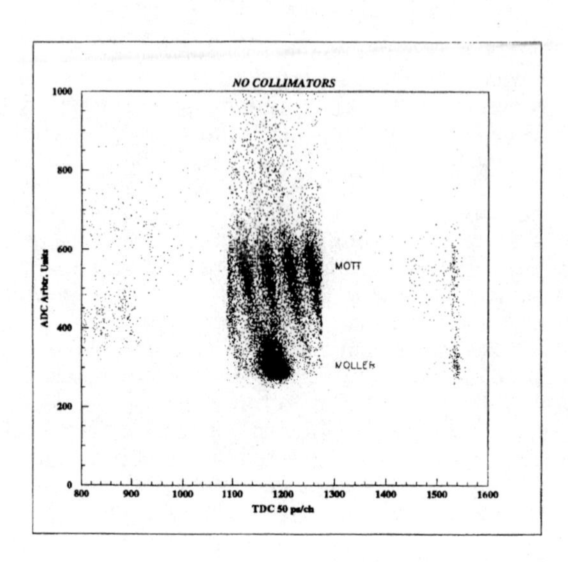

FIGURE 5. A plot of the pulse height versus the event time for one of the Møller electron detectors in time coincidence with the other electron detector, with the collimators fully open. Each scattering event is a dot on this plot. The pulse height is proportional to the detected electron energy. The incident electron beam had a 499 MHz time structure, so the beam bursts are 2 nsec apart, equal to 80 TDC channels. One can clearly see Møller scattered electrons in time coincidence with other Møller scattered electrons, and in accidental coincidence with Mott scattered electrons.

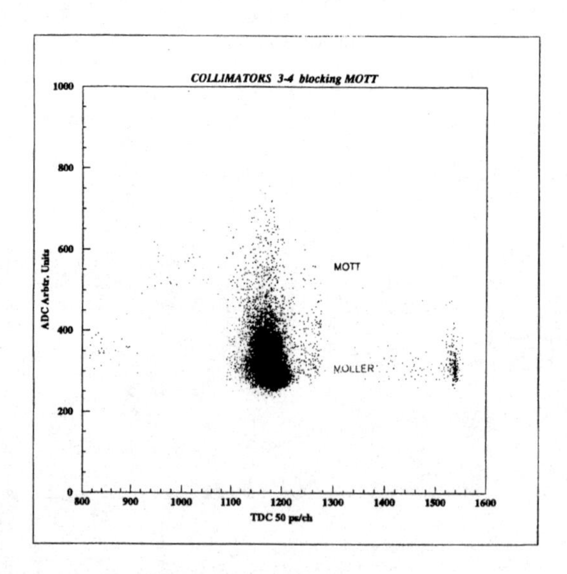

FIGURE 6. The same plot as in Figure 5, with the collimators moved to block the Mott-scattered electrons. The combination of pulse amplitude, time coincidence, and collimation is a powerful tool to separate Møller scattered electrons from backgrounds.

Møller scattering polarimeters had been in regular use for nearly two decades before a very important systematic effect was discovered and understood. This is the Levchuk effect, named after its discoverer. The effect is a result of the momentum carried by the atomic electrons in the Møller target. The tightly bound inner atomic electrons have quite large average momenta, while the loosely bound outer electrons have much smaller momenta. It is the outer atomic electrons which are polarized. The momentum of the atomic electrons broadens the laboratory angular distribution of the scattered electrons. For the case of the inner unpolarized electrons, this effect is quite significant. Depending upon the angular acceptance of the scattered electron detection setup, some beam electrons Møller scattered by the inner atomic electrons may be lost on collimators or fall completely outside the detector. Electrons scattered by the outer atomic electrons are much less likely to be lost by this effect. The net result is that, depending on the details of the detector and collimator arrangement, Møller scattering from the polarized atomic electrons is more likely to be detected than Møller scattering from the unpolarized electrons. This increases the effective analyzing power.

Values for electron beam polarization measured by Møller scattering prior to the understanding of the Levchuk effect are thus suspect. Since the change in the effective analyzing power is apparatus specific, no general statements can be made. The sign of the effect is always to increase the effective analyzing power, and thus decrease the true beam polarization from the measured value. Effects as large as 15% have been reported. Presently, the Levchuk effect is studied during the design stage of a Møller polarimeter by Monte-Carlo methods. The goal is to build a polarimeter in which the Levchuk effect is both small and sufficiently well-modelled that it is not a major contributor to the overall systematic uncertainty of the polarization measurement.

COMPTON POLARIMETERS

Compton polarimeters are the natural choice to measure the polarization of circulating beams in storage rings and stretcher rings, since they are "non-intercepting" devices; they do essentially no harm to the beam itself, so can be left on during accelerator operation. Their use is not restricted to ring applications, however. They have been successfully used with ordinary electron beams, most notably with the SLC at SLAC. In these polarimeters, polarized photons from a laser beam are backscattered by the beam electrons (or positrons). The backscattered photons are twice doppler-shifted, resulting in a laboratory backscattered photon energy distribution with a maximum photon energy given by:

$$E_{\max} = 4\gamma^2 E_\lambda \left(1 + 4\gamma E_\lambda / m\right)^{-1} \tag{6}$$

where E_λ is the energy of the laser photon, γ is the ratio of the electron beam energy to the electron mass m, and $E_{\max}$ is the maximum backscattered photon energy in the lab system. Typical electron beam energies give γ values from a thousand to very much greater. Thus, backscattering a few eV laser photon produces a continuous spectrum of gamma rays with a maximum energy of many MeV .

Compton scattering may be used to analyze either transverse or longitudinal electron polarization. To analyze transverse polarization, circularly polarized laser light is scattered off the polarized electrons. The backscattered gamma rate has a cosϕ dependence, where ϕ is the azimuthal angle of the backscattered gamma with respect to the polarization direction of the electron. This azimuthal dependence is usually measured by observing an up-down asymmetry, with respect to the scattering plane, in the backscattered counting rate. For longitudinal polarization analysis, circularly polarized laser light is scattered off the polarized electrons, and a difference is observed in the counting rate of the backscattered photons, depending on whether the laser and electron polarizations are parallel or antiparallel. The analyzing power for either transverse or longitudinal polarization depends strongly on the electron beam energy and the backscattered photon energy. High electron beam energies and high backscattered gamma energies give the largest asymmetries. This makes it desirable to use some form of energy discrimination in detecting the backscattered gammas.

In a storage ring, the electron and positron beams become transversely polarized (parallel or antiparallel to the magnetic guide field) by the Sokolov-Ternov effect. The ultimate polarization value is approached exponentially in time, and is exactly related to the polarization buildup time constant. This makes it possible to calibrate the analyzing power of a Compton polarimeter by measuring the asymmetry as a function of time. This is a very valuable characteristic of the beam polarization in storage rings.

The physics use of storage ring beam polarization requires longitudinal polarization. Some type of spin rotator must therefore be used in the storage ring to rotate the natural transverse polarization into longitudinal before the physics experimental interaction area, and back to transverse for transport around the ring. These spin rotators can be fairly complex from a beam optics standpoint, and may require a considerable amount of beamline space. In fact, although a number of Compton polarimeters have been built to measure the transverse polarization in storage rings, little use of the natural beam polarization has been made for lack of money and/or beamline space to install the necessary spin rotators. The major exception is the HERA storage ring. Spin rotators have been installed, a transverse Compton polarimeter is used to tune the storage ring for mazimum polarization, and a longitudinal Compton polarimeter is used to measure the beam polarization at the physics interaction point.

In stretcher rings, the beam does not stay in the ring long enough for the Solokov - Ternov effect to give significant polarization. Instead, one injects polarized electrons directly from a polarized source. Some form of "Siberian Snake" spin rotator is used to preserve the polarization in the ring. In this case, the beam polarization is measured with a Mott polarimeter before acceleration and injection into the ring, and the polarization in the ring is measured with a Compton polarimeter.

The laser and the electron beams collide in or very close to a "head on" geometry. The backscattering rate is related to the spatial and temporal overlap between the laser and electron beams. The diffraction limit of the laser beam determines the product of its diameter and divergence angle, and is often a real limitation on the maximum obtainable overlap of the two beams. The emittance of the electron beam is rarely a similar limitation. It is important to choose the interaction point between the laser and electron beams to be at a location where the storage or stretcher ring lattice parameters are optimized for the laser and detector used. For transverse polarimeters at high energies, where small position differences must be observed, this is essential.

Detectors for Compton polarimeters are very similar to detectors for Møller scattering. Counters sensitive to the total energy of the backscattered photon are used. These are usually lead-glass, lead-scintillator, or lead-lucite total absorption counters coupled to photomultipliers. Such counters give a signal amplitude proportional to the absorbed photon energy. It is common to add a "veto" scintillation counter in front of the total absorption counter, to assure that a neutral particle is detected. For measuring transverse polarization, position sensitive information is required. For this, one uses a thin sheet of high-Z material (lead or tungsten) to convert the photon into an electron-positron pair, followed by a high-resolution position-sensitive detector such as a drift chamber. For very high electron beam energies, where the maximum backscattered photon carries away a significant fraction of the incident electron energy, it becomes practical to detect the scattered electron rather than the backscattered photon. This allows a magnetic separation of the lower-energy scattered electron from the beam electrons. In principle one could detect the backscattered photon in coincidence with the scattered electron, though this is yet to be done in any Compton polarimeter.

The polarized photon "targets" provided by lasers are not dense. This fact, and the relatively small Compton scattering probability, lead to modest backscattering rates in Compton polarimeters. The principal backgrounds to Compton scattering are gamma rays produced by bremsstrahlung on the residual gas in the vacuum system, and x-rays produced by synchrotron radiation in magnets close to the interaction point. In designing Compton polarimeters, it is necessary take some care to ensure that the desired backscattering rate is large compared with the backgrounds. Two methods are used to accomplish this. In one, a low-duty-factor laser produces short duration optical pulses of high peak intensity. In this case, the backscattered photons arrive in a "burst", and the detector signal is integrated to obtain a measure of the total backscattered energy. Veto counters are not useful in this case. Q-switched and frequency-doubled Nd:YAG lasers are a common choice for this scheme. This method is essential for use with low-duty-factor electron beams, and may be selected for the case of continuous electron beams in a storage ring. The vacuum pressure in the electron-laser interaction region must be low (10^{-9} mbar or below is a typical requirement) so that bremsstrahlung gamma rays do not contribute significantly to the integrated counter signal. Similarly, the edges of nearby magnets must be magnetically "softened" with low field regions, or the detector must be shielded from line-of-sight to these magnets, to keep the synchrotron x-ray contribution small. These backgrounds may be measured by blocking the laser light.

Alternatively, a CW laser is used to continuously intercept the beam. If the beam current is low, as in accelerators like CEBAF, very high CW optical powers are required. Optical cavities with gains of 10^4 or greater have been proposed as a way to increase the CW optical power to yield adequate counting rates, though operation of these cavities in an accelerator environment is yet to be demonstrated. In this method, one detects individual backscattered photons or possibly scattered electrons. The vacuum requirements may be more demanding than in the pulsed laser case. The backscattered photon energy spectra is obtained, which is useful for discriminating between residual gas bremsstrahlung and Compton scattering. A variant of the CW scheme uses a cavity-dumped CW laser to produce a continuous train of moderate power optical pulses. The cavity-dumped laser choice is particularly appropriate for use with storage or stretcher rings, as the cavity dumping rate is well matched to the bunch revolution frequency in many rings. Argon-ion lasers are a common choice for the CW laser methods.

It is easy to reverse the sense of the circularly polarized light quite rapidly with a Pockels cell. This is an essential requirement for storage ring polarimeters, since the beam polarization is not reversible. However the circularly polarized optical beam is prepared, one must take care to keep the residual linear polarization components of the optical beam small, to avoid systematic effects. It is possible to slightly alter storage ring operating parameters to rapidly and completely depolarize the circulating beam. This latter trick also allows one to make very precise measurements of the storage ring beam energy. These techniques have been pushed to a point at the LEP storage ring where even the effects of Atlantic storms and electric railway currents are detectable!

SUMMARY AND SPECULATIONS

Three scattering processes have been used to date to measure electron beam polarization at accelerator energies. Mott scattering is uniquely suited to the measurement of beam polarization from polarized electron sources, and from the low-energy stages of electron injectors. Only transverse polarization can be measured. Fortunately, it is quite easy to rotate longitudinal polarization into transverse at low energy, so the longitudinal component can be measured fairly directly as well. While blessed with large analyzing power, Mott polarimeters, particularly at the lower energies, are subject to a number of difficult sources of systematic uncertainties. At energies of a few MeV, many of these uncertainties are greatly reduced. It appears possible to construct high-energy Mott polarimeters with well understood analyzing power and minimal systematic uncertainties, allowing high-quality polarization measurements to be made.

Møller polarimetry is the "work horse" instrument for polarization measurement at all fixed target electron accelerators. The effective analyzing power of all present Møller targets, based on magnetized foils, is low. Both longitudinal and transverse polarization can be measured, but the analyzing power for transverse polarization is smaller than for longitudinal by a factor of 7, making transverse measurements statistically challenging at best. With "easy" magnetization targets, there are significant systematic uncertainties in the knowledge of the effective target polarization. Such polarimeters are capable of giving polarization measurements with ~3–4% overall uncertainty. "Hard" magnetization targets have substantially lower systematic uncertainties, though they are moderately more expensive to construct and maintain. This type of polarimeter is capable of polarization measurements with better than 1% overall uncertainty

Compton polarimeters are universally used to measure the polarization of circulating beams in storage and stretcher rings, and are occasionally used in other circumstances. The analyzing power of Compton polaimeters is strongly beam energy dependent, growing with energy. Both transverse and longitudinal polarizations may be measured, with roughly equal ease. The laser "targets" are not dense, and the Compton scattering probability is not large, leading to a requirement for high-average-current electron beams and high-intensity lasers to obtain reasonable scattering rates. Compton polarimeters based on extremely high-gain CW laser cavities are under development for use in low-average-current CW accelerators. The time dependence of the polarization buildup in storage rings may be used to provide an absolute calibration for the analyzing power of Compton polarimeters.

By way of speculation, it has been suggested that the spin dependence of the intensity of synchrotron light may be developed into a circulating beam polarimeter for storage rings. Though very small, the effect has been observed at the VEPP-4 storage ring in Novosibirsk. The size of the effect increases with beam energy, so at higher-energy storage rings, such polarimeters become more practical. As with Compton polarimeters, the time dependence of the polarization buildup provides a calibration for the analyzing power.

Even more venturesome, some have suggested that it may be possible to detect the magnetic effects of a polarized beam directly by using SQUIDs. This certainly sounds exceptionally difficult with present technology. The magnetic fields due to the beam current itself are larger than those from the polarization, and there is presently not even a good SQUID based beam current monitor.

Møller polarimetry would be much more effective if one could develop a target with essentially 100% electron polarization. With much additional development, it might be possible to accomplish this with a jet of fully polarized atoms of hydrogen or helium. It is possible to completely polarize such atoms with optical pumping techniques. The difficulties arise in generating a sufficient target density of polarized atoms, and in avoiding dilution of the analyzing power with unpolarized atoms. Even more far out is the notion of using a low-energy electron beam of well-known polarization as a target for Møller scattering. Unlikely as these ideas seem, no doubt improvements in electron beam polarimetry will be developed in the future.

BIBLIOGRAPHY

General Sources

• *Polarized Electrons*, by J. Kessler, Springer-Verlag, Berlin, 1985, is a general monograph on polarized electrons and polarimetry, although the emphasis is on low-energy phenomena.

• The motion of the electron spin in traversing electromagnetic fields is given in full generality in V. Bargmann, L. Michel, and V. L. Telegdi, *Phys. Rev. Lett.*, **2**, p. 435 (1959).

• An International Symposium on High Energy Spin Physics has been held every second year since 1976. The proceedings of these symposia have been published in book form, and contain a wealth of information on polarized beams and targets, polarimetry, and the physics done with these beams. There are often workshops on various polarization and polarimetry topics held in association with these conferences. Electron Beam polarimetry has been addresses in several of these workshops.

• There is an established series of workshops held every other year, between the Spin Physics symposia above, known as the International Workshop on Polarized Gas Targets and Polarized Beams. The proceedings of these workshops are occasionally published, and although they deal primarily with the technology of polarized hadron beams, targets, and polarimeters, there is some information on polarized electron sources and polarimetry.

• The Particle Accelerator Conference, held every other year in the US and Canada, typically has one or two sessions dedicated to topics associated with the development and delivery of polarized beams and polarimeters. The proceedings of these conferences are published by the IEEE.

Mott Polarimetry

• There have been a number of calculations of the analyzing power of Mott scattering from a single atom, known as the Sherman function. The original paper is N. Sherman, *Phys. Rev.* **103**, 1601 (1956). More recent calculations are mentioned in the references below.

• T. J. Gay and F. B. Dunning, Rev. *Sci. Instrum.* **63**, p. 1635 (1992), is an excellent and thorough review of low-energy Mott polarimeters.

• T. J. Gay et al., *Rev. Sci. Instrum.* **63**, 114 (1992), presents a careful discussion of the systematic effects associated with target thickness extrapolations in low-energy Mott polarimetry.

• S. Mayer et al., *Rev. Sci. Instrum.* **64**, 952 (1993), describes a method for high precision calibration of low-energy Mott polarimeters.

• G. Mulhollan et al., in *Proceedings of the 1995 Particle Accelerator Conference*, Dallas, 1995, IEEE, p. 1043, describes an effort to cross-calibrate low-energy Mott polarimeters in use at a number of laboratories world-wide.

• D. Conti et al., in "Polarized Gas Targets and Polarized Beams," *AIP conference Proceedings*, **421**, AIP, New York, p. 326, is an excellent discussion of Mott polarimetry at high beam energies.

• J. S. Price et al., in *Proceedings of the 12th International Symposium on High Energy Spin Physics*, C. W. de Jager et al., eds. *World Scientific*, Singapore, p. 727, describes the development of a 5 MeV Mott polarimeter. Further work on this polarimeter is presented in J. S. Price et al., "Polarized Gas Targets and Polarized Beams," *AIP Conference Proceedings*, **421**, AIP, New York, p. 446.

Møller Polarimetry

• P. S. Cooper et al., *Phys. Rev. Lett.* **34**, p. 1589 (1975), is the original experiment which demonstrated that Møller scattering could be used to analyze the polarization of high energy electron beams.

• B. Wagner et al., *Nucl. Instr. Meth. A*, **294**, p. 541 (1990), is a description of the first "three axis" Møller polarimeter.

• L. G. Levchuk, *Nucl. Instr. Meth. A*, **345**, p. 496 (1994), is the original presentation of the Levchuk effect in Møller polarimetry.

• M. Swartz et al., *Nucl. Instr. Meth. A* **363**, p. 526 (1995), is a thorough experimental study of the Levchuk effect with a particular polarimeter.

• Both single-arm and coincidence Møller polarimeters have been employed at low duty-factor accelerators. Examples are described briefly by H. R. Band (single arm) and A. Feltham and P. Steiner (coincidence) in *Proceedings of the 12th International Symposium on High Energy Spin Physics*, C. W. de Jager et al., eds., World Scientific, Singapore, p.765 and p.782, respectively.

• The development of the "hard" magnetization Møller polarimeter for Jefferson Lab has been done by a number of students at the University of Basel. This work is briefly described in *Proceedings of the 12th International Symposium on High Energy Spin Physics*, C. W. de Jager et al., eds., World Scientific, Singapore, p. 768. Until this work is published in greater detail, the theses of M. Loppacher and S. Robinson contain much valuable material.

Compton Polarimetry

• The various asymmetries in Compton scattering of polarized photons from polarized electrons are worked out in full detail in F. Lipps and H. A. Tolhoek, *Physica* **20**, p. 85 (1954) and **20**, p. 395 (1954). The expression useful for longitudinal Compton polarimeters is given in S. B. Gunst and L. A. Page, *Phys. Rev.* **92**, p. 970 (1953).

• D. B. Gustavson et al., *Nucl. Instr. Meth.* **164**, p. 177 (1979), is the first demonstration of a Compton polarimeter, used to measure the transverse polarization of the circulating positron beam in the SPEAR storage ring.

• D. P. Barber et al., *Nucl. Instr. Meth. A*, **329**, p. 79 (1993), is a thorough discussion of the transverse compton polarimeter built at HERA.

• J. R. Johnson et al., *Nucl. Instr. Meth.* **204**, p. 261 (1983), describes the first precision measurement of storage ring beam energy using the resonance depolarization technique.

• The use of a Compton polarimeter for precision storage ring energy measurement has been fully exploited at LEP, as reported by A. Blondel et al., *Proceedings of the 12th International Symposium on High Energy Spin Physics*, C. W. de Jager et al. eds.,

World Scientific, Singapore, p. 267. Similar work at HERA is reported by F. Zetsche et al. in the same proceedings, p. 846.

• W. Lorezon, in Polarized Gas Targets and Polarized Beams, *AIP conference Proceedings*, **421**, AIP, New York, 1998, p. 181, describes the longitudinal compton polarimeter in use with the HERMES experiment at DESY.

• A proposed polarimeter employing a very high gain optical cavity is described by J. P. Jorda in *Proceedings of the 12th International Symposium on High Energy Spin Physics*, C. W. de Jager et al., eds, World Scientific, Singapore, p. 791.

Speculations

• A. V. Airapetian et al., in *Proceedings of the 12th International Symposium on High Energy Spin Physics*, C. W. de Jager et al., eds, World Scientific, Singapore, p. 762. This paper reports on a proposed synchrotron light polarimeter for HERA. The spin dependence of synchrotron light was originally observed in the VEPP-4 storage ring by Belomesthnykh et al., *Nucl. Instr. Meth.* **227**, p. 173 (1984).

Measuring the Proton Beam Polarization

Yousef Makdisi

Brookhaven National Laboratory, RHIC Project, Upton, NY 11973

Abstract. Polarimeters are necessary tools for measuring and maintaining the beam polarization during the acceleration process. They serve, as well, as a yardstick for performing spin physics experiments. In this paper, I will describe the principles of measuring proton beam polarization and the techniques that are employed at various energies. I will use as a guide the design work for the Polarized Proton Project at the Relativistic Heavy Ion Collider (RHIC) which is under construction at Brookhaven National Laboratory.

INTRODUCTION

A polarimeter is a tool that measures the beam polarization, that being the degree of alignment of the spins of an ensemble of protons in a beam with respect to a specific direction. This is generally done by sampling the spin projection of a large number of particles in a scattering process that is sensitive to the spin direction.

Polarmeters are used at various stages in an accelerator primarily as diagnostic tools to maintain the polarization of the beam during acceleration, storage, and beam transport. Of course, the goal is the physics measurement and polarmeters are often employed as a first stage of an experiment to measure the polarization of the beam impinging on the target. This normalizes the final result and the error in the polarization measurement has a direct statistical impact on the data. While the accuracy from electron beam polarimeters is now well below 5%, this represents a lofty goal for proton beam polarimeters and is the desired goal for the polarized proton project at RHIC.

BEAM POLARIZATION

The degree of beam polarization along a certain direction is defined as:

$$P = \frac{N_{\uparrow} - N_{\downarrow}}{N_{\uparrow} + N_{\downarrow}} \tag{1}$$

CP451, *Beam Instrumentation Workshop*
edited by R. O. Hettel, S. R. Smith, and J. D. Masek

where $N_\uparrow$ and $N_\downarrow$ are the number of protons with spins parallel and antiparallel to the desired direction. In an accelerator, the stable spin direction is usually transverse to the momentum vector and along the vertical. The task is to find proton-induced reactions that are sensitive to the spin alignment of the beam. This is often a nuclear reaction in a plane that is perpendicular to the beam polarization direction and is usually sensitive to the spin-spin or the spin-orbit interactions between the incoming proton beam particle and the target nuclei. The yield then depends on whether the beam is polarized up or down and is reflected in the number of scatters or events of particular interest measured in the apparatus.

Typically one tries to measure the number of events with the beam polarized up versus those with the beam polarized down, making sure that the experimental conditions remain unchanged. The resulting beam polarization is:

$$P = \frac{1}{A}\frac{n_\uparrow - n_\downarrow}{(n_\uparrow + n_\downarrow)} \tag{2}$$

where $n_\uparrow$ and $n_\downarrow$ refer to the number of scatters, properly normalized, with the beam polarization up and down respectively. A is a new term called the "analyzing power" of the reaction. This is a measure of the sensitivity of the reaction to the beam polarization, a number that varies between 0 and 100%. Of course the above equation assumes that one can flip the beam polarization without any other changes.

Similarly, an apparatus that can measure the number of scatters to beam-left, n_L, and beam-right, n_R, simultaneously will determine the degree of up and down beam polarization independently and one substitutes n_L and n_R in the above equation. In general both methods are utilized and combined in order to reduce systematic errors and potential dependence on geometrical effects.

The associated statistical error in these measurements is a function of both the analyzing power A and the total number of events N (($n_\uparrow + n_\downarrow$) or ($n_L + n_R$) or the sum of all four terms if a combined measurement is performed):

$$\Delta P = \frac{1}{A}\frac{1}{\sqrt{N}}. \tag{3}$$

As an example, for a 5% statistical precision in the beam polarization measurement and using a reaction with an analyzing power of 10%, one needs 4×10^4 events. It is therefore important to utilize reactions with large analyzing power as well as large cross section. The latter is a measure of the frequency of the reaction given a certain number of protons impinging on a specific target. In the design of a polarimeter, the quantity to optimize is the product:

$$N \times A^2. \tag{4}$$

THE SALIENT FEATURES OF A POLARIMETER

In general a beam polarimeter should serve as the following:

- A polarization monitor capable of several samples over a reasonable period of time.
- A diagnostic tool that can be used on demand with a turn key operation.

- A tool for machine tuning with measurements and feedback provided within a few minutes.

The beam polarimeter should have:

- A large dynamic range for dealing with the acceleration cycle (e.g., at RHIC the range is from 25 GeV at injection to 250 GeV at top energy).
- A large analyzing power, high cross section and low background contamination to the physics process.
- A reasonable cost.

It is not always possible to combine all these features and certain compromises are sometimes necessary.

Proton Beam Polarimeter Reactions, the Empirical Way

Unlike electromagnetic reactions (see the presentation by C. Sinclair on electron beam polarimeters in these proceedings), the analyzing power and cross section of these proton-induced nuclear reactions are not precisely calculable, especially at high energies. Thus one reverts to experimental results. These measurements are generally done with polarized proton targets the polarization of which is well-measured using NMR techniques and Masers. Until recently, these targets were not pure hydrogen, thus the target material presents undesired background, especially in inclusive measurements. However, this is not an impediment for exclusive reactions such as p-p elastic scattering or p-carbon scattering when both outgoing particles are detected to provide good kinematic constraints.

These methods have been applied at relatively low energies below beam momenta of 12 GeV/c with good results (3%–5%) and can in turn be used to measure the beam polarization to comparable statistical accuracy. This was done for the polarized proton project at the Argonne National Laboratory ZGS. At higher energies, the experiments are more difficult and the analyzing power becomes increasingly small, resulting in reduced accuracy.

At Brookhaven National Laboratory, an effort is underway to equip the RHIC with the capability to accelerate, store, and collide polarized proton beams at high luminosity. Polarimeters will be deployed to cover various energy ranges: 200 MeV, 4–24 GeV, and 24–250 GeV. I will discuss the choice of the physics processes and associated apparatus for each. But first, a few words about the RHIC project at BNL.

THE RELATIVISTIC HEAVY ION COLLIDER

The RHIC is a new project that is currently under construction at BNL with a target date of completion in June 1999. For heavy ions, the collider uses the Tandem-Booster-AGS as an injector to accelerate, store, and collide beams of various species from light ions to gold. The design luminosity for colliding gold beams is 2×10^{26} cm^{-2} sec^{-1}. The first round of experiments that are slated to receive beam include two large detectors, STAR and PHENIX, and two smaller detectors, BRAHMS and PHOBOS. The physics goal is to study a new state of matter under conditions of extreme temperature and pressure, the quark gluon plasma.

The polarized proton capability utilizes a polarized H^- source and the 200 MeV Linac-Booster as injectors to the AGS. The funding is provided in collaboration with the RIKEN, Japan, to equip the collider with the necessary hardware. This includes "Siberian Snakes," two in each ring, to preserve the beam polarization; two sets of spin rotators around the STAR and PHENIX experiments to orient the beam spins in the longitudinal or transverse directions depending on the physics demands; and beam polarimeters to tune the machine and measure the beam polarization during the 10-hour beam store. The goal is to study the contribution to the proton spin from the constituent quarks and gluons.

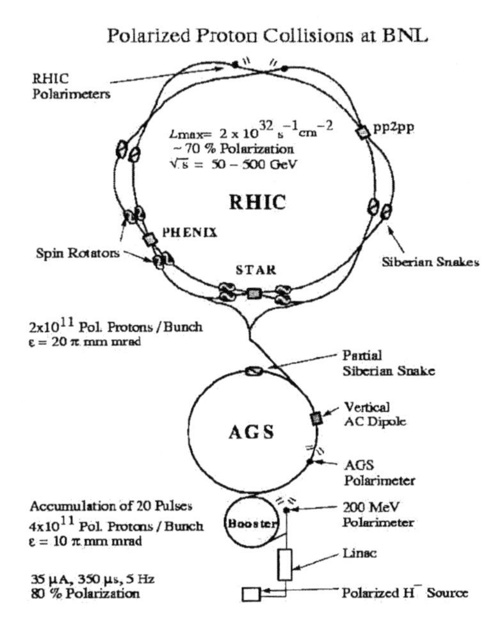

FIGURE 1. Polarized proton collisions at BNL.

The parameters that bear upon this discussion of polarimeters are the polarized beam intensities at the exit of the linac of 10–15 μA over a 300 μsec duration in the AGS, an approximately 2×10^{10} particle circulating beam, and a design goal of 2×10^{11} polarized protons per bunch in the RHIC, with a bunch spacing of approximately 100 nsec. A schematic of the complex is shown in Figure 1 in which the spin-related hardware and polarimeter locations are shown.

POLARIMETERS VERSUS ENERGY

The 200 MeV polarimeter

This polarimeter operates at the exit of the 200 MeV linear accelerator (1). At this energy, we have a large body of experimental data at our disposal. Experiments were carried out with polarized targets or using double scattering to infer the polarization of the outgoing particles. The cross sections are large and the analyzing power is appreciable, reaching unity in some cases (Figure 3). The task is easy and the apparatus is straightforward. The polarimeter of choice scatters polarized protons from a carbon target filament. The detectors are scintillators viewed by fast phototubes (Figure 2). The beam polarization is vertical and we measure the scattering in the horizontal plane. Two identical left/right spectrometers, subtending angles of 12 and 16 degrees in the laboratory frame on either side of the beam, measure the left/right scattering of the proton beam. A third spectrometer looks at 12 degrees in the vertical direction and serves two purposes: to measure any beam polarization component in the horizontal direction perpendicular to that plane, and to monitor intensity of the accumulated up versus down data. The polarization at the source is flipped on alternate pulses and the data of the two polarization states are accumulated simultaneously.

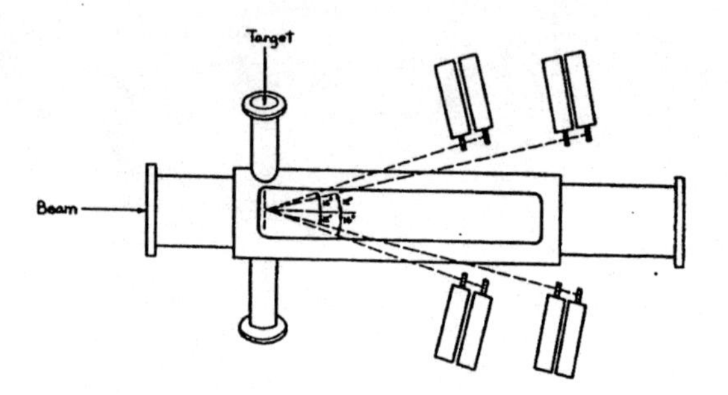

FIGURE 2. The 200 MeV polarimeter.

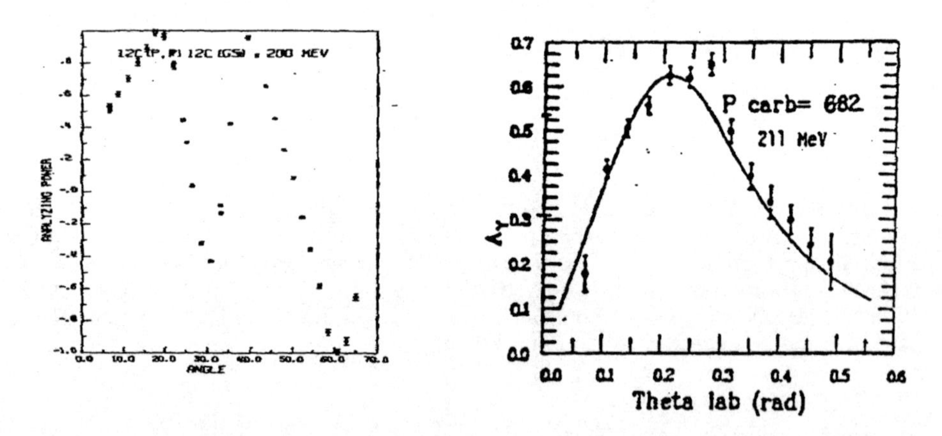

FIGURE 3. The analyzing power versus scattering angle in p-carbon elastic and inclusive production at 200 and 211 MeV, respectively.

The initial intention was to select the elastic scatters by inserting absorbers to uniquely define the scattered proton energies and angles. Later we settled on the inclusive measurement, at the expense of lower analyzing power (62% at 12° and 51% at 16°) but higher cross section. A 2% statistical measurement is achieved in about 2–3 minutes. The polarimeter was calibrated using the polarized beam at IUCF.

The AGS polarimeter

The polarimeter in the AGS (2) operates between the booster injection energy of a few GeV to the extraction energy at 24 GeV. The peak in the analyzing power in p-p elastic scattering at low four-momentum transfer values of 0.1–0.3 GeV^2/c^2 is utilized. As mentioned earlier, this has been measured precisely using polarized proton targets to better than 5% up to energies of 12 GeV. Beyond that, the cross sections are lower and the data get worse. The available measurements (Figure 5) were fitted and parameterized (3). Thus the asymmetry is known to about 10% and falls off with increasing beam energy as (~1/P). At the AGS energy the analyzing power ranges between 5 and 2%.

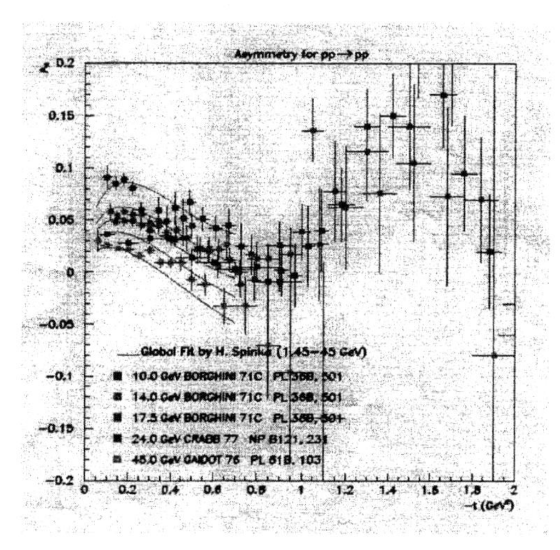

FIGURE 4. The analyzing power in p-p elastic scattering.

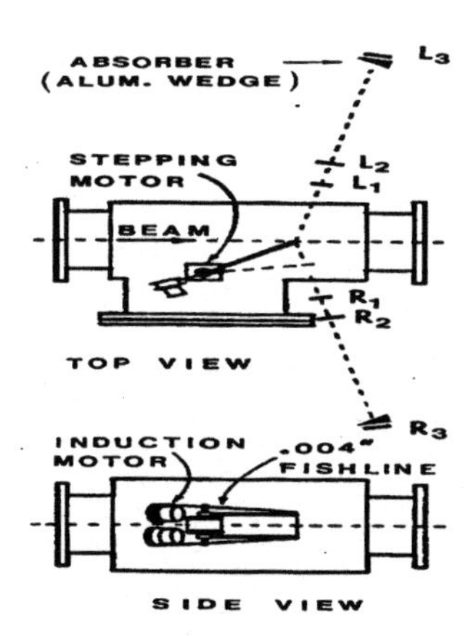

FIGURE 5. A schematic of the AGS polarimeter.

A typical p-p elastic polarimeter has two arms to simultaneously measure the forward scattered and recoil protons. However, the physical constraints in the AGS, the vacuum pipe and the 10 ft long straight sections, do not permit the placement of the forward arms that subtend a few degrees from the beam line. The measurement is done with the recoil proton at kinetic energy of approximately 500 MeV and scattering angle of approximately 76 degrees with respect to the beam direction (Figure 5). Time-of-flight, energy-range, and energy-angle correlation is utilized to select the p-p elastic scattering reaction from the multitude of inelastic scatters. The detectors are scintillation counters viewed by extremely fast phototubes. Similar to the 200 MeV polarimeter, this allows for relatively fast retrieval of the data with minimal on-line computer processing time. It is a counting experiment of the data that passed the trigger cuts.

The environment in the AGS is extremely harsh. The counting rates are high, necessitating debunching the beam during the polarization measurement. Pileup leads to accidental rates of the order of 10–20% depending on the energy. A nylon fish line target is used which is spooled at a rate of 100 cm/sec and flipped in and out of the beam in order to avoid damage due to heating or localized radiation. This requires a complicated computer controlled target mechanism. The loss of a target results in several hours downtime. A 1% statistical measurement takes from 5 minutes at low energy to 20 minutes at the higher energies. A second carbon target is periodically substituted for the fish line in order to measure and subtract the carbon contamination in the fish line. This apparatus serves as a relative polarimeter to tune the 50 imperfection and four intrinsic and depolarizing resonances in the AGS. For physics experiments using extracted AGS beams, this polarimeter is then calibrated against a separate external polarimeter that measures the fully constrained p-p elastic scattering process using a polarized proton target. Those in turn calibrate simple local polarimeters in each experiment.

Polarimeters at high energies (RHIC)

As the beam energy increases, the reach of the AGS-type polarimeter becomes increasingly difficult. Thus one has to revert to different reactions.

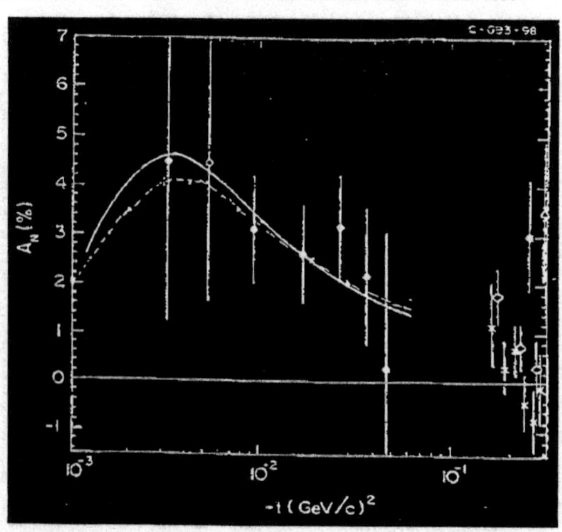

FIGURE 6. The CNI asymmetry in p-p elastic scattering.

An interesting possibility is to utilize the analyzing power in the Coulomb nuclear interference (CNI) region ($10^{-3} < t < 10^{-2}$) in p-p elastic scattering which arises primarily from the interference between the real electromagnetic helicity-flip amplitude and the imaginary hadronic helicity-nonflip amplitude. Unlike the previous process, this is calculable and the analyzing power is a respectable 5%, which appears to be energy independent. However, the hadronic interaction need not conserve helicity in the small t region and the inclusion of the single-flip hadronic amplitude may be relevant. The associated uncertainties with the latter render this measurement less certain. The precision of experimental data is good to approximately 15% (Figure 6).

Other calculable reactions are in the domain of proton-electron elastic scattering with both the proton and electron beams polarized. Polarimeters based on these processes, while experimentally challenging, may be quite feasible.

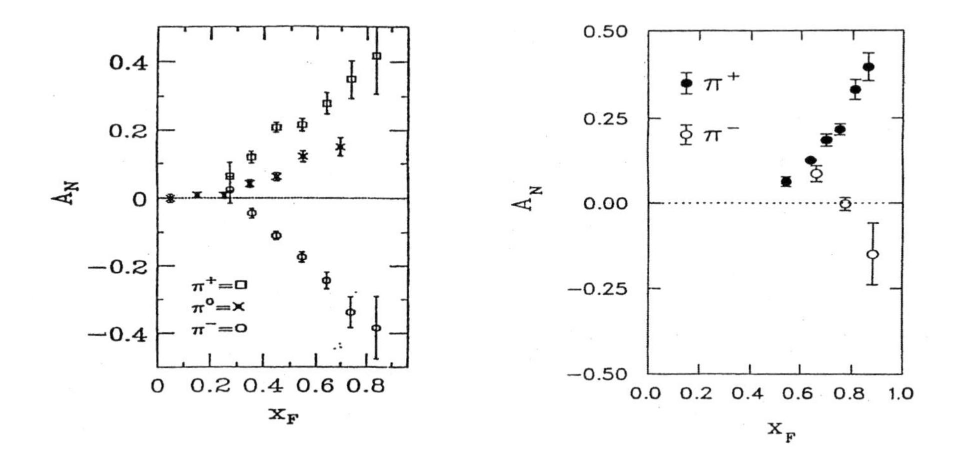

FIGURE 7. Asymmetries in pion production measured at Fermilab and the ZGS.

Experiments using polarized proton beam scattering from unpolarized hydrogen targets, carried out at 12 GeV/c at the ZGS (4) and at 200 GeV/c at Fermilab (5), observed large left-right asymmetries reaching over 30% in the inclusive production of pions (Figure 7). The copious production of pions and the associated large analyzing power seem to be ideally suited for the polarimeter application. The data and associated systematics provide a 10% absolute measurement which may be good for beam commissioning and the early physics results.

In what follows, I will sketch in some detail the steps in the design of this pion polarimeter for RHIC including the conditions that dictated the choice of parameters.

1. From the Fermilab data, one can parameterize both the A and the cross section data in terms of the kinematic variable x_f (~ the ratio between the scattered pion momentum to the beam momentum) and then optimize the value A^2N in order to determine the ideal conditions. For $p_t > 0.7$ GeV/c this resulted in x_f~0.5 and a respective analyzing power of 15%.
2. Choose an acceptable transverse momentum of the outgoing pion (p_t= 0.8 GeV/c), which determines the scattering angle, the geometry, and the production cross section. The scattering angle ranges from 64 mrad to 6.4 mrad as the energy is raised from 25 to 250 GeV.

3. The cross section also grows linearly with energy requiring an apparatus with variable acceptance. Having defined this, one then proceeds to calculate the time required for one measurement with a specified precision. For a 7% statistical measurement that is comparable to the experimental data at 15% analyzing power one requires 10^4 scattered pions in the detector.
4. The required accuracy in the measurement of the scattered pion momentum of ~1% sets the need on the magnet rigidity ~8 T-m. This and the desire to measure the left and right scattering dictates that the magnets and detectors fit within a 1-meter lateral dimension between the two RHIC beam pipes. This led to an ingenious 4 magnet toroidal, 30-meter design that satisfied these requirements. A $\pm 7^{\circ}$ wedge gap increased radially outwards which provided the variable acceptance. The magnets were independently powered to accommodate the continuous energy range. The magnet excitations for various beam energies are shown in the table below. The polarimeter layout is in Figure 8.

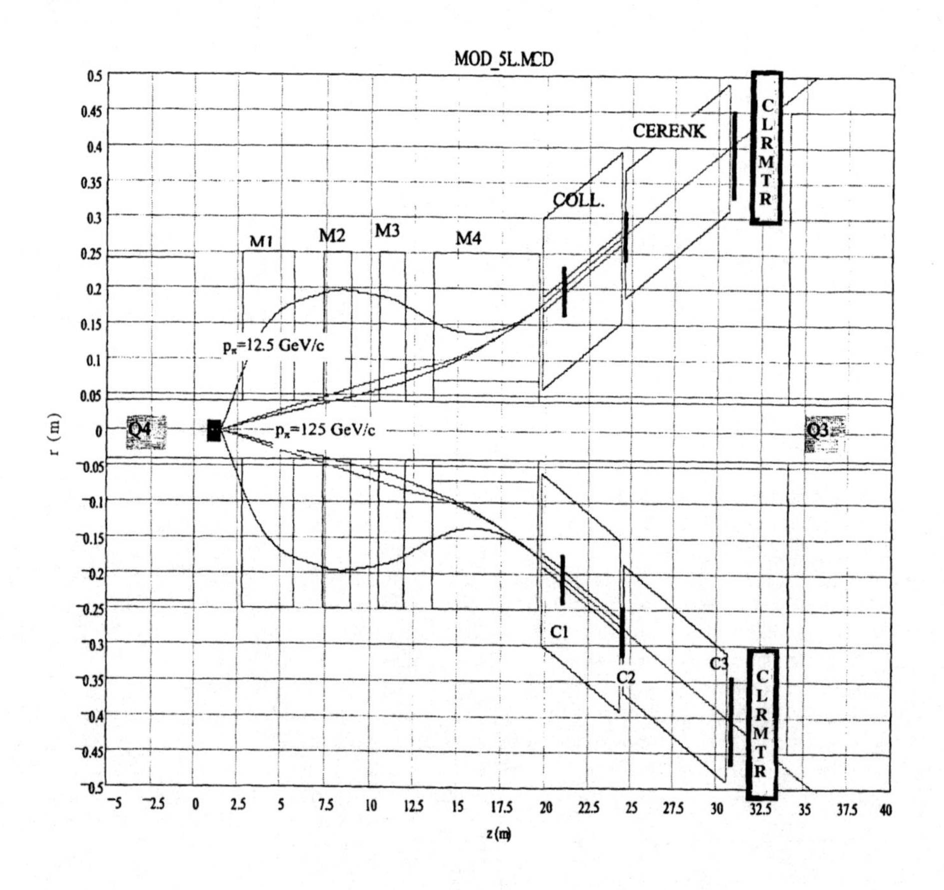

FIGURE 8. A schematic layout of the RHIC polarimeter and detectors.

TABLE 1. Polarimeter Element Lengths and Magnet Excitations

	TARGET	M1	M2	M3	M4	COLL	CRNKV
Length (M)	0	3	1.5	1.5	6	4.6	6
Position (m)	1.5	2.75	7.35	10.45	13.55	19.75	24.55
23 GeV/c		–1	–1	–0.859	+0.3743		
100 GeV/c		0.2	–1	–0.9576	+0.7706		
200 GeV/c		0	0	–0.305	+0.945		
250 GeV/c		0	0	+0.95	+0.95		

5. The carbon target heating was calculated, taking into account the beam intensity per bunch and the number of bunches in each ring (Figure 9):

$$T_{equ.} = \left(\frac{\varepsilon_h f_r N_p \dfrac{dE}{dx} \rho d}{8\pi\sigma_x \sigma_y \varepsilon_{rad} \sigma_{SB}} + T_0^4 \right)^{1/4} \tag{4}$$

where N_p = number of protons per fill, ε_{rad}= 0.8 (emissivity), f_r = 78 kHz (RHIC beam rev. freq.), σ_{SB}= Boltzmann const, dE/dx = 1.78 MeV/gm·cm^2 (energy loss), ε_h= 0.3 (heating efficiency), $\sigma_{x,y}(p)$= rms beam width at momentum p, and ρ =1.75 gm/cm^3.

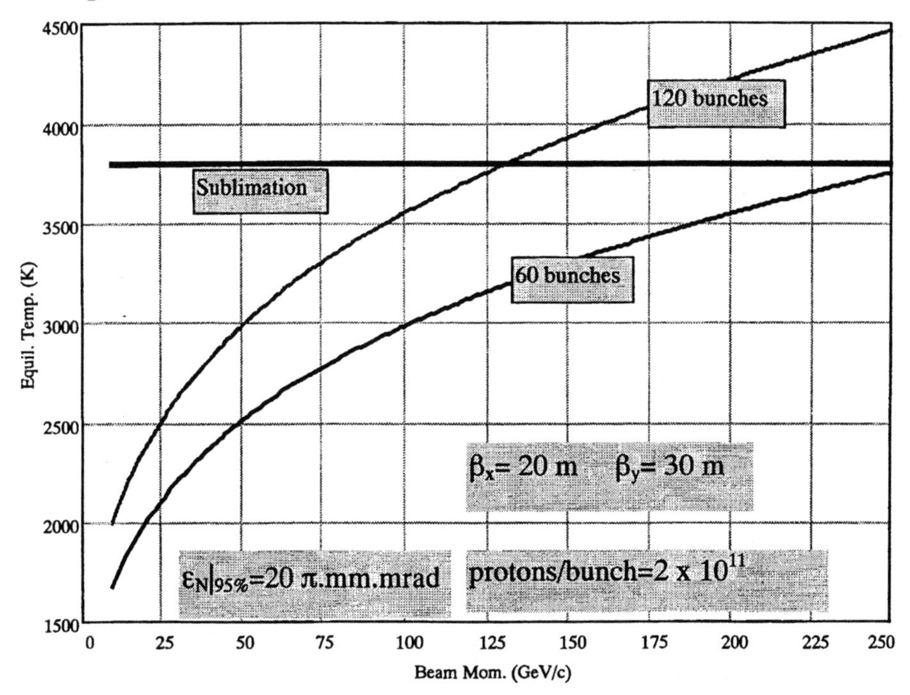

FIGURE 9. Heating of a 5-micron target in the RHIC polarimeter.

The 5-micron-thick target reaches the sublimation temperature at 125 GeV. This necessitates flipping the target in and out of the beam to allow cooling. A lighter filament, 20 μgm/cm², has been developed at the Indiana University Cyclotron Facility and is being regularly used as a target. Calculations indicate that it can survive beam heating at all energies (Figure 10). Calculations of the expected pion production rates for these two fiber configurations are shown in Table 2. The rates are quite reasonable even in the lighter target. One also needs to calculate the survival time for such a thin ribbon due to knock-out of its atoms. This appears to be a few hours of continuous beam bombardment. Thus the target design will allow for multiple ribbon configuration. The emittance growth in the worst case with a 5-micron target at 25 GeV energy is approximately 1 π·mm·mrad, which is acceptable.

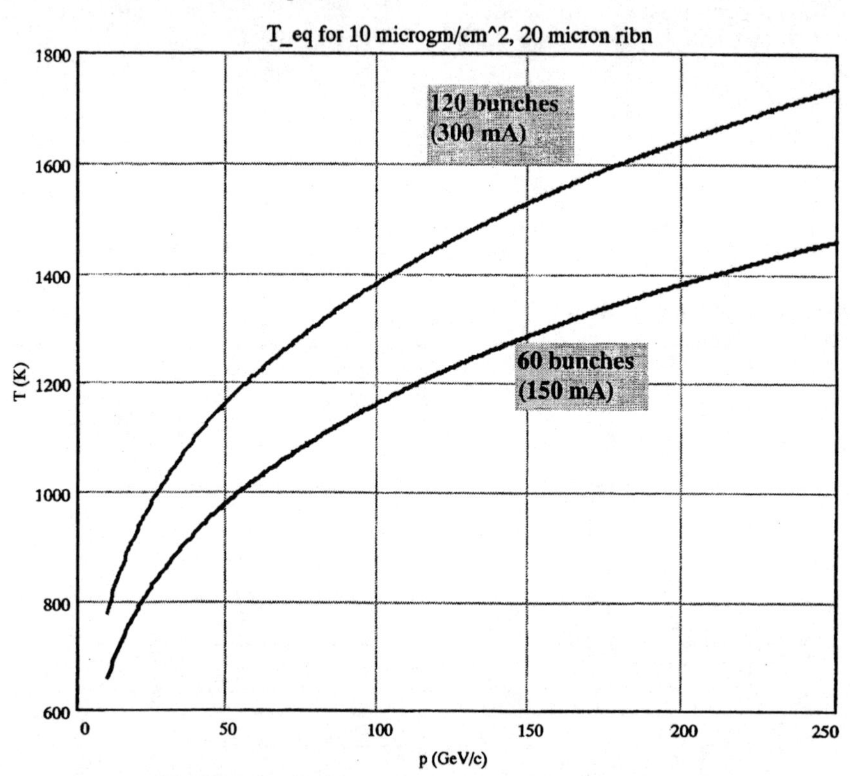

FIGURE 10. The expected heating of the ribbon target.

TABLE 2. Comparison of Expected Rates in the Kinematic Range: x_F=0.5, p_T=0.8 GeV/c, δx_F= +/–0.05

Target	P_{beam}(GeV/c)	Luminosity ($cm^{-2}s^{-1}$)	V (m/s)	Nπ/bunch	τ_{meas} (sec)	T_{max} (K)
5 μm fiber	250	3.9×10^{36}	4.5	4×10^{-2}	60	1700
	25	1.2×10^{36}	1.4	2×10^{-3}	130	1700
10 μg/cm², 20 μm wide ribbon	250	1.8×10^{35}	0	2×10^{-3}	0.8	1740
	25	5.6×10^{34}	0	6×10^{-5}	1.8	1000

From these simulations, it appears that the ribbon would reduce the rates per bunch crossing to quite comfortable levels. If technically feasible, a 10 μgm/cm², 20 μm-wide ribbon target would allow π^+ measurement with the present collimator arrangement.

6. The use of the carbon target presents another dilemma. Does the production from a nuclear target dilute the observed asymmetry from hydrogen? Is the asymmetry large enough at lower energies? The available negative pion data at lower energy is not similar to that at higher energy. The above questions led us to carry out a special measurement of the pion asymmetry, using the AGS polarized proton beam on a carbon target at 22 GeV. The preliminary results, shown in Figure 11, are quite promising. The numbers should be scaled (divided) by the average beam polarization, measured to be approximately 30% over the run. Thus it appears that using negative pion production is a strong possibility. This simplifies the polarimeter as no particle identification (Cerenkov counter) is needed with negative pions. The expected background is negligible.

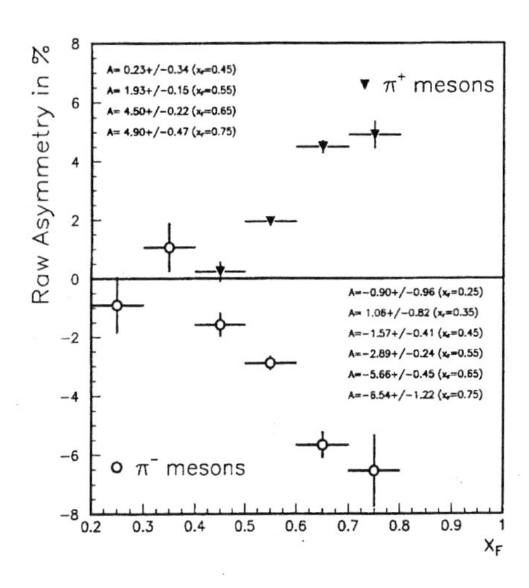

FIGURE 11. Preliminary results from AGS experiment E925 on asymmetries in charged pion production from a 22 GeV polarized proton beam on a carbon target.

7. This being the case, we turn our attention to computer simulations to determine the expected counting rate of charged particles from all sources (real pion scatters, scattering from the vacuum pipe, scattering from the magnet pole faces, neutral particle conversion, etc.) in the detectors. Ideally, and to avoid confusion, one would like no more than one particle in the detector per bunch crossing. This also determines the granularity of the detectors and the type that can be used.
8. Finally, the data acquisition and trigger systems have to be chosen to make sure a) that the correct scattering process is accepted, b) there is minimum dead time resulting after a trigger is accepted, and c) the data is processed and accumulated in a pipeline within the time spacing between bunches.

REALITY CHECK

The estimated cost of the above design for two polarimeters was approximately $1.5M. The current fiscal status of the project mandated a rethinking and simplifying of this concept. We are in the process of designing a single arm inclusive pion polarimeter for one ring using five existing conventional dipoles and scintillation detector hodoscopes from the E925 experiment at the AGS. This will serve during the beam commissioning phase, which we hope to carry out during the first year of RHIC running in FY 2000. The polarimeter described above is capable of measuring the absolute beam polarization to 10%. This is acceptable for Day-1 physics since luminosity will take some time to reach the design value. Currently we are also testing other polarimeter concepts such as p-carbon scattering in the CNI region. Should this prove viable, the detector is quite cheap can be easily configured to look at the same target of the pion polarimeter.

Beyond Day-1 and in order to reach the desired absolute polarization measurement of 5%, we need to calibrate these polarimeters using a polarized hydrogen jet target. The polarization of these jets can indeed be measured to the 5% accuracy.

ACKNOWLEDGMENTS

This work was performed under the auspices of the U.S. Department of Energy contract number DE-AC02-98CH10886 and grants from the U.S. National Science Foundation, and the Institute for Chemical and Physical Research (RIKEN), Japan. My colleagues G. Bunce, H. Huang, T. Roser, D. Underwood, H. Spinka, and A. Yokosawa provided major contributions to the development of these polarimeter concepts.

REFERENCES

For additional details on these polarimeters please consult the following and references therein:

[1] Rice, J. A., A 200 MeV Polarimeter, Masters thesis, Rice University, unpublished.

[2] Khiari, F. Z., et al., *Phys. Rev.* **D39**, (1989) p. 45.

[3] Spinka, H., et al., *Nucl. Instr. And Meth.* **211**, (1983) p. 239.

[4] Dragoset, W. H., et al., *Phys. Rev.* **D18**, (1976) p. 3939.

[5] Adams, D. L., et al., *Phys. Lett.* **B 264**, (1991) p. 462.

[6] Alekseev, I., et al., Design Manual for the Polarized Proton Collider at RHIC 1996, unpublished.

Cavity Beam Position Monitors

Ronald Lorenz

DESY Zeuthen, Platanenallee 6, D-15738 Zeuthen

Abstract. Beam-based alignment and feedback systems are essential for the operation of future linear colliders and free electron lasers. A certain number of beam position monitors with a resolution in the submicron range are needed at selected locations. Most beam position monitors detect the electric or the magnetic field excited by a beam of charged particles at different locations around the beam pipe. In resonant monitors, however, the excitation of special field configurations by an off-center beam is detected. These structures offer a large signal per micron displacement. This paper is an attempt to summarize the fundamental characteristics of resonant monitors, their advantages and shortcomings. Emphasis will be on the design of cylindrical cavities, in particular on the estimation of expected signals, of resolution limits and the resulting beam distortion. This includes also a short introduction into numerical methods. Fabrication, tuning, and other practical problems will be reviewed briefly. Finally, some resonant devices used for beam position diagnostics will be discussed and listed.

INTRODUCTION

A linear collider with a center of mass energy of 500 GeV could be used for detailed studies of the top quark system, for finding or excluding the Higgs boson and for the search of supersymmetric particles. Because a luminosity of about 10^{33} cm^{-2}s^{-1} is needed due to the small cross sections of the processes of interest, all linear collider studies worldwide consider flat colliding beams having vertical dimensions of 3–20 nm [16]. Beam-based alignment and correction schemes have to be adopted to bring these tiny beams into collision. A certain number of high-resolution beam position monitors (BPM) are needed at selected locations.

Free electron lasers (FEL) are under design [23] for which the power and the coherency of the radiation can be enhanced due to the bunching by emitted radiation in the first part of the undulator, a process known as self-amplified spontaneous emission. Very small beam emittances are essential to reach saturation within a reasonable undulator length. The position of the electron beam might vary inside the undulator, mainly because of field imperfections of the magnets. Since a precise overlap of electron and photon beams is essential for FEL operation, the position of the electron beam has to be measured with a high resolution and corrected along the undulator beamline.

CP451, *Beam Instrumentation Workshop*
edited by R. O. Hettel, S. R. Smith, and J. D. Masek

A BPM system consists of a 'transducer' close to the beam, transmission lines, electronics, and software. This paper concentrates mainly on the transducer. If a bunch of charged particles is centered in a circular, conducting beam pipe, then there is a uniformly distributed electro-magnetic field accompanying the beam. Its spectrum depends on the bunchlength σ_z, the bunch shape, and the spacing between two bunches. For a 'short' bunch spacing and processes having 'long' time constants, the spectrum contains discrete lines at harmonics of the bunching frequency. The fields of an off-center beam are not uniformly distributed. Many transducers used in BPM systems detect the electric or the magnetic field at different locations around the beam pipe. These signals are subtracted in the electronics or in a computer to measure the 'field distortion' [19].

In resonant cavities the beam excites special field configurations resonating at a certain frequency. The amplitude of these *modes* depends on the cavity orientation, the bunch charge, the beam position, and its spectrum.

Dipole modes are used for position detection, since their amplitude depends linearly on the beam position and is zero for a centered beam. The signals excited in the cavity are coupled into an external circuit, and the amplitude of this particular mode can be separated in the frequency domain. In principle, no additional subtraction is needed, the information about the position is given directly. Because of the large signal per micron displacement, these resonant monitors find application in situations where the beam signals are weak. This paper is an attempt to discuss the characteristics of cavity BPMs, and to summarize their advantages and the trade-offs. Emphasis will be on the design of circular cavities, including the expected signals and some numerical methods.

WAVEGUIDES AND CAVITIES

A waveguide can support waves of any frequency (above cut-off), the guide wavelength taking a value which causes the field patterns to satisfy the boundary conditions. These field patters can be divided into two basic sets of solutions or modes: 'transverse magnetic' or TM modes (no longitudinal magnetic field component) and 'transverse electric' or TE modes (no longitudinal electric field component).

The field components of the TM modes in a cylindrical waveguide can be derived from a Hertzian potential having a single component along the axis of the guide [5]. A *generating function* ψ_e, which satisfies the two-dimensional Helmholtz-equation, is then used to calculate these field components:

$$\mathbf{E}_t = \pm\Gamma\nabla_t\psi_e e^{(\pm\Gamma z)} \qquad E_z = -\nabla_t^2\psi_e e^{(\pm\Gamma z)} \qquad \mathbf{H}_t = \mp\frac{j\mu_0\varepsilon_0\omega^2}{Z_0\Gamma}\,\mathbf{a}_z \times \mathbf{E}_t \tag{1}$$

where k_c is the cut-off wave number, $\Gamma = k_c^2 - (\omega^2\mu_0\varepsilon_0)$ the propagation constant and $Z_0 = \sqrt{\mu_0/\varepsilon_0}$ the intrinsic impedance of free space. Usually, a solution of the form $\psi_e = U(u_1)U(u_2)$ may be found, where the functions U_1, U_2 depend on the geometry used. The allowed values of k_c^2 and Γ^2 are determined by the boundary conditions (E_z and $\mathbf{H}_t$ must vanish on the boundary for perfectly conducting walls). At a wavelength λ longer than the cut-off wavelength $\lambda_{c,mn}$ of the mnth mode the fields at a point z_0 from a source are attenuated by

$$\delta_{z_0} = \exp\left(-2\pi \cdot z_0 \cdot \sqrt{\lambda^{-2} - \lambda_{c,mn}^{-2}}\right) \qquad \lambda_{c,mn} = 2\pi \cdot \left(k_{c,mn}\right)^{-1} \tag{2}$$

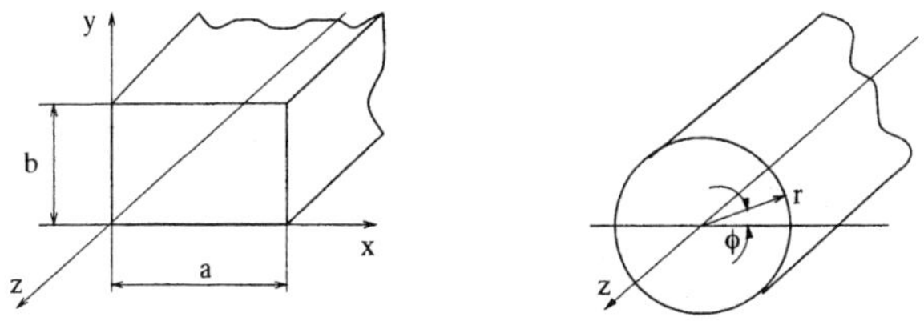

FIGURE 1. Coordinate systems for rectangular and for circular waveguides.

Rectangular waveguides

Following the coordinate system in Figure 1, the field generating functions of the TM modes and the cut-off wave numbers of rectangular waveguides are given [5] by

$$\psi_{nm} = \cos\left(\frac{n\pi x}{a}\right) \cdot \sin\left(\frac{m\pi y}{b}\right) \qquad k_{c,nm} = \left(\frac{n\pi}{a}\right)^2 + \left(\frac{m\pi}{b}\right)^2 \tag{3}$$

Circular waveguides

Using Figure 1, the field generating functions of the TM modes and the cut-off wave numbers of circular waveguides are given [5] by

$$\psi_{mn} = J_m\left(k_{c,mn}\right) \cdot \cos\left(m\phi\right) \qquad k_{c,mn} = \frac{a_{mn}}{r} \tag{4}$$

where a_{mn} is the n-th zero of the Bessel function J_m.

Resonant Cavities

If a waveguide is closed by two short-circuiting planes, boundary conditions will have to be met on these planes. This requires the planes to be separated by an integral number of half-wavelengths, and waves can be excited in the enclosure only at discrete frequencies. Such a resonating enclosure is called a *cavity* (but a cavity is not necessarily a waveguide section). In the following the indices for the different modes are neglected, where appropriate. For a more detailed discussion the reader is referred to [5].

Cavities — Equivalent Circuits

The energy in a cavity oscillates between pure electric and pure magnetic energy; the averaged stored electromagnetic energy is $W_s = \langle W_e \rangle + \langle W_m \rangle = 2 \cdot \langle W_e \rangle$. For a certain mode, this is equivalent to a circuit containing a capacitor and an inductor.[1] In an equivalent circuit for the whole cavity, there are many LC-circuits in parallel. Three cavity quantities that are independent of the field strength are needed to calculate the parameters R, L, and C for each mode. The resonance frequency is given by $\omega_r = (L \cdot C)^{-\frac{1}{2}}$. The definition for the shunt impedance following the circuit theory is

$$R = \frac{\left| \int_0^l E_z \cdot e^{j\omega t} dz \right|^2}{2P_d} = \frac{|V_{mn0}|^2}{2P_d}. \tag{5}$$

The integral in the numerator depends not only on the cavity shape and the chosen mode, but also on the integration path (here, the particle trajectory).

Practical cavities, made of metal, dissipate energy in their walls. In a cavity without any external coupling (free oscillation), the negative change in time of the stored energy is equal to the dissipated power P_d.

$$-\frac{dW_s}{dt} = P_d \quad \Rightarrow \quad W_s = W_0 \cdot \exp\left(-\frac{\omega_r}{Q_0} t\right) \quad \text{with} \quad Q_0 = \frac{\omega_r W_s}{P_d} = \frac{\omega_r \tau_r}{2} \tag{6}$$

Q_0 is the quality factor of the cavity and is independent of the field strength. Together with ω_r, it determines the decay time τ_r of the cavity.

Coupling to External Circuits

A cavity has also connections to the outer world, for feeding or to couple out the signals (e.g., for measurements). Such a connection can be realized by

1. Magnetic coupling, where the H-field passes through a loop and induces a voltage (as shown in Fig. 2a),

2. Electric coupling, where the E-field causes a current on a small dipole (Fig. 2c), and

3. Electromagnetic coupling by an aperture, where electric and magnetic fields penetrate through a hole in the common wall, corresponding to an electric dipole or magnetic dipoles in an external waveguide (Fig. 2b and 2d).

By choosing the antenna length, the form of the loop, and its position correctly the coupling factor β can be adjusted. The coupling is also explained in terms of the *external Q value* Q_{ext} and the *loaded* Q_{L}

$$Q_{\text{L}} = (Q_0 \cdot Q_{\text{ext}}) \,/\, (Q_0 + Q_{\text{ext}}) = Q_0 / (1 + \beta) \qquad \text{with} \qquad \beta = P_{\text{out}} / P_d. \tag{7}$$

---

1. Note: In real cavities, both energies are not concentrated in lumped elements!

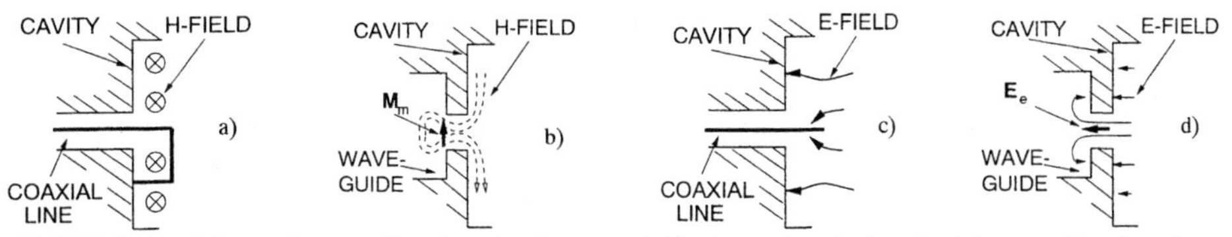

FIGURE 2. Magnetic coupling by a) a loop and b) through a hole; electric coupling by c) an antenna and d) through a hole.

Cavities in Particle Accelerators

Coupling to a Beam

One of the most common applications of a cavity in particle accelerators is the use as the source of the accelerating rf voltage. A charge q passing an empty cavity leaves a field behind it that can be represented by an infinite sum of modes (Fourier series). Each mode n can be described by ω_n, Q_n and the instantaneous voltage V_n along the particle path. The stored energy W_n in this mode will be proportional to the square of the charge. A fraction U_n of the total energy, the particle has lost, is radiated in mode n. It can be expressed by a decelerating voltage proportional to the induced voltage V_n. The statement that a charge passing a cavity sees exactly one-half of its own induced voltage (remaining after the passage) is called the *fundamental theorem of beam loading.* By conservation of energy, the stored energy in mode n must be equal to the energy radiated into this mode. Using Equation (5) one gets for V_n and for the *loss factor* k_n

$$V_n = 2 \cdot q \cdot k_n \qquad \text{with} \qquad k_n = \frac{V_n^2}{4 \cdot W_n} = \frac{\omega_n}{2} \cdot \left(\frac{R}{Q}\right)_n . \tag{8}$$

If we want the voltage V_n to be the actual voltage excited by the bunch in the cavity we have to take into account the change of the E-field during the passage time. Often this phase difference between the particle beam and the rf field is called the *transit time factor.* For a velocity of $v = c_0$ one finds

$$T_{tr} = \left(\int_{z1}^{z2} E_z \cdot e^{jkz} \cdot dz\right) / \left(\int_{z1}^{z2} E_z \cdot dz\right) = \frac{\sin\eta}{\eta} \qquad \text{with} \quad \eta = \frac{\pi \cdot l}{\lambda_{mn0}} . \tag{9}$$

Tuning of a Cavity

High-Q cavities having a small bandwidth can be detuned by small temperature changes. Fabrication tolerances and temperature drifts have to be tuned out for most applications to be exactly on resonance. Since the adjustment of an external impedance would also change the coupling, tuning is often done by deformations of the inner cavity surface ('screws'). This can be done in regions of high

- E-field by changing the length s of an E-field trajectory, corresponding to a change of the capacitance

$$(C + \Delta C) = \frac{Q}{E\,(s + \Delta s)}.$$

- H-field by changing the area A where the magnetic flux passes through (changes the inductance)

$$(L + \Delta L) = \frac{\mu H}{I}\,(A + \Delta A).$$

Both result in a new resonance frequency $\omega_r + \Delta\omega_r = [(L + \Delta L)(C + \Delta C)]^{-\frac{1}{2}}$.

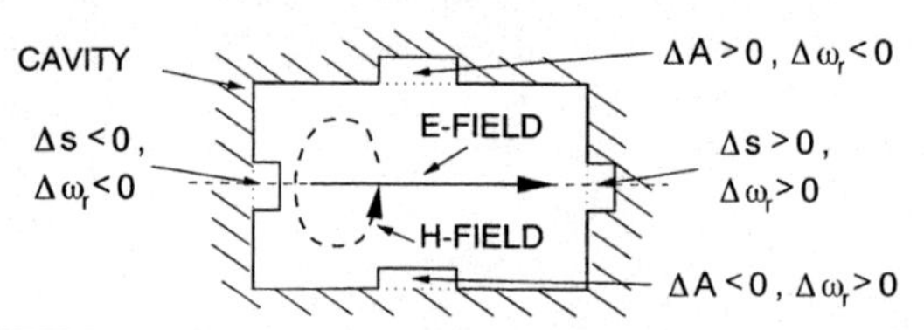

FIGURE 3. Tuning of cavity by changing the inner surface.

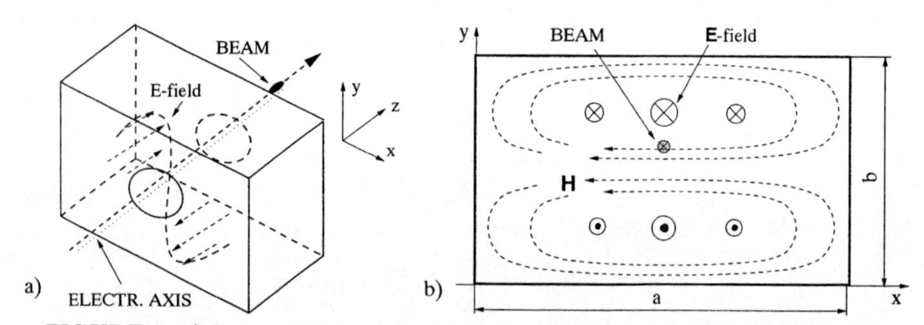

FIGURE 4. a) Beam-excited rectangular cavity; b) field pattern of the TM_{120} mode.

Example 1: Rectangular Cavities used as BPMs at SLAC.
Resonant monitors using rectangular cavities excited in the TM_{120} mode were built at SLAC many years ago [9]. The longitudinal E-field component of this mode and its resonant frequency are given by

$$E_z(x, y) = E_z^{max} \cdot \sin(\pi x/a) \cdot \sin(2\pi y/b) \qquad f_{120} = c_0/2 \cdot \sqrt{(1/a)^2 + (2/b)^2}. \quad (10)$$

The cavities were fabricated from OFHC copper plates and brazed. A resolution of 10 μm was reached for a beam current of about 100 μA by using a homodyne detection system. The cavity output power was about 50 μW/(mA^{2}mm^2).

CIRCULAR CAVITIES

The simplest microwave BPM structure is a circular cavity excited in the TM_{110} mode by an off-axis beam. The amplitude of this mode yields a signal proportional to the beam displacement and the bunch charge; its phase relative to an external reference gives the sign of the displacement. This signal is much stronger than the signal given by other monitors and is a linear function of the beam displacement.

Fields and Signals of the TM_{110} Mode

First, only unperturbed cavities are considered and the influence of the beam pipe and of coupling ports are neglected. The resonant frequency and the z-component of the mn0th TM mode electric field at a position δx are given by

$$E_{mn0} = C_{mn0} \cdot J_m \left(\frac{a_{mn} \delta x}{R_{\text{res}}} \right) \cdot \cos(m\phi) \cdot e^{-(j\omega z)} \qquad f_{mn0} = \frac{c_0 \cdot a_{mn}}{2\pi \cdot R_{\text{res}}} \tag{11}$$

(r, ϕ, z) is the position in cylindrical coordinates, t the time, and C_{mn0} the amplitude. The electric and magnetic fields of the first dipole mode are shown in Figure 5b. The Q value of this TM_{110} mode and the R/Q at the position $[r_{E_{max}} = 0.481 \cdot R_{\text{res}}]$ of its electrical field maximum are given by

$$Q_{110} = \frac{1}{2\pi} \cdot \frac{a_{11}}{1 + (R_{\text{res}} \cdot l^{-1})} \cdot \frac{\lambda_{110}}{\delta} \qquad \text{with} \qquad \delta = \sqrt{\frac{1}{\pi \cdot f_{110} \cdot \mu \cdot \kappa}} \tag{12a}$$

$$\left(\frac{R}{Q} \right)_{110} = \frac{(V_{110}^{\text{max}})^2}{2\omega_{110} \cdot W_{110}} = \frac{2 \cdot Z_0 \cdot l \cdot (J_1^{\text{max}})^2 \cdot {T_{tr}}^2}{\pi \cdot R_{\text{res}} \cdot J_0^2(a_{11}) \cdot a_{11}} \approx 130.73 \cdot \frac{l}{R_{\text{res}}} \cdot {T_{tr}}^2 \tag{12b}$$

where a_{11} is the first root of J_1, $J_1^{max} = J_1(1.841) = 0.582$ at its maximum.

Beam-Excited TM_{110} Signals

The voltage V_{110}^{in} induced in the TM_{110} mode by a charge q at $(\delta x, \phi = 90^o)$ can be calculated by using Equation (11) and the maximum voltage. $V_{110}(r_{E_{max}})$ can be expressed in terms of the loss factor k_{110} and the bunch charge q (Eqn. (8))

$$\frac{V_{110}^{in}(\delta x)}{V_{110}(r_{E_{max}})} = \frac{J_1\left(\frac{a_{11} \cdot \delta x}{R_{\text{res}}}\right)}{J_1^{max}} \qquad \Rightarrow \qquad V_{110}^{in}(\delta x) = \frac{2 \cdot k_{110} \cdot q}{J_1^{max}} \cdot \frac{a_{11} \cdot \delta x}{2 \cdot R_{\text{res}}} \tag{13}$$

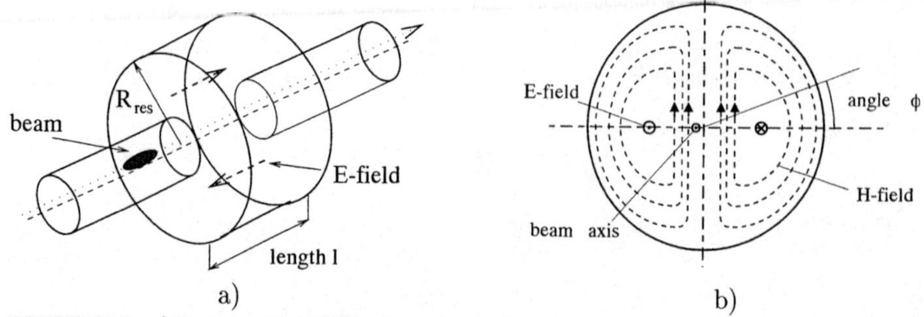

FIGURE 5. a) Beam-excited TM_{110} in a circular cavity; b) fields of the horizontal polarization.

The Bessel function $J_1(x)$ can be replaced by $x/2$ for small arguments,[2] and the voltage becomes linear proportional to δx. Together with Equation (12b) this gives

$$V_{110}^{in}(\delta x) = \left(\frac{R}{Q}\right)_{110} \cdot \omega \cdot q \cdot \left\langle \frac{a_{11} \cdot \delta x}{2 \cdot J_1^{max} \cdot R_{\text{res}}} \right\rangle = \delta x \cdot q \cdot \frac{l \cdot T_{tr}{}^2}{R_{\text{res}}{}^3} \cdot 0.2474 \left[\frac{\text{Vm}}{\text{pC}}\right]. \quad (14)$$

The term in the angle brackets is also called the *beam coupling coefficient*, M_b.

In the case of a cavity excitation by *multiple bunches* the cavity has to reach its steady state and an averaged position of the bunch train will be measured. With the length of the bunch train, τ_s, and with τ_r (Eqn. (6)) this results in $Q_{\text{L}} \ll \omega\tau_s$. The power coupled out of the cavity and the voltage into a 50-Ω system can be estimated using the averaged beam current $\langle I_b \rangle = 2 \cdot I_{DC}$

$$P_{out} = \left(\frac{R}{Q}\right)_{110} \cdot Q_L \cdot \langle I_b \rangle^2 \cdot M_b^2 \cdot \frac{\beta}{1+\beta} \qquad V_{110}^{out} = \sqrt{P_{out} \cdot 50\ \Omega}. \quad (15)$$

When a *single bunch* traverses the cavity it will excite a burst of rf that decays with τ_r. A proper Q_{L} has to be chosen so that the cavity is 'empty' when the next bunch comes (see also Eqn. (6))

$$V_{110}(t_b) = V_{110}(t=0) \cdot \exp\left(-\frac{1}{2} \cdot t_b \cdot \frac{\omega_{110}}{Q_{\text{L}}}\right) = \exp\left(-\frac{1}{2} \cdot \frac{t_b}{t_{r,l}}\right). \quad (16)$$

The rf current during such a burst will be $I_{rf} = q \cdot \tau_{r,l}/2$. With Equation (15) we get for the voltage excited by a single bunch and coupled into a 50−Ω system

$$V_{110}^{out}(\delta x) = \left(\frac{R}{Q}\right) \omega q \sqrt{\frac{50\ \Omega}{Q_{\text{L}}}} M_b \frac{\beta}{1+\beta} = V_{110}^{in}(\delta x) \left(\frac{R}{Q}\right)_{110}^{-\frac{1}{2}} \sqrt{\frac{50\ \Omega}{Q_{\text{L}}}} \sqrt{1 - \frac{Q_{\text{L}}}{Q_0}} \quad (17)$$

2. The error of this relation is less than 1% up to $\delta x = 0.14 \cdot R_{\text{res}}$.

Common-Mode Signals

Since the field maximum of the common modes is on the cavity axis, they will be excited much stronger than the TM_{110} by a beam near the axis. The excitation of the dominant TM_{010} mode at its own frequency can be estimated following [17]:

$$S_1 = \frac{V_{110}(\omega_{110})}{V_{010}(\omega_{010})} = \frac{1}{J_1^{\max}} \frac{\delta x \cdot a_{11}}{2 \cdot R_{\text{res}}} \cdot \frac{V_{110}^{\max}}{V_{010}} \approx \frac{5.4}{\lambda_{110}} \cdot \delta x \cdot \frac{k_{110}}{k_{010}}. \tag{18}$$

This gives the frequency-sensitive TM_{010} rejection, which has to be realized mainly in a band-pass filter. Assuming that both loss factors are identical, 69 dB of rejection are required to detect a beam displacement of 10 μm in a 1.52-GHz cavity.

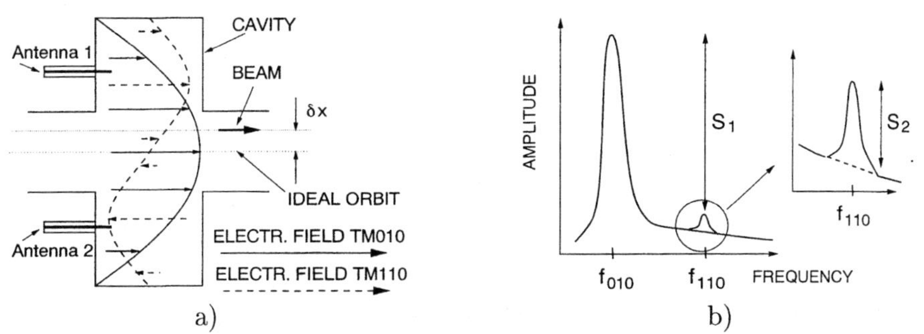

FIGURE 6. a) Excitation of the TM_{010} and the TM_{110}; b) signals in the frequency domain.

Due to their finite Q, all modes have field components even at the TM_{110}-mode frequency as sketched in Figure 6. The ratio of the spectral densities at ω_{110} was estimated in [17], leading to a position δx^{min} close to the electrical center where both signals are identical (smaller displacements can not be detected):

$$S_2 = \frac{v_{110}(\omega_{110})}{v_{010}(\omega_{110})} \approx S_1 \cdot Q_L \cdot \left(1 - \frac{\omega_{010}^2}{\omega_{110}^2}\right) \quad \Rightarrow \quad \delta x^{min} \approx \frac{k_{010}}{k_{110}} \cdot \frac{R_{\text{res}}}{2 \cdot Q_{\text{L}}} \tag{19}$$

This pessimistic estimation assumes a cw-excitation and a detection scheme which is not sensitive to the phase difference between the two voltages.

Effect of Beam Angle

Let us assume a beam, that goes through the cavity center with an angle x'. By integrating the E-field along this new particle trajectory, we get [14]

$$T_{tr} \cdot M_b = \frac{a_{11}^2 \cdot l^2 \cdot x'}{12 \cdot J_1(\rho_{01}) \cdot {R_{\text{res}}}^2} = 2.36 \frac{l^2 \cdot x'}{{R_{\text{res}}}^2} \qquad V_{110}(x') \simeq j \cdot A_3 \cdot q \cdot x'. \tag{20}$$

Together with Equation (14) we get an offset error of

$$\frac{\delta x}{x'} = \frac{2244.5}{\sin\left(\pi l \cdot \lambda^{-1}\right)} \cdot \frac{l^3}{{R_{\text{res}}}^2} \cdot \frac{1}{\lambda} \left[\frac{\text{m}}{\text{mrad}}\right] \tag{21}$$

Summary — Resolution Limits and Conclusions

As shown above, the cavity output signal at the frequency ω_{110} contains also other signals than a 'clean' TM_{110} voltage. This is summarized in Equation (22) (the A_n are constants) and illustrated in the right part of Figure 7b:

$$V_{cav}(\omega_{110}) = V_{110}^{out}(\delta x) + V_{0n0} + V_{110}(x') + V_n = A_1 qx + jA_2 q + jA_3 qx' + V_n. \tag{22}$$

The second term is caused by the common-mode leakage, the third term displays the beam angle, and the last one is the electronics noise.

Considering two opposing antennae (1 and 2 in the cavity of Fig. 6a), the TM_{010} and the TM_{110} fields have a phase difference of 180°. If these signals are combined in a hybrid or a Magic-T, the TM_{110} signal appears at the Δ-port. This effect is also sketched in Figure 11 (frequency domain). The rejection of common-mode components in such a field-selective filter — and thus the improvement of the 'center resolution' — is limited by the isolation between the Σ- and the Δ-port of the device used. Standard hybrids have an isolation of about 25 dB. Furthermore, the beam-angle signal and the common-mode signal are phase shifted by $\pi/2$ and can be suppressed by using a synchronous detector. If the remaining phase error in the system is very small, the resolution is mainly limited by the electronics noise.

Example 2: Cavity-BPMs installed at the Final Focus Test Beam (FFTB)

Often the amplitude of the beam position jitter is in the order of some microns and it is impossible to predict the beam trajectories. But at high energies the trajectories are 'straight' lines, and three BPMs can be used to measure their intrinsic resolution. A block of three C-band TM_{110} cavities was tested at the FFTB at SLAC [14], and the measured resolution was about 25 nm (Fig. 8b).

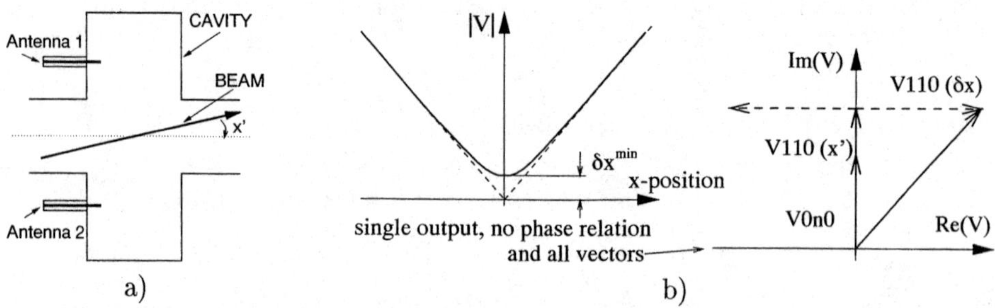

FIGURE 7. a) Effect of a beam angle; b) signals excited in the cavity (summary).

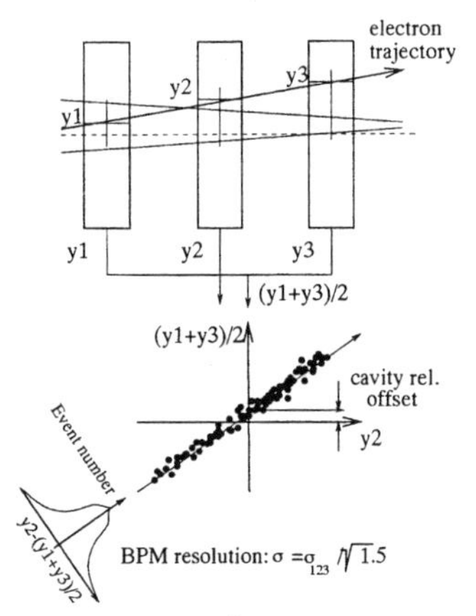

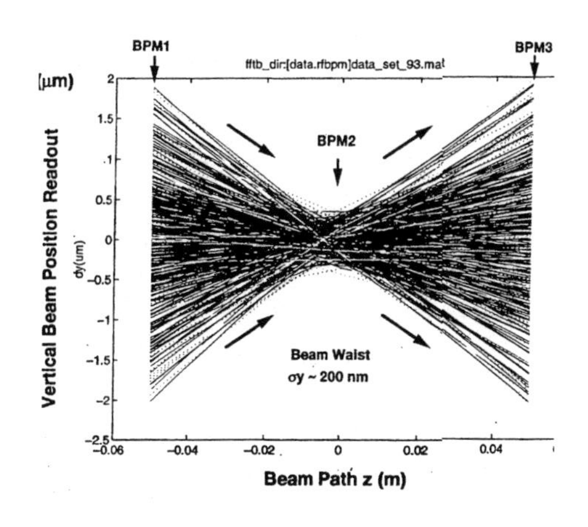

FIGURE 8. a) Resolution measurements at the FFTB using three cavities; b) pulse-to-pulse multiple beam trajectory traces [19].

Wakefields

Since a charged particle interacts electromagnetically with the surroundings (e.g., discontinuities in a metallic vacuum chamber), it generates an electromagnetic field. This so-called *wakefield* acts back on the motion of following particles within the same bunch or even within other bunches, leading to deflection or deceleration. The *wake potentials* are obtained by integrating the longitudinal or transverse components of the Lorentz force at a distance s behind the exciting charge moving on a straight path with constant velocity. The form of these potentials depends on the structure and on the bunch length.

Since impedances are also the Fourier transforms of the delta-function wake, there is a strong correlation between the signals coupled out of a structure ('pick-up' of any geometry) and the wakefield. Therefore, the high shunt impedance of a cavity BPM causes problems for many circular machines such as storage rings and light sources. Usually, the number of such devices adding a remarkable amount to the impedance budget has to be minimized.

An analytical approach for the wakefield of a cavity gap and very short bunches is the *diffraction model* [3]. It can be used to estimate the peak wake potential of a cavity of length g having an entrance aperture of radius R_0, and a Gaussian bunch with a rms length σ_z:

$$\hat{W}_{diff} = \frac{Z_0 c}{\sqrt{2}\pi^2 R_0}\sqrt{\frac{g}{\sigma_z}} = 8.1 \cdot \frac{1}{R_0} \cdot \sqrt{\frac{g}{\sigma_z}} \quad \text{(in units of } \frac{\text{V}}{\text{nC}}). \tag{23}$$

In most practical cases, the wake potentials must be calculated using computer codes. 2D codes such as ABCI [1] can be used for circular cavities if the effect of a coupling device is negligible. For more complex structures one has to use 3D codes such as MAFIA. The total loss factor k_{all} (all excited modes of a certain azimuthal symmetry) gives also the *power deposited* in the cavity, P_l:

$$P_l = k_{all} \cdot f_{av} \cdot q^2 \qquad \text{with} \qquad f_{av} = N_{bunches} \cdot f_{rep}. \tag{24}$$

Example 3: Diagnostic stations for the TESLA Test Facility Linac-FEL
Circular cavities were chosen for this purpose [13] because of the required single bunch resolution of 1 μm. Since V_{110}^{in} is proportional to the cavity size, the TM_{110} design frequency is 12 GHz (see Fig. 9b). With $q = 1$ nC, $\beta = 1$ and $Q_L = 1000$ we get for the voltages

$$V_{110}^{in} \approx 464 \text{ [mV/}\mu\text{m]} \qquad V_{110}^{out} \approx 9.9 \text{ [mV/}\mu\text{m]} \tag{25}$$

A complete monitor consists of two cavities separated in beam direction. The frequency shift due to the missing coupling in one plane is about 0.33%. The peak wake potential is about 20.7 V/pC for a bunch length of 50 μm.

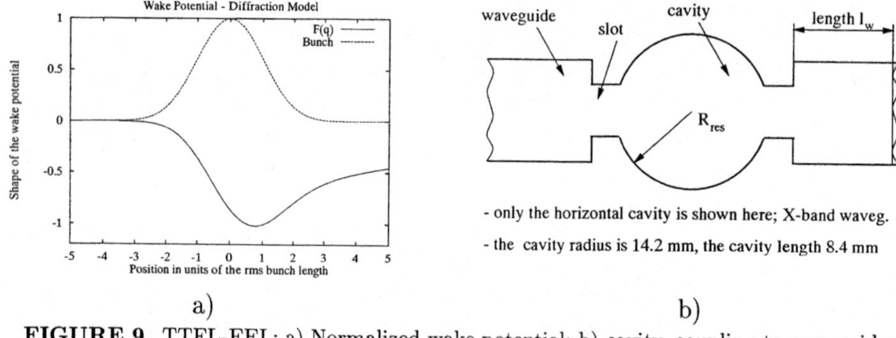

FIGURE 9. TTFL-FEL: a) Normalized wake potential; b) cavity, coupling to waveguides.

Design of a TM_{110} Cavity BPM

Cavities are narrowband devices and they have to be designed in the frequency domain. There are two main applications of cavity BPMs:

1. High resolution (in the nm-range), even for single bunches, and

2. Moderate resolution, but very low current (cw, about 1 nA).

TM_{110} Frequency and Cavity Size

There is no 'global' optimal TM_{110} frequency for each BPM application. A general rule is that it should be apart from the accelerating frequency in order to avoid interference of the position signals by leakage fields of the high rf power in the machine. In addition, f_{110} should be well below the frequency where the power density in the spectrum decreases by 3 dB. Otherwise, one has to take into account the bunch factor. This factor is about 1 for Gaussian bunches with $\sigma_z \leq 1$ mm and frequencies below 40 GHz. Beside that, the TM_{010} frequency should not be a harmonic of the bunching frequency (multibunching).

TABLE 1. Arguments for a higher or a lower resonance frequency (fixed beam pipe radius).

Parameter	Higher frequency	Lower frequency	Remark
$f_{TM_{020}}$ above f_c	+	-	common-mode rejection
costs	(-)	(+)	electronics, vac.-feedthroughs
signal level	+	-	better: re-entrant (!)
reference, phase	(-)	(+)	

For optimum common-mode rejection the resonant frequency should be high (Eqn. (18)). The cavity radius is limited by the field distortion due to the beam pipe: for cavities that are too small the electric field is no longer linear/constant along the gap. A lower limit (practical experience) is that the cavity should be larger than three times the beam pipe radius. Hence, it is impossible to get a TM_{020} frequency higher than the beam pipe cut-off without any distortion.

Concerning the costs, it is very important to chose the 'right' frequency, mainly because of the electronics (see Electronics section below) and existing hardware/software. Beside that, there are only a few vacuum feedthroughs on the market for frequencies above 10 GHz. Some arguments for a resonant frequency choice are listed in Table 1.

Influence of l/R_{res} on the TM_{110} Signal

For the detection of *single bunches* the voltage induced in the cavity (Eqn. (14)) should be high. Since T_{tr} depends on the ratio $\xi = l/R_{\text{res}}$, the function $\sin^2 \xi/\xi$ has to be optimized, yielding $\xi_{opt} = 0.6086$ for a cavity without beam pipes.

In the case of multibunching, the shunt impedance has to be optimized according to Equation (15). Since Q_0 and Q_{L} cancel and only a coupling term remains, this leads to an optimization of the transit time factor. The resulting value for ξ_{opt} is 0.742. A more effective way to optimize the shunt impedance is to make the cavity re-entrant ('nose-cones'). Such an optimization is usually done by using numerical codes.

Resonance Frequency for Multiple Bunches

Let us assume a cavity with $l = 0.742 \cdot R_{\text{res}}$ and $R_{\text{res}} = 3 \cdot R_o$, having two identical couplings of $\beta = 1$. Neglecting the ratio of the loss factors we get with equations

(12a), (7), and (19) for Q_L and the minimum resolution

$$Q_L \approx \sqrt{R_\mathrm{res}} \cdot \sqrt{\kappa} \cdot \frac{11.45}{(1+2\beta)} \qquad \delta x^{min} \approx \sqrt{R_\mathrm{res}} \cdot \frac{0.227}{\sqrt{\kappa}} \tag{26}$$

Coupling by Antennae or by Waveguides

The signals excited in the cavity have to be coupled into an external circuit (see discussion in previous section). There are two strong arguments for having two ports per plane: for symmetry reasons and to realize a common-mode rejection (see below, subsection entitled "Other Signals and Resolution Limits").

Type of Coupling

For frequencies below 10 GHz it makes sense to use antennae, mainly because of the size of waveguides in this frequency range (though these could be ridged). At higher frequencies it is both easier and more precise to use waveguides; antennae for such frequencies are so tiny that tolerances for length, thickness, angle, etc., are problematic.

Strong or Weak Coupling

A strong coupling leads to larger signals and is good for stabilizing the cavity (temperature drifts, etc.). But it lowers also Q_L (common-mode rejection, Equation (19)) and might lead to field distortion.

For the design it is important to determine the *coupling factor* of each port and to estimate the resulting *frequency changes.* The *perturbation method* can be used to predict the frequency change due to an antenna inserted in the cavity (Fig. 2c). First, a cavity is excited at its resonant frequency. A small object is then introduced, and the resulting frequency change is of the order of the volume ratio of the cavity and the object. In another theory by Bethe, the fields of small coupling holes are replaced by equivalent dipoles ($\mathbf{M}_m$ and $\mathbf{E}_e$ in Figure 2b and 2d). For more details of both methods the reader is referred to [5]. Another approach discussed below (see the "Numerical Methods" section) follows a method described in [22].

Number of Cavities

Both TM_{110} polarizations have to be measured to obtain the displacements in x and y, respectively. Single cavities are compact devices, and the distortion to the beam is smaller. A major problem is that of symmetry: the tuning and the coupling of one polarization effects also the other polarization. This additional asymmetry leads to a lower isolation between both polarizations ('cross-talk'). By using two cavities separated in beam direction, the unwanted polarization in each of the cavities can be detuned and/or additionally damped (thus reducing the wakefields and the decay time). The beam pipe acts as a waveguide below cut-off, the attenuation of a TM_{01} wave at a position z_0 can be calculated using Equation 2 and $\lambda_c = 2.613 \cdot R_{\text{res}}$. Table 2 summarizes some of the arguments for both structures.

TABLE 2. Comparison: two cavities versus a single cavity

Parameter	**Single cavity**	**Two cavities**
Isolation/Cross talk	about 25 dB	more than 40 dB
Tuning and Coupling	complex, less space	individual, better for symmetry
Coupling	possible, but less space	independent for both planes
Wakefields, losses	smaller	higher
Space	short	long
Fabrication and costs	easy/lower	complex/higher
Temperature stabilization	easier	complex, but independent

Numerical Methods

The equations for the calculation of fundamental cavity parameters are true only for cavities without any perturbation. Electromagnetic computer-aided design (ECAD) is needed, e.g., to estimate the impact of re-entrant parts, of beam pipe holes or of bellows on the resonance frequency. 2D codes such as SUPERFISH or URMEL/MAFIA-2D are good enough for the first step, since most of the structures have an axis symmetry. In the case of azimuthal asymmetries, 3D codes such as MAFIA [24] are needed.

An example for the latter is the insertion of waveguides or antennae for coupling purposes. In a method by Slater [22] the waveguide coupled to the cavity is shortened at a position l_w (Fig. 9b), thus forming a second resonator. This can be done for many positions l_w, and the resulting resonance frequencies of these systems of coupled resonators can be used to determine the external Q-value as well as the resonant frequency. This method is very useful for the numerical calculation of both quantities [11]. Besides these calculations in the frequency domain, the wake potentials and the total loss factor can be calculated in the time domain.

Mechanics, Fabrication, and Environment

Fabrication

Circular cavities can be built very precisely by turning. Most of the cavities used in particle accelerators are made of OFHC copper and their parts are brazed. In addition, aluminum and stainless steel were used for special applications [12]. For the cavities in Example 3, electro-discharge machining (EDM) was used to realize the waveguides and the coupling to the cavity. Stainless steel/copper/water interfaces should be avoided because of electrolytic erosion.

Since f_{110} depends mainly on the cavity radius, the effect of temperature changes on the resonance frequency scales roughly with the expansion coefficient α of the material: $\Delta f_{110} = f_{110} \cdot \alpha$. Close dimensional tolerances of less than 0.02% are unavoidable for the cavity radius. Every reduction in azimuthal symmetry leads to an unwanted coupling between both polarizations in the cavity (cross-talk). A major problem is therefore to maintain the symmetry even after brazing or welding. The isolation is often reduced to about 25 dB, whereas it reaches about 35 dB for 'ideal' cavities. All internal cavity surfaces may be 'polished' if necessary.

Sometimes it is impossible to use standard rf components because of the special vacuum requirements in particle accelerators (10^{-9} Torr), and vacuum windows or special feedthroughs (e.g., by KAMAN Corp. or KYOCERA) are required. These feedthroughs can be used to develop special coax-to-waveguide adaptors.

Cryogenic Environment

A cryogenic environment has a strong impact on the design and the required reliability of beam instrumentation. Since operation at low temperatures is best achieved inside evacuated vessels, access to the beam pipe is quite difficult and maintenance requires long time scales. *Heat losses* should be kept to a minimum.

Cool-down causes a *shrinkage* of the cavity and changes the Q-values (*conductivity*). Since active tuning systems are very expensive, the design frequency should be reached after each cool-down within a certain bandwidth. Many materials most commonly used are not appropriate. Often special procedures are required to get 'clean' surfaces, since extraneous particles may affect, for example, high-Q superconducting accelerating cavities. Finally, special *rf feedthroughs* and cables are required to work reliablly even at cryogenic temperatures.

Example 4: Cold monitors in the TESLA Test Facility Linac (TTFL)

Single circular cavities were built for steering correction at all superconducting quadrupoles, mainly because of the desired resolution, 10 μm in a cryogenic environment, and limited longitudinal space. The cavities are made of stainless steel to measure the position of single bunches at 1 μs spacing. A major problem in the mechanical design was to avoid asymmetries caused by welding. No active tuning system was allowed for this monitor.

Electronics

In most applications the resonance frequency is much higher than 1 GHz; consequently, rf-processing techniques are needed for signal detection. It is beyond

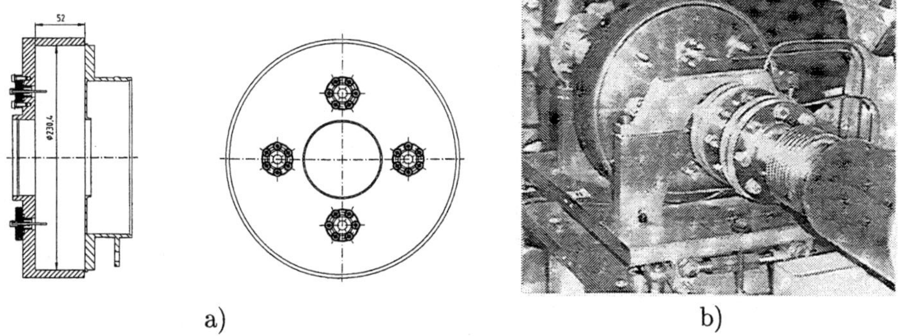

a) b)

FIGURE 10. a) Design of the cold TTFL-monitors; b) warm monitor installed in the TTFL.

the scope of this paper to review all existing electronics used for cavity BPMs, or to discuss the rf electronics in detail. But this fact leads to some general remarks since it results in a more complex R&D than for other (low-frequency) BPMs.

Often the electronics design depends on the availability of components — an additional argument for selecting the 'right' frequency. Some commercially used frequency bands are at 900 MHz, at 1.9 GHz (PCS), and around 11 GHz (TV-sat, DBS). Components developed for these applications are cheap and 'off the shelf' (COTS). Otherwise more sophisticated design tools are needed for the development of individual circuits, for which special care has to be taken in terms of shielding, reflections, etc.

Synchronous Demodulation

The signal coupled out of the cavity decays with the time constant τ_r and can be treated as an amplitude modulation of the TM_{110} resonance. Therefore, most of the electronics developed for a cavity BPM employ the superheterodyne receiving technique: the signal is demodulated by mixing it down to an *intermediate frequency* (IF) within one or more stages. A special case is the *homodyne receiver*, having an IF-frequency of $f_{IF} = 0$ and yielding the TM_{110} envelope. The phase between the reference signal and the TM_{110} signal has to be adjusted and stabilized. Sometimes *I/Q-mixers* are used [12], where the signals are mixed in-phase and in-phase-quadrature. This results in a coordinate system in which the vector of the output signals I and Q rotates with the IF frequency.

The dynamic range of a receiver is given by the maximum input signal which is amplified or mixed without distortion, and the noise. Signals of less than 10^{-19} W were detected at MAMI [7] by using a phase-sensitive synchronous demodulation scheme (lock-in amplifier at 100 kHz).

Reference Signal

A reference signal is needed for the normalization as well as for getting the starting phase of the resonating field. The latter gives the sign of the displacement: when the beam is on the right, the system can be set up to give positive video polarity.

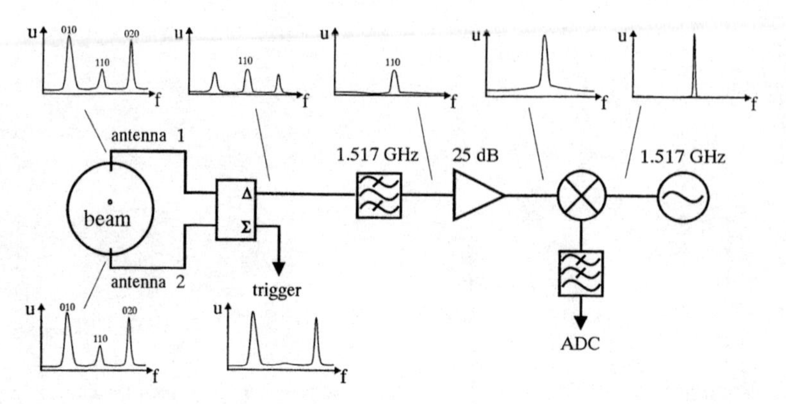

FIGURE 11. Block diagram of the homodyne receiver built for the detection of the beam-excited vertical TM_{110} polarization (TTFL-monitors, Example 4). The mixer is realized as an I/Q-mixer.

The signal changes the phase by 180° when the beam moves to the left, and for a centered beam it becomes zero. The phase difference relative to an external reference has to be measured approximately, however, at higher frequencies even this is not straightforward.

In many resonant monitors an additional circular cavity is used, excited in the TM_{010} mode. This cavity measures the bunch charge and yields a phase reference. Often its TM_{010} frequency is close to f_{110} in the BPM cavity. Another possibility is to use the *TM_{010} signal* of the BPM cavity. This signal — proportional to the bunch charge — is usually absorbed at the Σ-port of the field filter. In the third method, an external reference oscillator has to be phase-locked to the beam or to the timing system. Temperature drifts in long cables have to be corrected.

Measurements and Tests in the Lab

All cavity parameters — such as resonance frequencies, Q-values, coupling factors, and the effect of tuning systems — can be measured in the rf lab using a Vector Network Analyzer (VNA or generator). Usually, one can use the coupling ports of both polarizations, and no additional antenna is needed. The difference in the dielectric constants of vacuum and air causes a frequency offset of about 0.029 %.

The resolution and the precision of a BPM can be measured on a testbench by using a coaxial wire or an antenna. The cavity is excited by this antenna fed by a cw-source (VNA, Generator). For measuring the precision, the antenna is centered in one direction. Then the cavity is rotated by 180°, and the difference in the signal output yields the offset between the mechanical axis and the electrical center.

OTHER RESONANT STRUCTURES

Besides the simple circular cavity there are other resonant structures used in accelerators for beam position measurement (see, for example, [4]). One special example will be discussed below.

Higher Order Modes (HOM) in Accelerating Structures

Energy variations along the bunch train caused by wakefield effects lead to an increased projected emittance. This can be avoided in principle if the beam could be precisely centered in the accelerating structures to prevent the excitation of HOMs. Rather than relying on the BPMs and their good alignment with respect to the structures, a better approach to beam steering is to minimize the beam induced dipole signals directly [18]. Therefore, these signals occurring at frequencies higher than the accelerating mode[3] have to be measured.

A position measurement resolution of 12 μm was demonstrated for the dipole mode spectrum of the SLC-structures (4.14 to 4.35 GHz) by using amplitude and phase detection. The uncertainty in the determination of the absolute center position was larger.[4] Much better results were obtained during a test of a NLC prototype structure in the SLC linac. In these structures the lowest dipole mode band is Gaussianly detuned. Additional damping is provided by coupling the cells to four manifolds that run along the structure. These manifolds permit the measurement of the beam-induced dipole mode signals that originate throughout the structure. By these means it seems possible to realize an in situ straightness measurement of the structure.

CONCLUSIONS AND SUMMARY

Cavity BPMs offer many advantages, but they also have shortcomings:

- They provide very high transfer impedance (some kΩ) and high sensitivity, but also impose a large effect on the beam (wakefields, impedances).
- They offer good linearity over a wide position range and enable beam position measurement with a single output.
- The amplitude of the fundamental mode (the TM_{110} for rectangular and the TM_{010} for circular cavities) is proportional to the bunch charge or the current.
- The use of cavity BPMs saves longitudinal space compared to striplines, but costs transverse space.

3. For the properties of periodic structures, the reader is referred to [6].
4. Horizontal and vertical polarizations could not be separated; the couplers introduce asymmetries and the detected signals are a superposition from several structures.

Table 3 summarizes some principle arguments concerning the *cavity shape*. The signal levels are nearly the same for rectangular and for circular structures depending on the design.

TABLE 3. Arguments for the choice of the cavity shape.

Parameter	Circular Cavity	Rectangular Cavity	Remark
Fabrication	easy (turning)	standard waveguide	
Costs	low	moderate	
Precision	very high	moderate	welding, brazing
Tolerances	radius	waveguide length	most sens. parameter
Cooling	easy (copper tubes)	more complex	because of shape

Table 4 summarizes some of the existing cavities used for position detection; it is not a complete list of all resonant monitors.

TABLE 4. Existing cavities (∗) and new monitors (∘), beam-tested or under development.

Lab	Machine	Type	f_{110}	Q-values		Resolution	Excitation (min.)	
			[GHz]	Q_0	Q_L	[μm]	current (cw)	charge
Mainz	MAMI ∗	cyl.	2.449	9300	6100	50	1 nA	-
SLAC	SLC ∗	rect.	2.856			10	100μA	-
CERN	CLIC ∘	cyl.	30.000			0.1	-	1 nC
TJNAF	CEBAF ∘	cyl.	1.497		3500	10	1 nA	-
DESY	TTFL ∘	cyl.	1.517		1000	10	5 mA	8 nC
KEK	JLC ∘	cyl.	5.712		140	0.025	-	(1 nC)
Mainz	MAMI ∘	cyl.	9.795	7500	2500	(2)	1 μA	-
DESY	FEL ∘	cyl.	12.000		1000	10	-	1 nC
BINP	VLEPP ∘	cyl.	14.000			0.010	-	(?)

ACKNOWLEDGMENTS

I would like to thank all people who helped me in the preparation of this talk. Special thanks are extended to T. Shintake and H. Euteneuer for sending very useful information about their work at the FFTB and at Mainz.

REFERENCES

1. Chin, Y. H., *User's Guide for ABCI Version 8.7*, CERN SL/94-02 AP, 1994.
2. Balakin, V., et al., "Beam Position Monitor with Nanometer Resolution for Linear Collider," in *Proceedings of the European Particle Accelerator Conference (EPAC94)*, London, 1994, pp. 1539–1541.
3. Bane, K. L., M. Sands, "Wakefields of Very Short Bunches in an Accelerating Cavity," SLAC-PUB-4441, Nov. 1987.

4. Bossart, R., "Microwave Beam Position Monitor Using a Reentrant Coaxial Cavity," CLIC-Note 174, CERN, 1992.
5. Collins, R. E., "Field Theory of Guided Waves," New York: IEEE Press (1991).
6. Dome, G., "RF Theory," presented at the CERN Accelerator School, Exeter College, Oxford, 1991; also published as CERN 92-03, p. 1–96.
7. Euteneuer, H., Universitaet Mainz, private communication.
8. Doerk, T., Institut f. Kernphysik, Uni Mainz, Diploma Thesis, 1996.
9. Farkas, Z. D., et al., "Precision Energy Measurement Technique," SLAC-PUB-1970, 1977.
10. McKeown, J., "Beam Position Monitor Using a Single Cavity," in *IEEE Transactions on Nuclear Science*, Vol. 26, No. 3, 1979, pp. 3423–3425.
11. Kroll, N. M. and D. U. L. Yu, "Computer Determinations of the External Q and Resonant Frequency of Waveguide Loaded Cavities," *Part. Acc.* 34, pp. 231–250 (1990).
12. Lorenz, R., et al., "First Operating Experiences of Beam Position Monitors in the TESLA Test Facility Linac," contributed to the 1997 Particle Accelerator Conference (PAC97), Vancouver, May 1997.
13. Lorenz, R., et al., "Beam Position Measurement Inside the FEL-Undulator at the TESLA Test Facility Linac," presented at the DIPAC97, Frascati, 1997.
14. Mazaheri, G., et al., "Development of Nanometer Resolution C-Band Radio Frequency Beam Position Monitors in the Final Focus Test Beam," these proceedings.
15. Piller, M., et al., "1 nA Beam Position Monitor," unpublished manuscript.
16. Rossbach, J., "Options and Trade-Offs in Linear Collider Design," in *Proceedings of the 1995 Particle Accelerator Conference (PAC95)*, pp. 611–615 (1995).
17. Schnell, W., "Common-mode rejection in resonant microwave position monitors for linear colliders," CERN, CLIC Note 70 (1988).
18. Seidel, M., "Studies of beam Induced Dipole-Mode Signals in Accelerating Structures at the SLC," SLAC-PUB-7557, June 1997.
19. Shafer, R., "Beam Position Monitoring," in *AIP Conf. Proc. 212*, pp. 26–58 (1989).
20. Sladen, J. P. H., et al., "Measurement of the Precision of a CLIC Beam Position Monitor," CLIC Note 189, CERN (1993).
21. Sladen, J. P. H., et al., "CLIC Beam Position Monitor Tests," in *Proceedings of EPAC96* (conference held in Sitges, Barcelona, Spain, 10-14 June 1996), pp. 1609–1611 (1996).
22. Slater, J. C., *Microwave Electronics*, Bell Telephone Laboratory Series, New York: Van Nostrand (1950).
23. *A VUV Free Electron Laser at the TESLA Test Facility Linac — Conceptual Design Report*, DESY Hamburg, TESLA-FEL 95-03 (1995).
24. Weiland, T., "On the Numerical Solution of Maxwell's Equations and Applications in the Field of Accelerator Physics," *Part. Acc.* 15, pp. 245–292 (1984).
25. Vogel, V., "BPM for VLEPP," in *Proceedings of the ECFA-workshop on* e^+e^- *Linear Colliders (LC 92)*, workshop held in Garmisch-Partenkirchen, 1992.

Image Sensor Technology for Beam Instrumentation

R. Jung

CERN, CH1211 Geneva 23, Switzerland

Abstract. Beam monitoring using cameras has evolved from qualitative beam observation to precision measurement. After a description of the two main TV standards, various sensors including TV tubes (Vidicon), solid state sensors (Interline and Frame transfer CCDs, CMOS and CID X-Y matrices), and Fast Shutter/Intensifiers of the MCP type are reviewed. Comparative resolution measurements for the various sensors are given. The two types of sensor acquisition hardware, "frame grabbers" and "digital cameras," are described. Finally, special image processing requirements for beam instrumentation are reviewed, including radiation hardness, spectral sensitivity, fast acquisition, and enlarged dynamic range.

INTRODUCTION

Television cameras have been used since the early days of accelerators for beam observation. The main application was for a long time the observation of screens for beam steering purposes through transfer lines and for the first turn around circular machines. With the construction of lepton machines producing enough synchrotron radiation, and high intensity proton storage rings, the use of TV based monitors was extended to the measurement of beam dimensions for machine optimization and luminosity estimation. Instruments of this type were developed in several laboratories in the 1970s with tube TV cameras and included some form of digitization and numerical processing for beam size extraction. Despite the usefulness of these instruments, it took a certain number of years to have them accepted as precision instruments. The introduction of the CCD sensor, with its more than 100,000 cells having silicon-engraved precision of a few microns, came to maturity in the early eighties, just in time for LEP.

Camera-based beam monitors have now acquired a competitive position in beam instrumentation and are recognized as indispensable instruments in accelerators. The users now have maximum expectations from them.

The recent improvements of CMOS sensors and the interest of industry and universities in machine vision have induced enormous progress in the field and there is now a wide variety of hardware and software available, some of which is directly useful for beam instrumentation.

CP451, *Beam Instrumentation Workshop*
edited by R. O. Hettel, S. R. Smith, and J. D. Masek

TV STANDARDS

The original cameras were for TV broadcasting use. This application has imposed a certain number of features which are still applied in the field:

- the aspect ratio of 4/3 between horizontal and vertical image dimensions
- the interlace technique by which the full image is scanned and reconstructed with two frames, so as to limit the TV transmitter bandwidth
- the number of TV lines, which have to be an odd number, generated by a simple-to-build divider. The choices for the number of lines were:

$$525 = 3 \times 5^2 \times 7 \text{ [USA] and } 625 = 5^4 \text{ [Europe]}$$

- the frame frequency which is the mains frequency in order to maximize the noise rejection and satisfy flicker requirements acceptable to the human eye: 60 Hz in the USA and 50 Hz in Europe, and image frequencies which are half these values: 30 and 25 Hz
- the gamma correction for best visual contrast in monochrome reproduction, which states that the Object-to-TV monochrome image intensity relation is not linear but should follow a logarithmic law of the type:

$$I_{image} = k\,[I_{object}]^{\gamma}$$

  with $\gamma = 1.2$ for the whole chain. This value was determined by the movie film industry for black/white films to compensate the loss of color contrast. But for color TV, $\gamma = 1$ of course. This should also be the case for instrumentation. The cameras, including the focusing lens, tend to reproduce the spectral sensitivity of the eye, peaked around 550 nm, and spanning from 400 to 700 nm.
- the synchronization patterns for driving TV receivers.

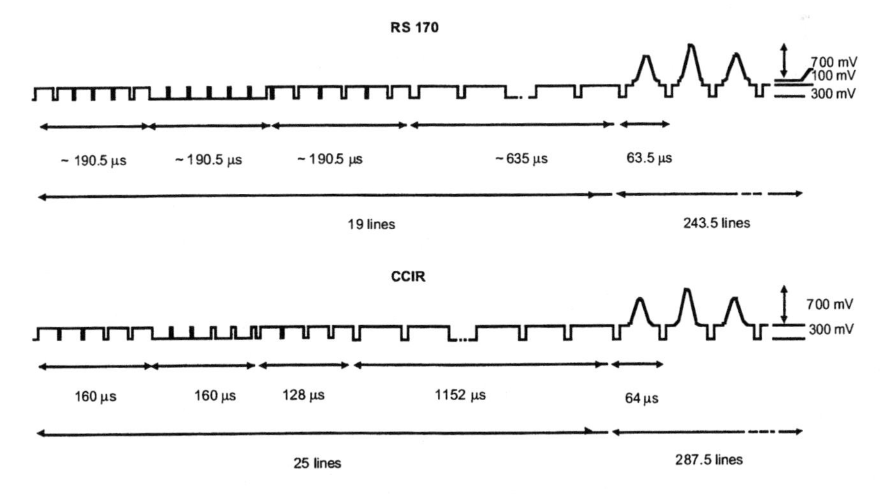

FIGURE 1. Timing diagrams and amplitude standards for an RS170 [USA] odd TV frame and a CCIR [Europe] even TV frame.

The previous features were translated into TV standards, RS170 for the USA and CCIR for Europe, similar to which all analog TV cameras comply. As can be seen in Figure 1, these standards are close to each other. The signal delivered by a TV camera is called the *composite video* signal. Only the so-called digital cameras do not comply to these TV standards.

A typical block diagram for a complete system is given in Figure 2.

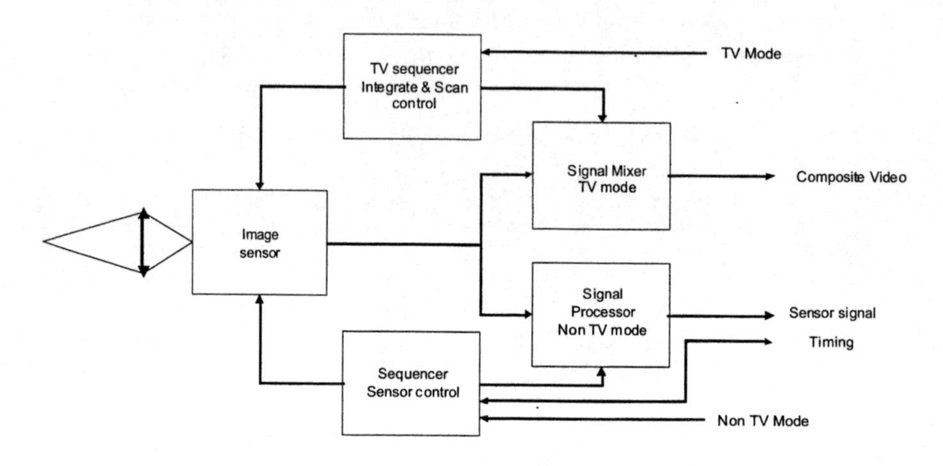

FIGURE 2. Block diagram of a complete camera system.

TUBE CAMERAS

The tube cameras were the first available image sensors. The only interest in them nowadays is for their radiation resistance, and hence the whole camera has to be made with tubes, including image sensor, amplifiers, and other active elements. Only the Vidicon (Figure 3), the most popular of them, and the SIT image sensors will be mentioned.

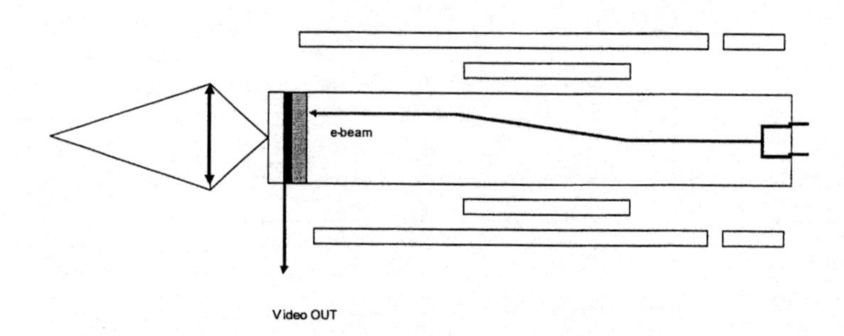

FIGURE 3. Schematic view of a Vidicon TV tube with deflection coils and lens.

The Vidicon sensor is an evacuated tube with a photoconductive target on which the scene of interest is imaged. The conductivity of the target is a replica of the scene illumination. The face of the target towards the input window is coated with a transparent metallic layer connected to the signal output of the tube. A low-energy electron beam orthogonally scans the inner surface of the target and the video output signal follows the illumination of the scene. The electron beam is focused with a solenoid coil and its deflection is controlled by magnetic fields generated by the deflection coils. The image is scanned in the usual raster scan every second line. The Vidicon has a sensitivity better than the eye in approximately the same spectral range, a gamma of around 0.7, and a rather long remanence of several frames. It has a very high radiation resistance.

The SIT (Silicon Intensified Target) is a two-stage tube. The first stage is an intensifier using a photocathode onto which the scene is imaged. The electrons emitted by it are accelerated to a silicon target made of tightly spaced p-n diodes where the accelerated electrons create electron-hole pairs. A scanning electron beam neutralizes the holes which are collected on the scanning side of the target and again generates the video signal, which is the replica of the scene. Due to the two stages, a gain of 1000 or more can be achieved with respect to the Vidicon. These tubes have a high sensitivity, a spectral range depending on the photocathode type, in general extended towards the infrared up to 800 nm, a low remanence, and a gamma of 1. The silicon target makes these tubes less radiation-resistant than the Vidicon.

Cameras of this type are used in high radiation areas where the integrated doses go above 10^4 Gy (10^6 rad). They suffer from external magnetic fields, which occur in pulsed machines and necessitate magnetic shielding. As for any standard TV cameras, they can be connected to TV monitors for direct observation and the composite video signal can be digitized for further processing by the standard “frame grabbers.” Due to the scanning mechanism, there is a time difference of one mains period between top and bottom of the same frame, and between two adjacent TV lines. This is in general not a severe limitation in usual instrumentation applications where time resolution is not a requirement. Their spatial resolution has been measured and is good (see Figure 15).

SOLID STATE CAMERAS

Solid-state sensors of the CMOS type appeared in the late sixties and CCDs in the early seventies. Despite its later appearance, the CCD sensor developed faster to a mature technology and is presently the leading technology. It is mainly the field of the large solid-state electronics companies and for the moment dominates scientific-type applications. CMOS is catching up rapidly. Because of its simpler technology, similar to that used for producing RAM, it is a field open to many companies, of all sizes, including university laboratory spin-offs. With this technology, it is possible to integrate, on the same chip, the image sensor and signal processing functions, including ADCs and digital processing.

Linear sensors exist mainly in CCD technology. They are a simplified version of the area sensor. They have many commercial applications and can be interesting in beam instrumentation applications where one-dimensional information is of interest and where readout speed is the main concern. Only area sensors will be further considered.

Both CCD and CMOS sensors are based on the photoelectric effect in silicon. When a photon of an appropriate wavelength (in general between 200 and 1000 nm) hits silicon, it generates an electron-hole pair. If an electric field is present, the electron and the hole are separated and charge can accumulate, proportional to the number of incident photons, and so reproduce the scene imaged onto the detector if a proper X-Y structure is present. Each basic element, defining the granularity of the sensor, is called a pixel (picture element). An image is composed of typically 400 × 300 pixels, with dimensions of the order of 10 to 25 μm. Two types of techniques can generate the pixel structure: a MOS capacitor or a p-n junction (see Figure 4).

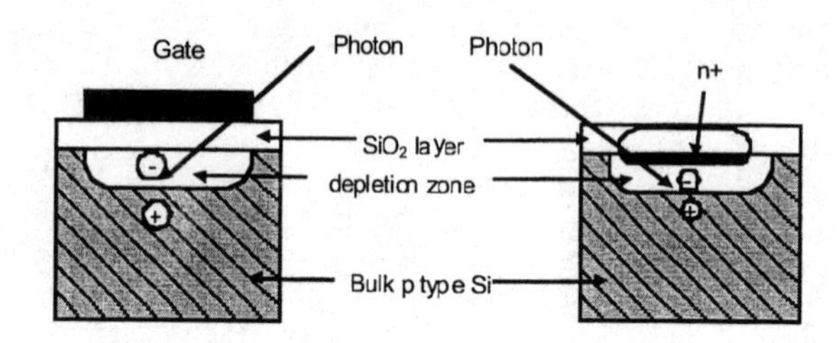

FIGURE 4. MOS capacitor and p-n diode structures for image sensors.

A positive bias applied onto the MOS gate creates a depletion region where the electrons are accumulated, whereas the holes disappear in the substrate. Electrons created in the substrate are either attracted to the depletion region or are lost, if they are created too far from it. This process is characterized by the Quantum Efficiency, having a maximum variance between 40% and 80%, depending on the technology. The electrons are accumulated for a certain length of time, in order to generate enough signal at read-out. In TV type applications, this integration length is limited by the mains frequency. Maximum accumulated charges are typically of the order of 300,000 electrons. As in any semiconductor device, there are also thermally generated electron-hole pairs which add to the signal and generate so-called dark current. In order to improve the sensitivity and reduce the noise, the sensors can be cooled, as the thermal current in silicon is divided by two every 8 K. The cooling ranges from tens of degrees, achieved with Peltier cells, down to cryogenic cooling in extreme cases.

The readout mode of the accumulated charges will define the type of sensor: CCD or X-Y matrix read-out. The solid-state sensors which have been used as examples in this tutorial are presented in Figure 5.

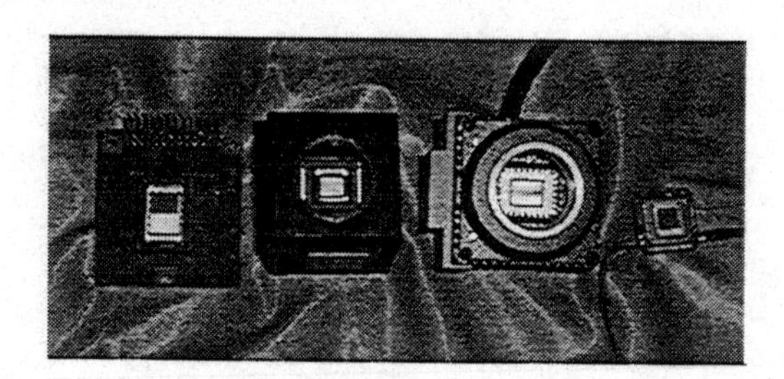

FIGURE 5. Solid state sensor cameras. From left to right: frame transfer and interline CCDs, CID, and CMOS XY sensors. The last one is a complete TV system.

CCD Sensors

Two types of CCD sensors are available: Interline (IL) and frame transfer (FT) sensors. Both work on the same readout principle, charge transfer, but the IL uses roughly half the silicon surface of that needed for a FT sensor and is hence more economical.

Interline Sensors

The structure of an IL sensor is given in Figure 6. The image-charge-to-vertical-memory transfer of low capacitance occurs in one clock period, which is fast, less than 1μs. The whole image is frozen at that given moment and all pixels have been exposed during exactly the same period. The integration period can easily be lowered to microseconds, if enough light is available, to create an "electronic shutter." From there onwards, the charges are shifted line-by-line towards the horizontal output register. Each line is then shifted out towards the unique output amplifier where the charges are converted into voltage by the "floating diffusion" capacitance. All these operations occur during one frame period. There are few individual elements in the readout chain, so the "fixed pattern" noise generated by these differences is kept small.

The light collection area is smaller than the chip because of the Vertical Memory columns. This is, in general, characterized by a filling factor. Because of this, the horizontal and vertical resolutions can be different. Some cameras have double the described structure for the best interlace resolution. They can work in "progressive scan" where the whole image, i.e. the two frames, are read together.

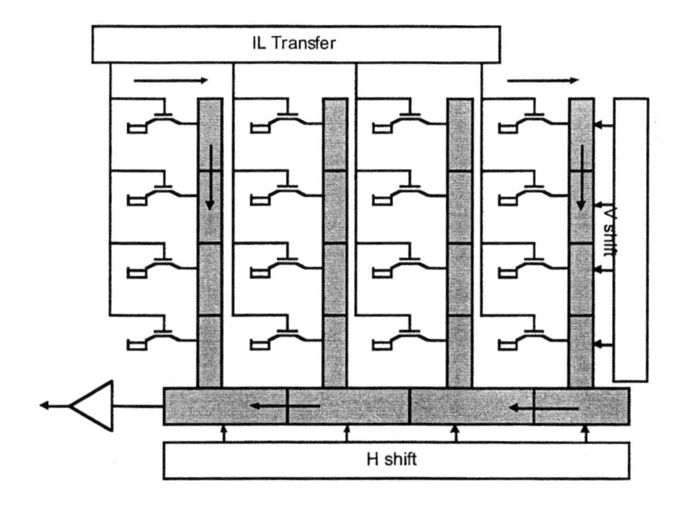

FIGURE 6. Interline CCD structure: after the integration period, the collected charges are all transferred to the column storage area, from there they are clocked down to the line storage register and finally clocked out to the Output amplifier, all during the next integration period. The storage registers are metallized (shaded area) to protect them from incident light.

Frame Transfer Sensors

The structure of a frame transfer CCD is given in Figure 7. The scene is imaged onto the Image Area which can be 100% available for integrating the incident light. After the integration period, the full Image Area is transferred to the Memory Area. The "image" is then shifted line by line to the output register, from where it is clocked out to the output amplifier at the chosen rate, which can be the standard TV rate.

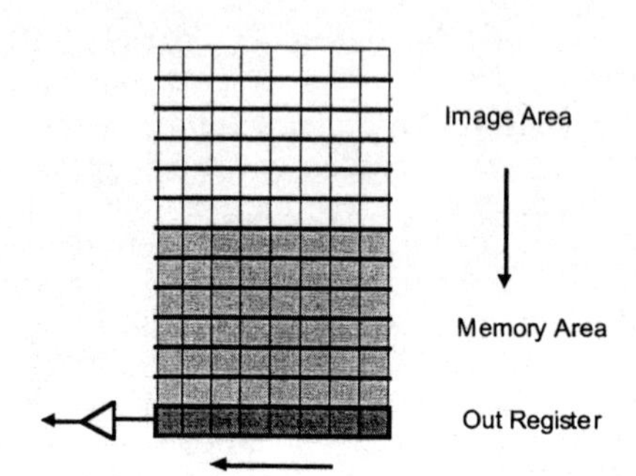

FIGURE 7. Frame transfer CCD structure: after the integration period, the charges of the Image area are shifted into the Memory area, and from there read out line by line into the out register from where they are clocked out into the amplifier.

The memory and output registers are protected from the ambient light by a metallic layer. The columns are defined by the silicon structure, with so-called stop-bands, whereas the lines are defined by electrodes deposited onto silicon oxide. These same electrodes will control the various shifts of the charges by well-defined timing sequences. This is shown in Fig. 8 for a four-phase system, corresponding to a column during integration and at four different moments of a charge shift sequence. The lower rectangle in each line corresponds to the silicon substrate. Above it is an SiO_2 layer, and above that, the polysilicon electrode structure, with a four-fold periodicity. The electrodes, which are positively biased, are represented in black. The first line corresponds to the integration period. The charges are collected under the first three electrodes. This defines the active pixel. The center-of-charge is below the central biased electrode.

Once the integration period is over, the bias voltage of the first two electrode sets is turned off, while the voltage of the third set is turned on. At the end of this transition, the charges have moved to the right, and so on, until the fifth line, when the charges have moved by one full period to the right. One full pixel shift takes place in approximately 1 μs. A full shift from Image to Memory area takes of the order of 300 μs. This is short compared to the 16.7 or 20 ms of one TV frame, but can induce a vertical "smearing" of the image in some cases.

For the next TV frame, the sensor has to produce an image, shifted by one line. This does not reproduce exactly the TV tube raster scan, as all charges are always collected. What is done instead is to shift the previous electrical pattern of the bias electrodes by two electrodes, the accumulation of charges being now centered under the white electrode of the first line in Figure 8. The center-of-charge is hence shifted by two electrodes or half a pixel, which is satisfactory for a TV monitor, and can be of some use for precision measurements.

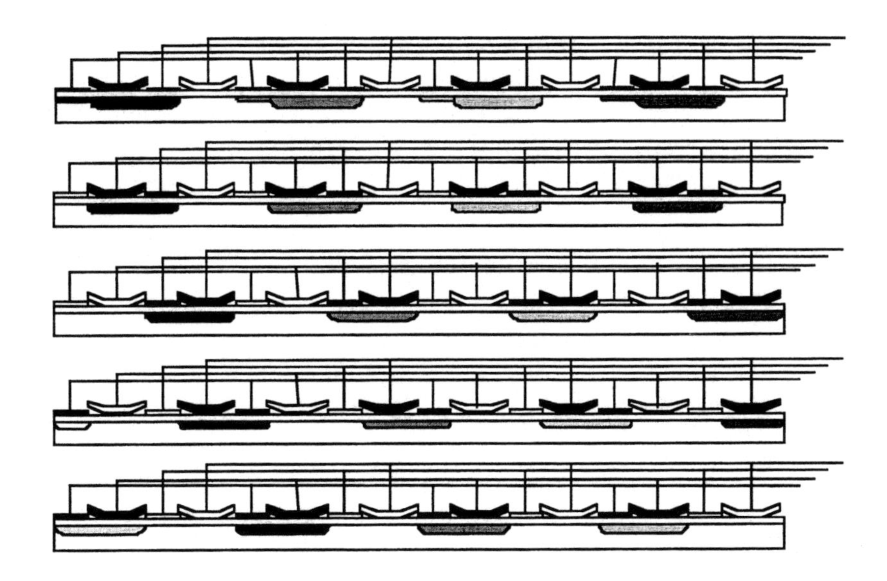

FIGURE 8. CCD structure and timing sequence for a four-phase sensor for the integration period and a complete shift of one pixel. The biased electrodes are in black, and the charge-collection areas are the shaded areas below the biased electrodes. During the integration, one electrode (gray shaded) is biased at a smaller voltage.

Frame transfer CCDs have, in principle, the best sensitivity, as the full silicon area is available for photon-to-charge conversion. They have for the same reason a good resolution for a given pixel size, a high output signal and a good uniformity. Their disadvantage comes from the many charge shifts, typically from 300 to 1000 from the closest to the farthest pixel. Figure 9 depicts a measurement of charge transfer efficiencies (CTE) made at different pixel readout frequencies. It should be pointed out that the individual shifts are very efficient: in this example from 99.97% to 99.99%.

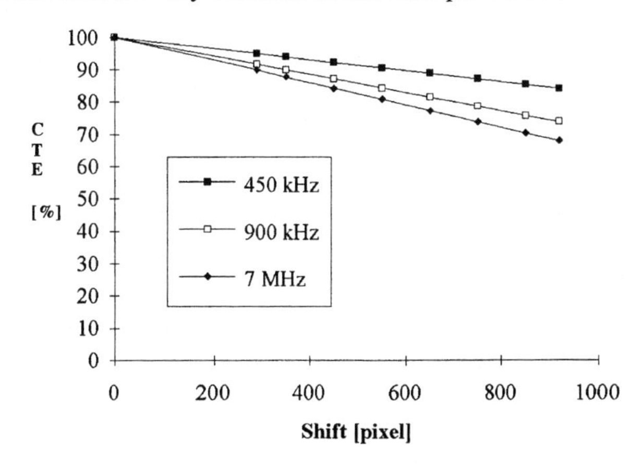

FIGURE 9. Measured Charge Transfer Efficiencies [CTE] as a function of readout frequency. The normal TV readout frequency is close to 7 MHz.

The lower spectral sensitivity of the CCD is limited by the electrode and protection window transmittances to wavelengths longer than 400 nm. To go below these wavelengths, either UV scintillator coatings or back-illuminated CCDs are used. The coated CCDs are a much cheaper solution, but their resolution is limited by the emission angle of the coating. The back-illuminated CCD has a normal CCD structure, illuminated from the back. The silicon has to be thinned so as to avoid the recombination of the photon-generated electrons with holes before reaching the collecting potential well. The efficiency can be the double of an ordinary CCD. This process results of course in a higher cost.

X-Y CMOS and CID sensors

In a CMOS sensor, each pixel is comprised of a photodiode with a MOS switch and is individually addressable in the following way. All gates on one line are connected to a horizontal register and all drains of the MOS transistors in one column are connected to a sense line connected through a transistor to the video output amplifier (see Figure 10). Each pixel is read sequentially once per frame.

Some more elaborate detectors of this type use several MOS transistors. Others, rather than sequentially reading the integrated charge, sequentially integrate the current generated by each photodiode for a given time. The most elaborate use a log amplifier in order to achieve the highest dynamic range.

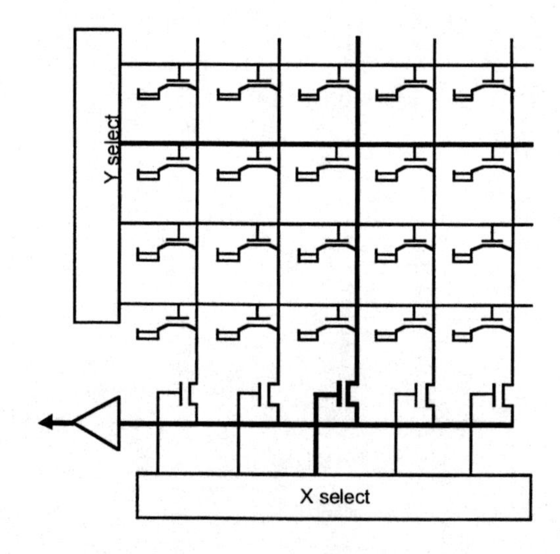

FIGURE 10. X-Y matrix image sensor: the Y selection register selects a pixel row, of which the X selection register selects a pixel at the intersection of both lines, which is connected to the output amplifier.

The major disadvantages of this architecture are a potential lower sensitivity due to the smaller photo-active surface available and a fixed-pattern noise due to the differences in the individual transistors. These disadvantages are disappearing rapidly as the technology progresses and as more signal processing functions are incorporated on the sensor chip. Such incorporation is easy in this technology. Recent sensors have a performance comparable to or even better than the best Frame transfer CCDs (see Figure 15). The possibility of integrating signal processing circuits on the silicon of the sensors is a big advantage. The suppression of the fixed-pattern noise by double correlated sampling is one of the important functions which are available. There can also be an automatic integration time adjustment for best dynamic range with fixed-diaphragm optics. This can be a disadvantage in pulsed-light operation, frequent in accelerator environments, unless precautions are taken. Some of the chips also incorporate an ADC, resulting in a true digital image sensor of small dimension. These components will probably be the dominant image sensors in the next generation.

The CID (Charge Injection Device) is also a sensor of the X-Y type. It is comprised of two MOS photogates. The charges are read non-destructively when they are shifted from one photogate to the other. Integration can be resumed after readout, in principle, for each pixel individually if the signal level is too low. The charges are discarded before a new integration period by "injection" into the substrate. They are the most radiation-resistant sensors. Unfortunately, these devices are single-sourced.

INTENSIFIERS

An intensifier is a vacuum tube optoelectronic device. It is comprised of a photocathode that emits electrons, a gain mechanism, and a screen transforming the electron flux back to photons. In first generation intensifiers, the gain was given by the electrostatic acceleration of the photoelectrons. It was moderate and the image distortions were not negligible. In the second generation, the gain is given by the multiplication of the electrons by secondary emission in a Multi Channel Plate (MCP) made of many little conductive glass tubes, typically 10 μm in diameter. One electron generates many electrons, which gives the amplification but also degrades the resolution of the device. There can be more than one plate. The principle is given in Figure 11.

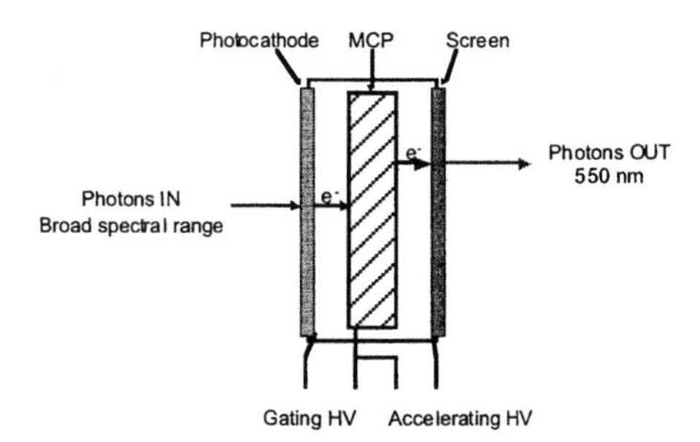

FIGURE 11. Operating principle of a MCP intensifier.

Third-generation intensifiers have a Gallium Arsenide photocathode in place of the usual multi-alkali photocathode of second generation intensifiers in order to increase the infrared response, which is not of concern in beam instrumentation in general. They will not be considered further here.

The intensifier can be used in DC or pulsed mode. Pulsed intensifiers are used to solve three different problems:

- not enough light available for a good Signal to Noise ratio
- time resolution by gating the intensifier
- wavelength shifting by proper choice of the photocathode and screen materials: the screen material is chosen for the best match with the sensor spectral sensitivity, and for a decay time compatible with the expected time resolution.

The intensifier is coupled to a CCD or another sensor either by lenses or by direct fiber optic coupling. The latter is the most compact and efficient method.

Intensifiers can produce very short gate times, down to a few nanoseconds, and work up to a certain repetition frequency; 10 kHz is often quoted by the manufacturers. The reality is in general less good. In the LEP synchrotron light telescopes, it was soon realized that the MCP caused two problems. First, it induced a broadening of the beam spot, offsetting completely the gain on diffraction-limited resolution by working in the UV (1); and second, it was not able to work for measurement purposes at the LEP revolution frequency of 11 kHz. The last limitation is the most serious one and is thought to come from the lack of electron replacement in the MCP tube. Systematic measurements were made on two types of MCPs: a standard one and a so-called "high strip current" type. In Figure 12 are given the normalized evolutions in amplitude and beam width for a 10 kHz repetition rate for different number of pulses per TV frame.

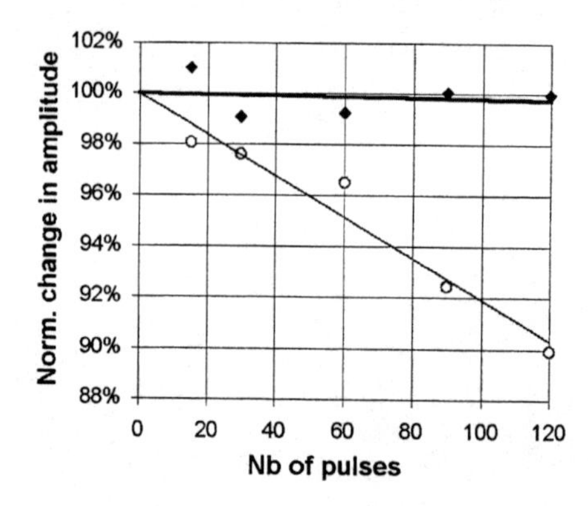

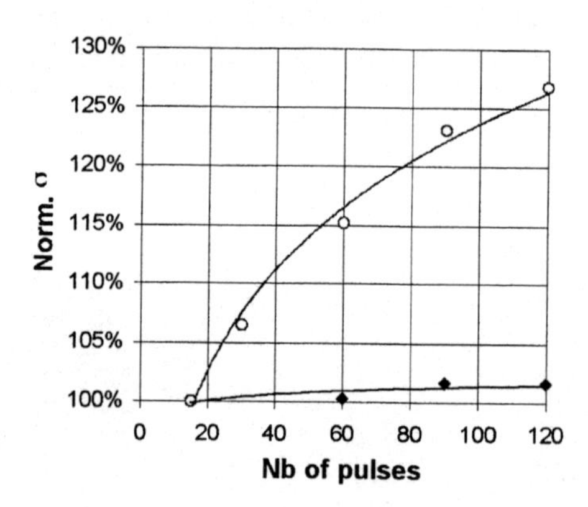

FIGURE 12. Comparison of normal (open circles) and "high strip current" (full diamonds) MCP intensifiers for an increasing number of pulses per TV frame. Left: normalized evolution of the signal level per pulse. Right: normalized evolution of the spot size under the same conditions. The measurements have been made at 10 kHz with a gain of 400.

The "high strip current" MCP has a better performance both in amplitude and for beam size conservation. The input signal level, gain, and repetition frequency also influence the performance.

DIGITAL IMAGE ACQUISITION

The direct observation of a beam spot on a TV screen has long been the principal use of camera monitors in beam instrumentation. With the possibility of digitizing the images, a new field was opened for these monitors: the evaluation and monitoring of beam emittances, replacing SEM-Grids in transfer lines and wire scanners in circular machines where light emission was available in sufficient quantity. There are two main ways to digitize an image: either by acquiring the standard TV signal with a "frame grabber," or by digitizing the pixels individually, which is done by the "digital" or "slow scan" cameras.

Frame Grabbers

A frame grabber takes the standard TV signal (RS170 or CCIR), uses the synchronization signals to start a frame recording and restart a new line, generates (in general) an internal clock, digitizes the video signal with a fast ADC, and stores the result in memory (Figure 13). There is sometimes a synchronization on the actual pixel clock and some pre-processing of the data stored in memory.

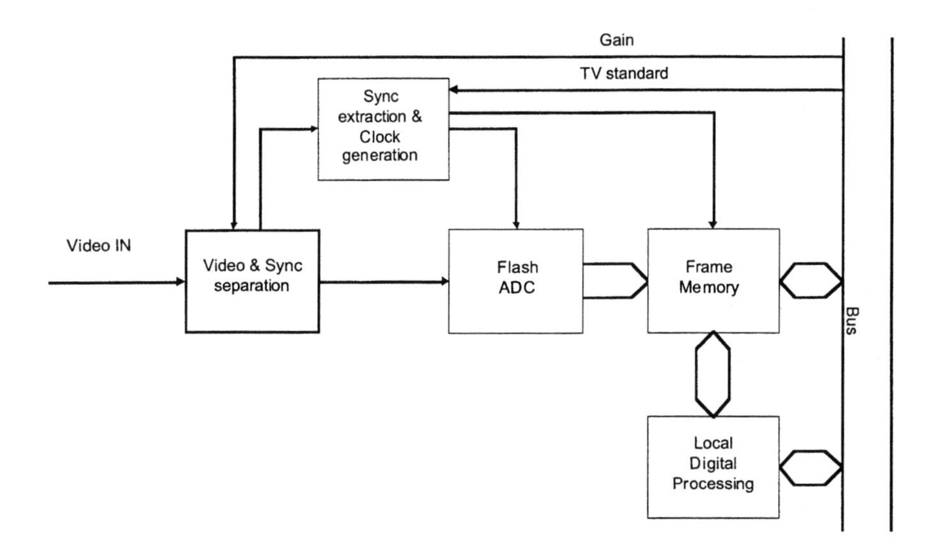

FIGURE 13. Block diagram of a frame grabber.

Most of the frame grabbers use an 8-bit flash ADC, which is adequate for the majority of applications. The digitizing rate is in general half the frame rate, i.e. 30 or 25 Hz. The interface with the digital world is either through a standard bus system or a serial link. The most popular ones are the PCI bus and the RS 232 or higher performance RS 422 interfaces. A few systems are compatible with the VME standard. The number of available PCI cards is growing rapidly.

Digital Cameras

A so-called digital camera is comprised of an image sensor and digitizing circuits, delivering to the outside world digital information which is or can be processed further. Being independent of the video standards, the system can be optimized for best performance. As the sensor quality is closer to 12 rather than to 8 bits, these systems start in general at 12 bits and go frequently up to 16 bits, i.e. 65,000 gray levels instead of 256! The price to pay is a slower digitizing frequency or a smaller digitized area, which is particularly well suited to beam instrumentation applications, where the beams are approximate ellipses located in a well known region. For CCD sensors, the digitizer is usually on a different card and takes full control of the sensor, i.e. integration start and stop, image-to-memory transfer and individual pixel read-out. The digitizer can also take a reference dark-level image to subtract from the real image in order to get rid of thermal and fixed pattern noise, and so increase the dynamic range. For CMOS sensors, the pre-processing and ADC can be incorporated on the sensor chip, increasing also the signal-to-noise and dynamic range. The recent progress in this field is remarkable. The data acquisition and bus interface board can contain many processing functions. A typical structure is illustrated in Figure 14.

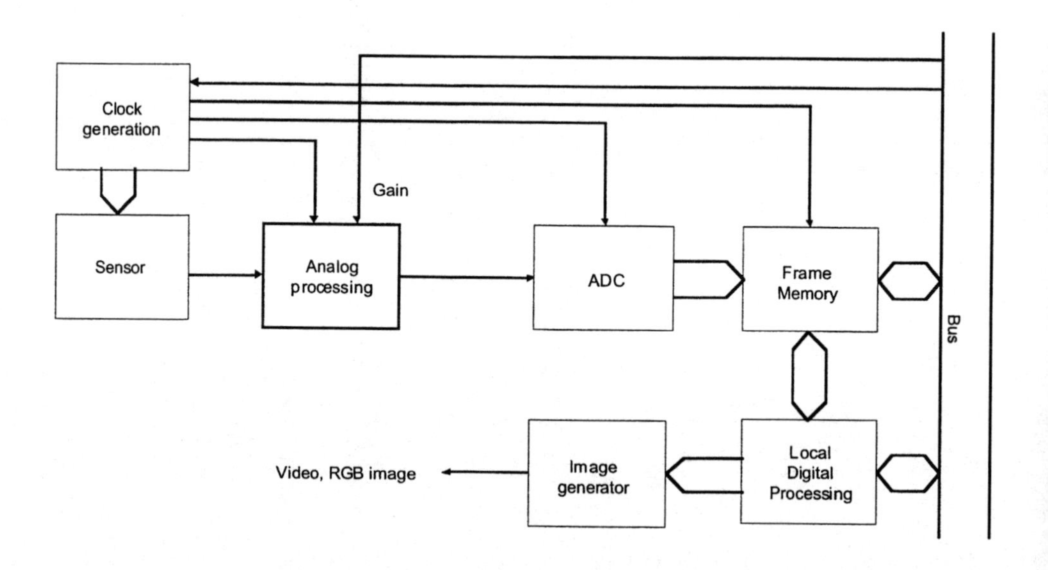

FIGURE 14. Block diagram of a digital camera.

The advantage of these cameras is their better digital data quality, their disadvantage being the lack of a direct video image to follow, in real time, the beam behavior. Some digital systems include image generators for that reason.

SPATIAL RESOLUTION MEASUREMENTS FOR THE VARIOUS SENSOR TYPES

Spatial resolution measurements have been made with a standard test target [USAF 1951] and a target to sensor magnification adjusted to have always a horizontal field of view of 10 cm, typical for screen observations. The images were acquired with a PCI frame grabber. The contrast C between the black and white bars of the test pattern in Figure 15, and the contrast transfer function (CTF), are defined as:

$$C = \frac{V_{\max} - V_{\min}}{V_{\max} + V_{\min}} \tag{1}$$

$$CTF = \frac{C_{image}}{C_{testplate}} \tag{2}$$

The results are summarized in Table 1 and Figure 15. The data for each camera are plotted as a function of the spatial frequency of the test plate, expressed in "line pairs per mm", together with the data of the FT CCD camera for easy comparison. The measurements have been performed with the sensors used around SPS and LEP. They are representative of the different technologies.

The best resolution is achieved by the modern X-Y CMOS matrix. The performance improvement is enormous compared to a previous sensor tested two years ago (2). The Frame transfer CCD, now ten years old, comes second. The interline camera comes next. The performance of the Vidicon camera is good, probably better than expected by most users, slightly better than the CID camera, and makes this type of camera a good solution for radiation areas.

The performance degradation of the FT CCD camera with MCP intensifiers has to be noticed (see Table 1). This degradation will increase with repetition rate. Nevertheless, the low resistance MCP has an honorable performance and can be used when fast gating is a necessity.

TABLE 1. Spatial Resolution Measurements Normalized to the Full Sensor Size

Sensor	Line pairs for 50% CTF	Line pairs for 10% CTF
X-Y CMOS	200	310
CCD Frame transfer	180	240
CCD Interline	140	230
X-Y CID	140	200
Vidicon	140	220
Low resist. MCP [1 Hz]	110	210
Normal MCP [1 kHz]	40	130

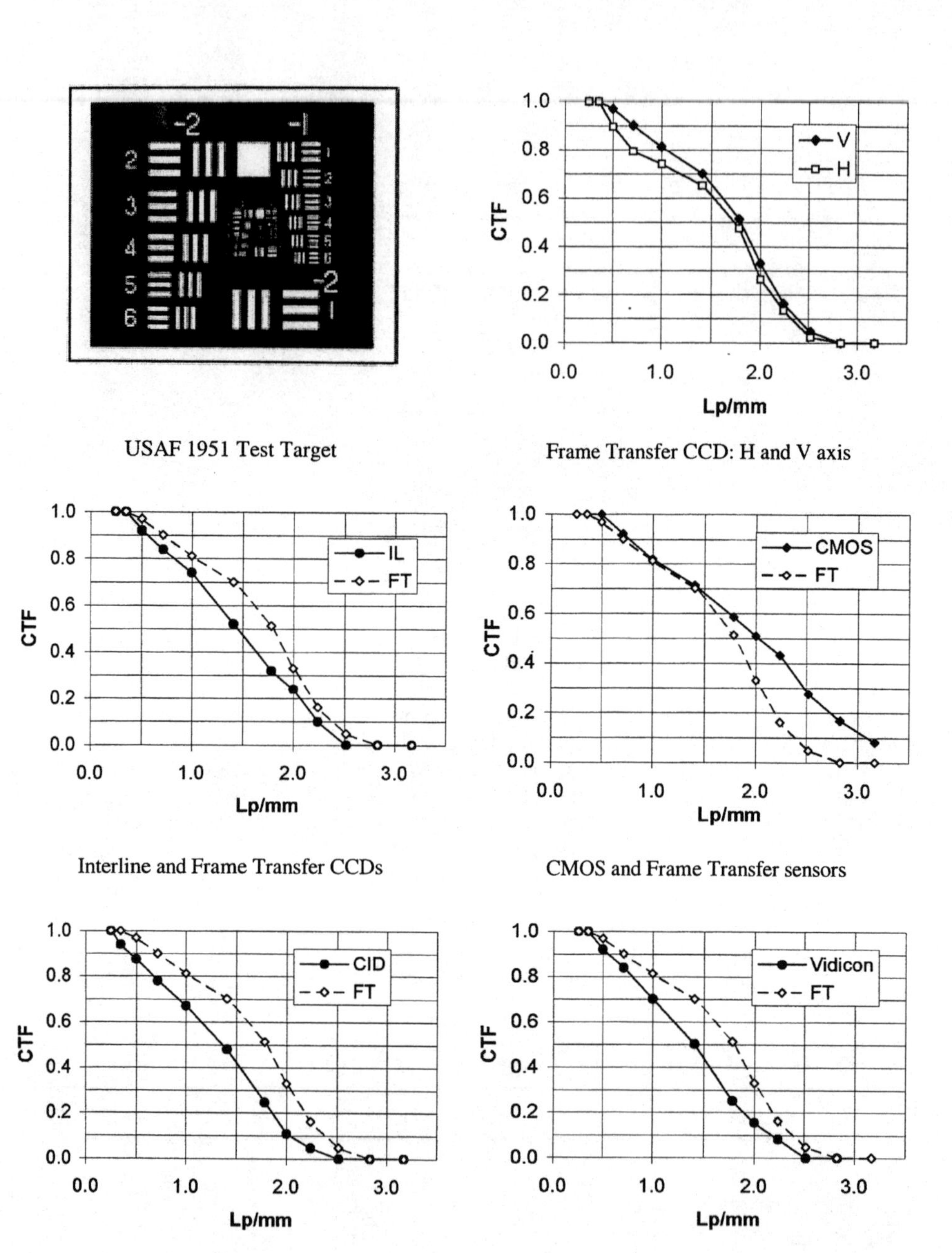

FIGURE 15. Comparative resolution measurements for various cameras. The measurements were performed with a USAF 1951 standard test target and a PCI frame grabber. The results are plotted as a function of the spatial frequency of the target, expressed in "line pairs per mm" for a 100 mm horizontal field of view.

SPECIAL FEATURES FOR BEAM INSTRUMENTATION

Beam instrumentation is a small field compared to the consumer and machine vision markets. Hence, commercially available devices are not specifically tuned toward this field and the specific demands for beam instrumentation have to be considered when choosing a system.

All of the instruments are in a radiation environment where they have to survive. The radiation dose and the lifetime to be achieved will, in general, determine the choice of sensor to be used: TV tube, CCD, CMOS, or their radiation hardened counterparts. Tubes make, of course, the most resistant cameras. At the CERN SPS such cameras are used in areas with annual doses greater than 10^6 Gy. Normal CCDs have shown damage at levels of 10 Gy. Radiation-hard CCDs and CIDs are claimed to resist from 10^3 to 10^4 Gy. The radiation causes an increase in dark current resulting in an overall decrease in well capacity, i.e. in contrast, which is clearly visible on the CTF (see Figure 16). The local radiation environment will also influence the location of the processing electronics. In some accelerators, this may mean that the processing takes place hundreds of meters away from the sensor. As a consequence, frame grabber acquisitions can be degraded with respect to pure digital acquisitions. This is visible in Figure 16 with the CTF measured for the same image sensor in video mode and in digital mode. One way to overcome this limitation is maybe to use fiber optic links, as was done at RHIC (3).

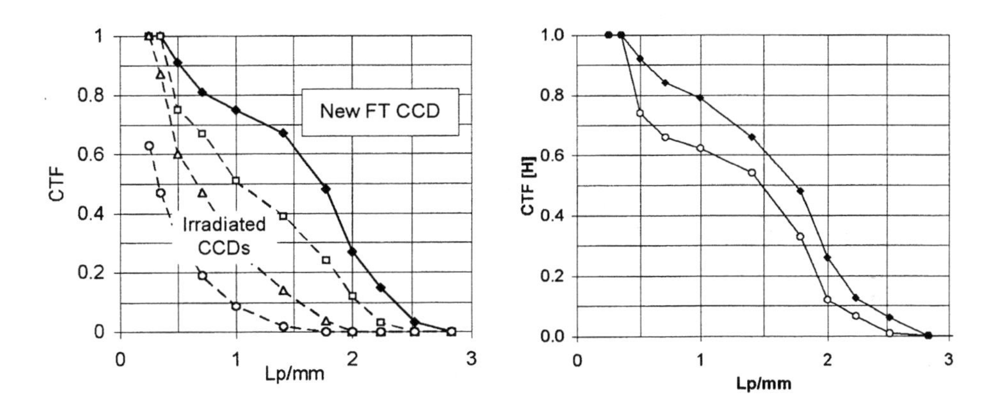

FIGURE 16. Contrast Transfer Functions. Left: for new (full diamonds) and irradiated (open figures) CCDs. Right: Digital (full diamonds) and frame grabber (open circles) acquisitions with 1000 m of cable between sensor and digitizer.

The spectral sensitivity of the sensor has to match the available spectrum or select the best slice for the precision of the instrument. Representative spectral sensitivities are given in Figure 17. Quite a number of cameras for visual applications include infrared filters, simulating the eye sensitivity for the best use of the commercial achromats which have been optimized between 450 and 650 nm. But a popular screen material, Al_2O_3 (Cr) has a peak emission around 700 nm, at the limit of the commercial IR filters, which has given surprises in some accelerators! In general, color cameras are of little or no interest in beam instrumentation.

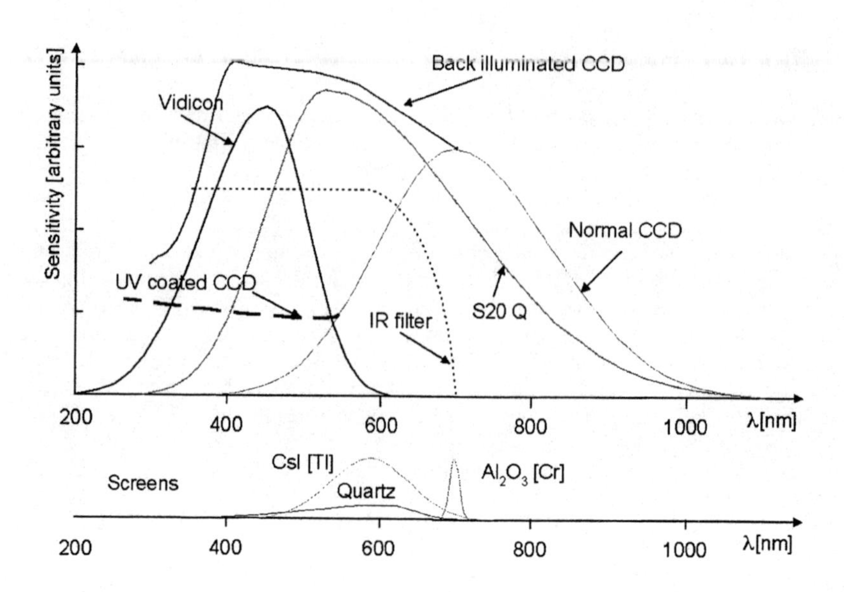

FIGURE 17. Relative spectral sensitivities of common image sensors and spectral emission of some widely used screens.

Fast turn-by-turn beam cross-section acquisition (2), beyond the mains frequency, is useful for beam instability or beam matching observations. The latter can be performed by measuring the profile variations over several turns after injection, but before filamentation. They can be made by using the CCD chip as an analog buffer memory. The principle is explained in Figure 18. It makes use of an intensifier used as a fast shutter and a CCD used as memory. It is called the "burst mode". The number of pulses stored on the chip can be doubled by using the full chip. The main limitation comes from the recovery time of the intensifier, limiting the acquisition frequency to around 10 kHz.

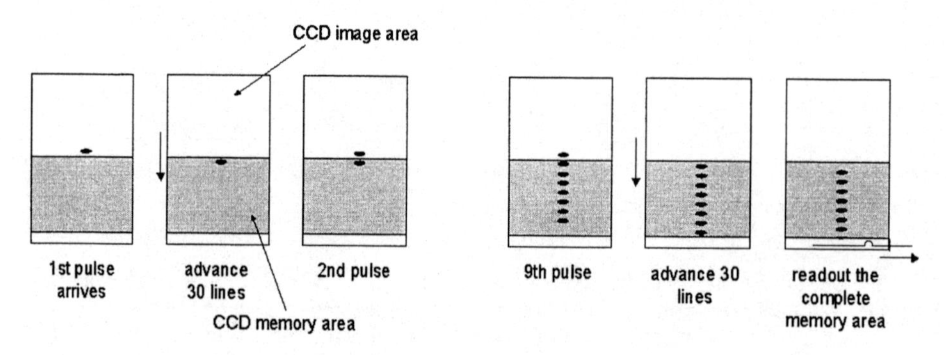

FIGURE 18. Principle of the "burst mode."

Projections are needed for emittance calculations and can be obtained in several ways. Most frame grabbers come with software, which in one way or another can perform projections of the data contained in a region of interest (ROI). Another possibility is to perform the projection by hardware-summing of the ROI in the frame grabber memory, which makes a 25 or 30 Hz rate possible. Finally, the projection can be achieved by summing the charges of a beam cross-section directly on a CCD, in a mode which is an extension of the "burst mode" described previously (see Figure 19). This time, the charges are not only stored on the CCD, but the CCD is used as a processing circuit, performing the summing operation on-chip. Here again, the limitation comes from the intensifier. With a perfect intensifier, the operation could be performed at a 30 kHz rate with a 12 bit ADC, 1000 times above the normal video image rate.

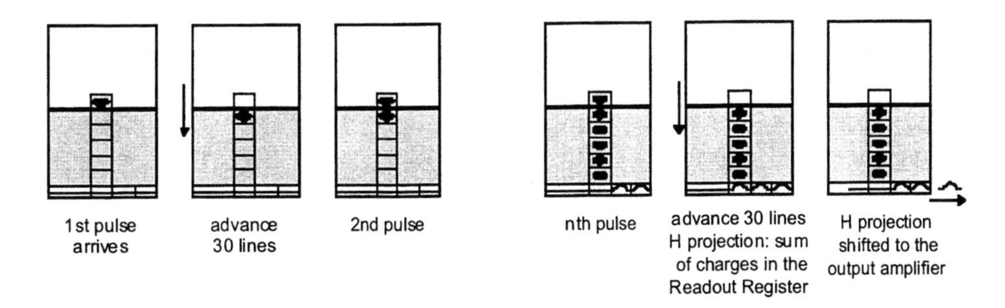

FIGURE 19. Principle of the "fast projection" mode.

Another field of interest is a large dynamic range to study non-gaussian beam tails. Whatever the light source used, be it synchrotron light or screens, there is, in general, little energy available in the region of interest. Another limitation is the radiation environment which makes it safer to keep most of the electronics away from the beam tunnel, increasing the distance between sensor and digitizer. Digitizing over 16 or more bits is in general not a solution, because of the Signal-to-Noise ratio. A possibility, which has been used at least in Argonne (4) and in LEP to extend the dynamic range to 16 bits equivalent, is to attenuate the dense part of the beam by a known amount and reconstruct from a 12-bit image a 16-bit equivalent beam profile (5).

Another possibility is to use the selective integration possibility of the CID sensors.

CONCLUSION

Beam monitors using image sensors are recognized now as precision instruments. Tube cameras are the most radiation-hard sensors. The Vidicon is the most widely used. It has good resolution and sensitivity. Silicon target tubes can be considered for higher sensitivity. The main drawbacks are the normal component lifetime, the volume, and the influence of external magnetic fields.

Solid state cameras are taking over where the radiation levels are lower.

The CCD is the mature technology, with the frame transfer type the most interesting for scientific uses. These have special features which make them interesting for accelerator diagnostics, such as the burst and fast projection modes. Interline CCDs are not interesting, except for microsecond shutter speed.

CMOS sensors are developing quickly. Their technology, similar to RAM, makes them cheaper to produce than CCDs, while yielding better performance than most CCDs. The machine vision and consumer market will push them to the front of the scene while pushing their prices down. Due to the on-chip processing capabilities, and considering the progress made over the past few years, it seems that they will be the image sensors of the future.

Software was not touched upon in this tutorial. A lot of it is already available commercially. Most of the frame grabbers come with sophisticated software, some of which is of direct interest in the field of beam instrumentation. Here again, it is the consumer and machine vision market which has stimulated development. It is estimated that half of the vision market, which amounts world-wide to hundreds of millions of dollars, is represented by software!

It is now possible to buy complete instruments comprised of sensor, digitizer, and software. But, because of the small market of accelerator beam instrumentation, the items available are not always adequate for these applications. Even worse, their specifications or descriptions are often misleading. One of the aims of this tutorial was to make the potential user aware of the various technologies and their performance. One must also be aware that the field is changing very rapidly.

ACKNOWLEDGMENTS

This tutorial condenses the experience accumulated over the years together with my colleagues on the optical monitors in the ISR, SPS and LEP. It is a pleasure to acknowledge their contributions, with a special mention for L. Robillard for his many sensor qualification measurements.

REFERENCES

[1] Burtin, G., et al., *Proc. of the 1993 IEEE Part. Acc. Conf.*, 1993, pp. 2495–2497.
[2] Colchester, et al., BIW'96, *AIP Conf. Proc.,* **390**, May 1996, pp. 215–222.
[3] Witkover, R. L., *Proc. of the 1995 IEEE Part. Acc. Conf.*, 1996, pp. 2589–2591.
[4] Lumpkin, A. H., M. D. Wilke, BIW 92, *AIP Conf. Proc.,* ***281***, 1993, pp. 141–149.
[5] Burkhardt, H., et al., SL-MD Note 238, CERN, May 1997.

INVITED PAPERS

Instrumentation and Diagnostics for PEP-II*

Alan S. Fisher

Stanford Linear Accelerator Center
Stanford University
MS 17, P.O. Box 4349, Stanford, California
94309

Abstract. PEP-II is a 2.2 km-circumference collider with a 2.1 A, 3.1 GeV positron ring (the low-energy ring) 1 m above a 1 A, 9 GeV electron ring (the high-energy ring); both rings are designed to allow an upgrade to 3 A. Since June 1997, we have had three runs totaling 14 weeks to commission the full HER, reaching a current of 0.75 A. Positrons were transported through the first 90 m of the LER in January 1998, with full-ring tests planned for the summer. This workshop provides a timely opportunity to review the design of the beam diagnostics and their performance, with an emphasis on what works, what doesn't, and what we're doing to improve it. This paper discusses: the synchrotron-light monitor, including both transverse imaging onto a CCD camera and longitudinal measurements with a streak camera; beam position monitors, with processors capable of 1024-turn records, FFTs, and phase-advance measurements; tune measurements with a spectrum analyzer, including software for peak tracking; measurements of both the total ring current and the charge in each bucket, for real-time control of the fill; and beam loss monitors using small Cherenkov detectors for measuring losses from both stored and injected beam.

INTRODUCTION

The PEP-II *B* Factory (1) is a 2.2 km-circumference, two-ring, e^+e^- collider under construction at the Stanford Linear Accelerator Center (SLAC) in the tunnel of the original PEP single-ring collider. The project is a collaboration with the Lawrence Berkeley and Lawrence Livermore National Laboratories (LBNL and LLNL). Its goal is the study of *CP* violation by tracking the decay of $B\overline{B}$ meson pairs produced with nonzero momentum in the lab frame. The design involves two rings at different energies; both rings require large currents for high luminosity. The 2.1 A, 3.1 GeV positron ring (the low-energy ring, or LER) runs 1 m above the 9 GeV, 1 A electron ring (the high-energy ring, or HER). At one interaction point (IP), the LER comes down to the height of the HER, and the two beams collide with zero crossing angle in the BABAR detector. Table 1 lists several of the parameters for PEP-II operation.

Because the HER reuses the PEP-I magnets (although with a new, low-impedance, vacuum chamber), it began commissioning first, in May 1997, and has accumulated 14

---

* Supported by the U.S. Department of Energy under contracts DE-AC03-76SF00515 for SLAC and DE-AC03-76SF00098 for LBNL.

CP451, *Beam Instrumentation Workshop*
edited by R. O. Hettel, S. R. Smith, and J. D. Masek

weeks of full-ring operation through the end of the January 1998 run. At that point, the maximum current reached 750 mA in 1222 bunches. Both horizontal and vertical feedback were running; longitudinal feedback had been commissioned but was not running since power amplifiers were out for repair. The full rf voltage (15 MV) was available, with low-level feedback to stabilize the output. This paper reviews the design of the beam diagnostics and discusses their performance during commissioning.

LER commissioning also began that month with the injection and transport of beam through the first 90 m of the ring. The first run of the complete LER is planned for July 1998, followed by high-current commissioning and colliding-beam studies in the fall. BABAR will be installed in early 1999.

TABLE 1. PEP-II Parameters

Parameter	HER	LER	Unit
Circumference	2199.318		m
Revolution frequency	136.312		kHz
Revolution time	7.336		μs
rf frequency	476		MHz
Harmonic number	3492		
Number of full buckets	1658		
Bunch separation	4.20		ns
Luminosity	3×10^{33}		$cm^{-2}\cdot s^{-1}$
Center-of-mass energy	10.58		GeV
Current	0.99 (3 max)	2.16 (3 max)	A
Energy	9.01 (12 max at 1 A)	3.10 (3.5 max)	GeV
rf voltage	14.0	3.4	MV
Synchrotron tune	0.0449	0.0334	
Betatron tunes (x,y)	24.617, 23.635	38.570, 36.642	
Emittances (x,y)	49.18, 1.48	65.58, 1.97	nm·rad
Bend radius in arc dipoles	165	13.75	m
Bend radius in SLM dipole	165	43.45	m
Critical energy in arc dipoles	9.80 (23.23 max)	4.81 (6.92 max)	keV
Critical energy in SLM dipole	9.80 (23.23 max)	1.52 (2.19 max)	keV

SYNCHROTRON-LIGHT MONITOR

Synchrotron radiation (SR) in the visible and near ultraviolet (600–200 nm) has been used to measure HER beam profiles in all three dimensions. The measurements are made in the middle of Region 7, a high-dispersion point in the middle of a HER arc. A second synchrotron-light monitor (SLM) for the HER was planned near the start of the arc, where the dispersion is low, but has not been built due to budget limitations. An SLM is also planned for the LER in Region 2. We first present the HER system, and then discuss the modifications needed for the LER.

HER Synchrotron-Light Monitor

The high current in each ring leads to a high SR power on the first mirror. The ring's design does not permit the solutions common in synchrotron light sources. Space in the narrow tunnel does not allow backing the mirror away to reduce the heat load. Access to SR is restricted, with few ports, all presenting little impedance to the beam. To reduce the power along the SR stripe, the beam is incident on the mirror at 4° to grazing, giving a maximum power (for nominal energy and 3 A) of 200 W/cm in Arc 7 of the HER (and 19 W/cm for the LER in interaction region 2 (IR-2)). We detail the arrangement for the HER.

The HER arcs are almost entirely filled by the 5.4 m dipoles, with a quadrupole, corrector and sextupole taking up much of the rest of the 7.6 m half cell (see Fig. 1).

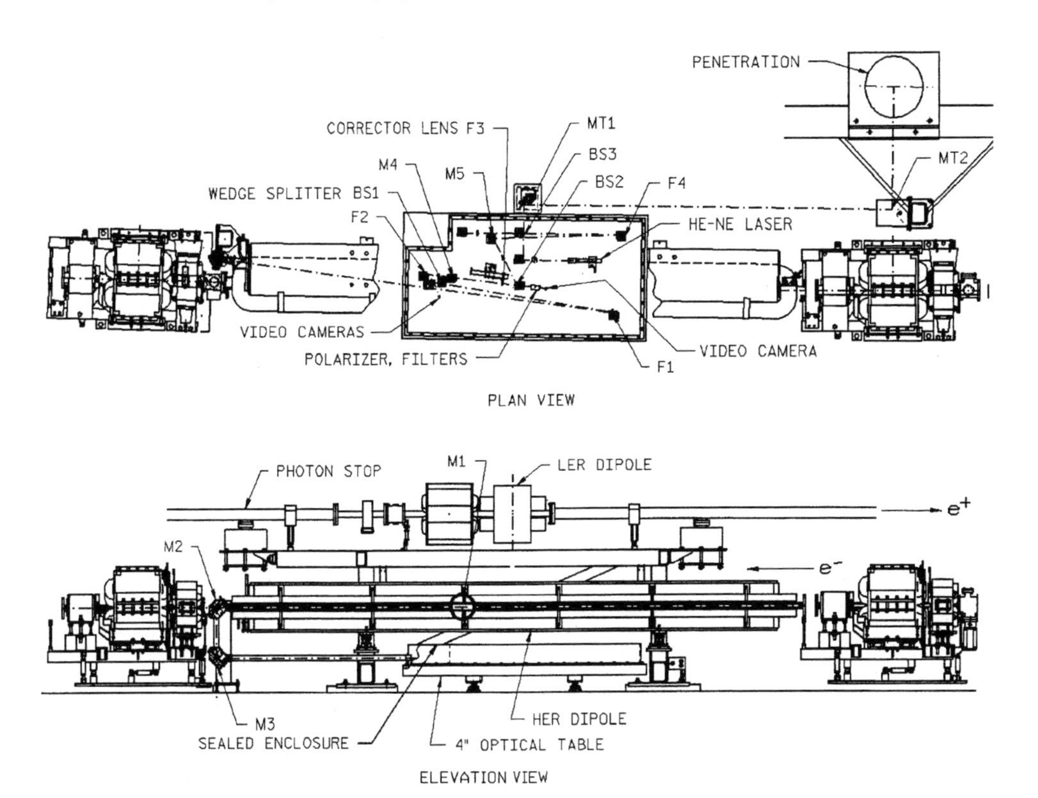

FIGURE 1. HER and LER beamlines in the middle of Arc 7 in (a) an elevation view and (b) a plan view, showing path of the HER synchrotron light going down to the enclosure on the optical table below the HER dipole. Part of the light continues from the table to the penetration leading up to the streak-camera lab.

The intense SR fan strikes the water-cooled outer wall of the chamber. The first mirror (Fig. 2), mounted in the vacuum chamber on the arc's outer wall, reflects the light horizontally across the chamber to the downstream inner corner. The mirror is slightly rotated to recess the upstream edge behind the opening in the chamber wall, so

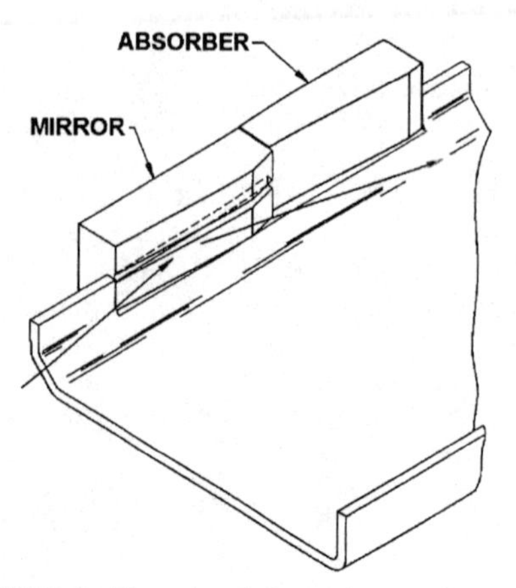

FIGURE 2. The slotted first mirror (M1) and the x-ray absorber, both mounted in the wall of the HER chamber.

that it does not receive power at normal incidence. The downstream edge sticks slightly into the chamber, shading the leading edge of the chamber as it resumes downstream of the mirror.

At 200 W/cm, the mirror cannot be cooled sufficiently to obtain adequate flatness for good imaging. Instead, note that the high-power SR fan at the critical energy is 15 times narrower than the visible fan we image. When the electrons travel on axis, a 4 mm-high slot along the mid-plane of the 7 cm long mirror passes the x-ray fan, while visible light reflects from the surfaces above and below. Because of grazing incidence, the x-rays never reach the bottom of the slot, which tapers from 0 to 5 mm in depth, but travel past the mirror to dump their heat into a thermally separate absorber (Fig. 2). The residual heat load of 1 W/cm^2, due largely to scattered SR, causes a temperature variation across the surface of less than 1° C.

However, the electron beam will not always be correctly positioned. Then we do not demand that the mirror be suitable for imaging, but only that it not exceed its yield strength. We can then steer the electrons back to their proper orbit and wait for the mirror to cool. Both the mirror and the dump are made of Glidcop® (copper strengthened with a dispersion of fine aluminum-oxide particles) with water-cooling channels, following techniques (2) developed for the Advanced Light Source. An ANSYS thermal analysis of a beam hitting the mirror 2.5 mm above the top of the slot shows that the temperature for a 3 A HER beam rises from 35° C to 160° C, and the stress rises to 90% of yield (3).

Two 45° mirrors, with a fused-silica window in between, transport the light from M1 to imaging optics in a nitrogen-filled enclosure on an optical table located below the HER dipole (Fig. 1(a)). This location was chosen to get good resolution from a short, stable optical path. The dipole itself provides radiation shielding. Both mirrors are motorized for remote adjustment, to correct for changes in the beam's orbit.

Figure 1(b) shows the imaging scheme, designed to compensate for the effect of the slot. In geometric optics, a slot or aperture placed in the plane of a lens (like a camera iris) serves only to restrict uniformly the amount of light reaching the image plane without otherwise affecting the image. Here, the first focusing mirror F1 images the slot onto the second focusing mirror F2, which then images the beam onto a CCD camera. Another camera images M1, so that we can center the SR on the slot. The third focusing element, a motorized lens F3, adjusts the focus for different beam orbits.

We also considered two effects of diffraction. Table 2 shows the loss in resolution due to the small vertical dimension of the beam. To reduce this effect, the light at all three emission points is taken near horizontally defocusing quadrupoles, where the beams are large vertically. For a point source, diffraction from the slot causes some narrowing of the full width at half maximum (FWHM) of the image and creates tails, but the effect is small for our 4 mm slot. However, when this pattern is convolved with a narrow ($\sigma_y/\sigma_d = 2$, where σ_d is the diffraction spot size) Gaussian electron beam, the tails disappear and the distribution broadens by 6% over the FWHM without a slot. The increase is less with PEP's broader beams.

TABLE 2. Resolution of the SLM for Measurements at 300 nm. Here, the diffraction spot size, $0.26(\rho\lambda^2)^{1/3}$, uses a larger experimental coefficient (from LEP) rather than the calculated value of 0.21. The image size is given by the quadrature addition of the source size and the diffraction size.

	HER Mid-Arc 7	LER Mid-Arc 7	LER IR-2
Radius of curvature in dipole [m]	165	13.75	43.45
Diffraction spot size σ_d [μm]	64	28	41
Electron/positron beam size σ_x [μm]	1000	622	2003
Electron/positron beam size σ_y [μm]	176	216	161
σ_y / σ_d	2.8	7.7	3.9
Image size σ_{image} [μm]	187	218	166
$\sigma_{image} / \sigma_y$	1.06	1.01	1.03

This system, installed for our September run, soon produced beam images. For example, at high current and without feedback, the SLM displayed strong oscillations on the electron beam. Bunch-by-bunch transverse (4) and longitudinal (5) feedback decreased the motion and the spot size (Fig. 3). We use a video digitizer to determine the transverse beam size from the images. As the beam current varies, we set the electronic shutter of the CCD (Pulnix TM-7EX) to adjust for the light level; color filters are not usually inserted.

Problems also became apparent early on. First, we found that a manufacturing error led to poor alignment between M1 and the axis of the exit tube. Some corrective bending of the chamber was needed in the clean room before installation. Afterward, a beam bump using horizontal corrector magnets let us position the electron beam to center the light reflected from M1 as it enters the exit tube.

More seriously, on the TV monitor we found a second, somewhat distorted, image of the electrons above the main one (the one we show in Fig. 3); the images are separated by about the height of the TV screen. By scanning the electrons vertically (in order to move the region of illumination above or below the slot in M1), we could see that the two images came from the two halves of the mirror. The shapes and separation of the two images did not vary with the beam current, suggesting that the distortion was not thermal in origin. We also were unable to reach best focus within the range of the motorized lens F3, again suggesting a mirror distortion.

We next tried to determine the cause of the distortion. The M1 assembly, with its internal cooling channels, was first brazed at SLAC before being sent out for nickel plating and optical polishing at SESO in Marseilles, France. To keep the mirror balanced during polishing, SESO asked us not to attach the long stainless-steel tubes that bring the water through the vacuum housing to the back side of the mirror. Instead, we attached short stumps of tubing before polishing, and welded the tubes to the stumps after the mirror returned. Additional complexity resulted from PEP's policy of forbid-

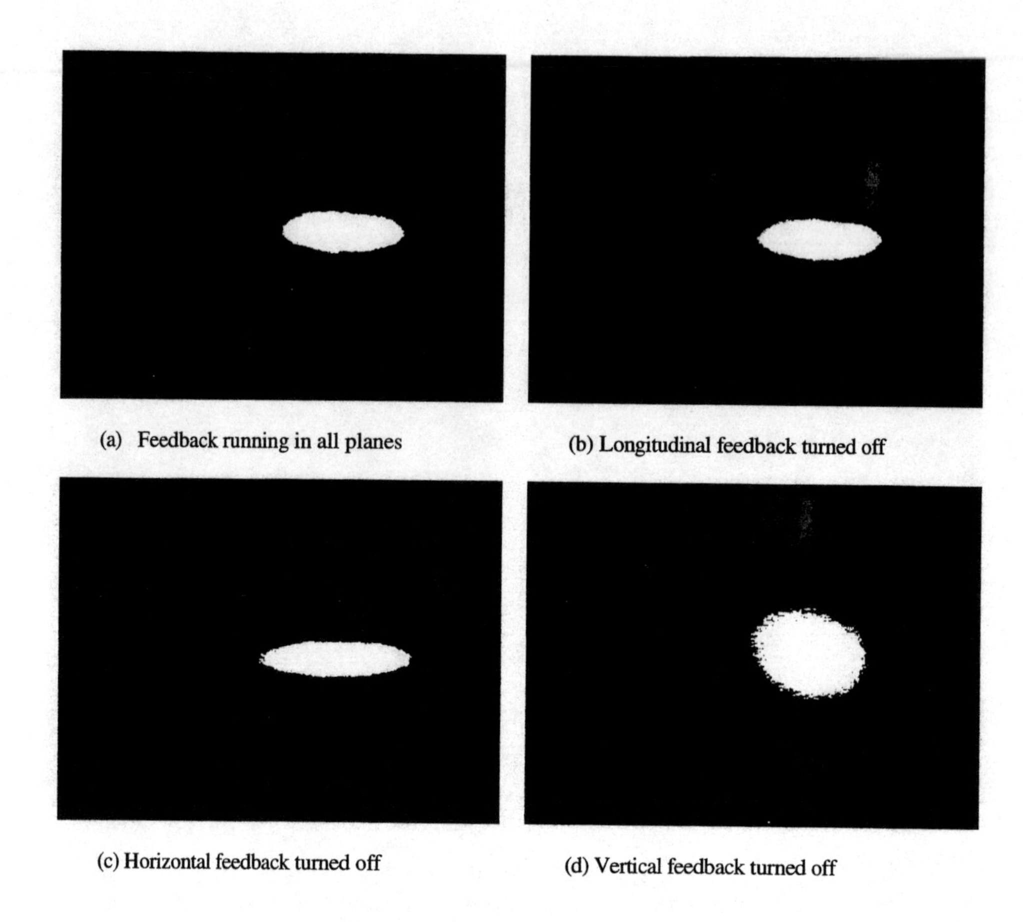

(a) Feedback running in all planes

(b) Longitudinal feedback turned off

(c) Horizontal feedback turned off

(d) Vertical feedback turned off

FIGURE 3. Transverse images of the HER beam.

ding welds or brazes between water and vacuum: an *air guard* (a volume connected to outside air) must surround the water weld.

In November, after the run, we removed the mirror assembly for inspection, and took it to LLNL to make an interferogram for comparison with one done at SESO. Although M1 had been flat to $\lambda/30$rms (using HeNe-laser red, 633 nm, and including both halves) after polishing, it was no longer so; the variations on one half alone were $\lambda/5$ rms. The worst deformation was on the front face opposite the insertion for a cooling tube on the back side. It appears that shrinkage while this weld was cooling deformed the mirror's front face. It also appeared that the mirror had folded slightly, so that the upper and lower halves were no longer on the same plane. When M1 is mounted inside the vacuum chamber (where we were unable to test it), this effect may be compounded by additional stress from the pressure of the mirror's support bracket against the mounting surface on the chamber, and from the weight and stiffness of the cooling tubes (although the tubes had been coiled to add flexibility).

After studying these problems, we remounted the mirror in December. A bellows just upstream of the source point was removed to place a survey telescope along the path of the synchrotron light, to re-survey and adjust the mirror's position. Two mirrors diverted the optical path into the aisle so that a reticle (focusing target) could be used as a source point. The CCD camera was repositioned to focus with F3 in mid-range.

The new camera position allowed us to focus the synchrotron light by scanning the focusing stage and finding a minimum spot size. We also bumped the orbit horizontally, to vary the distance along the tangent from the source point to M1. However, the minimum was at different settings for the vertical and horizontal planes, and for both, the measured beam size was larger than expected. The double image from the two mirror halves was still present.

For the long term, we have discussed two approaches. We might rebuild the mirror, with suitable modifications. The rear plate, where the cooling tubes attach, needs to be stiffer, and we must further relieve the mirror of the weight and stress of the long lines. The risk, of course, is that a new design carries the risk of new problems.

Alternatively, we are investigating the use of adaptive optics, a new technology in which a deformable "rubber" mirror inverts the wavefront errors measured by a wavefront sensor. The method was originally developed for the military and later for astronomy. Japan's KEK *B* Factory has devised its own version and tested it at the Photon Factory, in order to simplify the primary mirror of its SLM (6). (In addition, a special weak dipole was added to provide a source point with lower SR power.)

Now these components are about to become more commercial and less astronomical in both size and price. One firm (7) has a DARPA (Defense Advanced Projects Research Agency) contract to commercialize a 1 cm diameter deformable mirror with 325 actuators on a 0.5 mm grid, driven by a multichannel DAC and matched to a wavefront sensor. We are now discussing the possibility of SLAC participation in a "beta" test of this system for PEP's October 1998 run. The deformable mirror would be located at M1's image plane on the optical table, with the sensor just downstream to form a closed feedback loop. A PC can calculate the correction and control the loop directly, or, for greater bandwidth, the loop can be run by a digital signal processor (DSP) on the wavefront-sensor board. Since the deformation we see is largely independent of time and temperature, a static correction may be sufficient, but subtler, time-dependent effects as the beam current changes may become apparent after the static error is corrected.

LER Synchrotron-Light Monitor

LER measurements were originally planned (8) for the middle of Arc 7, with the LER's primary mirror directly above that of the HER. This arrangement would allow us to combine the optics for both rings on one optical table in the tunnel, and to transport the light from both up to the same optics room and streak camera. However, the aluminum vacuum chambers in the LER arcs are quite different from the copper chambers of the HER. In LER arcs, the SR diverging from the beam downstream of each dipole enters an antechamber; two-thirds of the photons strike a water-cooled photon stop 6 m beyond the bend (see Fig. 1(a)), while the remainder, emitted in the downstream part of the bend, continue to the next photon stop. To extract the photons, we would need a vertical slot in a photon stop; in addition, we would have to modify the design for the primary mirror M1 and its mount.

To economize, we are instead now planning to install the LER's SLM in IR-2. Because this is the straight section of Region 2, with the IP at the center, the LER has dipoles that steer the positrons down to the level of the electrons, then horizontally across the interaction point (IP), and finally back up. Downstream of the IP, one horizontal bend followed by a long drift has been selected for the SLM. The SR from the various bends makes the heat load in this straight larger than the others;

consequently, the chambers use the same octagonal copper extrusions as the HER arcs. We already have a copy of the HER's M1, originally built for the second HER SLM; this mirror will fit without modification into a chamber that will be fitted with a mirror-mounting flange and a light-exit port. The disadvantage of choosing IR-2 is that the LER system cannot share the HER's optical table, controls, or streak-camera room, and there is a much longer distance to any possible site for the streak camera.

The dispersion at the SLM location in IR-2 is low, as it also is in the original Arc-7 location. No second SLM had been planned for the LER since there is no suitable high-dispersion bend.

Streak-Camera Measurements

In addition to these transverse-profile measurements, the January 1998 run included longitudinal studies using a streak camera. The HER light is split in front of the camera; one half is transported through an 11 m penetration to a ground-level optics room, as shown in Figure 1(b). For single bunches, we measured bunch length versus current and rf voltage. Multibunch studies looked at longitudinal instabilities as we varied the current and fill pattern. These results are presented in a separate paper (9) at this Workshop.

BEAM POSITION MONITORS

The beam position monitor system (10) must provide a wide dynamic range, with the ability to measure both a multibunch beam in a full ring (up to 8×10^{10} $e^{\pm}$ with 238 MHz spacing) and a single bunch of 5×10^{8} during injection. To tune injection for a top-off fill, we plan to inject a single small bunch into the ion-clearing gap, and measure its position without interference from a 3 A stored beam. This requires a gated measurement and careful impedance matching to reduce round-trip reflections on the long cables to the processors. Table 3 compares the requirements to bench-test results.

TABLE 3. BPM resolution for single-bunch fills, with and without averaging over 1024 turns. Measurements were made on the bench and so do not include the beat-frequency effect discussed in the text.

Averaging [Turns]	Charge [$e^{\pm}$]	Required [μm]	Measured [μm]
1	5×10^{8}	1000	<100
1	10^{10}	100	<20
1024	10^{10}	15	<1

Each BPM uses four 15 mm-diameter pickup buttons, matched to 50 Ω and designed to contribute a low impedance to the ring. They are located at each quadrupole, with ≈300 sets per ring. To avoid synchrotron radiation, they are placed ±45° off the horizontal (or, in the octagonal HER arc chambers and the elliptical LER arc chambers, at points where 45° field lines terminate on the wall).

Except near the interaction and injection points, only one plane is measured—x only at QFs, y at QDs—in order to economize on cables and processors. In a filter-isolator box (FIB) next to the quad, the four button signals are filtered at $2f_{rf}$ (952 MHz) and combined in pairs (top and bottom, or left and right). Out-of-band power is matched into a load. For two-plane measurements, special FIBs filter but do not combine. An isolator on each FIB output limits a second pass of signal reflected from residual mismatch at the processor input.

The ring I&Q (RInQ) processor is a CAMAC module receiving one x and one y BPM signal from each ring. To limit cable length (for both cost and high-frequency loss), most RInQs are in the tunnel, in crates under the HER dipoles for shielding. Each channel has a 20 MHz bandpass filter, which sets the time resolution to 20 ns; a programmable input attenuator, to adjust for dynamic range; an in-phase and quadrature (I&Q) demodulator, to convert the signal to baseband cosine and sine components, while avoiding dependence on the phase of the rf reference; a track-and-hold with a gate opened once per turn and centered on any selected bucket; and a 14-bit digitizer. The wide bandwidth argued for direct conversion rather than using an intermediate frequency. A digital signal processor (a TI320C31 DSP) computes the position and records 1024 turns (either consecutive or every *N*th) for display, averaging, or a fast Fourier transform (FFT). Both the digital and rf circuits share a single printed-circuit board, made in two sections with different dielectrics.

To economize on cables and processors, the HER and LER signals are multiplexed for demodulation and digitizing. To measure a small charge in one ring while the other is full, a SPST switch is followed in series by a high-isolation DPST switch; the paths are separated and shielded upstream of the second switch. In retrospect, this choice was less economical than it appeared: the isolation was difficult to achieve, while cellular telephones have rapidly reduced the cost of components for this frequency range.

The RInQ includes two on-board, tunable, digitally synthesized frequency sources, clocked by the ring rf. One is used for the local oscillator (LO) for the I&Q; the other provides a calibration signal introduced into the signal path by a 10 dB directional coupler at the input. We calibrate the ADC pedestals, the gain ratios of the I and Q channels, their phase offset from 90°, and the top/bottom and left/right gain ratios. Since PEP must operate as a "factory," these calibrations must be performed with beam stored in the rings, by measuring during the gap. The calibration source is tuned slightly off the LO, many measurements are made, and the points are fitted to a sine at the beat frequency, in order to measure the gains and pedestals of both I and Q. Both the LO and the calibration source are tuned off 952 MHz to further isolate the calibration from the stored beam. The multibunch fill has a widely spaced spectrum, allowing a narrow-band measurement at any 238 MHz harmonic (such as 952), but the single-bunch case has lines spaced at 136 kHz, requiring a separate broadband calibration, in which the source sweeps through the channel's pass band.

Although this calibration process works well on the bench, many of the BPMs on the HER have been troubled by a false beam oscillation at the beat frequency between the LO and 952 MHz, or at twice this frequency. The oscillation's amplitude is equivalent to beam motion of 60 to 300 μm peak to peak. We normally program the LO to operate just off 952 MHz, typically from 21 Hz to 1 kHz off, because the beat frequency is a signature of a false oscillation, while a DC error would be indistinguishable from real beam position, and the error can be made to disappear by averaging the position over complete periods. The problem is related to the calibration; for example, as the LO phase slowly beats, the signal moves from cosine to sine, and an error in the I-to-Q gain ratio would show up at the second harmonic of the beat. DSP software changes have improved the situation, and recently there have been indications that the problem is, at least in part, related to the new power supplies for the PEP crates, which are different from those on the test bench.

During the January run, we commissioned a new BPM application to measure the phase advances and beta functions at BPMs around the ring. The beam is driven at the tune frequency in one plane, say x, using the tune system. All BPMs record the same 1024 turns. The DSP for the kth BPM then fits the responses x_{kn} for turn n to

$$x_k = A_{x,k} \cos(\nu_x \omega_{\text{rev}} t + \phi_{x,k}) \; , \tag{1}$$

giving the amplitudes $A_{x,k}$ and phase advances $\phi_{x,k}$ around the ring. Network traffic is kept low, since only one phase and amplitude per BPM are reported back to the Control System. Although the beta functions $\beta_{x,k}$ can be found from $\beta_{x,k} = A_{x,k}^2 / \varepsilon_x$, a better way (11) that is independent of BPM calibrations uses

$$\phi_{x,k} - \phi_{x,k-1} = \int_{s_{k-1}}^{s_k} \frac{ds}{\beta_x(s)} \; , \tag{2}$$

along with the known optics and measured phase advances from BPM k to BPMs k–1 and k+1. This method has allowed us to tune out beta beats with small changes in the IP quadrupoles.

TUNE MONITOR

The HER tune monitor, shown in Figure 4, takes signals from dedicated BPM-type pickup buttons, passes them through a downconverter and into a digital spectrum analyzer. The LER monitor will be a duplicate. (In addition, each ring has a second dedicated set of buttons reserved for special measurements, such as high-frequency spectra to examine bunch dynamics.)

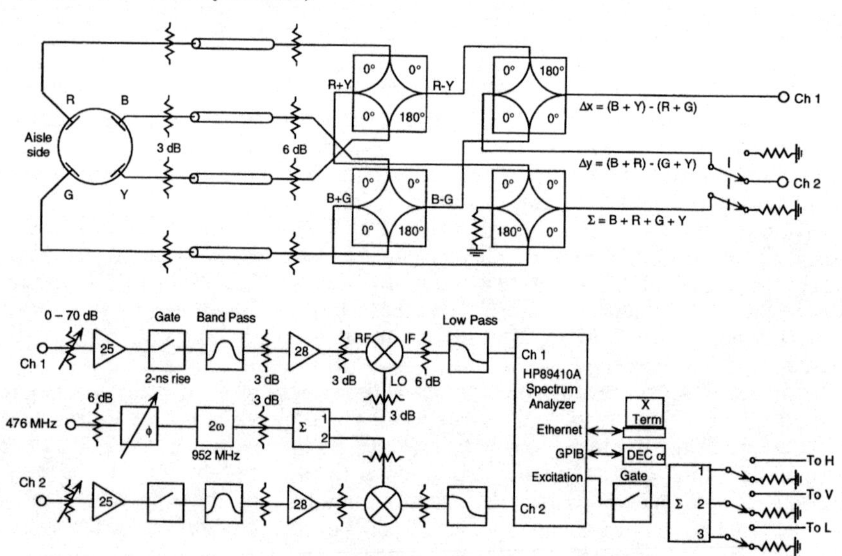

FIGURE 4. The PEP-II tune monitor (one per ring).

The button signals are combined with 180° hybrids to form a sum signal and horizontal and vertical difference signals; an rf switch selects two of these three. A computer-controlled step attenuator, followed by an amplifier, allows for a wide dynamic range, from a single bunch of 5×10^{8} electrons to 1658 bunches of 8×10^{10}, 4.2 ns apart (a full ring, except for a 5% ion-clearing gap).

Up to this point, the components are all broadband, to keep the pulses narrow. A fast GaAs switch (actually a pair of switches in series, to reduce leakage from other bunches) can then gate the signal from one or more bunches, or pass the signal from the entire ring. Several effects, such as the fast ion instability (12), may cause the tune to vary along the bunch train following the ion-clearing gap. The gate lets us measure these effects. Another important case will occur during collisions: the first and last bunches in the train will not experience one parasitic crossing (near-collision with a bunch from the other beam) 0.63 m (half the bunch spacing) away from the IP, because the other beam has a gap rather than a bunch as they approach or leave the IP. Compared to the other bunches, these will feel a different beam-beam kick and so will have different tunes. Another application of the gate is to measure the tune of a specific bunch while feedback is turned off for that bunch alone.

The two-channel spectrum analyzer (Hewlett Packard 89410A) uses digital signal processors (DSPs) and an FFT to compute high-resolution spectra from 0 to 10 MHz. To bring the signals from the pickups into range, the front end includes mixers at $2f_{rf}$.

The analyzer includes a tracking generator to excite the beam with a swept sine or broadband noise when needed. No separate excitation structures are needed. Instead, for transverse excitation, this signal is summed with the input to the power amplifiers for the stripline dampers of the transverse feedback system (13). For longitudinal excitation, the signal will be added to a signal sent by the longitudinal feedback system (14) to modulate the ring rf for control of low-frequency modes; this "sub-woofer link" was partly commissioned during the January run. The drive can be switched on or off separately for each plane, and also passes through another 2 ns gate, which can chop the excitation in order to drive all or part of the bunch train. To study multibunch instabilities, for example, we can drive one bunch while measuring the response of the following bunches. When the LER is complete, we can excite bunches in one ring and measure the corresponding bunches in the other.

In October, PEP multibunch spectra were filled with peaks as the current was raised. Much of this was due to noise imposed on the beam from the rf system, driving large synchrotron and synchro-betatron oscillations that were too big to be controlled by longitudinal feedback. The spectra cleaned up markedly in January after commissioning various feedback loops to control the outputs of the rf stations.

The spectrum analyzer interfaces through both GPIB—for data transfer and remote commands—and ethernet—for FTP data transfer and for an X-window display of an image of the front panel, showing the two traces and allowing control with a mouse. Its internal processor runs both built-in functions like peak finding and user programs in Instrument Basic; both can be executed on command from PEP Control-System software, with results returned to the control system. During January, we commissioned routines that record the tunes periodically for a history buffer, and that follow tune peaks during scans of chromaticity or xy coupling, to automatically make a correlation plot. To avoid reporting the wrong peak in a complex spectrum, the algorithm tracks the evolution of the two peaks (one x and one y) initially identified by the user. Future routines may control the beam-excitation signal to measure the peak with the minimum drive.

In July 1998, when the LER is commissioned, we plan to have our first run with both electrons and positrons. The tune monitor will be used to help bring the two beams into collision. We will excite one beam transversely with a sine wave, and look with the tune monitor for the coupling of this motion to the other beam as we scan the relative positions of the two beams at the IP. A correlation plot showing the amplitude of the response against position will find the best alignment for collisions. For good sensitivity, we will set the spectrum analyzer to a narrow resolution bandwidth (1 Hz). To get a larger amplitude for the driving or responding beam, we can choose to use the tune frequency of one ring, but we must avoid confusion with self-excited motion.

CURRENT MONITORS

A commercial DC current transformer (DCCT) (15) measures the total current in the HER with a 5 μA resolution over a 1 s integration time and a full-scale current of 5 A. (For comparison, a 1 A current with a 3-hour lifetime drops by 93 μA/s, and injecting 5×10^8 $e^{\pm}$ adds 11 μA.) A second unit is now being installed in the LER. Our DCCT housing places it outside the vacuum envelope, provides an electrical gap directing DC wall currents around the transformer core, and capacitively bypasses the gap for higher frequencies to present a low impedance to the beam. An integrating voltmeter (Keithley 2002) reads the DCCT output, and will use an input scanner to alternate between HER and LER.

The nominal bunch pattern has 1658 buckets, each separated by 4.2 ns (two rf periods) and filled to as much as 8×10^{10} $e^{\pm}$ for a 3 A beam. To balance the beam-beam kicks, the variation in charge per bunch within each ring must be held to ±2%. Such tight control requires a second current diagnostic, the bunch-current monitor (BCM) (16), to measure the charge in each of the 3492 rf buckets. A third system, the bunch-injection controller (BIC), then plans the filling sequence, while a fourth, the master pattern generator (MPG) implements it.

With a relative accuracy of 0.5%, the BCM in each ring updates measurements at 60 Hz. This rate compares to 40 Hz for interleaved injection into both rings (and 60 Hz for e^+ only, or 120 Hz for e^- only). In each ring we sum and filter the signals from a set of four BPM-type buttons, using a microstrip combiner with a 2-period comb filter at $3f_{rf}$ (1428 MHz). The filter is designed to avoid crosstalk from adjacent bunches. A mixer at $3f_{rf}$ brings this signal down to baseband, giving a DC to 1 GHz "video" output. The operating frequency is a compromise: a lower frequency would not allow the high video bandwidth and low adjacent-bunch crosstalk; at a higher frequency, the mixer output would be more sensitive to synchrotron oscillations, and some of the button signals would come from propagating modes in the beampipe.

The video goes to a VXI crate, where an 8-bit track-and-hold ADC clocked at f_{rf} digitizes every bucket at the same phase. This data stream is divided among 12 Xilinx field-programmable gate arrays. Even after this "decimation," the rate remains too high for processing. Each Xilinx downsamples by processing only one bucket out of 8 in each turn, so that it takes 8 turns to sample the entire ring. The data for each bucket is summed over 256 measurements in each 60 Hz interval, to improve the resolution and to average over many synchrotron oscillations, then is written into a table in a reflected (dual-port) memory. In addition, lifetime measurements of individual bunches require an accuracy of 0.05% in 1 s, to allow the detection of a lossy bunch (≤10 minute lifetime) at a rate useful for operator adjustments; consequently, the VXI processor maintains a second table with sums over 1 s intervals.

The BIC, in an adjacent VME crate, reads the memories of both rings, and also their DCCT voltages and exact measurement times using the Keithley's GPIB interface. It normalizes the individual bunch currents to the DCCT and calculates lifetimes. Over an

EPICS interface to the control system, it displays the DC currents, lifetimes, and bunch charges, and receives the user's desired fill pattern. User settings for the BCM go in the opposite direction, from EPICS to the BIC and through the reflected memory to the BCM. Based on the charge measurements, the BIC lists the injection sequence for the MPG, which controls the injector linac to fill the appropriate buckets in the rings with bunches of selected sizes. New charge is usually injected in a pattern of nine zones around the ring, so that the bunch can damp before more charge arrives in a zone. The complete fill-control system was commissioned during the January run.

BEAM LOSS MONITORS

A network of 100 beam loss monitors (BLMs) detects losses at selected points (collimators, septa, and selected quadrupoles) around the rings. The system covers a wide dynamic range, from high losses to well-stored beams, provides reasonable localization, and allows the measurement of injection loss. The output is used for machine tuning, for loss histories, and for the rapid detection of high losses requiring a beam abort.

We have chosen a Cherenkov detector, using a small (16 mm diameter) photomultiplier with 2 ns wide pulses (comparable to the bucket spacing). The Cherenkov radiator is an 8 mm diameter, 10 mm long, fused-silica cylinder placed against the fused-silica PMT window, with optical grease on the interface. The opposite end and the cylindrical surface are aluminized for internal reflection. The assembly is enclosed in 1 cm of lead to avoid synchrotron-radiation background, but remains small enough to be moved for commissioning and troubleshooting. Using the ring magnets as shielding, the BLMs can be placed for preferential sensitivity to HER or LER, or they can be exposed to losses from both rings. As the LER is installed, some BLMs are being moved from their initial placement on the HER.

BLM processors (BLMPs), ten-channel CAMAC modules distributed around the rings in the BPM crates, process each BLM signal through two input circuits that together provide a wide dynamic range. To measure low loss rates, the PMT pulses pass through a discriminator and are counted over 1 s or 8 ms intervals. A different procedure is needed for high losses, to avoid high count rates and pulse pile-up, and for injection losses, since a counter would record only a single count even if the injected bunch hits the wall by a BLM. For these situations, the PMT signal is integrated by a 10 μs (about one ring turn) RC filter, and a peak detector then saves the maximum for 8 ms. A multiplexer scans the channels and digitizes these lossiest-turn readings, which are available to the control system on request. The peak detectors are then reset.

If the integrated signal exceeds a programmable threshold, it is possible to abort one or both rings (determined by two programmable abort-enable bits) by firing a kicker on each ring to send the beam into a dump. The BLM processor then records the triggering channel and, through a daisy chain linking all the processors, causes all BLMs to freeze their most recent readings. Several other faults, such as a loss of rf or the closure of a valve, can also fire this abort system.

Another daisy-chain signal provides a 100 μs gate around injection time. During this interval, the BLM network is inhibited from aborting the stored beam, since faulty injection is a more likely source of a large loss (and stored-beam losses will persist after the gate). To measure injection loss, the multiplexer timing is restarted to digitize the outputs of the peak detectors within 1 ms after this inhibit interval. This scheme measures the loss not on the first turn, but on the worst turn, since the injected bunch may not scrape until a later turn, depending on its betatron phase at the obstacle. The control system then acquires the readings within 3 ms of injection.

ACKNOWLEDGMENTS

I would like to thank the many people from all three laboratories who have worked with me in developing PEP's diagnostics, as well as the many others involved in the building, installing, and commissioning the machine. The list is too long to repeat here, but you know who you are. I appreciate your help and the collegial and cooperative spirit in which it was offered. Of course, we still have to finish the LER, collide the beams, install the detector...

REFERENCES

[1] *PEP-II: An Asymmetric* B *Factory*, Conceptual Design Report, LBL-PUB-5379, SLAC-418, CALT-68-1869, UCRL-ID-114055, UC-IIRPA-93-01, June 1993.

[2] DiGennaro, R., and T. Swain, *Nucl. Instrum. Methods* **A291**, 313–318 (1990).

[3] Daly, E.F., A. S. Fisher, N. R. Kurita, J. B. Langton, "Mechanical Design of the HER Synchrotron-Light-Monitor Primary Mirror for the PEP-II B Factory," *Proc. IEEE Particle Accelerator Conf.*, Vancouver, BC, May 1997 (IEEE Press, Piscataway, NJ), in press.

[4] Barry, W., J. Byrd, J. Corlett, M. Fahmie, J. Johnson, G. Lambertson, M. Nyman, J. Fox, and D. Teytelman, "Design of the PEP-II Transverse Coupled-Bunch Feedback System," *Proc. IEEE Particle Accelerator Conf.*, Dallas, TX, May 1995 (IEEE Press, Piscataway, NJ, 1996).

[5] Teytelman, D., J. Fox, H. Hindi, C. Limborg, I. Linscott, S. Prabhakar, J. Sebek, A. Young, A. Drago, M. Serio, W. Barry, and G. Stover, "Beam Diagnostics Based on Time-Domain Bunch-by-Bunch Data,"; Prabhakar, S., Teytelman, D., Fox, J., Young, A., Corredoura, P., and Tighe, R., "Commissioning Experience from HER PEP-II Longitudinal Feedback," in these Proceedings.

[6] Mitsuhashi, T., S. Hiramatsu, N. Takeuchi, M. Itoh, and T. Yatagai, "A Design of Synchrotron Radiation Monitor for KEK B-Factory," *Proc. 11th Symp. Accelerator Technology and Science*, SPring-8, Ako, Hyogo, Japan, 21–23 Oct. 1997.

[7] MEMS Optical, Inc., Huntsville, Alabama.

[8] Fisher, A. S., D. Alzofon, D. Arnett, E. Bong, E. Daly, A. Gioumousis, A. Kulikov, N. Kurita, J. Langton, E. Reuter, J. Seeman, H. U. Wienands, D. Wright, M. Chin, J. Hinkson, D. Hunt, and K. Kennedy, "Diagnostics Development for the PEP-II B Factory," *Beam Instrumentation: Proceedings of the Seventh Workshop*, Argonne, IL, May 1996, AIP Conf. Proc. **390** (Amer. Inst. Phys., Woodbury, NY, 1997), pp. 248–256.

[9] Fisher, A. S., R. W. Assmann, A. H. Lumpkin, B. Zotter, J. Byrd, and J. Hinkson, "Streak-Camera Measurements of the PEP-II High-Energy Ring," in these Proceedings.

[10] Aiello, G. R., R. G. Johnson, D. J. Martin, M. R. Mills, J. J. Olsen, and S. R. Smith, "Beam Position Monitor System for PEP-II," *Beam Instrumentation: Proceedings of the Seventh Workshop*, Argonne, IL, May 1996, AIP Conf. Proc. **390** (Amer. Inst. Phys., Woodbury, NY, 1997), pp. 341–349. Smith, S.R.,

Aiello, G.R., Hendrickson, L.J., Johnson, R.G., Mills, M.R., and Olsen, J.J., "Beam Position Monitor System for PEP-II"; Johnson, R., Smith, S., Kurita, N., Kishiyama, K., and Hinkson, J., "Calibration of the Beam-Position-Monitor System for the SLAC PEP-II *B* Factory," *Proc. IEEE Particle Accelerator Conf.*, Vancouver, BC, May 1997 (IEEE Press, Piscataway, NJ), in press. Johnson, R., Smith, S., Aiello, G., "Performance of the Beam-Position Monitor System for the SLAC PEP-II B-Factory," in these Proceedings.

[11] Castro, P., et al, "Betatron Function Measurement at LEP Using the BOM 1000-Turns Facility," *Proc. IEEE Particle Accelerator Conf.*, Washington, DC, May 1993 (IEEE Press, Piscataway, NJ, 1993), pp. 2103–2105.

[12] Byrd, J., "An Initial Search for the Fast Ion Instability in the ALS," PEP-II AP-Note 95.49, 28 Aug. 1995.

[13] Barry, W., J. Byrd, J. Corlett, M. Fahmie, J. Johnson, G. Lambertson, M. Nyman, J. Fox, and D. Teytelman, "Design of the PEP-II Transverse Coupled-Bunch Feedback System," in *Proc. IEEE Particle Accelerator Conf.*, Dallas, TX, May 1995 (IEEE, Piscataway, NJ).

[14] Teytelman, D., J. Fox, H. Hindi, C. Limborg, I. Linscott, S. Prabhakar, J. Sebek, A. Young, A. Drago, M. Serio, W. Barry, and G. Stover, "Beam Diagnostics Based on Time-Domain Bunch-by-Bunch Data,"; Prabhakar, S., D. Teytelman, J. Fox, A. Young, P. Corredoura, and R. Tighe, "Commissioning Experience from HER PEP-II Longitudinal Feedback," in these Proceedings.

[15] Parametric Current Transformer, Bergoz Precision Beam Instrumentation, Crozet, France.

[16] Chin, M.J., J. A. Hinkson, "PEP-II Bunch-by-Bunch Current Monitor," *Proc. IEEE Particle Accelerator Conf.*, Vancouver, BC, May 1997 (IEEE Press, Piscataway, NJ), in press.

Techniques for Intense-Proton-Beam Profile Measurements*

J. D. Gilpatrick

Los Alamos National Laboratory
Los Alamos, NM 87545

Abstract. In a collaborative effort with industry and several national laboratories, the Accelerator Production of Tritium (APT) facility and the Spallation Neutron Source (SNS) linac are presently being designed and developed at Los Alamos National Laboratory (LANL). The APT facility is planned to accelerate a 100 mA H^+ cw beam to 1.7 GeV and the SNS linac is planned to accelerate a 1 to 4 mA-average, H^-, pulsed-beam to 1 GeV. With typical rms beam widths of 1 to 3 mm throughout much of these accelerators, the maximum average-power densities of these beams are expected to be approximately 30 and 1 MW-per-square millimeter, respectively. Such power densities are too large to use standard interceptive techniques typically used for acquisition of beam profile information. This paper will summarize the specific requirements for the beam profile measurements to be used in the APT, SNS, and the Low-Energy Development Accelerator (LEDA) — a facility to verify the operation of the first 20 MeV section of APT. This paper will also discuss the variety of profile measurement choices discussed at a recent high-average-current beam profile workshop held in Santa Fe, NM, and will present the present state of the design for the beam profile measurements planned for APT, SNS, and LEDA.

HIGH AVERAGE-BEAM-CURRENT ACCELERATORS

In partnership with industry and several other national laboratories, LANL is designing two high-average-current accelerators. The APT linac, which will be built at the Savannah River Site Laboratory, will accelerate a 100 mA cw H^+ beam to 1.7 GeV. The SNS linac, which will operate at the Oak Ridge National Laboratory, will accelerate a 1 to 4 mA-average H^- pulsed beam to 1 GeV.

The schematic representation in Figure 1 shows the APT accelerator general layout (1). During the first 211 MeV, beam is accelerated with room-temperature accelerating structures followed by a series of short superconducting accelerating structures (2). The

---

* Work supported by the U.S. Department of Energy.

CP451, *Beam Instrumentation Workshop*
edited by R. O. Hettel, S. R. Smith, and J. D. Masek

DC beam is initially extracted from the source at 75 keV. It is then bunched at 350 MHz and accelerated to 6.7 MeV with an 8 m-long radio frequency quadrupole (RFQ). A series of differently configured coupled-cavity drift-tube linacs (CCDTLs) and coupled-cavity linacs (CCLs) increase the beam's energy to approximately 100 MeV and 211 MeV, respectively. Up to an energy of 469 MeV, the superconducting linac (SCL) is composed of a series of cryogenic modules each of which is composed of either β=0.64 superconducting multi-cell cavity accelerators. Above 469 MeV, β=0.84 multi-cell-cavity accelerators are used. The beam is then transported by a high-energy beam transport (HEBT) to either a tune-up 2% duty-factor beam stop or the target/blanket assembly where the tritium is produced.

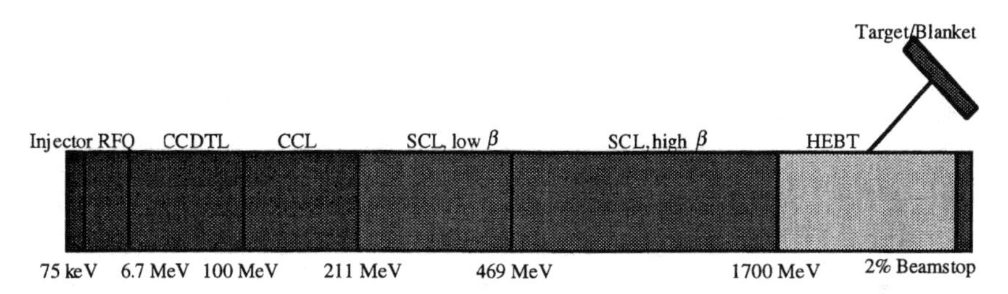

FIGURE 1. The APT accelerator is composed if a series of normal conducting (room temperature) and superconducting accelerator structures. After acceleration to 1700 MeV the cw beam is transported to the target/blanket facility where the tritium is produced.

At Los Alamos, we will be installing and commissioning the first 20 MeV (Fig. 2) of the APT accelerator in a facility called the LEDA (3). This facility is presently being assembled to perform the first experiments that verify the operation of the RFQ. The RFQ verification process will take place during three experiments: the first experiment will provide an initial indication of the RFQ by sustaining 8 hours of operation of a 100 mA cw beam, the second experiment will operate the RFQ for an extended period of time, and a third experiment will test the transverse-matching capability of the CCDTL lattice to the RFQ output beam. Later experiments will verify operation of the CCDTL.

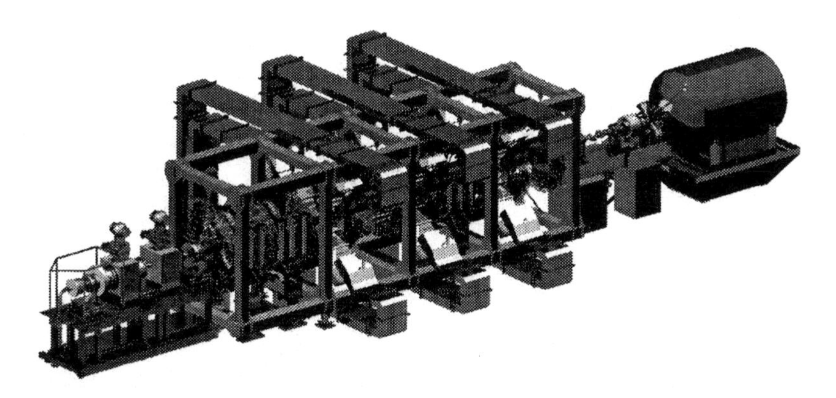

FIGURE 2. The first LEDA experiment configuration that includes the 75 keV injector, the 6.7 MeV RFQ, a short HEBT, and a 670 kW beamstop, is shown.

The SNS accelerator-facility concept is sketched in Figure 3. It consists of a 65 keV H^- injector, a 2.5 MeV RFQ, and a short medium-energy beam transport (MEBT) which matches the beam to a DTL that accelerates the beam to 20 MeV. The beam is then accelerated to 1 GeV by a CCDTL and CCL (4).

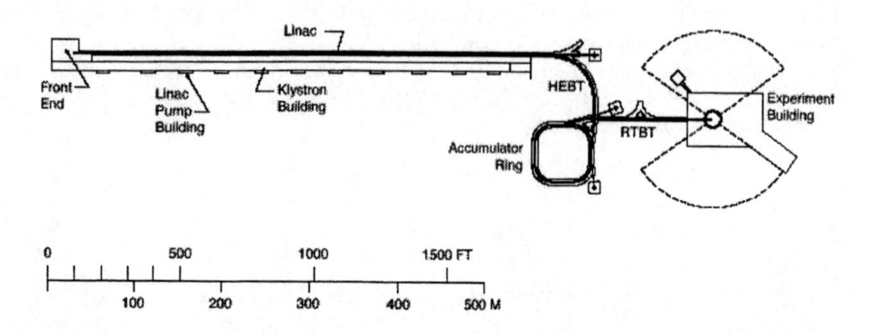

FIGURE 3. The SNS facility contains a room temperature linac consisting of an RFQ, DTL, CCDTL, and CCL. The 1 GeV beam is then transported to a ring where it is accumulated to a <1 μs pulse and transported to a neutron-production target. The short pulse of neutrons are used for diverse research applications.

Table 1 summarizes the beam commissioning and operational parameters of both facilities. While the APT cw average current is 25 to 100 times greater than the SNS average beam current, due to the SNS duty factor, its peak and average macropulse current is 1/4 and 1/7, respectively, of APT. Also note that the rms beam widths are similar for both facilities, and therefore, both facilities have very high beam-current densities.

TABLE 1. Summary of the APT and SNS Operational Beam Parameters

Beam Parameter	APT	SNS
Particle	H^+	H^-
Maximum average current (mA)	100	1, 2, 4
Maximum average macropulse current (mA)	100	18
Macropulse beam repetition rate (Hz)	cw, 1–10	60, or less
Macropulse length (ms)	cw, 1.0–0.1	1.04, or less
Chopped beam period (ns)	NA	841
Chopper transmission (%)	NA	65
Bunching frequency (MHz)	350	402.5, 805
Source energy (keV)	75	65
Output energy (GeV)	1.7	1.0
Transverse rms emittance, norm. (π mm-mrad)	0.16–0.19	0.14–0.17
Typical transverse rms widths (mm)	0.8 to 2.1	0.9 to 1.7

BEAM PROFILE MEASUREMENT MOTIVATION

The motivation for measuring beam profiles throughout these accelerators is primarily due to the requirement for minimizing beam loss in the accelerator. Lost beam will result in unacceptable radioactivation of the accelerator and beam line structures. For APT, reducing beam losses throughout the accelerator to ~0.1 nA/m or 0.1 mA total losses is an operational goal. Lost beam can be a result of a mismatch between the beam's Twiss parameters or betatron funtions and the accelerator's acceptance. Such a mismatch can cause the beam envelope to "flutter" (oscillate unevenly), producing beam transverse-distribution extent or "halo" growth and, in some cases, rms-width growth. Additional mismatch errors and magnet-fringe-field errors can further spread beam halo and increase beam losses. Figure 4 shows a typical representation of how these mismatches can cause a halo growth (5). The figure on the left shows a properly matched beam to a periodic transverse lattice. The figure on the right shows the result of a 50% mismatch; a halo has formed outside the beam core.

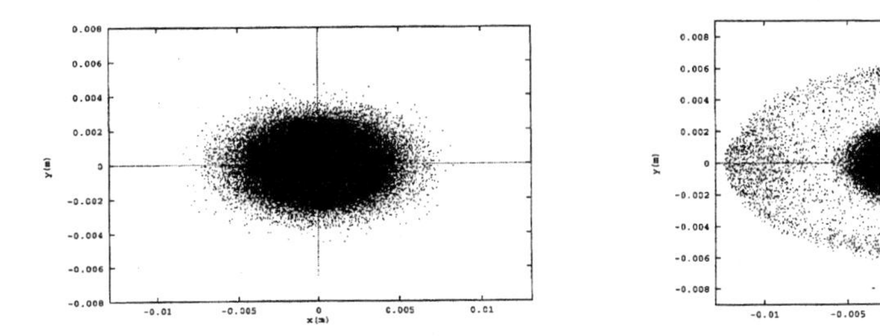

FIGURE 4. This figure shows a typical representation of the extent to which mismatches can cause halo growth. The figure on the right shows the transverse profile of a beam well-matched to a transverse periodic lattice. The figure on the right shows that a 50% mismatch can cause halo growth.

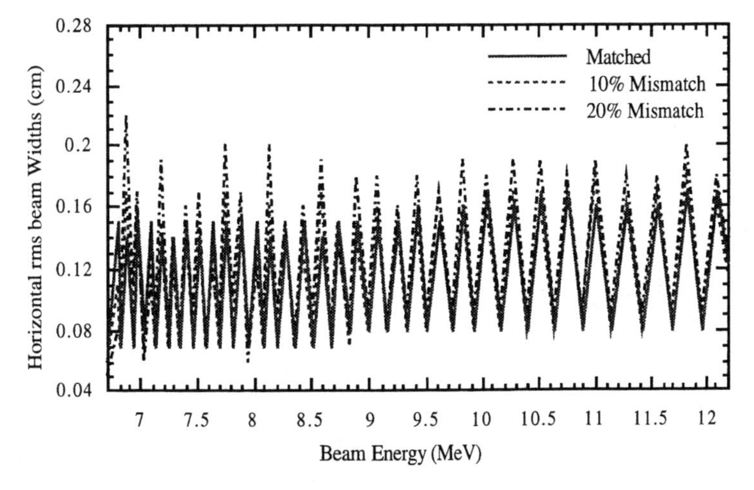

FIGURE 5. This graph shows the horizontal rms beam width as a function of beam energy in the CCDTL (or as a function of location in the CCDTL lattice). The three conditions plotted are a properly matched RFQ output beam, a 10% mismatched RFQ output beam, and a 20% mismatched RFQ output beam.

Another view of the mismatched beam condition can be seen in Figure 5. In this graph, the rms beam width is plotted versus beam energy within the APT CCDTL after the RFQ for three conditions: a matched beam, a 10% mismatched beam located at the RFQ/CCDTL interface, and a 20% mismatched beam at the same interface (6). Under the matched condition, the beam envelope oscillates <0.3 mm. Under the 10%- and 20%-mismatched conditions, the beam flutters <0.4 mm and <1 mm, respectively. The difference between these three conditions becomes less apparent after the beam is transported through 10 or more focus/defocus lattice periods (shown in the graph as several MeV of energy gain). At these lower energies, if the mismatch is sufficiently large, the rms beam width also increases after 10 or more lattice periods.

There are several methods for determining if the beam is mismatched. The method selected for APT measures a series of beam profiles at transverse-lattice locations in areas of the accelerator where beam mismatches are likely to occur. These likely locations for mismatched-beam include the region between the RFQ and CCDTL, and between the CCL and SCL. Multiple profiles can either be analyzed for minimum rms flutter after a mismatch or the profile rms widths can be fit in a least-squared-error sense to the simulated beam trajectory. Furthermore, if each profile is acquired over the range of about 2.5 rms widths, some small portion of the mismatched-beam-dependent halo may be directly sampled. Finally, in some of these locations within the accelerator, specific beam-halo measurements will be performed to acquire information on the far outer edges of the beam distributions (e.g. about 3 to 5 rms widths) (7).

PROFILE MEASUREMENT REQUIREMENTS FOR APT

The requirements for the APT profile measurements are summarized in Table 2. As previously stated, the "full" APT profile measurement, as defined by ~2.5 rms widths, must be measured because these "full" distribution measurements allow for the observation of slight mismatch conditions. Of course, there are the usual beam-line-location and space limitations that restrict the locations at which the profile measurements can be made. These measurements must be sufficiently robust and reliable. For example, each type of beam line device must have at least a one-year expected lifetime with little or no preventative maintenance required.

TABLE 2. Summary of the APT Profile Measurement Requirements

Measurement Parameter	Requirement
Maximum cw beam current density (mA/cm2)	~50 to ~2000
Beam energy (MeV)	0.75 to 1700
Beam current dynamic range (mA)	<10 to 110
rms width range (mm)	0.3 to ~5
rms width precision (mm)	1 % of rms width
rms width accuracy	~5% of rms width
"Full" distribution range	~2.5 rms widths
Pixel size	1% of distribution peak by 20% of rms width
Acquisition time (s)	≤ 60, prefer ≤1

SURVEY OF PROFILE MEASUREMENT CANDIDATES

At a recent Intense Beam Profile Workshop held in Santa Fe, NM, a variety of beam profile measurement techniques were discussed and analyzed for their application to the APT and SNS accelerator beam profile measurements. The profile measurement techniques discussed at this workshop included various types of interceptive and noninterceptive techniques and are reviewed in the latter half of this paper.

Traditional Interceptive Profile Measurements

Even during low-duty-factor pulsed-beam operation, APT beams are too intense for traditional interceptive techniques such as harps, slow wire scanners, or viewscreens. Figure 6 shows a calculation for the temperature of a 0.14 mm-SiC fiber placed in the APT linac beam. This calculation was generated for a 4 Hz, 0.15 ms, 100 mA, 1700 MeV beam with typical peak current densities of 1.6 A/cm^2 and the calculation assumes that fiber cooling only takes place through thermal radiation. Peak temperatures of >1600K are seen even during these very low duty factor conditions, and furthermore, during cw-beam operation, the fiber will vaporize within a few milliseconds. Is this pulsed-beam fiber temperature too high a temperature for a SiC fiber or wire? Figure 7 shows manufacturing data of SiC fiber strength versus fiber temperature. This graph suggests that the above pulsed beam fiber temperature is, indeed, too great. These manufacturing data were acquired in 1 atmosphere of air and argon, a harsher environment than what is expected for the APT and SNS evacuated beam lines, show that the SiC fiber's strength degrades by a factor of three at temperatures of ~2000°F (1400 K). Even though one expects the fiber strength degradation to be less in the evacuated beam line environments, the conservative approach for wire scanner and harp designs is to never allow SiC fibers reach temperatures greater than ~1400 K. The simulation in Figure 6 shows the fiber will periodically reach temperatures well past this safe operating temperature during even very-low-duty factor conditions, let alone cw–beam conditions.

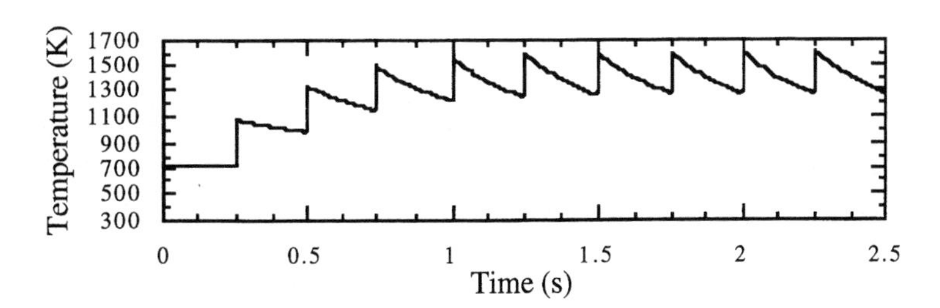

FIGURE 6. A calculation that shows the temperature of a SiC fiber placed in the APT 1.7 GeV beam. The beam is pulsed at 4 Hz with a 0.15 ms long pulse and a current density of 1.6 A/cm^2.

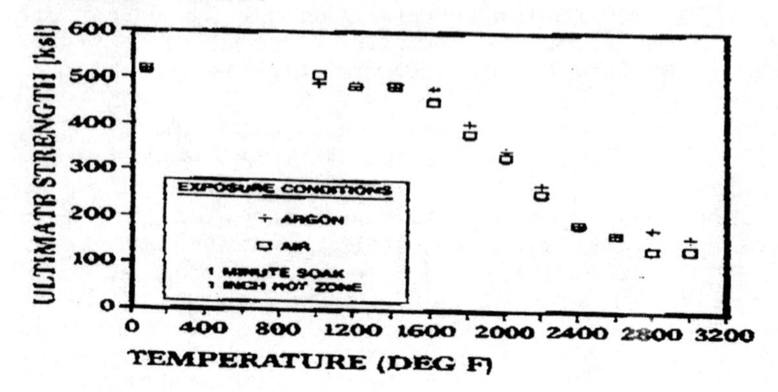

FIGURE 7. This graph depicts Textron System Division manufacturing data of 0.14 mm SiC fiber strength versus fiber temperature. These data were acquired in an environment of 1 atmosphere of air and argon (8).

Limitations of pulsed-beam operation are further reduced during the transverse lattice-tuning process in which beam widths will be smaller than nominal beam conditions. These smaller beams will increase the current densities and result in even higher peak fiber temperatures and likely fiber destruction. Therefore, under no cw-beam nominal-beam-width conditions and for a very limited number of pulsed-beam nominal-beam-width conditions can one expect to reliably operate slow-wire or harp-profile measurements. Furthermore, early in the linac where the beam energy is lower, the amount of beam energy deposited in the fiber is large. Table 3 shows the stopping power and range of protons in carbon. Note that the stopping power of a 6.7 MeV beam is ~32 times that of a 1700 MeV proton beam. The higher stopping power results in greater beam energy deposition into the fiber, higher fiber temperatures, and greater likelihood of fiber damage.

TABLE 3. Proton Beam Stopping Power and Range in Carbon

Beam Energy (MeV)	Stopping Power (MeV cm^2/g)	Stopping Power (MeV/cm)	Range (g/cm^2)	Range (cm)
0.75	> 500	> 1100	<0.002	<0.001
6.7	57.1	126.1	0.068	0.031
20	23.5	51.9	0.48	0.22
100	6.5	14.4	8.6	3.9
211	3.9	8.6	31.8	14.4
469	2.5	5.5	119	54
1700	1.8	4.0	741	336

While not as severe as APT, the SNS linac beam is also too intense for traditional interceptive techniques. For example, Figure 8 shows the SiC-fiber temperature calculated with a 18 mA, 1 GeV SNS beam. The pulsed-beam timing characteristics are a 0.15 ms pulse length and a 60 Hz repetition rate with typical current densities of 0.15 A/cm^2. As in Figure 6, the fiber temperature is well above the nominal 1400 K SiC fiber nominal operating temperature (i.e., shown here to be >1600 K).

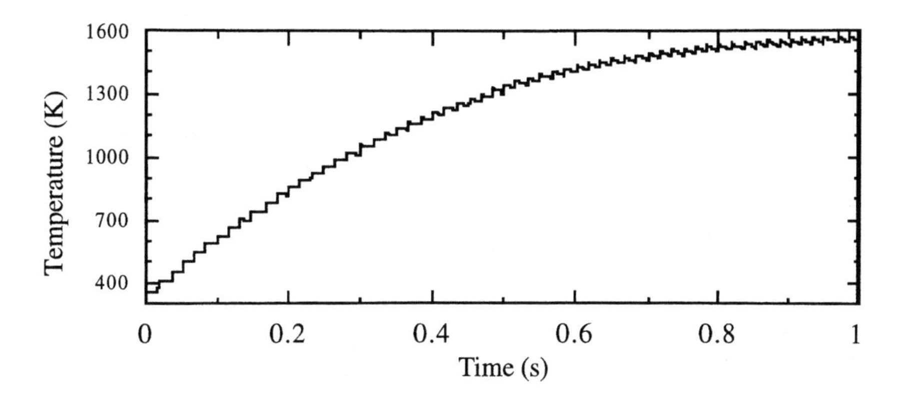

FIGURE 8. This simulation shows how the temperature of a SiC fiber increases to a steady state condition of >1600 K with the SNS beam.

Residual Gas Fluorescence Profile Measurements

At lower beam energies, where deposited beam energies in various materials can be very large, APT and SNS will use a minimally beam-interceptive profile measurement technique. This technique collects and integrates photons resulting from the beam/background-gas interaction. The resultant photon-flux density is dependent on three primary parameters: the beam current density, the beam energy, and the partial background-gas pressure. During the early 1980's several measurements were performed at LANL that characterized these three parameters for proton beams interacting with a nitrogen background gas (9,10). The authors reported that photon radiances of 91 mW/(sr cm^2) at wavelengths near ~400 nm are produced by a 100 mA/cm^2, 80 keV proton beam interacting with partial pressures of nitrogen in the 2X10^{-5} range. Also, this work produced a spectrum of the resultant photons from the beam/gas interaction (shown in Figure 9). The prominent spectral line groups are at ~391 and ~427 nm and are believed to be fast fluorescence process with single-transition excited-state lifetimes of less than 100 ns. These short lifetimes allow for sensing the beam's profile without any apparent beam profile broadening due to non-beam-related phenomenon, such as slower multiple-transition background-gas-ionization processes. During the early 1990's, similar experiences were found at the Ground Test Accelerator (GTA) and Accelerator Test Stand facilities (11). However, it is known that N_2 has metastable states at separate wavelengths which can cause apparent profile broadening. During the operation of GTA, profile broadening was observed as a broad background and thought to be the result of a slow metastable transition.

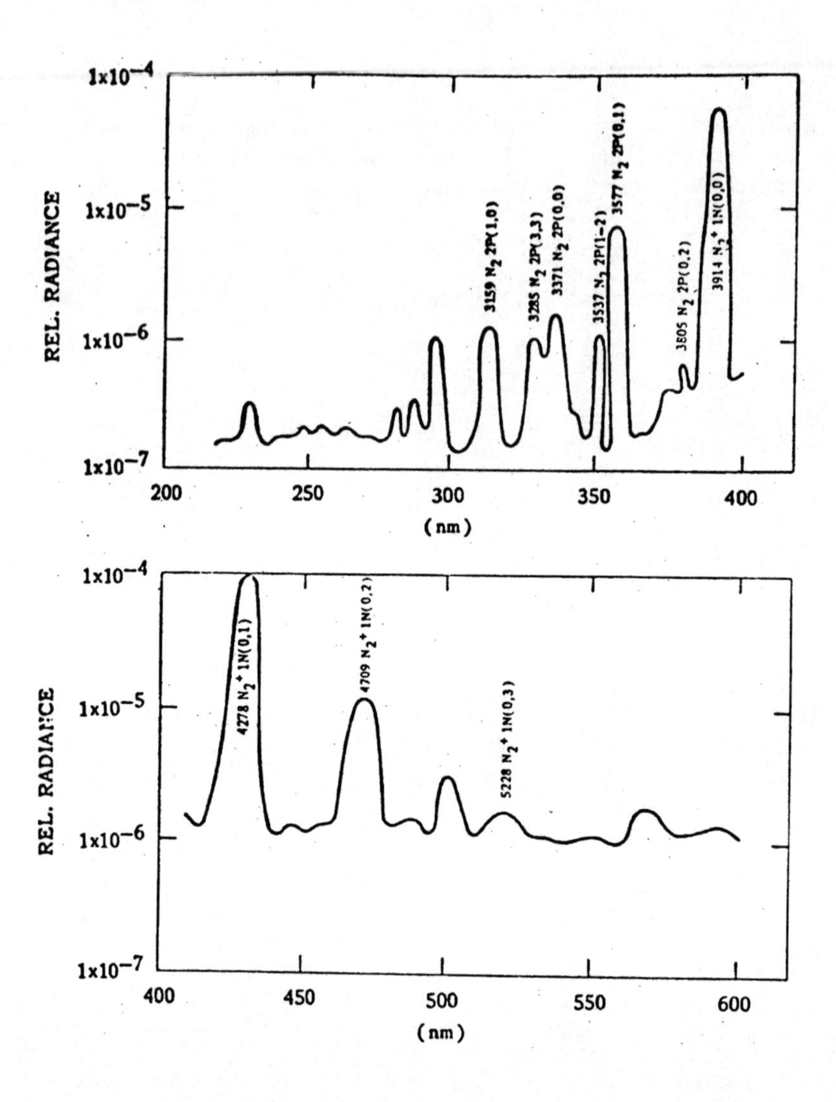

FIGURE 9. The graph is a spectrum of the fluorescence-based photons resulting from a 100 mA/cm^2, 80 keV proton beam interacting with a residual background nitrogen gas.

Table 4 lists calculated values of the viewable radiant power density (assuming a nominal partial pressure of 10^{-6} Torr), the average peak illuminance on a camera faceplate, and the resultant signal-to-noise under cw and 0.1 ms pulsed beam conditions. The cw beam illuminance numbers displayed assume that the NTSC-style cameras are unintensifed and the signal-to-noise numbers are calculated at the distribution peak. The pulsed beam illuminance numbers displayed assume that the NTSC-style cameras are intensified and the signal-to-noise numbers are calculated at the 5% to 10% threshold levels.

TABLE 4. Expected Radiant Power Densities, Average Illuminances, and the Resultant Signal-to-Noise Ratios for 100 mA Beams in 10^{-6} Torr Partial Pressures of Nitrogen

Beam Energy (MeV)	X rms width (mm)	Y rms width (mm)	Radiant Power Density (μW/cm²)	Faceplate Illum., cw (lux) [1]	S:N, cw [2]	Faceplate Illum., cw (lux) [3]	S:N, pulsed [4]
6.7	0.9	0.9	1.3	0.2	1000:1	570	16:1
100	1.2	1.2	0.09	0.01	71:1	31	4:1
211	2	1.5	0.019	0.0012	24:1	3.7	1.3:1
469	2.2	1.8	0.0089	0.00051	16:1	1.5	0.8:1
1000	2	1.5	0.00058	0.00038	13:1	1.1	0.7:1
1700	1.2	0.8	0.011	0.0013	25:1	3.5	1.3:1

1. Average faceplate illumination during cw beam operation at the distribution peak.
2. Signal-to-noise ratio during cw beam operation at the distribution peak.
3. Average faceplate illumination during 0.1 ms pulsed-beam operation at the distribution peak.
4. Signal-to-noise ratio during 0.1 ms pulsed-beam operation at the distribution peak.

Flying Wire Profile Measurements

The present baseline profile measurement technique for beam energies greater than ~200 MeV and rms beam widths greater than ~0.5 mm is a flying-wire profile measurement. This somewhat mechanically complicated technique sweeps a wire through the beam with wire velocities greater than a few meters per second. The relative position of the wire is acquired using either an encoder or a tachometer and the beam distribution information is acquired using either prompt ionizing radiation resulting from the wire/beam interaction (e.g., few MeV gammas) or collecting the secondary electrons emitted from the wire. Figure 10 shows the calculated 0.14 mm SiC fiber temperature as a function of time plotted on a logarithmic scale. As the fiber is flown at a 5 m/s velocity through the 100 mA, 1700 MeV, APT beam, the peak fiber temperature of ~1250 K is reached after the wire passes through the 0.7 × 1.5 mm wide beam. The peak temperature can be maintained within a safe operational temperature range as long as neither the wire velocity nor the beam widths is not reduced.

There are several facilities using this technique to acquire proton beam profiles. At Fermi National Laboratory (FNAL), Blokland rotationally translates a 0.033 mm carbon monofilament through their beam while attaining velocities between 2 and 5 meters per second (12). The 540° rotational scans appear to be repeatable to within 1% and the distributed charge is detected sensing prompt gammas with a scintillator material and a photomultiplier tube. CERN also uses several different configurations of the flying-wire technique (13, 14). Bovet reported that the flying-wire scanners used in LEP linearly translates a wire through their beams with speeds of 0.1 to 2 meters per second. CERN personnel have used a variety of wire materials including 0.05 mm Be, 0.036 mm C, 0.01 mm SiC, and 0.007 mm SiO_2. Extensive wire heating studies have been performed and reported in previous workshops and conferences (15). Another type of CERN flying-wire scanner rotationally translates a 0.03 mm carbon fiber through a beam with velocities of 10 to 20 meters per second. The beam charge distribution is detected by either sensing the depleted secondary electrons from the wires or by detecting nuclear interactions as a result of the beam/wire interaction. Finally, LANL has also partially developed and had limited experience with the flying wire profile measurements (16). In this case, 0.035 mm carbon fibers were rotationally translated through beam regions and peak wire velocities of 5 m/s were attained within the ~300° rotational angular range. The beam charge distributions were detected by sensing the captured secondary electrons. The monitor mechanism was repeatably tested at full speed on the bench to verify the velocity profiles and several crawling or slow-wire tests were performed on a 20 MeV electron beam to investigate background and charge sensing issues. However, no high wire-velocity beam tests were performed.

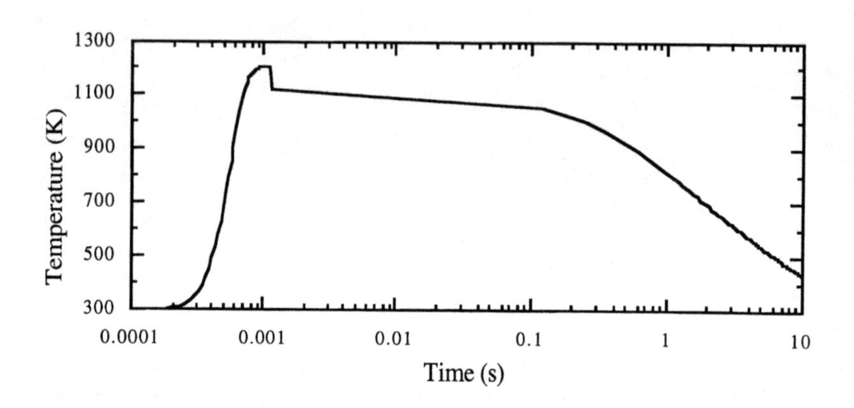

FIGURE 10. This graph plots the calculated temperature of a 0.14 mm SiC fiber or wire as it is flown at a 5 meter-per-second velocity through a 1700 MeV proton beam. The peak fiber temperature of ~1250 K is reached after the wire passes through the 0.7 × 1.5 mm wide beam.

Residual Gas Ionization Profile Measurements

Residual gas-ionization-profile measurement is another minimally interceptive technique used to measure proton beam profiles. As the beam transports through a evacuated beampipe and the beam's energy is deposited in the background gas, electron-ion pairs are created. An electric field is placed across the beam region so that either electrons or ions are electrostatically accelerated toward a collection device. The electrons or ions are then collected with a series of charge collectors typically in the form of either a multi-wire grid, a phosphor and video camera, or a micro-channel plate (MCP). The collected charge is integrated or averaged over time to reduce the random noise in the detected beam distribution.

The residual-gas-ionization profile-measurement technique has been successfully applied to low-intensity beam-profile measurements. However, for the APT and SNS high current beams, the beam space-charge will cause deep potential wells in the beam region. These potential wells can distort the trajectories of the electrons or ions, and therefore, limit the measurement spatial resolution. An additional magnetic field can focus the ionized electron trajectories and provide a finer profile spatial resolution. One drawback to this particular implementation of the residual gas-ionization-profile measurement is related to the required magnetic field strength. Due the to limited beam line space and the required magnetic field strength, required physically large magnets may not be placed in the linac, and hence, limit this particular implementation's APT and SNS usage.

One other concern is the quantity of free electron-ion pairs created by the interaction between the beam and background gas. The gas-ionization-cross section is dependent on the background gas partial pressures. If the vacuum is of high quality and background-gas partial pressures are low, the amount of free electron-ion pairs created within a reasonable amount of time can be very small. The lower electron currents require the charge collection detectors integrate the charge distributions over longer time periods to acquire adequate low-noise data. Unfortunately, the charge collection devices have leakage and dark currents that are also integrated with the signal currents. These extraneous noise or background currents limit the maximum integration times, and ultimately, limit the overall operational vacuum range of the profile measurement. One method of overcoming these noise problems as suggested by Jason, of LANL, is to initially discriminate and then count the collected electrons.

There are several facilities using this technique to measure beam profiles. Connolly, of BNL, reported, at the profile measurement workshop, on the gas-ionization profile measurement developed for RHIC (17). This particular implementation uses both a shaped 6.5 kV electric field and 0.2 T magnetic field for confine and accelerate the ionized electrons. The amplitude of the two fields reduces the electrons Larmor radius to <0.3 mm, therefore providing the spatial-resolution lower limit. Witkover, also at BNL, reported on a ionization profile measurement in the AGS that uses a shaped 45 kV electric field to confine and accelerate the collected ions. The collectors have 48 channels per plane with a 1.2 mm pitch. While this implementation is space charge limited for proton operation, it provides useful information for polarized protons and heavy ions.

Bovet at CERN, reported on two different implementations of a ionization profile monitor. K. Wittenburg developed a profile measurement that uses a shaped 22 kV electric field to collect electrons on to a MCP with a phosphor screen and T. Quinteros is developing a measurement that uses both electric and magnetic fields to collect electrons on to a MCP. Forck, of GSI, reported on a profile measurement for their heavy-ion synchrotron that uses a shaped 12.5 kV electric field to accelerate and confine the collected electrons. The collector is a 64 wire grid that uses 0.1 mm diameter wires with a spacing of 1 mm. Hahn, of FNAL, reported on a gas ionization profile monitor in the FNAL booster. This measurement uses a shaped 8 kV electric field that accelerates and confines the collected electrons. The collector used in this measurement is a MCP and typically collects profile information for 1.6 mm rms width beams. They are starting to look at adding a magnetic field in order to further confine or focus the electrons and improve the measurement's spatial resolution.

Other Less Applicable Profile Measurement Techniques

During the profile measurement workshop, Shafer, of LANL, gave a presentation that suggested other possible intense-beam profile measurement techniques and how applicable these other techniques for acquiring intense proton beam profiles.

The other traditional interceptive techniques such as phosphor screens, solid state detector arrays, IR imaging, etc. suffer from the same interceptive-related disadvantages as do slow-wire scanners discussed in the previous sections. All of these techniques are insufficiently robust for long-term usage with these very intense beams.

Various types of non-ionizing radiation-based light cannot be used because the proton beam is not very relativistic (e.g., $\gamma<3$). This means that synchrotron, transition (e.g., optical transition radiation), and Cherenkov radiation are not suitable for these very intense beams.

One method worth considering is measuring beam width by acquiring second-moment information on the beam from a multi-electrode beam-image-current device, such as a beam-position monitor. This method was first reported by R. Miller of SLAC in the Proceedings of the 1983 Conference on High Energy Accelerators. However, the moment method is very difficult to perform on APT and SNS because of the beam's low emittance and low aspect ratio, and provides only rms-width information (i.e. it provides little or no beam distribution information).

One method that may be of use for SNS is what one might call a laser beam probe (18, 19, 20). This method uses a laser of sufficient power, wavelength, and focused size to photodetach one of the electrons from the H^- ion. The residual neutral-particle beam produced can then be analyzed with a secondary-electron monitor. Various applications of these measurements have been used to measure an H^- beam's longitudinal and transverse phase space. The cross section for H^- beam is very high (i.e., ~40 Mb), however, this is not an applicable solution for APT profile measurements since the there are no electrons in the H^+ particle.

In a similar vein, low-energy ion beams have been suggested as a possible intense beam profile measurement solution and, in fact, CERN personnel are presently investigating this technique.

Electron-beam probes have been developed and reported. Paseru and Ngo performed limited tests with a 1 to 10 keV CRT electron gun and electron currents of 100s of nA (21). They observed deflections for 100 to 350 mA electron beams and showed that the beam positions were measurable. However, no beam size measurements were reported.

There have been several attempts to use various types of vapors and liquid metal beam profile measurements to detect intense beam profiles. CERN personnel reported using a sodium vapor curtain with some limited success (22). Hardek and Gray reported using a magnesium vapor ribbon (23, 24). However, the metal-vapor plates out onto insulating surfaces in vacuum systems, eventually resulting in the shorting out of the ion pumps. There have been reports of SLAC personnel developing a profile measurement that uses a 4 μm diameter pulsed-jet of low-melting point eutectic metal alloy. However, the liquid "wire" broke into beads after 10 or 20 cm, thereby limiting its useful range of operation. These special types of interceptive profile measurements using vapor or liquid metal materials to image the beam generally all suffer from lack of long-term robustness — they require a fair amount of "care and feeding."

CONCLUSIONS

Measuring the transverse beam-profiles of a 100 mA cw, 1 to 3 mm rms width, 1.7 GeV proton beam is a very difficult problem. While there are many profile measurement techniques used for various types of charged particle beams, few of these techniques are applicable to the APT intense-proton-beam profile measurement. There are slightly more applicable measurement techniques for the SNS H^- beam profile measurements (i.e., laser probe technique), however, it too is a very difficult problem. At the present time, there are three possible techniques we are pursuing for detecting the APT and SNS beam profiles. For low beam energies, the residual gas fluorescence technique is expected to provide a sensitive and reliable method of detecting the APT and SNS profiles. However, at higher beam energies the choices become less apparent. The two techniques we are presently pursuing are the flying-wire profile measurement and the residual-gas-ionization profile measurement.

ACKNOWLEDGMENTS

The author would like to thank the attendees and presenters of the Intense Beam Profile Measurement Workshop held in Santa Fe, NM on November 5–6, 1997. Their significant contributions to the workshop topic provided useful and helpful information for the further development of the APT and SNS beam profile measurements.

REFERENCES

[1] Lisowski, P. W., "The Accelerator Production of Tritium Project," presented at the PAC '97.

[2] Lawrence, G. P., and T. P. Wangler, "Integrated Normal Conducting/Superconducting High-Power Proton Linac for the APT Project," presented at the PAC '97.

[3] Schneider J. D., and K. C. D. Chan, "Progress Update on the Lower-Energy Demonstration Accelerator (LEDA)," presented at the PAC '97.

[4] Appleton, B. R., "The National Spallation Neutron Source," presented at the PAC '97.

[5] Simulation run by Robert Ryne, LANSCE-1, LANL.

[6] Simulation run by Edward Gray, LANSCE-1, LANL.

[7] Gilpatrick,J. D., et al., "LEDA and APT Beam Diagnostics Instrumentation,"

[8] Data from Textron Systems Division.

[9] Chamberlin,D. D., et al., "Noninterceptive Transverse Beam Diagnostics," *IEEE Transactions on Nuclear Science*, Vol. **NS-28**, No. 3 (PAC'81), pp. 2347, (June, 1981).

[10] Fraser, J. S., "Developments of Non-Destructive Beam Diagnostics," *IEEE Transactions on Nuclear Science*, Vol. **NS-28**, No. 3 (PAC'81), pp. 2137, (June, 1981).

[11] Sandoval, D., et al., "Fluorescence-Based Video Profile Beam Diagnostics: Theory and Experience," *AIP Conferences Proceedings* **319** (BIW'93), 1993, pp. 273.

[12] Blokland, W., "A New Flying Wire System for the Tevatron," presented at the PAC '97.

[13] Elmfors, P., et al., "Wire Scanners in Large Beam Emittance Accelerators," CERN/PS 97-04, (OP), March 7, 1997.

[14] Boser, J., and C. Bovet, "Wire Scanners for LHC," LHC Project Note 108, October, 1997.

[15] Camas, J., et al., "Observation of Thermal Effects on the LED Wire Scanners," CERN SL/95-20 (BI), May 1995, and *Proc. of the IEEE PAC'95*, Dallas TX (1996).

[16] Wilke, M., et. al., "Status and Test Report on the LANL-Boeing APLE/HPO Flying-Wire Beam-Profile Monitor," Los Alamos LA-12732-SR Status Report, UC-906, Issued: July, 1994.

[17] Connolly, R. C., et al., "A Prototype Ionization Profile Monitor for RHIC," presented at the PAC'97.

[18] R. C. Connolly, et al., "A Transverse Phase-Space Measurement Technique for High-Brightness H^- Beams," NIM, **A312**, pp. 415–419, (1992).

[19] V. W. Yuan, et al., "Measurement of Longitudinal Phase Space in an Accelerated H^- Beam using a Laser-Induced Neutralization Method," NIM, **A329**, pp. 381–392, (1993).

[20] Shafer, R. E., "Laser Diagnostic for High Current H^{-}Beams," presented at BIW '98, Stanford, California, June, 1998.
[21] Paseru, J. A., and M. T. Ngo, "Nonperturbing Electron Beam Probe to Diagnose Charged Particle Beams," *Rev. Sci. In.*, **63**, pp. 3027 (1992).
[22] Vosicki, B., and K. Zankel, "The Sodium Curtain Beam Profile Monitor of the ISR," *IEEE Transactions on Nuclear Science*, **NS-22**, No.3 (PAC'75), pp. 1475 (March, 1975).
[23] Hardek, T., et. al. "Very Low Intensity Storage-Ring Profile Monitor," *IEEE Transactions on Nuclear Science*, **NS-28**, No.3 (PAC'81), pp. 2219 (June, 1981).
[24] Cline, D. B., et. al., "Proton Beam Diagnostics in the Fermilab Electron Cooling Experiment," *IEEE Transactions on Nuclear Science*, **NS-26**, No.3 (PAC'79), pp. 3302 (June, 1979).

Real-Time Orbit Feedback at the APS

J. A. Carwardine and F. R. Lenkszus

Advanced Photon Source, Argonne National Laboratory, 9700 South Cass Avenue, Argonne, IL 60439, USA

A real-time orbit feedback system has been implemented at the Advanced Photon Source in order to meet the stringent orbit stability requirements. The system reduces global orbit motion below 30 Hz by a factor of four to below 5 μm rms horizontally and 2 μm rms vertically. This paper focuses on dynamic orbit stability and describes the all-digital orbit feedback system that has been implemented at the APS. Implementation of the global orbit feedback system is described and its latest performance is presented. Ultimately, the system will provide local feedback at each x-ray source point using installed photon BPMs to measure x-ray beam position and angle directly. Technical challenges associated with local feedback and with dynamics of the associated corrector magnets are described. The unique diagnostic capabilities provided by the APS system are discussed with reference to their use in identifying sources of the underlying orbit motion.

INTRODUCTION

The Advanced Photon Source (APS) is the foremost third-generation synchrotron light source in the United States, delivering intense x-rays to as many as 35 insertion device and 35 bending magnet beamlines. Orbit stability of the stored particle beam is critical to achieving optimum performance for the x-ray users. At the APS, the rms orbit motion is not to exceed 5% of the particle beam size, and with the design 10% x-y coupling, this translates to horizontal and vertical stability requirements of 17 μm rms and 4.5 μm rms, respectively.

This level of stability requires active feedback to reduce orbit motion from sources such as vibration (e.g., from ground motion) and from magnet power supply fluctuations that modulate the magnetic fields guiding the particle beam around the storage ring.

The APS orbit feedback system is designed to provide dynamic correction of low-frequency orbit disturbances, both on a global rms basis (long spatial wavelengths) and locally at each x-ray source point (local feedback). The global orbit feedback system has been in routine operation with APS users since June 1997 and corrects orbit motion up to approximately 50 Hz.

CP451, *Beam Instrumentation Workshop*
edited by R. O. Hettel, S. R. Smith, and J. D. Masek
1998 The American Institute of Physics 1-56396-794-4/98/$15.00

ORBIT CORRECTION PRINCIPLES

The principles of orbit correction have been covered extensively in the literature (1). A linear response matrix describes the relationship between small changes in chosen corrector magnet (dipole) fields and the resulting change in the particle beam orbit as measured at chosen beam position monitors (BPMs). This is described mathematically by the equation

$$\mathbf{R}\,\Delta\mathbf{c} = \Delta\mathbf{x}$$

where **R** is the so-called response matrix, $\Delta\mathbf{c}$ is a vector of corrector field changes, and $\Delta\mathbf{x}$ is the resulting change in orbit as measured at specific BPMs. This equation can be inverted mathematically to produce a relationship that maps orbit perturbations to changes in corrector magnet fields that would cancel those perturbations. For small orbit errors, the relationship is assumed to be linear and time-invariant. Different objectives, such as correcting the orbit in an rms sense or correcting specific locations, can be achieved by choice of BPM and corrector locations and by applying different weights to correctors or BPMs when computing the inverse response matrix.

At the APS, orbit correction is performed by two systems that operate independently but in parallel. A workstation-based system corrects the orbit at ten-second intervals using 80 correctors and more than 300 BPMs to compensate for slow changes in the global DC orbit and to return to a user-preferred orbit at the beginning of each fill (2). It is also used with different selections of BPMs and correctors to make local steering changes at the x-ray source points as requested by the users.

In addition to the workstation-based system, the real-time orbit feedback system that is the subject of this paper performs orbit corrections at a 1 kHz rate. It uses 160 BPMs and 38 correctors to correct dynamic orbit errors and incorporates a high-pass filter to roll off the response below 20 mHz. Orbit corrections below this frequency are performed by the workstation-based system.

Implementation of the Global Algorithm

Inspection of the orbit correction algorithm reveals that the calculation of each new corrector setpoint requires forming the vector dot product of the appropriate row of the inverse response matrix and the entire vector of measured BPM values (or 'errors' from the desired values). Since there are 38 correctors used in each plane, a total of 38 vector dot products must be computed to complete one correction step.

In the APS system, the computations are distributed between 20 VME crates (described in the next section), with each crate computing only the dot products associated with its correctors, as shown in Figure 1. The results of each vector dot product become the input to a control loop around each corrector.

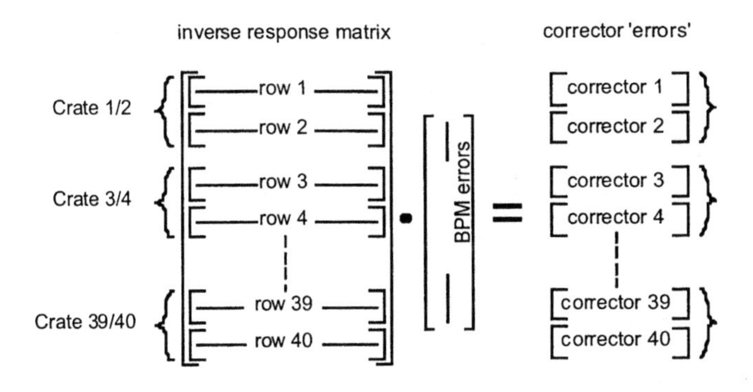

FIGURE 1. Separation of the global algorithm across 20 slave stations.

SYSTEM IMPLEMENTATION

The APS storage ring contains a total of 360 rf BPMs and 317 dual-function corrector magnets. All of these are accessible to the APS control system (through EPICS). Additionally, 320 of the BPMs and all of the correctors can be accessed by the real-time orbit feedback system. However, unlike the workstation-based system that uses the APS control system to access BPMs and correctors, the real-time system has dedicated data links to both BPM and corrector magnet power supply hardware.

The real-time system uses a total of 20 VME 'slave' crates distributed around the circumference of the 1.1 km storage ring. Each contains interfaces to 16 rf BPMs and 16 corrector magnet power supplies in each plane (3). Every slave crate also contains digital signal processors (DSPs) that implement the real-time algorithms. The overall layout of a typical slave crate is shown in Figure 2.

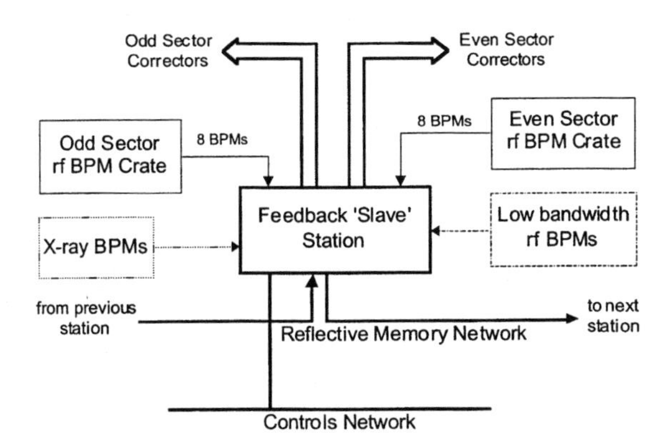

FIGURE 2. Real-time orbit feedback system layout.

A separate 'master' VME crate performs supervisory tasks and provides a central interface for data acquisition and control. All 21 VME crates communicate via a dedicated 'reflective memory' network that provides crate-to-crate data transfers at 29.6 Mbytes/second. The crates are also accessible over the APS controls network. Figure 3 shows the components associated with the master and with each slave crate.

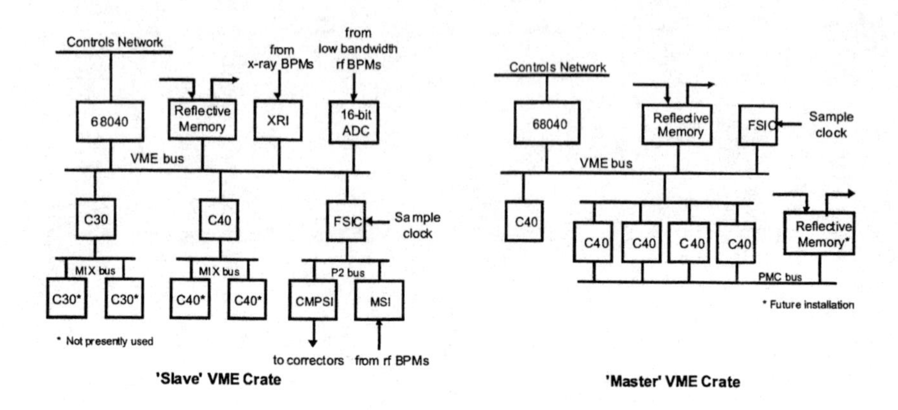

FIGURE 3. 'Master' and 'slave' crate components.

Custom hardware receives data from the rf BPM memory scanners (MSI) and delivers setpoints to the corrector magnet power supplies (CMPSI). There are also interfaces to x-ray BPMs and to new narrow-band BPM electronics that will provide, respectively, positional feedback from the x-ray beams and more stable measurements of the particle beam position at the insertion device source points.

Both master and slave crates contain a 68040 processor that runs EPICS core routines and provides access to the DSPs via the APS controls network. Each slave crate can contain up to six DSP processors. These are commercial VME boards based on the Texas Instruments TMS320 C30 and C40 DSP processors. Presently only two of the six processors shown in the slave crate are in use, implementing global rms orbit feedback. The additional processors will be added to implement 'local' feedback at the x-ray source points.

To date, the global algorithm has been implemented with a single C40 DSP performing orbit feedback in both planes at a 1 kHz rate. (The data presented here was collected with this configuration.) However, by sharing the global algorithm across two DSPs (one for each plane), we have upgraded the system to operate at 2 kHz. Operation at 2 kHz will be commissioned during the the APS "98-3" user run.

BPM Interface

The rf BPMs are installed at 360 locations around the storage ring, providing data every two turns. Raw BPM data flows along two separate data paths, allowing simultaneous access from the APS control system and from the real-time orbit feedback system. In order to band-limit the signal content provided to the real-time feedback system, raw BPM values are transmitted across a high-speed serial link to a 32-point boxcar averager in the orbit feedback slave crate. The averager was originally designed for orbit feedback sampling rates of 4 kHz, and has a –3 dB point around 1.85 kHz. Consequently, there is an aliasing issue with the data supplied to the orbit feedback system, the most noticeable impact of which is that synchrotron motion at 1.8 kHz is aliased to 200 Hz in the orbit feedback system. Fortunately this does not lie within the correction bandwidth of the system, but does influence certain measurements using the orbit feedback data path.

A new BPM filtering module is under development that will implement multi-stage decimation filters on an embedded DSP chip right at the data source. Filtered data will be available at several bandwidths and sampling rates consistent with their uses by the orbit feedback system and by the APS control system. The new module will also resolve some data integrity and reliability problems with the high-speed serial link that transfers the raw data values at 135 kHz to the orbit feedback system averager.

Corrector Interface

Each corrector power supply can receive setpoints from one of two sources, either from the APS control system as an EPICS process variable, or over a serial link from its associated orbit feedback slave crate. An electronic switch (controlled through EPICS) selects the source of the setpoints.

By throwing the switch to the EPICS source, the real-time feedback system can be completely isolated from the accelerator. This has proven very useful during commissioning and system debugging. An equally important feature is that the setpoint source can be changed 'on the fly' when beam is being delivered to the users. To do this, it has been necessary to guarantee that there are no transients on the output of the power supplies when the source is changed. When the orbit feedback control loops are opened, the original DC setpoint (read from EPICS) is automatically copied to the power supply by the orbit feedback system through its dedicated link. With the control loops closed, the DC setpoint is used as a bias that is added to the setpoint calculated in real time by the DSPs.

Originally setpoints were transmitted by the orbit feedback system hardware only once at the end of each feedback cycle (every 1 ms). However, it was found that transmission errors could cause a sufficiently large transient on the output of the power supply such that stored beam would be dumped. In addition to having improved transmission reliability, setpoints are now sent out repeatedly, at 64 μs intervals, regardless of whether a new setpoint has been calculated. While this in itself does not prevent the errors, it does limit the duration of the transient so that the effect on stored beam is minimal.

Data Sharing and Synchronization

The processes on the 21 feedback crates (master plus 20 slaves) must be synchronized for the system to operate properly. A feedback clock (currently 1 kHz, but soon to be 2 kHz), is generated by the APS timing system processor in the main control room and distributed via the APS timing event system (4) to each of the 21 locations. Each feedback crate receives these timing tics in a custom module containing a counter and a missing pulse detector. Each DSP waits for the counter to increment from zero, at which time it begins its algorithm. A counter value greater than one indicates the DSP is late. The missing pulse detector indicates a timing tic was late or missing. Both conditions are checked by the DSP and, if they occur, are reported to the control system.

Each DSP collects its local BPM data, computes BPM error values, and deposits them at the appropriate location in the BPM error vector in reflective memory. Each DSP needs the entire vector to compute its corrector values. This operation is synchronized via the reflective memory. After writing BPM error values to reflective memory, each DSP writes a one to an assigned word in a "data ready" vector, also residing in reflective memory. All DSPs spin-wait until this vector becomes all ones. When this occurs, each DSP reads the error vector and clears its "data ready" location in reflective memory. Processing then proceeds. Each DSP will spin-wait on the 'ready' vector for a limited amount of time. If the time expires, the DSP signals an error to the control system and returns to waiting for the next feedback clock tic.

The master DSP continuously monitors slave performance. It monitors the "data ready" vector and identifies to the control system slaves that do not report "data ready" within time. Also, each slave DSP increments a heartbeat location in reflective memory that is monitored by the master. Slaves that fail to increment their heartbeat locations on each feedback clock tic are reported to the control system. Needless to say, these conditions indicate a malfunction and are dealt with by the APS operators.

Global controls, such as loop on/off, filter breakpoint frequencies, filter on/off, etc. are distributed to the slaves by the master via the reflective memory.

Software Development

The DSP software is developed on a Unix workstation using Texas Instruments' TMS3203x/C4x Code Generation tools. While most of the code is written in the C language, procedures that perform operations on arrays or vectors are written in C-callable assembly language routines. This allows us to insure that TMS320 DSP features such as parallel and three-operand instructions are used. The latest compiler (v 5.00) however seems to be much more sophisticated in optimization than prior versions. Nevertheless, we still prefer to use the assembly language routines.

Optimized code produced by the compiler can cause unexpected behavior, particularly when the compiler rearranges code to minimize DSP pipeline conflicts. This can be difficult to diagnose. We've uncovered "features" of the DSP VME interface that were stimulated by code rearrangement that improved speed.

Code is downloaded to the DSP via the controls LAN from the control system file server using tools in SwiftNet, a product available from the DSP vendor (5). The tools run under VxWorks (6) on the VME controls processor located in the same crate as the DSP VME boards. The loader fetches executables via the LAN from a file server and transfers them over the VME bus to the DSP.

We have not made extensive use of vendor-supplied debugging tools, but have used reserved 'test' locations in dual-access RAM and in the reflective memories into which the DPSs write selected values. These locations are easily interrogated from a workstation via the VME controls processor. This method has little impact on the DSP algorithm and facilitates debugging at the normal sampling rate.

An important consideration in any real-time system is the time it takes for the code to operate. We've used two methods to measure this. The first method is to have each DSP write values to a VME DAC card, with a different value written after each major computation step. This allows the quick determination of where most time is consumed and is particularly useful with multiple DSP systems where each DSP accesses the VME bus, potentially causing bus contentions. Each DSP writes values to different DAC channels. System operation is easily observed on an oscilloscope connected to the DAC channels. The second timing method uses the DSP's on-chip timers to measure execution time. This method has minimal impact on execution, since no VME bus accesses are required, but does not easily lend itself to correlating the execution of multiple processors.

Operational Issues

A great deal of effort has been spent in establishing reliable operation and in streamlining routine tasks to reduce the burden on the APS operators. For example, we ensure that no detectable transients are generated when the feedback loops are opened or closed. Transients are minimized by closing the loops with regulator settings that produce fast settling times, with the settings gradually ramped to their normal values.

On occasion, it has been necessary to reconfigure the response matrix to remove misbehaving BPMs from the algorithm. Again, we have ensured that this can be performed with no impact on the users, other than having to open the control loops while the new matrix is downloaded via EPICS.

The system can also operate autonomously, such that the control loops are automatically opened when stored beam is lost, and re-closed when the beam returns.

Efforts in these areas have paid off in terms of the overall reliability. During the first nine months that the orbit feedback system has been in routine operation with users, availability has been greater than 99%, and there has been only one beam dump directly attributed to it (leading to 42 minutes of machine downtime).

SYSTEM PERFORMANCE

On-line machine diagnostics make it possible to measure orbit motion quickly and conveniently. Typically, we measure the orbit motion at 40 BPMs located in the insertion-device straight sections, with power spectral density measurements averaged over the 40 locations.

TABLE 1. Typical Measured rms Orbit Motion at Insertion Device Source Points

	Horizontal		Vertical	
	Feedback Off	Feedback On	Feedback Off	Feedback On
Required orbit stability (rms) (with 10% x-y coupling	17μm		4.5μm	
Orbit motion 0.016Hz–30Hz (rms)	18.4μm	4.4μm	3.1μm	1.8μm
Orbit motion 0.25Hz–500Hz (rms)	20μm	13.2μm	7.4μm	7.5μm
Typical beam size at source points (rms)[1]	335μm		18μm	
Beta at insertion device source points (design)	17m		3m	

[1]Inferred from measurements at S35BM with 100mA beam, stored in 81 bunches.

We routinely measure the orbit motion over two frequency bands, one sampled at 60 Hz (using the 'slow beam history' diagnostic (7)), the other at orbit feedback system sampling rates. The slow beam history diagnostic is used to measure orbit motion in the frequency range 0.016 Hz to 30 Hz, while the real-time feedback system is used to measure orbit motion from 0.25 Hz to 500 Hz. Typical results of measurements during the latter part of the APS "98-2" user run are shown in Table 1.

It should be noted that even though the orbit feedback system is sampling at 1 kHz, signal content up to 2 kHz is included in the 0.25 Hz-500 Hz measurement because of aliasing of the incoming BPM data.

Horizontal and vertical power spectra in the band from 0.01 Hz to 500 Hz are shown in Figure 4. The same data are shown in Figure 5 as cumulative rms motion from 0.01 Hz to 500 Hz. This data was collected at three different sampling rates using the real-time orbit feedback system.

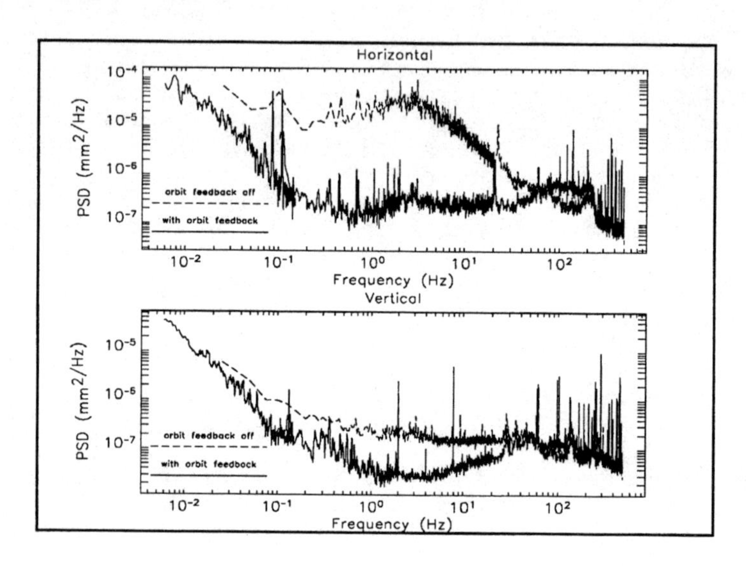

FIGURE 4. Orbit motion power spectra from 0.01 Hz to 500 Hz.

Horizontal orbit stability is well within the requirements in the band up to 500 Hz. In the case of the vertical plane, orbit stability below 30 Hz is within specification, but broad-band motion to 500 Hz is still too high, although to date this has not been an issue with the APS users.

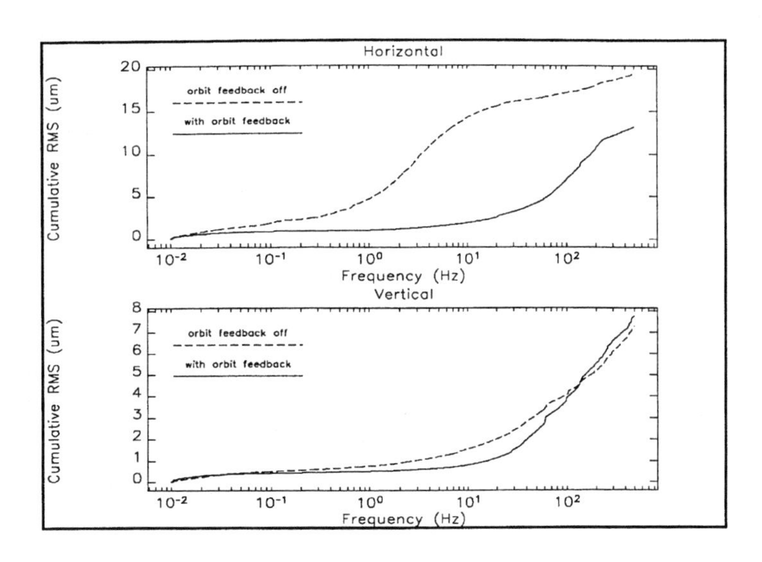

FIGURE 5. Cumulative rms orbit motion from 0.01 Hz to 500 Hz.

Horizontal orbit motion is dominated by power in the 0.1 Hz to 20 Hz frequency range. This is caused by small fluctuations in the current delivered to multipole magnets and corrector magnets in the storage ring (each magnet has a separate power supply). Fluctuations in the multipole magnet currents are converted into vertical AC dipole fields by the asymmetrical aluminum vacuum chamber. This effect is frequency dependent and peaks around 5 Hz. The evidence for this is clearly visible in the horizontal power spectrum shown in Figure 4 (the peak is shifted to around 3 Hz by the power spectrum of the power supply fluctuations). This contribution to orbit motion was unexpected, and was not taken into account when the power supply stability specifications were originally developed. Several narrow-band sources are visible in power spectra of both planes. These are caused by power converter ripple from the fast corrector power supplies that are not attenuated by the Inconel vacuum chamber. In most cases, their contribution to rms orbit motion is very small.

Two factors account for the fact that the orbit feedback system is more effective in reducing horizontal motion than vertical motion. First, the orbit feedback system is most effective in the frequency range that includes the horizontal motion between 0.1 Hz and 10 Hz. Conversely, since the vertical motion has a much flatter power spectrum, less of the power is included in the correction bandwidth. Second, the global algorithm itself is more effective in the horizontal plane, because more of the energy appears in spatial modes that are correctable by the feedback system. This has been shown in modeling where the maximum theoretical attenuation in the horizontal plane was found to be around 22 dB but only 15 dB in the vertical plane.

An important factor in achieving optimum performance over a wide frequency range is the tuning of the feedback regulator. At present we use a simple band-pass filter and PID. By adjusting the PID settings it is possible to trade off a fast rise-time (i.e., wide correction bandwidth) against an overshoot in the step response (that results in amplification of higher frequencies). Depending on the power spectrum of the underlying motion, this trade-off can have a significant impact on the rms orbit motion. The effects of regulator tuning on measured rms orbit motion up to 30 Hz and 500 Hz are shown in Figure 6. Notice that while there is a clear optimal tuning for the horizontal plane, the broad-band vertical motion only increases when the regulator gain is increased. This is primarily because the vertical power spectrum is much flatter so that a reduction in low-frequency motion is achieved at the expense of higher-frequency motion.

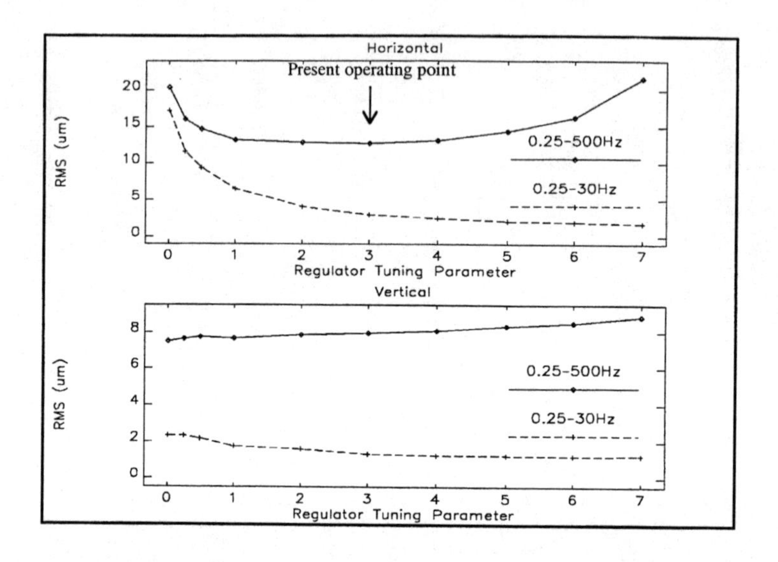

FIGURE 6. Measured rms orbit motion vs. regulator tuning.

Since the system does not currently allow different regulator settings for the two correction planes, the present regulator configuration was chosen as a compromise between the optimal setting for the two planes. As part of the upgrade to 2 kHz, we will increase the order of the regulator filter and will include separate filters for each correction plane.

CORRECTOR EQUALIZATION

All of the 320 dual-function corrector magnets in the APS storage ring can be made available for real-time orbit feedback. However, the dynamics associated with the magnets vary depending on their location in the lattice because of eddy-current effects in their associated vacuum chambers. Seven of the available eight corrector magnets in each sector are mounted on thick aluminum vacuum chambers, with the remaining corrector mounted on an Inconel bellows. Eddy-current effects in the aluminum chamber significantly impact field penetration to the particle beam by introducing a strong frequency dependence that slows the effective response of the corrector magnet. Eddy-current effects from the Inconel bellow have negligible impact and the response is dominated by power supply dynamics. In order to obtain the widest correction bandwidth, the global orbit correction system uses the 38 'fast' correctors available in each plane of the machine.

Figure 7 shows the step responses that were measured from the response of the closed orbit to a step change in setpoints to several of the magnets.

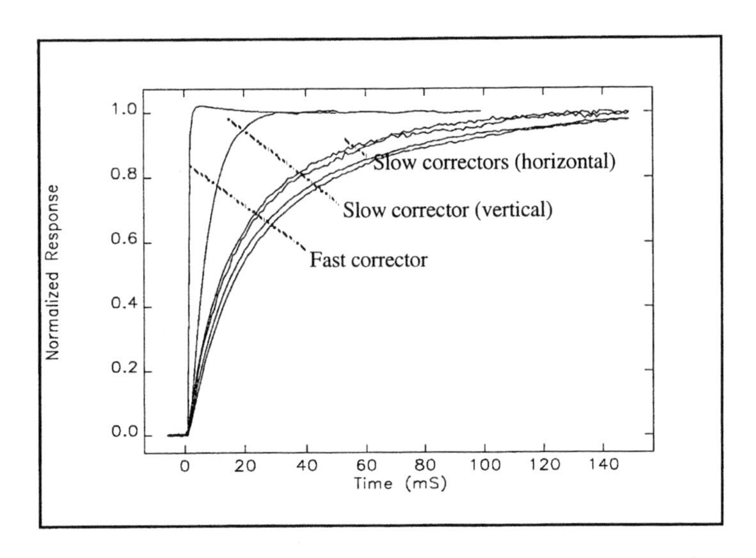

FIGURE 7. Dynamic orbit response to a step change in corrector setpoints.

Notice that there are significant differences between the step responses of horizontal and vertical windings for the 'slow' correctors. There are also slight differences in the responses of the four horizontal correctors that would be used to form a local bump around the insertion-device source point. The latter are caused by small differences in the vacuum chamber arrangements close to each of the magnets (e.g., there are pumping ports close to certain magnets). These differences in mechanical arrangements affect the eddy currents from the stray fields of each magnet and result in subtle differences in the dynamics of each corrector. The differences are most problematic when correctors are combined to form local bumps that might be used to steer the particle orbit through the x-ray source points. Small differences in the step responses of the magnets are magnified by the large bump coefficients, and the resulting bumps become far from closed (i.e., they are non-local) dynamically.

It is possible to use the orbit feedback DSPs to implement equalization filters such that the combination of filter and magnet are identical in each case. One complication in designing the filters is that the eddy current effects are nonlinear with frequency and cannot be exactly represented using a finite transfer function of poles and zeros. In modeling the responses of the correctors, we found that it is most effective to work in the time domain and to equalize measured step responses, rather than to equalize the frequency responses.

Even so, there are limitations to what can be achieved, particularly because of the time delay associated with the field penetration through the thick aluminum chambers. This means, for example, that it is not possible to equalize the slow correctors to give the same response as the fast correctors. So far, the most satisfactory results have been achieved by equalizing the correctors to a single-pole filter with a one-step time delay. Using a three-pole, three-zero discrete filter, it has been possible to equalize responses to within one part in a thousand. This is illustrated in Figure 8, where the equalized step response is shown together with original responses of the fast and slow correctors.

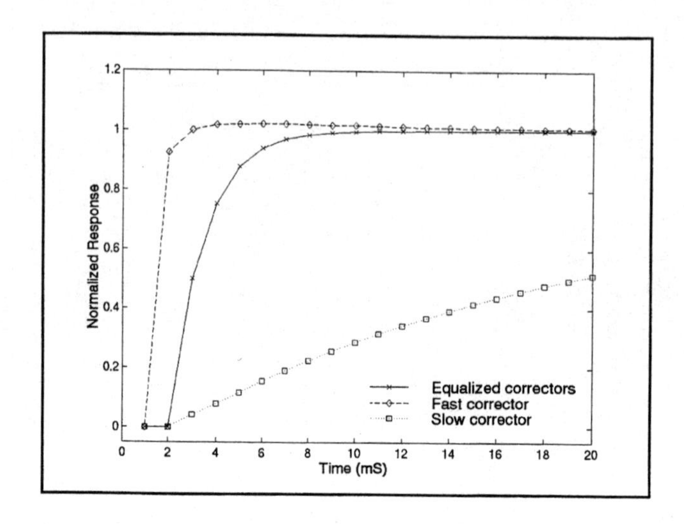

FIGURE 8. Equalized horizontal step response.

In all cases examined so far, the largest impulse applied to a corrector was a factor 20 higher than the desired step size. This is well within the dynamic range of the power supplies for any likely scenario.

While the equalized responses of the slow correctors are considerably faster than the original responses, there is some loss of bandwidth and increased phase delay with the fast correctors. If it is decided to implement a unified algorithm with both fast and slow (equalized) correctors, there will be a corresponding reduction in the maximum closed-loop bandwidth that can be achieved compared with a system that uses only fast correctors.

So far, the equalization filters have not been tested in the storage ring. It has been our experience that the optimization of the filter responses will be an iterative process because of contamination of the measured step responses by real orbit motion.

It should also be noted that the equalization filters have been designed for 1 kHz orbit feedback operation. The transition to 2 kHz should improve the overall result and may reduce the length of the time delay required in the equalized response.

ORBIT FEEDBACK AT X-RAY SOURCE POINTS

The APS real-time orbit feedback system was designed with both global and local feedback objectives in mind. However, implementation of the global feedback system was given priority over the local system since all users benefit from the improved global orbit stability. Indeed by using only global feedback, we have already met requirements for horizontal orbit stability. Nevertheless, local dynamic feedback will be implemented and will run in parallel with the existing global feedback algorithm.

Measurement of X-ray Beam Position

It has become clear that a usable local feedback system requires very stable high-resolution measurements of the x-ray source point or x-ray beam position. This has been most evident from experience with the workstation-based DC orbit correction system.

It was found that an algorithm that used many BPMs and relatively few correctors to correct the global rms orbit was more successful at stabilizing the x-ray source points than an algorithm that exactly corrected the rf BPMs around each source point. This is a consequence of current- and bunch pattern-dependence of the rf BPM offsets.

Photon BPMs are installed in each bending-magnet and insertion-device x-ray beamline and are specifically intended for use in local orbit feedback. By design, these offer much greater sensitivity and stability than the rf BPMs installed at the x-ray source points, and the photon BPMs installed in the bending magnet lines will shortly be integrated into the real-time feedback system. However, photon BPMs in the insertion device beamlines are subject to x-ray contamination problems, not only from adjacent bending magnet x-ray sources, but also from quadrupole and corrector magnets that are in direct line-of-sight of the BPMs. To date, no simple method has been found to eliminate the measurement ambiguities from these sources, and accuracy of the insertion-device x-ray BPMs is limited to about 20 μm. Techniques are being developed for masking the stray radiation by physically moving accelerator components. These are scheduled for testing during the latter part of 1998.

In order to provide more stable DC measurements of the particle beam position, new (narrow-band) rf BPM electronics are being installed at each insertion device source point. These offer significant improvements over the present rf BPM electronics in terms of long-term drift and beam-current dependence, and it is anticipated that the combination of these narrow-band BPMs together with the photon BPMs will provide the stability required for local orbit feedback.

To manage the limited dynamic range of the photon BPMs, a virtual "mapped" BPM will be created that maps each photon BPM to an equivalent rf BPM location. In the event that the photon BPM signal becomes invalid, the rf BPM value will be automatically used in its place, alleviating the need to reconfigure the orbit correction response matrices.

Local Feedback Algorithms

The classic approach to local feedback is to create three- or four-corrector closed orbit bumps, where the ratios of corrector strengths are chosen to minimize global orbit effects from changing the local bump strength. There are practical difficulties with this approach since the bump coefficients must be very well matched (and possibly be frequency dependent) to eliminate closure errors. However, the algorithm is straightforward to implement since it requires only local BPM information.

A great advantage of the APS orbit feedback system is that every BPM in the storage ring is already available to the slave crates via reflective memory. This means that algorithms requiring access to global orbit information can be considered.

An important consideration in selecting a local algorithm is that the correctors available for local feedback have significantly slower responses than those of the global correctors because of eddy-current effects in their respective vacuum chambers. As was previously discussed, digital equalization of the corrector responses is certainly possible, but at some cost to the usable bandwidth of the global correctors.

The concept that is applied to the implementation of global feedback will also be applied to the local feedback system. All local orbit feedback loops will be incorporated into one (globally oriented) matrix. Additional corrector feedback loops will be implemented in the same way as the global system; each corrector control loop operates independently, with its feedback 'error' signal being derived from the vector dot product of one row of the new 'local' inverse response matrix and the vector of measured BPM errors. By changing the inverse response matrix contents and the regulator configuration, it will be possible to explore several different algorithms that are either unified with the existing global algorithm or operate independently.

Two algorithm structures are currently under consideration, one decoupled from the global matrix in the frequency domain and the other decoupled in the spatial domain. The first option uses local feedback to reduce dynamic motion at the x-ray source points up to a few hertz, while retaining the existing global feedback system to reduce higher frequency orbit motion. This solution probably offers the widest correction bandwidth because the global feedback corrector response would not have to be compromised by equalization.

The second option is to digitally equalize the frequency responses of all (slow and fast) correctors and to integrate both local and global objectives into a single response matrix. Suitable weighting of the x-ray source points in the inverse response matrix would provide exact correction at the source points themselves, with global rms orbit minimization everywhere else. While this would certainly offer the best correction in spatial terms, it would come at the expense of some global feedback bandwidth because of the need to equalize all the correctors to the same response.

In either structure, depending on the stability of the source point measurements, the correction bandwidth could continue down to DC, or be rolled off at some low frequency so as to mesh with the existing workstation-based global correction algorithm. The possibility of incorporating different correction bandwidths for global and local correction within the unified algorithm is also being explored.

REAL-TIME BEAM DIAGNOSTICS

The reflective memory provides the means to share data and distribute global controls. The use of reflective memory to share the BPM error vector has already been mentioned. In addition, each slave DSP deposits BPM positions, computed corrector values, and x-ray BPM data in reflective memory at the feedback clock rate. This data is accessible to the master DSP for collection and analysis.

To take advantage of this data, a number of diagnostics have been incorporated into the orbit feedback system. These include running statistics of quantities such as rms corrector 'errors' and capture of waveforms from any of the signals deposited into reflective memory by each slave station. Typical uses include tracking the rms orbit motion, measuring response matrices using an 'AC-lockin' technique, and post-analyzing orbit motion that resulted in a beam dump. This system has also proven to be very useful for identifying BPM and corrector channels that are misbehaving.

'Dspscope'

Any of the signals deposited into reflective memory can be accessed at the feedback sampling rate and collected as a waveform. A total of 40 waveforms can be collected simultaneously, each with up to 4080 data points. Lower data collection rates are achieved by filtering and decimating the incoming data. This tool is used to routinely collect BPM and corrector 'error' signals for orbit motion measurements.

'AC Voltmeter'

Any of the signals can also be Fourier analyzed in real time with a sliding discrete Fourier transform, so that the time evolution of particular frequency components can be followed. This has been most useful for 'AC lock-in' measurements, where the orbit feedback system drives a chosen corrector at one frequency (typically 83.33 Hz) and measures the response at that frequency using the 'AC Voltmeter'. This technique has been successful in making rapid measurements of the response matrix and in identifying BPM gain errors.

Corrector 'Error' History Buffers

A corrector 'error' is defined as the result of the vector dot product of the measured positional errors (BPM 'errors') and the appropriate row of the inverse response matrix. As described in a later section, this signal is useful for localizing sources of orbit motion. The past 128 samples of corrector errors are stored in sliding buffers. When an unexpected beam dump occurs, the buffers are frozen and their contents automatically downloaded and archived with other information that can be used to trace the cause of the beam dump. These history buffers have proven useful in identifying the character of certain classes of beam dump and in localizing the origin of motion associated with it (a glitch on a magnet power supply, for example).

Sliding Algorithms

The AC Voltmeter is one of several algorithms that have been implemented as 'sliding' algorithms, where previously computed results are updated on every tick, as new data becomes available. In addition to the sliding Fourier transform, we have implemented statistical algorithms that estimate, for example, the mean and variance of the corrector error signals. The results are collected at periodic intervals and archived along with other machine data for subsequent analysis.

IDENTIFYING SOURCES OF ORBIT MOTION

The global orbit feedback algorithm uses 160 BPMs to determine corrections at 38 correctors in each plane. Since the forward response matrix is over-determined, the inverse response matrix is a least-squares solution that minimizes the rms of all BPM errors. The form of the matrix is band diagonal, meaning that each BPM influences only a subset of the correctors and that the response of each corrector is dominated by a subset of the BPMs. A practical implication is that for any given disturbance, orbit corrections are applied to magnets that are close to the source of the disturbance.

This feature has proved very useful in using orbit correction signals to localize strong sources of orbit motion. The 'dspscope' diagnostic in the orbit feedback system allows simultaneous data collection from all corrector drive signals or corrector 'error' signals in each plane at the full orbit feedback sampling rate. Taking the power spectral density of these signals, it is possible to generate a 'roadmap' of the sources of underlying orbit motion in the storage ring. An example is shown in Figure 9.

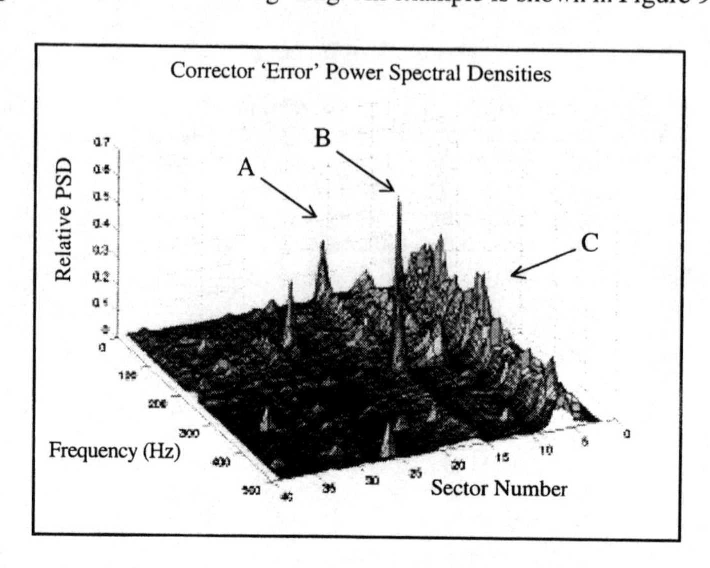

FIGURE 9. Example 'roadmap' of sources of horizontal orbit motion (see text for explanation of labeled features).

Three features are highlighted on the figure. Low-frequency, broad-band orbit motion in sector 16 ('A') was identified as a sextupole power supply with poor regulation. Narrow-band motion at 250 Hz, also in sector 16 ('B'), was identified as an oscillating corrector power supply. The wide-band orbit motion covering several sectors around sector 6 ('C') was identified as a bad BPM, and the indicated motion was not real. The bad BPM actually caused the orbit feedback system to generate motion, since the noisy BPM signal was interpreted as real beam motion. The spatial resolution of this technique for identifying sources of motion is actually better than the spacing between correctors. By examining patterns of adjoining correctors, it is possible to localize the source to about one third of a sector.

The number of corrector 'error' signals that are affected by a given source depends on the location of the source relative to adjacent feedback correctors. Patterns were measured by driving each of the ten quadrupoles in one sector with AC signals of equal magnitude and examining the response of the orbit feedback system. (Note that the AC modulation on the quadrupoles was converted to a vertical dipole field by the asymmetrical vacuum chamber.) Three separate patterns were identified and are shown in Figure 10. (More than three patterns can be identified if subtle differences are taken into account, although they have not proven to be useful in locating real sources of motion.)

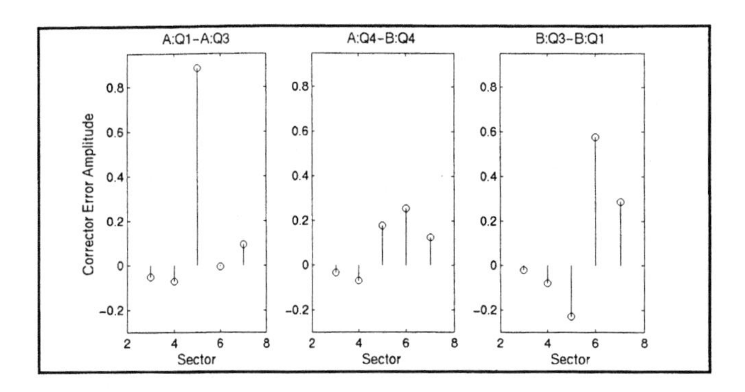

FIGURE 10. Horizontal corrector 'error' patterns for excitation of the ten different quadrupoles located in Sector 6.

The different corrector error patterns are associated with the location of the source within the sector in terms of betatron phase. The fact that we identified three separate patterns of correctors is consistent with there being two distinct changes in betatron phase advance through the lattice.

These beam-based techniques have proven useful in identifying a number of magnet power supplies with poor regulation and small oscillations on their outputs. However, in certain cases, strong sources have been localized but the sources themselves have not been identified. We have also found there is a more definitive indication of source locations when the measured response matrix is used instead of the modeled response matrix that is generally used for orbit feedback.

FUTURE PROSPECTS FOR ORBIT STABILITY AT THE APS

It was noted earlier that even with orbit feedback, the wideband vertical orbit motion is out of specification. This is accentuated by the fact that the APS is now to routinely deliver beam with 1% x-y coupling, a number that is likely to be further reduced in the future. If we are to maintain orbit stability within 5% of the beam size at such small x-y coupling levels, orbit motion will ultimately have to be reduced below 1 μm rms. This presents a tremendous challenge, not the least of which is our ability to measure and correct such small orbit motion.

For such small orbit motion requirements, an important consideration will be the frequency spectrum of the residual motion. While most users observe low-frequency orbit motion as fluctuations in x-ray intensity, higher-frequency motion is observed as an increase in the effective x-ray beam size. Consequently, it is reasonable to consider spectral shaping of the residual motion rather than to reduce the broad-band rms motion below certain absolute levels. (This is already being done to some extent with the choice of regulator tuning in the vertical plane.)

BPM Resolution and Stability

The need for stable and accurate measurements of the beam position has already been discussed within the context of local orbit feedback. The prospects for reducing global orbit motion to submicron levels are very much dependent on the quality of the beam position information provided by the rf and photon BPMs (8).

Provided that we are able to accurately measure the trajectory of the x-ray beams, implementation of the local feedback algorithms should reduce x-ray beam motion significantly below the present levels. The alternative is to continue to improve the performance of the global orbit feedback system, based on the premise that a large number of (relatively) poor sensors can provide more accurate information about the global orbit than a small number of (relatively) accurate sensors.

Global Orbit Feedback Improvements

Regardless of whether the local feedback system can be implemented, we will continue to work on improving the performance of the global algorithms.

The present algorithm contains 160 BPMs and 38 correctors, and will shortly correct the orbit at a 2 kHz rate (presently 1 kHz). However, there are many more BPMs and correctors available to the system, and with fixed processing power it is possible to trade off sampling rate against the number of elements (BPMs and correctors) used in the global algorithm. In terms of algorithm performance, the trade-off is in terms of correction bandwidth, signal-to-noise ratio, and number of spatial modes corrected.

The signal-to-noise ratio varies as the square root of the number of BPMs used. The spatial resolution also improves as the number of BPMs is increased, but there are diminishing returns as further BPMs are added. Modeling indicates that using four BPMs per sector (160 BPMs in total) gives an rms improvement of around 90% of the improvement obtained with all 360 BPMs.

The greatest impact on vertical orbit motion is expected from increasing the number of correctors in the global algorithm from 38 to 77 (two correctors per sector instead of one). This is estimated to increase the maximum attenuation of vertical motion from 15dB to 27 dB. While this would not increase the range of correction, it would further attenuate the orbit motion within the present correction bandwidth.

Increasing the correction bandwidth of the existing feedback system would also help reduce wideband rms orbit motion. Without increasing the sampling rate beyond 2 kHz, it is expected that the most significant improvements in correction bandwidth can be achieved by increasing the order of (and optimizing) the feedback regulator.

It should be noted that the VME bus backplane is rapidly becoming a bottleneck, with 50% of the total computation time being associated with accessing the BPM data from reflective memory. At this point, it would only be possible to increase the sampling rate significantly above 2 kHz by reducing the number of BPMs in the global algorithm.

Corrector Quantization

The present corrector magnet power supplies have 16-bit resolution and a maximum kick of ±1 mrad that generates a peak orbit deviation of ±10 mm at the x-ray source points. Statistically, quantization errors caused by the finite resolution can be considered a white noise source with values uniformly distributed between ±0.5 least-significant bits. This has an rms value of 0.29 bits, with energy equally distributed over the entire frequency band from DC to half the sampling rate. On the assumption that the effective resolution is 15 bits, then rms orbit motion caused by quantization errors of the 38 global correctors is estimated to be around 1.1 μm.

In order to reduce these quantization errors, a program is underway at APS to increase the resolution of the corrector power supplies. An alternative would be to reduce the dynamic range of the power supplies, but this approach is less desirable, since it also reduces flexibility in steering the DC orbit.

Reducing the Magnitude of the Underlying Orbit Motion

Clearly the most effective method of reducing the rms orbit motion is to remove the underlying sources of the motion. The program of identifying strong sources of orbit motion is continuing, and work is also underway to improve the stability of the magnet power supplies that are presently dominating the orbit motion power spectra.

CONCLUSIONS

The real-time orbit feedback system implemented at the APS successfully reduces horizontal orbit motion below the specification of 5% of beam size. Vertical orbit motion up to 30 Hz is also within specification, although broad-band vertical motion is still out of specification. To date, however, this has not been an issue with the APS users.

Local feedback is presently not implemented. Several options are available for its implementation in combination with the existing global orbit feedback algorithms. Initial evaluations of these options on the APS storage ring are anticipated to occur during the latter part of 1998.

There are several diagnostic capabilities built into the orbit feedback system that provide the ability not only to measure orbit motion and track system performance, but also to localize the sources of orbit motion.

Long-term plans for the APS include reducing the vertical rms orbit motion to sub-micron levels. This provides a tremendous challenge to our ability to measure and correct very small orbit motions. It is expected that, with continuing improvements, the system will meet this challenge.

ACKNOWLEDGEMENTS

The authors wish to thank J. Galayda for his continued support and encouragement in the development of this system. We also wish to thank M. Borland, G. Decker, and L. Emery for their encouragement and assistance with orbit correction and physics issues; and C. Doose, A. Hillman, D. McGhee, R. Merl, and A. Pietryla for their assistance with technical issues.

REFERENCES

[1] Winick, H., ed., *Synchrotron Radiation Sources - A Primer,* pp. 344-364, World Scientific (1994).

[2] Emery, L., M. D. Borland, "Advances in Orbit Drift Correction in the Advanced Photon Source Storage Ring," *Proceedings of the 1997 Particle Accelerator Conference*, Vancouver BC, Canada, May 1997 (to be published).
[3] Carwardine, J. A, F.R. Lenkszus, "Architecture of the APS Real-Time Orbit Feedback System," *Proceedings of the 1997 International Conference on Accelerator and Large Experimental Physics Control Systems*, Beijing, China, November, 1997 (to be published).
[4] Lenkszus, F. R., R. Laird, "The Advanced Photon Source Event System," *Proceedings of the 1995 International Conference on Accelerator and Large Experimental Physics Control Systems,* Vol. 1, pp. 345–350 (1996).
[5] Wind River Systems, Alameda, CA 94501.
[6] Pentek, Upper Saddle River, NJ 07458.
[7] Lenkszus, F. R., et al., "Beam Position Monitor Data Acquisition System for the Advanced Photon Source," *Proceedings of the 1993 Particle Accelerator Conference* (Washington DC, USA, May 1993), pp. 1814–1816 (1993).
[8] Decker, G. D., J. A. Carwardine, O. V. Singh, "Fundamental Limits on Beam Stability at the Advanced Photon Source," these proceedings.

RHIC Instrumentation

T. J. Shea and R. L. Witkover

Brookhaven National Laboratory
Upton, NY 11973 USA

Abstract. The Relativistic Heavy Ion Collider (RHIC) consists of two 3.8 km circumference rings utilizing 396 superconducting dipoles and 492 superconducting quadrupoles. Each ring will accelerate approximately 60 bunches of 10^{11} protons to 250 GeV, or 10^9 fully stripped gold ions to 100 GeV/nucleon. Commissioning is scheduled for early 1999 with detectors for some of the 6 intersection regions scheduled for initial operation later in the year. The injection line instrumentation includes: 52 beam position monitor (BPM) channels, 56 beam loss monitor (BLM) channels, 5 fast integrating current transformers and 12 video beam profile monitors. The collider ring instrumentation includes: 667 BPM channels, 400 BLM channels, wall current monitors, DC current transformers, ionization profile monitors (IPMs), transverse feedback systems, and resonant Schottky monitors. The use of superconducting magnets affected the beam instrumentation design. The BPM electrodes must function in a cryogenic environment and the BLM system must prevent magnet quenches from either fast or slow losses with widely different rates. RHIC is the first superconducting accelerator to cross transition, requiring close monitoring of beam parameters at this time. High space-charge due to the fully stripped gold ions required the IPM to collect magnetically guided electrons rather than the conventional ions. Since polarized beams will also be accelerated in RHIC, additional constraints were put on the instrumentation. The orbit must be well controlled to minimize depolarizing resonance strengths. Also, the position monitors must accommodate large orbit displacements within the Siberian snakes and spin rotators. The design of the instrumentation will be presented along with results obtained during bench tests, the injection line commissioning, and the first sextant test.

OVERVIEW OF RHIC

Upon completion in 1999, the Relativistic Heavy Ion Collider (RHIC) at Brookhaven National Laboratory will accelerate and collide protons, polarized protons, and heavy ions (1) (2). Heavy ion collisions up to Gold on Gold at 100 GeV/u beam energies will produce extended nuclear matter with energy densities an order of magnitude greater than that of the nuclear ground state. This should result in temperatures and matter densities that prevailed a few microseconds after the origin of the universe. It is also believed that these extreme conditions could produce a phase transition to a quark-gluon plasma. The Spin Physics program at RHIC will utilize polarized protons at up to 250 GeV and 70% polarization. The primary goal is to study the spin structure function of the proton.

The collider consists of two rings separated horizontally by 90 cm in a tunnel 3.834 km in circumference. Collision points are provided in six insertion regions that

CP451, *Beam Instrumentation Workshop*
edited by R. O. Hettel, S. R. Smith, and J. D. Masek
1998 The American Institute of Physics 1-56396-794-4/98/$15.00

are connected by six arcs. The total complement of magnets for both rings include 288 arc dipoles, 276 arc quadrupoles, 108 insertion dipoles, and 216 insertion quadrupoles. Additional magnets include 72 trim quadrupoles, 288 sextupoles, and 492 correctors. All ring magnets are superconducting.

The Alternating Gradient Synchrotron (AGS) complex is the injector to RHIC. Beam is extracted through the H-10 septum magnet into the AGS-to-RHIC (AtR) transfer line as single bunches during a flat-top of the AGS magnet cycle. The AGS cycle is repeated until each of the two RHIC rings are filled. The 1900-foot-long AtR line consists of several sections, starting as the "U-line", becoming the "W-line," the branch into the "V-line" to the g-2 experiment. A switch magnet at the end of the W-line directs beam into either the "X-line" or "Y-line" to the counter-rotating RHIC rings.

RHIC construction officially began in 1991. In the Fall of 1995, the first part of the AtR line was commissioned up through the W line. In February, 1997, beam was transported to a temporary dump at the end of the first sextant. Collider commissioning will begin in early 1999 with project completion scheduled for June of that year. Relevant parameters are listed in Table 1.

TABLE 1. RHIC Parameters

Parameter	Value
Kinetic Energy, Injection-Top (each beam)	Gold: 10.8 – 100 GeV/u Protons: 28.3 – 250 GeV
Luminosity, Au-Au at 100 GeV/u	$2 \cdot 10^{26}$ cm^{-2} s^{-2}
Operational lifetime, Gold	10 hours
No. of bunches/ring	60
No. of particles/bunch	Gold: 10^9 (at times, a few 10^7) Protons: 10^{11} (upgrade, $3 \cdot 10^{11}$)
Bunch length	from 20 ns down to 1 ns
Normalized Emittance, gold	Injection: 10π mm mrad After 10 hr: 40π mm mrad
Filling mode	Bunch to bucket, 30 Hz peak rate
Filling time	1 min, each ring
Acceleration time	75 s
Revolution frequency	About 78 kHz
rf harmonic number	Acceleration system: h = 360 Storage system: h = 2520
Beta at crossing, H,V	During injection: 10 m Low beta insertion: 2 m
Transition energy	$\gamma_T = 22.89$
Circumference	3833.845 m
Beam tube i.d. in arcs	69 mm
Vacuum, warm beam tube sections	$7 \cdot 10^{-10}$ mbar
Operating temp., helium refrigerant	< 4.6 K

INSTRUMENTATION

A list of the major instrumentation systems is provided in Table 2. Most of the transfer line systems have already been installed and tested during the 1995 AtR commissioning and the 1997 Sextant Test (3) (4) (5). All of the listed systems are expected to be available for collider commissioning in 1999, although some will provide only a subset of potential functionality.

TABLE 2. Table of Instrumentation

System	Quantity
Position monitors	52 measurement planes in transfer line 667 planes total in collider rings
Ionization profile monitors	One horizontal and one vertical per ring
Wall current monitors	One per ring
Transverse feedback	Two kicker units per ring (each provides horizontal and vertical deflection)
Schottky cavities	One per ring (each provides horizontal, vertical and longitudinal signals)
Transfer line intensity monitors	Five integrating current transformers
Loss monitors	120 ion chambers in the transfer line 400 ion chambers in the collider tunnel
Collider ring current monitor	One DCCT per ring

ATR BEAM PROFILE MONITORS

Video Profile Monitors (VPMs) are used in the AtR line (6). A total of 12 VPMs were installed, any four of which can be viewed on a single transfer via a 4 × 16 video multiplexer. Low mass phosphor screens minimize scattering and allow four profiles to be acquired on a single bunch for an emittance measurement. VME-based image processing electronics are used to acquire and process the data.

Gadolinium Oxy-sulfide doped with Terbium (Gd_2O_2S:Tb) phosphor was chosen because it has low mass and can be deposited in a thin (0.002") coating. In order to see individual profiles on successive transfers the light had to decay fully in 33 ms. Chromium doped aluminum oxide (Chromox) has too long a decay time and is not available in such thin sheets. Tests have shown Gd_2O_2S:Tb to fully decay faster than the 30 Hz camera frame rate. The screen is mounted at 45° and pneumatically inserted. The drive, viewing window and two lamp ports are on the same 8" conflat flange, simplifying alignment. A 45° first-surface mirror transfers the image to the camera over a typical 300 cm path length. Screen durability has not been established but none have been damaged in the low-intensity gold beam runs. Aluminum oxide screens are used at the two upstream locations where intense proton beams are transported to the g-2 experiment.

Pulnix TM-7cn CCD cameras are used in all locations except in the AGS ring tunnel. Because typical CCD camera lifetime is 100 to 700 Rad, a Cidtek (3710D) charge injection device (CID) camera, rated at 20 kRad, was used there, but failed after one week. The dose was estimated to be consistent with the expected life. Other alternatives are to use an MRad version of the Cidtek camera or a Dage radiation hardened vidicon camera. At most locations in the transfer line the cameras are mounted in 14" diameter tubing inserted into the tunnel wall to provide gamma shielding. Since the light lasts only a millisecond, the camera must either acquire the full frame at once (Cidtek) or have an overlap period in which the odd and even fields are both sensitive (Pulnix). The camera must be synchronized with the beam in this way or every other line on the vertical display will be blank. More than 3 decades of light intensity is expected, but a typical lens will only adjust over a 150:1 range. Lenses with a graded neutral-density center spot can cover 3 decades but are very expensive and require a motor drive interface. A simple device using solenoids to insert up to four neutral density filters between the camera and the lens, providing a 20,000:1 range has worked well. A 500 mm f/5.6 reflector lens is used at most locations, but 1000 mm f/8 reflector lens and 300 and 400 mm f/5.6 refractors are also used. About half of the cameras were rotated 90° to best match the beam aspect ratio with the orientation re-established in the computer-generated display. The lens, camera, and filter array are aligned on an optical rail using commercial optical mounting hardware. The rail sits on a sliding tray which uses tapered pins for precision location, allowing rapid replacement. An inexpensive "leveling" laser substituted for the camera and lens was used to align the optics.

Phosphor screen grain size, air waves and mechanical motion are not significant limitations on resolution for the AtR beam sizes, which range from several mm to several cm at 2.5 sigma. With the screen at 45°, depth of field is a factor which can be significant if the beam is well off center or vertically large. Camera resolution is limited by the number of pixels and transmission and electronics bandwidths. Wideband (6 MHz) analog fiber-optic links are used to preserve resolution over the longest (1700 ft) run to the centrally located VME-based image processing electronics. The resolution was calculated to be better than 0.25 mm using measurements and manufacturer's data for the cameras and lenses. Measurements of the fine fiducial marks were consistent with the estimate.

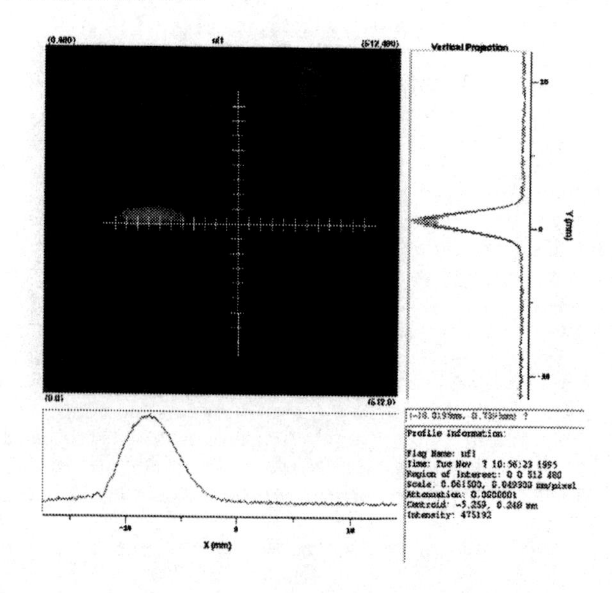

FIGURE 1. Video Profile Monitor display.

The acquisition and processing of the video data uses Imaging Technology Inc. VME modules running under VxWorks. These include four IMA-VME-4.0 boards with 4 AMVS-HS acquisition modules, 2 CMCLU-HS Convolver arithmetic modules, and 2 CMHF-H histogram/feature extractor modules. Forty-eight 512×512 frames, plus base frames for background subtraction, and computational results can be stored on-board. A 128×128 subset of the full frame can be generated either from a 4×4 convolution or a region of interest (ROI), which can be pre-selected or dynamically determined by a threshold setting. The full RHIC fill can be stored for these reduced data sets and later sent over the LAN for higher level processing. On-line real-time computations available include: pixel-by-pixel base frame subtraction of full frame data, centroid of the full frame, H and V projections, and sum of all pixels. So far, only the full 512×512 frame data has been used. High-level code was written to control and display the beam profiles and calculate emittance. Figure 1 shows a typical profile display.

ATR BEAM INTENSITY MONITORS

Beam intensity is monitored at five junction points in the AtR line using Integrating Current Transformers (ICT) and electronics manufactured by Bergoz (7). For the RHIC sextant test, an ICT from the left injection arc was moved to the end of the string of superconducting magnets. The design of the ICT provides passive pre-integration of a fast current pulse, reducing the effect of core losses. The slower risetime signal is then integrated and held for acquisition by the RHIC MADC (8). Initially noise from the AGS extraction kicker, which is coincident with the beam, interfered with the signal, but passing multiple turns of the tri-axial signal cable through a ferrite core significantly reduced the pick up. Reliable signals were then obtained even for 10^7 gold ions.

ATR BEAM LOSS MONITORS

The Beam Loss Monitors (BLMs) are ion chambers mounted on the vacuum flanges downstream of each magnet. To limit the number of electronic channels while providing complete coverage, signals from the 120 detectors were grouped into 56 channels. The electronics were located in four equipment houses. The BLMs, designed to be sensitive one decade below the nominal 10^9 gold ion intensity, were able to monitor the losses for beam intensities from 10^6 to 2×10^7 Au79 ions.

The BLMs are Tevatron ion chambers (9) modified by using an isolated BNC to break the ground loop formed by the signal and HV cable shields. Rexolite is used rather than PTFE for the insulators in the BNC and SHV connectors to improve the radiation hardness. The ion chamber (10) consists of a 113 cc glass bulb filled with argon to about 725 mTorr. Each chamber is calibrated using a cesium-137 source. The mean sensitivity in the middle of the plateau (1450 V) is 19.6 pA/R/h, with 95% within (1.5 pA/R/hr of the mean. Where multiple detectors were used on a single channel, they were grouped by calibration with the average value used for the group.

The ionization current from the detector is fed to a low-leakage (less than 10 pA), gated integrator and read out using the standard RHIC MADC. The integrator input pre-integrates the electron signal with a millisecond time constant (comparable to the ion collection time), greatly reducing noise from the kicker magnets which are time coincident with the beam. To take full advantage of the pico-Amp sensitivity of the detector and electronics, it was necessary to use non-tribo-electric cable (Belden 9054) to reduce noise caused by mechanical motion. These features resulted in the noise level of about 10 pA observed during the AtR commissioning.

POSITION MONITORS

Position Monitor Electrodes

The beam position monitor (BPM) electrode assemblies for the collider ring and the AtR line share a common mechanical design (11). Nearly all of the collider assemblies operate at 4.2 Kelvin while the all of the AtR assemblies operate at room temperature. They contain 23 cm long, shorted striplines with a carefully controlled 50-ohm impedance. These large striplines couple enough power to allow accurate measurement of low-intensity pilot bunches. The shorted design requires electronics with a low return loss in order to control beam coupling impedance, but the static cryogenic load is reduced by a factor of two over the more expensive back-terminated designs. The assemblies are constructed of 316L stainless steel and are copper brazed in a hydrogen reducing atmosphere. The resulting assembly is fully annealed and therefore mechanically stable under extreme temperature variations. The feedthrough is a coaxial, glass-ceramic design supplied by Kaman Instrumentation. Unlike conventional ceramic-kovar feedthroughs, these units can be reliably thermally cycled from a 300-degree Celsius bakeout temperature down to a 4.2-Kelvin operating temperature while providing excellent microwave performance. Over 1400 feedthroughs have been tested and installed with no insulator failures. The cryogenic installation requires a special cable to bring the signal to room temperature feedthroughs mounted on the cryostat. These .141" diameter cables have a Tefzel insulator that provides increased radiation hardness over the standard Teflon. To optimize thermal performance, the jacket is made of stainless steel instead of copper and the central portion of the cable is thermally stationed at the 55-Kelvin heat shield.

Most of the assemblies contain two opposing striplines and are oriented horizontally or vertically such that the position measurement is made in the plane having the larger beta function. These are designated Type 1 monitors; a cutaway view of one is shown in Figure 2. In critical areas, particularly in the insertion regions, the monitors allow simultaneous measurement of horizontal and vertical position by including four striplines. In order to match the expected beam size, they are constructed with either large (Type 3) or small (Type 2) apertures. The collider ring contains 480 electrode assemblies plus additional units for the spin rotators, Siberian snakes, and the beam dumps. The AtR line contains 39 assemblies.

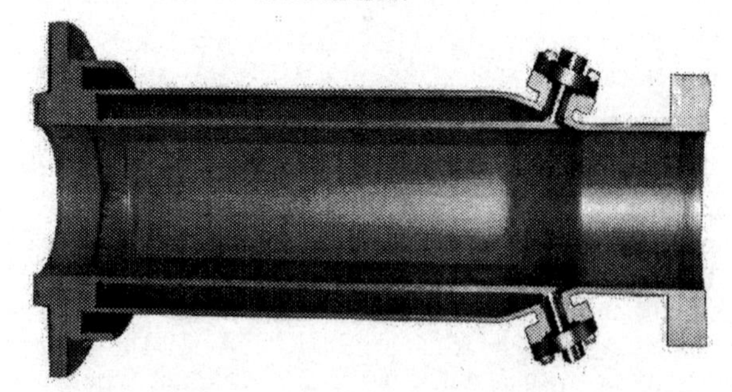

FIGURE 2. Position monitor electrode assembly.

For ease of commissioning, the offset between electrical centers of the position monitors and the magnetic centers of the quadrupoles must be accurately characterized. Even tighter constraints are imposed during polarized proton acceleration, which requires accurate orbit control in order to avoid depolarizing spin resonances. Because the main arc quadrupoles do not have individual trim supplies, the position monitor offset characterization cannot be easily performed online. Therefore, this offset was surveyed during magnet assembly by performing the following procedure (12):

1. The magnet center is measured relative to cryostat fiducials. For early units, this measurement was made with a magneto-optical technique (13). For later units, a measurement coil was used.
2. An antenna is inserted into the electrode assembly. The position of the antenna is optically surveyed relative to the cryostat fiducials.
3. An rf signal is injected into this antenna and the signal amplitudes at the position monitor ports are measured. This provides a measurement of the offset between the electrical center and the antenna.
4. The offset between the position monitor electrical center and the magnet's center is calculated. The estimated tolerance of the offset measurement is about 100 μm rms.

Position Monitor Electronics

Analog Sampler

Each channel of position monitor electronics employs a broadband sampler (14) depicted in Figure 3. This circuit has evolved since publication of the 1995 reference, but the basic design remains. All measurement planes are treated independently. Therefore, dual-plane electrode assemblies (Type 2 and Type 3) are connected to two independent electronic channels. Two channels are contained within the beam position monitor module that will be described in the next section. As shown in Table 2, the AtR line requires 52 channels and the collider rings require a total of 667 channels. All AtR electronics are rack mounted in equipment buildings that are accessible during operations. These signals are transported on 3/8-inch, solid, shielded coax between the tunnel and the equipment building. Where possible, the collider ring channels are cabled in a similar manner with 1/4-inch coax. Channels in the insertion regions, injection area, and dump area are all cabled this way with up to 150-meter-long runs. All other modules are located in the tunnel and cabled to the appropriate electrodes with shorter, 2-meter-long cables. These channels are all in cryogenic regions. The low vacuum should minimize beam-gas scattering and the resulting radiation field at the electronics. In all cases, the signal first passes through an attenuator and a coaxial, 135 MHz low-pass filter before reaching the analog sampler. The filter rejects high-frequency, high-voltage signals from short bunches while the attenuator provides improved return loss. An upgrade to higher cost diplexers will be considered if required in the future.

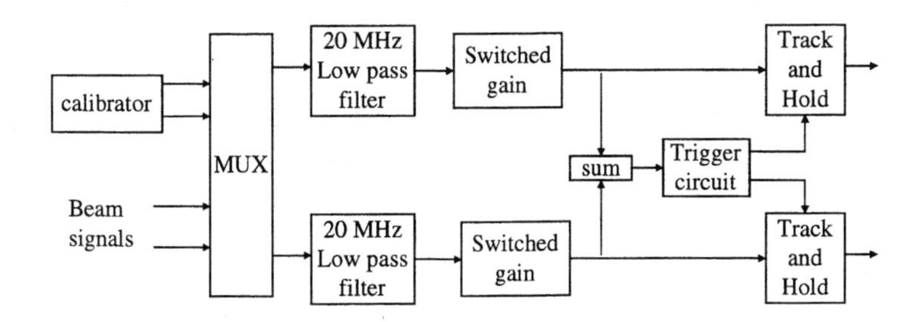

FIGURE 3. Block diagram of the BPM analog sampler.

Referring again to Figure 3, a multiplexer (MUX) at the input to the sampler selects between beam signals and an on-board calibration pulser. This MUX can also swap the input signals to allow observation of imbalance in the following electronics. Signals then pass through matched 20 MHz, lowpass filters. These are Bessel filters with good transient response. A high impedance sum circuit provides the input to a self-trigger circuit. The trigger threshold is adjustable and the self trigger can be completely disabled to allow external clocking. When the self trigger is enabled, an external gate is applied to select a particular bunch. Independent delays are adjusted to assure that each signal is sampled precisely at the peak. Track-and-holds with 14-bit linearity are used to sample the signals. The output is digitized by a 16-bit ADC. In the AtR line every transported bunch is sampled. This leads to a maximum acquisition rate of 30 Hz. In the collider ring modules, a selected bunch is sampled turn by turn at the revolution frequency of about 78 kHz.

Data Acquisition

Each analog sampler resides on a circuit board that also contains the data acquisition hardware. Each board includes two 16-bit digitizers, a fixed point digital signal processor (DSP) subsystem, an in-system programmable gate array, a beam synchronous timing interface, and an IEEE1394 Serial Bus (Firewire) interface (15). Two boards are packaged together in each module, but each board functions as a completely independent position channel. The control system communicates with the channels via shared memory in a VME/Firewire interface board. As shown in Figure 4, up to 12 channels are connected to each interface board via the Firewire Serial Bus. In the AtR line, a data record containing all the positions of all transported bunches is sent to the shared memory on every AGS cycle. In the collider rings, the channels can operate in different modes. During injection, a turn-by-turn record for each injected bunch is written to shared memory. For the rest of the collider cycle, the channels will periodically send a turn-by-turn record for a particular bunch and simultaneously stream signal averaged position data at 10 Hz.

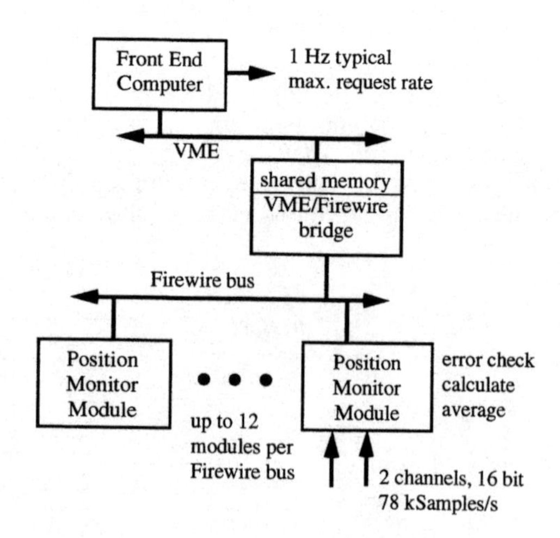

FIGURE 4. Data flow from the BPM system.

Test results

Results from several tests are summarized in Table 3. To ease comparison, all position amplitudes are normalized to the nominal 69 mm aperture of the RHIC arcs. The bench tests were made on production prototypes of the final modules. All beam tests were performed with similar but earlier prototype samplers.

TABLE 3. Summary of Position Sampler Tests

Test	Result	Comment
Minimum measurable bunch intensity	$5 \cdot 10^8$ charges	Bench measurement
Maximum measurable bunch intensity	10^{12} charges	Bench measurement
Noise in turn by turn measurement	2 × μm rms	Bench measurement, max. intensity
	1 mm rms	Bench measurement, min. intensity
	<10 μm rms	Measured proton bunch in Tevatron during store
Accuracy	±100 μm	Bench measurement, $5 \cdot 10^8 <$ bunch intensity $< 10^{12}$
	< 1 mm	Orbit correction results from AtR test. Includes other error sources.
Scatter in single pass, single bunch measurement	20 μm rms	Vertical scatter of bunches extracted from AGS
Drift over 5 hours	±5 μm	Calibration pulse injected into channel during AGS extraction
	±25 μm	Measured vertical orbit of bunches extracted from AGS

RING BLM SYSTEM

BLM System Design

The primary function of the RHIC Ring BLM system is to prevent a beam-loss quench of the super-conducting magnets. It will also provide quantitative loss data for tuning and loss history in the event of a beam abort. The system uses 400 AtR style ion chambers, but since the AtR is single pass, different electronics are employed. It has been estimated that the RHIC superconducting magnets will quench for a fast (single turn) loss > 2 mJ/g or a slow (100 ms) loss > 8 mW/g. This is equivalent to 78.3 krad/s at injection (49.3 krad/s at 100 GeV/c) for uniform loss over a single turn and 4.07 rad/s at injection (0.25 rad/s at 100 GeV/c) for slow losses. This will yield a range of signal currents from 5.5 mA for the injection fast loss level to 17.6 nA for a slow loss quench at full energy. Allowing for studies results in a dynamic range of 8 decades in detector current. The amplified signal is digitized at 720 Hz and continually compared to programmable fast and slow loss levels which can cause a beam abort. This will halt data acquisition, providing a 10-second history of the pre-abort losses. BLM parameters are adjusted during injection, magnet ramp and storage phases to set gains, fast and slow loss thresholds, and abort mask bits on specific RHIC Event Codes.

The Detectors

The detectors and cable are as in the AtR line. Half of the ion chambers (198) are mounted between the two RHIC Rings on the quadrupole cryostats using stainless steel "belly bands." This will not provide equal sensitivity to losses from each ring, but if more precision is required a second detector can be added on the outer ring and either the outputs of the two BLMs connected in parallel or the number of channels doubled. Ninety-six BLMs are placed at insertion region quads. In the warm regions, 68 detectors are mounted on the beam pipe at expected sensitive loss points. In addition, 38 BLMs are available as movable monitors. Since the Ring BLM system is used for quench prevention, redundancy is provided by separate HV power supplies for the two cables which provide the bias voltage to alternate detectors. Further redundancy is not required since the system is not to be used for personnel protection.

Electronics

The analog circuitry is packaged in an 8-channel module. A micro-controller module manages up to 8 analog boards independent of the crate front-end computer (FEC), once the write list values have been set through high level code. This insulates the real-time operations from the control system I/O, allowing the BLM system to operate during a controls link failure. Commercial digital I/O and DAC modules are used to control the HV power supplies. The electronics will be located in service buildings at 2,4,5,7,8,10 and 12 o'clock, allowing access during beam storage. Standard VME crates were modified for the special needs of the BLM electronics. Tests indicated that the standard ±12 V switcher power supplies were too noisy for the high sensitivity analog circuitry. While DC-DC converters might have been used, due to limited real estate for the converters and filters and the possibility of oscillator noise, it was decided incorporate a separate linear ±15 V supply into the crate. A piggy-back board across the last nine P2 connectors provided a dedicated bus between the micro-controller module and the 8 analog modules and a means of supplying the ±15 V.

The Analog Module

Figure 5 shows a simplified schematic of the analog section for one of the eight channels on the Analog Module. An input low-pass RC filter, matched to the magnet thermal time constant, integrates the fast loss impulse, greatly reducing the dynamic range while providing a sufficiently fast rising signal to protect against a single turn loss quench. Back-biased, matched, low-leakage diodes (DPAD-5) protect the amplifier input from high voltage spikes. The low-current amplifier (OPA627AU) is rolled off to a 10 μs rise time. To allow for BLM shielding differences, jumpers can set two alternate gains. The front-end op-amp output is applied to a second amplifier with programmable gains of 1 or 10 prior to signal acquisition. The data is read at 720 Hz by a RHIC VME MADC configured for ±10 V, 13 bits and stored in a 1 Mbyte on-board memory. An optional off-board 360 Hz anti-aliasing filter is available. Readings can be taken at additional times as required for specific applications. For the nominal jumper setting and a buffer gain of 10, one LSB represents 12.5 pA, comparable to the noise observed in beam tests. Offsets, typically a few LSBs, are not adjustable since these can be removed in the higher level processing.

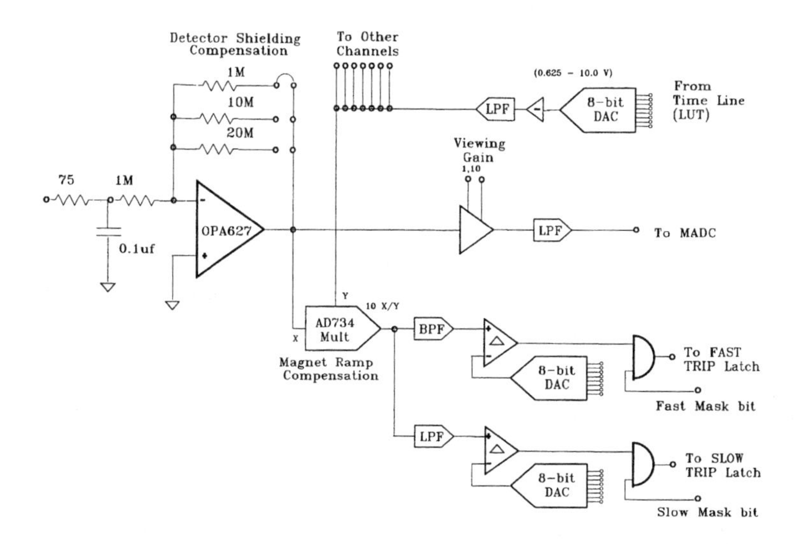

FIGURE 5. Schematic of a typical channel of the BLM Analog Module.

The front-end op-amp output also goes to an AD734BN analog multiplier which provides a gain to compensate for the increased magnet quench sensitivity with current. An 8-bit DAC sets the gain for all multipliers on a board. A high-pass (100 μs) and low-pass (20 ms) filter direct the signal to respective fast- or slow-loss comparators with independent programmable references. The 8-bit reference DACs are sufficient due to the magnet current compensation provided by the analog multiplier. Each comparator can be masked to prevent a bad BLM from inhibiting the beam or to allow special conditions. The gains, mask bits and trip levels may be changed by Events on the RHIC Event Link. Any trip latches the states allowing the location to be determined. An Altera 7128 chip performs all logic functions and communication with the BLM Micro-controller module via the dedicated bus on the VME P2 backplane.

The Micro-Controller Module

The RHIC Control System talks to the BLM micro-controller (16) which controls the BLM analog module. This was necessary due to the large number of set-point changes, particularly during injection, acceleration and transition. The micro-controller, once in possession of the write-list, completely controls the BLM analog module, freeing the FEC and allowing the BLM system to continue to provide beam loss quench protection even in the event of a controls failure. A Microchip PIC16C64 micro-controller services the 256 byte registers on the BLM analog modules. A 64k × 16 bit memory holds the write-lists for 256 RHIC Event codes each associated with up to 255 address/data values. On detection of a specific Event, the corresponding write-list is sequentially executed with the data (gain, fast/slow trip level...) going to a particular register. Altera 9320 and 7128 gate arrays are used on the board.

BLM Test Results

Figures 6a and 6b show the losses from an 8 bunch transfer at 30 Hz for 2×10^{12} protons, 20 times RHIC intensity. The top trace is the analog output showing the signal rising rapidly due to the loss from each bunch transfer. The signal then decays with the 100 ms front-end filter time constant. The middle trace of Figure 6a is the slow-loss filter output with the fast losses rejected. The bottom trace is the comparator output for a 2 V reference. The middle trace of Figure 6b is the fast loss filter output, showing only the electron component of the signal. The bunches did not have equal losses so the comparator does not trip for every transfer. It is clear that the circuit has the ability to discriminate between fast and slow losses.

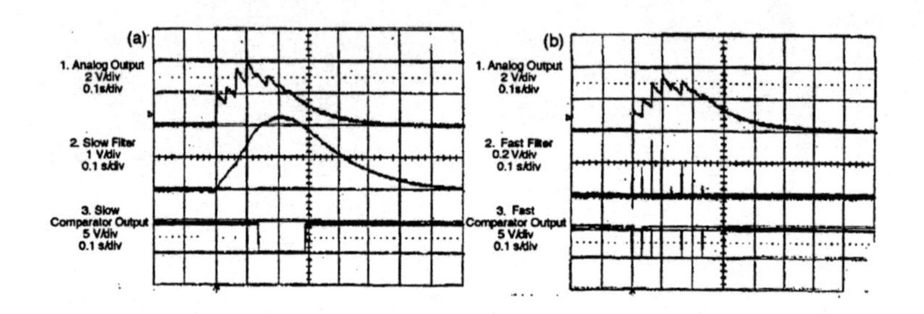

FIGURE 6. Eight bunch 2×10^{12} proton transfer. a) Slow filter and comparator output. b) Fast filter and comparator.

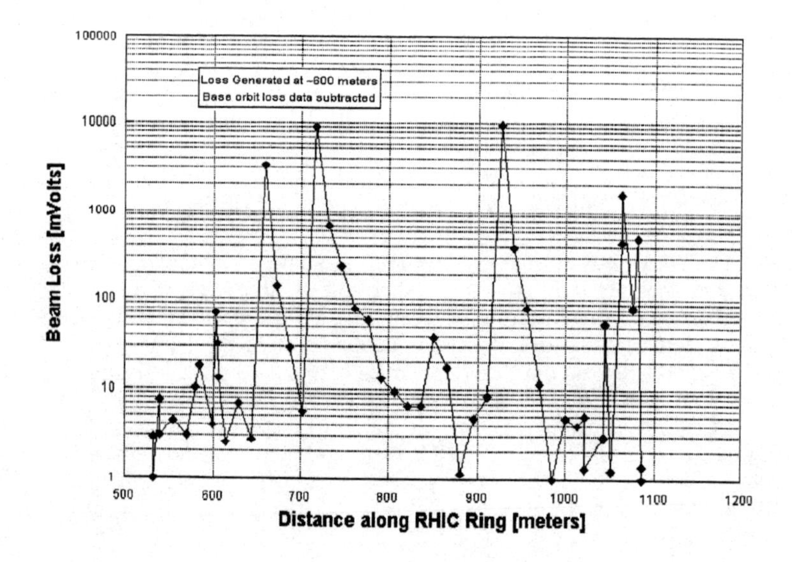

FIGURE 7. Loss data from the Sextant Test.

In January 1997, an Au^{+79} beam was injected into one ring of the first sextant of superconducting magnets. BLMs were installed at their final locations, which for these tests were on the inside of the Ring carrying the beam. Since the beam was single pass and sufficient Ring electronics were not yet available, AtR electronics (integrators) were used. The results of an average of two runs normalized to 10^8 ions is shown in Figure 7. A large loss was purposely created at 600. With the BLMs located at the quads (about 15-meter interval) there is sufficient spatial resolution and dynamic range to determine the direction of the beam causing the loss. In other runs, larger losses were purposely generated requiring the integrator gain to be lowered and gate width to be reduced. The observed integrator noise and offsets, of the order of a few LSBs, are similar to that observed with the prototype Ring BLM electronics installed in the AtR line. These tests indicted that the BLM system will have sufficient range to meet the design requirements.

THE RHIC RING BCM SYSTEM

System Requirements

The ring beam current monitor (BCM) must cover the range shown in Table 4. Because studies intensities may be an order of magnitude lower, the intensity may range from 1×10^9 to 1.2×10^{13} charges or 12.5 μA to 150 mA. RHIC is a storage accelerator so the BCM must be able to measure DC current yet be able to observe bunch-stacking at 30 Hz. Since the beam in RHIC will cross transition, the BCM bandwidth must allow intensity changes over several turns to be observed, although at lower resolution. Stability should be ±10 μA over the 10-hour storage time.

TABLE 4. RHIC Intensity and Current

Number of Bunches	Gold (+79) [Intensity current]	Protons [Intensity current]	Pilot protons [Intensity current]
1	7.9×10^{10} 0.99 mA	1×10^{11} 1.25 mA	1×10^{10} 0.125 mA
57 (Nominal)	4.5×10^{12} 56.2 mA	5.7×10^{12}/ 71.1 mA	5.7×10^{11} 7.11 mA
120	9.5×10^{12} 118 mA	1.2×10^{13} 150 mA	1.2×10^{12} 15.0 mA

The Beam Current Monitor

A DCCT for each ring was purchased from Bergoz (17). The unit has remotely switchable 50 and 500 mA maximum current ranges. Modulator noise is less than 1 μA when integrated over 30 ms. The long term stability has not yet been measured, but sensitivity to temperature is consistent with the 5 μA/ °C quoted by the vendor. An RTD mounted on the sensor will be used to correct for this effect. The unit was specified with 75-meter-long cables to allow front-end electronics to be removed from the RHIC tunnel. However, preliminary tests indicate that the modulator noise is more than an order of magnitude greater than with the standard 3-meter cables. This noise is not a problem when viewed on the electronics low-pass output (~4 kHz high-order filter), or when the wide-band output is integrated over 30 ms or more, but higher frequency measurements will be affected. Because the modulation is regular, the high over-sampling makes it is possible to digitally filter much of this noise. Certainly the result will be better with the shorter cables which may be used for the RHIC commissioning.

The BCMs will be located in the warm region of the 2 o'clock sector, which will be baked to 150° C. The BCM housing has been designed to insulate the transformer core from the heated beam pipe and prevent it from exceeding 60° C. Thermocouples on the detector will be used to interlock the heater blanket. The outer shell of the housing will provide the bypass path for the wall current around the transformer.

BCM Data Requirements

Beam intensity information will be used in a number of ways which set different requirements on the data as summarized in Table 5.

TABLE 5. BCM Data Requirements

Measurement	Read Rate	Resolution	Data
High Resolution (Decay Rate)	1 Hz	20 bits	Display each reading Display 1000 point sliding avg
Medium Resolution	78 kHz	16 bits	Non real-time digital filtering Display 1000 point sliding avg
Low Resolution (Tuning)	720 Hz	13 bits	Average of last 72 readings With 10 Hz display
Low Resolution (Logging)	720 Hz	13 bits	Each reading recorded In 10-second sliding memory
Injection	30 Hz	16 bits	Average over 33.3 ms All bunch display

The long term beam decay rate will be monitored by a very high resolution, slow update digital multimeter (DMM) for each Ring. Since loss rates of less than 10 μA must be detected, acquisition of 1 μA or better will be needed. This 20-bit resolution will be provided by a Keithley Model 2000 with a IEEE-488 interface, which can be programmed to provide a rolling average, lessening the load on the FEC. Measurement of the intensity at injection or around transition will require a faster acquisition at medium resolution. A 16-bit, 16-channel, 100 kS/s/channel ADC with 4 Mbytes on-board memory will be used to read at the 78 kHz revolution frequency. This data, suitably averaged, will also be used at injection to track the bunch stacking. The modulation noise on the signal is highly periodic and amenable to digital filtering to obtain sub-millisecond non real-time intensity information. A standard RHIC MADC will provide the 720 Hz low-resolution (13-bits plus sign). The low-pass output BCM signals will be stored in a 10-second deep memory which will be available in the event of a beam abort. An average of 72 MADC readings will be used to provide a display at 10 Hz for tactile feel tuning.

WALL CURRENT MONITOR

The wall current monitor system incorporates ferrite loaded pickups based on the design by Weber (18). One pickup is installed in each ring. The ferrite has been selected to attain a flat frequency response down to 3 kHz with a transfer impedance of 1 ohm. The response extends to 6 GHz, which is well above pipe cutoff. Interfering modes will be attenuated by microwave absorber installed on either side of the pickup. A calibration port has been included.

The signal from the pickup will be digitized by a LeCroy LC584AL oscilloscope. This scope has bandwidth of 1 GHz and will digitize in 8 Gsa/s bursts at a trigger rate of up to 30 kHz. The scope is controlled and read out over GPIB by a computer running LabVIEW. This software is based on a similar application developed at Fermilab by Barsotti (19). The RHIC control system communicates with this application via shared memory on a VME/MXI interface board. The entire system is event driven and synchronized by the RHIC beam synchronous event system. Functions provided by the system are summarized in Table 6.

TABLE 6. Wall Current Monitor Functions

Function	Features
Injection and acceleration bucket fill pattern	Reports integrated charge within each of the 360 buckets, and total charge
Store bucket fill pattern	Reports integrated charge within each of the 2520 buckets, and total charge
Bunch profile and beam centroid	Mountain Ranges
Calculated bunch parameters	length, peak current, area...
Spectral waterfall	Time resolved frequency domain view

IONIZATION PROFILE MONITORS

The ionization profile monitors (IPM) collect electrons that are produced as a result of beam-gas interactions (20). Two monitors will be installed in each ring, one horizontal and one vertical. Because the dispersion is non-zero at the location of the horizontal IPM, the measured beam width will be affected by both the transverse emittance and the momentum spread. A desirable future upgrade will be the addition of a horizontal monitor in an area of high dispersion.

The strong space charge field of the RHIC beam affects the both the ions and electrons that are produced from the residual gas interactions. However, the electrons are easily confined to a small Larmor radius by a weak magnetic field. Therefore, a permanent magnet dipole is installed over the vacuum chamber. The confined electrons are swept out of the beam in a few nanoseconds by an applied electric field. Meanwhile, the ions are slowly accelerated in the opposite direction where they pass through an electron suppression grid near the opposite wall. The extracted electrons are amplified by a two stage chevron microchannel plate. The resulting charge is collected on 64 striplines that are spaced 0.6 mm center to center. Special preamplifier hybrids designed by the BNL Instrumentation Division are used to integrate, shape and buffer the signal before transmitting it out of the tunnel.

Gold beams will give single-bunch profiles, while proton beam profiles will be generated by integrating the signals from all bunches for several turns. With 10^9 gold ions/bunch, beam width measurement will be accurate to ±3%. To keep the MCP from saturating with gold beams (signal from proton beams is too small to saturate MCP), the sweep field will be turned on only during data collection. There will be two measurement modes:

1. The profile of a single bunch will be measured on every turn. The sweep field will be off during the passage of all other bunches.
2. Every bunch will be measured for one complete turn. The sweep field will be left off for 100-1000 turns for the MCP to recover.

The system block diagram is shown in Figure 8. All timing is controlled by the beam synchronous event system. The 10 Msample/s, 12 bit ADCs consist of 8 channel VME boards with 128 ksamples of memory behind each channel. These digitizer boards and the timing board reside in the control system front end computer.

A prototype IPM was successfully tested during the 1997 sextant test. Single-pass profile measurements of bunches containing 10^8 Gold ions were made. Beam widths agreed at the few percent level with those measured by the VPM system.

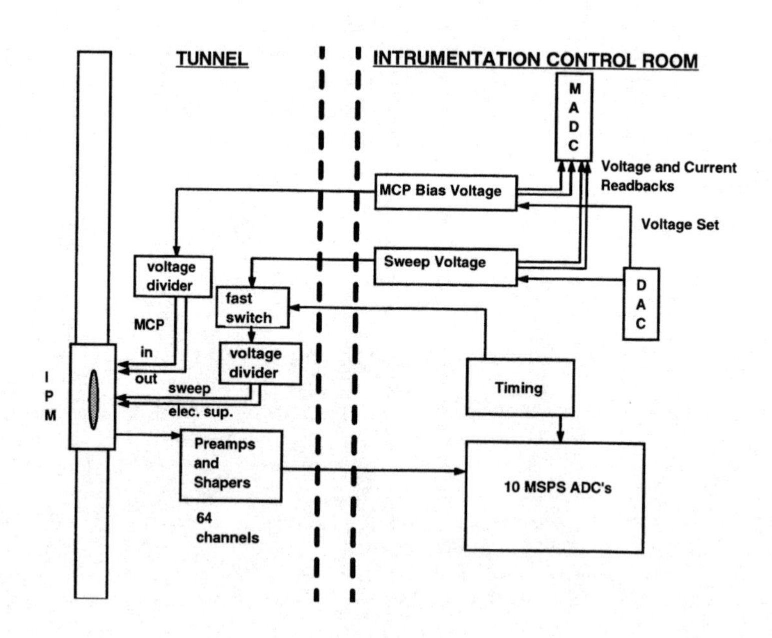

FIGURE 8. Block diagram of the Ionization Profile Monitor system.

TRANSVERSE FEEDBACK SYSTEM

The transverse feedback system will provide the following functions.

- Excitation of coherent betatron motion for diagnostics using one of the following modes:
  1. Single kick (50 μm amplitude at storage energy, expected decoherence time of a few hundred turns)
  2. Random sequence of kicks (larger betatron line, more emittance growth)
  3. Swept frequency
- Phase Lock Loop tune tracking
- Damping of injection errors
- Damping of transverse instabilities

The single kick and random kick modes will be provided for tune measurement during commissioning. All other functionality will be developed as experience is gained during operations.

The kicker system employs 50-ohm stripline kickers. Each unit is 2 m long and has four electrodes thus providing both horizontal and vertical deflection. Each ring has two units that will be wired in series for early operations. To provide large deflections at reasonable cost, the kickers are driven by solid state, switched pulsers capable of delivering up to 3 kV, 50 ns pulses at the revolution frequency of 78 kHz. Therefore, a selected bunch can be kicked turn by turn. After operational experience is gained, wideband linear amplifiers may be employed to drive one kicker unit per ring. This will allow development of a phase lock loop tune tracker and bunch by bunch feedback for damping potential instabilities.

The digital electronics consists of a Motorola VME based processor board and a Technobox PMC module for digital I/O. The VME board contains a 300 MHz PowerPC that will process the of 78 ksample/s turn by turn data stream while the PMC module contains an Altera gate array that will handle the 9.4 Msample/s bunch by bunch data stream. For tune measurements during early operations, turn by turn data from the standard position monitor channels will be used. Later, dedicated position monitor electronics will provide low noise, bunch by bunch measurements.

SCHOTTKY SYSTEM

A high frequency cavity from Lawrence Berkeley National Laboratory will be used to detect high frequency Schottky signals. Although somewhat harder to interpret than signals at lower revolution harmonics, these high-frequency signals suffer less contamination from coherent power (21). The cavity's transverse modes of interest are the TM_{210} and the TM_{120} at about 2.1 GHz. These two modes have a measured Q of about 4700 and are separated by 4 MHz. A longitudinal mode is at 2.7 GHz.

The signals will be carried from the tunnel on 7/8" solid shield coax to a digital signal analyzer located in the instrumentation control room. This analyzer has a 10 MHz bandwidth. During commissioning, the system will be locally operated. As certain data proves useful, it will be made available to the control system through a shared memory interface.

ACKNOWLEDGMENTS

The following are among the many who have contributed to RHIC beam instrumentation systems: P. Cameron, P. Cerniglia, B. Clay, R. Connolly, J. Cupolo,

C. Degen, A. Drees, M. Grau, L. Hoff, D. Kipp, W. MacKay, J. Mead, R. Michnoff, R. Olsen, V. Radeka, W. A. Ryan, T. Satogata, H. Schmickler, D. Shea, R. Sikora, G. Smith, S. Tepikian, E. Tombler, N. Tsoupas, J. Weinmann, P. Zhou, P. Ziminski, E. Zitvogel.

This work was supported by the U. S. Department of Energy.

REFERENCES

[1] *RHIC Design Manual*, April 1998.
[2] Peggs, S., "RHIC Status," presented at the 1997 Particle Accelerator Conference, Vancouver, BC, May 1997.
[3] Shea, T. J., et. al., "Beam Instrumentation for the RHIC Sextant Test," *Proc. of the 4th European Particle Accelerator Conference*, 1994, pp. 1521–1524.
[4] MacKay, W. W., et. al., "AGS to RHIC Transfer Line: Design and Commissioning," *Proc. of the 5th European Particle Accelerator Conference*, 1996, pp. 2376–2378.
[5] Shea, T. J., et. al., "Performance of the RHIC Injection Line Instrumentation Systems," presented at the 1997 Particle Accelerator Conference, Vancouver, BC, May, 1997.
[6] Witkover, R. L., "Design of the Beam Profile Monitor System for the RHIC Injection Line," *Proc. 95 Particle Accel. Conf*, 1995, p. 2589.
[7] Bergoz, 01170 Crozet, France.
[8] Michnoff, R., "The RHIC General Purpose Multiplexed Analog-to-Digital Converter System," *Proc. 95 Particle Accel. Conf*, 1995, p. 2229.
[9] Shafer, R. E., et al., "The Tevatron Beam Position and Beam Loss Monitoring Systems," *Proc. The 12th Int'l. Conf. High Energy Accel.*, 1983, pp.609.
[10] Troy-onic Inc., 88 Dell Ave, P.O. Box 494, Kenvil, NJ 07847.
[11] Cameron, P. R., et. al, "RHIC Beam Position Monitor Characterization," *Proc. 95 Particle Accel. Conf.*, 1995, p. 2458.
[12] Trbojevic, D., et. al, "Alignment and Survey of the Elements in RHIC," *Proc. 95 Particle Accel. Conf.*, 1995, p. 2099.
[13] Goldman, M. A., et. al., "Preliminary Studies on a Magneto-Optical Procedure for Aligning RHIC Magnets," *Proc. 93 Particle Accel. Conf*, 1993, p. 2916.
[14] Ryan, W. A. and T. J. Shea, "A Sampling Detector for the RHIC BPM Electronics," *Proc. 95 Particle Accel. Conf.*, 1995, p. 2455.
[15] Shea, T. J., et. al., "Evaluation of IEEE 1394 Serial Bus for Distributed Data Acquisition," presented at the 1997 Particle Accelerator Conference, Vancouver, BC, May, 1997.
[16] Michnoff, R., see "http://www.rhichome.bnl.gov:80/Hardware/lossmon1.htm"
[17] Loc. Cit. Reference [7]
[18] Weber, R. C., "Longitudinal Emittance: An Introduction to the Concept and Survey of Measurement Techniques Including the Design of a Wall Current Monitor," *AIP Conf. Proc.* **212**, 1989, p. 85.
[19] Barsotti, Jr., E. L., "A Longitudinal Bunch Monitoring System Using LabVIEW and High Speed Oscilloscopes," *AIP Conf. Proc.* **333**, 1995, p. 466.
[20] Connolly, R. C., et. al., "A Prototype Ionization Profile Monitor for RHIC," presented at the 1997 Particle Accelerator Conference, Vancouver, BC, May, 1997.
[21] Goldberg, D. A. and G. R. Lambertson, "A High Frequency Schottky Detector for Use in the Tevatron," *AIP Conf. Proc.* **229**, 1991, p. 225.

A Cryogenic Current Comparator for the Absolute Measurement of nA Beams

Andreas Peters*, Wolfgang Vodel†, Helmar Koch†, Ralf Neubert†, Hannes Reeg*, Claus Hermann Schroeder*

*Gesellschaft für Schwerionenforschung, Darmstadt, Germany
†Institut für Festkörperphysik, Friedrich Schiller Universität, Jena, Germany

Abstract. A new type of beam transformer, based on the principle of a Cryogenic Current Comparator (CCC), was built to measure extracted ion beams from the SIS, the heavy ion synchroton at GSI. A current resolution of 0.006 – 0.065 $nA/\sqrt{Hz}$, depending on the frequency range, could be achieved allowing us to measure ion beams with intensities greater than 10^9 particles per second with high accuracy.

Numerous investigations were carried out to study the zero drift of the system which shows a strong exponential slope with two time constants. In addition, the influence of external magnetic fields was measured. Furthermore the microphonic sensistivity of the system was studied by measuring noise spectra of the detector's vibration and the output signal.

Measurements with neon and argon beams will be presented and compared with signals emitted from Secondary Emission Monitors (SEM). Another measuring function of the CCC-detector aims at the analysis of the beam's time structure to get information about beam spill fluctuations. With an extended bandwidth (0–20 kHz) of the detector system it is now possible to compare simulations of extracted beams from synchrotons with measurements of the CCC.

MOTIVATION

The main research topics of GSI are mid-range nuclear physics of heavy ions, atomic physics of highly-charged ions, biophysics (especially tumor treatment with carbon ions), and radioactive beams. Three different sources produce the ion beams for the UNILAC, the UNIversal Linear ACcelerator (Fig. 1). The highest energies at the UNILAC in the order of 10–20 MeV/u allow several experiments at the Coulomb barrier in the older experimental hall. One of these experiments aims at the production of new super-heavy elements, the last that was found is the element 112.

For higher energies up to 1–2 GeV/u a heavy ion synchroton (SIS) was added in 1989. High-energy beams can be kicked to the the experimental storage ring (ESR) for further atomic physics experiments using different cooling techniques.

CP451, *Beam Instrumentation Workshop*
edited by R. O. Hettel, S. R. Smith, and J. D. Masek

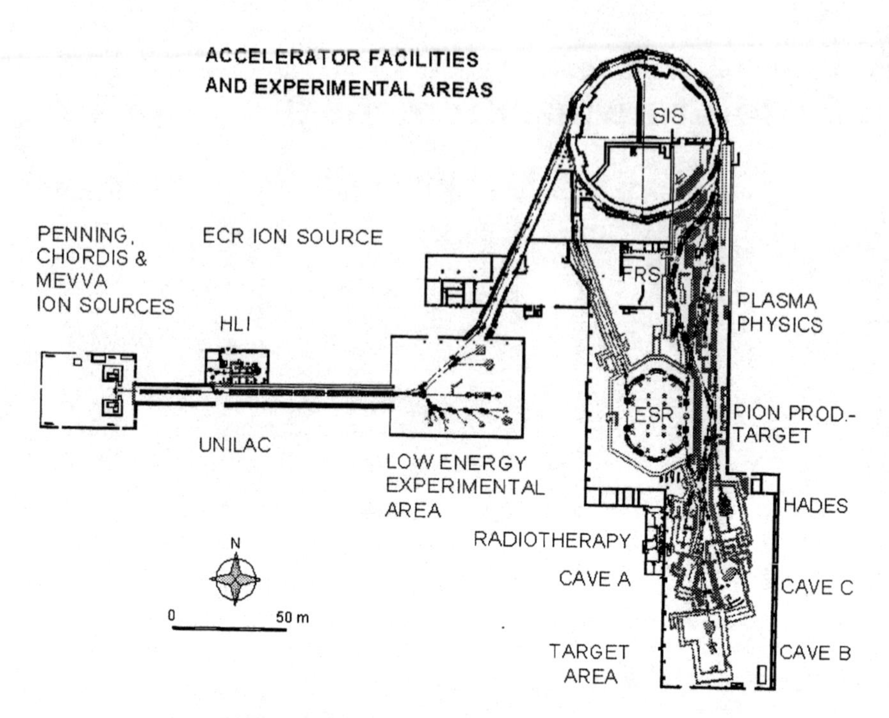

FIGURE 1. The GSI Accelerator Facilities.

On the other hand a lot of experiments use a resonant extracted beam for target radiation, e.g., to produce radioactive beams in the fragment separator (FRS). All these experiments will need higher beam intensities in future because of interest in measuring more exotic nuclei of smaller cross sections.

The GSI-accelerators UNILAC and SIS will be improved in the next three years so that the planned intensities of the SIS (defined by the incoherent space charge limit) can be reached [1]. To reach this aim a new injector consisting of RFQ- and IH-cavities will be installed instead of the old Wideroe-section. After the upgrade 2×10^{11} light ions (e.g., ^{12}C, ^{16}O, or ^{20}Ne) or 4×10^{10} heavy ions (e.g., ^{197}Au, ^{208}Pb, or ^{238}U) will be accelerated in the synchrotron corresponding to beam currents of about 300–500 mA at 1 MHz circulating frequency. Due to extraction times of 1–10 s the currents in the high-energy beam transport system could extend to some hundreds of nanoamps.

Besides new detectors for the UNILAC and the SIS, an expansion of the diagnostics for "high-intensity beams" in the beamlines to the new target hall was necessary. Possible detectors with relevant properties are as shown in Table 1.

Most detector principles are based on the interaction between the beam and material (e.g., scintillators, ionization chambers), so the beam is more or less distorted

TABLE 1. Detector Types and Properties

Detector principle	on-line	non-destruct.	radiation resistent	absolute calibration	suitable for vacuum
Nucl. trace counting	–	–	–	+	–
Faradaycup	+	–	+	+	+
Ionization chamber	+	–	+	–	–
Scintillation counter	+	–	–	+	–
Diamond detector	+	o	+	+	+
Sec. emission monitor	+	o	+	–	+
Residual gas monitor	+	+	+	–	o
Beam transformer	+	+	+	+	+

+ = yes/possible; o = possible only under favorable conditions; – = no/not possible

by scattering or energy loss. Diamond detectors represent a new and very interesting development in this field [2]. But all required properties are fulfilled only by beam transformers. Because the function of a DC transformer is based on both the modulation of a magnetic material and the observation of a shift of the hysteresis curve in the presence of a beam current through the pick-up coil, the limiting factor is the Barkhausen noise produced by wall-jumps of magnetic regions. To avoid this phenomenon no modulation of the core material is allowed.

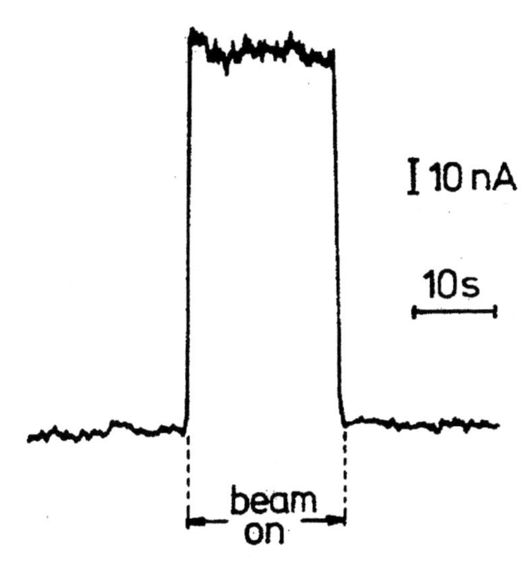

FIGURE 2. An electron beam current measured by a CCC at the PTB Berlin in 1977.

To reduce other noise effects, a cooling of the detector is often useful. Using low temperatures, modern superconductive detector principles could be taken into account as done in the early 70s by research groups in national standards laboratories. The first Cryogenic Current Comparator (CCC) was developed by I. K. Harvey in Australia 1972 [3]. A few months later a similar CCC was built at the Physikalisch-

Technische Bundesanstalt (PTB) in Berlin [4]. This group was the first which used a CCC based on SQUIDs for measuring beam currents. An 108 nA electron beam was generated by a Van de Graaff accelerator at 2.5 MeV (see Fig. 2 for their best measurement). For the beam current of about 100 nA the corresponding magnetic field is only 0.2 pT at a distance of 10 cm.

Because the accelerator at the PTB was shut down and the problems with the self-made SQUIDs were large the development of this device was stopped. But the first measurements showed that a CCC with SQUIDs works in an accelerator environment with rf background and large magnetic stray fields. Further developments with modern techniques seemed to be promising.

THE FUNDAMENTAL PRINCIPLES OF A CRYOGENIC CURRENT COMPARATOR

Figure 3 shows the construction principle of a CCC [5]. Two currents counterflow through a superconducting tube and introduce a differential current on the surface of this tube which can be measured outside the tube via the magnetic field by a SQUID, a highly sensitive magnetic sensor (see below).

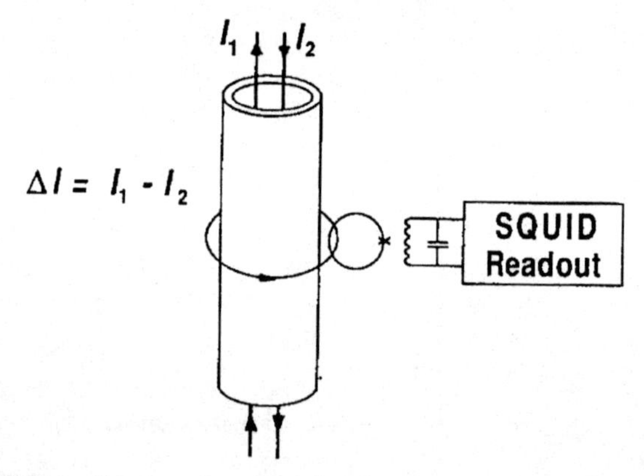

FIGURE 3. Construction principle of a cryogenic current comparator.

The ideal coupling is only obtained for an infinitely long shield cylinder but in practice this arrangement has to be replaced by inverting the shield cylinder into a toroidal geometry (see Fig. 4). To avoid a short circuit a little slit in the shielding is necessary. It is easy to recognize that only the azimuthal field of a current passing through the hole of this toroid can induce a magnetic field inside. Because of the geometry, all other field components are strongly weakened!

The mathematical treatment of the attenuation of different shield arrangements [6] gives the solutions shown in Table 2 (for geometries, see Fig. 5):

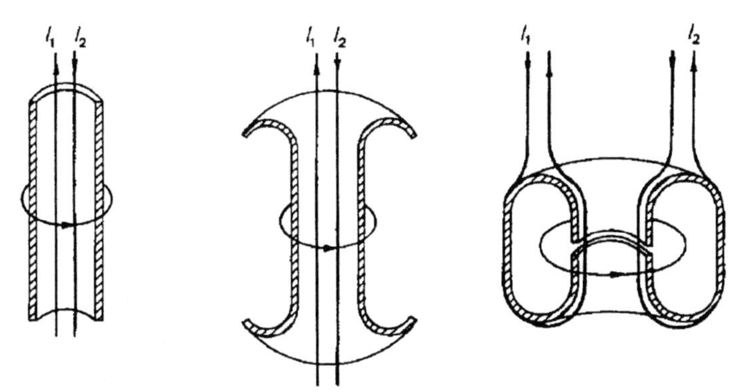

FIGURE 4. Transition of a cylinder geometry to a toroid geometry for a CCC.

TABLE 2. Shield Types and Their Attenuation Properties

Type	Attenuation
Coaxial cylinder	$A_{CC} = exp\{-\frac{2}{1+r_i/r_o} \cdot \frac{l}{r_a}\}$ $(r_a/r_i \leq 1,5)$
Simple cylinder	$A_{SC} = exp\{-1,84 \cdot \frac{l}{r_c}\}$
Ring cavity, type I	$A_{RC1} = \{\frac{r_i}{r_o}\}^2$
Ring cavity, type II	$A_{RC2} = 1$

r_i, r_o are inner and outer radii for coaxial cylinders and ring cavities of type I; r_c is the radius of the cylinder for simple cylinders; l is the length of the gap.

CCCs made of long (coaxial) cylinders have the advantage of high attenuation factors but they have only a small inner radius. A CCC consisting of ring cavities allows a larger inner radius which is necessary for proper functioning. The disadvantage of a smaller attenuation in this geometry can be compensated by using more cavities.

In addition to the magnetic shielding, a highly sensitive magnetic sensor is necessary to measure fields in the fT and pT ranges. Superconducting QUantum Interference Devices are sensors consisting of a superconducting ring with two weak links for the DC type. Two superconducting effects should be mentioned to under-

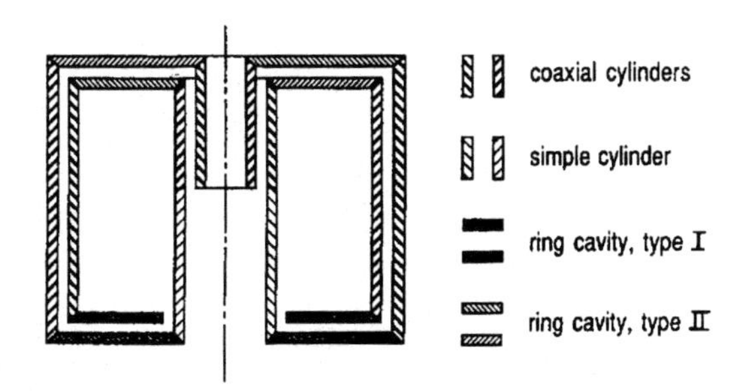

FIGURE 5. Classification of superconducting shielding geometries.

stand the function of a DC-SQUID:

- In a superconducting ring a quantization of the magnetic flux is observed:
  $\Phi_{mag} = n \cdot \Phi_0, \qquad n = 0, 1, 2, \ldots \qquad \text{with } \Phi_0 = \frac{h}{2e} = 2.07 \times 10^{-15} Vs$
- In the presence of a weak link the so-called "DC Josephson Effect" appears (Figure 6):

  The affiliated wave functions of the Cooper pairs are:

$$\Psi_1(r) = |\Psi_1(r)| e^{i\varphi_1(r)}$$
$$\Psi_2(r) = |\Psi_2(r)| e^{i\varphi_2(r)}$$

  The two wave functions are coupled to each other due to the possibility of tunneling through the barrier. The solution of this coupled equations is:

$$I(t) = I_0 \cdot sin(\tfrac{2eV_0}{\hbar} t + \varphi_0)$$

  For zero voltage across the junction there is a DC current $I = I_0 \cdot sin\varphi_0$ which can assume any value between $-I_0$ and $+I_0$.

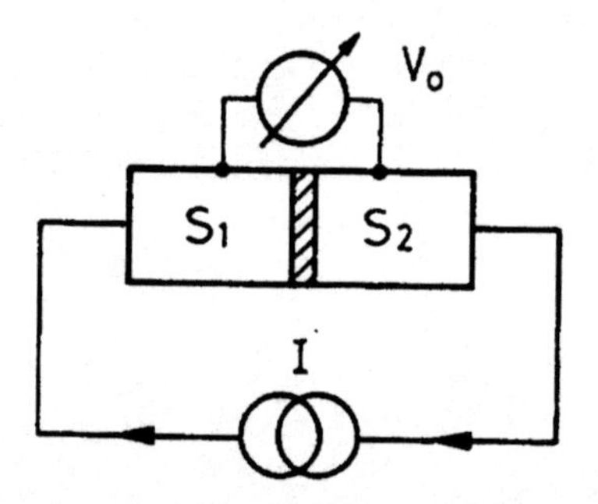

FIGURE 6. Schematic arrangement of a Josephson junction.

A superconducting loop consisting of two Josephson junctions in parallel exhibits interference phenomena. In the DC case the total current is

$$I = I_a + I_b = I_0 (sin\varphi_a + sin\varphi_b).$$

When a magnetic flux Φ_{mag} threads the area of the loop, the phases differ by

$$\varphi_a - \varphi_b = \tfrac{2e}{\hbar} \oint \vec{A} \cdot d\vec{s} = \tfrac{2e}{\hbar} \Phi_{mag}.$$

With $\varphi_0 = (\varphi_a + \varphi_b)/2$ the current I is

$$I = I_0 \cdot sin\varphi_0 \cdot cos(\tfrac{e}{\hbar} \Phi_{mag}).$$

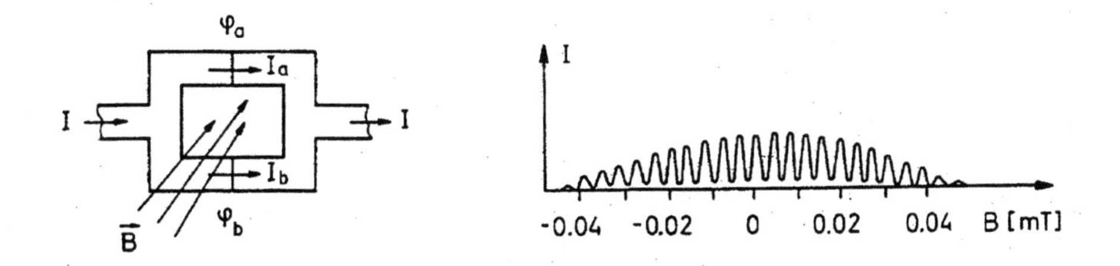

FIGURE 7. SQUID current dependence on an external magnetic field.

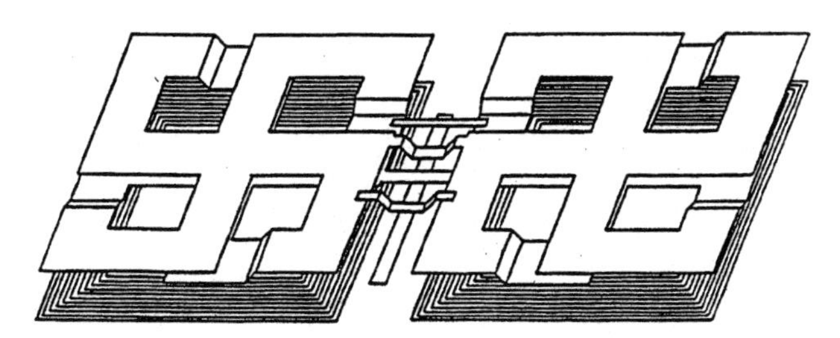

FIGURE 8. Modern SQUID layout (schematically).

Figure 7 shows the result of this function which is the same as the interference pattern of a double slit in optics. The output voltage of the SQUID is directly coupled to the magnetic flux Φ_{mag} caused, for example, by a beam passing a pickup coil of a CCC.

Today DC SQUIDs are fabricated like micro chips with multilayer techniques. Figure 8 shows a simplified drawing of the layout of the DC SQUID UJ 111 designed and manifactured by the Institute of Solid State Physics at the Friedrich Schiller University, Jena, Germany. To suppress the influence of external magnetic fields a gradiometric design was chosen [7].

To read out the signals generated by a SQUID chip, special electronics are required [8]. This consists of a low-noise preamplifier, a bias current source, a lock-in detector (with a modulation frequency of 125 kHz) followed by an integrator and filter module. The scheme of the DC SQUID electronics is shown in Figure 9.

THE DESIGN OF THE GSI PROTOTYPE-CCC

At the project's start in 1992, the following requirements for the detector were defined:

- Cryostat:
  - Working temperature: 4.2 K (liquid helium temperature)
  - Boil-off rate: $\leq$ 5 l liquid helium per day
  - A "warm hole" for the passing ion beam with 100 mm aperture

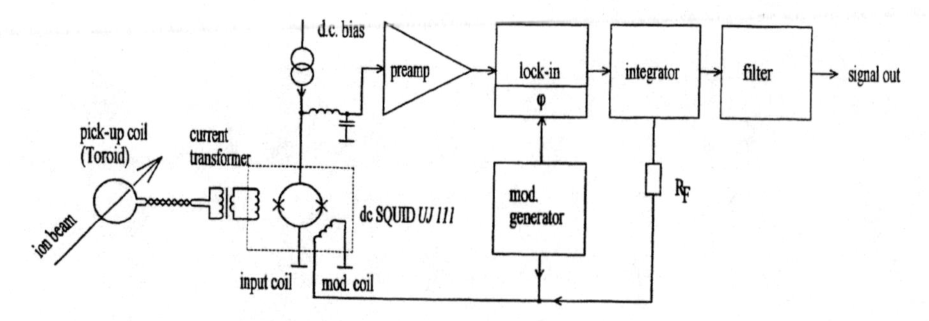

FIGURE 9. Scheme of the DC SQUID electronics.

- Magnetic shielding highly efficient against stray fields
- Resolution of the whole system: $\leq 1\ nA/\sqrt{Hz}$
- Current zero drift: $\leq$ 1nA/min
- Maximum bandwidth of the output signal: 0–5 kHz
- Maximum range: 10 μA
- Accuracy of the whole measurement system: $\pm$ 5 %

To meet these requirements a special liquid helium bathcryostat with a "warm hole" of Ø 100 mm for the passing ion beam was designed. The outer radiation shield consists of a superinsulated copper vessel cooled to about 40 K by a Gifford-McMahon refrigerator. Figure 10 shows the mechanical setup of the cryostat which is nearly 1.2 m high and has a diameter of about 0.66 m.

In Figure 11 the design of the magnetic shielding and the input coil is shown. The input coil — a single winding formed as a toroid with a VITROVAC core — is made from niobium while the meander shape shielding is produced from lead plates and tubes insulated by teflon foil. An attenuation factor of 2×10^{-9} for external background fields (non-azimuthal) was calculated. The whole detector is mounted on a carrier tube with a ceramic gap to prevent parasitic wall currents.

The DC SQUID head designed and manifactured by the Friedrich Schiller University, Jena, Germany (see Fig. 12) is directly attached to the magnetic shielding whereas the other parts of the system — the preamplifier and the SQUID controller — are mounted outside the helium dewar.

MEASURED PARAMETERS

Cryogenic Performance

Several steps are necessary to cool down the cryostat and to fill it with liquid helium. Because the LHe-container including the detector system weighs about

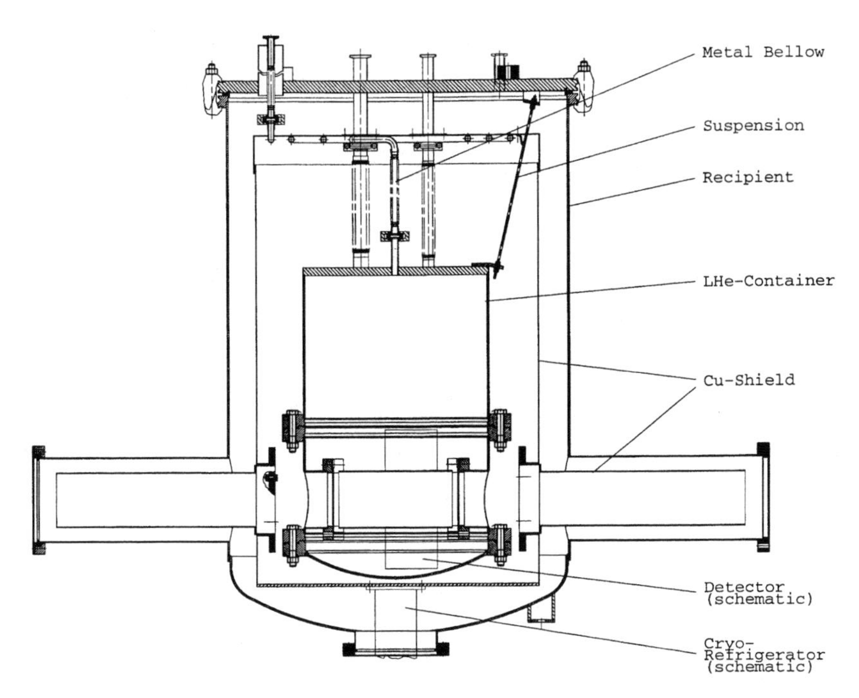

FIGURE 10. Design of the special bath-cryostat for the CCC.

100 kg this process takes about three days. As a result of continuous measurements of the helium level a minimum boil-off rate of 5.6 l LHe/d was observed. This corresponds to a heat loading of the LHe-vessel of 170 mW and is in good agreement with the theoretical calculated power consumption of about 125 mW.

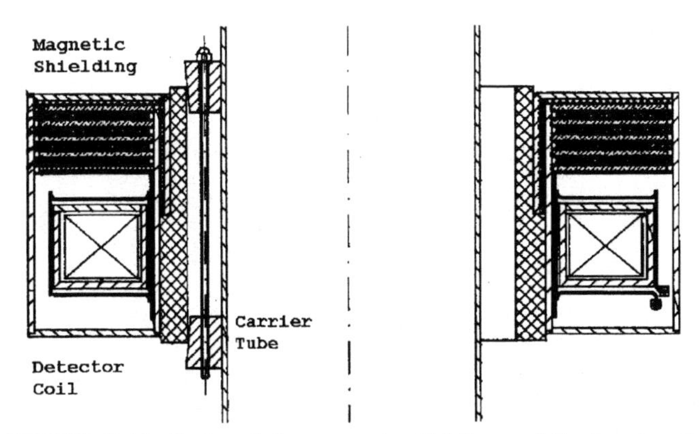

FIGURE 11. Design of the magnetic shielding and the input coil.

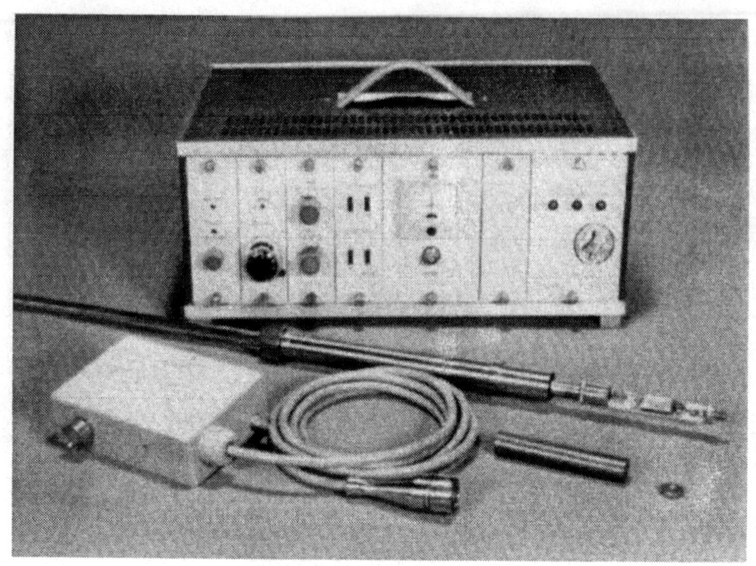

FIGURE 12. DC SQUID system model 4.

Current Sensitivity

A first output signal was observed by feeding a 10-nA test pulse of a calibrated current source (Keithley model 261) into the calibration winding through the pickup coil (see Fig. 13). For this measurement a bandwidth of 500 Hz was chosen.

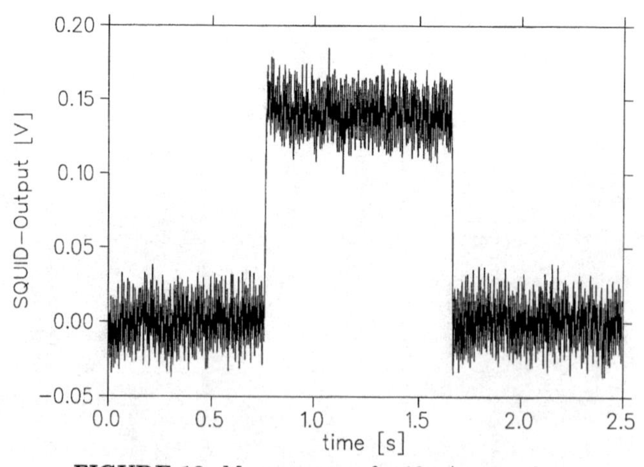

FIGURE 13. Measurement of a 10 nA test pulse.

In Figure 14 the plotted output voltage of the SQUID electronics as a function of the calibration current is presented. The current sensitivity of the detector system was determined to 181.0 nA/ϕ_0, the linearity error is smaller than 0.5 %

(1 ϕ_0 corresponds to an output signal of 2.5 V in the most sensitive range of the SQUID system). During the measurements the refrigerator was switched off to avoid microphonic effects and magnetic interference (see below).

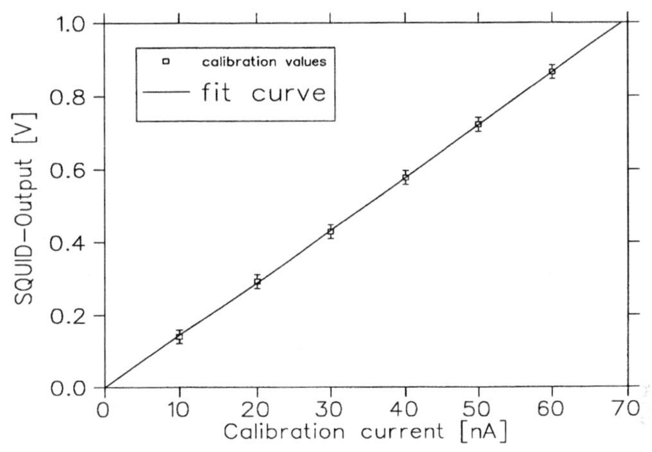

FIGURE 14. Calibration curve of the CCC.

Noise Level and Current Resolution

To determine the current resolution a noise spectrum (with open input) was taken. The noise level of 0.08-0.9 $mV_{RMS}/\sqrt{Hz}$ corresponds to a current resolution of 0.006-0.065 $nA/\sqrt{Hz}$, depending on the measurement frequency (see Fig. 15). The cut-off frequency of the system was found at about 10 kHz (small signal mode).

Current Zero Drift

At the beginning of each test the output signal of the CCC shows a strong zero drift. It was supposed that residual flux in the core of the coupling coil fades slowly away because of imperfect superconducting contacts in the input circuit. To study this effect the detector was cooled down to 4.2 K over a period of about 100 hours. The results of the long-term measurement are shown in Fig. 16. The curve is fitted by the sum of two exponential functions where the time constants are 3.3 hours and 30.8 hours. The shorter time constant is caused by a thermal effect. After a cooling time of about 100 hours the zero drift drops to values under 0.5 mV/s ($\cong$ 35 pA/s current drift), during another test run a minimum value of 0.23 mV/s ($\cong$ 16 pA/s current drift) was achieved.

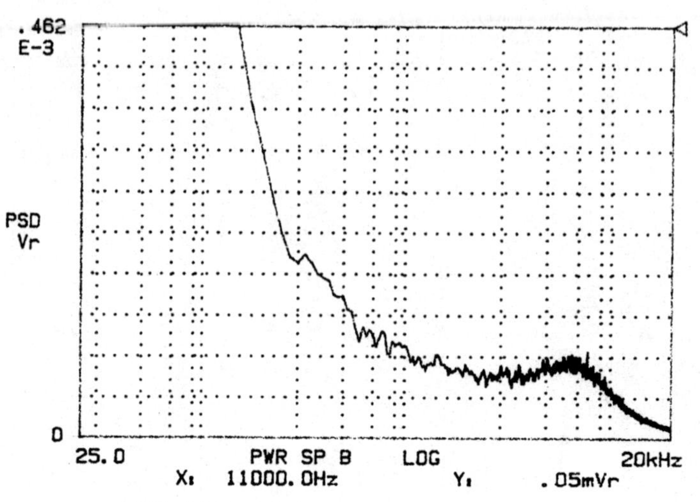

FIGURE 15. Noise spectrum of the CCC.

Influence of External Magnetic Fields

Further measurements were carried out to study the influence of external magnetic fields. A field of 10^{-5} T produced by Helmholtz coils yields the following apparent currents:

$\vec{B} \parallel \vec{I}$	3.3 nA
$\vec{B} \perp \vec{I}$	22 nA

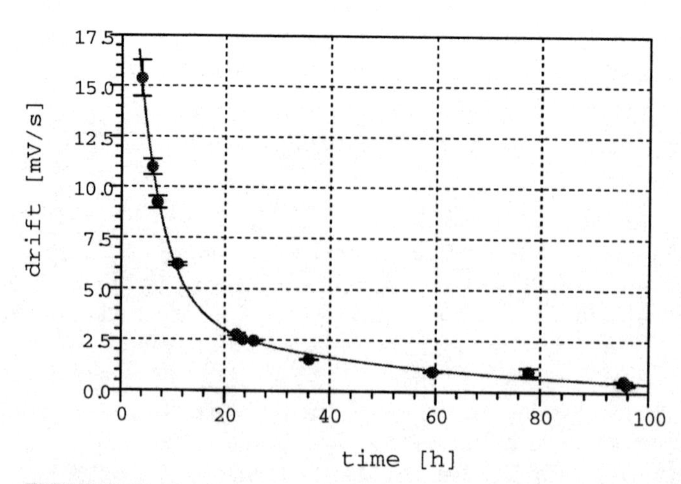

FIGURE 16. Long-term measurement of the current zero drift.

These values are 1–2 orders of magnitude higher than expected, but small enough to allow tests under real conditions in the beam line of the beam diagnostics test stand.

Influence of External Vibration

Numerous investigations with an accelerometer were carried out to study the microphonic sensistivity of the detector system. For this test the Gifford-McMahon

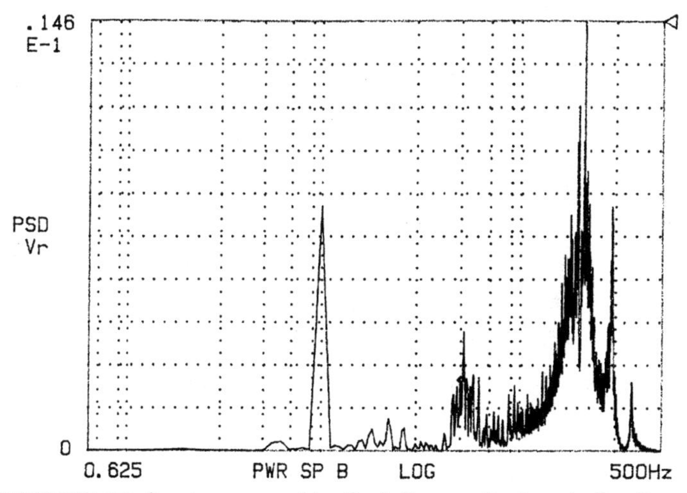

FIGURE 17. Spectrum caused by the influence of external vibrations.

refrigerator was switched off because of the strong mechanical influences. If the vacuum pumps close to the detector are switched on during measurements, an unavoidable mechanical influence in an accelerator environment besides vibrations of the building, the noise spectrum (see Fig. 17) shows several characteristic lines. Besides the disturbing electrical interferences caused by the mains (50, 100, and 150 Hz) there are some other peaks in the range around 70 Hz and corresponding harmonics at higher frequencies caused by mechanical resonances of the dewar.

Measurement results were taken into account for the design of the vibration-insulated installation of the detector in the test beam line. The device is mounted on three rubber bearings, and vibrations of the beam pipe are damped by metal bellows on each side of the CCC. Figure 18 shows the CCC at the beam diagnostics test bench.

MEASUREMENTS OF NEON BEAMS

First measurements were carried out in May 1996 with a $^{20}Ne^{10+}$ beam at 300 MeV/u. About 4×10^{10} particles per machine cycle were accelerated in the SIS

FIGURE 18. Cryostat with detector system at the beam diagnostics test bench (Photo: Achim Zschau, GSI).

and were extracted to the beam diagnostics test bench with a transmission of about 50%. A typical measurement is shown in Figure 19. The extracted ion beam shows a strong modulation with current peaks up to 35 nA while the average current is about 12 nA.

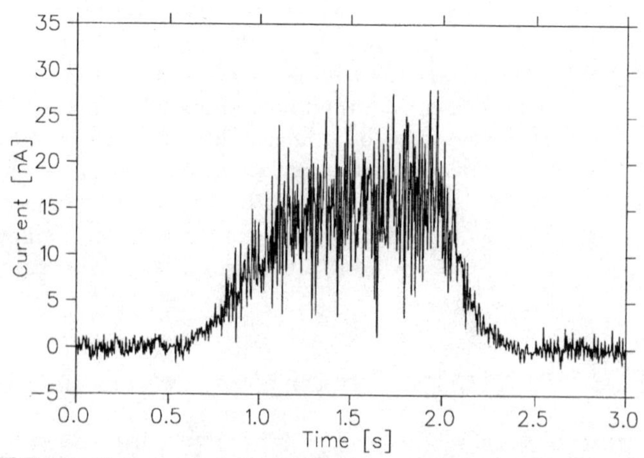

FIGURE 19. Measurement of a 300-MeV/u neon beam extracted of the SIS.

Calibration of SEMs versus CCC

A secondary electron monitor [9] made of three aluminum plates is mounted closely behind the CCC to provide a comparable measuring device. Again for a $^{20}Ne^{10+}$ beam at 300 MeV/u the spill rate was determined by a SEM and the CCC detector. Within a 5% error of each measurement system this plot shows a good linear correlation of the datas (see Fig. 20).

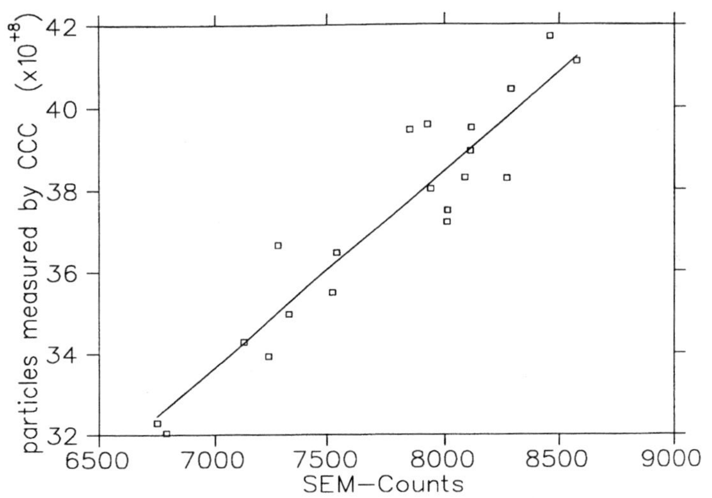

FIGURE 20. Calibration curve, CCC versus SEM data.

For the determination of the particle rate measured by the SEM the energy loss of the ions inside the Al material is calculated using the Ziegler formalism [10]. For the specific yield — the amount of detectable secondary charge per unit energy loss — a formula was used that shows a slight dependence on the nuclear charge of the incoming ions [11]. Further comparative measurements with various ion species are necessary to verify this behaviour.

FIRST ANALYSIS OF THE STRUCTURE OF EXTRACTED ION BEAMS

With a broadened bandwidth of 20 kHz more details of a Neon beam can be observed. Figure 21 shows the "burst" structure of the extracted ion beam.

M. Pullia studied this behaviour of the resonant extraction process [12]. Simulations assuming a monochromatic beam with a ripple on the power supplies showed a similar spill structure compared to the CCC measurement (see Fig. 22).

To find out the main accelerator components responsible for the ripples, a frequency analysis was calculated which is shown in Figure 23. The strongest peaks at

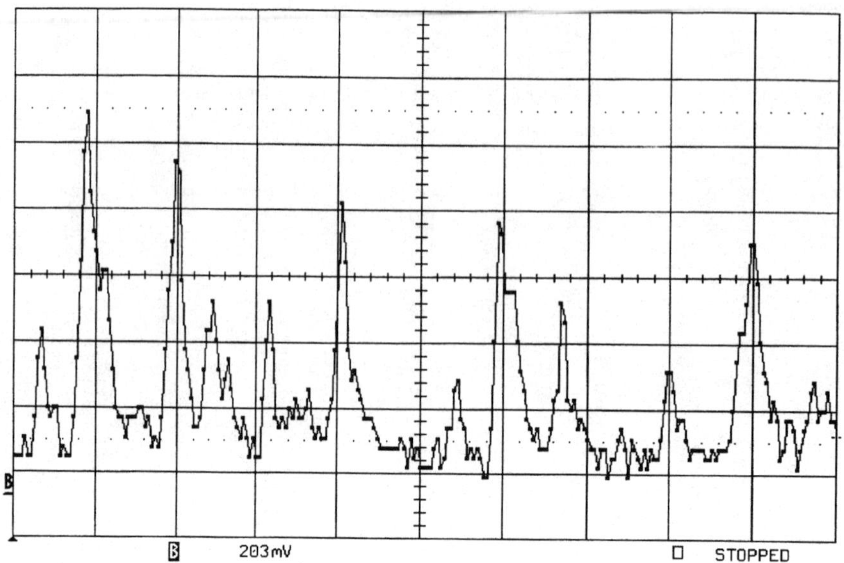

FIGURE 21. Detailed view of an extracted neon beam (x-axis: 0.5 ms/div.; y-axis: 0.21 V/div. = 15.2 nA ion current per div.).

1200Hz and 600 Hz (and the corresponding harmonics) are correlated to the characteristics of the 12-way rectifier bridges of the dipole and quadrupole power supplies

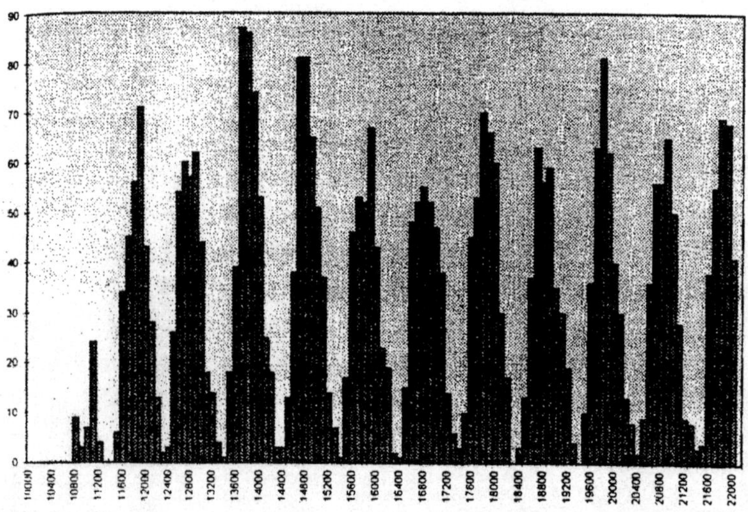

FIGURE 22. Simulation of a resonant extracted beam from a synchroton by M. Pullia.

in the SIS. The sextupole power supplies are possible further sources responsible for the ripples, especially in the lower frequency range. Additional investigations are necessary to minimize these effects.

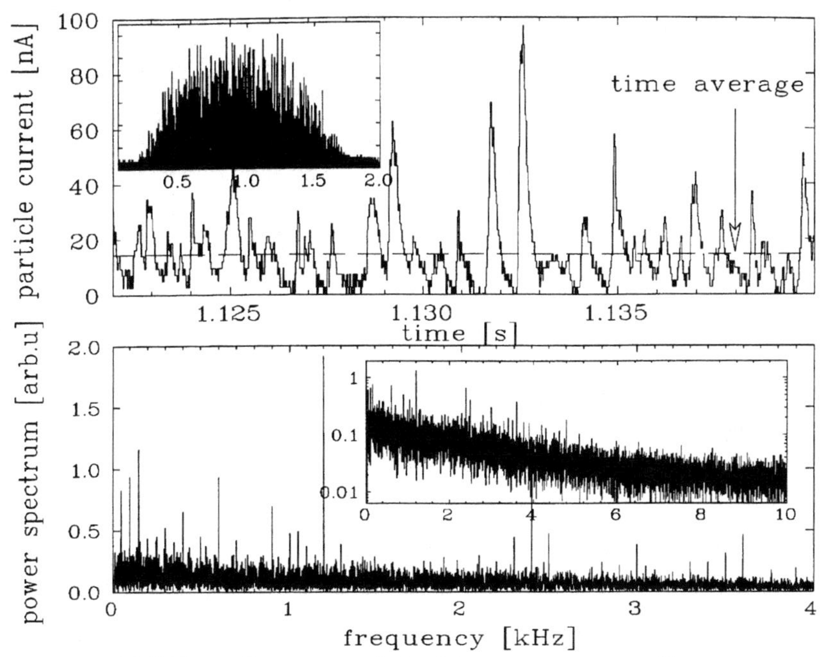

FIGURE 23. Frequency analysis of the resonant extracted beam.

FURTHER DEVELOPMENTS AND OUTLOOK

Further improvements will be done in the near future:

- A higher dynamic range of the DC SQUID electronics is needed. Until now the slew rate is limited to about 5000 ϕ_0/s ($\cong$ 1 nA/μs). If the current rise exceeds this value the feedback circuit of the CCC electronics becomes unstable and negative spikes can be observed. To avoid these instabilities a new version of the SQUID electronics is in preparation. The development aims at a higher slew rate combined with an increased bandwidth.
- In addition, an automatic offset correction will be installed to minimize the zero drift during a typical measurement interval of 1–10 s.
- Because of the strong vibrations of the Gifford-McMahon refrigerator, a prototype of a pulse tube cooler was used which produces a factor of 1000 less disturbance [13].
- For an autonomic detector, a small and cheap helium reliquifier is needed.

CONCLUSION

A new type of beam transformer using the measurement principle of a CCC and based on a high performance DC SQUID was developed and successfully tested in the extraction beamline of the SIS at GSI. Beam currents down to 1 nA can be measured nondestructively. Because of the high resolution of the CCC the beam structure can be analyzed within a bandwidth of 10 kHz.

In addition to the device's extremely high measurement sensitivity the most important advantage is greatly reduced effort for absolute calibration.

REFERENCES

1. Blasche, K., "Status Report on SIS-ESR," in *Proceedings of the Fourth European Particle Accelerator Conference (EPAC94)*, p. 133, London (1994).
2. Moritz, P., E. Berdermann, K. Blasche, H. Stelzer, and F. Zeytouni, "Diamond Detectors with Subnanosecond Time Resolution for Heavy Ion Spill Diagnostics," these proceedings.
3. Harvey, I. K., "A precise low temperature dc ratio transformer," *Rev. Sci. Instrumen.*, vol. 43, p. 1626 (1972).
4. Grohmann, K., D. Hechtfischer, J. Jakschik, and H. Lübbig, "A cryodevice for induction monitoring of dc electron or ion beams with nano-ampere resolution," in: *Superconducting quantum interference devices and their applications*, Walter de Gruyter & Co., p. 311 (1977).
5. Gutmann, P., and H. Bachmair, "Cryogenic Current Comparator Metrology," in *Superconducting Quantum Electronics*, V. Kose, Berlin, Heidelberg, New York: Springer-Verlag (1989).
6. Grohmann, K., D. Hechtfischer, J. Jakschik, H. Lübbig, "Field attenuation as the underlying principle of cryo current comparators," *Cryogenics* 16, pp. 423–429 (1976).
7. Vodel, W., and K. Mkiniemi, "An ultra low noise SQUID system for biomagnetic research," *Measurement Science and Technology*, Vol. 3, No. 2, pp. 1155–1160, December 1992.
8. Koch, H., W. Vodel, and T. Döhler, "DC-SQUID CONTROL 4 Instruction Manual," Jena, Germany (1990).
9. Forck, P., P. Heeg, and A. Peters, "Intensity measurement of high energy heavy ions at the GSI facility," in *Proceedings of the Seventh Beam Instrumentation Workshop*, workshop held at Argonne, May 1996; AIP Conference Proceedings 390.
10. Ziegler, F. J., *The Stopping Power and Ranges of Ions in Matter*, Vol. 5, Pergamon Press (1980).
11. Junghans, A. and K. Sümmerer, GSI, private communication.
12. Pullia, M., "Time Profile of the Extracted Spill," in *Proton-Ion Medical Machine Study*, CERN-Report, September 1996.
13. Häfner, H.-U., Leybold Vakuum GmbH, Cologne, Germany, private communication.

CONTRIBUTED PAPERS—TALKS

DAΦNE Beam Instrumentation

A. Ghigo, C. Biscari, O. Coiro, G. Di Pirro, A. Drago, A. Gallo, F. Marcellini, G. Mazzitelli, C. Milardi, F. Sannibale, M. Serio, A. Stecchi, A. Stella, G. Vignola, M. Zobov

INFN Laboratori Nazionali di Frascati
00044 Frascati (Roma), Italy

Abstract. DAΦNE, the Frascati Φ-Factory, is now under commissioning. The accelerator complex is composed of a linac, an accumulator-damping ring, and two separate main rings, one for electrons and the other for positrons, with two interaction regions in which the experiments will be placed. In order to achieve the luminosity goal, high performance instrumentation and beam diagnostics have been installed. Some of the relevant beam measurements performed are: beam emittance, transverse and longitudinal dimensions, beam positions and tunes, overlap in the interaction points, and luminosity. An overview of the diagnostic instrumentation of the accelerator complex is given together with measurement examples and discussion of operational experiences.

INTRODUCTION

DAΦNE, the Frascati Φ-Factory (1) now under commissioning, is an electron-positron collider designed to produce very high luminosity at 1020 MeV center of mass. The main aim of the machine is to permit the observation of CP violation through the measurement of ε'/ε in the K^0 decay with the KLOE detector; a smaller detector for the spectroscopy of Lambda hypernuclei will be installed in the second DAΦNE interaction region.

In order to achieve the high luminosity needed for precision measurements (about two orders of magnitude larger than the highest luminosity achieved so far at this energy), the strategy adopted was to increase the collision frequency, operating in multibunch mode, maintaining the beam dimensions and the tune shift parameter at reasonable values. This operation mode implies storing very high current distributed in many bunches (up to 120). In order to minimize the parasitic crossings, two separate rings, one for electrons, the other for positrons, with two common interaction regions have been built. In the IR the two beams collide at a full horizontal angle of 25 mrad in the waist of a vertical low beta function (flat beam). Table 1 summarizes the DAΦNE main rings parameters.

CP451, *Beam Instrumentation Workshop*
edited by R. O. Hettel, S. R. Smith, and J. D. Masek
1998 The American Institute of Physics 1-56396-794-4/98/$15.00

TABLE 1. DAΦNE Main Rings Design Parameters

Parameter	Value
Energy (GeV)	0.51
Maximum luminosity [$cm^{-2}s^{-1}$]	5.3×10^{32}
Single bunch luminosity [$cm^{-2}s^{-1}$]	4.4×10^{30}
Trajectory length (each ring) [m]	97.69
Emittance, $\varepsilon_x/\varepsilon_y$ [mm·mrad]	1/0.01
Beta function, β_x/β_y [cm]	450/4.5
Transverse size σ_x/σ_y [mm]	2/0.02
Beam-beam tune shift, ξ_x/ξ_y	0.04/0.04
Crossing angle, θ_x [mrad]	± 12.5
Betatron tune, ν_x/ν_y	5.09/6.07
rf frequency, f_{rf} [MHz]	368.25
Number of bunches	120
Minimum bunch separation [cm]	81.4
Particles/bunch [10^{10}]	8.9
rf voltage [MV]	0.250
Bunch length σ_z [cm]	3.0
Synchrotron radiation loss [keV/turn]	9.3
Damping time, τ_ε/τ_x [ms]	17.8/36.0

To achieve and maintain high average luminosity, a very efficient full energy injector, able to top up the beam current, was realized. The injector consists of a 550MeV (e^+)/800MeV (e^-) linac and an accumulator-damping ring.

The electron and positron beams travel from the linac to the accumulator and, after damping and extraction, to the main rings. A sizable fraction of the transfer lines are traversed in opposite directions by the linac and accumulator beams.

The main purpose of the DAΦNE beam instrumentation (2, 3) is to measure the beam parameters, to help to understand the machine behavior during the commissioning phase and to maintain the luminosity performances during operation (see Table 2).

TRANSVERSE DIMENSIONS

The Synchrotron Light Monitor (4) is used extensively in this first part of commissioning to measure the transverse beam sizes. The synchrotron radiation in the visible range produced from the beam passing through a dipole magnet is extracted from the vacuum chamber by a 45° tilted aluminum mirror through a fused-silica window. The beam image is focused with a lens system, after slit selection, on a commercial Philips VCM6250 CCD camera and processed by a frame grabber (Spiricon LBA-100A).

TABLE 1. DAΦNE Beam Instrumentation Summary Table

Type	Tr. Lines	Storage Rings Acc.	Storage Rings e^+/e^-	Interaction Regions Day One	Interaction Regions KLOE	Interaction Regions FINUDA
Secondary Emission (SEM) Hodoscope	1					
Faraday Cup	1					
Fluorescent Flag	18	2	1/1			
Slit/Scraper	4		3/3			
Toroidal Current Monitor	9	1				
Wall Current Monitor		1	1/1			
DC current Monitor		1	1/1			
Beam Position Monitor - Stripline	23	4		8		
Beam Position Monitor - Button		8	33/33	26	10	10
Beam Position Monitor - Special			3/3			
Transverse Kicker - Stripline pair		2	2/2			
Synchrotron Light Monitor		2	1/1			
Transverse Tune Monitor/Tr. Feedback		2	2/2			
Synchr. Tune Monitor/Long. Feedback		1	1/1			
Beam Loss Monitor		1	8/8	8	4	4
Luminosity Monitor				2	1	1

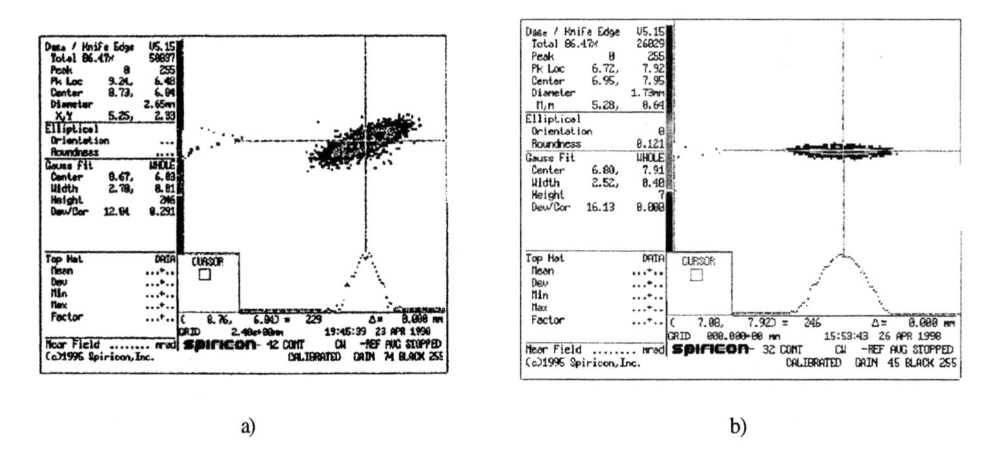

FIGURE 1. Beam images from the synchrotron light monitor a) before, b) after correction of coupling.

The beam transverse emittances were directly evaluated from horizontal and vertical dimension measurements since the position of the point source is in a zone with vanishing value of the dispersion function.

The coupling and the rotation due to the beam's vertical displacement in the sextupoles, to the off axis passage in the interaction regions quadrupoles and to tilted quadrupoles was measured and corrected to the design value of 0.01, as shown in Figure 1. The beam-beam blow-up is easily observable.

BUNCH LENGTH

The bunch length in DAΦNE is measured by processing the beam signal from a broad band button electrode (5), connected with a low attenuation cable (Andrew FSJB-50B), 9 m long, to a sampling oscilloscope Tektronix 11801A, equipped with a sampling head SD-24 with a rise time of 17.5 psec and an equivalent bandwidth of 20 GHz. Stability of the waveform, even in the presence of longitudinal oscillations, has been achieved by using the signal from a stripline electrode as trigger. The waveform is sent via a GPIB interface to the control system for storage and off-line reconstruction after correcting for the (small) cable distortion and the pick-up transfer impedance.

Figure 2 shows the comparison between the bunch length vs. bunch current and the numerical simulations at two different rf voltages. From these bunch lengthening measurements it has been possible to evaluate the vacuum chamber impedance. We find $Z/n \approx 0.6\ \Omega$ (below the design value of 1Ω). the bunch length at the design bunch current is $\sigma_z < 3$ cm.

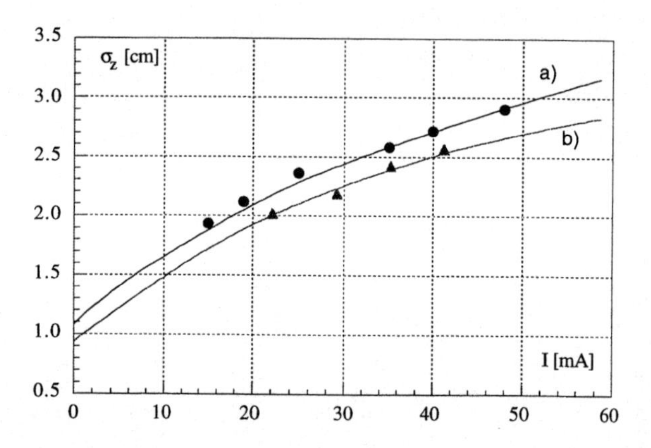

FIGURE 2. Bunch length vs bunch current in DAΦNE Main Ring; measurement results (dots) and numerical simulation (solid line) a) V_{rf} = 150 KV, b) V_{rf} = 200 KV.

TRAJECTORY AND CLOSED ORBIT

The Beam Position Monitor (BPM) is the most important diagnostic system in DAΦNE (6). Two different kinds of BPM are used in the accelerator complex: striplines, which have higher sensitivity and are used in single-pass measurements in the transfer lines, and during the accumulator and main ring injection; and button monitors, which are more numerous and permit accurate measurements on the stored beam with the advantage of little contribution to the machine impedance.

In order to measure the stationary closed orbit in the Main Rings, 35 button BPM have been installed in each ring and 26 in the interaction regions. Several different monitor configurations have been realized because of the differing vacuum chamber shape along the accelerator. For each monitor configuration, a calibration procedure with numerical and bench measurements has been accomplished, to obtain an accurate determination of the beam position.

The BPM detector was developed by Bergoz Beam Instrumentation System for DAΦNE: it consists of a superheterodyne receiver which converts the 240th harmonic (twice the rf frequency) of the beam induced signals down to an intermediate frequency if = 21.4 MHz before amplitude detection. At the circuit output, two voltages, which are software processed to obtain horizontal and vertical beam positions, are provided.

The acquisition system has been developed in the VME standard. The signals are multiplexed and measured with HP E1352A FET Multiplexers and HP E1326B Digital Multimeters controlled by dedicated CPUs.

The higher level of DAΦNE control system collects from these peripheral units the position data to be used in the high-level orbit reconstruction and analysis programs. The whole closed orbit of the main rings is acquired at a rate of 5 per second.

The main rings striplines have been very useful during the commissioning phase, indeed, the high sensitivity permits the injected beam to follow turn-by-turn and they are used to measure and correct the orbit in the first turns and help in the injection kicker setting. One optimization criterion was to achieve the number of turns permitted by the spiralization due the synchrotron radiation loss with the rf cavity off; a hardcopy of the striplines signal at injection, digitized by an oscilloscope (Tektronix 644B), is shown in Figure 3. The peak intensity decreases because the debunching effect broadens the pulse shape; when the rf cavity is switched on, the previously injected beam is stored in the ring.

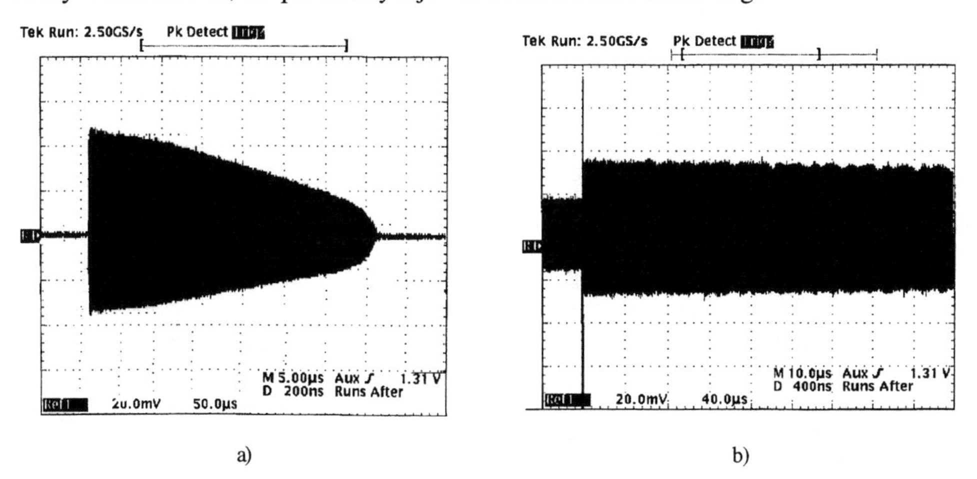

FIGURE 3. a) The injected beam in the main ring detected by a stripline monitor with rf cavity off b) the stored beam with rf cavity on.

TUNE

The fractional parts of the horizontal and vertical betatron tunes are measured by exciting the beam at rf frequency with transverse stripline kickers and measuring the beam response in the excitation plane with a transverse pick-up.

Two different sets have been adopted to perform the tune measurements. In the first, the Network Analyzer HP 4195A (10 Hz–500 MHz) rf output, amplified up 100 W by class A amplifiers, provides the sweeping excitation. The horizontal and vertical coherent beam response is picked-up by stripline pairs. The signal is combined in hybrid junctions and detected with the Network Analyzer. In the second system the other beam is excited with white noise and the oscillation signal is extracted by broad-band button electrodes and sent to the spectrum analyzer (HP 70000 system) operating in detector (zero span) mode. The spectrum analyzer if output is down-converted with an HP 89411A module and processed by a real time FFT anlyzer, HP model 3587S.

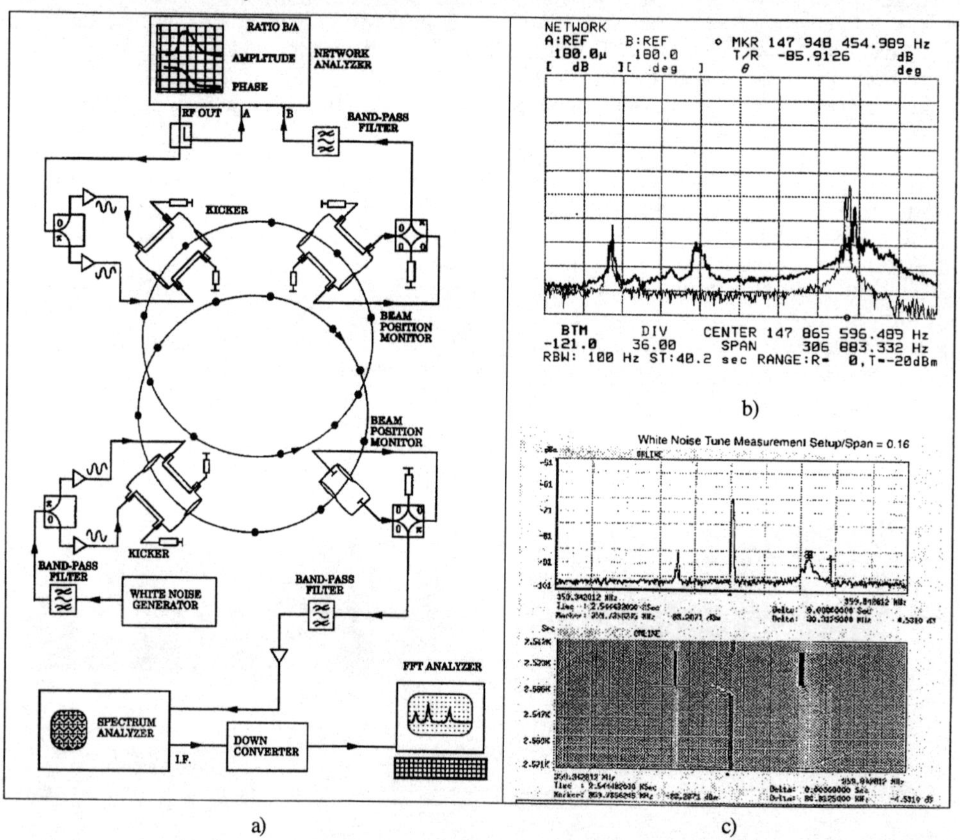

FIGURE 4. a) Tune measurement systems layout; b) swept measurement on the electron beam: lighter line out of collision, darker line during collision; c) white noise measurement on positron beam during collisions, below the spectrogram representation.

In Figure 4 a layout of the two measurement sets, together with the results of two simultaneous tune measurements in the two rings are shown. The horizontal and vertical signals are combined so that they appear in the same trace, horizontal at lower frequency and vertical at the higher. In the upper picture, the swept measurement performed on the electron beam shows clearly the tune split on the horizontal tune (the two peaks at lower frequency) during the beam collisions (darker line), in comparison with the tune measurement out of collision at low current (lighter line).

Below (Figure 4c), the measurement with the white noise performed on the positron beam is shown, using a spectrum representation, in which the tune split appears in the horizontal and vertical planes and the spectrogram, in which the time evolution (downwards) of the tune peaks shows two distinct phases: no interaction and collisions.

The two measurement systems have been largely used during this commissioning phase to tune the machine working point and to observe the beam-beam effects during the collisions.

BEAM CURRENT

The direct measurement of the beam current throughout the DAΦNE accelerator complex is crucial in order to maintain the best integrated luminosity.

In the injection system nine toroidal integrating current transformers by Bergoz have been placed to measure the transfer efficiency from the linac to the accumulator and from the accumulator to the main rings. In the accumulator a toroidal current monitor is used to measure the injected current in each linac pulse.

In the accumulator, the stored current is measured by a DC current monitor (made by Bergoz). A control program, based on this measurement, permits the injection to stop at a pre-set current value, ready for extraction into the main rings, in order to equalize the bunch currents.

Similar DC current monitors have been also installed in each main ring. The usual ceramic gap has been shunted by an array of parallel resistors in order to strongly reduce the resonant impedances of the monitor structure. The voltage developed across the resistive by-pass is picked up at four locations and used as a longitudinal monitor. The current value is continuously acquired and stored by the control system in order to perform lifetime measurements and to keep a log of the integrated current.

LUMINOSITY

The DAΦNE luminosity monitor (7) is based on the measurement of the photon production in the single bremsstrahlung electromagnetic reaction at the interaction point during the collisions. The luminosity value is given by the single bremsstrahlung photon-counting rate multiplied by the reaction cross section.

Because of the high counting rate of the single bremsstrahlung, the monitor has a fast response. This feature proved very useful during the machine tune-up.

Two monitors are placed at both ends of the two interaction regions, where the common vacuum chambers are split into separate rings by means of split-field magnets. At these positions, thin aluminum windows permit the photons to escape the chambers, hitting the detectors.

The detector is a proportional counter composed of a sandwich of lead-scintillating fiber with photomultiplier read-out. The integral of the signal coming from the detector is proportional to the incoming photon energy. The energy analysis and photon counting is provided by NIM-VME electronics.

The most important contribution to the measurement error is the photon production in gas bremsstrahlung reactions; this background is subtracted measuring the counting rate with the two beams kept separated.

The evaluations of the luminosity from the tune split measurement, described before, at the nominal machine parameter values and the direct measurements with the luminosity monitor are in good agreement.

ACKNOWLEDGMENTS

The authors wish to thank C. Marchetti and D. Pellegrini for highly professional and continuing technical support and P. Possanza for editing this paper.

REFERENCES

[1] Vignola, G., and the DAΦNE Project Team, "DAΦNE, the First Φ-factory," *Proceedings of the Fifth European Particle Accelerator Conference*, Sitges, Spain, June 1996, p. 22.

[2] Serio, M., DAΦNE Team, "Operational experience with the DAΦNE beam diagnostic system," *Proceedings of the 3RD European Workshop on Beam Diagnostics and Instrumentation for Particle Accelerators DIPAC97*, Frascati, Italy, October 1997, p. 138, LNF-97/048.

[3] Biagini, M. E., et al., "Overview of DAΦNE Beam Diagnostic," *Proceedings of the Fifth European Particle Accelerator Conference London*, UK, July 1994, p. 1503.

[4] Ghigo, A., Sannibale F. and M. Serio, "Synchrotron Radiation Monitor for DAΦNE," *AIP Conference Proceedings No. 333 of the Beam Instrumentation Workshop*, Vancouver, Canada, October 1994, p. 238.

[5] Marcellini, F., Serio M., A. Stella and Zobov M., "DAΦNE Broad-Band Button Electrodes," *Proceedings of the 3RD European Workshop on Beam Diagnostics and Instrumentation for Particle Accelerators DIPAC97*, Frascati, Italy, October 1997, p. 106, LNF-97/048.

[6] "Beam Position Monitor System of DAΦNE," these Proceedings.

[7] Di Pirro, G., et al., "The DAΦNE Luminosity Monitor," these Proceedings.

Laser Diagnostic for High Current H^- Beams*

Robert E. Shafer

Los Alamos National Laboratory
Los Alamos, NM 87545

Abstract. In the last 5 years, significant technology advances have been made in the performance, size, and cost of solid-state diode-pumped lasers. These developments enable the use of compact Q-switched Nd:YAG lasers as a beam diagnostic for high current H^- beams. Because the threshold for photodetachment is only 0.75 eV, and the maximum detachment cross section is 4×10^{-17} cm^2 at 1.5 eV, A 50 mJ/pulse Q-switched Nd:YAG laser can neutralize a significant fraction of the beam in a single 10 ns wide pulse. The neutral beam maintains nearly identical parameters as the parent H^- beam, including size, divergence, energy, energy spread, and phase spread. A dipole magnet can separate the neutral beam from the H^- beam to allow diagnostics on the neutral beam without intercepting the high-current H^- beam. Such a laser system can also be used to extract a low current proton beam, or to induce fluorescence in partially stripped heavy ion beams. Possible beamline diagnostic systems will be reviewed, and the neutral beam yields will be calculated.

INTRODUCTION

Laser systems have been in use at the Los Alamos LAMPF 800 MeV proton linac and on various low-energy H^- beamlines since about 1980 to do research or diagnostics on the accelerated H^- beam. The basis for these systems is that the threshold for photodetaching an electron is about 0.75 eV, and the photodetachment cross section rises to about 4×10^{-17} cm^2 for photons of about 1.5 eV (800 nm).

A Q-switched laser, when triggered, fully discharges in a few ns. Thus a small Q-switched laser with, say 50 mJ pulse energy and 10 ns pulse length, has the instantaneous power of 5 MW. Furthermore, a 50 mJ pulse at 1064 nm wavelength contains over 2×10^{17} photons. Because the photodetachment cross section is substantial, a significant fraction of the beam can be neutralized during the laser pulse. The Q-switched laser beam can either be focused to select a thin slice of the transverse beam profile, or defocused to nearly uniformly illuminate the entire beam. Real-time

---

* Work supported by DOE.

CP451, *Beam Instrumentation Workshop*
edited by R. O. Hettel, S. R. Smith, and J. D. Masek

measurements can be made on the extracted neutral H^0 beam to determine parameters of the H^- beam. These parameters include transverse beam profiles, beam current, and even emittances.

In the photodetachment process, the neutralized beam maintains nearly the original phase-space parameters of the H^- beam from which it was extracted. This is because neither the laser photon nor the recoiling photodetached electron transfer significant momentum to the H^0 atom. Thus the transverse spatial profile, transverse divergence, emittance, energy, energy spread, and phase spread characteristics of the H^0 and H^- beams are the nearly identical. Furthermore, because the neutralized beam will not be deflected by either electric or magnetic fields, the H^- beam parameters can be deduced from measurements on the drifting neutral beam, even after it is separated from the H^- beam by magnetic fields. Measurements on the neutral beam, even if totally destructive, will have no impact on the H^- beam from which it was extracted.

For high current H^- beams, such as the one being planned for the 1 MW Spallation Neutron Source at Oak Ridge National Laboratory, laser photodetachment of H^- ions provides a way to measure beam parameters that is neither disruptive to the primary beam, nor destructive to the beam diagnostic.

THEORY

Photodetachment Cross Section

A plot of the photodetachment (stripping) cross section vs. photon energy, in the rest frame of the H^- atom, is shown in Figure 1 (1–3).

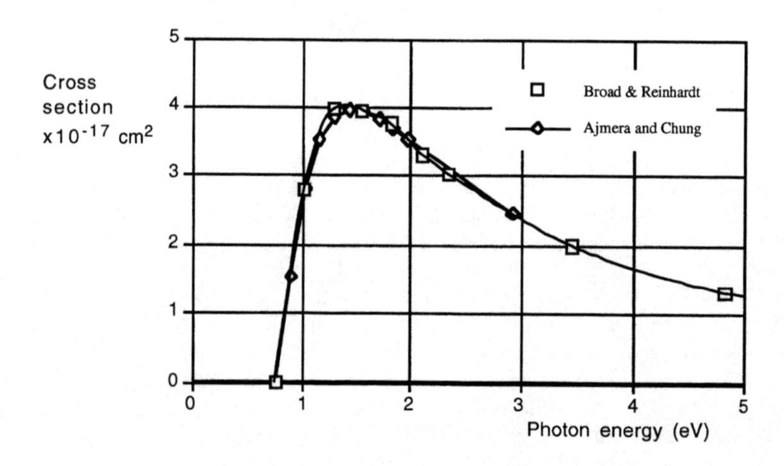

FIGURE 1. Photodetachment cross section of H^- vs. photon energy in the H^- rest frame. The threshold is at 0.75 eV (1650 nm). The maximum cross section occurs at about 1.5 eV.

The threshold is at about 0.75 eV and the peak cross section, 4×10^{-17} cm^2, is at about 1.5 eV. Because the binding energy of the remaining 1s electron in the neutral hydrogen atom is 13.6 eV, it will not be stripped by the laser.

Lorentz Transformation

Because H^- beams can be accelerated to energies of 1 GeV or more, there is a very sizable relativistic shift of the laser photon energy to higher energies in the H^- rest frame, often referred to as a "Lorentz boost". The photon energy E_{CM} in the H^- rest frame is related to the laser photon energy E_L by the equation

$$E_{CM} = \gamma E_L\left[1 - \beta \cos\left(\theta_L\right)\right] \tag{1}$$

where β and γ are the Lorentz parameters of the H^- beam, and θ_L is the laboratory angle of the laser beam relative to the H^- beam.

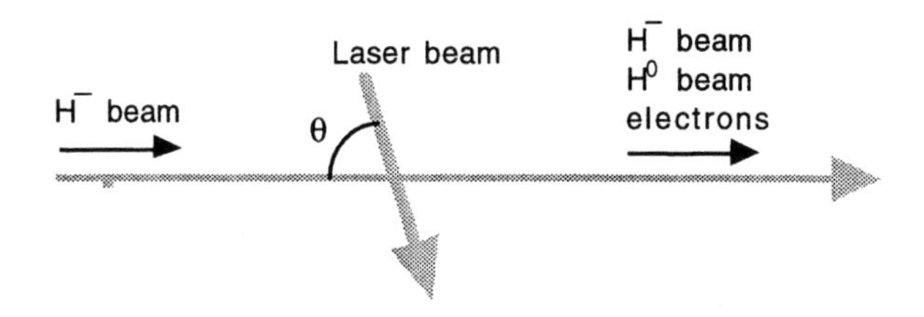

FIGURE 2. Geometry for laser photodetachment.

Photodetachment Yield

For a Gaussian-profile laser beam with N_L photons intercepting a Gaussian-profile H^- beam of current I_b at an angle θ_L, the yield Y_1 (number of neutral hydrogen atoms produced per laser-H^- beam crossing) is given approximately by (4)

$$Y_1 = \frac{I_b N_L}{e\beta c}\frac{1-\beta\cos\theta_L}{\sin\theta_L}\frac{\sigma_N(E_{cm})}{2\pi\sigma_b\sigma_L}\int_{-\infty}^{\infty}\exp\left(\frac{-x^2}{2\sigma_b^2}\right)\exp\left(\frac{-x^2}{2\sigma_L^2}\right)dx$$

$$= \frac{I_b N_L}{\sqrt{2\pi} e\beta c}\frac{1-\beta\cos\theta_L}{\sin\theta_L}\frac{\sigma_N(E_{cm})}{\left(\sigma_b^2+\sigma_L^2\right)^{1/2}} \tag{2}$$

where σ_b and σ_L are the transverse rms sizes of the H^- and laser beams normal to the plane of incidence, and σ_N (E_{cm}) is the photodetachment cross section at photon energy E_{cm} in the H^- rest frame.

The yield of photodetached H atoms for a 50 mJ 1064 nm Nd:YAG laser pulse on a 50 mA, 1 GeV H^- beam, using Eqs (1) and (2) and the following parameters:

- $\theta_L = 85°$ (chosen to optimize both cross section and mirror angle)

- $E_{cm} = 2.22$ eV (Lorentz-boosted photon energy in rest frame of H^-)
- $\beta c = 0.875 \times 3 \times 10^{10}$ cm/s (beam velocity)
- $N_L = 2.68 \times 10^{17}$ (photons per laser pulse)
- σ_b and $\sigma_L = 0.2$ cm (rms width of laser and H^- beams)
- $\sigma_N(E) = 3 \times 10^{-17}$ cm^2 (photodetachment cross section at energy E_{cm}),

is $Y_1 = 1.25 \times 10^8$ H^0 atoms per laser pulse (single crossing).

This technique can also be used for low-energy (< 10 MeV) H^- beams because the detachment cross section (Fig. 1) is 3.5×10^{-17} cm^2 at 1.17 eV (1064 nm). In fact, because the yield is inversely proportional to β, the yield is larger for low-energy beams. In the above example, if the beam energy is lowered to 2.5 MeV, the yield increases to 1.6×10^9 atoms.

Yield Enhancement

A variety of mirror configurations for reflecting the laser beam through the H^- beam many times are possible. The simplest configuration is two parallel front-surface mirrors. Another configuration is an internally-reflecting cylindrical mirror with its axis aligned along the beam. Also, an elliptical shaped internally-reflecting mirror (integrating sphere) is possible. In theory, if the reflection coefficient is 100%, all the photons could be stored in the integrating sphere until they were either absorbed by the beam or exited through the entrance aperture. The actual reflectivity of the mirror limits the number of reflections to a few tens or hundreds of times. Another practical limit is that to take advantage of the temporal resolution of a very short Q-switched laser pulse, which is useful in maximizing signal to noise, the effective photon lifetime in the mirror should not exceed a few ns. An effective lifetime of 10 ns corresponds to a photon path length of about 300 cm, which represents about 30 reflections inside a 10 cm diameter mirror assembly. Thus the optimum mirror assembly needs to reflect the laser beam through the H^- beam only about 30 times, an easily achievable number even with modest reflectivities.

Specifically, if the laser beam passes through the H^- beam N times, the fractional yield F_N of H^0 current is related to the fractional yield F_1 of a single crossing by the relation

$$F_N = 1 - (1 - F_1)^N \cong NF_1, \tag{3}$$

where the approximation is true when the depletion of the H^- beam is not significant.

In the first example above, the average H^0 beam "current" during a 10 ns Q-switched laser pulse is about 2 mA, or 4% of the H^- beam current. Thus with N=10 mirror reflections, the total yield is about 17 mA (34% of the H^- current).

Backgrounds

There are two sources of background uniquely associated with H^- beams. They are magnetic stripping and residual gas stripping. If not controlled, these stripping mechanisms can contaminate the signal obtained by laser stripping. For high current, high energy H^- beams, these loss mechanisms can also contribute to a significant amount of activation. A beam loss of one watt per meter at 1 GeV can lead to activation levels in the range of tens of mrad/hr.

A relativistic H^- beam can be stripped by the Lorentz-transformed magnetic field of a typical beamline magnet. The theory of electric and magnetic field stripping of H^- beams is discussed by Sherk(5) and by Jason(6). As an example, the stripping loss rate of a 1 GeV H^- beam in magnetic fields of 0.3 T, 0.35 T, and 0.4 T is 0.12, 7.4, and 164 ppm per meter respectively.

A relativistic H^- beam can also be stripped by inelastic collisions with residual gas atoms. The cross sections for this process have been evaluated by Gillespie(7). As an example, the cross sections for stripping a 1 GeV H^- beam in hydrogen and nitrogen gas are about 1.2 and 8.9×10^{-19} cm^2/atom, and scale approximately as $1/\beta^2$. For a 1×10^{-7} torr (273 K) vacuum, these cross sections represent stripping losses of about 0.08 and 0.6 ppm per meter respectively.

It is relevant here to mention the possibility of using solid state laser diodes, whose instantaneous light output power is a few watts, for beam diagnostics applications. The Q-switched Nd:YAG laser example above has an instantaneous output power of 50 mJ $\div$ 10 ns = 5 MW, and yields peak photodetachment currents of a few percent of the H^- beam. Thus backgrounds of 1 ppm/meter from either magnetic or residual gas stripping will create significant interference with any laser whose peak output is only a few watts. Thus although solid-state laser diodes may be useful as a device for extracting very small average currents from a H^- beam, they are probably not suitable as a beam diagnostic.

EXPERIMENTAL APPLICATIONS

Characteristics of Commercially Available Q-Switched Lasers

Inexpensive shoebox-sized Q-switched Nd:YAG lasers can produce 10 ns long, 50 mJ, 1064 nm pulses (or harmonics) at 60 Hz. These units are totally enclosed, and can be installed directly on a beamline. The 1064 nm line is nearly ideal for general diagnostics on H^- beams, because of its proximity to the peak in the photodetachment cross section. The 10 ns pulse width is adequate for many applications where good temporal response is required, and this can be improved if necessary by using external polarizers and Pockels cells.

Possible Experimental Layouts and Measurements

A generic layout for a laser diagnostic is shown in Figure 3. In Figure 3, the Q-switched laser beam intercepts the H^- beam at an angle θ_L. A mirror assembly produces multiple passes of the laser beam. A dipole magnet separates the neutral beam from the H^- beam. If a dipole magnet, such as in a bend, is not possible, then a weak dipole field will deflect the detached electrons, which can be detected. After the neutral beam emerges from the dipole magnet, it may be foil-stripped to produce a proton beam. A variety of beam diagnostics for characterizing the resultant proton beam are possible. Because the proton beam is low power, the diagnostic may totally intercept the protons.

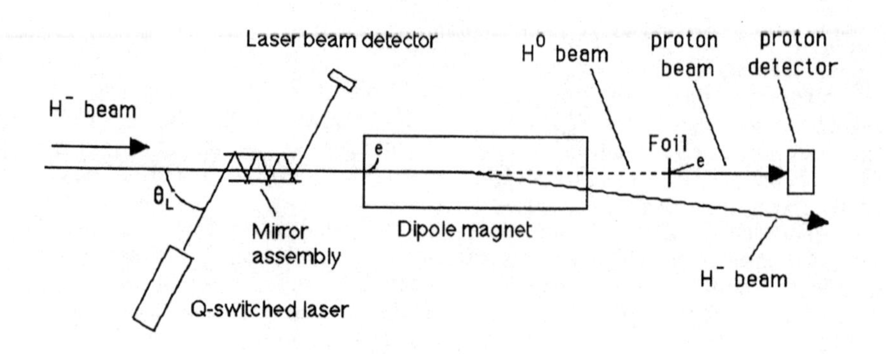

FIGURE 3. A generic arrangement for laser beam diagnostics.

A laser beam for transverse beam diagnostics can either be a thin "laser wire", neutralizing only a thin slice of the incident H^- beam, or intercept the entire beam(8–10). The width of the laser wire can be of the order of 0.2 or 0.4 mm. If used in a high dispersion region, it may be possible to measure the H^- energy spread. For measuring the proton yield, possible proton diagnostics include phototube-scintillator assemblies, Faraday cups, secondary-emission monitors, etc. Because the photodetachment yield is higher at low energies, lasers may be a good substitute for intercepting wire scanners which are particularly hard to use in low-energy, high dE/dx beams.

It may also be possible to measure the H^- phase spread by detecting the rf phase of secondary-emission electrons from the stripper foil (11). In this application, the secondary emission electrons, which maintain the temporal response of the incident charged particles, can be accelerated into a rf deflector synchronized to the rf bunch structure and through a slit to determine their rf phase to perhaps 15 or 20 ps.

Phase spread of H^- beams can also be measured using cw mode-locked lasers (12). In this method, very short laser pulses, synchronized with the beam microstructure, are used to neutralize 20 ps temporal "slices" of the H^- microbunches. These laser systems usually require special clean rooms for the laser oscillator, amplifier, and pulse compressor.

A very specific application in the proposed Spallation Neutron Source project is to measure the beam current in a 1.18 MHz, 280 ns wide, beam chopper gap, which must be less than about 0.3 μA (about 1×10^{-5} of the 28 mA, H^- beam). The laser system can extract a neutral current of about 0.10 μA from this gap for 10 ns, equivalent to about 6200 particles. This can be measured using either charge or scintillator pulse detection techniques to determine the cleanliness of the gap. The very high dynamic range and charge sensitivity required for the beam-in-gap measurement is also useful for exploring the halo region of the primary beam. This is a difficult measurement to make with normal beam profile diagnostics.

When measurement of the photodetached H^0 atom or proton is difficult, measurement of the photodetached electron is possible. The electron has about $1/1840^{th}$ of the proton rigidity, and is easily deflected into detectors by weak magnetic fields. This technique has been used in photodetachment experiments. The photodetached electron is easily deflected by space charge forces in high current H^- beams, however, so the electron signal cannot be analyzed for obtaining accurate H^- beam emittance information.

Resonances in the photodetachment total cross section near the n=2 threshold (10.953 eV) have been used to measure H^- beam momentum and momentum spread (13). In this experiment, a 50 mJ Q-switched Nd:YAG laser operating at 266 and 355 nm was used. Both the Feshbach resonance (10.926 eV, width 30 μeV) and the shape resonance (10.975 eV, width 25 meV) can be used for this measurement, although the widths and strengths of these resonances are not optimum.

Under certain conditions, the H^- beam itself will fluoresce. Laser-induced fluorescence of a 50 MeV H^o beam has been observed (14). If a frequency-quadrupled Nd:YAG laser (266 nm wavelength) intercepts a 1 GeV H^- beam at 99.2°, the photon energy in the H^- rest frame is 10.975 eV. This is the energy for exciting the n=2 shape resonance, with the H^0 final state being either the 2s or 2p state. The cross section for the 2p final state is about 4.5×10^{-17} cm^2 (15). The lifetime of the 2p-1s transition is about 1.6 ns (13), which Lorentz-transforms into a decay length ($\beta c\gamma\tau$) of about 87 cm in the laboratory (16). The cross section for the metastable 2s final state is about 1×10^{-17} cm^2, and can be quenched by using a magnetic field to Stark-mix it with the 2p state (14). The wavelength and angular distribution of the 121.6 nm 2p-1s fluorescence must be Lorentz-transformed back into the laboratory reference frame. Detection of this laser-induced fluorescence may provide another alternative to detection of the photodetached electrons or protons for beam diagnostics applications.

CONCLUSION

Laser photodetachment can be used on high-current, high-energy H^- beams to carry out a wide variety of beam diagnostic measurements parasitically during normal operation, without having to operate the facility at either reduced current or duty cycle. Suitable Q-switched laser systems are inexpensive, small, and can be mounted on or near the beamline. Most of the proposed laser-based diagnostics techniques have already been demonstrated. Photodetachment can be measured by detecting either the detached H^o, the detached electron, or H^o fluorescence. This method appears to be suitable for low-energy (< 10 MeV) as well as high-energy (1 GeV) H^- beams.

REFERENCES

[1] Broad, J. T., and W. P. Reinhardt, "One and Two-electron Photoejection from H^- Atoms," *Phys. Rev. A,* **14**, pp. 2159–72 (1976). See Table IV.

[2] Ajmera, M. P., and K. T. Chung, "Photodetachment of Negative Hydrogen Ions," *Phys. Rev., A* **12**, pp. 475–79 (1975).

[3] Daskhan, M., and A. S. Ghosh, "Photodetachment Cross Section of the Negative Hydrogen Ion," *Phys. Rev. A,* **28**, pp. 2767–69 (1983).

[4] S. M. Shafroth and J. C. Austin, Eds.; *Accelerator-Based Atomic Physics Techniques and Applications,* A.I.P. Press (1997). Chapter 6 has a comprehensive discussion of photodetachment kinematics and yields, as well as a complete review of many laser experiments on H^- beams.

[5] Scherk, L., "An Improved Value for the Electron Affinity of the Negative Hydrogen Ion," *Can. J. Phys.*, **57**, p. 558 (1979).

[6] Jason, A. J., D. W. Hudgings, and O. B. VanDyck, "Neutralization of H^- Beams by Magnetic Stripping," *IEEE Trans. Nucl. Sci.,* **NS-28**, pp. 2704–6 (1981).

[7] Gillespie, G. H., "Electron-Loss Cross Sections for High Energy H^- Collisions with Low and High Z atoms," *Phys. Rev. A,* **15**, pp. 563–573 (1977), and A16, pp. 943–950 (1977).
[8] Swenson, D. R., E. P. MacKerrow, and H. C. Bryant, "Non-Invasive Diagnostics for H^- Ion Beams using Photodetachment by Focused Laser Beams," Proc. 1993 Beam Instrumentation Workshop (Santa Fe), pp. 343–52, *A.I.P. Conf. Proceedings,* **319** (1994).
[9] Connolly, R. C., et al., "A Transverse Phase Space Measurement technique for High Brightness H^- beams," *Nucl. Inst. and Meth. A*, **312**, pp. 415–419 (1992).
[10] Sandoval,D. P., "Non-Interceptive Transverse Emittance Measurement Diagnostic for an 800-MeV H^- Transport Beam," 1994 Beam instrumentation Workshop (Vancouver). *A. I. P. Conference Proceedings,* **333**.
[11] Witkover, R. L., "A Non-Destructive Bunch-Length Monitor for a Proton Linear Accelerator," *Nucl. Inst. and Meth.*, **137**, pp. 203–11 (1976).
[12] Yuan, Y. W., et. al., "Measurement of Longitudinal Phase Space in an Accelerated H^- beam using a Laser Induced Photoneutralization Method," *Nucl. Inst. and Meth., Vol. A*, **329**, pp. 381–392 (1993).
[13] MacKerrow, E. P. et al., "Laser Diagnostics For H– Beam Momentum And Momentum Spread," Reference 8, pp. 226–35 (1994).
[14] Sander, R. K. et al., "Laser-Induced Fluorescence Measurement Of A 50-Mev Hydrogen Atom Beam," *Rev. Sci. Instr.* **62**, pp. 1893–98 (1991).
[15] M. Cortes and F. Martin, "Photodetachment of H^- with Excitation to N=2," *Phys. Rev. A,* **48**, pp. 1227–38 (1993). See Figures 3–5, and associated text.
[16] Bethe, H. A. and E. E. Salpeter, "Quantum Mechanics of One and Two-Electron Atoms," Springer-Verlag (1957), see p. 266.
[17] See chapters 17 and 18 of Ref. 4 for further discussion of laser-induced fluorescence in atomic beams.

Linac-Beam Characterizations at 600 MeV Using Optical Transition Radiation Diagnostics*

A. H. Lumpkin, W. J. Berg, B. X. Yang, and M. White

Advanced Photon Source, Argonne National Laboratory
9700 South Cass Avenue, Argonne, Illinois 60439 USA

Abstract. Selected optical diagnostics stations were upgraded in anticipation of low-emittance, bright electron beams from a thermionic rf gun or a photoelectric rf gun on the Advanced Photon Source (APS) injector linac. The upgrades include the installation of optical transition radiation (OTR) screens, transport lines, and cameras for use in transverse beam size measurements and longitudinal profile measurements. Using beam from the standard thermionic gun, tests were done at 50 MeV and 400 to 650 MeV. Data were obtained on the limiting spatial ($\sigma \sim 200$ μm) and temporal resolutions (300 ms) of the Chromox (Al_2O_3 : Cr) screen (250 μm thick) in comparison to the OTR screens. Both charge-coupled device (CCD) and charge-injection device (CID) video cameras were used, as well as a Hamamatsu C5680 synchroscan streak camera operating at a vertical deflection rate of 119.0 MHz (the 24th subharmonic of the S-band 2856 MHz frequency). Beam transverse sizes as small as $\sigma_x = 60$ μm for a 600 MeV beam and micropulse bunch lengths of $\sigma_\tau < 3$ ps have been recorded for macropulse-averaged behavior with charges of about 2 to 3 nC per macropulse. These techniques are applicable to linac-driven, fourth-generation light source R&D experiments, including the APS's SASE FEL experiment.

INTRODUCTION

An increased interest in diffraction-limited light sources for the next-generation sources and the implementation of prototype or scaling experiments has been evident since the Fourth Generation Light Source Workshop, held in Grenoble in January 1996 (1). At the Advanced Photon Source (APS), a research and development effort had been underway for several years to use the injector linacs with a low-emittance electron beam source (2-4). More specifically, an rf thermionic gun would be used for injection into the 100 to 650 MeV linac subsystem based on the existing 200 MeV electron linac and 450 MeV positron linac. The low, normalized emittance beams ($\varepsilon_n \approx 5\ \pi$ mm mrad)

---

* Work supported by the U.S. Department of Energy, Office of Basic Energy Sciences, under Contract No. W-31-109-ENG-38.

CP451, *Beam Instrumentation Workshop*
edited by R. O. Hettel, S. R. Smith, and J. D. Masek
1998 The American Institute of Physics 1-56396-794-4/98/$15.00

require an upgrade to the existing Chromox viewing screens for characterization of beam transverse size and bunch length. We are in the process of testing optical transition radiation (OTR) screens at selected positions in the beamline to provide sub-100 μm spatial resolution and sub-ps response times (5, 6). In order to compensate for the reduced brightness of this conversion mechanism, both gated, intensified cameras and streak cameras are being used to measure the beam properties. Initial tests with beam at 550 to 650 MeV (but generated by a conventional thermionic gun) have been done with a charge-injection device (CID) camera, a charge-coupled device (CCD) camera, and a streak camera. Additionally, the feasibility of using coherent transition radiation (CTR) and diffraction radiation (DR) based techniques will be evaluated. The diagnostics will be used to characterize, optimize, and monitor the bright beams needed to support self-amplified spontaneous emission (SASE) scaling experiments at $\lambda \sim$ 120 nm and a beam energy of 400 MeV as described separately (7).

EXPERIMENTAL BACKGROUND

Linac

The APS facility's injector system uses a 250 MeV S-band electron linac and an in-line S-band 450 MeV positron linac. The electron gun is a conventional thermionic gun in standard operations. For the alternate configuration, an rf thermionic gun, designed to generate low-emittance beams (<5 π mm mrad) and configured with an α-magnet, injects beam just after the first linac accelerating section (4). Then both in-line linacs can be phased to produce 100–650 MeV electron beams when the positron converter target is retracted.

The rf-gun's predicted, normalized emittance is about an order of magnitude lower than that of the conventional gun. Consequently, much smaller beam spot sizes are produced ($\sigma_{x,y} \approx 100$ μm). The standard intercepting screens are based on Chromox of 0.25 mm thickness, with a 300 ms decay time (8). Previous experiences on the Los Alamos linac-driven free-electron laser (FEL) with a low-emittance photoelectric injector (PEI) support the applicability of optical transition radiation screens in this case (9). A summary of projected beam properties is given in Table 1.

TABLE 1. APS Linac Beam Properties in the Low-Emittance Mode (rf Gun)

Parameter	Specified Value
rf frequency (MHz)	2856
Beam energy (MeV)	100–650
Micropulse charge (pC)	350
Micropulse duration (ps)	3–5 (FWHM)
Macropulse length (ns)	30
Macropulse repetition rate (Hz)	1–20
Normalized emittance (π mm mrad)	~5 (1 σ)

However, initial tests of a Ti foil used as an OTR screen, the transport of the OTR out of the tunnel, and the CCD camera and streak camera setup have been done with a "surrogate" beam from the conventional gun. These were done at a beam energy of 650 MeV, 2–5 nC in the macropulse, and 25–30 pC in each of 80 micropulses.

Beam Characterizations

A general description of the proposed techniques for beam characterization is given in Reference 2. A subset of those, based on optical techniques, now in the installation and testing stage, are presented here.

Transverse Characterizations

The transverse beam sizes and profiles are key to evaluating the beam emittance and its preservation throughout the accelerator and transport lines. At the 50 MeV station, two additional OTR screens are being installed. Although their axial spacing is less than 1 meter, beam quality will be initially checked at this point using the two-screen-beam-size-measurement technique, as well as the beam-size-versus-quadrupole-field-strength scan technique.

Another key station is at the end of the linac in the transport line, nominally the 650 MeV station. At this point, an optical transport line has been installed to bring the OTR light to an optical table outside of the linac tunnel. This table will provide an experimental base for measurements with a streak camera and a gated, intensified charge-coupled device (ICCD) camera (Stanford Computer Optics, Quik-05A). With the microchannel plate (MCP)-based shutter, samples 5 ns wide from the beam macropulse are possible. The gain factor of the MCP also allows for imaging of defocused spots during a quadrupole field scan for an emittance measurement. The tests of these cameras have already been done on the APS positron accumulator ring (PAR) and the booster synchrotron using synchrotron radiation from ~1 nC charge passing through a dipole.

The PAR bypass transport line provides a unique opportunity with its 10 m drift space to perform a three-screen emittance measurement. As described in Reference 7, the center screen is 4.8 m from the two end screens. Relay optics will bring the images to a lead-shielded ICCD camera. A fourth screen may be used for OTR interferometer experiments in conjunction with the center screen.

Longitudinal Characterizations

Because longitudinal beam brightness is related to evaluations of SASE gain, the measurement of bunch duration and profile are also critical in this program. At the 50 MeV station, one of the OTR screens and one part of the beamline cross will be configured to send the far-infrared (FIR) coherent transition radiation, generated by the few-ps or mm-long bunches, to a FIR Michelson interferometer. An optical autocorrelation technique will be evaluated as a bunch duration diagnostic (10).

The baseline technique will use a Hamamatsu C5680 dual-sweep streak camera viewing the incoherent OTR signal from the 650 MeV station. The transport of OTR to the optics table outside the tunnel has facilitated these experiments. The most useful vertical sweep plug-in has been a synchroscan unit phase-locked to 119.0 MHz, the 24th subharmonic of the 2856 MHz linac frequency. Low jitter of the synchronous sum of beam bunches is advantageous in dealing with the very low charge in a single micropulse. Because the S-band micropulse spacing is much smaller than the 119.0 MHz period, the sequence of micropulses will best be displayed using the dual-sweep technique *if* light levels are sufficient. This particular 119.0 MHz unit has been successfully phase-locked to an rf source at the Duke Storage Ring FEL facility, which is injected by an S-band linac (11). At the APS, a low-jitter countdown circuit has been built using Motorola ECLIN PS logic to generate the 24th subharmonics. It has been tested with a 0.7 ps (rms) jitter pulse generator, and the total jitter was observed to be 1.1 ps. Bandpass filters on the output result in a clean 119.0 MHz sine wave to be used with the synchroscan unit (12). The initial results are given in the next section.

RESULTS

An initial test of OTR source strength has been done using 1–3 nC of beam in a macropulse from the conventional gun, at energies of 580 and 650 MeV, with an in-tunnel camera (13). Since the Ti foil was placed over only half of the Chromox screen at this station, the e-beam could be steered and focused on the Chromox first and then steered onto the OTR foil.

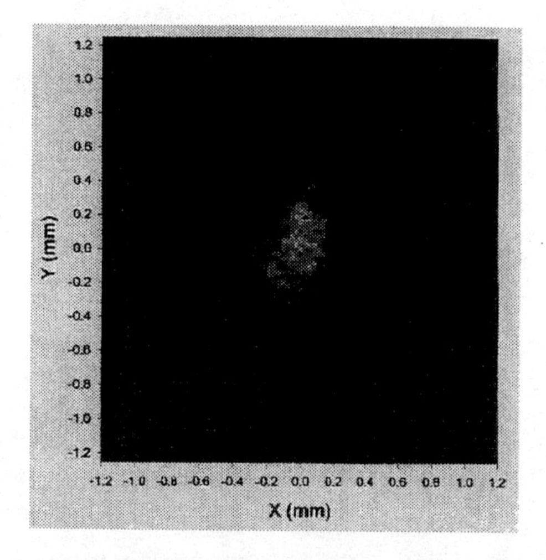

FIGURE 1. One of the first OTR images of APS linac beam at 650 MeV and ~3 nC in a macropulse from the *conventional* thermionic gun. A four-frame average was used to improve the statistics.

Figure 1 shows a sample beam image using the new optical transport to bring the OTR outside the tunnel to the optics table. Figure 2 shows the horizontal and vertical profiles with Gaussian fits. Focused spots (~0.5 mm, FWHM) were readily imaged with the CCD camera with about 3 nC in a macropulse. However, normal transport conditions usually have a larger spot size at this location and are seen much more readily with the Chromox screen. This baseline measurement supports the OTR screen choice because the increased beam image size from the Chromox screen implied that it had a 200 μm (σ) resolution limit under these conditions.

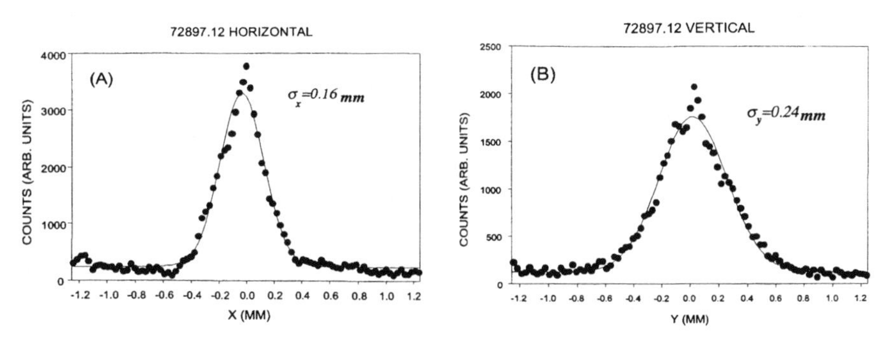

FIGURE 2. The horizontal (a) and vertical (b) spatial profiles for the OTR image in Figure 1. The observed sizes are smaller than those from the Chromox screen with its 200 μm limiting resolution.

We have also performed initial bunch-length measurements using the OTR conversion mechanism and the streak camera operated in synchroscan mode. As noted, the beam was generated by the conventional thermionic gun. Our configuration of rf BPM electronics limited us to 100 mA in a macropulse. This corresponded to only about 25–30 pC in each of 80 micropulses, separated by 350 ps. Although this charge is lower by an order of magnitude than was projected for rf-gun operations, we were able to obtain streak images by using 8- or 16-event averages in the digitizing system. We then performed the synchronous summing of micropulses from the same section in the macropulse.

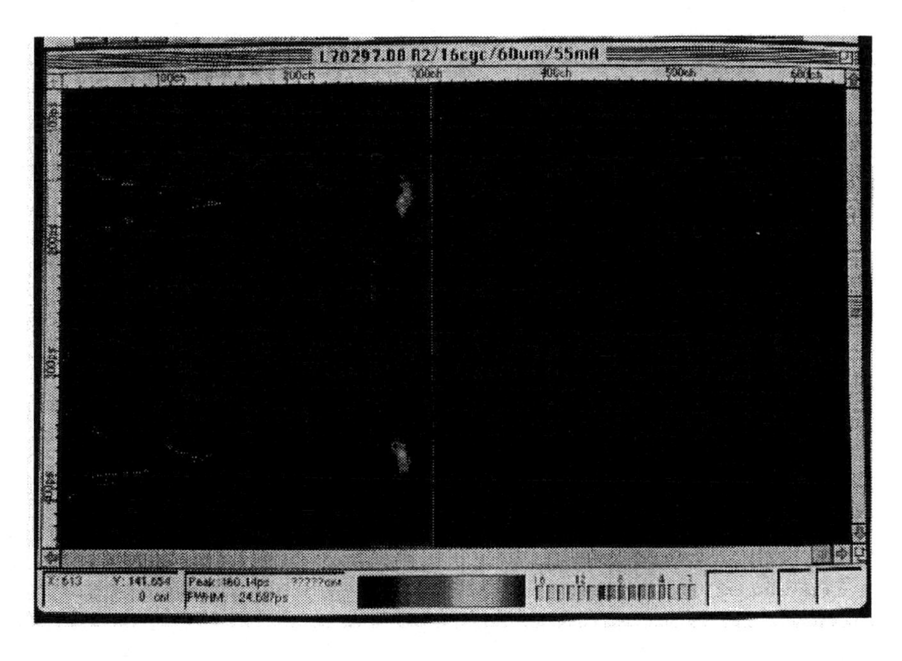

FIGURE 3. Synchroscan streak image summing over several micropulses showing a *y-t* tilt similar to a head-tail wakefield effect. These data were taken without a bandpass filter so the total bunch length was σ = 10 ps or 24 ps (FWHM).

Images on four streak ranges were obtained. Figure 3 shows an example from range #2 (R2) that spans ~480 ps. Due to the S-band repetition frequency of the microbunches, more than one micropulse is displayed with the 119 MHz sweep rate. An intriguing feature of the image is the curvature in *y-t* space displayed. The data are reminiscent of a head-to-tail transverse kick on the submicropulse timescale, perhaps due to transverse beam position offsets while transiting the linac accelerator structures. The displacement of the spatial profile centroid from the early to late part of the micropulse was about 200 μm with the observed bunch length of 10 ps (σ). As the peak current is quite low for this case, further data are needed. A short time later, the e-gun was observed to be arcing, and this resulted in noticeable differences in micropulse arrival time. The data were taken without a bandpass filter, so some temporal dispersion effects are involved.

In Figure 4 the streak camera focus mode profile (a) is shown to give a limiting resolution of about 2.6 ps, while the micropulses bunch length averaged over 4 micropulses for 4 macropulses is about 3.6 ps in Figure 4(b). The 550 × 40 nm bandpass filter was used to reduce the chromatic dispersion effects. These data involved a streak speed three times slower than the fastest range, so 1 ps bunch lengths are addressable.

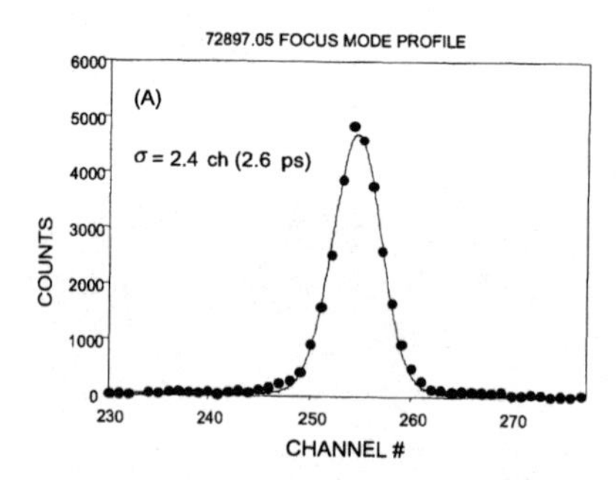

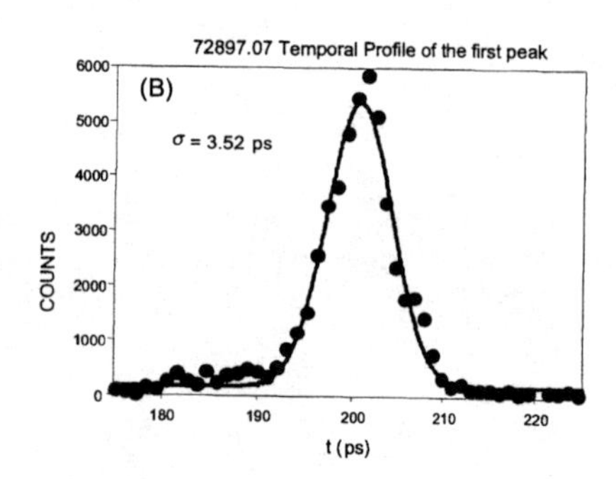

FIGURE 4. Streak camera images taken with a 550 × 40 nm bandpass filter for the focus mode (a) and the R2 streak range (b). The total observed bunch length is about 3.6 ps (σ) from the conventional gun and buncher system.

In the two cases above, an intercepting OTR foil is used. For nonintercepting bunch length measurements, coherent DR is a possible way to extend the Michelson interferometer technique (14, 15). In the streak camera case, a bend in the transport line, a special few-period diagnostics wiggler, or the final prototype wiggler for the SASE experiments are potential nonintercepting sources of optical radiation for a bunch length measurement.

SUMMARY

In summary, the adjustments of optical diagnostic techniques in preparation for low-emittance beams are well underway in the APS linac. Tests of some techniques (e.g., OTR, gated cameras, and the synchroscan (119.0 MHz) streak camera) have already been done with alternative particle beam sources. Further tests will be done with the conventional injector, and the initial tests with rf-gun injected beam are expected in 1998.

ACKNOWLEDGMENTS

The authors acknowledge the foresight of John Galayda (Accelerator Systems Division) in keeping the options open for an undulator test line and Stan Pasky for supporting linac studies time for the OTR tests.

REFERENCES

[1] Laclare, J., (Ed.), *Proceedings of the Workshop on Fourth Generation Light Sources*, Grenoble, France, January 22–25, 1996.

[2] Lumpkin, A. H., S. Milton, and M. Borland, "Proposed Particle-Beam Characterizations for the APS Undulator Test Line," *Nucl. Instr. and Methods A* **341**, pp. 417–421 (1994).

[3] Lumpkin, A. H. et al., "Diagnostics for the APS Undulator Test Line," *Proc. of the Fifth Beam Instrumentation Workshop*, Santa Fe, NM, Oct. 20-23, 1993, *AIP Conf. Proc.* **319**, pp. 211–219 (1994).

[4] Borland, M., "An Improved Thermionic Microwave Gun and Emittance Preserving Transport Line," Proc. 1993 Particle Accel. Conference, Washington, DC, May 17–20, 1993, pp. 3015–3017 (1993).

[5] Rule, D. W., R. B. Fiorito, A. H. Lumpkin, R. B. Feldman, and B. E. Carlsten, *Nucl. Instr. and Methods A* **296**, p. 739 (1990).

[6] Lumpkin, A. H., R. B. Fiorito, D. W. Rule, D. Dowell, W. C. Sellyey, and A. R. Lowrey, *Nucl. Inst. and Methods A* **296**, p. 150 (1990).

[7] Milton, S. V. et al., "The Advanced Photon Source Low-Energy Undulator Test Line," Presented at the 1997 Particle Accelerator Conference, Vancouver, B. C., May 12–16, 1997, (Proceedings in press).

[8] White, M., and A. Lumpkin, Argonne National Laboratory, private communication, Feb. 15, 1997.

[9] Lumpkin, A. H., "Advanced, Time-Resolved Imaging Techniques for Electron Beam Characterization," Presented at the 1990 Workshop Accelerator Instrumentation, Batavia, IL, Oct. 1–4, 1990, *AIP Conf. Proceedings* **229**, p. 151, and references therein (1991).

[10] Lihn, H. C. et al., *Phys Rev. E* **53(b)**, p. 413 (1996).

[11] A. H. Lumpkin et al., "Initial Application of a Dual-Sweep Streak Camera to the Duke Storage Ring OK-4 Source," Presented at the 1997 Particle Accelerator Conference, Vancouver, B.C., May 12–16, 1997, (Proceedings in press).

[12] Laird, R. and F. Lenkszus, Argonne National Laboratory, private communication, April 1997.

[13] Lumpkin, A. H., W. J. Berg, and B. X. Yang, "Planned Optical Diagnostics for the APS Low-Energy Undulator Test Line," Presented at the 1997 Particle Accelerator Conference, Vancouver, B.C., May 12–16, 1997, (Proceedings in press).

[14] Barry, W., "Measurements of the Sub-picosecond Bunch Profiles Using Coherent Transition Radiation," Presented at the 1996 Beam Inst. Workshop, Argonne, IL, May 6–9, 1996, *AIP Conf. Proc.* **390**, p. 173 (1997).

[15] Shibata, Y. et al., *Phys Rev.* **E 52**, p. 6787 (1995).

A High Resolution Electron Beam Profile Monitor and its Applications

W. S. Graves, E. D. Johnson, and S. Ulc

Brookhaven National Laboratory, Upton, NY 11973 USA

Abstract. A beam diagnostic to measure transverse profiles of electron beams is described. This profile monitor uses a cerium-doped yttrium:aluminum:garnet (YAG:Ce) crystal scintillator to produce an image of the transverse beam distribution. The advantage of this material over traditional fluorescent screens is that it is formed from a single crystal, and therefore has improved spatial resolution. The resolution is ultimately limited by the diffraction of visible light to approximately 1 micron. The application of these scintillators in a very compact three-screen emittance monitor is also described.

INTRODUCTION

The high-brightness electron beams now being produced for short wavelength FELs and high-energy colliders have focused sizes as small as a few microns. Resolution below this is necessary for accurate reproduction of the transverse beam profiles. Fluorescent screens based on phosphors have been widely used to measure the transverse profiles of electron beams at high-energy accelerator facilities. Traditional fluorescent screens [1], such as ZnS, produce a bright image but have relatively poor resolution. Phosphors have an individual grain size of 50–100 microns. Internal scattering of the fluorescence limits the image resolution to the same scale. A newer technology with improved resolution is the use of optical transition radiation (OTR) [2] produced by the electron beam passing from vacuum through a material such as a thin carbon or aluminum foil. This technique has excellent spatial resolution but suffers from very weak light output (more than two orders of magnitude less intense than phosphors in the visible spectrum). The YAG:Ce scintillators described here produce an image as bright as a good phosphor screen with a resolution comparable to that of an OTR screen.

Several devices based on these crystals have been built at the NSLS. They will be the main profile monitors installed in the Source Development Lab, a new free electron laser laboratory. A periscope pop-in monitor [3] now uses them for high precision beam profiles inside a small-gap undulator installed at the BNL Accelerator Test Facility. The crystals are also used in a compact three-screen emittance

CP451, *Beam Instrumentation Workshop*
edited by R. O. Hettel, S. R. Smith, and J. D. Masek

Table 1. Summary of YAG:Ce properties

YAG:Ce Physical Properties	
Density	4.55 gm/cm^3
Rise Time	5 ns
Fast Decay Time	90 ns
Slow Decay Time	300 ns
Photon Yield	18,000 photons/MeV
Peak Fluorescence Wavelength	560 nm

monitor installed in the Smith-Purcell experiment at ATF. This monitor is described in detail below.

SCINTILLATOR PROPERTIES

The crystals used to date have been cut as cylinders 0.5 mm thick by either 6 or 10 mm in diameter. They are available in a wide range of sizes and shapes. The thinnest standard size we have found is 0.2 mm, available from Preciosa Crytur. Crystals 0.1 mm thick are available on special order from the same manufacturer.

The photon yield is dependent on the type of exciting particle [4]. The yield is approximately 45% of NaI(Tl) [4] at 1.8×10^3 photons/MeV. This is more than two orders of magnitude brighter than screens utilizing optical transition radiation.

There is a finite rise time of the scintillator light of about 5 ns which may be due to a time constant of the lattice-to-activator energy transfer. The decay time is marked by two exponentials which depend on the method of excitation [5]. The fast time component has a decay constant of 90 ns and accounts for about 80% of the emitted light. The slow decay has a time constant of 300 ns and accounts for the remaining 20% of the light.

We initially coated the crystals with a 60%/40% Au/Pd metal layer a few tens of nm thick. This is to drain adsorbed charge from the electron beam. However, light reflections from flaws in the coating caused intensity variations in the beam image. Subsequent tests of an uncoated crystal using an electron microscope beam indicated that the bare YAG:Ce is adequately conducting, and we now routinely use uncoated crystals in high-energy beams.

The beam at BNL's Accelerator Test Facility has been used to test damage resistance. Several thousand pulses of charge of 0.5 nC each were run through the crystal in spot sizes ranging from tens of microns to 1 mm. There has been no measurable deterioration in light output. Visual inspection of the crystal also indicates no damage. The crystals have also been tested with large amounts of charge [6] (40 nC/pulse, 200,000 nC integrated charge) at the Argonne Wakefield Accelerator facility and have shown no damage.

Linearity of the light output with beam charge is good [7]. Results from Argonne [6] also show good linearity for larger beam charges.

RESOLUTION LIMITS

The ultimate spatial resolution is set by both the minimum object size that can be produced by the crystal and by the optical transport of visible light. The minimum object size for high-energy electrons is limited by multiple Coulomb scattering of the beam through the crystal and by the generation of bremsstrahlung. The multiple scattering angle through the material is given approximately by [8]:

$$\theta_{ms}(z) = \frac{13.6\mathrm{MeV/c}}{p}\sqrt{z/Z_0}\,[1 + 0.20\ln(z/Z_0)] \tag{1}$$

where z/Z_0 is the material thickness in radiation lengths and p is the electron momentum. The radiation length for YAG is 5.2 cm. For a 50 MeV beam traveling through a 0.5 mm thick crystal, $\theta_{ms} = 1.9$ mrad, yielding a minimum possible spot size of

$$x_{ms} = \int_0^{0.5\mathrm{mm}} \theta_{ms}(z)dz = 0.6\mu\mathrm{m}. \tag{2}$$

The critical energy, where energy losses due to ionization and radiation of bremsstrahlung are equal, is about 50 MeV. Above this energy the x-rays generated by electrons passing through the material will also cause scintillation, limiting the minimum measurable beam size. The characteristic emission angle of the x-rays is $1/\gamma = 10$ mrad at 50 MeV. Thus $x_{brem} = 2.5$ μm at 50 MeV.

The minimum spot size limits due to either multiple scattering or bremsstrahlung are both dependent on the electron beam energy. The limits are reduced at higher energies (Fig. 1).

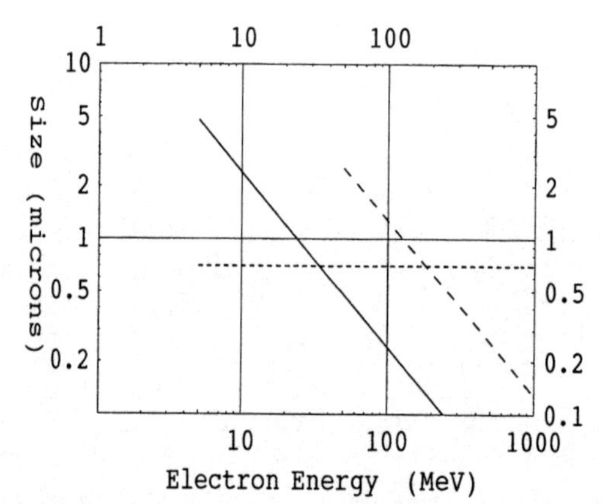

FIGURE 1. Effects that limit minimum spot size are plotted as a function of electron beam energy for a 0.5 mm thick crystal. Solid line is multiple scattering, dashed line is bremsstrahlung, dotted line is diffraction limit.

In addition to the intrinsic limits set by the electron beam passing through the crystal, the final image quality is also determined by the transport of visible light through the optical focusing system. The diffraction-limited diameter of a lens focus is $d_0 = 2.44\lambda(f/D)$, where the wavelength, λ, of the light is 560nm, f is the lens focal length, and D is the lens diameter. This is the diameter of the zeroes of the Airy disk. For reasonable working distances of a few centimeters, the lowest diffraction-limited f/D that can be achieved is about 3. Therefore the optics of visible light limit our resolution to beam spots of 4 microns diameter, or 0.7 microns rms size. Figure 1 summarizes the rms spot size limits due to multiple scattering, bremsstrahlung, and diffraction.

Spherical aberration, chromatic aberration, and depth of field are the most important lens effects that will distort the image and prevent diffraction-limited performance. The crystals are transparent so that the beam image is created along the entire depth of the crystal. This produces an extended longitudinal object that must be accurately focused at the plane of the camera. The crystals should be as thin as possible to reduce the depth of field and the amount of electron beam scattering. A lens system that is telecentric (magnification is the same for all conjugate planes, see Figure 2) will reduce the depth of field problem.

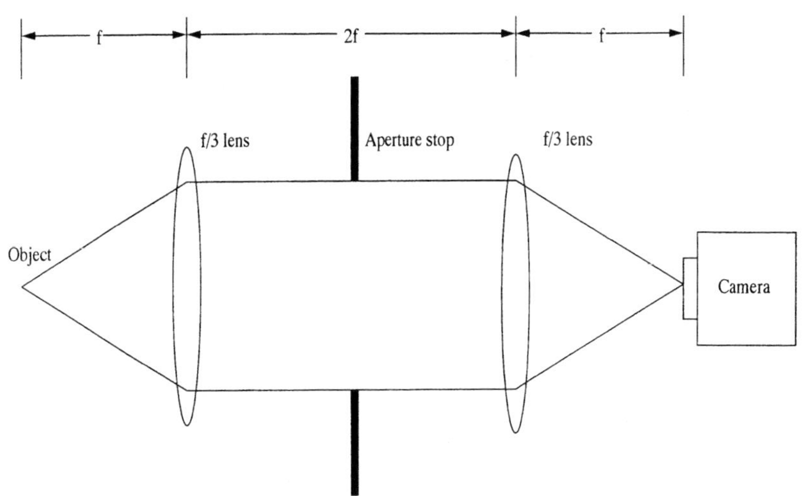

FIGURE 2. A telecentric lens arrangement reduces distortion due to depth of field and has minimum distortion if doublet achromatic lenses are used.

Spherical and chromatic aberration may then be reduced by choosing achromatic doublets for the lenses, which can be made diffraction-limited over a range of wavelengths.

THREE-SCREEN EMITTANCE MONITOR

The layout of the three-screen emittance monitor is shown in Figure 3. It is installed in beamline 1 of the BNL Accelerator Test Facility in an experiment to

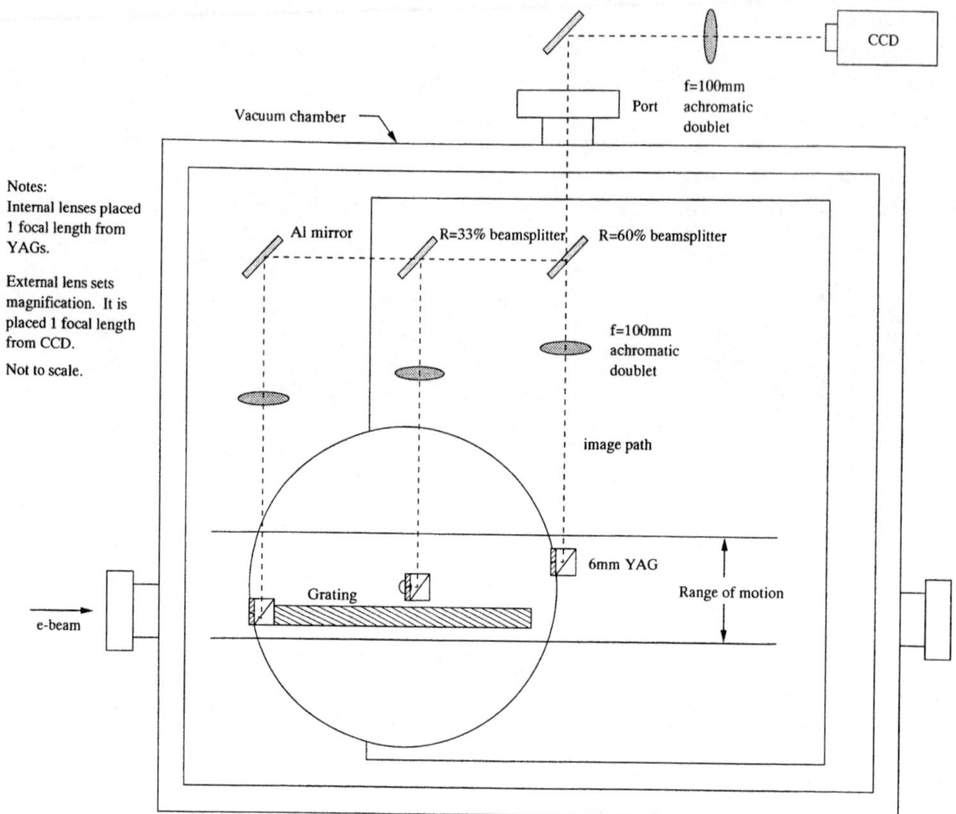

FIGURE 3. Layout of optics of three-screen emittance monitor. Spacing between screens is just 6 cm.

measure Smith-Purcell radiation from a relativistic electron beam [9]. The infrared Smith-Purcell radiation is transported by optics that are not shown. In this experiment the YAG:Ce crystals act purely as an electron beam diagnostic and do not affect the generation of Smith-Purcell radiation. The table on which the crystals are mounted translates across the beam direction. This enables the table to be placed so that the beam skims the surface of the grating or intercepts any one of the three screens. The table also rotates so that the grating may be made parallel to the beam. The YAG:Ce crystals facilitate the precise positioning of the grating with respect to the beam because the distances from each screen to the grating and to each other are known with high accuracy. Thus, position measurements allow the beam to be placed just off the surface of the grating and parallel to it. The crystals also indicate when the upstream focusing magnets are properly tuned to produce a beam waist at the grating center. Finally, they are used to measure the emittance using the three-screen technique.

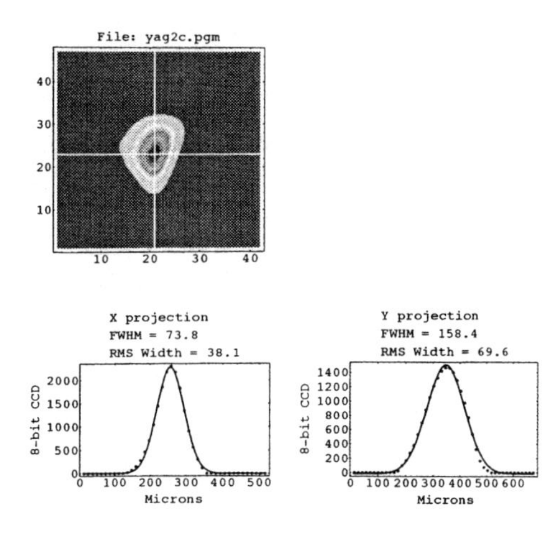

FIGURE 4. Image of the electron beam on the center profile monitor in Smith-Purcell experiment.

For this experiment the resolution requirements are not difficult (beam size greater than 40 μm). The doublets were arranged to be approximately telecentric. This arrangement proved useful for decoupling the lens focus from the translation table position. The vacuum window was used as an aperture stop. All lenses were 25 mm diameter and 100 mm focal length. The beam splitter reflectivities shown in Figure 3 were chosen to transport approximately equal amounts of light intensity from each screen to the camera. A Cohu 4910 CCD camera and a PC-based framegrabber were used to record the image [7]. Note that the short decay time of the scintillator (much less than a camera frame time) requires that the camera be synchronized to the beam arrival. Alignment and focusing of the optical transport was done by placing a pinhole successively at each screen until it was in focus at each and the images overlapped to with 1 or 2 pixels. With this arrangement, illuminating the apparatus with an external light allows all three screens to be seen on the CCD simultaneously. The grating shown in the illustration is 10 cm long, and the total spacing from the first crystal to the last is just 12 cm.

To accurately measure emittance, the beam's phase space ellipse should be sampled via profiles at equal intervals of betatron phase advance [10] as it rotates through 180 degrees. If three screens are available, then the ideal spacing is 60 degrees of phase advance between screens. This requirement is met when there is a beam waist at or near the center screen *and* the beam is a factor of two larger

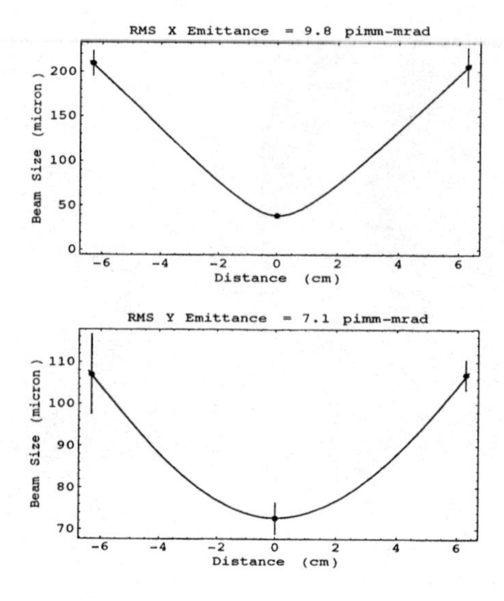

FIGURE 5. Horizontal and vertical emittance measurements using three screens in the Smith-Purcell experiment.

at each of the outer screens. Note also that if more than three screens are used, or the quadrupole-tuning method of single screen emittance measurement is employed, that the requirement that the screens be spaced at equal intervals of phase advance implies that the screens (or quadrupole tuning points) are not equidistant from each other.

A typical beam image and its horizontal and vertical projections are shown in Figure 4. This image is of a beam waist at the center screen of the experiment. Figure 5 shows horizontal and vertical profiles for all three screens and gives the calculated normalized rms horizontal emittance as 9.8π mm-mrad and the vertical emittance as 7.1π mm-mrad. The betatron advance between screens is not ideal for either transverse plane. The beam was tuned to give a minimum horizontal waist at the grating center as the best condition for producing Smith-Purcell radiation.

CONCLUSIONS

The performance of YAG:Ce scintillator crystals was described. These crystals are bright, have linear response, are highly damage resistant, have a fast decay time, and have excellent spatial resolution.

A three-screen emittance measurement using a very compact arrangement of these crystals was described. The novel aspects of this beam diagnostic are its compact size and placement directly in another experiment that is critically dependent on accurate knowledge of the beam size, position, angle, and divergence.

An important aspect of beam profile monitors based on these crystals is the low system cost. The crystals themselves are very inexpensive and the visible light produced is well matched to current silicon-based detectors.

ACKNOWLEDGMENTS

We are grateful to Vitaly Yakimenko of the ATF for producing the measured beams, and to Harold Kirk of the Physics Dept at BNL for assistance with the installation and mounting of the profile monitor at the ATF. This work performed under the auspices of the US Department of Energy Contract DE-AC02-76CH00016.

REFERENCES

1. Seeman, J. T., V. Luth, M. Ross, and J. Sheppard, *Beam Tests of Phosphorescent Screens*, Technical Report, SLAC, 1985. Single-Pass Collider Memo CN-290.
2. Jackson, J. D., *Classical Electrodynamics*, p. 685, Wiley (1975).
3. Johnson, E. D., W. S. Graves, and K. E. Robinson, "Periscope Pop-in Beam Monitor," in these proceedings (1998).
4. Moszynski, M., T. Ludziejewski, D. Wolski, W. Klamra, and L. O. Norlin, "Properties of the YAG:Ce Scintillator," *Nuc. Inst. Meth*, A 345:461–467 (1994).
5. Ludziejewski, T., M. Moszynski, M. Kapusta, D. Wolski, W. Klamra, K. Moszynska, and L. O. Norlin, "Investigation of Some Scintillation Properties of YAG:Ce Crystals," *Nuc. Inst. Meth*, A 398:287–294 (1997).
6. Power, J. G., N. Barov, M. E. Conde, W. Gai, R. Konecny, and P. Schoessow, "Initial Characterization of the YAG Crystal," AGN #32, Technical Report, Argonne National Lab (1997).
7. Graves, W. S., E. D. Johnson, and P. G. O'Shea, "A High-Resolution Electron Beam Profile Monitor," in *IEEE Particle Accelerator Conference* (1997).
8. Particle Data Group, editor, *Particle Properties Data Booklet*, North-Holland (1990).
9. Woods, K., J. Walsh, R. Stoner, H. Kirk, and R. Fernow, "The Smith Purcell Experiment," *Phys. Rev. Lett.*, 74:3808–3811 (1995).
10. Emma, P., private communication.

First Multi-GeV Particle-Beam Measurements Using a Synchroscan and Dual-Sweep X-ray Streak Camera†

Alex H. Lumpkin and Bingxin Yang

Advanced Photon Source
Argonne National Laboratory
Argonne, IL 60439 USA

Abstract. Particle-beam characterizations of a multi-GeV storage ring beam have been done for the first time using a synchroscan and dual-sweep x-ray streak camera at the Advanced Photon Source (APS). The hard x-rays (2–20 keV) from a bending magnet source were imaged using an adjustable pinhole aperture. Both the horizontal size, $\sigma_x \sim 190$ µm, and bunch length, $\sigma_t \sim 28$ ps, were measured simultaneously. The Au photocathode provides sensitivity from 10 eV to 10 keV covering the three orders of magnitude in wavelength from the UV to hard x-rays.

INTRODUCTION

As interest grows in the development of diffraction-limited x-ray sources, the challenges for particle beam diagnostics and the need for demonstration of measurement capability have been identified (1, 2). The general concept of a diffraction-limited source at 1 Å and the corresponding target particle beam parameters of $\sigma_{x,y} \sim 10$ µm, $\sigma_{x',y'} \sim 1$ µrad (for $\beta = 10$m), and $\sigma_t \sim 1$ ps to 0.1 ps have been noted. Fortunately, the present third-generation synchrotron radiation facilities can provide a test bed for some of these parameters in one plane by the use of low-vertical coupling (3). We have imaged the 7 GeV stored positron beam at the Advanced Photon Source (APS) using hard x-ray synchrotron radiation (XSR), an adjustable pinhole aperture, and a unique synchroscan/dual-sweep x-ray streak camera. Both transverse beam size and bunch length have been determined. Much as the visible single-sweep streak cameras were extended to the soft x-ray regime in the 1970s (4), the synchroscan and dual-sweep fea-

† Work supported by the U.S. Department of Energy, Office of Basic Energy Sciences, under Contract No. W-31-109-ENG-38.

CP451, *Beam Instrumentation Workshop*
edited by R. O. Hettel, S. R. Smith, and J. D. Masek
1998 The American Institute of Physics 1-56396-794-4/98/$15.00

tures of the late 1980s visible streak cameras (5,6) have now been extended to the x-ray regime. Potentially, the XSR can provide better spatial resolution than visible radiation by a reduced contribution from the diffraction limit. The potential for characterizing the smaller beam sizes of lower emittance beams in the picosecond regime can assist R&D with third-generation sources towards the next-generation sources.

EXPERIMENTAL BACKGROUND

The APS is a third-generation hard x-ray synchrotron radiation facility. Its main parameters include a 7 GeV beam energy, 100 mA stored beam current and a 8.2×10^{-9} m rad natural emittance. Other parameters are listed in Table 1. Measuring and monitoring the beam quality is one of the missions of the Diagnostic Group of the Accelerator Systems Division. To this end, one of the forty sectors in the ring is dedicated to particle-beam diagnostics using both optical synchrotron radiation (OSR) and XSR (7). In the past, we have measured head-tail kicks in a micropulse (8), bunch length versus single-bunch current (9), and other beam dynamics issues (10). As we push towards lower vertical coupling in the facility to enhance x-ray beam brilliance over the original 10% coupling baseline, the vertical particle beam size becomes less than 50 μm at 1–2% coupling and even smaller at 0.1% coupling, the long-range target. Due to diffraction effects, it is very difficult to image these beam sizes using OSR. The diagnostic sector was planned to support these long range developments. A part of this plan is the development of a beamline that transports undulator radiation and dipole radiation from both a dispersive and nondispersive point in the lattice. In this experiment the x-ray streak camera was positioned 27 m from the dipole source as schematically indicated in Figure 1. The adjustable pinhole, based on a remotely controlled four-jaw aperture, was located 10.1 m from the source point, resulting in a magnification of 1.67 at the photocathode (PC) of the x-ray streak camera.

TABLE 1. Accelerator Parameters for Diagnostics

Parameter	Storage Ring
Energy (GeV)	7
rf freq. (MHz)	351.93
Harmonic no.	1296
Min. bunch spacing (ns)	2.8
Rev. period (μs)	3.68
No. of bunches	1-1296
Design max. single bunch current (mA)	5
Bunch length (2σ) (ps)	35-100
Damping times $\tau_{h,v}$ (ms)	9.46
Tunes ν_h, ν_v	35.22, 14.30
Damping time τ_s (ms)	4.73
Synch. freq. f_s (kHz)	1.96

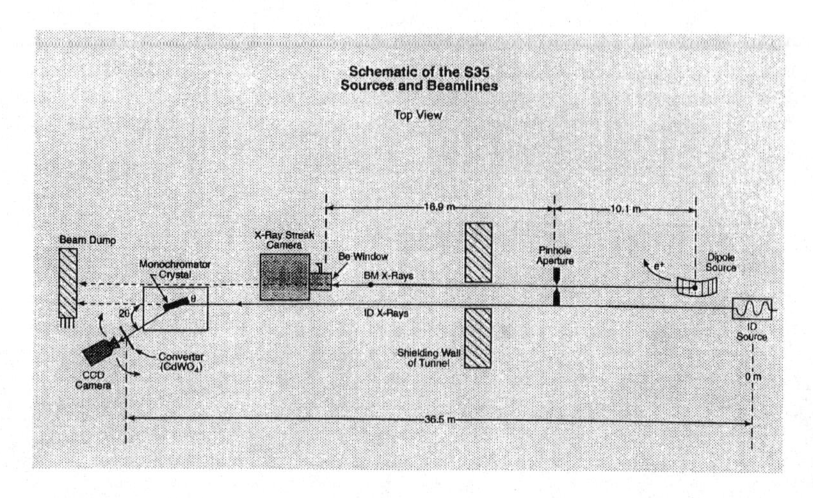

FIGURE 1. A schematic of the Sector 35 ID and dipole sources and the locations of the x-ray streak camera and adjustable pinhole.

An important practical aspect of the set-up was the remote-controlled positioning of the camera using three stacked translation stages as shown in Figure 2. These stages were under the mainframe and provided x-, y-, and $z(t)$-axis motion. This facilitated the alignment of the Au strip photocathode (6 mm (H) $\times$ 50 μm(V)) to the x-ray pinhole image field. Coarse alignment was done with the standard "burn paper" mounted on the front flange of the camera.

FIGURE 2. A photograph of the x-ray streak camera mounted on the three translation stages for x-, y-, and $z(t)$-axis motion. The front window is a Be. The streak tube vacuum is obtained with the external ion pump.

The Au photocathode allows detection of radiation from at least 10 eV to 10 keV and has some efficiency at energies up to 20–25 keV (see Figure 3). The soft x-rays are strongly attenuated by the Be windows in the transport. The x-ray tube is housed in a mainframe compatible with the plug-in units of the Hamamatsu C5680 series (11). For these experiments the synchroscan plug-in provided the vertical (fast time) sweep and the M5679 unit the horizontal (slow time) sweep. Although x-ray streak tubes with single fast deflections have been used for years in other fields (4), to our knowledge these are the first data taken on an accelerator or synchrotron radiation facility with the vertical sweep of the x-ray tube locked to the accelerator's rf frequency subharmonics, with repetition speeds greater than 100 MHz (117.3 MHz) and with jitter less than 1 ps. A second, orthogonal set of deflection plates in the tube was also used for a dual-sweep demonstration down to 200 μs, which was limited by photon intensity in this configuration. The use of x-rays for imaging also reduces the resolution limit from diffraction effects compared to optical (visible) synchrotron radiation (OSR) imaging, so that some beam dynamics issues may be addressed at smaller spot sizes.

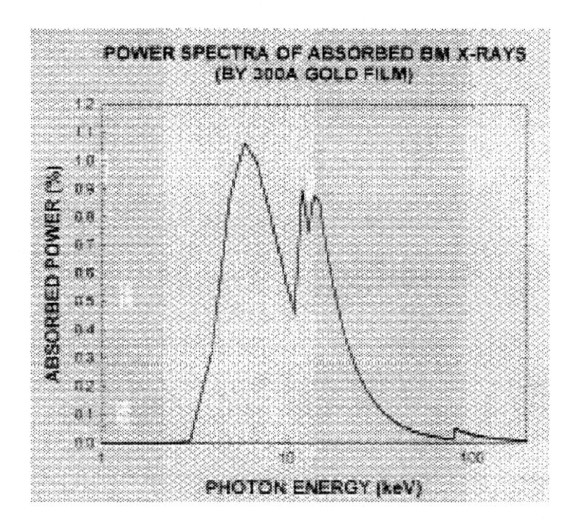

FIGURE 3. An estimate of the absorbed power in percent by the Au photocathode of the x-rays spectral distribution (in keV) from the bending magnet source. The attenuation of the soft-x-rays by the beamline exit Be window and the x-ray tube's entrance Be window has been included.

RESULTS

In this section we will address some of our first results in using the x-ray camera to image the 7 GeV particle beam. From these images both spatial and longitudinal information were obtained. Since the bending magnet source point was from an essentially nondispersive location in the lattice, no energy spread effects contribute to the horizontal or vertical beam size.

The first practical problem was to properly position the streak camera's slit photocathode in the sub-mm-sized x-ray beam. After a coarse mechanical height adjustment, better alignment was done with the standard "burn paper" mounted on the front flange of the camera. A Hg lamp was used then to illuminate the photocathode via a window and internal mirror to verify all voltages were applied and the PC was intact. As shown in Figure 4(a), the UV light irradiated the slit photocathode, and the image indicated a basically uniform response. The static spread function of the tube for UV light is found to be 4.5 channels (FWHM) and supports our previous sub-ps resolution test using a laser (12, 13). We then set the x-ray pinhole apertures at $50 \times 50\ \mu m^2$ and scanned the vertical plane using the remotely-controlled translation stage. As shown in Figure 4(b), the effective static spread function in the hard x-ray field resulted in a y-profile size of 9.1 ch (FWHM). The larger size is due to the physical limitations of focusing the few-eV energy spread of the photoelectron emitted from the Au's interaction with few-keV photons (4). Figures 4(a) and 4(b) graphically illustrate one of the issues for obtaining ps-resolution for an x-ray streak camera. For a vertical deflection speed of 0.31 ps/ch on the fastest range, the UV time resolution is about 1.5 ps (FWHM) or 0.6 ps (σ), and the hard x-ray resolution is estimated at 2.8 ps (FWHM) or about 1.2 ps (σ). The latter resolution is still sufficient to characterize the presently estimated limits of future storage ring beam bunch lengths.

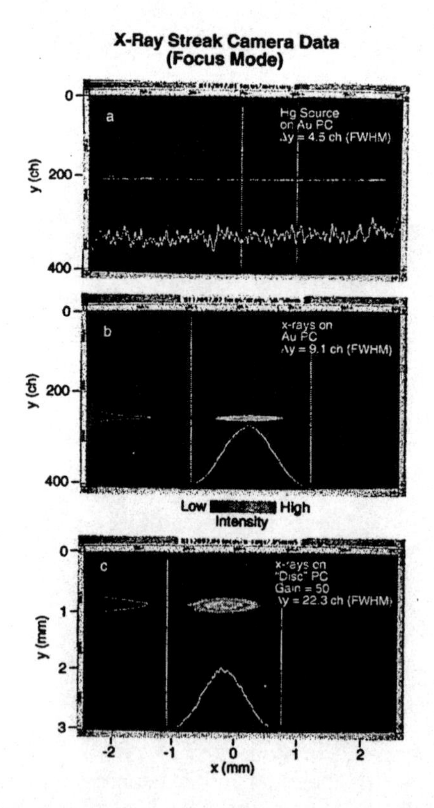

FIGURE 4. Images from the x-ray streak tube in focus mode: (a) the Hg UV light source, (b) the bending magnet x-ray source on the Au PC, and (c) the x-ray incident off the y-location of the Au PC.

During the course of vertical scanning for the Au PC we observed a much weaker, full image when the streak camera was located about ±1 mm from the strip photocathode position. By increasing the external microchannel plate (MCP) gain (×100), we obtained what appears to be the full spatial pinhole images of the particle beam. In Figure 4(c), the y-profile size of 22.5 ch (FWHM) corresponds to about 100 μm (FWHM) at the source point or 43 μm (σ). The x-profile is about 190 μm (σ) at the source point. These sizes are consistent with low vertical coupling (1–2%) in the ring.

The synchroscan streak ranges were next exercised. Three of the four were tested with x-rays. The phase adjustment was done using the z-translator instead of an rf phase-delay unit. A validation of the camera calibration was also executed by acquiring data at different $z(t)$ locations from the source. The centers of the streak images were determined for each $z(t)$ position, and the calibration factors were found to be within 5% of the factory calibration performed with a different method. On a subsequent run with the aperture at 100 × 100 μm^2, the focus mode image is shown in Figure 5(a), and in Figure 5(b) the synchroscan image with the observed bunch length of 65 ps (FWHM) or 28 ps (σ_{est}) is seen. On this streak deflection range the camera resolution is estimated from the focus mode image size (static spread function) to be σ_{res} ~ 4 ps. The fastest range available is about four times faster.

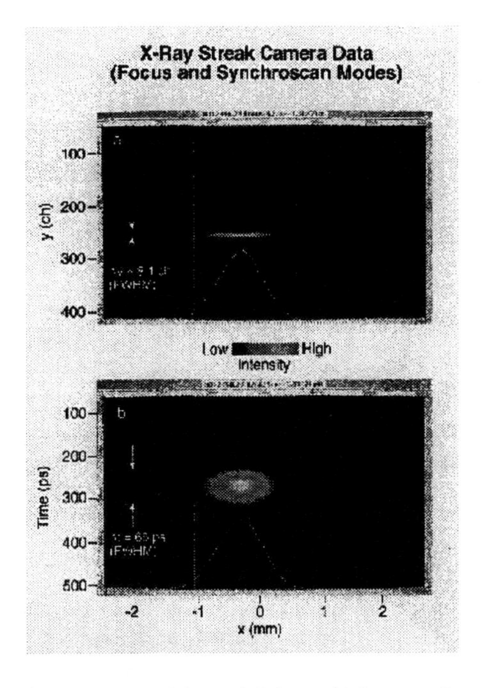

FIGURE 5. X-ray streak camera images of the particle beam (a) focus mode, and (b) synchrotron streak mode with 4 ps resolution. The observed bunch length is σ_t = 28 ps.

In Figure 6 we show a test of (a) the single horizontal sweep function and (b) the dual-sweep with a 10 ms horizontal by 500 ps vertical coverage. In this latter case the phase tilt upward going from left to right is an instrumental alignment effect (since it is also in 6(a)). The modulation of the bunch intensity on the few-ms timescale is attributed to vertical beam motion on and off the slit PC or a longitudinal effect in the beam. Source strength issues need to be evaluated, but shorter time samples might be done with an undulator source instead of a single dipole magnet and by using less magnification to increase the signal per pixel. The present case involves the nominal 80 mA stored positron beam. We have obtained images with a 100 times smaller pinhole aperture so bunch length measurements of less than 1 mA should be possible. This would address the realm of single-bunch, multiple-turn data in our ring.

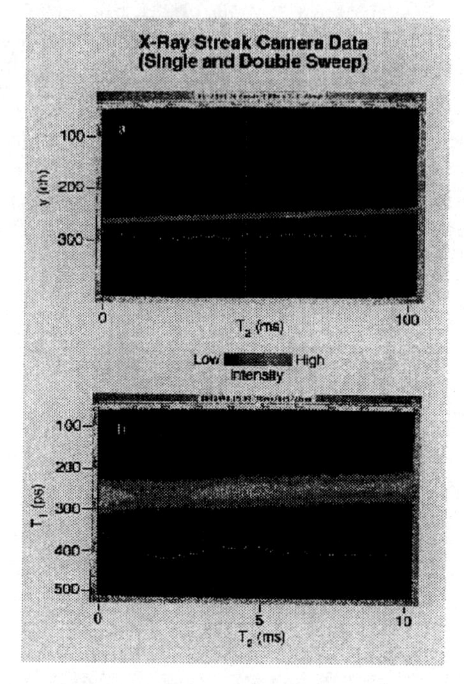

FIGURE 6. Examples of x-ray streak images: (a) a focus mode with single, 100 ms horizontal sweep on and (b) a dual-sweep with 10 ms horizontal time axis coverage and 500 ps vertical coverage (synchroscan). Some modulation of the intensity on the few-ms timescale in (b) is evident.

SUMMARY

In summary, we have performed initial measurements of a multi-GeV beam with hard XSR (2-20 keV) using a unique x-ray streak camera. Both beam transverse size and bunch length measurements are possible simultaneously. This capability should extend studies of beam dynamics in small beams. Additionally, this technique addresses the need identified for ps-domain x-ray detectors for time-resolved experiments using 10 eV to 10 keV radiation for user experiments (14).

ACKNOWLEDGMENTS

The authors acknowledge the support of John Galayda and Glenn Decker of the APS Accelerator Systems Division and the assistance of William Cieslik of Hamamatsu Photonics Systems.

REFERENCES

[1] Lumpkin, A. H., "On the Path to the Next Generation of Light Sources," *Nucl. Instrum. Methods in Phys. Res. A* **393**, pp. 170–177 (1997).

[2] Lumpkin, A. H., "Beam Diagnostics Challenges for Future FELs," *Proceedings of the SPIE'97 FEL Challenges Conference*, SPIE Vol. 2988, p. 70 (1997).

[3] Lumpkin, A.H. and Bingxin X. Yang, "Measurements of Near-Prototypical Fourth Generation Light Source Beam Parameters," submitted to *Phys. Rev. Letters*, April 1998.

[4] Stradling, G., "Soft X-ray Streak Cameras," *Proc. of the 18th International Conference on High Speed Photography and Photonics*, SPIE Vol. 1032, p. 194 (1988).

[5] Lumpkin, Alex H., "Advanced Time-Resolved Imaging Techniques for Electron-Beam Characterization," *Proc. of the Accelerator Instrumentation Workshop,* AIP Conf. Proc. **229**, p. 151 (1991).

[6] Rossa, E., C. Bovet, L. Disdier, F. Madeleine, and J. J. Savioz, "Real-time Measurement of Bunch Instabilities in LEP in Three Dimensions using a Streak Camera," *Proc. of the Third European PAC,* Vol. 1, pp. 144–146 (1992).

[7] Yang, B. X., A. H. Lumpkin, G. A. Goeppner, S. Sharma, E. Rotela, I. C. Sheng, and E. Moog, "Status of the APS Diagnostics Undulator Beamline," *Proc. of the 1997 Particle Accelerator Conference* (held May 12–16, 1997, Vancouver, BC, Canada), to be published.

[8] Lumpkin, Alex H., "Time-Domain Diagnostics in the Picosecond Regime," *Proc. of the Microbunches Workshop,* Upton, NY, Sept. 1995, AIP Vol. **367**, p. 327 (1996).

[9] Lumpkin, A. H., B. X. Yang, and Y. C. Chae, "Observations of Bunch-lengthening Effects in the 7-GeV Storage Ring," *Nucl. Instrum. and Meth. A* **393**, pp. 50–54 (1997).

[10] Yang, B. X. et al., "Characterization of the Beam Dynamics in the APS Injector Rings Using Time-Resolved Imaging Techniques," *Proc. of the 1997 Particle Accelerator Conference,* (held May 12–16, 1997, Vancouver, BC, Canada), to be published.

[11] Hamamatsu Photonics, C5680 Universal Streak Camera Data Sheet, May 1993.

[12] Lumpkin, A. H., B. Yang, W. Gai, and W. Cieslik, "Initial Tests of the Dual-Sweep Streak Camera System Planned for the APS Particle-Beam Diagnostics," *Proc. of the 1995 PAC* (Held Dallas, TX, May 1–5, 1995), Vol. 4, p. 2476 (1996).

[13] Lumpkin, A., *Nucl. Instr. and Methods A* **375**, p. 460 (1996).

[14] Knotek, M. (Ed.), Presented at the Workshop on Scientific Opportunities for Fourth-Generation Light Sources, Argonne National Laboratory, October 27–29, 1997.

Beam Diagnostics Based on Time-Domain Bunch-by-Bunch Data*

D. Teytelman, J. Fox, H. Hindi, C. Limborg, I. Linscott, S. Prabhakar, J. Sebek, A. Young

Stanford Linear Accelerator Center P.O. Box 4349 Stanford, CA 94309, USA

A. Drago, M. Serio

INFN - Laboratori Nazionali di Frascati, P.O. Box 13 I-00044 Frascati (Roma), Italy

W. Barry, G. Stover

Lawrence Berkeley National Laboratory, 1 Cyclotron Road Berkeley, CA 94563, USA

Abstract. A bunch-by-bunch longitudinal feedback system has been used to control coupled-bunch longitudinal motion and study the behavior of the beam at ALS, SPEAR, PEP-II, and DAΦNE. Each of these machines presents unique challenges to feedback control of unstable motion and data analysis. Here we present techniques developed to adapt this feedback system to operating conditions at these accelerators. A diverse array of techniques has been developed to extract information on different aspects of beam behavior from the time-domain data captured by the feedback system. These include measurements of growth and damping rates of coupled-bunch modes, bunch-by-bunch current monitoring, measurements of bunch-by-bunch synchronous phases and longitudinal tunes, and beam noise spectra. A technique is presented which uses the longitudinal feedback system to measure transverse growth and damping rates. Techniques are illustrated with data acquired at all of the four above-mentioned machines.

INTRODUCTION

A bunch-by-bunch feedback system has been developed by a multi-laboratory collaboration for control of coupled-bunch longitudinal motion at ALS, PEP-II, and DAΦNE. The architecture of the system has been described in detail in earlier publications (1), (2), (3). DSP-based design allows synchronized real-time data acquisition in conjunction with feedback processing.

Table 1 summarizes the parameters of different machines on which the feedback system has been used. The feedback system is configured in each case to maintain a constant ratio between the bunch sampling frequency and the synchrotron frequency. Downsampling matches the feedback processing rate to the longitudinal oscillation frequency and results in a significant reduction in the computational load on the DSP array

* Work supported by DOE contract No. DE-AC03-76SF00515

CP451, *Beam Instrumentation Workshop*
edited by R. O. Hettel, S. R. Smith, and J. D. Masek

as compared to the non-downsampling approach. A table-driven programmable down-sampler module allows operation on the machines with widely different numbers of bunches and downsampling factors.

TABLE 1. Machine Parameters

	ALS	DAΦNE	PEP-II	SPEAR
Number of bunches	328	120	1746	280
Bunch crossing rate, MHz	500 MHz	368 MHz	238 MHz	359 MHz
Revolution frequency	1.5 MHz	3 MHz	136 kHz	1.28 MHz
Synchrotron frequency	11 kHz	36 kHz	6 kHz	28 kHz
Growth time	2 ms	90 us	5 ms	16 ms
Downsampling factor	22	14	6	14
Bunch sampling rate	68 kHz	214 kHz	22 kHz	91 kHz
Growth time, samples	130	20	110	1500
Processors	40	60	80	40
Bunches/processor	9	2	22	7
Samples/bunch in a transient record	1008	4032	661	2016

DIAGNOSTIC TECHNIQUES

A large number of diagnostic techniques based on the time-domain transient and steady-state data have been developed. Transient data is used for measurements of growth and damping rates and injection transients. From steady-state data one can extract information on the system noise floor and a set of bunch-by-bunch parameters such as currents, synchronous phases and synchrotron frequencies.

The different accelerators listed in Table 1 vary significantly in the growth times of the unstable modes. For example, at SPEAR, growth time is comparable to the number of samples stored by the DSP. Techniques have been developed to facilitate growth and damping rate measurement in such cases. For weakly unstable modes positive feedback is used to speed up the growth. In cases when the damping rates of naturally stable modes are to be determined, the external excitation method is used (4).

Records of steady-state bunch motion provide a wealth of information about the beam and the performance of the feedback and rf systems. By capturing bunch motion while in negative feedback mode, one can quantify the residual noise level due to quantization, as well as determine frequencies and amplitudes of driven motion. Such measurements of driven motion were used during the PEP-II HER commissioning to characterize the performance of the rf system (5).

From steady-state records one can also extract information about bunch currents and synchronous phases. To measure bunch currents, we detect the level of low-frequency driven motion, e.g., line frequency harmonics, in the signal of each bunch. Bunch-by-bunch synchronous phase information can be extracted from the DC levels of different bunches. Bunch currents and synchronous phases can be used to measure machine impedance at the revolution harmonics (6).

EXPERIMENTAL RESULTS

From steady-state measurements, synchronous phase and synchrotron frequency per bunch can be extracted. In PEP-II, bunches are driven by baseband rf noise. Within the bandwidth of the synchrotron resonance, this noise has relatively flat spectrum. Consequently, baseband driven motion of the bunches has spectral characteristics of a damped oscillator excited by white noise. To obtain bunch-by-bunch synchrotron frequency, a second order oscillator response is fitted to the power spectrum of the time-domain motion of each bunch. In PEP-II, there is a significant gap transient resulting from the low-revolution frequency. This transient is characterized by the synchronous phase and synchrotron frequency variation as illustrated in Figure 1. Significant tune shift of the first group of bunches after the gap provides a possible explanation of the phenomenon observed in PEP-II HER, using a synchroscan streak camera (7), in which the head of the bunch train does not participate in unstable motion. Due to the tune shift, bunches in the head of the train are effectively decoupled from the rest of the bunches.

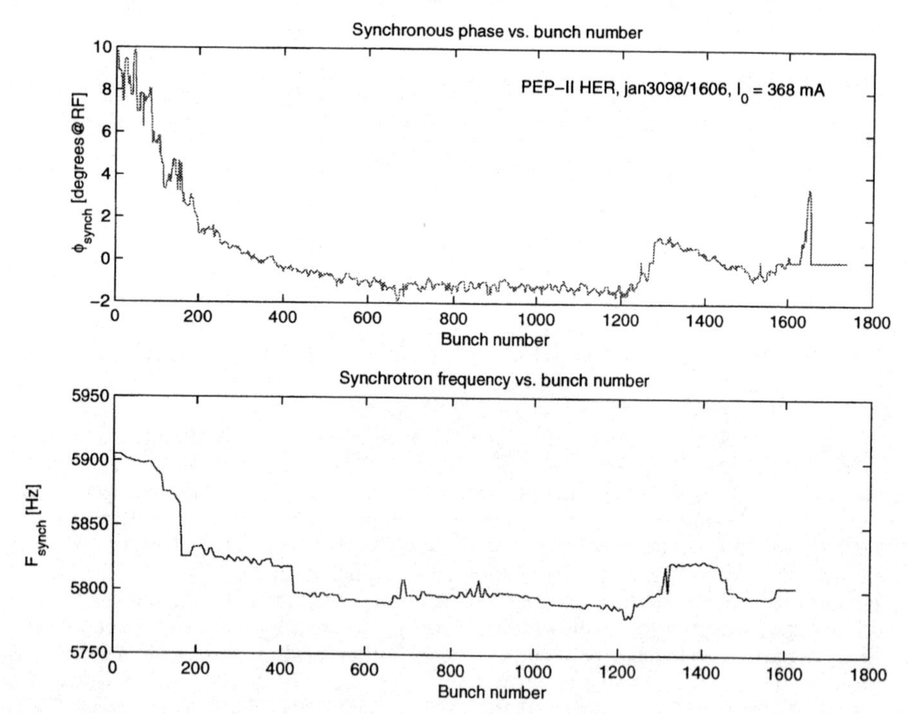

FIGURE 1. Synchronous phase (upper graph) and synchrotron frequency (lower graph) transient in the PEP-II HER. Tune shift between the bunches in the head and the middle of the train is greater than 100 Hz.

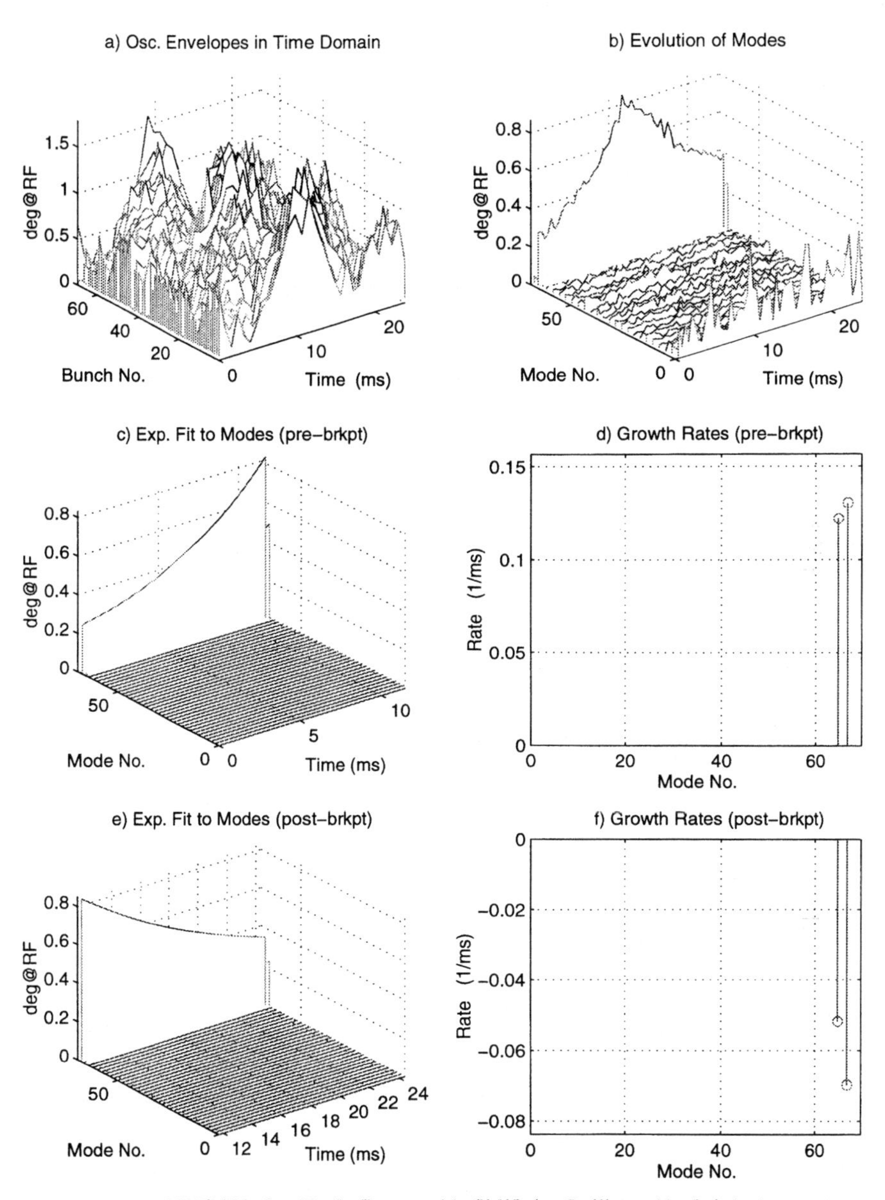

aug0197/1454: Io= 29mA, Dsamp= 14, ShifGain= 5, Nbun= 69, Gain1= 1,
Gain2= 1, Phase1= −120, Phase2= 60, Brkpt= 1065, Calib= 4.15cnts/mA–deg.

FIGURE 2. Grow/damp transient from SPEAR. The system starts in the negative feedback mode, controlling unstable motion. On software trigger, the sign of the feedback is reversed (goes to positive feedback) and after a predetermined hold-off period, recording starts. At t=12 ms, the system returns to negative feedback and the damping transient is recorded.

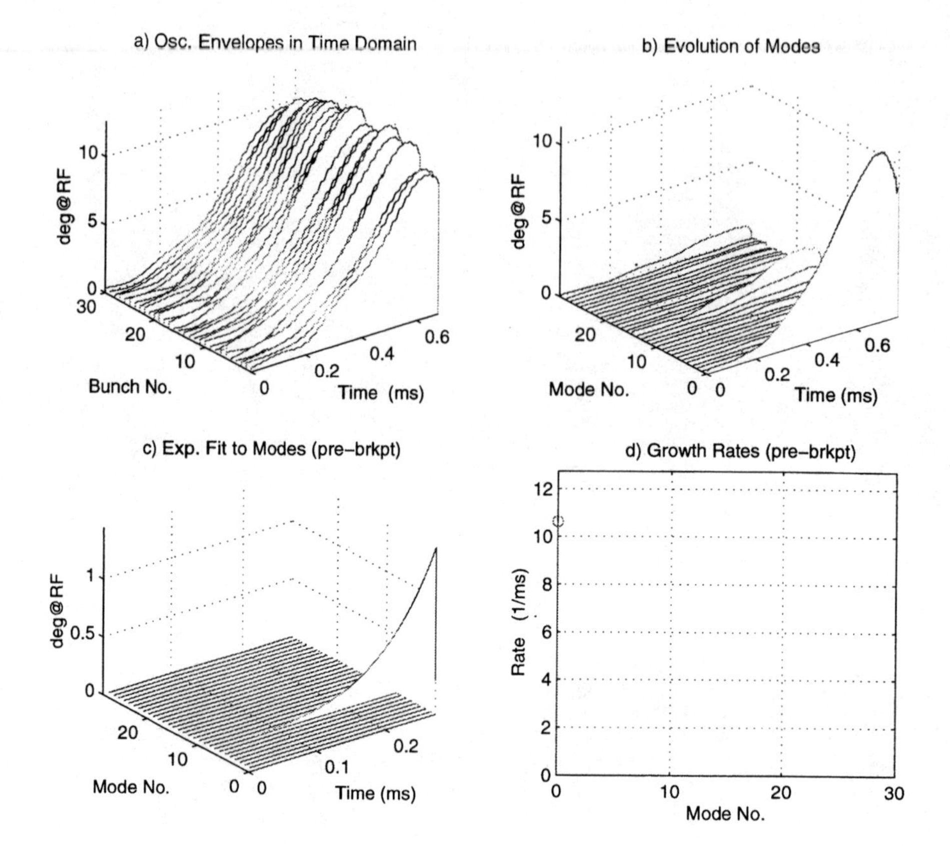

DAFNE/mar2598/0124: Io= 100mA, Dsamp= 14, ShifGain= 6, Nbun= 30, Gain1= 0,

Gain2= 1, Phase1= 50, Phase2= 165, Brkpt= 60, Calib= 1.166cnts/mA–deg.

FIGURE 3. Grow/damp transient from DAΦNE. At the t=0 feedback is turned off and oscillations of the bunches are recorded. In this transient, the growth rate is 10.8 ms^{-1} and motion reaches the fullscale of the phase detector (15 degrees at rf) in 500 μs.

As discussed earlier, the time scale of transient events differs significantly between different machines. Figure 2 shows a grow/damp transient from SPEAR. At this beam current, the growth rate of the unstable modes is very low and positive feedback is used to speed up the growth. Two modes are excited in this transient and their growth and damping rates are measured. In the case of DAΦNE, the growth rates are an order of magnitude higher. A growth transient from the positron ring is shown in Figure 3.

Using the feedback system as a triggered recorder, it is possible to capture transverse grow/damp transients. Downsampling aliases betatron tunes to lower frequencies. However, in this process phase information is retained, so the coupled bunch mode amplitudes can be reconstructed. Since the envelope of motion is of interest here, downsampling does not affect the measurement of growth and damping rates. Figure 4 shows such a measurement from PEP-II. In this case, the A/D converter was connected to the baseband vertical monitor output of the transverse feedback system to obtain bunch-by-bunch vertical positions. A mixer was used to open and close the vertical

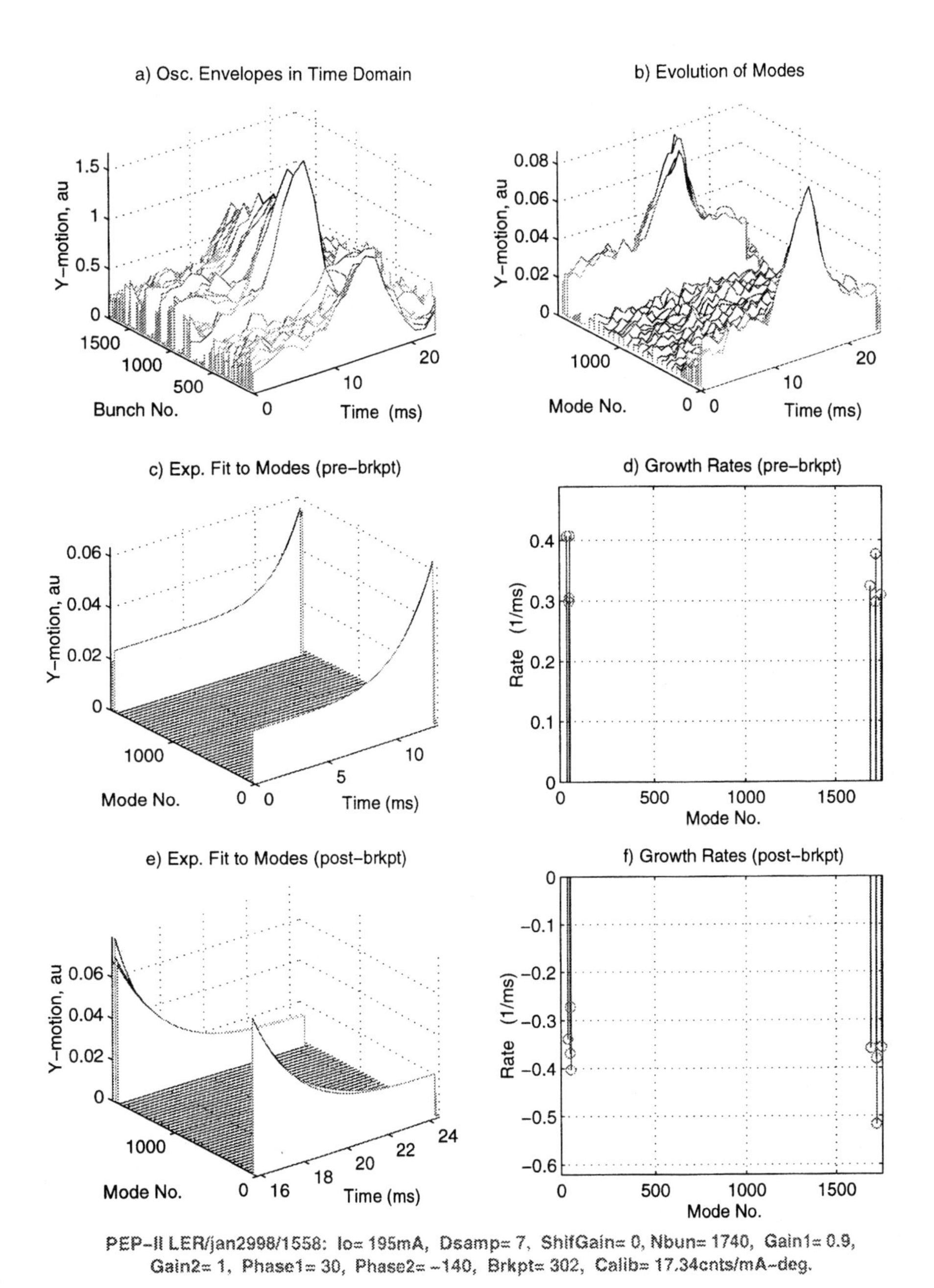

FIGURE 4. Transverse grow/damp transient from PEP-II. Two groups of modes participate in unstable motion. Exponential fits to the growth and damping portions allow to measure growth and damping rates for a large number of modes in a single transient.

feedback loop under control of an external trigger. The same signal was utilized to trigger recording in the longitudinal system. In this measurement, vertical feedback is turned off at t=7 ms. Bunch oscillations grow until t=11 ms, at which point the feedback is turned on. In the modal domain, we observe motion at the upper sidebands in two regions: the low and high-numbered modes in the spectrum. This corresponds to the upper and lower sidebands of the low revolution harmonics, which are driven by the resistive wall impedance.

SUMMARY

Transient and steady state diagnostics based on the bunch-by-bunch time-domain data provide a versatile tool for study of longitudinal and transverse beam dynamics. DSP-based architecture and tight synchronization of the longitudinal feedback system support transient measurements in a wide range of beam conditions. Open software architecture allows us to quickly develop and integrate new diagnostics.

ACKNOWLEDGMENTS

The authors thank B. Cordova-Grimaldi, J. Hoeflich, J. Olsen, and G. Oxoby of SLAC and J. Byrd, J. Corlett, and G. Lambertson of LBL for numerous theoretical and technical contributions. We also thank ALS, SPEAR, PEP-II, and DAΦNE operations groups for their support of the measurements.

REFERENCES

[1] Teytelman, D., et al, "Operation and Performance of the PEP-II Prototype Longitudinal Damping System at the ALS," presented at the 16th IEEE Particle Accelerator Conference (PAC 95) and International Conference on High Energy Accelerators, Dallas, Texas, May 1–5, 1995.
[2] Claus, R., et al, "Software Architecture of the Longitudinal Feedback System for PEP-II, ALS and DAΦNE," presented at the 16th IEEE Particle Accelerator Conference (PAC 95) and International Conference on High Energy Accelerators, Dallas, Texas, May 1–5, 1995.
[3] Young, A., et al, "RF and Baseband Signal Processing Functions in the PEP-II/ALS/DAΦNE Longitudinal Feedback System," presented at the 3rd European Workshop on Beam Diagnostics and Instrumentation for Particle Accelerators (DIPAC 97), Frascati (Rome), Italy, October 12 - 14, 1997.
[4] Fox, J. D., et al, "Observation, Control and Modal Analysis of Longitudinal Coupled-Bunch Instabilities in the ALS via a Digital Feedback System," presented at the 7th Beam Instrumentation Workshop, Argonne, IL, May 6–9, 1996.
[5] Prabhakar, S., et al, "Low-Mode Longitudinal Motion in the PEP-II HER," SLAC-PEP-II-AP-NOTE-98-06, March 1998.
[6] Prabhakar, S., et al, "Calculation of Impedance from Multibunch Synchronous Phases: Theory and Experimental Results," SLAC-PEP-II-AP-NOTE-98-04, February 1998.
[7] Fisher, A., et al, "Instrumentation and Diagnostics for PEP-II," this conference.

Characterizing Transverse Beam Dynamics at the APS Storage Ring Using a Dual-Sweep Streak Camera

Bingxin Yang, Alex H. Lumpkin, Katherine Harkay, Louis Emery, Michael Borland, and Frank Lenkszus

Advanced Photon Source, Argonne National Lab
9700 South Cass Avenue, Argonne, IL 60565

Abstract. We present a novel technique for characterizing transverse beam dynamics using a dual-sweep streak camera. The camera is used to record the front view of successive beam bunches and/or successive turns of the bunches. This extension of the dual-sweep technique makes it possible to display non-repeatable beam transverse motion in two fast and slow time scales of choice, and in a single shot. We present a study of a transverse multi-bunch instability in the APS storage ring. The positions, sizes, and shapes of 20 bunches (2.84 ns apart) in the train, in 3 to 14 successive turns (3.68 μs apart) are recorded in a single image, providing rich information about the unstable beam. These include the amplitude of the oscillation (~0 mm at the head of the train and ~2 mm towards the end of the train), the bunch-to-bunch phase difference, and the significant transverse size growth within the train. In the second example, the technique is used to characterize the injection kicker-induced beam motion, in support of the planned storage ring top-up operation. By adjusting the time scale of the dual sweep, it clearly shows the amplitude (±1.8 mm) and direction of the kick, and the subsequent decoherence (~500 turns) and damping (~20 ms) of the stored beam. Since the storage ring has an insertion device chamber with full vertical aperture of 5 mm, it is of special interest to track the vertical motion of the beam. An intensified gated camera was used for this purpose. The turn-by-turn *x*-*y* motion of a single-bunch beam was recorded and used as a diagnostic for coupling correction. Images taken with uncorrected coupling will be presented.

INTRODUCTION

The streak camera has been successfully used for accelerator diagnostics in the past two decades (1–7). Originally developed for the diagnostics of ultra-fast laser pulses, the camera's temporal resolution, from several ps down to less than 1 ps, is particularly suitable for measuring the length and longitudinal phase of particle bunches in modern accelerators. The addition of scanning capability in the second, horizontal axis enabled

CP451, *Beam Instrumentation Workshop*
edited by R. O. Hettel, S. R. Smith, and J. D. Masek
1998 The American Institute of Physics 1-56396-794-4/98/$15.00

recording of successive images of the beam and study of its longitudinal dynamics (6,7). Many measurements also make use of one transverse dimension in the image (1–5), which amounts to taking the side or top view of the traveling particle bunch. This technique enabled the visual demonstration of dynamic effects such as intra-bunch wakefield interaction in the linac (1), strong head-to-tail instability in a high-energy storage ring (2,3), etc. When the horizontal beam size is dominated by the particle energy spread, found in high dispersion regions, the top view of the bunch was used to image the longitudinal phase-space distribution during longitudinal damping (5). In these measurements, the longitudinal beam dynamics were studied.

In this work, we present a streak camera technique that emphasizes recording the evolution of the front view of the particle beam, either from multiple particle bunches or from multiple turns of a single bunch in a circular accelerator. We will show that it is a powerful and flexible technique for studying transverse beam dynamics.

Basic Technique

The basic technique is illustrated in Figure 1. In a streak tube, photoelectrons generated by the instantaneous light pulse of the particle bunch on a photocathode are accelerated and focused on to a micro-channel plate (MCP), and the intensified electronic image is converted to light by the phosphor screen. When the vertical sweeping field ramps rapidly, the image scans down the screen, leaving a trace proportional to the length of the particle bunch. If, however, the sweeping field ramps slowly and the images from the head and tail of the bunch are almost overlapping, the resulting image is a good approximation of the front view of the particle bunch.

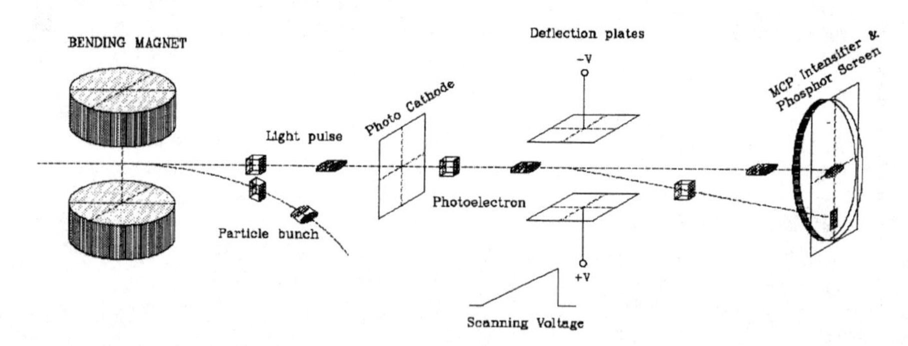

FIGURE 1. Schematic for recording the front view of the bunch with streak cameras. The accelerating and focusing electro-optics, as well as the horizontal deflection plates, are not shown for simplicity.

When the next event, the next bunch, or the next turn of the same bunch comes after a time period much longer than the bunch length, its image will be placed on a different location of the phosphor screen, preferably not overlapping with the previous one. In this manner a "fast sweep"—normally in the vertical direction and triggered by a timing pulse—records a chain of successive events. In a dual-sweep streak camera, when the horizontal sweep is also enabled, a second vertical sweep, triggered after a chosen time period, records a second series of images that are shifted from the first ones horizontally. In this manner, a matrix of beam images can be recorded. As we will illustrate below with examples, these images reveal rich information about beam dynamics. A few points are important for the application of this technique:

- The photocathode entrance slits need to be wide open for a good field of view.
- The imaging optics need to be carefully configured to maintain a balance between good spatial resolution and an adequate field of view.
- Because the bunch length is used to determine the exposure time of each image, the interbunch time (or turn period) needs to be much longer than the bunch length, a requirement often satisfied on many accelerators.
- The vertical time scale needs to be chosen to record an adequate number of events.
- The structure of the vertical timing pulse train and the horizontal scan time needs to be adjusted to suit the time scale of the dynamic phenomenon studied.

Multi-turn Images of a Bunch Train in the APS Storage Ring

The experiment to study the effect of chromaticity on bunch stability was performed at the APS UV/visible diagnostics beamline (8) with a Hamamatsu C5680 streak camera. The results are shown in Figure 2. We started with a high chromaticity setting. The 12 mA stored beam, in the train of 20 bunches filling successive buckets, is stable and the matrix of images are evenly spaced (Figure 2A). Each spot is the front view of a passing positron bunch. The vertical spacing between two adjacent spots is determined by the interbunch spacing of 2.8 ns, with the head of the train placed at the top of the image. The vertical scan is triggered every turn (3.7 μs). The horizontal scale is chosen to fit 14 columns of images on the screen without overlap.

As the horizontal chromaticity of the lattice was lowered, the last trailing bunch became unstable and its size grew horizontally (Figure 2B). Gradually, the latter half of the train began to execute a coherent oscillation, likely induced by interbunch coupling (Figure 2C). Eventually the horizontal size and oscillation amplitude of the trailing bunches got so large (Figure 2D) that they started to lose charge quickly.

From these beam images, one can obtain the center coordinates, horizontal and vertical size of each bunch in the train, and their turn-by-turn progression. This amount of information can put strong constraints on the theoretical model used to explain the instability. Without going into details, we can make some general observations: (1) The shape of the bunch train almost repeats itself every five turns, corresponding well with the fractional horizontal betatron tune of ~0.2 at the time of experiment. Hence, we conclude that the instability is transverse in nature and exciting betatron oscillations. (2) The amplitude of the oscillation increases quickly from the head of the train (~0 mm) to its tail (~2 mm), suggesting that (near-neighbor) interbunch coupling is feeding the oscillations through a wakefield. (3) The "wavelength" of the train shape in the coherent oscillation (Figure 2C) is determined by the phase of such couplings. It may provide clues about which wakefield component is responsible for the instability and what its bandwidth is.

Figure 3 shows two images taken with a test orbit after the recent installation of an insertion device chamber with 5 mm vertical aperture. Horizontal instability was observed on the left panel as a small-amplitude, coherent bunch-train oscillation in the horizontal direction. After we changed the current of sextupole magnets, the horizontal chromaticity was increased and the oscillation stopped. The vertical chromaticity, however, was simultaneously reduced and a vertical instability set in, as shown by the increased size of the last three bunches in the train on the right panel. Since these instabilities do not cause significant movement of the charge center of the six-bunch train, it is difficult for an electronic pickup device to detect. This example shows that the time-resolved transverse imaging technique may be used to quickly diagnose and visualize subtle instabilities and aid the search for a stable operating parameter space.

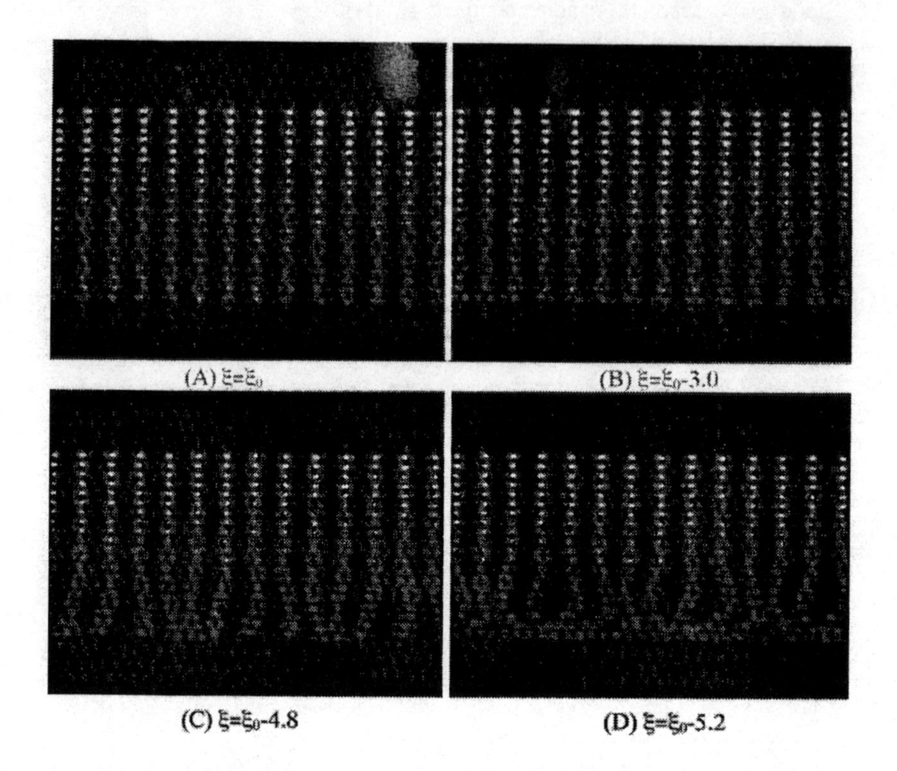

FIGURE 2. Image of a long bunch train in the APS storage ring with different chromaticity settings (in arbitrary units). The vertical scale is 100 ns and the horizontal scale is 50 μs.

FIGURE 3. Images of a short bunch train in a test orbit in the APS storage ring after a 5 mm insertion device chamber was installed. The vertical scale is 50 ns and the horizontal scale is 50 μs. Current of a sextupole (S4) was increased by 4 amperes (< 3%) to produce the image on the right.

Multi-turn Images of a Single Bunch in the APS Storage Ring

To support the top-up mode of operation, one needs to understand the transient beam motion after injection. We combined the streak camera with a gated intensified camera (Stanford Computer Optics QUIK05) to record the transient beam motion. Since the ring is equipped with insertion device chambers with vertical apertures as small as 5 mm, we are particularly interested in the amplitude of the vertical motion.

The initial experiment was performed with single-bunch stored current (~4 mA) in an orbit with the vertical coupling uncorrected. The injection kicker was fired at a 2 Hz rate to simulate the injection event. Figure 4 shows the streak camera image of the front view of the bunch in successive turns. Before the injection kicker was fired, a regular array of beam images was obtained (left panel). After the kicker was fired, the beam started to move in the horizontal direction. The motion quickly (in a single turn) coupled to the vertical direction and the beam entered into a large amplitude betatron oscillation in both directions. The bunch quickly decohered and its size increased dramatically. A number of streak images were taken and appeared to be identical to Figure 4. This indicated that the kicker operated reliably and reproducibly. We switched to a gated intensified camera for its higher spatial resolution and larger field of view.

FIGURE 4. Effect of injection kicker on a single bunch in the APS storage ring. The full scale of the vertical sweep is 100 μs and the horizontal one is 5 ms. The vertical sweep was triggered every 6 ms. Before the injection kicker was fired (left panel), and after the kicker was fired (right panel).

Figure 5 shows the images taken with the gated camera. Calibration was done with a target located at the bending magnet source point, which has four 1 mm holes arranged in a 3 mm × 4 mm rectangle. The fifth hole shown is used to indicate the upper-outboard corner of the target. From this image, we infer that the total field of view is 6 mm × 6 mm and that the image is tilted by 4.5° by the transport optics. A sixth hole, 127 μm in diameter, located at the center of the pattern, was used to estimate the rms system resolution, which is 120 μm(H) × 90 μm(V).

The center coordinates of the bunch can be extracted from the turn-by-turn images by fitting their integrated profiles. The horizontal motion follows a simple sinusoidal curve, indicating a well-behaved free betatron oscillation after the initial kick (Figure 6, left panel). The vertical betatron oscillation started one turn later (Figure 6, right panel), and its amplitude does not remain constant, as expected from tracking simulations.

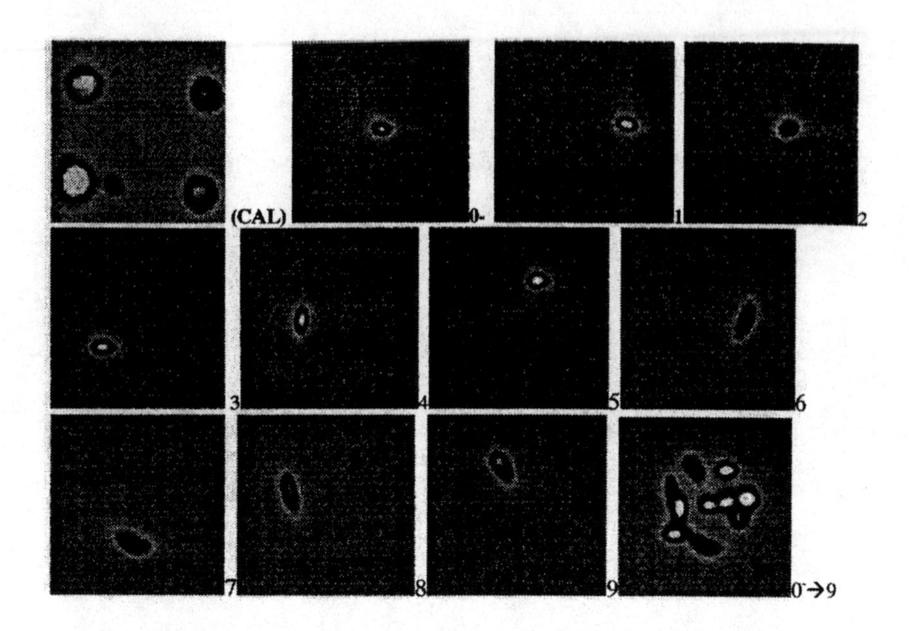

FIGURE 5. Image of a single bunch in the APS storage ring. Image (CAL) is the image of the calibration target. Images labeled 0 through 9 were taken on 0th through 9th turn after the kicker was fired. The last image is the integrated image from No. 0 through No. 9, taken with long exposure time.

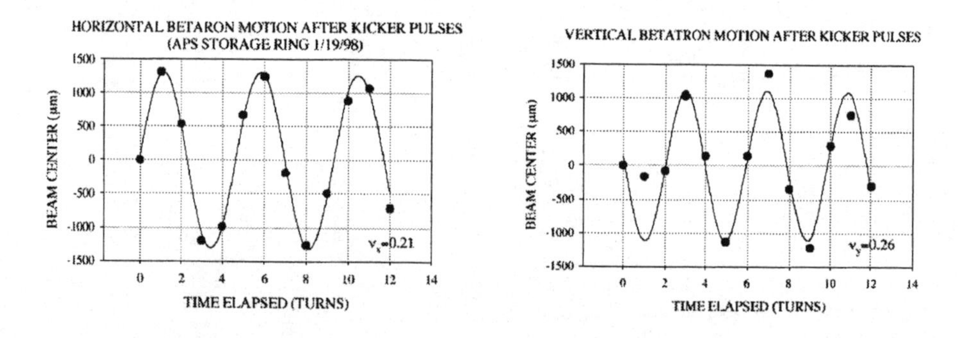

FIGURE 6. Coordinates of the electron bunch center as a function of time after the kicker was fired.

The betatron period is about four to five turns, and the integrated image of the first 10 turns (in Figure 5) gives a good average of beam motion over all betatron phases. The vertical extent of the image is now routinely used as a direct visual aid when vertical coupling of the injection orbit is corrected.

Figure 7 show the same series of single-pass images extended further in time. The bunch decoherence before and damping after 500 turns can be seen clearly. The width of bunch profiles may be extracted from these images and give information about transverse damping (Figure 8). These data compare well to the theoretical damping time of 9.5 ms in either transverse plane.

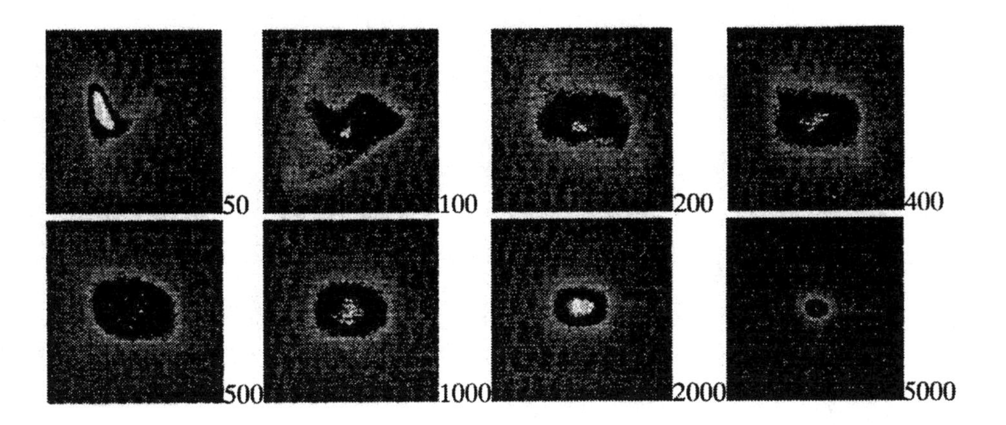

FIGURE 7. Image of a single bunch in the APS storage ring. Label for each panel shows the time of recording (in number of turns) after the kicker event.

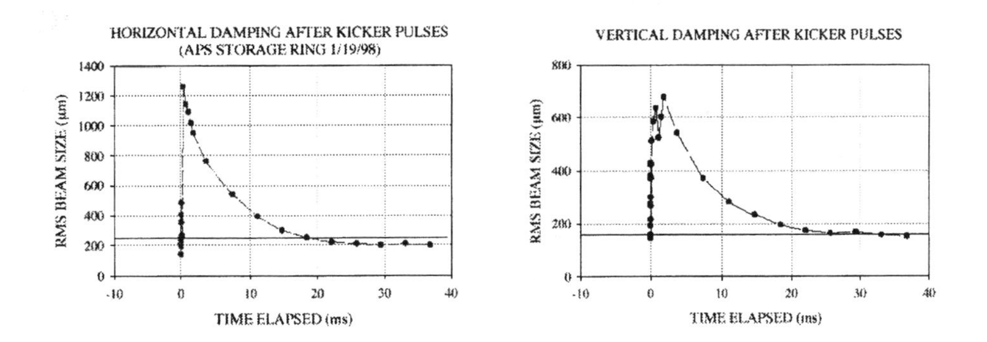

FIGURE 8. The rms bunch profile width as a function of time after the injection kicker is fired.

CONCLUSION

The timing structure of rf accelerated particle beams enables us to make a series of front-view images of the bunches in a train. This extension of the dual-sweep technique makes it possible to display non-repeatable beam transverse motion in two fast and slow time scales of choice, all in a single shot. It is a powerful and flexible technique for studying beam dynamics.

ACKNOWLEDGMENTS

We acknowledge J. Galayda and G. Decker for their foresight and strong support of the diagnostics beamline project.

REFERENCES

[1] Lumpkin, A. H., "Advanced, Time-Resolved Imaging Techniques for Electron-Beam Characterizations," Accelerator instrumentation, *AIP Conf. Proc.* **229**, 1991.

[2] Rossa, E. et al., "Real Time Measurement of Bunch Instabilities in LEP in Three Dimensions Using a Streak Camera," *Proc. of 3rd European Particle Accelerator Conference*, **1**, p. 144, 1992.

[3] Lumpkin , A. H., "Time-Domain Diagnostics in the Picosecond Regime," Micro Bunch Workshop, *AIP Conf. Proc.* **367**, 1995.

[4] Rossa, E., "Real time single shot three-dimensional measurement of picosecond bunches," BIW 94, *AIP Conf. Proc.* **333**, 1995.

[5] Yang, B. X., et al., "Characterization of Beam Dynamics in the APS Injector Rings Using Time-Resolved Imaging Techniques," presented at PAC 97, (Proceedings in press).

[6] Hinkson, J., et al., "Commissioning of the Advanced Light Source Dual-Axis Streak Camera," presented at PAC 97, (Proceedings in press).

[7] Fisher, A. S., et al., "Streak-Camera Measurements of the PEP-II High-Energy Ring," these proceedings.

[8] Yang, B. X. and A. H. Lumpkin, "Initial Time-Resolved Particle Beam Profile Measurement at the Advance Photon Source," SRI 95, *Rev. Sci. Instrum.* **67**, 1996.

Fundamental Limits on Beam Stability at the Advanced Photon Source

Glenn Decker, John Carwardine, Om Singh

Advanced Photon Source, Argonne National Laboratory,
9700 South Cass Avenue, Argonne, IL 60439 USA

Abstract. Orbit correction is now routinely performed at the few-micron level in the Advanced Photon Source (APS) storage ring. Three diagnostics are presently in use to measure and control both AC and DC orbit motions: broad-band turn-by-turn rf beam position monitors (BPMs), narrow-band switched heterodyne receivers, and photoemission-style x-ray beam position monitors. Each type of diagnostic has its own set of systematic error effects that place limits on the ultimate pointing stability of x-ray beams supplied to users at the APS. Limiting sources of beam motion at present are magnet power supply noise, girder vibration, and thermal timescale vacuum chamber and girder motion. This paper will investigate the present limitations on orbit correction, and will delve into the upgrades necessary to achieve true sub-micron beam stability.

INTRODUCTION: POWER SPECTRAL DENSITY

An essential tool in the study of beam stabilization is the power spectral density. Simply put, the power spectral density is the mean square signal per unit frequency, whether the signal is measured in volts, microns, or furlongs per fortnight. Upon integration (or summation) over the available frequency band represented in a given data set, one arrives at the mean square signal in that band, the square root of which yields the rms signal. Shown in Equation (1) is the definition of a discrete Fourier transform pair for a time-sampled data set $\{x_n\}$ containing N samples, while Equation (2) is a statement derived from Parseval's theorem, forming the basis for the definition of power spectral density (1):

$$X_k \equiv \frac{1}{N}\sum_{n=0}^{N-1} x_n e^{2\pi ink/\mathrm{N}} , \qquad x_n = \sum_{k=0}^{N-1} X_k e^{-2\pi ink/\mathrm{N}} , \tag{1}$$

$$\sigma_x^2 \equiv \frac{1}{N}\sum_{n=0}^{N-1} (x_n - \langle x\rangle)^2 = \langle x^2\rangle - \langle x\rangle^2 = \sum_{k=1}^{N-1} |X_k|^2 \; ; \quad \left(\langle x\rangle \equiv \frac{1}{N}\sum_{n=0}^{N-1} x_n\right) . \tag{2}$$

CP451, *Beam Instrumentation Workshop*
edited by R. O. Hettel, S. R. Smith, and J. D. Masek
1998 The American Institute of Physics 1-56396-794-4/98/$15.00

The key point to notice in Equation (2) is that the sums, whether taken in frequency or time domain, produce an invariant quantity, namely the mean square signal amplitude. The k=0 component of the sum over frequency (i.e., k) in Equation (2) has been moved to the left-hand side of the equation and is none other than the square of the mean, or DC value for the data set $\{x_n\}$, as can easily be verified by inserting k = 0 into the first of Equations (1). Power spectral density is quantitatively defined to be $2|X_k|^2/\delta f$, (k = 0...N/2), where $\delta f = 1/(N\ \delta t)$ is the frequency discretization interval, δt is the sampling period, and N is the number of samples. Values of $|X_k|$ from k = N/2+1 to N-1 are just a mirror image of those from k = 1 to N/2−1, explaining the factor of 2. "Integrating" the power spectral density from zero up the the Nyquist frequency $f_{Ny} = (N/2)\delta f = 1/(2\delta t)$ yields the mean square signal defined in Equation (2).

One impediment to the understanding of spectra displayed in other than power spectral density units is the fact that one must carefully account for the sampling rate and duration in order to be able to infer rms noise amplitude. For example, a plot of peak FFT amplitude vs. frequency, while convenient for picking off the amplitude of a pure sine wave, can be difficult to interpret in the case of purely random noise. A white noise source with constant power spectral density P volts^2/Hz, for example, corresponds to a signal having an rms amplitude of simply $(P\ \Delta f)^{1/2}$, where Δf is the frequency band of interest, measured in Hz. In the context of beam stability, a useful measure for power spectral density is (mm^2/Hz).

Keep in mind that information is unavoidably discarded when a continuous signal is sampled at discrete time intervals. In particular, changes in a continuously variable quantity, i.e., $x(t)$, that occur more rapidly than the sample period δt will be "aliased" (1) to lower frequencies when the variable is represented as a set of discrete numbers $x_n = x(n\ \delta t)$. In spite of this, Equation (2) will provide an accurate value for the rms noise in the continuous signal $x(t)$, provided that the number of samples N is large enough, even though the shape of the spectrum X_k vs. k will not be an accurate representation of the spectrum of $x(t)$.

SOURCES OF BEAM MOTION

Shown in Table 1 are the quantities of rms beam motion contained in four frequency bands: 10^{-6} to 0.017 Hz (drift), 0.017 to 30 Hz (jitter), 30 to 500 Hz (high frequency), and 500 Hz to 135 kHz (very high frequency). These numbers must be combined in quadrature to obtain the totals listed in the last row. At the APS, low-frequency drift motion is controlled using workstation-based orbit correction software, while intermediate-frequency motion (jitter) is corrected using real-time digital feedback operating at a 1 kHz update rate (2). Correction bandwidth presently extends to about 50 Hz. The dominant sources of beam motion are indicated in the table for each frequency band.

Shown in Figure 1 is a plot of beam position power spectral density, averaged over the forty zero-dispersion straight sections at the APS, corresponding to the locations of insertion devices (IDs) (35 of the forty are available for IDs, the other five are used for injection and rf hardware).

TABLE 1. Beam Stability Performance to Date (5/98)

Frequency Band	Horizontal Motion ΔX (microns rms)	Vertical Motion ΔY (microns rms)	Limitations
Low-frequency drift (10^{-6}–0.017 Hz)	$< \sigma_x^*$	2.5 to 20	BPM electronics intensity dependence, mechanical and electrical thermal effects, rf BPM bunch pattern dependence, ID x-BPM stray radiation
Jitter (0.017–30 Hz)	< 4.5, or 1.3% σ_x^*	< 1.8, or 10% σ_y^*	High bandwidth corrector availability, power supply stability, ID x-BPM stray radiation
High frequency (30–500 Hz)	< 12.4	< 7.5	Power supply stability rf voltage stability
Very high frequency (0.5–135 kHz)	5	6	rf voltage stability, multi-bunch Instabilities
Broadband TOTAL	14.1 + drift	10.1 to 22.3	Long-term drift

* Beam sizes σ_x = 335 microns, σ_y = 18 microns @ 1% coupling

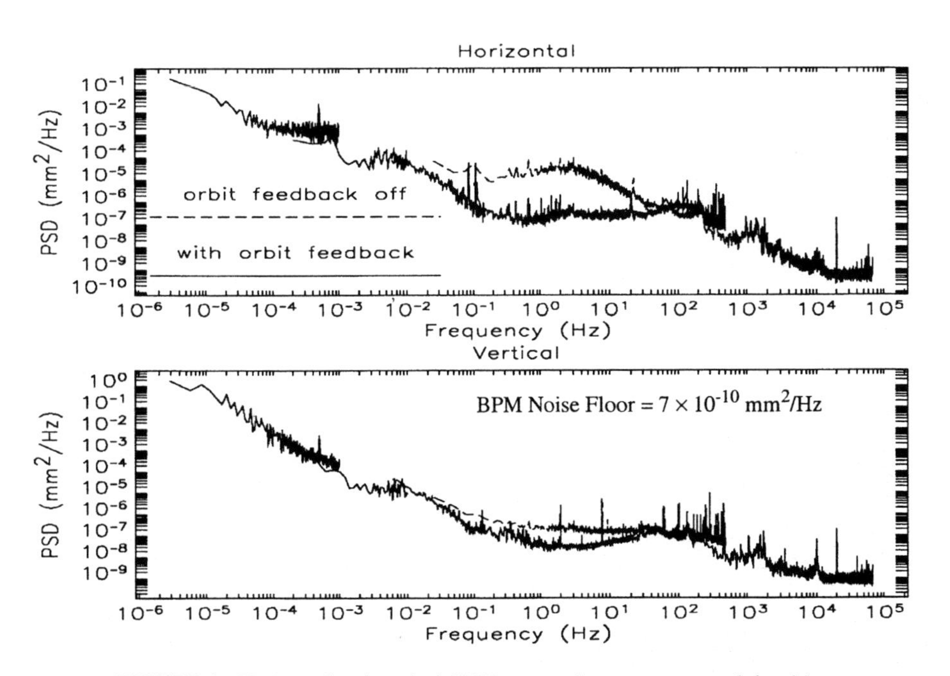

FIGURE 1. Horizontal and vertical APS beam motion power spectral densities.

This plot shows ten decades of frequency, extending from effects occurring on a turn-by-turn basis down to long-term drift effects of several days. The data shown was acquired using the same diagnostic, namely the broadband rf BPM system (3) that employs amplitude-to-phase conversion of 100 ns long, 352 MHz tone bursts derived from individual bunches, and sampled on a turn-by-turn basis (270 kHz sampling rate).

Features to note are a spectral line at approximately 5×10^{-4} Hz, corresponding to a one-half hour water temperature cycle, and two lines near 10^{-1} Hz, one caused by the "DC" orbit corrections that occur every ten seconds or so, and a second caused by the BPM intensity-dependent offset correction that occurs in the same time frame. A significant bump occurs in the open-loop horizontal motion, centered about 1 or 2 Hz, that is caused principally by power supply rumble, ground motion, and other electromagnetic interference. The thick aluminum vacuum chamber provides an excellent shield from stray electromagnetic fields above a few Hz. Vertical spectral lines at 2, 4, and 8 Hz are the result of the injector synchrotron's 2 Hz duty cycle. The synchrotron tune generally lies near 1.7 kHz and is observable in the horizontal plane. All of the storage ring power supplies except the main dipole are powered independently by pulse-width-modulated DC-DC converters that switch at a 20 kHz rate. The resulting 20 kHz spectral line and harmonics at 40 and 60 kHz are easily identified; however, they are most likely picked up in the BPM data acquisition stage and largely do not correspond to real beam motion. Although not apparent in Figure 1, if one were to show the horizontal spectrum measured at a high dispersion point in the lattice, a spectral line at 360 Hz and harmonics thereof would be seen coming directly from the rf system high-voltage power supplies that cause amplitude and phase ripple, inducing energy oscillations.

BEAM POSITION MEASUREMENT LIMITATIONS

Aliasing

The primary diagnostic used for orbit measurement and control is the monopulse receiver system that records turn-by-turn information and provides hardware averaging (up to 2048 samples to EPICS, 32 turns for feedback). This averaging is presently a "fundamental performance limitation." Data is sampled on every turn, alternating between horizontal and vertical, yielding an effective sample rate of 135.8 kHz in each plane, i.e., half the APS revolution frequency. Thus the available bandwidth on a single turn is the Nyquist frequency, about 68 kHz. The bandwidth available at the output to the 2048-sample "boxcar averager" is thus 68 kHz/2048 = 33.1 Hz; out of the 32-sample averager it is 2.1 kHz. The maximum realistic sampling rate supported by the EPICS controls system is 10 Hz, meaning that a significant aliasing problem exists. A similar issue is present for the feedback system that was originally intended to operate at a 4 kHz rate but is presently sampling at 1 kHz. Thus everything between the 500 Hz feedback sampling Nyquist frequency and the 2.1 kHz signal bandwidth gets aliased down below 500 Hz. An upgrade to the digital processing is in the works that will eliminate these aliasing issues for both the real-time and EPICS sampling rates.

Noise Floor

A careful measurement indicates that the monopulse receivers generate the equivalent of about 7 microns of rms noise on a single measurement, i.e., in a 68 kHz bandwidth. Since this noise scales as the square root of bandwidth, an equivalent statement is that the BPM noise floor is about 27 nanometers/$\sqrt{\text{Hz}}$, or 7.3×10^{-10} mm^2/Hz (see Fig. 1). It is desirable for the beam to be stable to within about 5% of its size, and since the vertical beam size with 1% coupling is about 20 microns rms, the tightest stability specification at the APS presently is 1 micron rms. Based on the noise-floor measurements, we are able to measure this level of motion in bandwidths of 1.4 kHz or less, another fundamental limit. Motion faster than a few tens of Hz for most x-ray experimenters appears as an effective small increase in beam size. They are thus more tolerant of beam motions in this band provided that the effective beam size is stable at lower frequencies.

Intensity and Bunch Pattern Dependence

While the noise floor dictates the high-frequency performance of the the monopulse receivers, more serious limitations occur at very low frequencies as a result of systematic intensity dependence. As the beam decays, position monitor readbacks have been found to vary by some tens of microns in the absence of real beam motion. The effect can be modeled as an intensity-dependent offset and is quantified at the beginning of each operational period. Corrections are calculated and applied as the beam current decays. Reproducibility of this effect is at best a few microns, giving another stability limitation.

Another serious effect is a sensitivity of the monopulse system to changes in fill pattern. The APS injectors provide single pulses at a 2 Hz rate, and the timing system is adjusted to direct the injected bunch at different buckets as a fill proceeds, resulting in small changes in fill pattern from one fill to the next as injection efficiency varies. An upgrade to the monopulse system's rf front end is underway to address this issue (3).

To overcome both the intensity and bunch pattern dependence, new electronics based on narrow-band switched heterodyne receivers (4) have been purchased and are undergoing early commissioning. These receivers are to be placed immediately upstream and downstream of all insertion device x-ray source points, with pickup electrodes fixed to the small-aperture vacuum chambers (most have an 8 mm vertical full aperture). These small-aperture chambers result in a position sensitivity that is greater than for the standard vacuum chambers by a factor of 3 vertically and 6 horizontally. If the electronics described in Reference 4 indeed achieve micron-scale resolution as stated with a 40 mm aperture chamber similar to the standard APS chamber, then in principle at least, long-term stability at the 100- to 300-nanometer rms scale should be attainable (relative to the vacuum chamber location, of course).

MECHANICAL STABILITY

In spite of the potential for beam position monitor electronics to detect submicron beam motions, the lack of an absolute mechanical datum and mechanical component stability dominate our ability to stabilize the orbit at very low frequencies. For example, our small-aperture insertion device vacuum chambers are rigidly supported 1.4 meters above the floor. The air and water temperature in the tunnel is generally stabilized to within ± 0.3 °C rms. Given that thermal expansion coefficients tend to be on the order of 1×10^{-5}, this translates into a vertical chamber motion of order $1.4 \times 0.3 \times 1 \times 10^{-5} = 4.2$ microns rms. This will almost always be the case; mechanical components typically cannot be stabilized to better than 5 to 10 microns rms owing to one's ability to regulate temperature. Worse yet, as the beam is injected and decays away, the thermal load on the water-cooling systems varies and vacuum chamber shape distortions inevitably result (5). Chamber motions at the APS are typically smaller than ± 2 microns, a consequence of careful masking of all vacuum chamber surfaces by discrete water-cooled radiation absorbers.

Another source of beam motion is associated with mechanical component vibrations (6). Typically, ground motion in the APS accelerator enclosure amounts to a few tens of nanometers rms from 2 to 50 Hz, while the magnet motion is in the range of 80 to 100 nanometers rms in the same frequency band. This is not yet a limiting source of beam motion since power supply noise dominates at these frequencies.

X-RAY BEAM POSITION MONITORS

One significant exception to the limitation on mechanical stability is the support structure for the x-ray BPM (x-BPM) photoemission blade monitors (7) where special attention has been given to thermal insulation and vibration damping. The x-ray BPMs are placed inside the accelerator enclosure, some 15 to 20 meters from the x-ray source points, with two units installed for each beamline. On bending magnet beamlines, a two-blade design is used with one upper and one lower blade for vertical position detection only. The insertion device beamlines employ four-blade designs, with the blades arranged in such a manner that the upstream set of blades does not cast shadows on the downstream unit's blades. The electronics presently in use consist of a set of micro-ammeters, the outputs of which are processed to generate delta over sum-type position signals. The bending magnet x-BPMs, owing to their superior mechanical support design and simple radiation geometry, provide the most believable data on long-term vertical beam stability. Shown in Figure 2 is a set of data, collected over a (good) four-day period, indicating vertical beam motion of ± 2.7 microns rms.

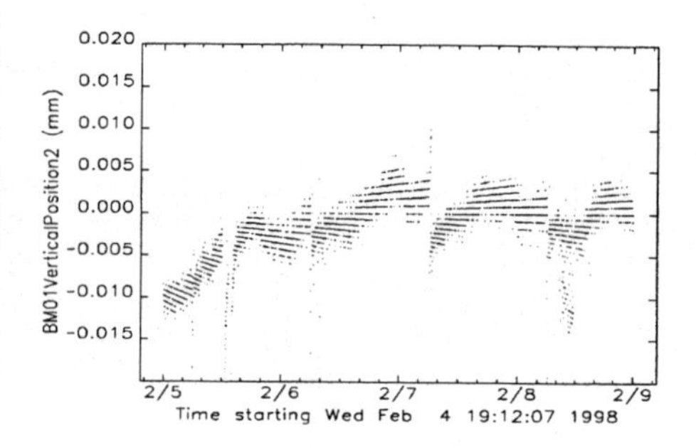

FIGURE 2. Bending magnet x-BPM data.

The insertion device x-ray BPMs, although similar mechanically and electrically to their bending magnet counterparts, have some critical differences that complicate their use. The most important difference is that the radiation travelling along the beamline is composed not only of the insertion device photons of interest, but also of stray radiation emanating from upstream and downstream dipole magnet fringe fields, from steering correctors, and from sextupoles and quadrupoles with offset trajectories. The effect of the stray radiation is to induce an offset in the measured ID beam position. Because the relative proportion of stray radiation to insertion device radiation changes when the insertion device gap is varied, an apparent "gap dependent" offset is observed. While significant progress has been made using look-up tables derived from translation stage scans to compensate for this effect (8), performance to date is at the 10- to 20-micron level. This limitation most likely arises from the fact that the stray radiation effects are sensitive to small trajectory changes taking place outside of the insertion device.

A research effort presently underway to address this issue involves the introduction of a chicane into the accelerator lattice to steer the stray radiation away from the x-BPMs (Figure 3). A horizontal parallel translation of the insertion device allows only ID photons and radiation from two nearby correctors to travel down the beamline, simplifying the radiation pattern considerably. Stray radiation is displaced by up to 2 cm horizontally at the x-BPM's location so that it should be easily masked.

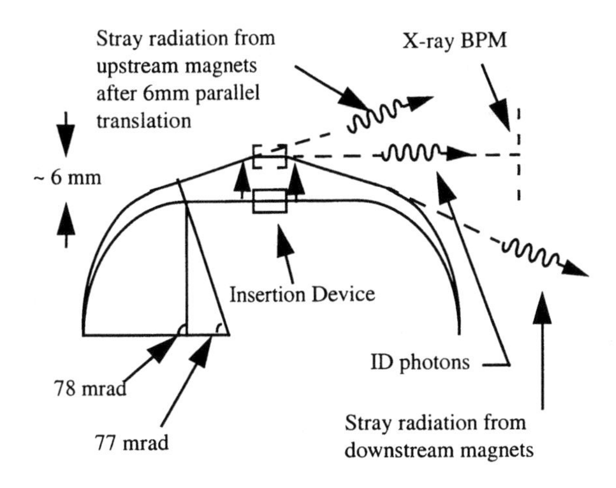

FIGURE 3. Concept to eliminate x-BPM stray radiation.

Orbit Correction Algorithms

The result shown in Figure 2 demonstrates that excellent orbit stabilization is achievable in spite of significant systematic effects such as BPM intensity dependence. The orbit correction algorithm at the APS makes use of as many of the 360 BPMs as possible together with only 80 of the available 317 correctors in each plane. A least-squares algorithm is used to minimize errors detected at the BPMs. This implementation is insensitive to short spatial wavelength (presumed unphysical) orbit motions, such as those caused by intensity dependence, that vary essentially randomly from unit to unit. Additionally, a nonlinear "de-spiking" algorithm is used to replace suddenly misbehaving BPM readbacks by the average of neighboring units, adding to the robustness of the correction.

The alternative to correcting long-wavelength orbit errors is to use so-called "local" control that exactly corrects the orbit at rf BPMs straddling each x-ray source point. The few-micron errors that accumulate over hours or days would result in microradian-scale pointing direction errors with local control. Multiplying by the 20-meter lever arm out to the second x-BPM of Figure 2, this amounts to 20 microns of beam motion instead of the 2.7 microns achieved. The use of an explicitly global correction algorithm that takes advantage of the statistics of having a large number of BPMs is what makes this possible (9). Local steering for individual beamlines is conducted infrequently and only upon beamline user request.

ACKNOWLEDGMENTS

The authors would like to thank many individuals in the APS Accelerator Systems Division for their support, including John Galayda, who has always been a strong supporter of beam instrumentation for many years. Michael Borland generated the majority of high-level software necessary to create graphics such as Figure 1, and both he and Louis Emery have spent countless hours in the control room implementing algorithms, the results of which are reported here.

REFERENCES

[1] Proakis, J., D. Manolakis, "Digital Signal Processing," New York, Macmillan Publishing Company (1992).
[2] Carwardine, J., "Real-Time Orbit Feedback at the APS," these proceedings.
[3] Lill, R., "Advanced Photon Source Monopulse RF Beam Position Monitor Front-End Upgrade," these proceedings.
[4] Unser, K., "New Generation Electronics Applied to Beam Position Monitors," *Proc. 1996 Beam Instrumentation Workshop*, AIP **390**, pp. 527-535 (1997).
[5] Safranek, J., O. Singh, L. Solomon, "Orbit Stability Improvement at the NSLS X-ray Ring," *Proc. 1995 PAC*, IEEE 95CH35843, p. 2711 (1996).
[6] Decker, G., Y. G. Kang, S. Kim, D. Mangra, R. Merl, D. McGhee, S. Sharma, "Reduction of Open-Loop Low Frequency Beam Motion at the APS," *Proc. 1995 PAC*, IEEE 95CH35843, p. 303 (1996).
[7] Shu, D., J. Barraza, H. Ding, T. M. Kuzay, and M. Ramanathan, "Progress of the APS High Heat Load X-ray Beam Position Monitor Development," *Proc. 1997 SRI Conference*, AIP CP417, pp. 173-177 (1997).
[8] Shu, D., T.M. Kuzay, "Smart x-ray beam position monitor system for the Advanced Photon Source," *Proc. 1997 SRI, Rev. Sci. Instr.*, **67** (9), p. 3367 (September 1996).
[9] Emery, L., M. Borland, "Advancements in Orbit Drift Correction in the Advanced Photon Source Storage Ring," *Proc. 1997 PAC*, to be published.

Alignment Measurement of an X-Band Accelerator Structure Using Beam Induced Dipole Signals

Chris Adolphsen

Stanford Linear Accelerator Center
Stanford, California

Abstract. Precise beam-to-structure alignment is critical for the acceleration of small emittance beams in linear accelerators. For the Next Linear Collider (NLC), a prototype accelerator structure has been developed in which the beam induced dipole mode signals can be readily accessed and processed to extract alignment information. In a test in the SLC linac, we used these signals to measure the internal alignment of the structure and to steer the beam in an attempt to minimize its wakefield. We used a second bunch to directly measure the wakefield and inferred from the results that a better than 40 micron alignment had been achieved. In this paper, we review these results and describe how we want to implement this alignment scheme for the approximately ten thousand structures in the NLC.

Paper not available for publication.

CP451, *Beam Instrumentation Workshop*
edited by R. O. Hettel, S. R. Smith, and J. D. Masek

From Narrow to Wide Band Normalization for Orbit and Trajectory Measurements

Daniel Cocq, Giuseppe Vismara

CERN, Geneva, Switzerland

Abstract. The beam orbit measurement (BOM) of the LEP collider makes use of a narrow-band normalizer (NBN), based on a phase processing system. This design has been working fully satisfactorily in LEP for almost 10 years. Development work for the LHC, requiring beam acquisitions every 25 ns, has led to a new idea of a so-called "wide-band normaliser" (WBN), which exploits most of the pickups differentiated pulse spectrum. In the WBN, the beam position information is converted into a time difference between the zero-crossing of two recombined and shaped electrode signals. A prototype based on the existing NBN unit has been developed and tested to prove the feasibility of this new idea. For this the bandpass filters and the 90° hybrids are replaced by lowpass filters and delay lines.

INTRODUCTION

Beam position measurements have greatly evolved in the last few years with demand for better performance in both dynamic range and bunch-to-bunch time resolution (< microsecond). Since the number of BPMs is increasing with the size of the machines and the spacing among bunches is getting smaller, acquisition systems become very complex and expensive, unless a normalization process is employed.

A normalization process will produce the analog ratio Δ/Σ required for the position measurement, which largely simplifies the digitization process. The beam orbit measurement (BOM) (1) of the LEP accelerator is a good example of a normalization process based on the "phase normalization" principle, associated with an 8-bit FADC.

The main advantages of this technique are:

- Normalization using only passive components
- Wide dynamic range (>50 dB) on the input signal without gain change
- 10 dB compression of the position dynamic range for 1/3 of the normalized aperture
- Higher sensitivity for centered beam, due to the arctangent transfer function
- Reduction by a factor two of the number of digitization channels
- Simplicity and reliability
- Low cost electronics

CP451, *Beam Instrumentation Workshop*
edited by R. O. Hettel, S. R. Smith, and J. D. Masek

The limiting point of this system concerns the minimum bunch spacing (>100 ns) due to the restricted bandwidth of the resonant filter, the 90° Hybrid and the settling time of the transmission path. In order to reduce the digitization time (25 ns for LHC), one has to increase the transfer function bandwidth from the filter up to the ADC. In the framework for the LHC development, a new normalization idea called a "wide band normalizer" (WBN) has emerged (2). The idea is based on an evolution of the "phase normalizer" principle where the "phase" is replaced by the "time" and the applied signal has a single oscillation period.

FROM NARROW-BAND (PHASE) TO WIDE-BAND (TIME) NORMALIZATION PRINCIPLE

The position measurement is determined by the normalized difference of signals on opposite electrodes

$$\text{Position} = K_x * (V_a - V_b) / (V_a + V_b) = K_x * \Delta / \Sigma = K_x * CR \qquad (1)$$

where $CR = \Delta/\Sigma$ is the normalization and K_x is the scale coefficient.

In phase normalization (3) (Figure 1), the two electrode signals are applied, after proper filtering, to the inputs of a 90° Hybrid. Each signal is shifted by 90° and then added to the opposite in-phase signal. The resulting phase difference (Φ) of the two outputs corresponds to their normalization

$$\Phi = \text{Arc-tangent}\,(V_a / V_b) \qquad (2)$$

This relation is valid for a continuous wave or a single oscillation period after proper settling time, which depends on the bandwidth of the processing chain. For a fixed frequency f_0, the 90° phase shift phase can be realized by delaying (in the time domain) the signals using $\lambda_0/4$ length coaxial cables (4). Since all components are wide band, there is no difference between a repetitive signal and a single oscillation (or a bipolar pulse). Figure 2 shows the block diagram of a delay normalizer.

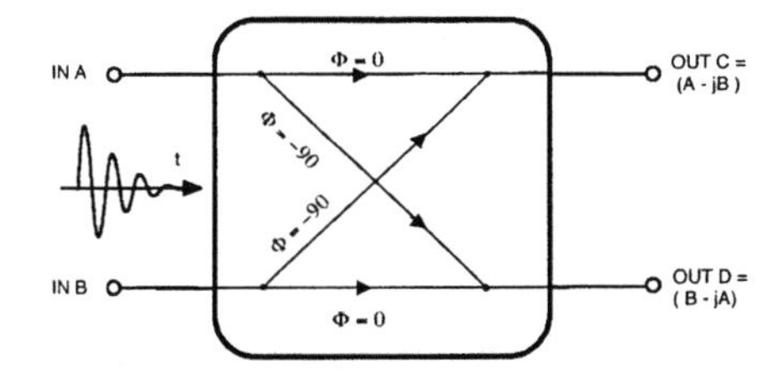

FIGURE 1. Phase normalizer schematic.

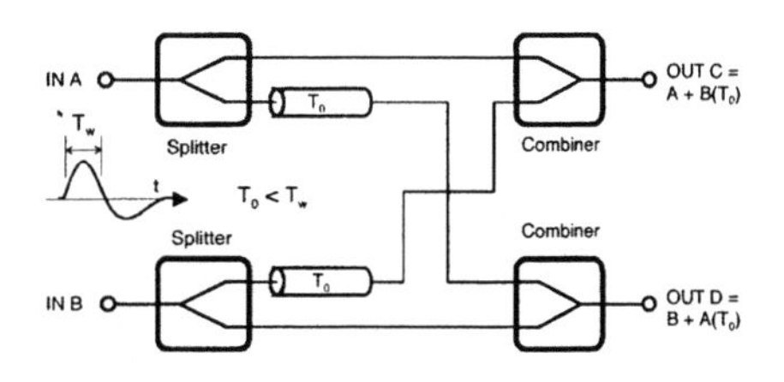

FIGURE 2. Time normalizer schematic

The induced signals on the button pickup are differentiated by the electrode capacitance and the resulting signal is a bipolar pulse (Figure 4). For simplification, let us assume a linear pickup. Under this assumption, the sum of the electrode signals is therefore constant with respect to the position. The signals from both electrodes of a plane are split in two and one branch is delayed by the time T_0. The delayed signal of one channel is then added to the direct signal of the other channel, and vice versa.

At output C, the time of passage through zero can vary up to a maximum of T_0 depending on the relationship of the amplitudes on the pickup. At output D, the same variation can be observed but in the opposite sense (Figure 3 right), hence the maximum relative variation between C and D is therefore $2T_0$. The transfer function, measured as the Δt between the two outputs for the negative zero crossings, is almost linear and the sensitivity is proportional to the delay. In the example described, T_0 is fixed at 1.5 ns but can be any value shorter than the signal width T_W. When approaching T_W, the transfer function gets more and more distorted and is no longer exploitable.

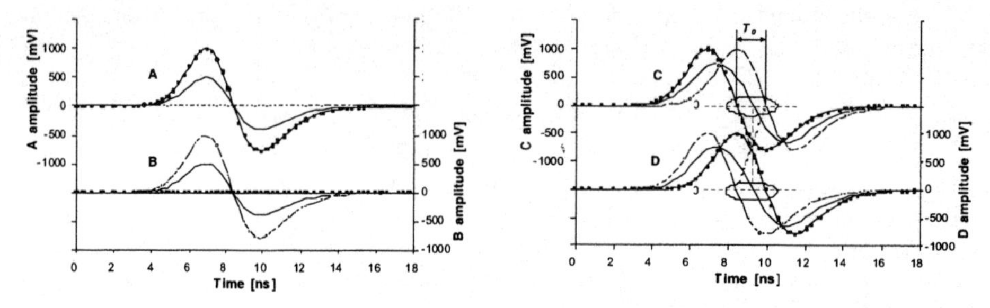

FIGURE 3. Signals at Normalizer inputs (left) and outputs (right) for three different beam positions.

Two fast comparators detect the zero passage and activate a flip-flop that generates a pulse T_{out} where the width will vary up to $2T_0$. By integrating this pulse, the time variation is transformed into amplitude and can be read by an A/D converter. As has been shown, this solution offers all the advantages of the phase normalization and extends the bandwidth of the transfer function by at least one order of magnitude.

SIGNAL ANALYSIS

The following considerations are general but examples are related to the LHC applications. The induced signal has a Gaussian distribution and the pulse width can change during the accelerating cycle. For the LHC, it varies from 1.0 to 0.5 ns and the Fourier transformation shows the content difference of the two spectrums (Figure 4A).

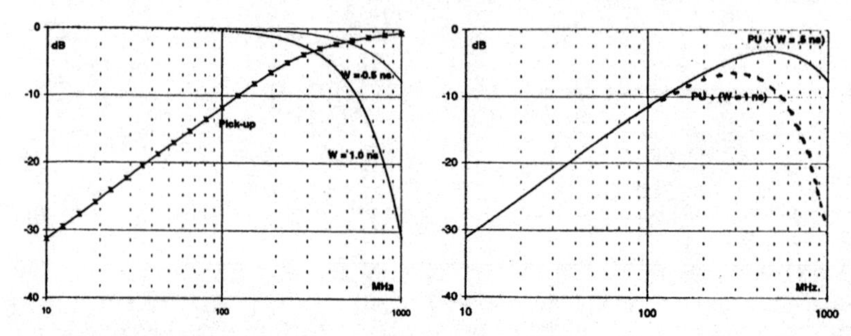

FIGURE 4. A: frequency spectrum and pickup response B: pickup output spectrum.

The capacitive monitors (buttons, couplers, etc.) have, by definition, a low-frequency cutoff which is determined by the geometrical dimensions and the resistive load.

The LHC buttons show a 397 MHz low-frequency cutoff. The estimated high-frequency cutoff is over 3 GHz (Figure 4A "Pick up"). When combining the beam spectrum content and the button response, the output signal has a response similar to a band-pass filter, as can be seen in Figure 4B. As a consequence, all induced signals will be differentiated and the average value is zero; all processing chains are submitted to these restrictions. The transmission path between the electrode and the signal treatment unit should be free of reflections for a correct measurement and to avoid false triggering; an acceptable figure but quite difficult to obtain would be ≈1% (40 dB). The interconnecting cables should have a length such that the reflection comes after the measurement instant and before the next pulse. For the LHC application, cable lengths between 5 to 10 ns will be used. The filter at the receiving end should be matched over the whole frequency spectrum.

LOW-PASS FILTER

In order to obtain an output signal whose shape is independent of the beam dimensions, the high frequencies should be limited to a value where the spectrum content does not vary. For easy signal treatment, the pulse response should correspond to a bipolar pulse of a single oscillation having nearly symmetric amplitudes.

Choice of the cutoff frequency

Several contradictory criteria should be taken into account:

1. *The output pulse shape should be independent of the pulse length.* For the LHC beam this corresponds to frequencies less than 210 MHz for a signal amplitude difference smaller than <10%.
2. *The minimum pulse length should be used to obtain a good S/N ratio and a linear transfer function*, which can easily be handled by the active electronics. Since a logic (ECL lite) time jitter is ≈3 ps rms. a minimum full-scale time excursion of 3 ns is required, when using a 10-bit digitizer. This requires a 1.5 ns delay line which implies that the output spectrum central frequency (f_0=4/T_0) should not exceed 166 MHz.
3. *The maximum pulse duration should be used* for unaffected double-pulse measurement. The BW and more strongly the filter response determine the residual amplitude. A Gaussian or linear phase filter is recommended. The LHC specifications ask for a residual amplitude of <0.2% for a bunch spacing of 25 ns.

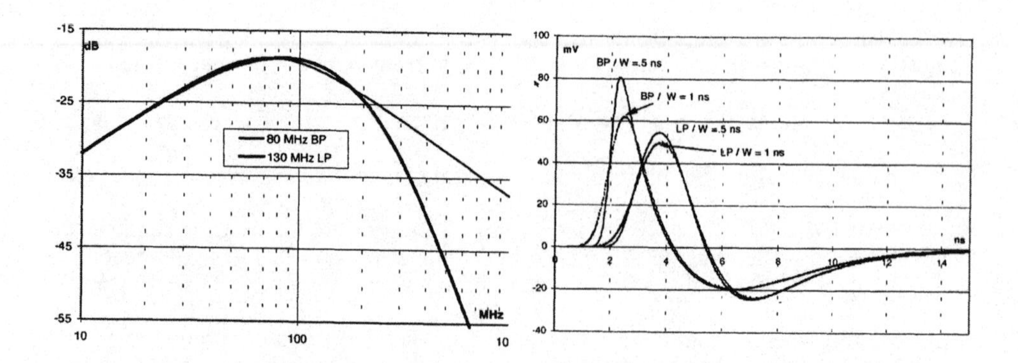

FIGURE 5. Filters frequency response.

FIGURE 6. Pulse response for a pilot bunch.

Filter

The simplest filter that can be implemented is a single resonator, which can be realized by simply adding a serial inductance on the button (Figure 5, 80 MHz BP). However such a scheme does not filter properly the high frequency spectrum contribution. The result is an asymmetric bipolar pulse which still depends on the pulse length (Figure 6, BP/W=0.5/1.0 ns). The best compromise is to add a Gaussian low-pass filter to cut high frequencies sharply. The global transfer function still has the same central frequency (≈80 MHz), while the LP filter is computed for 130 MHz (Figure 5, 130 MHz LP). The global bandwidth corresponds to greater than 120 MHz. The amplitude gets lower (≈20%) but the pulse symmetry and the variations versus beam dimensions are much better ($\Delta V_+ < 10\%$) (Figure 6, LP/W=0.5/1 ns).

NORMALIZATION

While keeping the same normalization principle, a simplified version making use of an inverting transformer (Figure 7) has been developed for the realization of the prototype. One of the two input signals is delayed by T_0, inverted and applied to the two negative inputs of a dual differential amplifier. The second one is directly applied to one of the positive inputs, then delayed by $2T_0$ and applied to the second positive input. The resulting output signal from the amplifier is the same as for the original idea but the signal has twice the amplitude.

Since the delay T_0 can be part of the interconnection cable and the dual amplifier is an integral part of the comparator circuit, the normalizer components are reduced to an inverter transformer and a coaxial cable ($2T_0$).

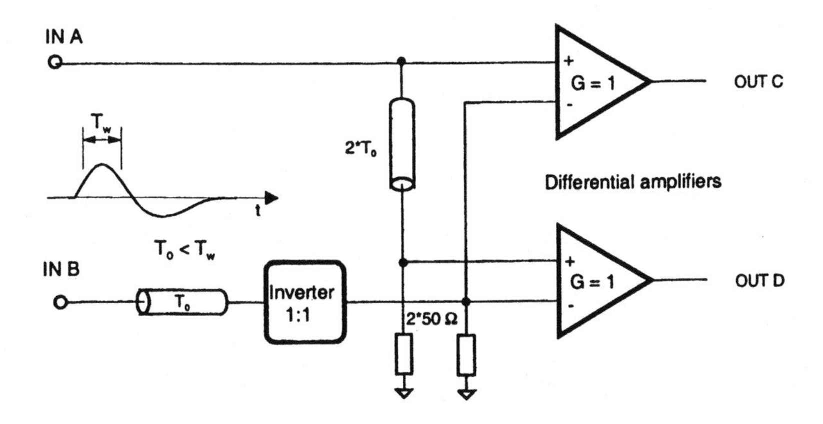

FIGURE 7. Inverter Normalizer.

ZERO-CROSSING DETECTOR

A dual ultra-fast comparator (AD96687) as in the BOM NB acquisition system provides the zero-crossing action. The input stage is a differential amplifier which realizes part of the normalization action (recombination of the delayed signals). The latch inputs are biased to generate a small hysteresis (10 mV) which will determine the threshold of the comparator. Since the comparator gain is over 60 dB, a positive-going signal above 10 mV will switch the output of the comparator to its high state.

The input offset is trimmed to bias the comparator just at the lower edge of the hysteresis cycle, in order to obtain a real zero-crossing and to be totally independent of the amplitude. The output pulse length depends on the selected LP filter frequency and on the input signal ratio. In the LHC development it varies from 3 to 5 ns.

The useful input dynamic range is >50 dB, and corresponds to the ratio between the maximum differential input (>3 V) and the threshold (10 mV). The comparator input noise is

$$\frac{3nV}{\sqrt{Hz}} \tag{3}$$

over a bandwidth >200 MHz, which results in a total input noise of 50 µV (rms.). The signal to noise ratio (S/N) corresponding to the pilot beam is >46 dB.

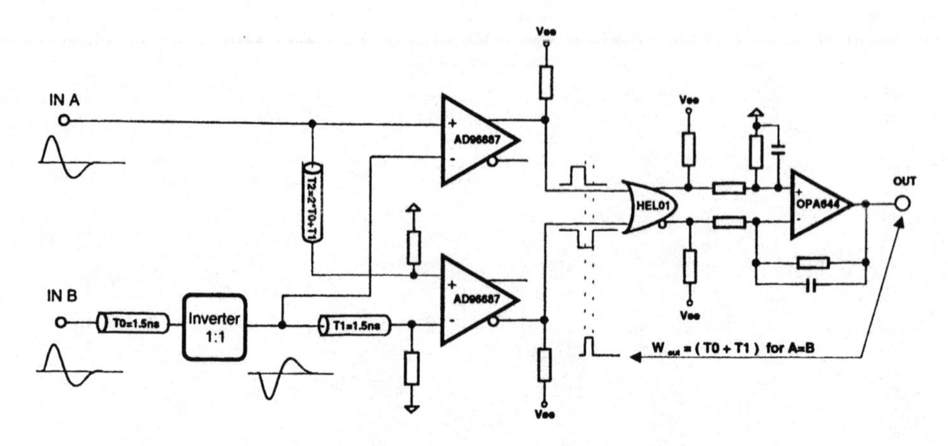

FIGURE 8. Wide band normalizer block diagram.

The propagation delay is the most critical point, since it depends on the signal overdrive and can vary up to 1 ns. It is very important that the tracking between channels is similar; previous experience has shown an average figure of 30 ps over a 40 dB dynamic range, which corresponds to an error of 1% of the full scale. Most of this error is present on the first 6 dB overdrive from threshold.

PRESENT STUDIES ON DIGITIZATION

The position information is the time difference between the trailing edges of the two-comparator outputs. A simple NOR function of the comparators Out1 and the complementary Out2 ($\overline{Out2}$) signals produces an output pulse width equal to the time difference. To avoid non-linearity problems due to overlapping of signal transitions the minimum pulse width should be ~1.5 ns. This is obtained by inserting a coaxial line between the two comparator inverting inputs and adding an extra 1.5 ns on the other branch. The NOR output pulse width will range from 1.5 ns up to 4.5 ns. Since it is difficult to digitize a time of a few ns with a resolution in the picosecond range, the NOR output pulse charge is digitized instead. To keep the circuit as simple as possible the two outputs are filtered by a 50 MHz Gaussian low-pass filter and applied to a video amplifier-buffer. The signal present at the ADC input will vary both in amplitude and time.

Two digitizing solutions have been explored:

a) Low resolution (8 bit) and high sampling rate (>500MS/s) FADC.

A minimum of 12 samples per measurement is taken in a window synchronous with the pickup signal. The data is averaged to improve the resolution by a factor 3 to 4 according to the sampling rate. It requires ultra fast FADC and RAM, and a fast processor to treat the data. The clock can be synchronous with the signal, which offers a stable data acquisition and linearity <±0.1% or asynchronous with data deviation <±0.1% and excellent linearity. This solution is quite complex and expensive.

b) High resolution (10 bit) and moderate sampling rate (>40 MS/s) ADC.

The ADC sampling time should be fixed relative to the PU signal and delayed by ~1 ns after the pulse peak. The amplitude varies as function of the delay, hence a calibration is required. The aperture uncertainty for a 10-bit ADC should be <10 ps in order to keep at least 9 ENB. Under these conditions the linearity error is ≈±1%. This figure can be can be improved by a factor x5 using a polynomial fit.

Both solutions require deeper investigation but the results are very promising.

MEASUREMENTS

Since the phase and time normalizer concepts are quite similar and they require the same basic components, it has been quite simple to adapt an existing BOM NB normalizer circuit to the WB version.

The LP 130 MHz circuit has replaced the BOM 70 MHz BP filter. On the normalizer unit, an inverter transformer and delay cable have replaced the 90° Hybrid. On the "zero crossing detector and the TAC" hybrid, a pin-to-pin compatible OR chip has replaced the Ex-OR. The video LP filter is set to 50 MHz instead of 20 MHz and the video amplifier replaced by a faster and pin to pin compatible OPA644.

Set-up

A bipolar signal filtered at 130 MHz is applied to a common attenuator (employed for stability measurements), split into two branches by a 0° Hybrid and applied to the unit under test via two high-resolution (0.25 dB) attenuators. The attenuators have been measured for a propagation delay difference <±10 ps and an attenuation accuracy better than two parts per thousand. From the attenuation ratio between the two branches (A/B) we obtain the normalized ratio CR=(1–A/B)/(1+A/B). The measurements are done with a digital oscilloscope sampling at 1GS/s in a 20 ns window.

Output Stability versus Input Level

The variation on the output signal for different common attenuator values is normalized to the full-scale aperture. The measurement has been done for 3 different conditions corresponding to a centered beam and two symmetric offset positions. From twice the threshold level up to 45 dB overdrive, the normalized output stays within ±1% of the full aperture.

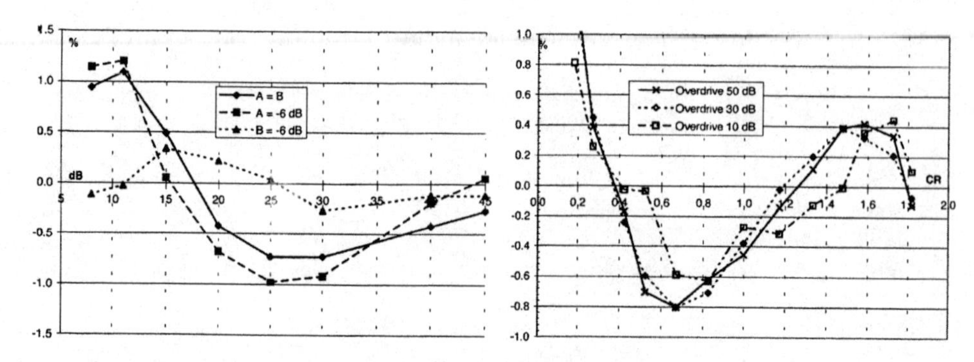

FIGURE 9. Normalized out variation vs. overdrive

FIGURE 10. Linearity errors at different overdrive.

Linearity

Figure 10 shows the linearity errors relative to a straight line versus the normalized aperture (over 80% range) for different overdrives above the comparator threshold.

The 10 dB overdrive corresponds roughly to the pilot beam; its slope is 3% less sensitive than the average. From 20 up to 50 dB overdrive the sensitivity remains constant within a fraction of one percent. The linearity error is <0.6% rms and is repetitive at different input levels. A second-order polynomial fit will reduce the error to the range of 0.1% rms.

Noise

Figure 11 shows the sigma of the output signal, normalized to the full-scale aperture versus the input signal overdrive. A pilot beam corresponds to a +10 dB overdrive.

The measurement has been done for a centered beam ($A=B$) and for an offset beam ($B=-25$ dB of A).

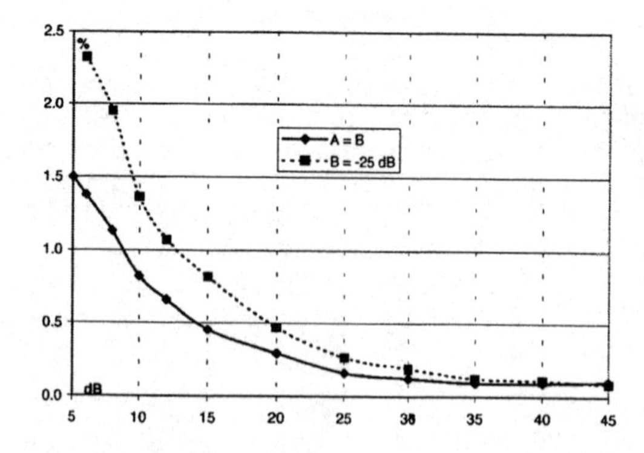

FIGURE 11. Normalized noise vs. input level above threshold.

EXTENDED APPLICATIONS

The principle can be extended to a wide variety of beam length and pickup types by properly filtering the induced signals and scaling the delay lines. A beam length ratio up to a factor of 10 seems to be easily attainable. The coupler electrodes associated with a proper LP filter can also be used, therefore taking advantage of their larger sensitivity.

For particle accelerators having a large signal length variation during the acceleration cycle or for accelerator bunches with different length (i.e., transfer lines PS to SPS), the LP filter and the cable delay should be designed to work properly for the longest bunch.

A general circuit that could be exploited by a large number of particle accelerators should be designed to the have removable coaxial delay and a large set of Gaussian LP filters. If wide band requirements are not needed, integration over long time periods can be simply obtained by increasing the output filter time constant.

CONCLUSIONS

The prototype performances fully satisfy the LHC specifications.

The building blocks of this system are identical to the LEP NB normalizer where 500 BPMs have been operating for almost 10 years with an excellent reliability and requiring a minimum number of human interventions. This expertise will facilitate the LHC development and avoid basic errors.

The WB normalizer solution offers all the advantages reported for the NB normalizer. Furthermore the realization of the LP filter requires less critically matched-paired units. The simplicity of the electronics, the reduced number of digitization channels associated with the reliability and the low cost per channel make the WBN solution ideal for many BPM acquisition systems.

ACKNOWLEDGMENTS

We would like to thank Dr. Hermann Schmickler for his invaluable contribution and support during the development. Special thanks go to Mr. Jean-Luc Pasquet for the realization of the prototype and part of the tests.

REFERENCES

[1] Borer, J., D. Cocq "The second generation of optimized Beam Orbit Measurement (BOM) system of LEP: Hardware and performance description," CERN-SL-95-60 BI, Paper presented at DIPAC'95—2nd European Workshop on Beam Diagnostics and Instrumentation for Particle Accelerators 28–31, May 1995.

[2] D. Cocq "The Wide Band Normalizer —a new circuit to measure transverse bunch position in accelerators and colliders," CERN-SL-97-74 BI, submitted to and accepted by NIM.

[3] R.E. Shafer, R.C. Webber et al. "An rf beam position measurement module for the Fermilab energy doubler" *IEEE Transactions on Nuclear Science*, **NS-28**, No.3, June 1981, p. 2323.

[4] Shafer, R.E. "A large dynamic range amplitude comparator," rf digital connection, September 1986, p. 74.

Improvement of the Noise Figure of the CEBAF Switched Electrode Electronics BPM System*

Tom Powers

Thomas Jefferson National Accelerator Facility
12000 Jefferson Ave., Newport News, VA 23606

Abstract. The Continuous Electron Beam Accelerator Facility (CEBAF) is a high-intensity continuous wave electron accelerator for nuclear physics located at Thomas Jefferson National Accelerator Facility. A beam energy of 4 GeV is achieved by recirculating the electron beam five times through two anti-parallel 400 MeV linacs. In the linacs, where there is recirculated beam, the BPM specifications must be met for beam intensities between 1 and 1000 μA. In the transport lines the BPM specifications must be met for beam intensities between 100 nA and 200 μA. To avoid a complete redesign of the existing electronics, we investigated ways to improve the noise figure of the linac BPM switched-electrode electronics (SEE) so that they could be used in the transport lines. We found that the out-of-band noise contributed significantly to the overall system noise figure. This paper will focus on the source of the excessive out-of-band noise and how it was reduced. The development, commissioning and operational results of this low noise variant of the linac style SEE BPMs as well as techniques for determining the noise figure of the rf chain will also be presented.

INTRODUCTION

The switched-electrode electronics beam position monitor (SEE BPM) system, which was developed and installed in 1993–1994, was designed to operate in the CEBAF accelerator linacs where the designed beam intensity range is 1 μA to 1000 μA (1). In most of the remainder of the machine, which is collectively known as the transport lines, the nominal beam intensity range is between 100 nA and 200 μA. The exception is the Hall B transport line where beam intensities as low as 200 pA are used. Because these beam intensities are below the limits of the existing BPM systems used at CEBAF, two new systems have been developed. The system which is used at the very low-beam intensities, known as the 1 nA beam position monitoring system, is a cavity-based system which makes use of lock-in amplifiers to provide synchronous detection of the position sensitive signals. This system is fully described in (2). The second system,

* Supported by DOE Contract #DE-AC05-84ER40150

CP451, *Beam Instrumentation Workshop*
edited by R. O. Hettel, S. R. Smith, and J. D. Masek

which will be described in this paper, is known as the transport line switched-electrode electronics system (TL-SEE). It is a low-noise variant of the linac style system which makes use of GaAs switches and variable gain amplifiers to cover the required dynamic range while maintaining a precise gain balance between the plus and minus signals of the X and Y electrode planes.

SYSTEM PERFORMANCE

The system performance for the linac style and transport line style SEE BPM system is summarized in Table 1. The dynamic range is defined as the range of currents for which the system operates within all other specified limits. The lower end of the dynamic range is limited by thermal noise and the detectivity of the if down-conversion circuit. The high end is limited by signal compression in the electronics.

TABLE 1. Summary of Linac Style and Transport Line Style Performance Specifications

Performance Specifications	Linac Style	Transport Line Style
Dynamic range	700 nA – 2000 μA	70 nA – 200 μA
Nominal measuring rate out of control system	1 meas./s	1 meas./s
Beam position range	$\lvert x \rvert$, $\lvert y \rvert \leq 5$ mm	$\lvert x \rvert$, $\lvert y \rvert \leq 5$ mm
Resolution (rms fluct. at nominal meas. rate)	≤ 0.1 mm	≤ 0.1 mm
Current dependence	≤ 0.1 mm	≤ 0.1 mm
Multipass capability	"snake" pulse	Not required
if bandwidth	1 MHz	50 kHz
Analog bandwidth	70 kHz	7 kHz
Maximum measurement rate	114 kS/s	7.1 kS/s

SYSTEM DESCRIPTION

Both types of SEE systems use VME based if, acquisition, and control modules. The VME crates are located in the service buildings approximately 10 meters above the beam line. Each channel consists of a BPM detector, an rf module located in the tunnel and an if module located in the VME crate. The beamline sensor has four wire-type stripline antennas previously described (3). Each pair of opposing electrodes is time-domain multiplexed into one-signal conditioning chain at the front end of the rf module (X+X–X+X–... and Y+Y–Y+Y–...). This is done to insure a balanced gain independent of gain variations in the remainder of the system. These signals are amplified and down-converted to 45 MHz before transmission to the if module. In the if module, the signal is amplified using a three-stage amplifier which has digitally controlled gain with a dynamic range of 84 dB. The signal is filtered, then down-converted to base band, filtered again and further multiplexed before it is transmitted to a commercial data acquisi-tion module as X+Y+X–Y–X+Y+... data. Each crate has a timing module which is used to synchronize the system to the accelerator and to generate specific timing signals required by the if and rf modules. The VME crate also contains three commercial data acquisition modules and a single board microcomputer which are used to acquire and process the position signals prior to transmitting them to the machine control system.

The system has two fundamental software interfaces. The first is a once-per-second average position readout which provides the average position and its rms noise to the machine operators. The second is a beam oscilloscope program which provides time domain and frequency domain measurements of the beam position and intensity modulation at any location of the machine. Additionally, as shown in Figure 1, the program provides a measure of energy "jitter" based on beam position readings in any of five different high dispersion regions of the accelerator. The measurement shown in Figure 1 was made using five transport line BPMs whose data was acquired synchronously and processed off line using machine optics information from the CEBAF model server (4). This energy modulation was due to a problem with the phase control loop of one of the 330 superconducting cavities which are the accelerating elements at CEBAF. The problem was quickly rectified once it was identified by the BPM system.

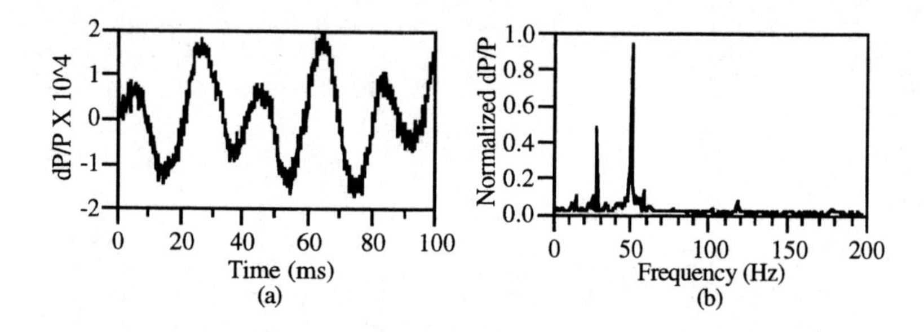

FIGURE 1. (a) Time domain and (b) frequency domain plots of the energy "jitter" in the Hall C transport line. The 50 Hz and 29 Hz frequency content was due to an SRF cavity with an open phase control loop.

BACKGROUND ON NOISE FIGURE CALCULATIONS

The noise factor of a device is the signal-to-noise ratio at the output of the device divided by the signal-to-noise ratio at the input. The noise figure is the noise factor expressed in units of dB. The noise figure is the specification that is normally provided in the component data sheet. One of the standard ways to define the noise factor (F) and noise figure (NF) is:

$$F = \frac{S_i/N_i}{S_o/N_o} = \frac{N_a + kT_oBG_a}{kT_oBG_a} \tag{1}$$

$$NF = 10 \times \log(F) \tag{2}$$

Here N_a is the noise power introduced by the device, G_a is the power gain of the device, T_0 is 290 K, B is the bandwidth of the system, and k is Boltzmann's constant. Typical noise figures for inexpensive room temperature amplifiers are 1.5 to 10 dB. Amplifiers with noise figures as low as 0.9 dB are commercially available. The noise factor of a passive two port device ($G_a < 1$), which does not contain noise sources other than thermal noise, is equal to the loss of the device, $L = 1/G_a$.

When two or more devices are cascaded, including passive devices, the noise factor for the network is given by the following equation:

$$F = F_1 + \frac{F_2 - 1}{G_1} + \frac{F_3 - 1}{G_1 G_2} + \cdots \quad (3)$$

Here F_1, F_2 and F_3 are the noise factors for the first, second and third stages respectively and G_1, G_2, and G_3 are the available gains of the first second and third stages respectively. Thus, once the first gain stage is accomplished the noise figure contributions to the overall noise figure by subsequent devices is reduced. This is the reason that the most important noise figure is that of the first stage amplifier and that the passive devices before the first amplifier stage should avoided if possible. However, if the front end amplifier gain is not high enough, additional losses may be such that subsequent devices may contribute significantly to the overall noise figure.

There are four contributions to the noise figure when a mixer is used as a down converter (5). The first is equal to the conversion loss of the mixer. The second is the effect of the local oscillator phase noise. The third is the noise generated by the electronic devices, (i.e., FETs for square law mixers and diodes for typical switching-mode mixers). Typically the "electronic" noise adds one or two dB to the noise figure of the circuit. The fourth contribution is the noise at the image frequency of the rf signal. Consider the signal shown in Figure 2. If there is equal noise power at the image frequency, the output noise power at the if frequency is equal to the sum of the noise power at the signal frequency plus that at the image frequency. This image frequency noise adds 3 dB to the system noise figure independent of the gain of the previous stages. In this case the noise power is added, because the two noise signals are uncorrelated.

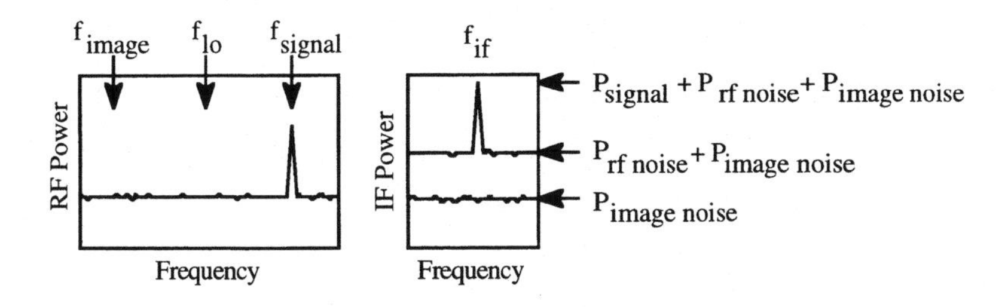

FIGURE 2. Contribution of image noise to if noise at the output of a mixer.

Measurement Techniques

There are several ways to measure the noise figure of a device or system. The first is to plot the output noise power with a 50 ohm source impedance at different temperatures (typically 300 K and 77 K)(6). The P-intercept of a straight-line plot of measured output power as a function of temperature is the value of the noise power N_a of the device under test. The second approach is to use a calibrated noise source, a low-noise amplifier (LNA) and noise figure meter, or spectrum analyzer with a noise figure personality module. In this technique, the spectral characteristics of the noise source and LNA are measured, the device under test (DUT) is inserted between them, and the instrument calculates the gain and noise figure of the DUT. The third method requires an LNA, a spectrum analyzer and a 50 ohm termination. In the latter two techniques the LNA is used to increase the sensitivity of the measuring instrument in order to increase the measured noise voltage above the noise floor of the instrument. The last technique has the added advantage that "man-made" signal sources become apparent when the measurement is being made and one may make compensations, while the first two systems provide only a noise figure at specific frequencies.

The noise factor of an amplifier can also be written as:

$$F = \frac{P}{kTBG} \tag{4}$$

Here P is the output noise power with the input terminated, T is the absolute temperature of the termination, k is Boltzmann's constant, and G is the amplifier gain. For a 50 ohm system with a preamplifier inserted between the DUT and the spectrum analyzer this equation may be written as:

$$NF(\text{dB}) = 10 \log \frac{V^2}{R} - 10 \log B - 10 \log (G_a + G_p) - 10 \log kT \tag{5}$$

Here G_P is the gain of the preamplifier, V is the rms voltage reading, and B is the measurement bandwidth. The reading is done using the voltage scale in order to increase the resolution. This equation can be further reduced to:

$$NF(\text{dB}) = 20 \log V - 10 \log BW - 10 \log (G_a + G_p) - 187.27 \text{ dB} \tag{6}$$

Here BW is the 3 dB bandwidth of the spectrum analyzer, and 187.27 is a result of the sum of four numbers: –10 log 50 ohms, –10 log kT, –10 log 1.2 (an approximate correction factor to go from noise power bandwidth to Gaussian 3 dB bandwidth), and +1.05 dB (detector correction factor) (7). In order to make accurate voltage measurements, strong video averaging is applied with typical averaging factors of 1000. Additionally, one must be careful to keep the peak signals, without averaging, below the saturation levels of the instrument by adjusting the vertical scale with averaging off and turning averaging on before making a measurement.

IMPROVEMENT TO THE NOISE CHARACTERISTICS

Four fundamental changes were made to the system to improve the low current operations. Three changes were made to the rf module which improved the noise characteristics at lower beam intensity by 8 dB or a factor of 2.5. Improvements were made to the if module which increased the sensitivity by 13 dB, or an additional factor of 4.4 lower in beam intensity. These improvements should have extended the dynamic range at low currents by a factor of 11, from 700 nA to 64 nA. As will be shown, the actual increase in sensitivity was only a factor of 10, or down to 70 nA.

Intermediate Frequency Module Improvements

The function of the if module is to amplify the if signal provided by the rf module and down-convert it to a baseband signal that can be digitized with a commercial ADC module. A simple schematic diagram of the linac style if module is shown in Figure 3. A detailed description of the function of the linac style if module may be found in (1). Three changes were made to improve the noise characteristics of the system: the 1 MHz BW at 45 MHz LC filter was changed to a 50 kHz BW at 45 MHz crystal filter; the 860 kHz low-pass filter was changed to a 100 kHz low-pass filter; and the integration time of the gated integrated filter was changed from 3.2 μs to 30 μs. This last change was done because the switching clock frequency was reduced from 248 kHz to 14.2 kHz so that the plus-to-minus modulated if signal has settled sufficiently before the integration process is initiated. The settling time for the 1 MHz BW filter is 500 ns while the settling time of the 50 kHz BW filter is 25 μs. The changes which were made to the baseband section of the if module improved the noise characteristics of the output signal, while the low-current limitation of the system remained the ability of the video detector to lock on the if frequency in the presence of noise. Thus the only improvement which impacted the low-current limit was the reduction of the bandwidth of the 45 MHz band pass filter.

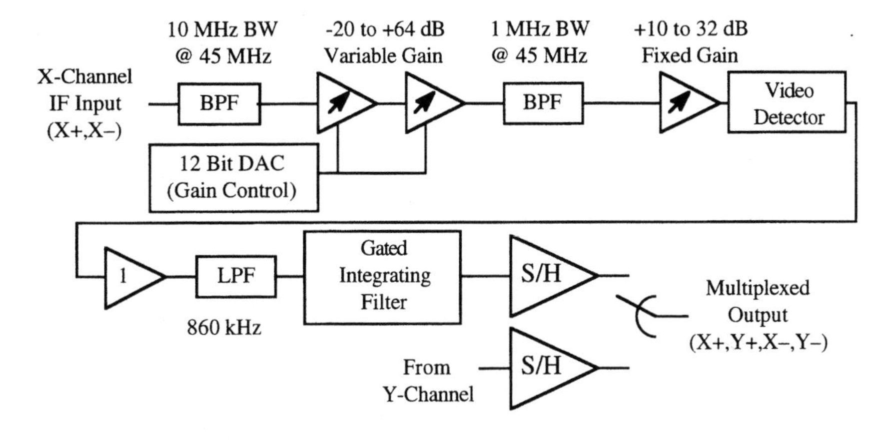

FIGURE 3. Simplified schematic diagram of the linac style if module.

Radio-Frequency Module Improvements

A detailed functional description of the rf module can be found in (1). The rf chains for the two systems are shown in Figure 4. Three changes were made to the linac-style system in order to improve the low-current performance. The first change is that the 1497 MHz band pass filter was relocated to just before the mixer in order to improve the overall noise figure by 6.8 dB. For the second change, the two-stage rf amplifier was changed from two MAR-6 amplifiers to one MAR-6 amplifier and one ERA-3 amplifier. This increased the rf gain from 23 dB to 32 dB without degrading the noise figure. Thirdly, the gain was adjusted on the output stage if amplifier to provide an overall gain of 38 dB in contrast with the 25 dB setting used in the linac style system. This, along with increased overall gain prior to this section, had the additional benefit of decreasing the overall noise figure by another 1.5 dB. The range of input power levels for the linac style rf module is –77 dBm to –4 dBm, which equates to an intensity range of 700 nA to 2 mA. The range of inputs for the transferline-style rf module is –97 dBm to –27 dBm, which equates to an intensity range of 70 nA to 200 μA.

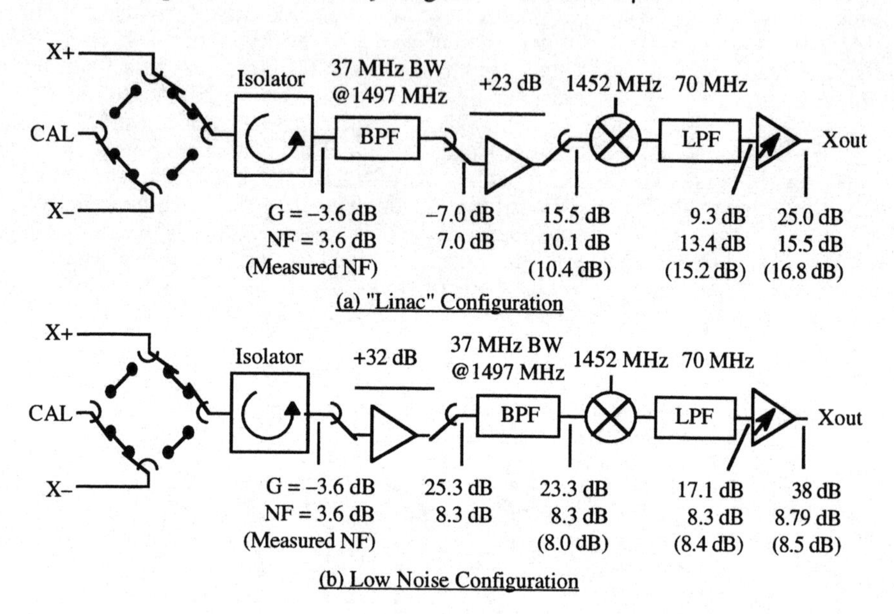

FIGURE 4. Switched-electrode electronics rf chain schematic diagrams showing the linac configuration and the low-noise, transport line, configuration. The noise figures within parentheses are measured values while the other noise figures, as well as the gains (G), were calculated based on typical measured values.

Relocating the filter had two expected and one unexpected effects. The first effect is the reduction of the overall noise figure by 2 dB (the in-band attenuation of the filter) because the filter was moved to a point after the initial gain stage. Thus the contribution to the noise figure due to the loss was reduced by $(F_{filter}-1)/G_i$ where G_i is the combined gain of the switches, the circulator and the two-stage amplifier. The second effect was the reduction of the overall noise figure by 3 dB because the filter rejected the noise at the image frequency of 1542 MHz. The third effect was one of the two more subtle improvements associated with the changes.

When performing the initial system noise figure measurements, the circuit, at the mixer output, had 2 dB of unaccounted-for noise figure. Figure 5 shows the normalized noise power as a function of frequency at the input of the mixer for both circuits. The noise power at the image frequency for the circuit with the filter prior to the amplifier is 1.8 times the power level at the signal frequency. By adding this to the in-band noise you would get a noise figure increase of 5 dB as opposed to the 3 dB that is expected from an SSB down conversion with uniform noise in both the image and signal frequency bands. Further investigation of the characteristics of the band pass filter indicated that the filter does reject signals at the image frequency by more than 30 dB. However, the output port of the filter is almost totally reflective at the image frequency. A reflective source will reflect any noise power coming out of the input port of the amplifier circuit back into the amplifier for amplification. The second subtlety is the noise figure of the AD603 variable gain amplifier, which is the last stage of the circuits shown in Figure 4. The first page of the specification sheet indicates that the input noise spectral density is 1.3 nV/Hz$^{1/2}$, which corresponds to a noise figure of 9.5 dB for the amplifier bandwidth of 50 MHz and a gain of 32 dB. However, the variable gain function of this amplifier was implemented by an R-2R ladder network followed by a fixed-gain amplifier. This ladder network is, in effect, a voltage-controlled attenuator. Thus, as the overall gain is reduced by 10 dB, the noise figure of the AD603 increases by 10 dB. By increasing the gain of the if stage from 18 dB to 23 dB, the noise figure of the if stage was reduced from 21 dB to 16 dB and the overall noise figure of the rf module was reduced by 0.85 dB.

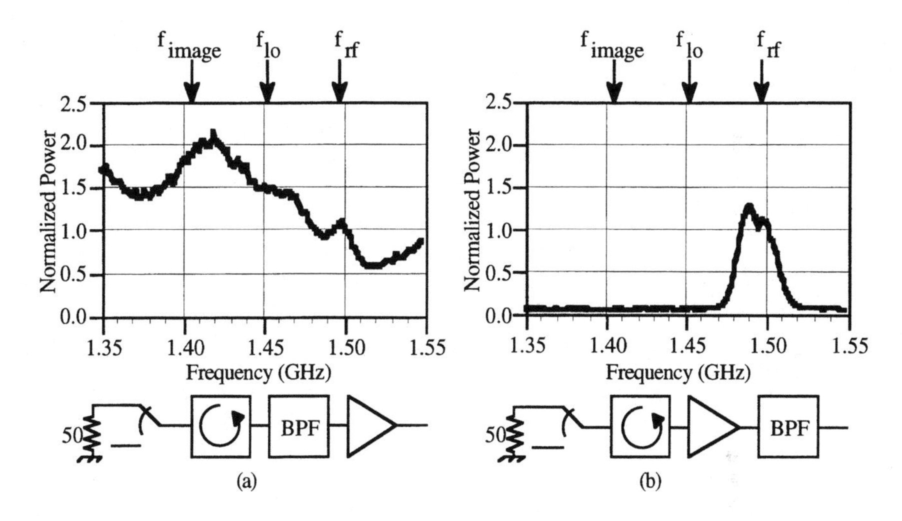

FIGURE 5. Normalized noise power as a function of frequency for two different circuit topologies. The first is the topology used in the linac style system while the second is that used in the transport line system.

RESULTS

During the production testing the low-current capability of all of the rf modules was determined. For practical reasons, this low-current limit is defined by the rms value of the position noise. As is shown in Figure 6, the low-current beam position is valid (within 200 µm of the maximum value) when the rms value of the instantaneous beam position readings is less than 1 mm. Further analysis of the production test data shows that position readings at the minimum current value on all of the channels used to generate Figure 6(b) were within +/– 120 µm of the maximum value. An additional 12% reduction in beam current was required before any of the modules varied by more than 200 µm from the maximum position. Figure 7 shows histograms of the results of this analysis. Only two of the transport line modules that were analyzed had a measured low current limit above 70 nA while most of the modules madc it down to 60 nA. This data is shown in Figure 7(b). An analysis of the production data from the linac style modules indicates that the low-current limit for this style is 700 nA with most modules making it down to 550 nA. This data is shown in Figure 7(a).

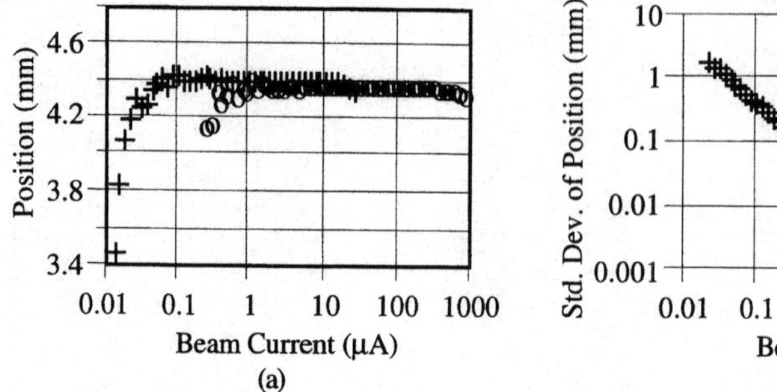

FIGURE 6. Typical plots of (a) beam position as a function of beam intensity and (b) standard deviation of the 140 μs beam position readings as a function of beam intensity. Both sets of data were taken using transport style electronics in the laboratory. The data indicated with a "+" is for the high gain setting and an "o" is for the low gain setting.

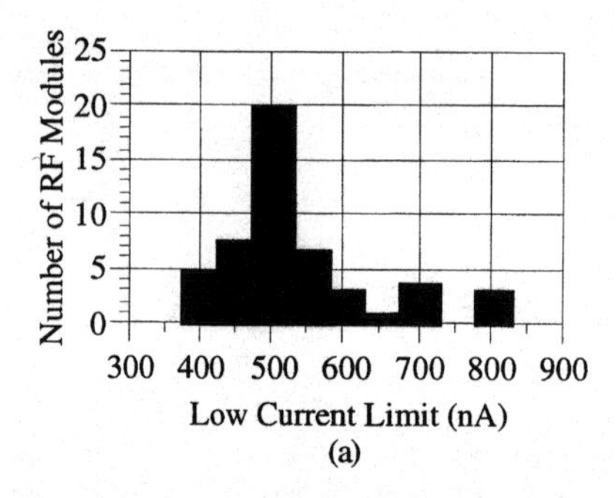

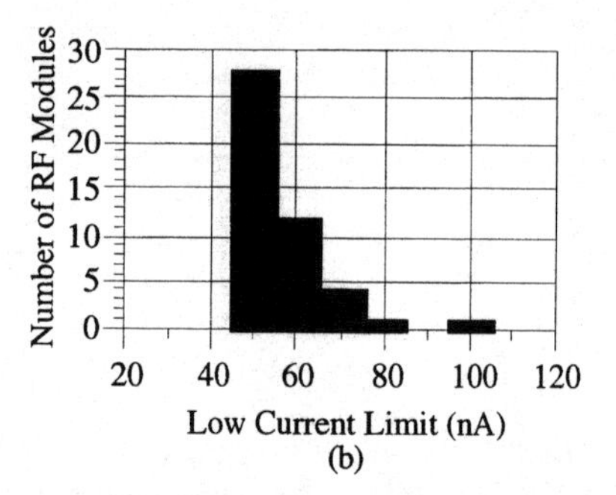

FIGURE 7. Low current limitations of (a) 48 linac-style rf modules and (b) 46 transport line rf modules as measured during production testing.

CONCLUSION

A description of how the noise figure is measured as well as the basic formula for calculation of the noise figure of a multi-stage system have been presented. Application of these principles to improve the low current operation of the SEE BPM system was also described. By changing the bandwidth of the if section, the judicious swapping of circuit elements in the rf section, and the adjustment of the gains in the rf module, the low-current capabilities have been improved from 700 nA to below 70 nA. This was accomplished without doing a major redesign of the system packaging, digital interface, or software.

REFERENCES

[1] Powers, T., et al., "Design, Commissioning and Operational Results of Wide Dynamic Range BPM Switched Electronics," *Proceedings of the Seventh Beam Instrumentation Workshop*, p 257.

[2] Piller, C., et. al, "1 nA Beam Position Monitor," presented at this conference.

[3] Barry, W., *Nucl Instrum Meth A*, **301**, pp. 407–416, 1991.

[4] Van Zeijts, J., et al., "Integrated On-Line Accelerator Modeling at CEBAF", *Proceedings of the 1995 Particle Accelerator Conference*, p. 2181.

[5] Hayward, W., *Introduction to Radio Frequency Design*, American Radio Relay League, Inc. 1994, pp. 202–246.

[6] Gardiol, F. E., *Introduction to Microwaves*, Artech House, Inc. 1984, p. 362

[7] "Spectrum Analysis, . . . Noise Analysis" Hewlett Packard Application note 150–4, April 1974.

Studies of Beam Position Monitor Stability[1]

P. Tenenbaum

Stanford Linear Accelerator Center, Stanford University, Stanford, California 94309

Abstract. We present the results from two studies of the time stability between the mechanical center of a beam position monitor (BPM) and its electrical/electronic center. In the first study, a group of 93 BPM processors was calibrated via a test pulse generator once per hour, in order to measure the contribution of the readout electronics to offset drifts. In the second study, a triplet of stripline BPMs in the Final Focus Test Beam, separated only by drift spaces, was read out every six minutes during one week of beam operation. In both cases offset stability was observed to be on the order of microns over time spans ranging from hours to days, although during the beam study much worse performance was also observed. Implications for the BPM system of future linear collider systems are discussed.

INTRODUCTION

One of the most ubiquitous and critical tuning elements of future linear colliders is the beam position monitor system. The proposed NLC design, for example, calls for a BPM to be installed in the bore of each quadrupole, with a total of 3000 such units. Each "Q" BPM is expected to have a single-pulse resolution of 1 micron, an *ab initio* installation accuracy (magnetic to electrical center) of 200 microns, and a 24-hour stability of the electrical center of 1 micron [1,2].

Previous experiments have demonstrated the required BPM resolution for bunch charges comparable to the NLC's [3], and other experiences indicate that the installation accuracy required can also be achieved [4]. We report on two experiments which seek to quantify the time stability of state-of-the-art SLAC stripline BPMs, in order to assess the achievability of the NLC specification for electrical center drift.

CALIBRATION PULSER EXPERIMENT

High-resolution single-pulse BPM processing electronics are used at SLAC [5]. The signals from a pair of striplines (T/B or L/R) are amplified in a two-channel

1. Work supported by Department of Energy contract DE-AC03-76SF00515.

CP451, *Beam Instrumentation Workshop*
edited by R. O. Hettel, S. R. Smith, and J. D. Masek
1998 The American Institute of Physics 1-56396-794-4/98/$15.00

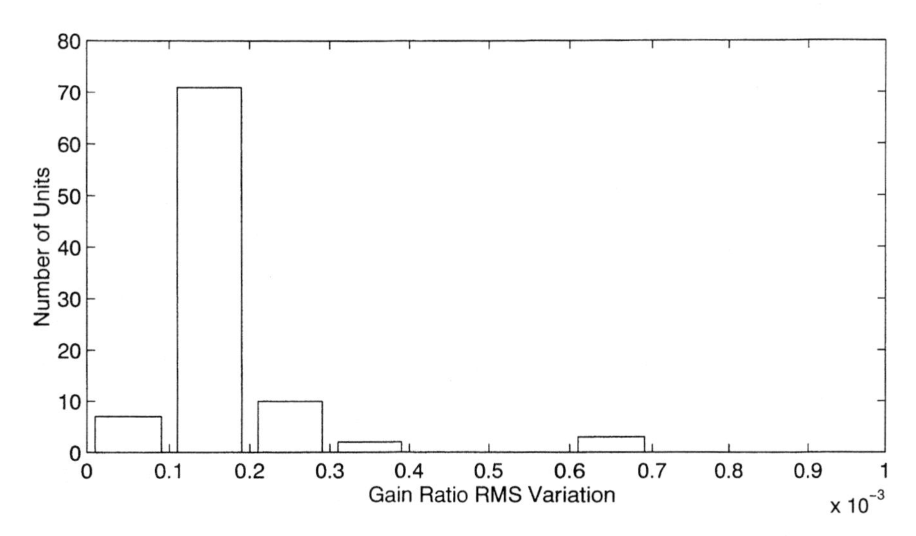

FIGURE 1. The rms gain ratio variations of 93 BPM processors when calibrated 30 times in a five-minute period.

amplifier, then digitized by a two-channel 16-bit track-and-hold (the "NiTnH"). The resulting digital words are then converted to position via the formula:

$$x, y = \frac{a}{2} \frac{(V_{R1} - P_1) - M(V_{R2} - P_2)}{(V_{R1} - P_1) + M(V_{R2} - P_2)}, \tag{1}$$

where a is the BPM radius, V_{R1} and V_{R2} are the two raw digital signals, P_1 and P_2 are the pedestals of channels 1 and 2, respectively, and M is the gain ratio between the two channels. The values of P_1, P_2, and M are determined via calibration. P_1 and P_2 are the digital words generated when the NiTnH is triggered in the absence of signal and M is measured by generating a test pulse and injecting it simultaneously into both channels of the head amplifier. By ramping the test pulse amplitude, the system gain as a function of input signal is measured for both channels and fit to a straight line for each channel. M is the ratio of the slopes.

Let us assume that after the calibration described above is performed, the pedestals and gain ratio change to $P_1 + S_1$, $P_2 + S_2$, and $M(1 + \epsilon)$, respectively, where $S_{1,2} \ll V_{R1,2} - P_{1,2}$ and $\epsilon \ll 1$. If a beam position is decoded from the raw signals using the old calibration, the error in the position determination to lowest order is:

$$dx, dy \approx \frac{a}{4} \left[\frac{MS_2 - S_1}{M(V_{R2} - P_2)} - \epsilon \right]. \tag{2}$$

The calibration described above was executed once per hour on a total of 93 BPM processors: 41 in the Next Linear Collider Test Accelerator (NLCTA) equipment area, and 52 in the Final Focus Test Beam (FFTB) instrumentation shacks.

The experiment lasted for one week, allowing long-term drifts to be assessed on a meaningfully large population of processors.

Resolution of the Method

In order to assess the resolution of the method, the calibration procedure was executed 30 times in rapid succession (less than five minutes was required). The rms drift of the gain ratio and the pedestals over five minutes gives an estimate of the resolution of the system.

Figure 1 shows the distribution of rms gain ratio variations measured in this procedure. Note that nearly all processor gain ratios were stable to within 2×10^{-4} in this procedure, which is taken to be the resolution of the system. Similarly, the pedestals were found to be stable to within one count, which is taken to be the resolution of the pedestal variations.

Results of the Calibration Pulser Experiment

Figure 2 shows the distribution in rms drifts of pedestals over one week. Typical units were stable at the level of two counts. Considering Equation 2, and assuming $M(V_{R2} - P_2) \approx 16,000$, a variation in pedestals of two counts would result in a shift in the measured BPM center of roughly 0.25 microns for a BPM with 6 mm radius.

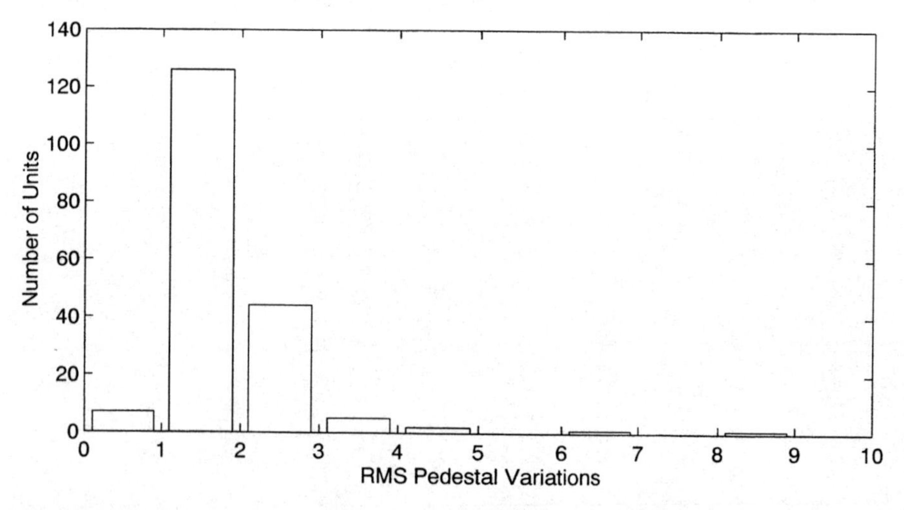

FIGURE 2. The rms pedestal variations of 93 BPM processors when calibrated once per hour over one week.

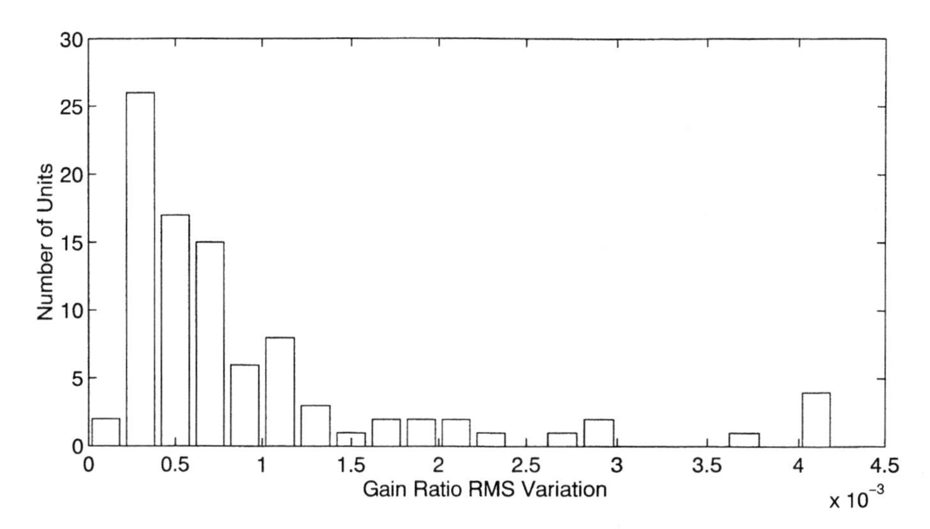

FIGURE 3. The rms gain ratio variations of 93 BPM processors when calibrated once per hour over one week.

Figure 3 shows the distribution in rms drifts of gain ratios over one week. Only 88 processors are represented; 5 of the 93 units displayed discontinuous "jumps" in gain ratio or other pathologies which indicated probable electronic failure of the processor, and were eliminated from the study. Most units were stable to within 1.2×10^{-3} of their mean values, equivalent to an rms offset drift of 1.8 microns for a BPM with 6 mm radius. Furthermore, the drifts were found to be highly correlated to the temperature of the crate containing the NiTnH (up to 85% correlation). When the temperature-correlated portion of the gain ratio drift is subtracted from each processor, the resulting distribution in rms gain ratios is as shown in Figure 4: 64 out of 88 units are stable to within 6×10^{-4} of their mean gain ratios, resulting in an offset drift of less than 1 micron. Note also that the tail of the distribution in Figure 4 is less extended than that in Figure 3. Typical values of the gain ratio/temperature slope were from from $-5.0 \times 10^{-4}/°\text{C}$ to $+5.0 \times 10^{-4}/°\text{C}$.

BEAM POSITION MONITOR TRIPLET EXPERIMENT

The Final Focus Test Beam [6] includes a diagnostic region in which three consecutive BPMs are separated by drift spaces. In this region, the betatron functions are relatively small, and therefore potential issues of beam scraping near the striplines are minimized. Let us consider a set of such BPMs in which the distance from the first to the second and from the first to the third are L_2 and L_3, respectively, and in which the offsets of the BPMs are d_1, d_2, d_3. If the measured BPM readings are given by x_1, x_2, x_3, then the relationship between the measured positions and the

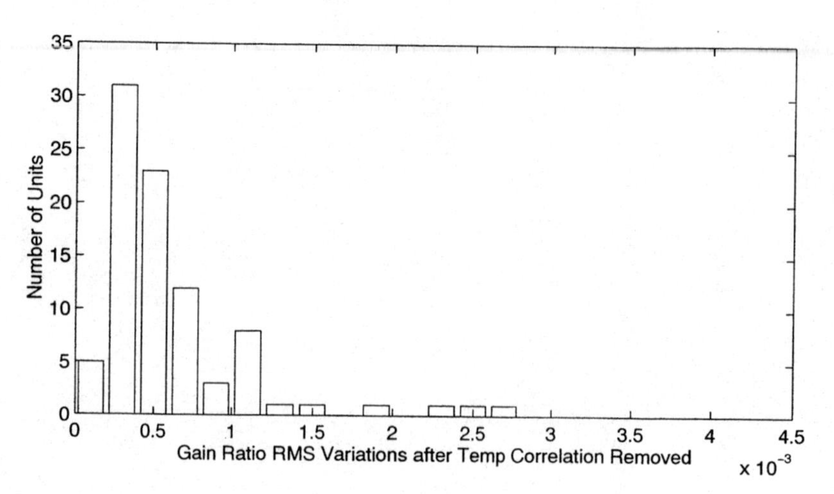

FIGURE 4. The rms gain ratio variations of 93 BPM processors after temperature-correlated drifts are subtracted off.

BPM offsets is given by:

$$x_3 - \frac{L_3}{L_2}x_2 + \left(\frac{L_3}{L_2} - 1\right)x_1 = d_3 - \frac{L_3}{L_2}d_2 + \left(\frac{L_3}{L_2} - 1\right)d_1. \tag{3}$$

If we define $X \equiv x_3 - \frac{L_3}{L_2}x_2 + \left(\frac{L_3}{L_2} - 1\right)x_1$, and we assume that the BPM offsets in the three BPMs are varying incoherently with time with an RMS variation of σ_{BPM}, then we can expect that:

$$\sigma_{BPM} = \sigma_X \left[1 + (L_3/L_2)^2 + (L_3/L_2 - 1)^2\right]^{-1/2}. \tag{4}$$

During the FFTB run of May 1997, the quantity X was read out and stored once every six minutes, for horizontal and vertical planes. For each stored value of X, four pulses were averaged; consequently the expected contribution to σ_{BPM} from BPM signal-to-noise limitation is 0.5 microns. While the calibration pulser experiment concentrated on the readout electronics, this experiment measures the contributions of all parts of the BPM system from the stripline to the main control computer.

Results of the BPM Triplet Experiment

Figure 5 shows the value of $Z \equiv X\left[1 + (L_3/L_2)^2 + (L_3/L_2 - 1)^2\right]^{-1/2}$ in the horizontal plane as a function of time. Several "fliers" have been removed from the dataset, which are believed to result from massively mis-steered pulses from the linac producing copious spray in the FFTB apertures. The rms incoherent offset drift implied by Figure 5 is 17 microns. Note also that, due to data acquisition errors, the data in the first few days of the run was saved at a much lower frequency than the 10 measurements per hour desired.

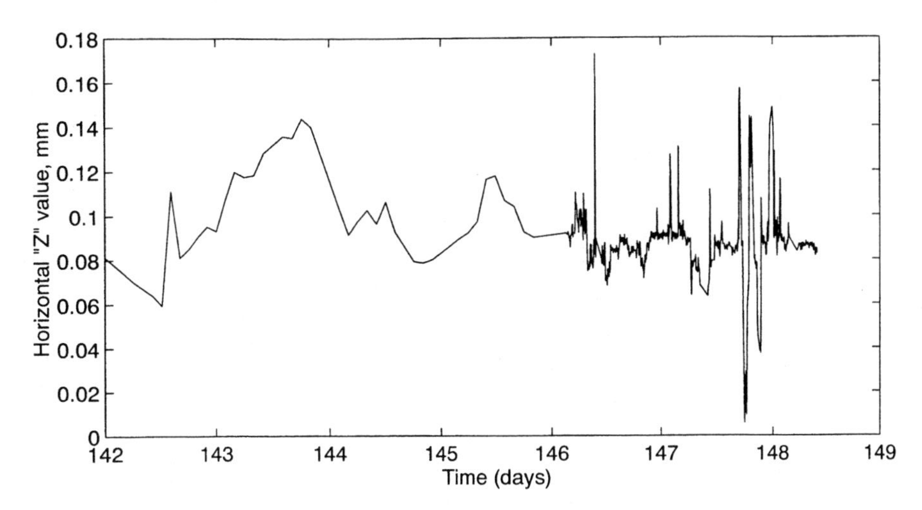

FIGURE 5. The value of Z for the horizontal plane during the May 1997 FFTB run. The implied rms offset drift per BPM is 17 microns.

Figure 6 shows the value of Z in the vertical plane, again with "fliers" suppressed. Here the implied rms drift is 4.2 microns, with several periods of extremely stable conditions during which drifts as small as 1.5 microns were observed for up to half a day. It is believed that the smaller drifts in the vertical plane result from the smaller vertical normalized emittance during the run (3.6 versus 36 mm/mrad), and also from the fact that the horizontal plane is the bend plane of the FFTB and thus synchrotron radiation and low-energy tails will primarily affect the horizontal measurements.

It is worth noting that the NLC beam position monitor has an aperture roughly half that of the FFTB unit (6 mm versus 11.5 mm in radius). If the drifts mentioned above are due primarily to effects in the cables and the feedthroughs, then the offset drifts for the NLC could be as small as 2 microns for a similar quality installation. Furthermore, there is no way to determine what fraction of the 4.2 microns measured in the vertical plane can be eliminated with further improvements in beam quality.

CONCLUSIONS

In both the experiment with the calibration pulser and the experiment with the BPM triplet, we see that BPM offset stability on the order of a few microns over

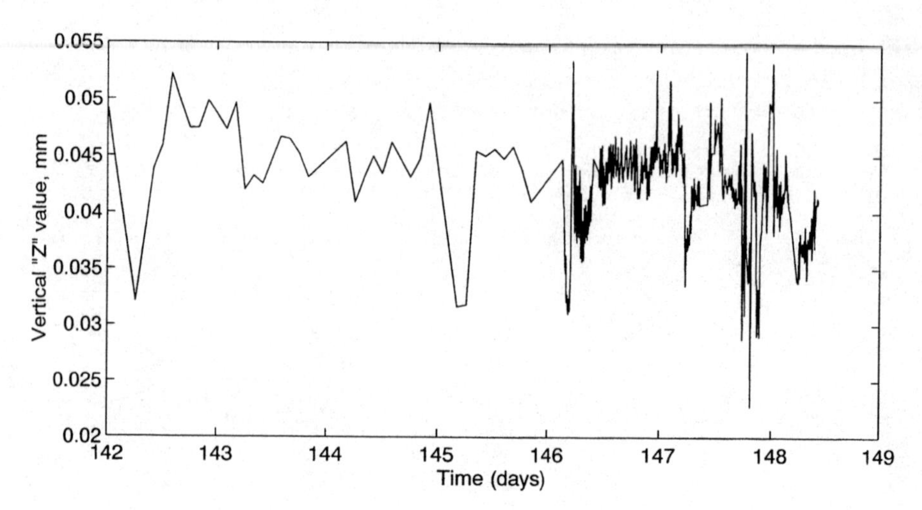

FIGURE 6. Value of Z for vertical plane during May 1997 FFTB run. The rms offset drift per BPM is 4.2 microns.

time periods up to one week can be reasonably achieved with present-day technology at future linear colliders. Higher beam quality may yield some improvements and a system which automatically calibrates the BPM processors continually (rather than the present scheme of calibration-on-demand) seems warranted. The value of temperature stability is also evident.

While the present systems described are not grossly inadequate to meet the NLC specifications, it remains to be demonstrated that reasonable improvements in temperature control of cables, electrical isolation of processors, etc., can reduce the slow offset drifts to the level required for such a future collider.

Future experiments may provide further insight into the various sources of BPM offset drift. These include running both the triplet and the calibration pulser experiment simultaneously and adding bunch charge to the set of variables read out by the triplet data acquisition system in order to measure and suppress any charge-position correlations in the BPM system.

ACKNOWLEDGMENTS

The authors wish to thank Steve Smith for many ideas and insights into the issues discussed herein, and Karey Krauter for authoring the triplet BPM acquisition software.

REFERENCES

1. NLC Design Group, *Next Linear Collider Zeroth-Order Design Report*, Stanford: SLAC, p. 440 (1996).
2. Smith, S., unpublished.
3. Tenenbaum, P., *Expanded Studies of Linear Collider Final Focus Systems at the Final Focus Test Beam*, Stanford: SLAC, 1995, p. 108.
4. Williams, S., unpublished.
5. Hayano, H., et al., *Nucl. Inst. Methods* **A320**, p. 47 (1992).
6. Oide, K., *Proceedings of the 1989 IEEE Particle Accelerator Conference*, Piscataway: IEEE Press, p. 1319 (1989).

Experiences of the QSBPM System on MAX II

Peter Röjsel

MAX-lab, Lund University, Lund, Sweden

Abstract. The MAX II is a third-generation synchrotron radiation source. The first beamline is in operation and several others are in the commissioning phase. The storage ring is equipped with a Quadrupole Shunt BPM (QSBPM) system for *in situ* calibration of the button-pickup BPM system. The calibration system uses switchable shunts on the combined-function quadrupole-sextupole magnets to find their magnetic centers. The BPM system has a time constant of several seconds, so a switched system is the only possibility. Each BPM pickup head and its corresponding electronics have been calibrated with the QSBPM system. The system has been in operation for about two years and operational experience, together with the technique itself, is discussed. The quadrupole shunts that are a part of the QSBPM system are, together with a spectrum analyzer and a tracking generator, also used to measure the beta functions individually in all main quadrupoles of the machine.

INTRODUCTION

The BPM system of MAX II is a multiplexed button pickup system of a type similar to that used at many third generation synchrotron radiation sources. The light source has a high brightness and small beam dimensions, which sets very high requirements on the stability of the beam and consequently on the BPM system. The QSBPM system makes it possible to do an *in situ* calibration of the BPM system's zero position reading using the main quadrupole magnets as the position reference.

Error sources in a BPM system

A number of sources generate errors in the BPM readings. Apart from the obvious mechanical positioning errors of the BPM pickups, we have a number of electrical "positioning" errors. The main error source in the MAX II BPM system is the unequal attenuation of the rf signals in the signal path of the BPM system, where the four rf signals from the buttons in one BPM head travel through different cables, connectors, filters and different inputs of the rf-multiplexer at the input of the BPM electronics. From the multiplexer's common point onwards, the BPM-button signals travel through and are processed by common circuits, multiplexed in time.

CP451, *Beam Instrumentation Workshop*
edited by R. O. Hettel, S. R. Smith, and J. D. Masek

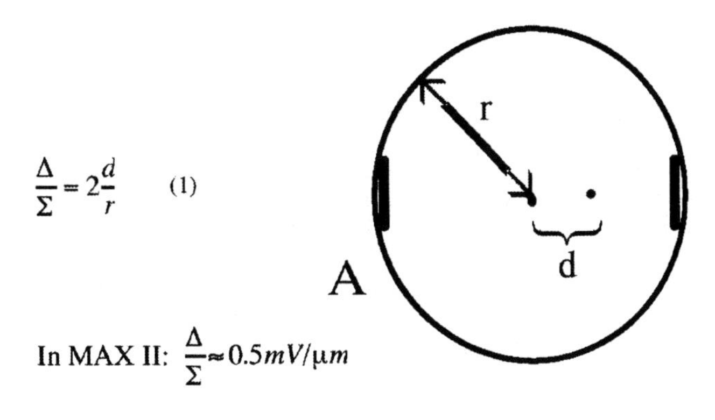

FIGURE 1. Signal in a BPM head.

Normal tolerances of attenuation imbalance in the electronic rf components can result in significant amounts of position error. Here is an example: A difference in attenuation between the channels A and B, connected to a BPM head of ±0.3 dB results in an error in the absolute beam position reading from that BPM head of ±0.5 mm. How can this be?

Let us, in our example in Figure 1, take d=0 and the voltage output from our BPM electronics to be 15 V for button A, a perfect signal, and the signal from button B to be (15–0.0005)V, as a less-perfect signal. The difference between A and B in dB is then: 0.5µm = 0.0003 dB and 1mm = 0.58 dB. The input multiplexer of the MAX II BPM system has a guaranteed maximum unbalance of 0.3 dB between the four input channels. This amounts to approximately 0.5 mm in offset!

An SMA connector exhibits a change in through attenuation in the range of 0.006 dB when being unplugged and then reconnected (and tightened using an SMA torque wrench), which corresponds to 10 µm offset. It should be noted that SMA connectors are among the type of connectors exhibiting the best data for this kind of application.

We clearly have a need to calibrate the BPM systems position offset *in situ* since even the unplugging and reconnection of high-quality connectors will affect the offset.

QSBPM PRINCIPLE

The basic idea of the QSBPM is quite old. It is based on the principle that, when modulating the strength of a single quadrupole magnet in a circular machine holding a stored beam, the closed orbit will move if the beam is not centered in the modulated quadrupole magnet. The modulation can be made in a variety of ways. In MAX II we use individually switchable shunts on each quadrupole magnet in the machine since our BPM system has a time constant of several seconds and a faster modulation would not be registered by the BPM system. The QSBPM system measures the BPM pickups offset with respect to the quadrupole magnets' magnetic centers in the machine by shunting off some current from the quadrupole nearest to the BPM head where the position offset is to be measured. The QSBPM system requires a BPM system with rather high resolution to read the beam position changes. We shall see later that some tricks can be used to increase the amount of data available, which, via averaging, increases the precision.

MAX II combined quadrupole-sextupoles

In MAX II the main focusing quadrupole magnet also has the main sextupole component integrated (4).The sextupole component is determined by the shape of the magnet's iron yoke so the sextupole component can only be slightly adjusted with a backleg winding. To make things even more complex, the yoke is slightly saturated when the machine is ramped to its full energy of 1.5 GeV. The saturation results in the magnetic center of the magnet being moved horizontally. We compensate for this with backleg windings. When a shunt on such a magnet is activated, the saturation is reduced and the magnetic center is moved about 3 mm horizontally. This could be compensated with a backleg winding on that magnet either with an additional shunt that is made variable or by providing an individual power supply for the backleg winding. In either case, we would rely on the magnetic field maps provided by the manufacturer and a model, which would reduce the precision in the BPM offset measurements since the exact magnetic center of the magnet will not be well known. This saturation effect makes it difficult to use the MAX II magnets for QSBPM purposes at full energy. Fortunately the beam in MAX II shows a long lifetime, even at the injection energy of 500 MeV where the magnets are far from saturation and the backleg windings can be shut off completely. Now we have a well-known magnetic center for the focusing magnets. Since the energy of the machine does not affect the BPM offsets we decided to measure the BPM offsets at the injection energy of 500 MeV.

The Shunts

The shunts on the quadrupole magnets are quite simple. They consists of a power resistor, a power MOSFET as the switch, and a photovoltaic optocoupler as the galvanic isolator and drive circuit for the powerMOS switch transistor. The schematic can be seen in Figure 2. There is one shunt on each quadrupole magnet. Each shunt can be controlled individually from the accelerator's control system, which also runs the BPM program and the QSBPM calibration program. The shunts in MAX II take 2 – 3 % of the magnet current.

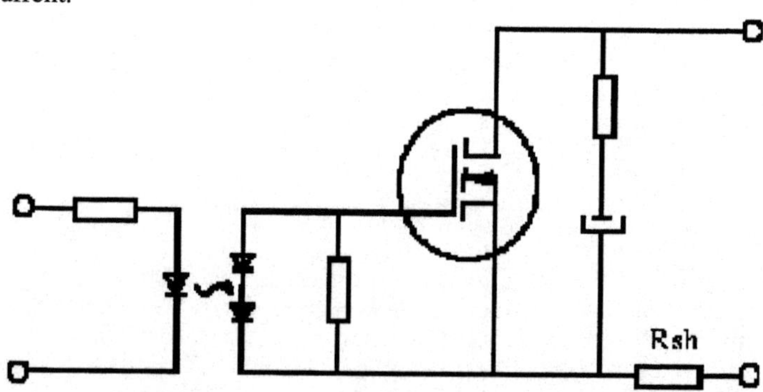

FIGURE 2. Schematic of a quadrupole shunt.

The BPM System

The BPM system must be able to measure the orbit shift induced by the quadrupole shunts. To do this with enough precision to create a BPM offset table that has the same or better precision as an ordinary BPM measurement, we have to use some tricks. One is to use data from all BPMs when determining the offset in one BPM. We can do this since only the amount of beam movement is interesting here.

The BPM system in MAX II has a time constant of a few seconds. The time for the reading to settle to its final precision is about 30 seconds. However the reading of one BPM takes almost as long as the reading of all 30 due to the parallelism used in the readout system.

Method Used to Find the BPM Offsets

We use a method that is rather easy to program so that the process can be automated, since it is rather lengthy. The example below is from MAX II but should, with adaption, be useful at most storage rings.

1. Start with beam in the machine, and a reasonably corrected, closed orbit.
2. Choose a BPM to be calibrated: e.g. Cell # 4 BPM #1 horizontal.
3. Set the nearby correction magnet to its current value –500. (The range is ±2047).
4. Measure the closed orbit and store it in a buffer.
5. Activate the shunt on the quadrupole closest to the BPM.
6. Measure the closed orbit again and subtract "buffer" from the readings.
7. Release the shunt.
8. Apply the formula given in Equation 2, below, to the shunt-induced difference position. Now, you have one point in Figure 3.
9. Increase the value of the correction magnet by +200.
10. Go to 4 if 5 points are not measured.
11. Do a least-squares fit of the five data points and find the minimum point. The correction value corresponding to the minimum point puts the beam in the middle of the shunted quadrupole.
12. Set the calculated correction value and read the BPM to be calibrated. This reading is the offset value since the beam now is in the center of the BPM.

$$f(I_{corr}) = \frac{1}{30}\sum_{i=1}^{30}\left(x_{BPMi}\right)^2 \tag{2}$$

As we can see in Figure 3 the curve minimum is below zero. A little disturbing. We found after some experiments that a fourth-order fit, which is the exact curve fit for five points actually gives a more accurate minimum although it is mathematically wrong. When this minimum is used, we get a better BPM offset reading. This is empirically confirmed, so we use it.

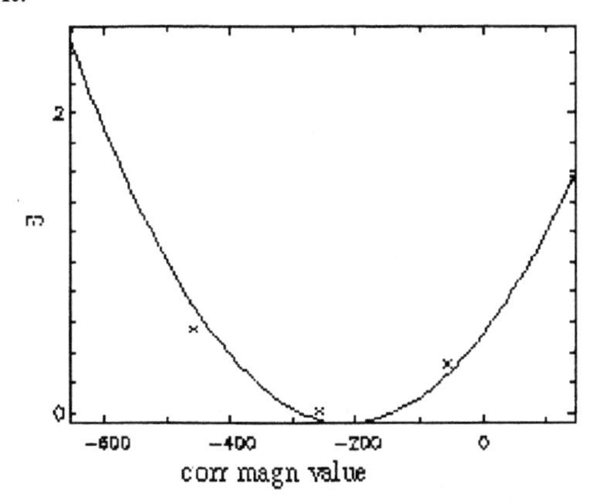

FIGURE 3. Plot of the fitted function of a real BPM calibration.

RESULTS FROM MAX II

Several BPM calibration runs using the QSBPM principle have been performed on MAX II. The results from a series of calibrations can be seen in Figures 4 and 5.

The vertical calibration runs in Figure 5 show an offset, randomly spread around the machine, with a mean value of zero. The horizontal calibration runs in Figure 4 have an offset of roughly 0.5 mm. This is an effect of the integrated sextupole component in the quadrupole magnets. If we change the backleg winding current or just use a different magnetization cycle we will change the horizontal offsets. We have thus defined a specific backleg winding current of zero and a specific magnetization cycle for the quadrupoles. This will also keep the closed-orbit reproducibility in control.

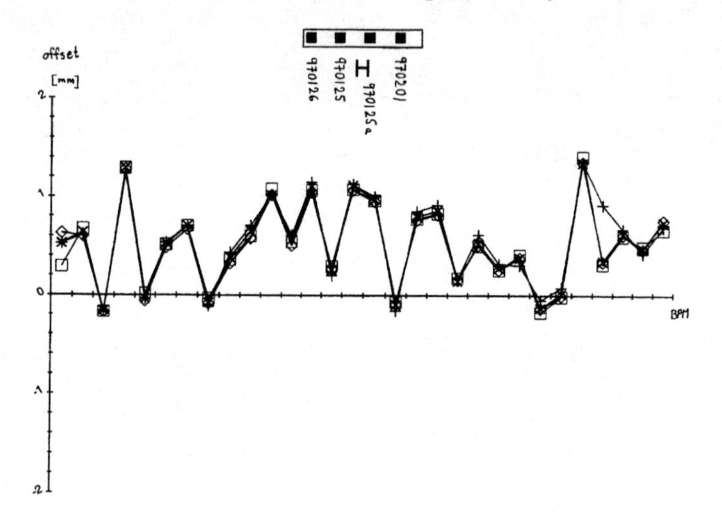

FIGURE 4. Calibration runs in horizontal, showing fairly good reproducibility.

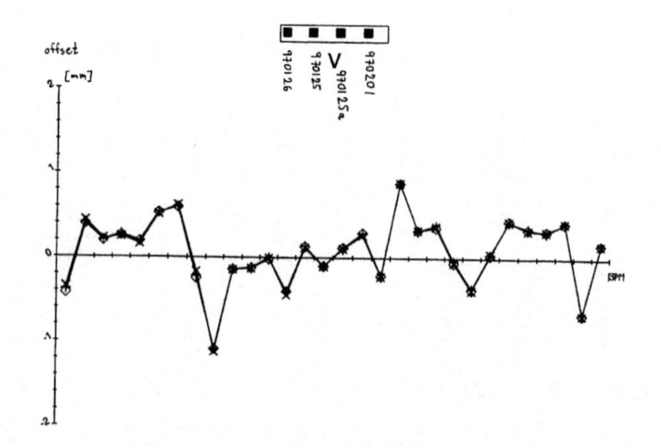

FIGURE 5. Calibration runs in vertical, showing very good reproducibility.

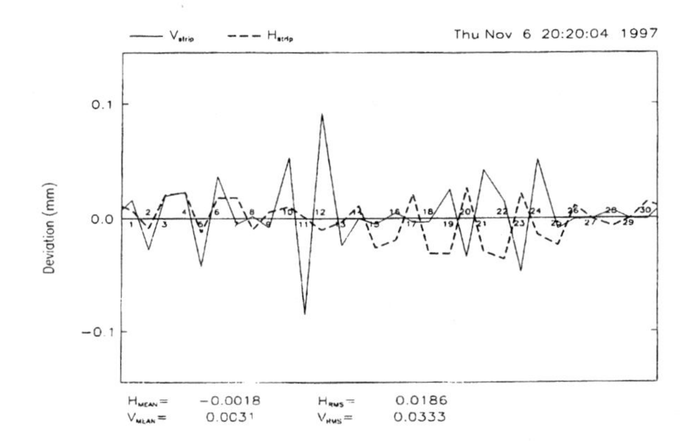

FIGURE 6. The best possible orbit correction without offset calibration.

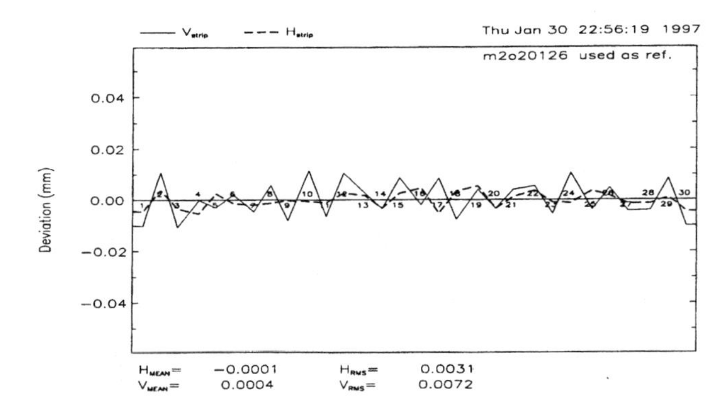

FIGURE 7. Final orbit correction after BPM calibration with the QSBPM method.

TABLE 1. BPM Offsets Measured at One QSBPM Run on MAX II

h:	BPM1	BPM2	BPM3	v:	BPM1	BPM2	BPM3
cell 1	0.559	0.627	–0.098	cell 1	–0.528	0.316	0.130
cell 2	1.387	0.011	0.611	cell 2	0.188	0.127	0.456
cell 3	0.780	–0.058	0.462	cell 3	0.516	–0.271	–1.203
cell 4	0.640	1.054	0.640	cell 4	–0.250	–0.176	–0.086
cell 5	1.128	0.271	1.186	cell 5	–0.493	0.067	–0.174
cell 6	0.993	–0.064	0.840	cell 6	0.021	0.258	–0.330
cell 7	0.890	0.186	0.617	cell 7	0.763	0.243	0.225
cell 8	0.258	0.393	–0.034	cell 8	–0.145	–0.417	–0.068
cell 9	0.121	1.401	0.446	cell 9	0.362	0.261	0.222
cell 10	0.734	0.479	0.802	cell 10	0.334	–0.716	0.074

As can be seen in Table 1, the BPM offsets are up to 1.4 mm horizontally and up to 1.2 mm vertically.

BETA FUNCTION MEASUREMENTS

The shunts can also be used to determine the local beta functions in the shunted quadrupole magnet. If we observe the tune shift induced by the activated shunt from every quadrupole magnet, we can plot the machine beta functions. This gives lattice information that can be compared to the theoretically calculated values.

CONCLUSIONS

The QSBPM system for BPM calibration has improved the quality of the delivered synchrotron radiation and shown that it can be implemented even on machines with rather "exotic" magnets such as MAX II.

QSBPM systems have been implemented in the two storage rings at MAX-lab and the technique is also used at many other labs. (2,3)

REFERENCES

[1] Röjsel, P. "A beam position measurement system using quadrupole magnets magnetic centra as the position reference," *Nucl. Instr. and Meth. A* **343** (1994) 374–382.

[2] Corbett, W. J., R. O. Hettel, and H. D. Nuhn. "Quadrupole Shunt Experiments at SPEAR" (SLAC-PUB-7162), presented at the 7th Beam Instrumentation Workshop (BIW 96), Argonne, IL, May, 1996.

[3] Barrett. I., et al., "Dynamic Beam Based Calibration of Orbit Monitors at LEP," presented at the 4th International Workshop on Accelerator Alignment, Tsukuba, Japan, December 1995, KEK Proc. 95–12 and CERN-SL/95–97 (BI).

[4] Eriksson, M., et. al., Design report for the MAX II Ring, MAX publications ISRN LUNTDX/NTMX--7019--SE (1992).

The Measurement of Chromaticity Via a Head-Tail Phase Shift

D. Cocq, O.R. Jones and H. Schmickler

CERN, CH-1211 Geneva 23, Switzerland.

Abstract. The most common method of measuring the chromaticities of a circular machine is to measure the betatron tune as a function of the beam energy and then to calculate the chromaticity from the resulting gradient. Even as a simple difference method between two machine energies this technique does not allow instantaneous measurements, for instance during energy ramping or beta squeezing. In preparation for the LHC, a new approach has been developed which uses the energy spread in the beams for chromaticity measurements. Transverse oscillations are excited with a single kick and the chromaticity is calculated from the phase difference of the individually sampled head and tail motions of a single bunch. Using this method the chromaticity can be calculated using the data from only one synchrotron period (about 15-50 msec in the case of the LHC). This paper describes the theory behind this technique, two different experimental set-ups, and the results of measurements carried out in the SPS.

INTRODUCTION

The tight tolerances on beam parameters required for successful LHC operation implies a good knowledge of the chromaticity throughout the ramping cycle. However, many of the methods currently used to measure chromaticity in circular machines (see (1) and references therein) are likely to be incompatible with LHC high intensity running. For example, the most common method, of measuring the betatron tune as a function of beam energy, might be difficult to implement due to the tight tolerances imposed on the betatron tune itself. Chromaticity can also be calculated from the amplitude of the synchrotron sidebands observed in the transverse frequency spectrum. However, this method suffers from resonant behaviour not linked to chromaticity, and the fact that the low synchrotron tune of the LHC would make it difficult to distinguish these sidebands from the main betatron tune peak. The width of the betatron tune peak itself, or the phase response of the beam transfer function, also give a measure of chromaticity, but require a knowledge of how the momentum spread in the beam changes with energy.

In this paper we describe a method that allows the chromaticity to be calculated from the turn-by-turn position data of a single bunch over one synchrotron period, after the

CP451, *Beam Instrumentation Workshop*
edited by R. O. Hettel, S. R. Smith, and J. D. Masek

application of a transverse kick. It will be shown that the chromaticity can be obtained by determining the evolution of the phase difference between two longitudinal positions within the bunch. This technique does not rely on an accurate knowledge of the fractional part of the betatron tune and, for a machine operating well above transition, the calculated chromaticity is virtually independent of beam energy.

SINGLE PARTICLE DYNAMICS IN A BUNCHED BEAM

The longitudinal motion of a particle within a bunch can be described in terms of the synchrotron frequency and the difference between the particle's momentum and the nominal momentum. This can be expressed as

$$\tau = \hat{\tau}\cos(\omega_s t + \phi) \tag{1}$$

where τ is the variation with time t of the particle from the bunch center, ω_s is the angular synchrotron frequency and ϕ is the initial longitudinal phase of the particle (Fig. 1).

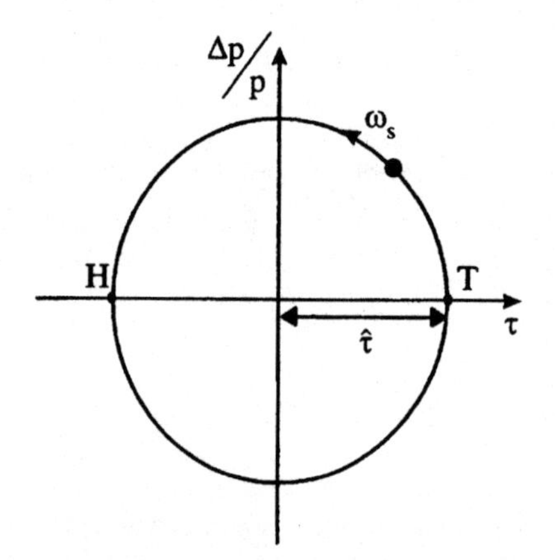

FIGURE 1. Longitudinal phase-space.

During this longitudinal motion the particle also undergoes transverse motion, which can be described by the change in the betatron phase, $\theta(t)$, along the synchrotron orbit. The rate of change of this betatron phase is given by

$$\dot{\theta} = Q\omega = (Q_0 + \Delta Q)(\omega_0 + \Delta\omega) \tag{2}$$

where Q_0 is the nominal transverse tune, ΔQ is the change in tune due to the momentum spread, ω_0 is the nominal angular revolution frequency and $\Delta\omega$ is the change in this frequency due to the momentum spread.

Now $$\frac{\Delta\omega}{\omega_0} = -\eta\frac{\Delta p}{p}, \quad \frac{\Delta Q}{Q_0} = \xi\frac{\Delta p}{p}, \quad and \quad \dot{\tau} = -\frac{\Delta T}{T_{rev}} = \frac{\Delta\omega}{\omega_0} \tag{3}$$

where:

ξ = relative chromaticity (Q'/Q_0 where Q' is the chromaticity)
η = $1/(\gamma)^2 - \alpha$ (where $\alpha = 1/(\gamma_{tr})^2$ is the momentum compaction factor)
T_{rev} = revolution period

Hence $\Delta Q = -\frac{\xi}{\eta}Q_0\dot{\tau}$ and $\Delta\omega = \omega_0\dot{\tau}$, which on substitution into (2) gives

$$\dot{\theta} = \left(Q_0 - \frac{\xi}{\eta}Q_0\dot{\tau}\right)\left(\omega_0 + \omega_0\dot{\tau}\right) = Q_0\omega_0\left(1 - \frac{\xi}{\eta}\dot{\tau}\right)\left(1 + \dot{\tau}\right) \approx Q_0\omega_0\left(1 + \dot{\tau}\left(1 - \frac{\xi}{\eta}\right)\right) \tag{3}$$

Integrating one obtains $\theta(t) \approx Q_0\omega_0(t+\tau) - \omega_\xi\tau + \theta_0$ where $\omega_\xi = Q_0\omega_0\frac{\xi}{\eta}$ and is known as the chromatic frequency. Since τ is usually much smaller than t, one can write:

$$\theta(t) \approx Q_0\omega_0 t - \omega_\xi\hat{\tau}\cos(\omega_s t + \phi) + \theta_0 \tag{4}$$

where θ_0 is the initial betatron phase of the particle.

If we assume that the displacement due to the kick is much larger than the betatron oscillations performed by the particles in the unperturbed bunch, then the initial betatron phase can be set such that all the particles have the same initial position in the transverse plane; i.e. at t = 0 a particle with a longitudinal position described by $\hat{\tau}_i$ and 0_i will have an initial betatron phase given by

$$\theta_0 = \omega_\xi\hat{\tau}_i\cos(\phi_i) \tag{5}$$

Hence the position of a particle in the bunch, y_i, at any time t is given by

$$y_i(t) = A\cos(\theta(t)) \tag{6}$$

where A is the maximum amplitude of the betatron motion.

Extracting the Chromaticity by Considering Head-Tail Evolution

Consider a system where the particles are distributed along a single synchrotron orbit, and where the particle position is measured as a function of time at the tail of the bunch (point T in Fig. 1). The initial synchrotron phase, ϕ_i, of the particle at the tail after a given time t is given by $\phi_i = -\omega_s t$, and therefore the corresponding initial betatron phase, $(\theta_0)_i$, can be written as $(\theta_0)_i = \omega_\xi \hat{\tau}_i \cos(-\omega_s t)$. Hence the transverse amplitude at the tail evolves as

$$\begin{aligned} y_T(t) &= A\cos(Q_0\omega_0 t - \omega_\xi\hat{\tau}_T\cos(\omega_s t + \phi) + \omega_\xi\hat{\tau}_T\cos(-\omega_s t)) \\ &= A\cos(Q_0\omega_0 t + \omega_\xi\hat{\tau}_T(\cos(-\omega_s t) - 1)) \end{aligned} \qquad (7)$$

Measurements are carried out turn by turn, which corresponds to time steps of $t = 2\pi n/\omega_0$, where n is the turn number. Hence the signal at the tail of the bunch for a given turn, n, can be written as

$$y_T(n) = A\cos\left(2\pi n Q_0 + \omega_\xi\hat{\tau}_T\left(\cos(2\pi n Q_s) - 1\right)\right) \qquad (8)$$

where Q_s is the synchrotron tune, with $\omega_s = Q_s\omega_0$. A similar expression can also be derived for the head of the bunch

$$y_H(n) = A\cos\left(2\pi n Q_0 - \omega_\xi\hat{\tau}_H\left(\cos(2\pi n Q_s) - 1\right)\right) \qquad (9)$$

The phase difference between head and tail as a function of turn number is therefore given by

$$\Delta\psi(n) = -\omega_\xi\left(\hat{\tau}_H + \hat{\tau}_T\right)\left(\cos(2\pi n Q_s) - 1\right) \qquad (10)$$

This equation is a maximum when $nQ_s = ½$, i.e. after half a synchrotron period, giving

$$\Delta\psi_{MAX} = -2\omega_\xi\left(\hat{\tau}_H + \hat{\tau}_T\right) = -2\omega_\xi\,\Delta\tau \qquad (11)$$

The chromaticity can therefore be written as

$$\xi = \frac{-\eta\,\Delta\psi(n)}{Q_0\,\omega_0\,\Delta\tau\left(\cos(2\pi n Q_s) - 1\right)} \quad or \quad \xi = \frac{\eta\,\Delta\psi_{MAX}}{2Q_0\,\omega_0\,\Delta\tau} \qquad (12)$$

Since the revolution frequency and total tune of the machine are known to a high degree of accuracy, and if one considers a machine operating well above transition, the chromaticity will depend only on the maximum phase difference attained between two regions of the same bunch, separated longitudinally by a known time.

EXPERIMENTAL SET-UP

In order to illustrate the principle of this technique, head-tail phase measurements were performed on proton bunches in the SPS. Each acquisition involved the measurement of 100 to 200 turns of data (around one SPS synchrotron period) after the application of a single transverse kick. An exponential transverse coupler was used to provide the transverse position data as a function of the longitudinal position within the bunch. In the first experiments the signal from the coupler was analyzed using two sample-and-hold amplifiers. This then provided information on the transverse position of two specific locations within the bunch (one positioned near the head of the bunch and the other, near the tail). Later experiments relied on a digital oscilloscope with fast-frame capabilities, from which a single trace gave transverse position information for the whole bunch as a function of longitudinal position.

Head-Tail Phase Measurements using Head and Tail Samplers

The experiments using the head and tail samplers were performed on the Machine Development (MD) cycle of the SPS, which provided a single bunch of ~2×10^{10} protons at an energy of 26 GeV and with a length of ~3.8 ns.

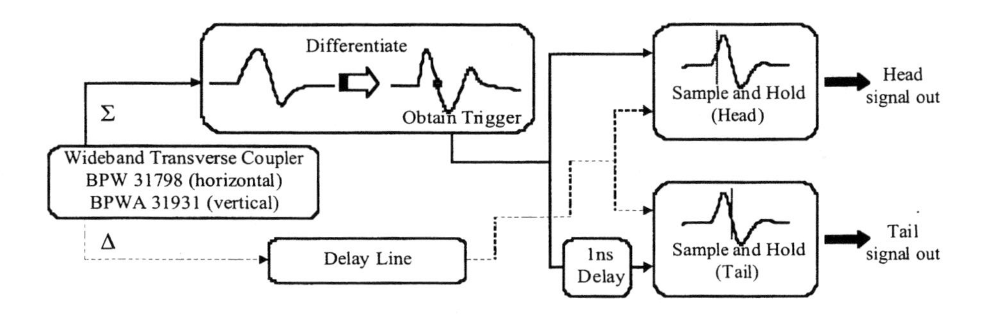

FIGURE 2. Sampler acquisition schematic.

Figure 2 shows a schematic of the acquisition system. The coupler sum signal was used to auto-trigger two Avtech AVS-105 variable gate width sample-and-hold amplifiers. The trigger of one sampler was delayed by ~1 ns with respect to that of the other, to allow the transverse bunch position to be sampled at two longitudinal positions separated in time by this delay. Each sampler had a gate width of ~200 ps and provided an output pulse of 500 ns duration with an amplitude proportional to the input amplitude. The output from both samplers was stored on a turn-by-turn basis on two channels of a Le-Croy 9354, 500 MHz, digital oscilloscope. Each acquisition, which consisted of around 200 turns of information, could then be transferred to a computer for analysis.

Head-Tail Phase Measurements using a Tektronix 784A Oscilloscope

By using a Tektronix 784A digital oscilloscope, with an analog bandwidth of 1 GHz and a maximum sampling rate of 4 GS/s, the complete coupler signal of a single bunch could be reconstructed. Additionally, a fast-frame acquisition setting meant that the signal could be recorded on consecutive SPS machine turns. This removed the necessity of having individual sample-and-hold amplifiers. When using the Tektronix 784A, the turn-by-turn acquisition was triggered directly from the SPS rf system.

Experiments using this acquisition system were again performed on the SPS MD cycle, this time with 20 bunches in the machine, each bunch having the same characteristics as described previously. The internal memory of the oscilloscope provided enough space to record up to 160 turns of data, which could be transferred directly to a computer for further analysis.

RESULTS

The raw data obtained from the head-tail samplers give graphs such as those of Figure 3(a & b). Here one sees the variation in the transverse position of two segments of the bunch separated by ~1ns, after applying a kick to the beam.

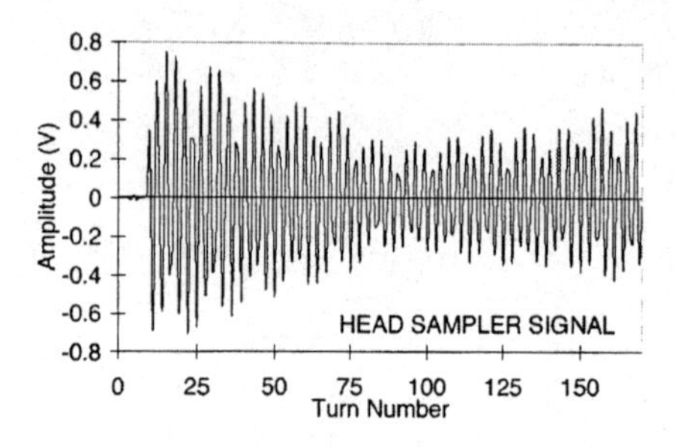

FIGURE 3(a). Head sampler signal.

FIGURE 3(b). Tail sampler signal.

In order to calculate the phase relationship between the head and tail signals, a sweeping harmonic analysis was performed. This involved working out the phase by harmonic analysis for each turn, using the current data point and 10 points on either side. Since the same calculation was performed on both signals, the phase difference for each turn could be found by simply subtracting the individual phases. The results of such a harmonic analysis sweep is shown in Figure 4. Here one clearly sees the de-phasing and re-phasing of the two positions within the bunch during one synchrotron period, with the maximum phase difference occurring after half a synchrotron period.

Once the phase difference evolution is known, it is possible to calculate the chromaticity, using Equation (12), on a turn-by turn basis. The result of such a calculation is shown in Figure 5. Due to the cosine term in the denominator of Equation (12) the error in the calculation of the chromaticity is very large at the start and end of the synchrotron period, while it is at a minimum at half the synchrotron period. In this particular case, there was sufficient signal to enable the chromaticity to be calculated again during the second synchrotron period. The average value of the chromaticity in this particular case, indicated by the dashed horizontal line, was calculated from the 40 turns around the center of both synchrotron periods.

A typical 'mountain range' display result obtained by capturing the coupler signal using the Tektronix 784A can be seen in Figure 6(a). The signal was sampled at 4 Gs/s and automatically interpolated to give output data points every 20 ps. The second, smaller, out-of-phase oscillation seen in the figure is the reflection of the main signal from the exponential coupler. In order to determine the chromaticity, we require knowledge of the phase difference evolution of two separate locations within the bunch. Hence, two cuts in the longitudinal plane were performed to obtain the evolution of two points located to either side of the bunch center (Fig. 6(b)). By performing a sweeping harmonic analysis on these two data sets, as previously described, the phase difference evolution and, hence, the chromaticity could be extracted on a turn-by-turn basis.

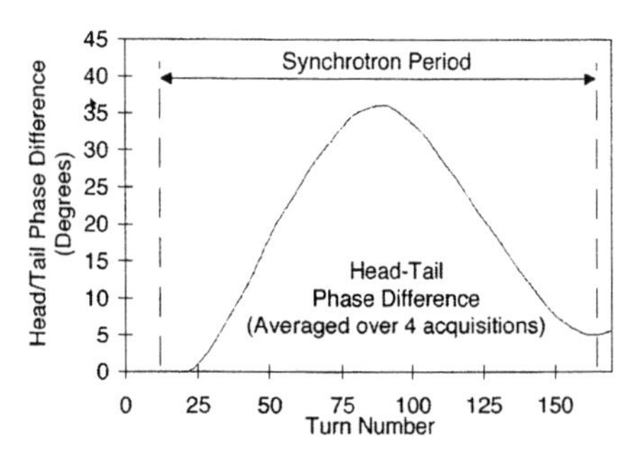

FIGURE 4. Phase difference evolution.

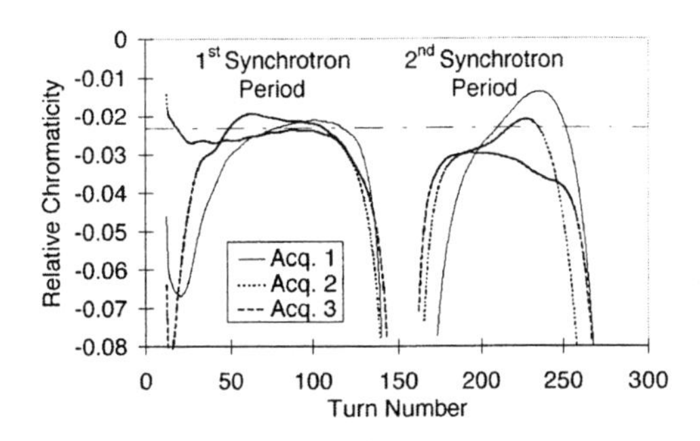

FIGURE 5. Turn-by-turn chromaticity.

This phase difference evolution for two consecutive acquisitions is plotted in Figure 7. The measurements were found to be reproducible, with the calculated value for the chromaticity in agreement to within 10%.

Finally, Figure 8 shows the vertical chromaticity, calculated by the head-tail phase difference, as a function of the vertical chromaticity machine trim. The results are what one would expect, with a linear relationship found between the trim and chromaticity.

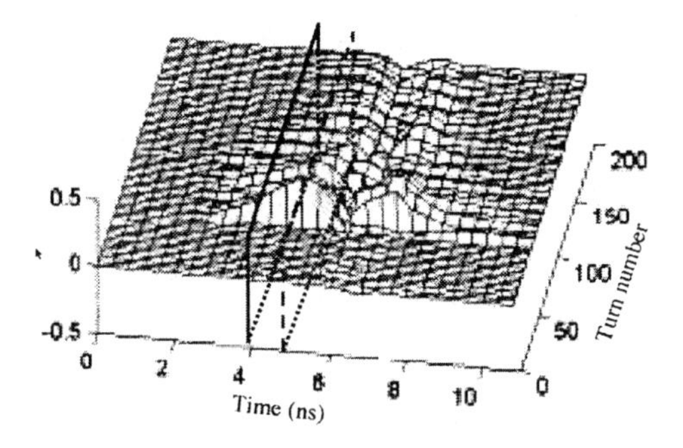

FIGURE 6(a). Tektronix 784A output.

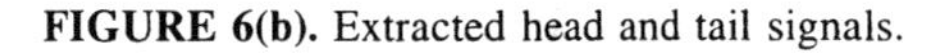

FIGURE 6(b). Extracted head and tail signals.

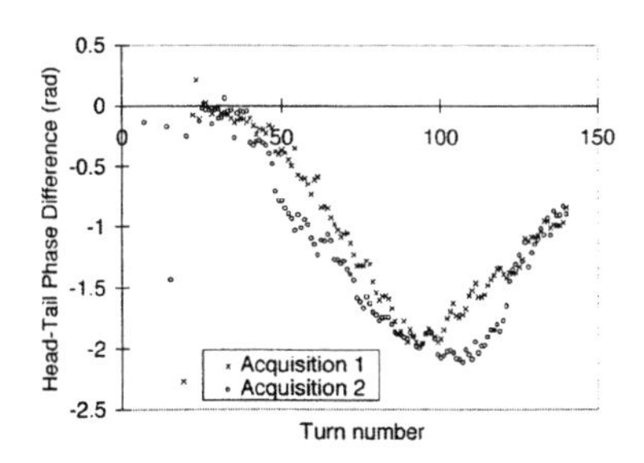

FIGURE 7. The time evolution of $\Delta\psi$.

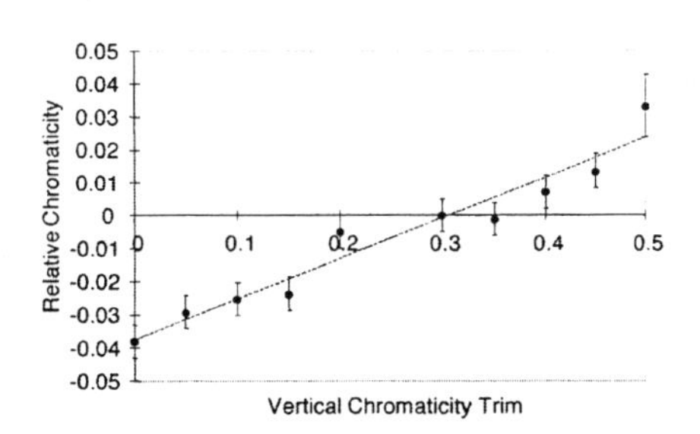

FIGURE 8. Variation of ξ with machine trim.

CONCLUSION

It has been demonstrated that the chromaticity of a circular machine can be calculated from the evolution of the phase difference between the head and tail of a single bunch after the application of a transverse kick. Using this technique, a measure of chromaticity can be obtained from a single synchrotron period. The sensitivity of the head-tail method relies on the fact that a sufficiently large phase difference can be achieved between the head and tail signals. Since, for a given chromaticity, this phase difference is proportional to the head-tail separation, the sensitivity is ultimately governed by the bunch length. Hence, this technique should work in proton machines where the bunch lengths are relatively large, but may prove difficult to implement in lepton machines, where the small bunch lengths imply a very short maximum head-tail separation.

Further studies are planned to investigate the effect of higher order fields, such as octupolar fields, which will be of consequence for the LHC, and also to determine the signal-to-noise ratio required for accurate measurements. This latter point will ultimately determine the emittance blow-up caused by repeated measurements, which is again of great importance for any proton machines.

REFERENCES

[1] Schmickler, H., "Diagnostics and Control of the Time Evolution of Beam Parameters" (CERN-SL-97-68), presented at the 3rd European Workshop on Beam diagnostics and Instrumentation for Particle Accelerators (DIPAC 97), Frascati, Italy, October 1997.

CONTRIBUTED PAPERS—POSTERS

An Automated BPM Characterization System for LEDA*

R. B. Shurter, J. D Gilpatrick, J. Ledford, J. O'Hara, J. Power

Los Alamos National Laboratory
Los Alamos, NM 87545

Abstract. An automated and highly accurate system for "mapping" 5 cm-diameter beam position monitors (BPMs) used in the Low Energy Demonstrator Accelerator (LEDA) at Los Alamos is described. Two-dimensional data is accumulated from the four micro-stripline electrodes in the probe by sweeping an antenna driven at the LEDA bunching frequency of 350 MHz in discrete steps across the aperture. These data are then used to determine the centroid, first- and third-order sensitivities of the BPM. These probe response coefficients are then embedded in the LEDA control system database to provide normalized beam position information to the operators. A short summary of previous systems we have fielded is given, along with their attributes and deficiencies that had a bearing on this latest design. Lessons learned from this system will, in turn, be used on the next mappers that are currently being designed for 15 cm and 2.5 cm BPMs.

BACKGROUND

We have implemented several BPM characterization systems for previous projects at Los Alamos. The first system we fielded was a basic "taut-wire" fixture whereby a wire antenna was stretched through the BPM between two parallel plates mounted on micrometer-driven X-Y linear stages. The antenna was driven at the accelerating cavity frequency while the signals at the output ports were monitored with a power meter and manually recorded as the stages incrementally translated the antenna across the aperture. Some of the drawbacks to this system were:

1) A high chance for error in the manual positioning of the stages and data recording of the outputs of all four ports for each increment. This slow and

* Work funded and supported by the U.S. Department of Energy.

CP451, *Beam Instrumentation Workshop*
edited by R. O. Hettel, S. R. Smith, and J. D. Masek
1998 The American Institute of Physics 1-56396-794-4/98/$15.00

tedious process also precluded obtaining measurement accuracy statistics through multiple measurements.

2) The antenna rigidly anchored at both ends made assembly and disassembly of the fixture for insertion and removal of the probes difficult.
3) The determination of the antenna position relative to the probe fiducials was problematic.
4) Creating a continuous signal return path necessitated using gold-plated rf "finger stock" which slid on the surface of the gold plated end plates. With repeated operations, the plating became worn, changing the resistance. The fingers were also broken or were easily bent over time.
5) The frictional drag of the rf fingers created positional inaccuracy through torquing of the fixture or exacerbating backlash.
6) It was determined that perturbation of the beam-generated fields surrounding the probe lobes affected the measurement.
7) The dynamic range of the power meter that measures the output signal from the probe can be a limiting factor, becoming proportionally worse with probe diameter.
8) "Over-running" of the stages often caused antenna breakage.

Addressing probe-mapping problems has been an iterative process. Through successive implementations of BPM mapping systems for several projects, many of the above problems have been addressed, with the present LEDA mapper representing a culmination of the best approaches we have determined thus far (1).

THE LEDA MAPPER

Table 1 lists the accuracy requirements for the 2" LEDA BPM (2).

TABLE 1. LEDA BPM Requirements

24 mm radius Probe	(mm)	(dB)	(mv)
Alignment fiducial fixture	0.075	0.101	5
Alignment scale reading	0.025	0.034	2
Probe offset elect. calibration repeatability with test fixture	0.074	0.100	5
Probe sensitivity (at 10% radius) elect. cal. repeat. w/ TF	0.148	0.200	10
Probe 3rd order (at 30% radius) elect. cal. repeat. w/ TF	0.037	0.050	2
Differences between test fix. and beamline	0.026	0.035	2
Test fix. probe mounting repeatability	0.026	0.035	2
Through lobe transmission errors	0.074	0.100	5
SOS Total (1 std. dev. or 34%)	0.205	0.276	13

The LEDA mapping system is comprised of a dual-axis, four-inch travel stage driven by stepper motors with absolute encoder feedback controlled by a Compumotor

4000 controller commanded by a PC running Labview software. The stages support a pair of "paddles," within which a probe with "spool pieces" is supported stationarily on a table. The top paddle, vertically positioned by an additional motor-driven stage, holds the antenna connector and vertical-alignment assemblage. The bottom paddle contains the antenna tensioner assemblage.

Under computer control, the antenna, driven with rf energy at the LEDA bunching frequency, is stepped in a serpentine fashion across the probe aperture. At each point, the mapper pauses and the rf signal transmitted to the four lobes is measured by a power meter and uploaded to the computer. The computer program then manipulates and stores the data as a text file. Afterward, the file is opened using macros within the Excel spreadsheet application to reduce the 2-dimensional data to 1-dimensional and do third-order fits to verify that the offset, sensitivity and third-order terms match expected results (3). Later, an inverse fit is done on the 2-dimensional data and the coefficients are used in the accelerator control system to transform the in-situ probe data into corrected beam position information (4).

FIGURE 1. LEDA BPM Mapper.

To better understand the LEDA mapping system, we shall look at the mapper as being divided somewhat arbitrarily into subsystems: the antenna positioning system, mechanical probe support/alignment system, computer control, and the stimulus/

measurement system. Table 2 delineates the various tests and results that were obtained from the mapper characterization.

TABLE 2. LEDA Mapper Error Summary

Test #	Test Name	Offset σ	Sensitivity σ	3rd Order σ	Note
		mm	dB/mm	dB/mm³	General: Averaged Std. Dev. of Both Axes
1	step sizes	N/A	N/A	N/A	Test done to confirm steps are correct
2	encoders' vs. lasers' wire zero	0.0000	N/A	N/A	
3	RF leakage w/ coax probe	N/A	N/A	N/A	–41 dBm max.
4	RF leakage w/ NF probe	N/A	N/A	N/A	–82 dBm max.
5	power vs. paddle height	0.0244	0.0000	0.0000	These errors include extreme height values
6	offset vs. shorting gaps	0.0198	N/A	N/A	±0.08 dB total power change max.
7	offset vs. terminations	0.0291	0.0000	0.0000	These errors include out-of-spec. termination values
8	offset vs. cal. switch	0.0074	N/A	N/A	HP's repeatability spec. in dB xformed to s in mm
9	offset indep. of pwr.head	0.0220	0.0000	0.0000	
10	map vs. input power	0.0164	0.0025	0.0000	+10 & +20 dBm only
11	map same for A,B ports	0.0244	0.0000	0.0001	
12	offset vs. mult. probe insert's	0.0181	0.0019	0.0000	
13	map long-term repeatability	0.0163	0.0019	0.0000	
14	laser repeatability w/ R&R	0.0000	N/A	N/A	±3.94E–06 mm
15	alignment plug repeatability	0.0000	N/A	N/A	Axis 3–1: ±1.38E–05 mm Axis 4-2: ±1.97E–05 mm
16	verify wire perpendicularity	N/A	N/A	N/A	during set-up
17	verify orthogonality of x&y axes	N/A	N/A	N/A	verified w/ lasers and dial indicators
18	verify lasers parallel to axes	N/A	N/A	N/A	part of above
19	frequency dependence	0.0348	0.0029	0.0001	350± 50 Mhz
20	RSS total of errors	0.0710	0.0047	0.0001	

N/A = not applicable; R&R = remove, replace

Mechanical Probe Support/Alignment system

The BPM is precisely located in the fixture using two of the four beam-line alignment monuments incorporated into the probe. The lobe-one and lobe-two monuments were chosen as the primary and secondary alignment points with the primary being pinned to control rotation. A spring-loaded ball screw, located 45 degrees from the centers and on the opposite side of the monuments, firmly presses their faces against fence blocks. The mechanical features of the probe were measured with respect to the four monuments to an accuracy of ± 2.5 μm at the time of manufacture.

Two 152.4 mm long dual-flanged cylinders, or "spool-pieces," are bolted to each end of the probe to contain and create uniform rf fields around the probe. The flanges at the opposite end of each spool-piece create a capacitively coupled return path for the antenna current. The impedance of this coupling is given by:

$$C = \varepsilon_0 \frac{A}{d} \tag{1}$$

$$Z = X_C = j\omega C^{-1} \tag{2}$$

where Z is the impedance (purely reactive), X_c is the capacitive reactance, A is the area of the 114.3 mm diameter flange face (8,516 mm^2), d is the separation distance of the flange from the paddle (0.508 mm), ω is $2\pi f$ (f = 350MHz), and the permitivity of air $\varepsilon_0 = 1$, yielding:

$$Z = 0.008\Omega \tag{3}$$

The narrow gap (0.254–0.508 mm) between the flanges and the top and bottom paddles also serves as an rf choke to keep the rf energy inside, precluding possible interference with the power sensors.

The antenna, a 0.102 mm stainless steel (SS) wire, is secured at the SMA input rf connector, guided through a 0.1143 mm tapered-bore, sapphire insulating bushing. A 4.76 mm diameter ball bearing with a drilled 0.203 mm hole is attracted by a samarium-cobalt magnet in the bottom paddle which tensions and terminates the antenna in a slip-fit Macor guide. To deploy the antenna, the top paddle is first spaced at a predetermined position established by running the computer program. One end of a length of the SS wire is then soldered into the bearing, using resistance heating. The ball bearing is lowered into the Macor insert in the bottom paddle, where it is held in contact with the magnet. The wire is then inserted into the sapphire guide and exits at the top of the SMA connector assembly. While the wire is held taut, the nylon antenna capture screw is screwed in until the ball bearing is pulled free and suspended above the magnet.

Antenna alignment with respect to the probe's mechanical center is accomplished through several relationships. On initial assembly of the BPM mapper, an alignment disk with a 1.5875 mm dowel pin extending 25.4 mm on each side, through the center, coupled with a fixture block, is used to permanently position the X and Y axis probe-alignment fence blocks. Henceforth, for each new probe map, the antenna is located at the mechanical center of the probe by first "zeroing" two Keyence laser micrometers on the center of the dowel pin. The stages are then manually actuated to position the antenna at zero. The tolerance stack-up is given in Table 3.

TABLE 3. Alignment Tolerances

Initial Setup	Table hole/Alignment disk	±0.0254 mm
	Alignment disk/ dowel pin	±0.0127 mm
	Alignment dowel pin/Fixture block	±0.0381 mm
	Fixture block/Fence block	±0.0000 mm
Test Tolerances	Fence block/BPM centerline	±0.0127 mm
Total Tolerance Stackup		±0.0889 mm

Several different approaches to determining the wire position were assessed, including the above-mentioned rf (and previously DC) edge detection, triangulation-laser, video and laser-shadow imaging. A commercial scanning laser micrometer was found to be most practical (5).

Table 4 lists the mechanical specifications of the commercial components.

TABLE 4. Mechanical Specifications

Component	Accuracy	Resolution	Repeatability
Daedal stage[1]	± 20.8 μm/mm		± 1.27 μm
Motor/driver[2]		4.1 μm /motor step	
Absolute encoder[3]		3.05 μm /encoder step	
Laser micrometer[4]	±0.20 μm	Set to ±2.54 μm	±0.31 μm

1. Daedal #106042P0E-LH.
2. Compumotor NEMA 23 motor, Zeta57-83-MO; Zeta4 driver (set to 50,000 steps/rev)
3. Compumotor AR-C absolute encoder (set to 16,384 steps/rev)
4. Keyence LS-5041/LS-5501

The Positioning System

There were several problems concerning the controller, motor drivers, and encoders. We attempted to overcome the previously mentioned antenna breakage (caused by the stages over-running limits) by beveling the ball-guide in the bottom paddle, using absolute-encoders, employing new more sophisticated motor drivers, and using a new controller (with packaged Labview software) which would allow complete coordination between the controller and the computer program. As the new Compumotor AT6400 controller does not support our absolute encoders, we were unable to use it on this mapper but will employ it on the 15 cm BPM mapper currently under construction.

Computer Control

A suite of Labview applications needed to be developed to do the main mapping functions as well as individually controlling and monitoring the subsystems and functions: antenna alignment, paddle positioning, arbitrary X-Y positioning, and power meter testing.

At program initiation, the operator puts the probe radius and map-grid step size into the main mapping program, running on a PC, which then computes and graphically displays the map circumference and the map points. The top paddle then moves into position, which lowers the antenna into the measurement position and, if necessary and selected by the operator, goes into "antenna zeroing" mode.

When mapping is begun, the antenna is stepped in a serpentine fashion, across the probe aperture, pausing at each point, while the rf signal transmitted to the four lobes is measured by a power meter and uploaded to the computer. The computer program then manipulates and stores the data as a text file.

Stimulus/Measurement System

An HP 8640B signal generator provides about 20dBm of CW rf power to the antenna.

Much effort went toward fulfilling the requirement of obtaining an rf power meter with sufficient dynamic-range accuracy. We decided to use a Boonton 4300 six-channel power meter with a dynamic range of –70 dBm to +20 dBm. Four channels were used to read the probe outputs, unlike previous mappers that used two channels alternately multiplexed by rf relays to the opposing ports.

During off-frequency testing of the stimulus/measurement system, we found power measurement anomalies at frequencies other than 350 MHz. Figure 2 shows four channels from the calibrated power meter as the (unleveled) frequency was swept.

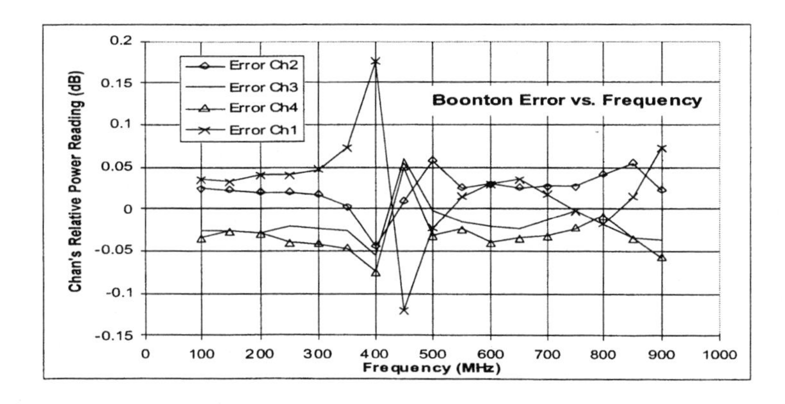

FIGURE 2. Power meter channel deviations from average.

It is apparent that the errors, in excess of 0.25 dB (p-p), are beyond the allowable errors for the beam position measurement (Table 1). Although the actual map is done only at 350 MHz, these deviations could emerge with even very small frequency variations, causing non-repeatable results. After reviewing various possible causes for the power meter errors, we finally determined that the underlying cause lay with problems in the power meter chassis, channel boards, and possibly in the power-head calibration. With spare power heads and channel boards, we were able to match combinations which had similar characteristics, such that the errors were within the LEDA BPM requirements. Figure 3 shows the results of differential measurements using these matched pairs. The power meter was sent back for warranty repair.

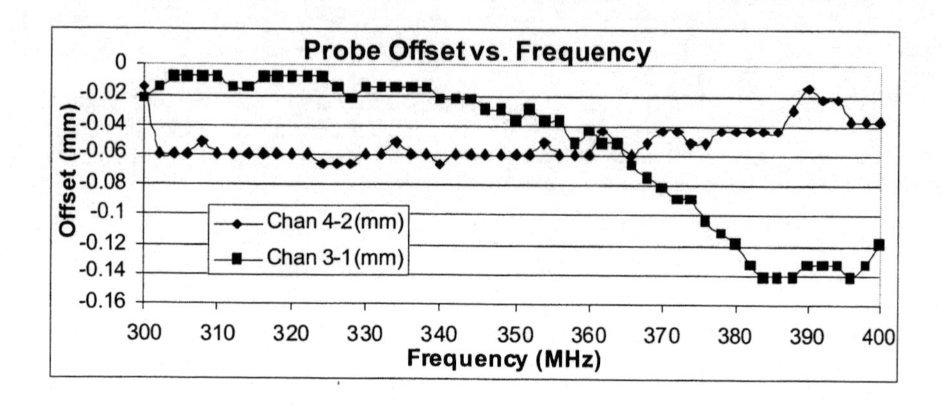

FIGURE 3. Paired channel differential errors.

ACKNOWLEDGMENTS

Much of the alignment strategy is owed to Rick Wood and Dale Schrage. Chris Rose and Dick Martinez also were invaluable in the early stages of development.

REFERENCES

[1] Shurter, R. B., J. D. Gilpatrick, "Tuned-Antenna Driver for Microstrip Probe Sensitivity Testing," Second Annual Accelerator Instrumentation Workshop, Batavia, IL, 1990.

[2] Gilpatrick, J. D., "LEDA Beam Position Tolerance Budget, v1.5," Internal Document, Los Alamos National Laboratory, January 1998.

[3] Power, John F., et al., "Characterization of Beam Position Monitors in Two-Dimensions," 1992 Linear Accelerator Conference Proceedings, Ottowa, August 1992, (in proceedings, pp. 362–364).

[4] Gilpatrick, J. D., "Derivation of the LANSCE BPM Nonlinear Beam Position Response," Internal Technical Note LA-CP-97-21, Los Alamos National Laboratory, February 1997.

[5] Keyence Corporation of America, Phone: 201–930–1400, Fax: 201–930–0088

500 MHz Narrowband Beam Position Monitor Electronics for Electron Synchrotrons

I. Mohos and J. Dietrich

Forschungszentrum Jülich GmbH
Institut für Kernphysik
Postfach1913, D-52425 Jülich, Germany

Abstract. Narrowband beam position monitor electronics were developed in the Forschungszentrum Jülich-IKP for the orbit measurement equipment used at ELSA Bonn. The equipment uses 32 monitor chambers, each with four capacitive button electrodes. The monitor electronics, consisting of an rf signal processing module (BPM-RF) and a data acquisition and control module (BPM-DAQ), sequentially process and measure the monitor signals and deliver calculated horizontal and vertical beam position data via a serial network.

INTRODUCTION

The beam position monitor system at ELSA Bonn (1) consists of 32 monitor chambers, each having four capacitive button electrodes. For position measurements, the narrowband signal chain tuned to the fundamental harmonic of the bunch frequency is used. The position data are proportional to the quotient of the difference and sum signals and to a coefficient depending on the monitor geometry (K_x, K_y). The electronics described in this paper uses the sequential processing method (2,3). In each acquisition cycle the button signals ($V_{LU}, V_{RU}, V_{LD}, V_{RD}$) are measured first, then the position data are computed as follows:

$$P_X = K_X * \frac{(V_{RU} + V_{RD} - V_{LU} - V_{LD})}{(V_{LU} + V_{RU} + V_{LD} + V_{RD})} \tag{1}$$

$$P_Y = K_Y * \frac{(V_{LU} + V_{RU} - V_{LD} - V_{RD})}{(V_{LU} + V_{RU} + V_{LD} + V_{RD})} . \tag{2}$$

CP451, *Beam Instrumentation Workshop*
edited by R. O. Hettel, S. R. Smith, and J. D. Masek

MEASUREMENT ELECTRONICS

The rf section (Fig.1), consisting of narrowband superheterodyne rf electronics, processes the fundamental bunch frequency (number of bunches along the accelerator ring, multiplied with the revolution frequency) component of the button signals. At the input, low-pass filters reject the higher harmonics, and a GaAs analog rf multiplexer scans sequentially the output of the filters. A low-noise narrowband preamplifier (*B*=5MHz) amplifies the selected low-level button-signal. For very high signal levels, a 30dB attenuator can be inserted. A GaAs mixer transposes the desired frequency range to the intermediate frequency, where narrowband filters reduce the if bandwidth to 220kHz and if amplifier with controlled gain enhances the signal level appropriately for demodulation.

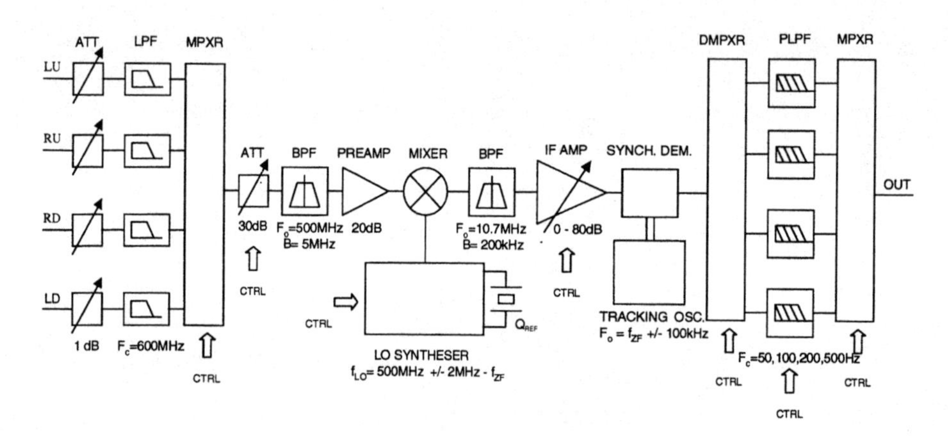

FIGURE 1. Block diagram of the rf signal processing module.

An on-board synthesizer generates the LO signal applied to the mixer. Its frequency determines the band-center frequency of the signal processing. Due to the low particle mass the revolution frequency is nearly constant in the acceleration ramp. Small frequency changes during the ramp within the if bandwidth will be automatically tracked by the demodulator in real time. Band-center frequency adjustments for special measurement purposes can be achieved by synthesizer remote control in the range of 500 MHz ± 2 MHz with 50 kHz steps.

The output signal of the highly linear synchronous demodulator is proportional to the rms value of the fundamental component of the amplified button signal and carries amplitude changes with frequencies up to 500 Hz. The overall demodulation bandwidth can be selected by means of switched filters to frequencies between 0.1 Hz and 500 Hz in 13 steps. The gain control range of the processing chain is about 100 dB. Signal levels between −80 dBm and +10 dBm are allowed. Scan timing and the step gain control are synchronized; four button signals will be measured in each cycle with the same gain, so consistent data can be used for position computing. The scan sequence of the button measurements is programmable.

The data acquisition module (Fig. 2) consists of an 8-bit microcontroller with an 8 kbyte EPROM and 32 kbyte RAM and built-in timer, a half-duplex 1 Mbit/s asynchronous serial interface, with a galvanic isolated twisted-pair transceiver for data communication, a 12-bit ADC for digitizing of the demodulated electrode signals, a 12-bit DAC for gain control, several bits for timing and bandwidth control, and a 3-wire serial interface for synthesizer control.

The timer of the microcontroller controls the rf multiplexer and the timing phases of the acquisition. After digitizing of the button signals, depending on the acquisition parameters, the microcontroller filters by means of a digital lowpass filter (0.1–500 Hz) and/or averages the signal values for the preset number (1–4096) of measurement cycles. Although the acquisition frequency is always 1 kHz, the transfer rate can be set by remote command to 1–256 ms, corresponding to the selected lowpass filter. Data transfer begins after the averaging is finished.

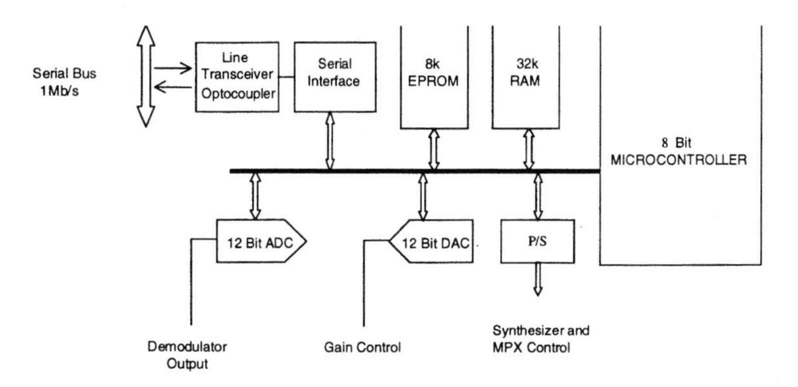

FIGURE 2. Block diagram of the data acquisition and control module.

DATA COMMUNICATION

A control module and an rf module form a BPM station. A group of eight stations is connected to the host computer via a serial bus with 100 m maximum length. The whole installation consists of four such groups, each with eight stations. The host initiates the data transfer; data collision is not allowed.

CONCLUSIONS

Compared to the parallel acquisition method (producing a sum and difference signal and processing them simultaneously with separate electronics), the sequential method measures the monitor signals with the same electronics. The time resolution of the monitor electronics described here is 1 ms; the position resolution is 0.5μm (@ B_{res}= 1 Hz, P_{in}= –40 dBm, K_{BPM}=14.5 mm). Advantages of the sequential data acquisition include:

- the minimization of differential measurement errors by using a single signal processing chain; and
- cheaper measurement hardware due to the single signal processing chain.

Possible disadvantages include:

- a 25% signal processing speed, compared with a parallel signal chain system having the same rf bandwidth; and
- pseudo beam position changes at rapidly changing signal levels (filling/fast extraction), due to the time delay between the four measurements within one cycle.

Proper additional circuits and algorithms prevent the theoretical disadvantages, and so sequential acquisition has proven to be the more accurate and cheaper solution in beam position electronics.

REFERENCES

[1] Schillo, M., "Das Strahldiagnosesystem für ELSA," PhD thesis at the University Bonn, Germany (October 1991).

[2] Biscardi, R., and J. W. Bittner, "Switched Detector for Beam Position Monitor," *Proc. of the 1989 IEEE Particle Accelerator Conference*, Chicago, IL, **3**, 1516 – 1518 (1989).

[3] Hinkson, J., "Advanced Light Source Beam Position Monitor," *AIP Conference Proceedings on Accelerator Instrumentation*, **252**, Newport News, 21–41 (1991).

The Calibration of BEPC Beam Position Monitors

K. Ye, L. Ma, and H. Huang

Division 10, IHEP, Beijing 100039, China

Abstract. The basic requirement for the BEPC beam position monitor is the measurement of the beam orbit with 0.1 mm precision near the collision point. To improve the measurement accuracy, the response of the beam position monitor pickups was mapped in the laboratory before they were installed in the BEPC ring. The microcomputer-controlled test set consists of high frequency coaxial switches to select each pickup electrode, a movable antenna to simulate the beam, a signal source, a spectrum analyzer to measure the pickup signals, and analysis software. The signal source operates well below the –3 dB cutoff frequency of the pickups (buttons). We believe that the low-frequency measurement yields the same information as the real beam. The button signals were clear. This calibration technique is satisfactory for BEPC operation.

INTRODUCTION

Four-button type beam position monitors (BPM) are used in the BEPC 2.2 GeV storage ring. The BPM assembly is an electrostatic type with four disk electrodes and BNC vacuum feedthrough connectors. The beam pipe is a cylinder. The buttons are rotated 45 degrees off vertical and horizontal axes to avoid the fan of synchrotron radiation. A tooling ball located on the top of the vacuum chamber is used as the fiducial mark for survey and alignment of the BPM assembly, relative to an adjacent quadrupole magnet. Electrical differences in the buttons and the mechanical installation tolerances cause the BPM to report beam offsets that are not real. We measured these offsets in our test set.

The button disk is made of stainless steel. The disks are welded to the center conductor of the BNC feedthroughs. The feedthroughs themselves are welded into the vacuum pipe. The disks are flush with the vacuum pipe wall. Button signals can reach a few GHz. A spectrum analyzer was used to measure button response.

The BPM quality was measured with three different tests. First, we measured the button response to a fast pulse on an antenna in the center of the chamber. This measurement gave the button sensitivity to beam current. In the second test, we excited the antenna with a low-frequency field with the antenna centered. This gave us the BPM offset due to electrical and mechanical errors. If the buttons had equal capacitance and were perfectly installed relative to the center of the beam pipe, there would be no offset error and all buttons would produce identical signals. The third test we performed was

CP451, *Beam Instrumentation Workshop*
edited by R. O. Hettel, S. R. Smith, and J. D. Masek

an evaluation of the BPM sensitivity to antenna position. Figure 1 shows button signals from a typical antenna scan.

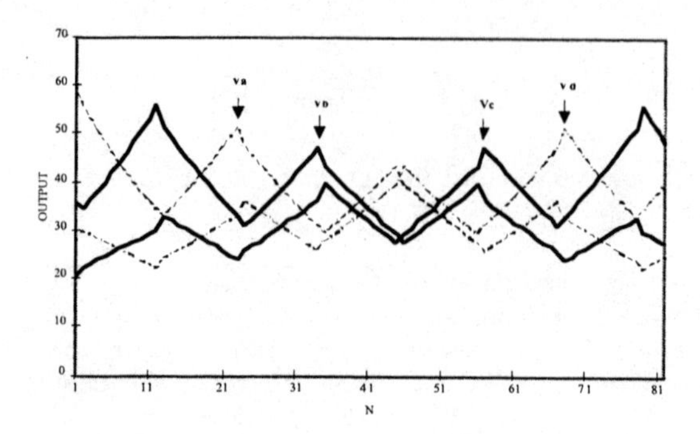

FIGURE 1. Button outputs over 10 × 10 mm scan.

The measuring setup is shown schematically in Figure 2. The antenna may be moved transversely inside the monitor along the *X*- and *Y*-axes. A SP6T rf switch is used to select signals from each button. The insertion loss of the switch was tested and corrections were made to the data. The attenuation of all button cables were measured. The mechanical error of the stepping motor was considered while the antenna moved. A personal computer (PC) controls the equipment via a PC I/O board, the RS232 port, and a GPIB board. The measurement is completely automatic.

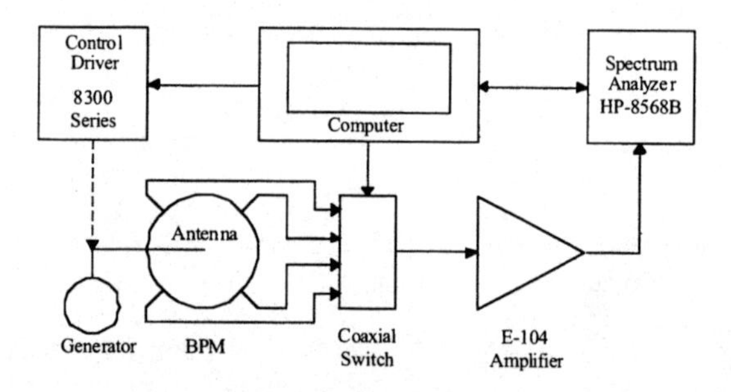

FIGURE 2. BPM calibration method schematic.

The button electrode faces a beam, and the button senses an image current Ib:

$$I_b = \frac{dQ}{dt} \propto \frac{a^2}{b} \frac{d\rho}{dt} \tag{1}$$

where ρ is the linear charge density, a is the radius of the button, and b is the radius of the duct (beam pipe).

The self-capacitance of the button to the wall of the beam duct is C_b. A load resistor R, in shunt with C_b, will produce a frequency-dependent coupling impedance to the beam that acts like a high-pass filter:

$$Z = \frac{R}{1 + i\omega RC_b} \tag{2}$$

The button output signal is as follows:

$$V = I_b Z = I_b \frac{R}{1 + i\omega RC_b} \tag{3}$$

$$V \propto \frac{a^2}{b}\left(\frac{i\omega}{\beta C}\right)\left(\frac{R}{1 + i\omega RC_b}\right)I(\omega) \approx \frac{a^2}{b}\left(\frac{i\omega}{\beta C}\right)RI(\omega) \tag{4}$$

where β is the beam velocity relative to light. Here C_b is very small, only 10 pF. With R equaling 50 ohms the –3 dB high-pass cutoff frequency is 318 MHz, given by

$$f_{3dB} = \frac{1}{2\pi RC_b}. \tag{5}$$

In our case, the antenna radiates a 4 MHz signal to simulate the beam in the beam duct. In this case, we believe that the low-frequency measurement yields the same information as the real beam using the usual difference/sum algorithm.

The antenna is moved by the stepper motors over a 20 by 20 mm area in 2 mm steps. The PC gives a conversion mapping of the normalized electrical position (H, V) to the mechanical position (X, Y) at the experimental spots. At the center of the BPM the response is linear. At a large antenna displacement the data show pincushion distortion. Fourth order polynomials are used to determine actual antenna position from the measured data:

$$X = \sum_{i=0}^{4}\sum_{j=0}^{i} A_{i-j,j} H^{i-j} V^j$$
$$Y = \sum_{i=0}^{4}\sum_{j=0}^{i} B_{i-j,j} H^{i-j} V^j \tag{6}$$

The $A_{0,0}$ $B_{0,0}$ terms show the offset of the electrical signal center from the geometric center.

Curve-fitting in Mathcad was used to extract polynomial coefficients from the experimental data using the least square method. The monitor has X–Y symmetry, so major distortion contributions come from the terms of $A_{1,0}$, $A_{1,2}$, $A_{3,0}$ for X, and $B_{0,1}$, $B_{2,1}$, $B_{0,3}$ for Y. The other terms are negligible.

CALIBRATION

Each BPM is calibrated before it is installed into the accelerator. Two kinds of calibration are necessary to convert the four button signals into beam position. One is obtaining the mapping diagram from the antenna position. The other calibration involves the antenna-setting error which we know from mechanical measurements. From its original position the antenna is moved step-by-step over the desired area in the center of the BPM assembly. At 4 MHz, the coupling impedance of the buttons to the antenna is low, so pre-amplifiers are used to boost the signals and to improve the measurement noise figure. The bipolar signal from these amplifiers is not easily measured on an oscilloscope to the required accuracy, so an HP8568B spectrum analyzer is used to measure the button signal amplitude. The signal from the analyzer screen is clear with a good signal-to-noise ratio.

Conversion between the signals from four buttons to the actual beam position is done by using polynomial expressions fitted to the BPM mapping. X and Y are nonlinear functions of H and V. Given the signal amplitudes at the four buttons, H and V are found by the usual difference over sum algorithm

$$H = \frac{V_A + V_D - V_B - V_C}{V_A + V_B + V_C + V_D}$$
$$V = \frac{V_A + V_B - V_C - V_D}{V_A + V_B + V_C + V_D} \tag{7}$$

where V_A, V_B, V_C, V_D are the output signals from corresponding button electrodes. Figure 3 shows the H and V values plotted on a rectangular grid.

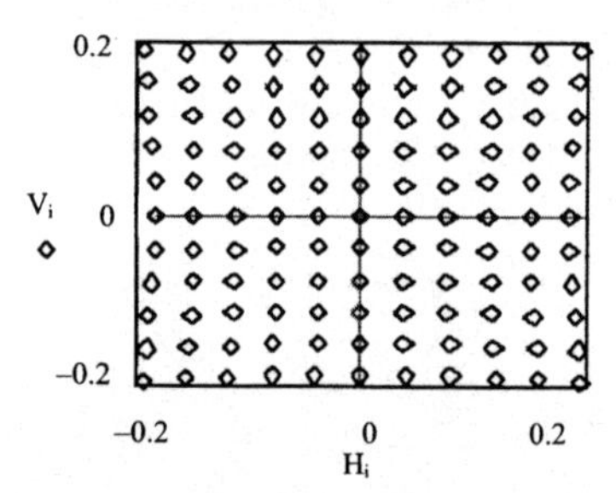

FIGURE 3. The H and V distribution from four buttons.

In Figure 3 pin cushion distortion is seen at large antenna offset from center. From these data the BPM sensitivity in %/mm may be found at any location in the measurement area. X' and Y' are found from

$$X' = X + \Delta X$$
$$Y' = Y + \Delta Y \tag{8}$$

where X and Y are geometrical coefficients and chamber electrical offsets, and ΔX and ΔY are offsets of the chamber mechanical center referred to the magnetic center of an upstream quadrupole magnet which is used as the primary reference point for beam position measurement. Figure 4 shows the calculated X and Y position found from above. Note that the pin cushion distortion has been eliminated.

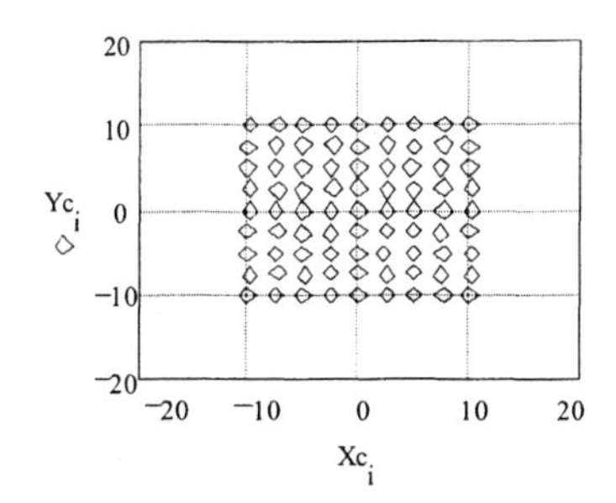

FIGURE 4. Using the fourth order polynomials to reconstruct X,Y from H,V.

During actual closed-orbit measurements we have obtained false position data from some BPMs mainly due to poor contact in switches. We have determined that we can obtain beam position from only three buttons. Below we show how to obtain beam position from buttons B, C, and D. Figure 5 shows the highly distorted H and V.

$$H = \frac{V_D - V_C}{V_C + V_D}$$
$$V = \frac{V_B - V_C}{V_B + V_C} \tag{9}$$

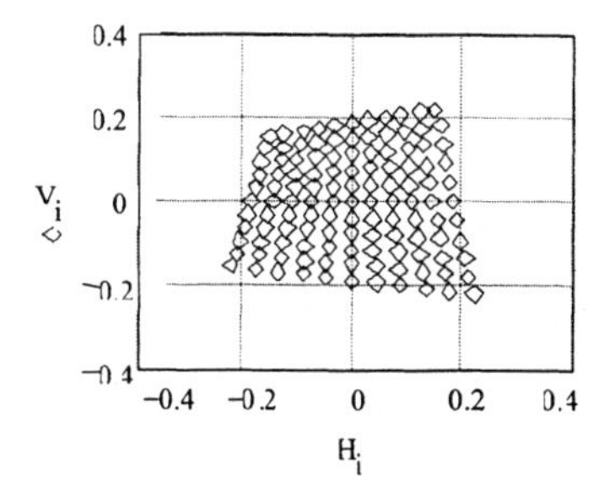

FIGURE 5. H and V calculated from buttons B, C, and D.

When all button signal paths function correctly we find the same beam position calculated in five ways. We compare signals from (a,b,c,d), (a,b,c), (b,c,d), (c,d,a), and (d,a,b). These calculations agree to 0.02 mm in the laboratory calibration. In cases where a large error is found the bad data are rejected. If the difference from five calibrations is larger than 0.4 mm, the data are considered bad. This is a good way for us to determine if we have a defective channel. Figure 6 shows X and Y for signals obtained from buttons B, C, and D.

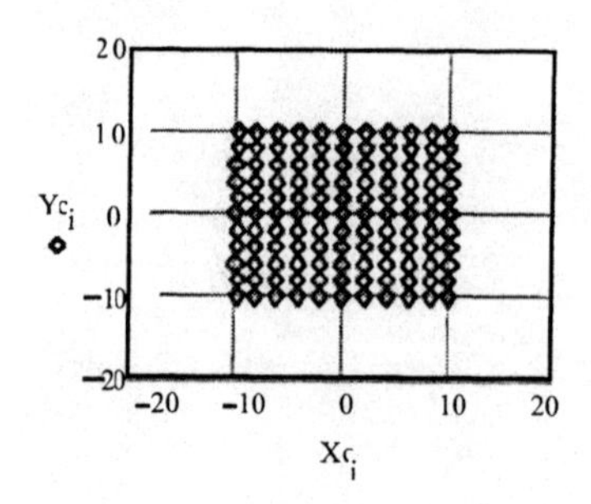

FIGURE 6. *X* and *Y* calculated from 3 buttons. It is essentially the same as Figure 4.

ERROR ANALYSIS

In the BPM calibration the maximum fitting error and rms errors are found as follows:

$$\sigma_{X_p} = \sqrt{\frac{\sum_{n=1}^{N}(\overline{X}_P - X_P(i))^2}{N}} \tag{10}$$

$$\sigma_{Y_p} = \sqrt{\frac{\sum_{n=1}^{N}(\overline{Y}_P - Y_P(i))^2}{N}} \tag{11}$$

where N is the number of measurements taken at each point, $\overline{X}_P$ is the X coordinate average value of pth point, and $\overline{Y}_p$ is the Y coordinate average value of the pth point. If we measure 25 points, the steps are 5 mm, and the fitting error is 0.05 mm. If we measure 81 points, the steps are 2.5 mm, and the error is about 0.03 mm.

The relative accuracy for a selected button signal measurement is

$$\frac{\Delta V}{V} = \sqrt{\frac{1}{4}\sum_{j=1}^{4}\frac{1}{P}\sum_{k=1}^{P}\frac{1}{N-1}\sum_{i=1}^{N}\frac{\left(\frac{V_{ijk}}{\sum_{j=1}^{4}V_{ijk}} - \frac{1}{N}\sum_{i=1}^{N}\frac{V_{ijk}}{\sum_{j=1}^{4}V_{ijk}}\right)^2}{\frac{1}{N}\left(\sum_{i=1}^{N}\frac{V_{ijk}}{\sum_{j=1}^{4}V_{ijk}}\right)}} \tag{12}$$

where in $1 \leq k \leq P$, P is the number of measured points; $1 \leq j \leq 4$ is a sum over the four buttons; and in $1 \leq i \leq N$, N is the number of measurements at every point.

In this case we tested every point 100 times. The relative error is less than 0.1%.

CONCLUSION

We have shown a method to test the BEPC BPM pickups in the laboratory. All error terms are compensated. We have demonstrated a technique for eliminating the geometric distortion inherent in all BPM pickups of this type. In addition we have shown how we can obtain good beam position readings even if one of four signal channels has become defective.

In the future, the antenna will be improved. We are considering working at a higher frequency to determine whether our assumption that low-frequency measurements are adequate is correct.

ACKNOWLEDGMENTS

The authors are grateful to Cai Yixing for doing the mechanical design for this system and J. Hinkson for his many helpful suggestions and discussions. We also thank our colleagues who contributed to the tests, and J. Sebek and R. Schlueter offered who helped us with our work.

REFERENCES

[1] Cheng, Y., G. Decker, "Beam Position Monitor Calibration for the Advanced Photon Source," presented at BIW '94.
[2] Hinkson, J., Advanced Light Source Beam Position Monitor, Berkeley.
[3] Sebek, J., R. Hettel, "SSRL Beam Position Monitor Detection Electronics," presented at BIW '94.
[4] Ieiri, T., H. Ishil, Y. Mizumachi, A. Ogata, J.-L Pellegrin, and M. Tejima, "Performance of the beam position monitor of the TRSTAN accumulation ring," IEEE Trans. NS-30 (1983) 2356.
[5] Smith, S. R., "Beam Position Monitor Engineering," BIW '96, *ATP Conf. Proc.* **390**, 1997, p. 350.
[6] Meller, R. E., C. R. Dunnam, "Beam Position Monitor for the CESR Linac", *Proc. of the 1989 IEEE PAC.*
[7] Shintake, T., M. Tejima, "Sensitivity Calculation of Beam Position Monitor Using Boundary Element Method," *Nuclear Instruments & Methods in Physics Research*, **A254** (1987).

Optimization of Four-Button BPM Configuration for Small-Gap Beam Chambers*

S. H. Kim

Advanced Photon Source
Argonne National Laboratory
9700 South Cass Avenue
Argonne, Illinois 60439 USA

Abstract. The configuration of four-button beam position monitors (BPMs) employed in small-gap beam chambers is optimized from 2-D electrostatic calculation of induced charges on the button electrodes. The calculation shows that for a narrow chamber of width/height (w/h) >> 1, over 90% of the induced charges are distributed within a distance of 2h from the charged beam position in the direction of the chamber width. The most efficient configuration for a four-button BPM is to have a button diameter of (2–2.5)h with no button offset from the beam. The button sensitivities in this case are maximized and have good linearity with respect to the beam positions in the horizontal and vertical directions. The button sensitivities and beam coefficients are also calculated for the 8 mm and 5 mm chambers used in the insertion device straight sections of the 7 GeV Advanced Photon Source.

INTRODUCTION

Circular button electrodes are commonly used as beam position monitors (BPMs) in a variety of particle accelerators (1, 2). For highly relativistic filamentary beams of electrons or positrons, the Lorentz contraction compresses the electromagnetic field of the charged beam into the 2-D transverse plane. This results in the induced currents on the beam chamber wall having the same longitudinal intensity modulation as the charged beam. When the wavelength of the beam intensity modulation is large compared to the button diameters, the calculation of the induced charges on the buttons may be simplified as a 2-D electrostatic problem. In the insertion device (ID) straight sections of the 7 GeV positron storage ring for the Advanced Photon Source (APS), beam chambers 8 mm and 5 mm in height are used to optimize the ID magnetic parameters. In this paper the configuration of four-button BPMs in a small-gap beam chamber is optimized, and BPM sensitivities and coefficients are calculated assuming that the button electrodes are flush with the chamber wall.

---

* Work supported by the U.S. Department of Energy, Office of Basic Energy Sciences under Contract No. W-31-109-ENG-38.

CP451, *Beam Instrumentation Workshop*
edited by R. O. Hettel, S. R. Smith, and J. D. Masek
1998 The American Institute of Physics 1-56396-794-4/98/$15.00

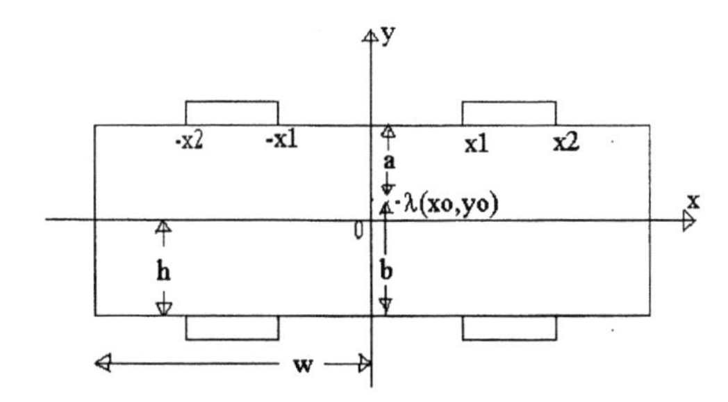

FIGURE 1. Cross section of a beam chamber with a height of 2h and width of 2w. The diameter of the four button electrodes for the BPM system is $(x_2 - x_1)$, and $b = h + y_o$, $a = h - y_o$.

IMAGE CHARGES

Assuming that the width of the beam chamber in Figure 1 is much larger than the height (w >> h), the induced charges are calculated by the method of image charge. The beam chamber is also assumed to have a high electric conductivity and is grounded. Then the vertical positions of the positive and negative image charges of a charge λ at (x_o, y_o) are given by

$+\lambda$ at $y = 2m(a+b) + y_o = 4mh + y_o$, $(m = -\infty, 0, \infty)$

and $-\lambda$ at $y = 2a + 2m(a+b) + y_o = 2a + 4mh + y_o$(for integer m, $-\infty<m<\infty$) (1)

with $a = h-y_o$ and $b = h+y_o$. For ease of calculation the vertical position for $(-\lambda)$ is shifted by $2h$ so that $y' = y-2h = 4mh - y_o$. (In the 3-D geometry the charge λ is a line-charge density along the z direction.) Then the electrostatic potential distribution $\Phi(x,y)$ within the chamber may be calculated from

$$\frac{-\lambda}{2\pi\varepsilon_0}\ln\frac{|z-z_o|\prod_{m=1}^{\infty}|(z-z_m)(z-z_{-m})|}{|z'-z'_o|\prod_{m=1}^{\infty}|(z'-z'_m)(z'-z'_{-m})|} = \Phi(x,y) = \frac{-\lambda}{2\pi\varepsilon_0}\ln\frac{\prod_{m=-\infty}^{\infty}|z-z_m|}{\prod_{m=-\infty}^{\infty}|z'-z'_m|}, \qquad (2)$$

where ε_0 is the permittivity in free space, $z = x + i\,y$, $z' = x + i\,y'$, $z_{\pm m} = x_o + i\,(\pm 4mh+y_o)$, and $z'_{\pm m} = x_o + i\,y' = x_o + i\,(\pm 4mh-y_o)$. Equation (2) may be simplified to a closed form

$$\Phi(x,y) = \frac{-\lambda}{2\pi\varepsilon_0} \mathrm{Re}\left\{ \ln \frac{\sin\pi(\frac{z - x_o - iy_o}{4hi})}{\sin\pi(\frac{z' - x_o + iy_o}{4hi})} \right\}$$

$$= \frac{-\lambda}{4\pi\varepsilon_0} \ln \frac{\cosh\pi\frac{x - x_o}{2h} - \cosh\pi\frac{y - y_o}{2h}}{\cosh\pi\frac{x - x_o}{2h} + \cosh\pi\frac{y - y_o}{2h}} \tag{3}$$

where y' is shifted back to $y + 2h$ in the final expression.

The induced charge densities per x/h in the top and bottom surfaces of the chamber, σ_t and σ_b, are calculated from $[-\varepsilon_o\, d\Phi/dy]_{y=\pm h}$

$$\sigma_t = -\frac{\lambda}{4} \frac{\cos py_o}{\cosh p(x - x_o) - \sin py_o},$$

$$\sigma_b = -\frac{\lambda}{4} \frac{\cos py_o}{\cosh p(x - x_o) + \sin py_o}. \tag{4}$$

Here $p = \pi/2$ and by setting $h = 1$ the coordinate system is normalized to the half height of the chamber. By adding up the induced charges in the top and bottom surfaces in Equation (4), the total induced charge, which should be proportional to the sum signal for a typical four-button BPM system of Figure 1, is given by

$$Q_s = Q_s(x_2) - Q_s(x_1) = \int_{x1}^{x2} (\sigma_t + \sigma_b)dx + \int_{-x2}^{-x1} (\sigma_t + \sigma_b)dx$$

$$= -\lambda \frac{1}{2} \int_{x1}^{x2} \{ \frac{\cos py_o \cosh p(x - x_o)}{\cosh^2 p(x - x_o) - \sin^2 py_o} + \frac{\cos py_o \cosh p(x + x_o)}{\cosh^2 p(x + x_o) - \sin^2 py_o} \} dx. \tag{5}$$

The induced charges proportional to the signals for the vertical and horizontal positions of the charged beam, Q_y and Q_x, may be calculated from Equation (4):

$$Q_y = Q_y(x_2) - Q_y(x_1) = \int_{x1}^{x2} (\sigma_t - \sigma_b)dx + \int_{-x2}^{-x1} (\sigma_t - \sigma_b)dx, \tag{6}$$

$$Q_x = Q_x(x_2) - Q_x(x_1) = \int_{x1}^{x2} (\sigma_t + \sigma_b)dx - \int_{-x2}^{-x1} (\sigma_t + \sigma_b)dx. \tag{7}$$

Here Q_y and Q_x are the differences in the induced charges between the top and bottom, and right and left buttons, respectively. As one expects from beam position measurements, Q_y is an odd function in y_o and even in x_o, and Q_x is the opposite. After Taylor expansions up to the third order in the charged beam position (x_o, y_o), indefinite integrals of Equations (5–7) are given by

$$\frac{Q_s(x)}{-\lambda} = \frac{1}{p}\tan^{-1}[\sinh px] + (y_o^2 - x_o^2)\frac{p\sinh px}{2\cosh^2 px} + y_o^2 x_o^2 \frac{p^3}{4}(\frac{\sinh px}{\cosh^2 px} - \frac{6\sinh px}{\cosh^4 px}), \quad (8)$$

$$\frac{Q_y(x)}{-\lambda} = y_o[\tanh px + x_o^2 p^2 \frac{-\sinh px}{\cosh^3 px}] + y_o^3[p^2 \frac{\sinh px}{3\cosh^3 px}$$

$$+ x_o^2 p^4 (\frac{2\sinh px}{3\cosh^3 px} - \frac{2\sinh px}{\cosh^5 px})], \quad (9)$$

$$\frac{Q_x(x)}{-\lambda} = x_o[-\mathrm{sec}h(px) + y_o^2 p^2\{\frac{1}{2}\mathrm{sec}h(px) - \mathrm{sec}h^3(px)\}] + x_o^3[\frac{p^2}{6}\{2\mathrm{sec}h^3(px)$$

$$-\mathrm{sec}h(px)\} + y_o^2 p^4\{2\mathrm{sec}h^5(px) - \frac{5}{3}\mathrm{sec}h^3(px) + \frac{1}{12}\mathrm{sec}h(px)\}]. \quad (10)$$

To the first order in y_o/h and x_o/h, calculations of the induced charges from $x_1 = 0$ to $x_2 = \infty$ in Equations (8–10) give $Q_y = -\lambda y_o/h$, $Q_x = -\lambda x_o/h$, and the total induced charge $Q_s = -\lambda$ as expected. The derivatives of $Q_s(x)$, $Q_y(x)$, and $Q_x(x)$ with respect to x/h may be called "the effective induced charge densities for the sum, vertical, and horizontal signals." The first terms of the charge densities and Equations (8–10) are plotted in Figure 2.

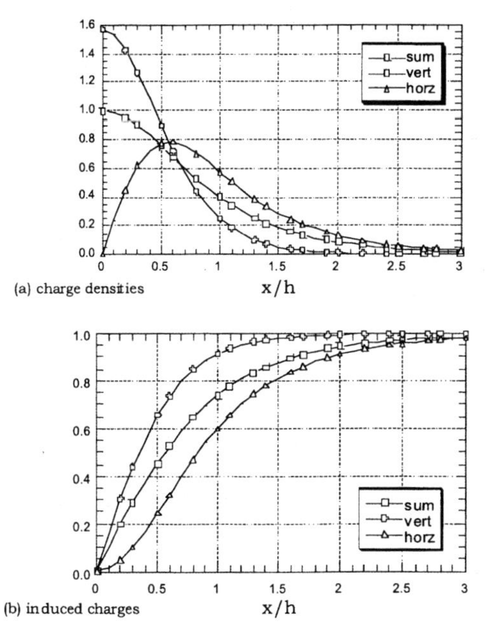

FIGURE 2. (a) Induced charge densities and (b) induced charges integrated from 0 to x/h. The induced charges and densities corresponding to Q_s, Q_y, and Q_x in Equation (8) are denoted as sum, vert, and horz in the legends, respectively, with units of $-\lambda$, $-\lambda y_o/h$, and $-\lambda x_o/h$.

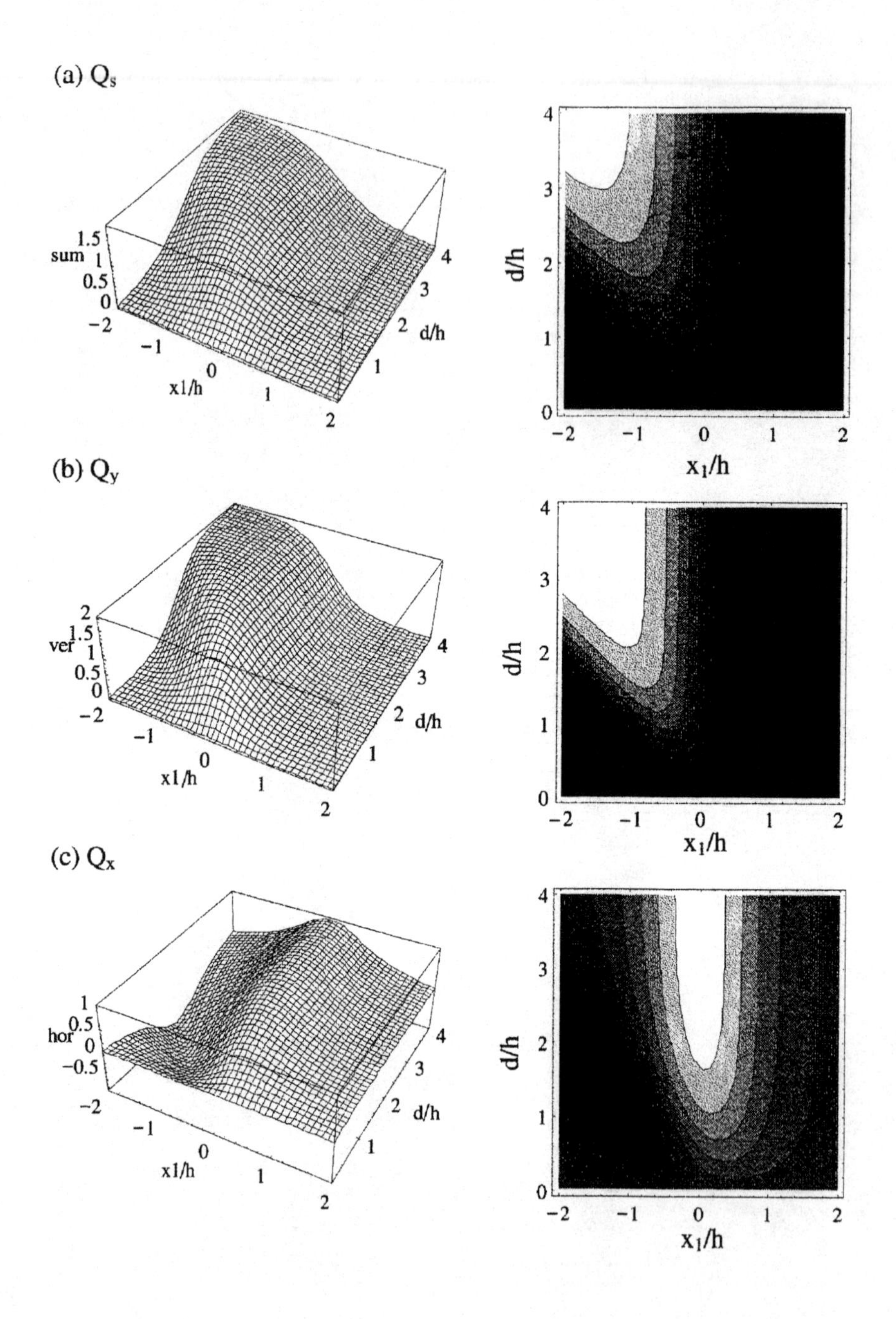

FIGURE 3. 3-D plots of the induced charges for Q_s, Q_y, and Q_x as functions of normalized button offset x_1/h and button diameter d/h on the left side, and their contour plots on the right side. The respective units for Q_s, Q_y, and Q_x are $-\lambda$, $-\lambda y_o/h$, and $-\lambda x_o/h$.

For small buttons (e.g., $x/h < 0.5$), when the beam is located near the origin, the horizontal beam displacement is not as sensitive to changes in the distances between the beam and the buttons as the vertical beam displacement. This makes the density distribution for Q_x broad with the peak near $x/h = 0.6$. The density for Q_y, on the other hand, has its peak at $x/h = 0$. This implies that, when the measurements of vertical displacements are critical for a beam chamber of small height, the location of the buttons should include the range of small x/h. For buttons located in the range of $x = 0 - 2h$ with button diameter of $2h$, for example, over 94%, 99%, and 91% of the available sensitivities for sum, vertical, and horizontal can be registered on the buttons. Therefore, any buttons located more than $2h$ (one chamber height) from the beam position in the horizontal direction would be very inefficient.

Shown in Figure 3 are 3-D plots and their contours for Q_s, Q_y, and Q_x. The negative position of x_1 is possible by rotating the four-button system with respect to the vertically symmetrical axis. For $x_1 = 0$ and a button diameter d larger than $2h$, it is seen that Q_s, Q_y, and Q_x do saturate as already expected from Figure 2. When the buttons are extended to both sides of the x-axis by rotating the four-button system and the diameter is larger than $4h$, the values of Q_s and Q_y increase by a factor of 2 because most parts of the buttons are still located within $x/h < 2$ where the charge densities are high. On the other hand, Q_x decreases because the charge density for Q_x is asymmetric with respect to x. Therefore, a four-button system with a button diameter of approximately (2~2.5)h and a button offset of $x_1 = 0$ would collect nearly all the induced charges and be the most efficient.

BUTTON SENSITIVITIES

With $x_1 = 0$ and $d = 2h$, where the button diameter d is $(x_2-x_1)h$, Q_s, Q_y, Q_x, and their normalized values to Q_s are calculated from Equations (5–7). The results give an optimized BPM configuration and are plotted in Figure 4 as functions of the normalized beam position (y_o/h, x_o/h). The button sensitivities and coefficients for y_o/h and x_o/h for the optimized configuration are calculated from Equations (8–10).

Optimized configuration:

$$Q_s = 0.945[1.0 + 0.07143\ \{(y_o/h)^2 - (x_o/h)^2\} + 0.0842\ (x_o/h)^2(y_o/h)^2],$$

$$Q_y = 0.9963[\{1.0 - 0.01836\ (x_o/h)^2\}(y_o/h) + \{0.00612 + 0.0295\ (x_o/h)^2\}(y_o/h)^3,$$

$$Q_x = 0.9137[\{1.0 + 1.4649\ (y_o/h)^2\}(x_o/h) - \{0.4883 + 2.7354\ (y_o/h)^2\}(x_o/h)^3],$$

$$Q_y/Q_s = 1.0542[\{1 + 0.0531\ (x_o/h)^2\}(y_o/h) - \{0.0653 + 0.0631\ (x_o/h)^2\}(y_o/h)^3],$$

$$Q_x/Q_s = 0.9669[\{1 + 1.3935\ (y_o/h)^2\}(x_o/h) - \{0.4169 + 2.6159\ (y_o/h)^2\}(x_o/h)^3]. \quad (11)$$

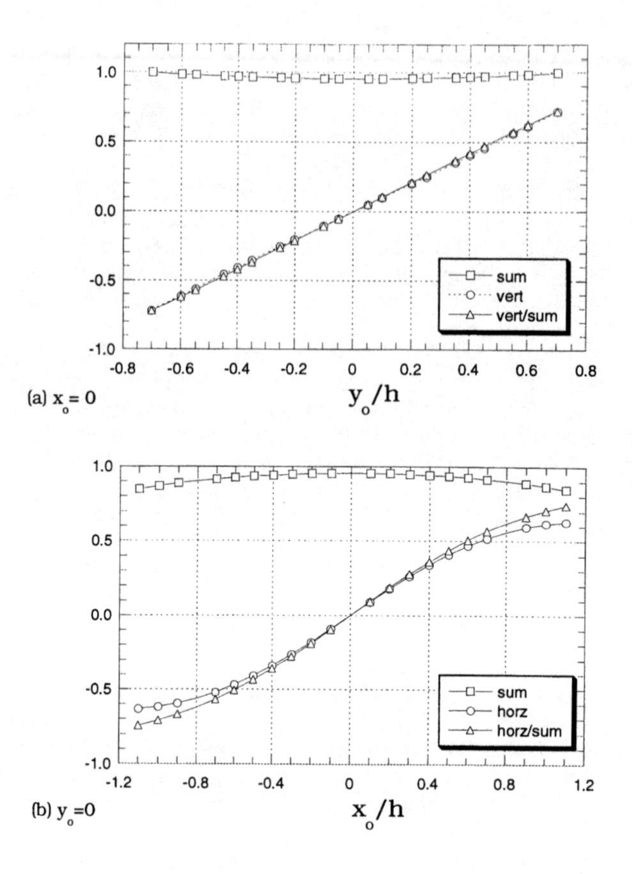

FIGURE 4. For the optimized BPM configuration, normalized button positions of $x_1/h = 0$ and $x_2/h = 2$ (normalized diameter $d/h = 2$), variations (a) Q_s, Q_y, and Q_y/Q_s are plotted as a function of normalized vertical beam position y_o/h for $x_o = 0$, and (b) Q_s, Q_x, and Q_x/Q_s as a function of normalized horizontal beam position x_o/h for $y_o = 0$. Here Q_s, Q_y, and Q_x are denoted as sum, vertical, and horizontal and their respective units are $-\lambda$, $-\lambda y_o/h$, and $-\lambda x_o/h$.

Figure 4(a) and Equation (11) show that the vertical signals Q_y and Q_y/Q_s within $\pm 0.7 y_o/h$ have excellent linearity in y_o/h and x_o/h. This is particularly important since vertical measurements are generally critical in a small chamber height. The horizontal signals Q_x and Q_x/Q_s, on the other hand, are less linear compared to those for the vertical as seen from Figure 4(b) and the coefficients of y_o/h and x_o/h in Equation (11).

In the APS storage ring, beam chambers with relatively small chamber heights are used for the IDs in the straight sections (3). Several four-button BPMs with button diameters of 4 mm and button-center separations of 9.65 mm have been installed for chamber heights of 8 mm ($h = 4$ mm, $x_1 = 0.7075h$, $x_2 = 1.7075h$, diameter $= 1.0h$) and 5 mm ($h = 2.5$ mm, $x_1 = 1.132h$, $x_2 = 2.732h$, diameter $= 1.6h$). One can see from Figure 2 that these buttons are located at relatively inefficient positions compared to the optimized case of $x_1 = 0$ and $x_2 = 2h$. The button sensitivities and y_o and x_o coefficients for the two chambers are:

APS ID chamber (2h = 8 mm):

$$Q_s = 0.3178[1.0 + 0.0529\ \{x_o^2 - y_o^2\} + 0.00778\ x_o^2 y_o^2],$$

$$Q_y = 0.0465[\{1.0 + 0.2199\ x_o^2\}\ y_o + \{-0.0733 + 0.00228\ x_o^2\} y_o^3,$$

$$Q_x = 0.1144[\{1.0 - 0.00738\ y_o^2\}\ x_o + \{0.00246 + 0.00827\ y_o^2\} x_o^3],$$

$$Q_y/Q_s = 0.1464\ [\{1 + 0.1669\ x_o^2\}\ y_o + \{-\ 0.2033 + 0.00502\ x_o^2\}\ y_o^3],$$

$$Q_x/Q_s = 0.360\ [\{1 + 0.0456\ y_o^2\}\ x_o + \{-0.5049 + 0.00229\ y_o^2\}\ x_o^3]. \qquad (12)$$

The smallest aperture APS chamber (2h = 5 mm):

$$Q_s = 0.1957[1.0 + 0.1817\ \{x_o^2 - y_o^2\} - 0.0104\ x_o^2 y_o^2],$$

$$Q_y = 0.02205[\{1.0 + 0.7246\ x_o^2\}\ y_o - \{0.2415 + 0.1285\ x_o^2\} y_o^3,$$

$$Q_x = 0.1205[\{1.0 - 0.1509\ y_o^2\}\ x_o + \{0.0503 + 0.0136\ y_o^2\} x_o^3],$$

$$Q_y/Q_s = 0.1127\ [\{1 + 0.5429\ x_o^2\}\ y_o - \{0.0598 + 0.00858\ x_o^2\}\ y_o^3],$$

$$Q_x/Q_s = 0.6156\ [\{1 + 0.0308\ y_o^2\}\ x_o - \{0.1314 + 0.0146\ y_o^2\}\ x_o^3]. \qquad (13)$$

As seen from Equations (12) and (13), the most critical signals Q_y for 8 mm and 5 mm chambers are only 0.046 and 0.022 of the unit $-\lambda y_o/h$. Compared to Q_y, the horizontal signals Q_x are over 0.11 of the unit $-\lambda x_o/h$ for both chambers. Even if the normalized signals are not too small (because of the small values of Q_s), one should expect that the noise/signal ratios for Q_y/Q_s and Q_x/Q_s in Equations (12) and (13) are relatively large compared to those in Equation (11).

ACKNOWLEDGMENTS

The author would like to thank Glenn Decker for his numerous suggestions and useful discussions concerning this work.

REFERENCES

[1] Shafer, R. E., "Beam Position Monitoring," presented at the First Accelerator Instrumentation Workshop, Upton, NY, Oct. 23–26, 1989, *AIP Conference Proceedings* **212**, 26-58 (1989).
[2] Barry, W. C., "Broad-Band Characteristics of Circular Button Pickups," *Proceedings of the Fourth Accelerator Instrumentation Workshop*, Berkley, CA, Oct. 27-30, 1992, *AIP Conference Proceedings,* **281**, 175–184 (1993).
[3] Lumpkin, A. H., "Commissioning Results of the APS Storage Ring Diagnostics System," *Proceedings of the Seventh Accelerator Instrumentation Workshop, Argonne*, IL, May 6-9, 1996, *AIP Conference Proceedings,* **390**, 152–172 (1997).

Advanced Photon Source Monopulse RF Beam Position Monitor Front-End Upgrade*

Robert M. Lill and Glenn A. Decker

Advanced Photon Source, Argonne National Laboratory
9700 South Cass Avenue, Argonne, Illinois 60439 USA

Abstract. This paper will describe and analyze the rf beam position monitor (RFBPM) front-end upgrade for the Advanced Photon Source (APS) storage ring. This system is based on amplitude-to-phase (AM/PM) conversion monopulse receivers. The design and performance of the existing BPM front-end will be considered as the base-line design for the continuous effort to improve and upgrade the APS beam diagnostics. The upgrade involves redesigning the in-tunnel filter comparator units to improve insertion loss, return loss, and band-pass filter-matching that presently limit the different fill patterns used at APS.

INTRODUCTION

The Advanced Photon Source (APS) is a third-generation synchrotron x-ray source that provides intense x-rays for basic and applied research. The stability of the x-ray beam is imperative for the operation of the APS. The storage ring beam stability must be less than 17 microns rms horizontally and 4.5 microns rms vertically in the frequency range from DC to 30 Hz. This beam stability is largely dependent on the quality and accuracy of the BPM system. The rf beam position monitor (RFBPM) system provides single turn capabilities for commissioning and diagnosing machine problems. The RFBPMs also provide the input to the beam feedback systems and the beam position limit detector (BPLD). The specification of the APS storage ring RFBPM system is listed in Table 1.

The RFBPM upgrade will provide improved signal strength to the input of the receiver. This will enable us to utilize the top end of the receiver's dynamic range. The upgrade will also reduce VSWR problems and band-pass filter side lobes that will allow the accelerator to be operated with less dead time between bunch trains. The upgrade will also address the calibration and maintainability of the system.

* Work supported by U.S. Department of Energy, Office of Basic Energy Sciences under Contract No. W-31-109-ENG-38.

CP451, *Beam Instrumentation Workshop*
edited by R. O. Hettel, S. R. Smith, and J. D. Masek
1998 The American Institute of Physics 1-56396-794-4/98/$15.00

RFBPM BASE LINE IMPLEMENTATION

The measurement of the APS storage ring beam position is accomplished by 360 RFBPMs located at approximately 1-degree intervals around the 1104 m ring circumference (1). The RFBPM processing electronics are located above the tunnel in 40 VXI crates with 9 channels per crate (2). The RFBPM signal processing topology used for the APS storage ring is a monopulse amplitude-to-phase (AM/PM) technique for measuring the beam position in the *x*- and *y*-axes. A logarithmic amplifier channel measures the beam intensity.

The RFBPM system provides the following capabilities:

- Measures beam position both during injection at 2 Hz and with stored beam.
- Provides single-bunch tracking around the ring.
- Measures position of different bunches at each BPM turn-to-turn.
- Measures position at each turn (3.68 μs revolution period).
- Provides average beam position for higher accuracy.
- Provides 32,768 samples of the beam history for each BPM.

TABLE 1. Specification of the Present APS Storage Ring RFPBM System

Parameter	Specified Value
First turn, 1 mA resolution/accuracy	200 μm / 500 μm
Stored beam, single or multiple bunches resolution/accuracy @ 5 mA total	25 μm / 200 μm
Stability, long term	±30 μm
Dynamic range, intensity	≥40 dB
Dynamic range, position, standard configuration	±20 mm
Dynamic range, position, 5 mm aperture chamber	±2 mm

Analysis of the Filter Comparator

The design and performance of the existing BPM front end will be considered as the baseline design for the continuous effort to improve and upgrade the APS beam diagnostics. A block diagram of the filter comparator is shown in Figure 1. The primary function of the filter-comparator unit is to convert the voltage impulse from the buttons into pulse modulated signals at 351.93 MHz, the ring's rf frequency. It also compares the four rf signals to create a beam intensity signal and two deviation signals, one for the *x*-axis and one for the *y*-axis.

The original filter-comparator design shown in Figure 1 uses 6 dB pads to match the button outputs and 2 dB pads to help match the input of the band-pass filters. The filters, hybrid comparator, and pads add up to a total insertion loss of 15 dB. This reduces the in-band power into the receiver to less than –8 dBm @100 mA with the standard fill pattern. The standard fill pattern is 10 mA in a cluster of 6 bunches followed by 90 mA in 25 triplets. The RFBPMs are presently configured to sample the 10 mA bunch of 6, or target cluster. A considerable dead time of hundreds of ns is necessary prior to the arrival of the target cluster to avoid the effects of time-domain side lobes and small reflections. The present goal is to fill the entire ring with singlets or triplets evenly spaced around the ring with as little as approximately 100 ns dead time between bunch trains.

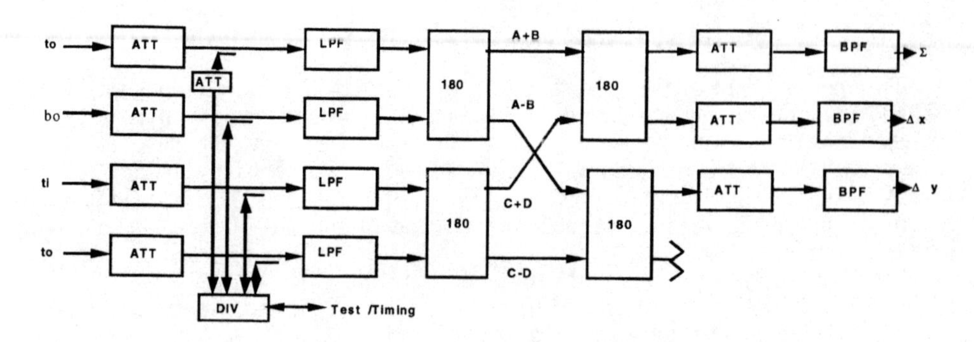

FIGURE 1. Filter-comparator block diagram (original design).

It is desirable to operate the system such that the maximum receiver input (+5 dBm) is realized in order to minimize the noise. The output noise of the receiver can be described as:

$$\Delta/\Sigma \text{ sensitivity} = 1 \text{volt} / 90 \text{ degrees} \tag{1}$$

$$\text{Phase jitter } \Delta\theta = 1 / \sqrt{\text{SNR}} \text{ rads} \tag{2}$$

$$\text{Receiver output noise} = \text{Phase jitter } x \; \Delta/\Sigma \text{ sensitivity} \tag{3}$$

The thermal noise power (kTB) for the 20 MHz bandwidth is –91 dBm. Since there are two channels, the noise is noncoherent and will sum for a total equivalent noise of –88 dBm.

The other problem with this design is the time-domain side lobe caused by the phase response of the band-pass filters (27 dB down) specified at 60 dB. The side lobes and reflections become a problem when the storage ring is completely filled and there is minimum dead time between bunches.

There are other problems with maintaining a system that is partitioned with the receiver front end located in the tunnel. It becomes very difficult to troubleshoot and isolate problems between the buttons and the receiver that arise during run periods.

Upgrade Design Approach

The upgrade involves redesigning the in-tunnel filter-comparator units to improve insertion loss, return loss, and band-pass filter impulse response that presently limit the different fill patterns used at APS. The design improvements will be delineated into two phases. The first phase involves improving the signal strength and matching the output of the button electrodes into 50 ohms. The second phase will replace the existing filter comparator with improved components to minimize allowable cluster spacing.

Reviewing Figure 1, we notice 8 dB of insertion loss due to the attenuators. These attenuate standing waves between the filter comparator and the button. The new design (Figure 2) will eliminate the need for the pads by carefully matching the components and, most importantly, the source.

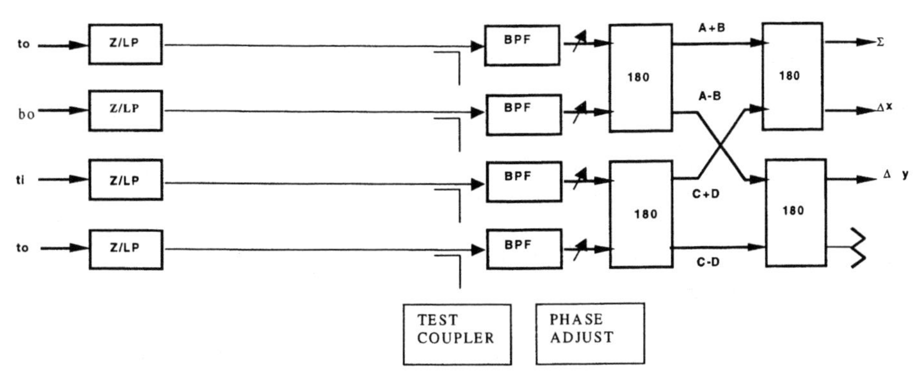

FIGURE 2. Filter-comparator block diagram (upgrade design).

The buttons have a very poor return loss when measured from the feedthrough side, which results in the reflection of 97% of the power. The button impedance is principally reactive with a small resistive component. To match the button's impedance, an inductor is placed in parallel with the capacitive electrode. This effectively creates a parallel resonate circuit, driven by the image current source. This technique of resonating the capacitive pickup provides a controllable response in a compact electrode design (3). The total impedance (*Ztotal*) and voltage developed by the button (*Vbutton*) and can be described as follows:

$$Ztotal = 1/(1/Zres + 1/Zcoax) \tag{4}$$

$$Vbutton = Ztotal \times Iimage \tag{5}$$

As the equivalent impedance of the resonant circuit, *Zres*, increases, the total impedance, *Ztotal*, looks more like *Zcoax* or 50 ohms.

The matching network will also include a low-pass filter that will provide an additional 46 dB of filtering at the second harmonic (704 MHz). The overall bandwidth will be maximized to 100 MHz at the –3dB power point in order to minimize the effect of button-to-button differences in capacitance and to ensure no interaction of the downstream band-pass filter. The implementation of the button-matching network will be considered phase 1 of the front-end upgrade and will be used together with the existing filter-comparator units.

The second phase will involve replacing the 180-degree hybrid comparators and band-pass filters. The hybrid comparators will be implemented using a rat-race bridge topology, either laid out with mini coax or stripline. The rat-race hybrid will provide predictable and stable performance over the required 50 MHz bandwidth and will have less than 0.7 dB insertion loss and a 30 dB return loss at 352 MHz.

The most difficult and critical part of phase 2 will be the implementation of the band-pass filters. The filters must be matched in phase and amplitude to ensure the vector addition and subtraction of the input signals. They must be phase matched to within 5 degrees over 20 cycles and amplitude matched better than 0.2 dB (100 μm) across the passband. They must also have time-domain side lobe rejection of 50 dB minimum at 100 ns or greater.

Tests using band-pass filters with single-pole quarter-wave cavity resonators have had good success. The filters exhibit very low loss and can be matched to within 0.1 dB of each other with minimum effort. The one problem encountered with the coaxial cavity is that we require the filter to ring down 50 dB in 100 ns so there is no interaction between bunches. Experiments have been conducted that trade off bandwidth for ring down time. At this time we have determined empirically that a 16 MHz bandwidth is the limit, due to the required data acquisition time of 100 ns. Ideally we need 100 ns of continuous wave 352 MHz output to the receiver with reflection and side lobes down a minimum of 60 dB.

Another implementation of the band-pass filter that is being investigated is the transversal filter shown in Figure 3.

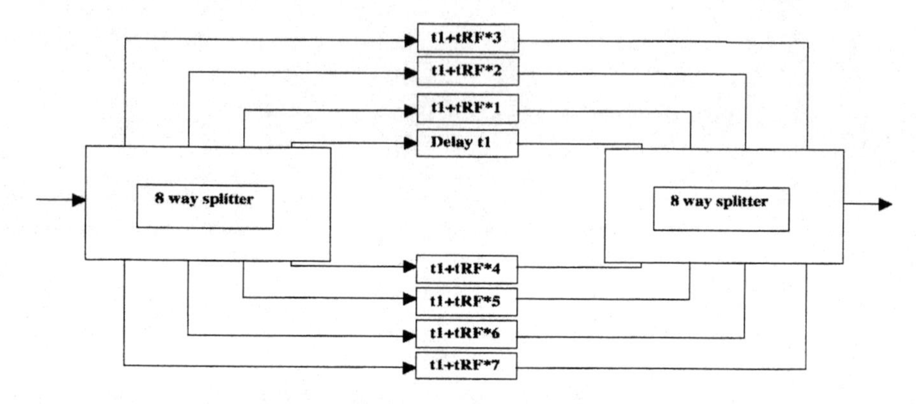

FIGURE 3. Transversal filter block diagram.

This device is basically a pulse repeater that delays each pulse or bunch of pulses by multiples of 2.84 nanoseconds with respect to each other. The result is a (SIN F/F)^2 response that has good matching characteristics unit to unit and good cancellation over all frequencies and times. The filter must be designed such that a minimum of 24 pulses, or 24×2.84 ns, are generated, which will yield a 68 ns pulse train. This is the minimum number of pulses required due to the 50 ns integration time and timing considerations. It is desirable to increase the number of pulses, provided that the overall size requirements can be maintained. A prototype set of filters has been constructed, using coaxial cable delays, with favorable results. Presently we are trying to find a cost-effective way of implementing this design on standard low-cost microwave board materials. We are also investigating implementing the delay using lumped components in order to minimize the cost and size of the filter assembly.

An alternative transversal filter implementation under investigation employs surface acoustic wave (SAW) band-pass filters. These filters exhibit high stability and reliability with good performance and no adjustments. This technology is available in the frequency and bandwidth required and has the same (SIN F/F)^2 performance as the transversal filter described above. The insertion loss is typically 4 dB for a filter similar to those meeting our requirements. Such filters are commonly used in many receiver band-pass filter applications.

Another important consideration of the upgrade will be the improvement of self-test capability, maintainability, and calibration of the RFBPM system. The test couplers will provide access to the input signals to aid in the troubleshooting and isolation of problems occurring during run periods. They will also be used as injection inputs for a self-test module that will provide input stimulus to an entire sector of BPMs during maintenance periods for calibration.

Preliminary Performance for Upgrade

The upgrade described in Figure 2 was implemented in the APS storage ring with 16-MHz cavity band-pass filters and the M/A-COM H-9 monopulse comparator network located above the tunnel.

The matching network improves the button's return loss to the specified value at 352 MHz of 30 dB and results indicate better than 34 dB. The Smith chart in Figure 4 depicts the button and matching network after installation; it shows significant improvement over the button alone. Figure 5 gives the output of one button impedance matching network when driven by 6 bunches at 1.67 mA/bunch.

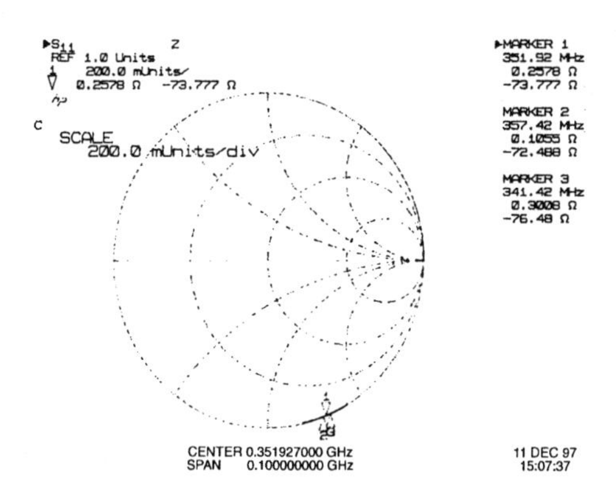

FIGURE 4. Measured impedance of a typical button electrode with matching network.

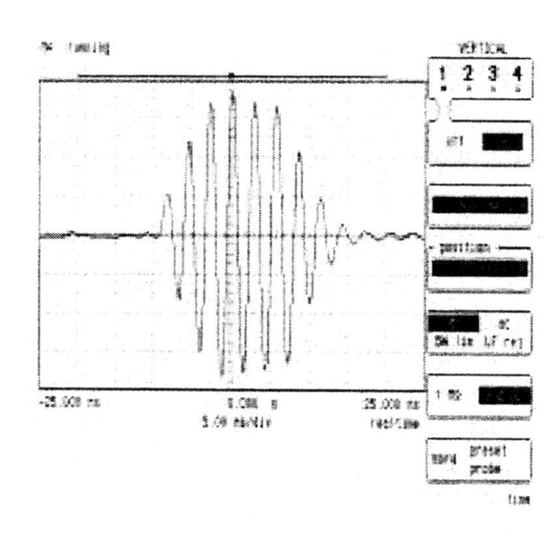

FIGURE 5. Output of impedance matching network with 6 bunches at 1.67 mA/bunch.

Figure 6 shows the sum of the 16 MHz BW cavity resonators into the receiver. The receiver output is seen in Figure 7 with the top trace being the Σ and the lower trace Δ/Σ. An overall improvement of 22 dB was realized from the original design. This is due to a 13.5 dB improvement of insertion loss from the filter comparator (15 dB–1.5 dB) and a gain from the buttons of 8.5 dB.

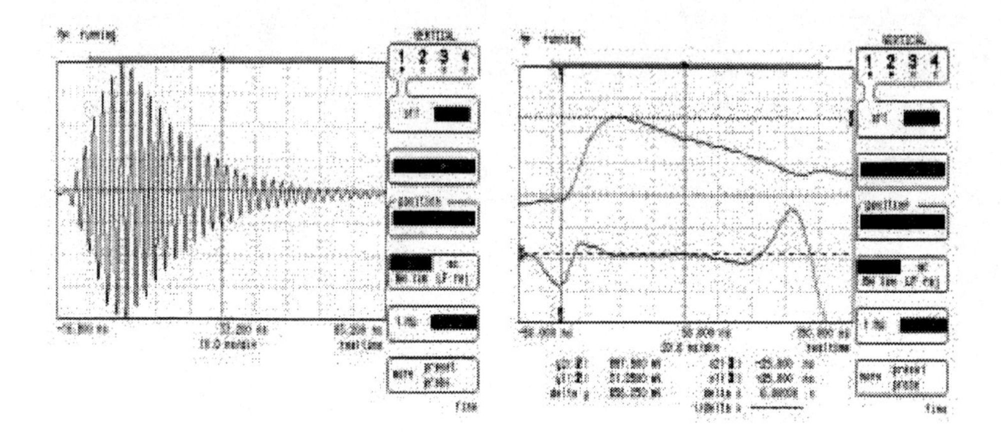

FIGURE 6. Sum of the cavity band-pass filter with 6 bunches at 1.67 mA/bunch.

FIGURE 7. Receiver output with 6 bunches at 1.67 mA/bunch.

CONCLUSION

The testing to date is very encouraging, and most of the system specifications are satisfied. Phase 1 implementation is planned for the July 1998 shutdown for 9 sets of BPMs. Data logging studies over long periods are planned to prove stable operation. The final band-pass filter and comparator configuration implementation will follow early next year.

REFERENCES

[1] Kahana, E., "Design of Beam Position Monitor Electronics for the APS Storage Ring," *Proceedings of the 3rd Accelerator Beam Instrumentation Workshop*, AIP Conference Proceedings **252**, pp. 235–240 (1992).

[2] Lenkszus, Frank R., Emmanuel Kahana, Allen J. Votaw, Glenn A. Decker, Young-joo Chung, Daniel J. Ciarlette, Robert J. Laird, "Beam Position Monitor Data Acquisition for the Advanced Photon Source," *Proceedings of the 1993 Particle Accelerator Conference*, pp. 1814–1816 (1993).

[3] Glen Lambertson, "Dynamic Devices-Pickups and Kickers," AIP Conference Proceedings **153**, Volume 2, pp. 1413–1442 (1987).

Calibration of an Advanced Photon Source Linac Beam Position Monitor Used for Positron Position Measurement of a Beam Containing Both Positrons and Electrons*

Nicholas S. Sereno

Advanced Photon Source, Argonne National Laboratory
9700 South Cass Avenue, Argonne, Illinois 60439 USA

Abstract. The Advanced Photon Source (APS) linac beam position monitors can be used to monitor the position of a beam containing both positrons and electrons. To accomplish this task, both the signal at the bunching frequency of 2856 MHz and the signal at 2 × 2856 MHz are acquired and processed for each stripline. The positron beam position is obtained by forming a linear combination of both 2856 and 5712 MHz signals for each stripline and then performing the standard difference over sum computation. The required linear combination of the 2856 and 5712 MHz signals depends on the electrical calibration of each stripline/cable combination. In this paper, the calibration constants for both 2856 MHz and 5712 MHz signals for each stripline are determined using a pure beam of electrons. The calibration constants are obtained by measuring the 2856 and 5712 MHz stripline signals at various electron beam currents and positions. Finally, the calibration constants measured using electrons are used to determine positron beam position for the mixed beam case.

INTRODUCTION

The APS linear accelerator is used to accelerate a positron beam to 450 MeV and deliver it to a positron accumulator ring (PAR). The positron beam is produced by irradiating a tungsten target with a 1.7A, 200 MeV electron beam (1). Low-energy positrons and electrons generated in the pair-production process are captured by the S-band (2856 MHz) linac rf wave and accelerated together to the end of the linac. Positron

* Work supported by the U.S. Department of Energy, Office of Basic Energy Sciences under Contract No. W-31-109-ENG-38.

CP451, *Beam Instrumentation Workshop*
edited by R. O. Hettel, S. R. Smith, and J. D. Masek
1998 The American Institute of Physics 1-56396-794-4/98/$15.00

position measurement using traditional log-ratio or difference-over-sum techniques breaks down for the mixed species beam because the S-band signal of each BPM pickup contains information about the electron beam current and position. As previously described (2) the first harmonic signal (2×S-Band or 5712 MHz) of each BPM pickup can be used along with the S-Band signal to determine the position of both the electrons and positrons in the mixed species beam. This paper describes a calibration procedure developed for an APS linac BPM using a pure beam of electrons. The calibration constants so determined for each BPM stripline are then used to determine electron and positron position for the mixed species beam used for injection into the PAR during normal operations.

BPM STRIPLINE SIGNALS FOR A MIXED SPECIES BEAM

The output of a pair of APS linac BPM striplines at an integer multiple *n* of the bunching frequency can be written as (2)

$$V^S(\omega_n) = \xi^S(\omega_n)\{I_p(1 \pm \eta^S(\omega_n)y_p) - (-1)^n I_e(1 \pm \eta^S(\omega_n)y_e)\}, \tag{1}$$

where I_p, y_p, I_e and y_e denote the positron and electron beam current and position, respectively, and $\omega_n = n\omega_o$, where ω_o is the (S-Band) bunching frequency 2856 MHz. In Equation (1), the + sign is used for the top ($S = T$) stripline and the – sign is used for the bottom ($S = B$) stripline (3). $\xi^S(\omega_n)$ is a parameter that depends on the bunch length, BPM stripline length, detection electronics bandwidth, cable attenuation, and number of positron or electron bunches in a linac pulse. $\eta^S(\omega_n)$ is the BPM sensitivity that depends on the azimuthal angle subtended by the stripline and the stripline distance from the center of the beam pipe. Ideally, $\eta^S(\omega_n)$ is a constant independent of frequency but in practice can depend on frequency due to the fact that at high frequencies one cannot neglect the effect of the stripline connection to the output transmission line.

Using Equation (1), the BPM stripline signals can be written as

$$W_n^S = (I_p - (-1)^n I_e) \pm \eta_n^S(I_p y_p - (-1)^n I_e y_e), \tag{2}$$

where the sign convention is the same as in Equation (1) and the shorthand notation

$$W_n^S \equiv \frac{V^S(\omega_n)}{\xi^S(\omega_n)} \tag{3}$$

$$\eta_n^S \equiv \eta^S(\omega_n) \tag{4}$$

$$\xi_n^S \equiv \xi^S(\omega_n) \tag{5}$$

is used. For n=1,2 Equation (2) represents four equations that can be solved for the electron and positron positions as functions of the measured signals and calibration parameters. The algebra is tedious but straightforward, and the result is given here in the following list of equations and definitions:

$$U_n^{\pm} \equiv \eta_n^B W_n^T \pm \eta_n^T W_n^B \tag{6}$$

$$\chi_n^- \equiv \frac{U_n^-}{2\eta_n^T \eta_n^B} \tag{7}$$

$$\theta_n^- \equiv \chi_n^- - \frac{\eta_n^\Delta U_n^+}{2\eta_n^\Sigma \eta_n^T \eta_n^B} \tag{8}$$

$$\eta_n^\Delta \equiv \eta_n^B - \eta_n^T \tag{9}$$

$$\eta_n^\Sigma \equiv \eta_n^B + \eta_n^T \tag{10}$$

$$y_{p,e} = \eta_1^\Sigma \eta_2^\Sigma \left(\frac{\theta_1^- \pm \theta_2^-}{\eta_2^\Sigma U_1^+ \pm \eta_1^\Sigma U_2^+} \right) \tag{11}$$

$$I_{p,e} = \frac{1}{2}\left(\frac{U_1^+}{\eta_1^\Sigma} \pm \frac{U_2^+}{\eta_2^\Sigma} \right) \tag{12}$$

In Equations (11) and (12) the + sign is used for the positron position and the – sign is used for the electron position. Equation (11) reduces to the familiar difference over sum result for the case where η_n^S is constant.

CALIBRATION OF APS LINAC BPM STRIPLINES USING AN ELECTRON BEAM

The calibration procedure for a pair of BPM striplines consists of using an electron beam to measure the calibration constants ξ_n^S and η_n^S by changing the electron beam position using a corrector for various beam currents. For each position and beam current the n=1 signal at 2856 MHz and the n=2 BPM stripline signal at 5712 MHz are detected using an HP8562A spectrum analyzer connected to the striplines by approximately 30 m of quarter-inch heliax. The spectrum analyzer was set up to detect pulsed rf with rf bandwidth (RBW) = 1 MHz, video bandwidth (VBW) = 3 MHz, and a span of 16 MHz. The linac was set to produce 30 ns beam pulses at a 30 Hz rate. Each beam pulse consisted of 86 electron and positron bunches separated by half an rf wavelength at 2856 MHz (2). A side benefit of using a high-energy electron beam to calibrate the striplines is that scraping is minimized because of the low beam emittance.

From Equation (1), the power output in dBm of the striplines as a function of electron beam position and current can be written as

$$P_n^S = 20\log\left(\frac{\xi_n^S I_e}{\sqrt{50\Omega \bullet 10^{-3}}}\right) \pm \left(\frac{20}{\ln 10}\right)\eta_n^S y_e, \tag{13}$$

$$P_n^S = P^S(\omega_n), \tag{14}$$

where the + sign refers to the top stripline (S=T) and the – sign refers to the bottom stripline (S=B). Equation (13) shows that the slope of a linear fit to the power output for a given frequency labelled by n is proportional to η_n^S. In addition, by measuring the power as a function of position for various beam currents, the gain parameter ξ_n^S is determined by a linear fit to the position intercept of Equation (13) as a function of beam current.

The position of a 580 MeV electron beam was changed over a range of 5 mm in the vertical plane for various beam currents using a corrector upstream of the test BPM striplines. The corrector excitation current was calibrated for the beam position by recording the electron beam position using the BPM striplines connected to standard processing electronics. Each stripline was bench calibrated to determine its sensitivity in dB/mm using a wire strung down the center of the BPM excited at the bunching frequency 2856 MHz and moved to various transverse positions (4). This means that the calibration constants determined in this procedure ultimately derive from the original BPM bench calibration measurements. Each position scan was repeated for various electron beam currents from 5 to 31 mA. The beam current was measured using a wall current monitor adjacent to the BPM striplines under calibration.

TABLE 1. Calibration Parameters Determined Using a Pure Electron Beam

Stripline (S)	**Frequency Index n**	$\frac{\xi_n^S}{\sqrt{50\Omega \bullet 10^{-3}}}$ **mA^{-1}**	$\left(\frac{20}{\ln 10}\right)\eta_n^S$ **dB/mm**
Top - (T)	1	2.63×10^{-3}	0.84
Top - (T)	2	0.12×10^{-3}	1.11
Bottom - (B)	1	2.78×10^{-3}	0.88
Bottom - (B)	2	0.40×10^{-3}	0.40

Table 1 summarizes the calibration results. The stripline sensitivity parameter η_n^S is given in terms of the average slope from Equation (13) found for all beam currents. The gain parameter ξ_n^S is determined from the best fit line to the position intercept from Equation (13) vs. the beam current. One can see by comparing the parameters in the table that this pair of striplines is quite well matched at 2856 MHz compared to 5712 MHz. When designing a pair of striplines specifically for the purpose of detecting the beam positions of particles in a mixed species beam, some care should be taken to try to match the striplines at both frequencies of interest.

POSITRON AND ELECTRON POSITION MEASUREMENT FOR A MIXED SPECIES BEAM

The calibration constants determined in the previous section are now used to measure the position of electrons and positrons in the mixed species beam used to inject into the PAR. The HP8562A spectrum analyzer was set up as in the calibration procedure except for the span. At each frequency, the spectrum analyzer was set to zero span so that fluctuations in each stripline signal could be observed over time and averaged. Zero span proved helpful when measuring the 5712 MHz signal, which was nearly at the noise floor of –60 dBm. Nominal beam energy was measured to be 395 MeV for the positrons and 380 MeV for the electrons by using a spectrometer located at the position of the BPM striplines under test.

Figure 1 shows the derived position of the electrons and positrons as a function of corrector current. The corrector was the same one used in the calibration procedure described in the previous section. The figure clearly shows the electrons and positrons moving in opposite directions under the influence of the corrector field. The ratio of the absolute value of the electron-to-positron slopes from the best fit line in the figure is 0.98±0.13, which agrees with the energy ratio 0.96 to within measurement error.

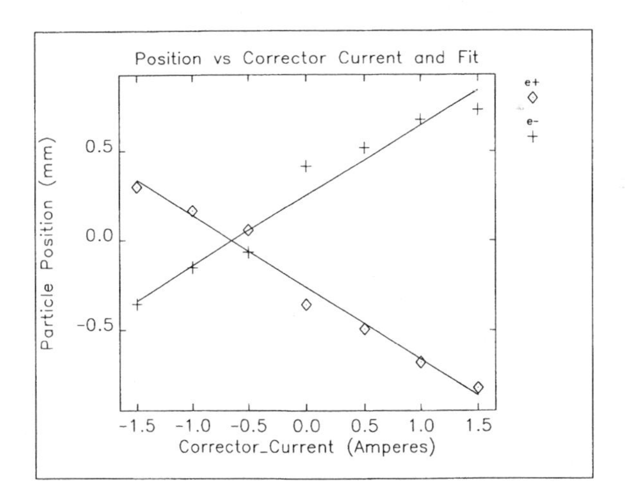

FIGURE 1. Positron and electron position vs. corrector current.

Figure 2 shows the electron and positron beam current for each corrector current. The figure indicates there is probably some amount of scraping occurring at the extreme values of corrector current, and hence, the beam position. The average positron and electron currents from the figure are seen to be about 15 and 11 mA, respectively. This is about a factor of two higher than the values recorded at the Faraday cup current monitors located downstream of the spectrometer. The difference is most likely due to the relative calibration between the Faraday cup and the wall current monitor used in the electron beam calibration procedure.

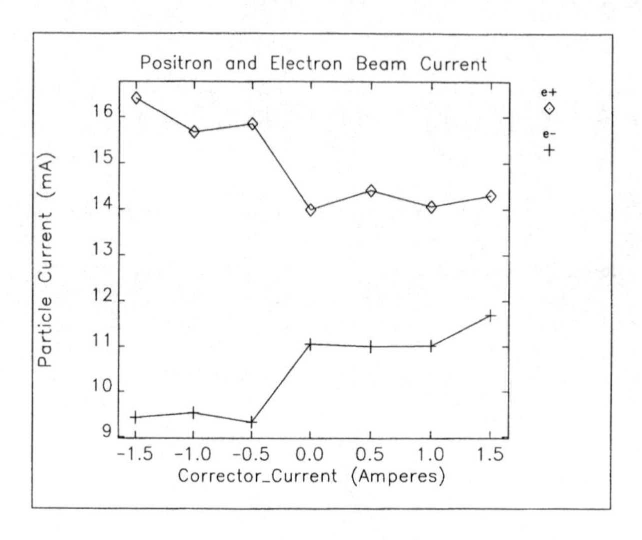

FIGURE 2. Positron and electron beam current vs corrector position.

CONCLUSION

This paper has described a calibration procedure used to calibrate a pair of APS linac striplines to measure the particle position of a mixed species beam consisting of electrons and positrons. The calibration procedure consisted of measuring the stripline signals at 2856 and 5712 MHz for various positions and beam currents for a pure electron beam. The calibration constants determined were then used to derive the electron and positron beam positions for the mixed species case. The final calibration of the stripline ultimately derives from the bench calibration used for the single particle beam case (an electron beam for the APS linac). The noise apparent in Figure 1 is mostly due to the fact that the 5712 MHz signal for both striplines was near the noise floor at –60 dBm. A practical design of a BPM system used to detect the position of a mixed species beam should include optimized striplines at 5712 MHz rather than the bunching frequency (2856 MHz). This should not be a problem since the fundamental beam signal is naturally large in any case.

ACKNOWLEDGMENTS

The author would like to thank M. Borland, G. Decker, L. Emery, R. Fuja, J. Galayda, and M. White for providing valuable insight and comments regarding the ideas and analysis presented in this work. The author is grateful for the assistance of M. White in tuning the APS linac to provide the beams used in these measurements as well as her assistance with data taking. The author would like to thank C. Gold for assistance in setting up the spectrum analyzer used to record stripline signal power.

REFERENCES

[1] White, M. et al., "Performance of the Advanced Photon Source (APS) Linear Accelerator," *Proceedings of the 1995 Particle Accelerator Conference,* Dallas, *TX*, pp. 1073–1075 (1996).

[2] Sereno, N. S. and R. Fuja, "Positron Beam Position Measurement for a Beam Containing Both Positrons and Electrons," *Proceedings of the Seventh Beam Instrumentation Workshop, Argonne, Il, AIP Conference Proceedings* **390,** pp. 316–323 (1997).

[3] Shafer, R. E., *AIP Conference Proceedings,* **249**, p. 608 (1992).

[4] Fuja, R., private communication.

Beam Position Monitors for the Fermilab Recycler Ring

E. Barsotti, S. Lackey, C. McClure, R. Meadowcroft

Fermi National Accelerator Laboratory
P.O. Box 500, Batavia, Illinois 60510

Abstract. Fermilab's new Recycler Ring will recover and cool "used" antiprotons at the end of a Tevatron store and also accumulate "new" antiprotons from the antiproton source. A wideband rf system based on barrier buckets will result in unbunched beam, grouped in one to three separate partitions throughout the ring. A new beam position monitor system will measure position of any one partition at a time, using low-frequency signals from beam distribution edges. A signal path including an elliptical split-plate detector, radiation-resistant tunnel preamplifiers, and logarithmic amplifiers, will result in a held output voltage nearly proportional to position. The results will be digitized using Industry Pack technology and a Motorola MVME162 processor board. The data acquisition subsystem, including digitization and timing for 80 position channels, will occupy two VME slots. System design will be described, with some additional emphasis on the use of logamp chips.

INTRODUCTION

Fermilab's Recycler Ring will be an 8 GeV storage ring constructed of permanent magnets (1). It will increase Tevatron collider luminosity in two ways. First, "used" antiprotons at the end of a store will be "recycled" and cooled in the Recycler. Also, accumulator high antiproton production rate will be maintained by periodically sending its stack to the Recycler for storage. The entire project was a late addition to the Main Injector project, funded with contingency money. The resulting time and money constraints played a role during subsystem development.

Besides the task of orbit measurement, the Beam Position Monitor system will be used for ion clearing. In fact, this dominated the decision to use two BPM detectors per half-cell, or 414 BPMs for the 3320-meter ring. The associated beamlines have an additional 32 BPMs. A cost-saving move briefly considered was to instrument only half the detectors.

CP451, *Beam Instrumentation Workshop*
edited by R. O. Hettel, S. R. Smith, and J. D. Masek

The Recycler will have up to three partitions of beam at a time: cooled, warmer, and injected beam. These partitions will be separated by six "barrier buckets"created by the wideband rf system. No other rf will be used to modulate the beam, so the roughly 1 μs beam distribution edges at the barrier buckets will produce the only signals for BPM processing. This necessitates good low-frequency sensitivity, which, in turn, requires a high termination impedance and preamplifier near the detector. The rest of the processing chain is based on logarithmic amplifiers, or "logamps," before a timing module and commercial digitizers.

TUNNEL ELECTRONICS

The preamp and its associated circuitry reside in a metallic rf box, mounted on the base of the detector stand at the tunnel ceiling. Short coaxial cables and an additionalial grounding strap connect the box and the detector. The 0.3 m-long elliptical split-plate detectors match the Recycler beam pipe shape, with axis dimensions 96 mm by 44 mm. Particularly over the inner half-aperture, the vertical and horizontal detectors provide a nearly linear response to beam displacement, 0.32 and 0.7 dB/mm, respectively. Orthogonal transverse plane dependence is insignificant.

The parallel combination of detector and cable capacitance and termination resistor determine the low-frequency corner in a high-pass filter model of the detector (2). Setting that frequency (and resistor) involved a tradeoff. It must be high enough so that the signal from one beam partition decays before the next, to prevent interference. The anticipated gap between beam partitions is 1-2 μs. Too high a corner, however, decreases signal strength and decay rate unnecessarily. A simulation program was written to determine the optimal low-frequency corner of 300 kHz. Signal-to-noise optimization set a lowpass filter corner at 2 MHz. A shunt 50 MHz low-pass filter to 100 Ω was added to improve beam impedance at higher frequencies. Figure 1 shows actual detector and preamp response to a simulated beam signal. The analog processing focuses on the decaying signal, so matching the falltimes of the two plate signals is important.

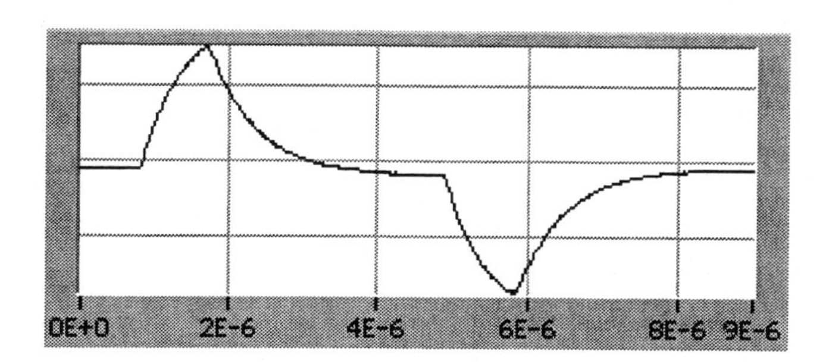

FIGURE 1. Response of a detector and preamp module to simulated beam signals.

A capacitive detector model is used here, since the detector is terminated with high impedance at the preamp. Above the corner frequency, signal strength on the detector is determined by:

$$V_c(t) = \frac{l\, I_b(t)}{2C_{tot}\ \beta_b\, c}, \quad C_{tot} = C_{\det} + C_{cable}\ , \qquad (1)$$

for detector capacitance $C_{\det}$= 60 pf, cable capacitance C_{cable}= 40 pf, detector length l = 0.3 m, $\beta_b \approx 1$ for relativistic beam, and beam current $I_b(t)$= 10 mA for accumulator transfers of 10^{11} antiprotons. This represents the lower limit for beam current, with a maximum roughly 40 times higher. Position variation within a practical region of detector aperture causes signal ratios within ±10 dB. Adding the effect of the 300 kHz corner frequency, the preamp will handle a peak input voltage range of 21 mV to 2.4 V.

While radiation resistance is not a major concern throughout most of the ring, local "hot spots" due to the underlying Main Injector accelerator dictate the choice of preamp circuitry. These locations will reach up to 5 MRad/year, but this dosage is reduced by an estimated factor of 5 by shielding from the Recycler Ring itself (3). For 15 months, four types of high-frequency, single-ended current buffers (CLC115, AD9620, AD9630, and HFA115) and their associated passive components were rad-tested in the Fermilab Booster. After a dosage of 12 MRad, none of the CLC115, AD9620, or AD9630 circuits had degraded. The quad-packaged CLC115 was selected as the only active component, while metal film resistors and mica capacitors were used whenever possible (4).

A study of the accelerator noise within the 300 kHz-2 MHz passband indicated that the standard solution of coaxial cable may not be best in this case. A balanced, shielded twisted-pair cable with an isolation transformer was chosen for the long runs (150' to 1300') from preamp box to service building racks. Both signal twisted-pair cables were bundled together in a custom cable with another shielded twisted pair for preamp DC power. This "reduced" the total cable to 350,000', reducing the huge installation costs. The cable losses, 0.8 dB/100' @ 1 MHz for the signal cables and 6.2 Ω/1000' for the power cable, were acceptable.

To check for radiation damage and general system operation, test signals can be applied to both channels just before the preamp chip. The amplitude of each test signal is controlled in a scan available on the console application program. The data, however, is not used as a calibration at this time. The signals are created upstairs and coaxially daisy-chained to all preamp boxes in that sector. When test signals are turned off during beam measurement, a DC voltage is turned on to bias on the overvoltage protection diodes in the preamp boxes. This reduces possible noise coupling over the daisy-chain.

Two ion-clearing high voltages, up to ±1000 V, are applied to the detector plates through the preamp box. The HV supplies will be interlocked and separately controlled. Again, a pair of daisy-chains connects all preamp boxes within a sector, but megaohm coupling resistors and shunt capacitors reduce noise coupling concerns.

SERVICE BUILDING ELECTRONICS

Accuracy specifications, schedule, and previous experience were major factors in selecting a method to process the preamp signals. A moderate accuracy in the ±1 mm range was requested, especially within the inner half aperture, over a 44 dB range of beam intensity and cable loss. A high-speed digitizer and digital signal processing were considered because of a similar concurrent project in another group. A more familiar system based on logamps was chosen, however, partially because of the lack of available research and development time.

Logarithmic Amplifiers

The response of an ideal split-plate detector is given by:

$$\frac{A-B}{A+B}=\frac{x}{b} \quad \rightarrow \quad \frac{A}{B}=\frac{1+x/b}{1-x/b}, \tag{2}$$

where A and B are the signal intensities on the two plates, x is position, and b is detector radius. Converting the ratios to decibels, taking the Taylor expansion, and rearranging the first term gives:

$$20\log(A/B)=\frac{40}{\ln 10}\left[\frac{x}{b}+\frac{1}{3}\left(\frac{x}{b}\right)^3+\frac{1}{5}\left(\frac{x}{b}\right)^5+\ldots\right]\rightarrow\frac{x}{b}=\frac{\ln 10}{2}(\log A-\log B). \tag{3}$$

Implementing the difference of logs can be simpler than difference over sum at many BPM processing frequencies. Recent commercial advances have made possible logarithmic demodulating amplifiers with high dynamic ranges and bandwidths (5). The ideal response of these integrated circuits is given by:

$$V_{out}=k\ \log\left|V_{in}/V_x\right|, \tag{4}$$

where k and V_x determine slope and intercept, respectively. At a CW beam signal frequency f, the resulting output includes baseband and the (typically filtered) higher order even harmonics, each with logarithmic gain compression.

Analog Devices produces several integrated circuits that approximate the ideal log function of Equation 4. The "log nonconformance" error has periodicity and amplitude dependent on the gain of discrete internal gain stages, and this varies among the chip set. This error directly translates into position measurement error and varies according to specific plate intensity ratios. In addition to a log output, these chips have a limiter output for other applications.

The AD640 has 120 MHz bandwidth and 0.3 dB (measured) maximum error, from its 10 dB gain stages. A cascaded pair is required to get 60 dB dynamic range at its higher frequencies. It was used in the 1996 design of the Main Injector 8GeV beamline BPM system to process batches of 53 MHz bunched beam, a similar application to the work investigated at the SSC (6). It has also been used here at low frequencies. The AD606 is a better limiter choice than the AD640, but its logamp output error is larger. Its limiter function has been investigated in an AM/PM circuit, in addition to a 53 MHz synchronous detector with the AD831 active mixer.

The AD608 chip combines an active mixer and a logamp/limiter into one "receiver if subsystem" package. The logamp section has 30 MHz input bandwidth and 0.6 dB (measured) maximum error, from its 16 dB gain stages. A dynamic range of over 60 dB can be achieved with a single chip, for about 10 percent of the price of cascaded AD640s. Figure 2 shows the log nonconformance of several chips, with individual slopes and intercepts normalized.

The low signal bandwidth of Recycler BPM signals and the cost made the AD608 a candidate, but a trick to reduce the log nonconformance enabled its use (6). Taking advantage of the somewhat sinusoidal, 16 dB periodic error function, each input was applied to two separate AD608s. By attenuating the second chip's input by half the error period (8 dB) and adding the outputs, the error is significantly reduced.

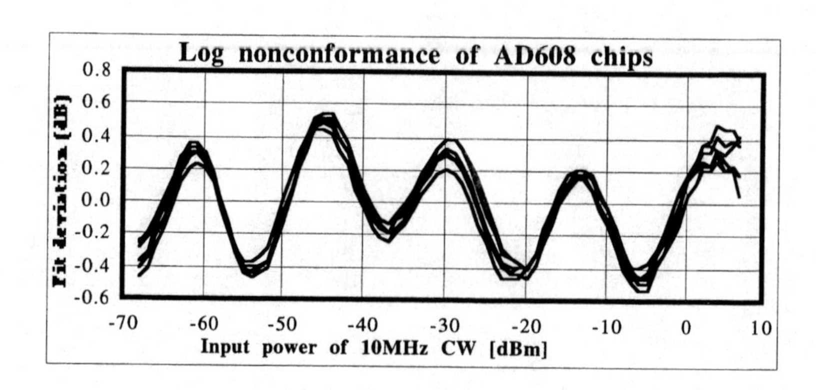

FIGURE 2. Deviation from the ideal logamp response for several AD608 chips.

Analog Processing Module

The balanced inputs to the analog processing module are converted to single-ended with an isolation/balun transformer. These signals are lowpass filtered and applied to the logamps. The four logamp outputs (A_{0dB}, A_{-8dB}, B_{0dB}, B_{-8dB}) are combined into $\log A - \log B$ using AD830s, which perform differencing and summing with excellent common mode rejection and no resistor matching. An external TTL timing signal controls the AD783 track-and-hold, with its 200 ns 0.1% acquisition time and $0.02\ \mu V/\mu s$ droop rate. A final gain and offset stage prepares the held signal for digitization.

The signal at the preamp output is reduced primarily by a back termination and cable attenuation. The range of signal strength at the logamp inputs is estimated between 2 mV and 1V, a good match for the AD608. The exponentially decreasing input (Figure 1) results in a nearly linear output decay (Figure 3), with a slight ripple according to the chip's error characteristic. The difference between two such signals with equal falltimes is ideally flat. Sampling of the position signal occurs early into the decay.

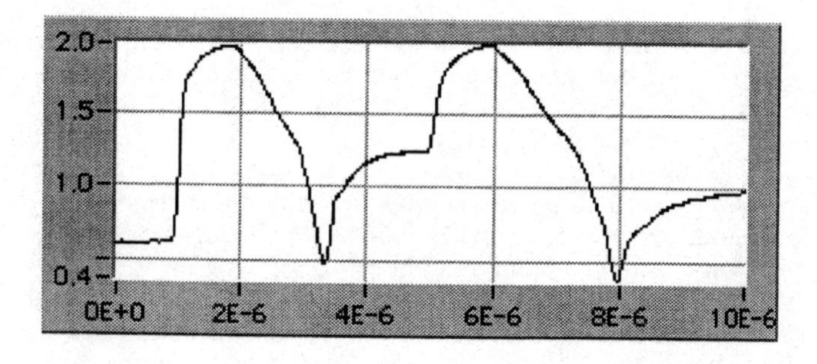

FIGURE 3. Logamp output (33 mV/dB) to preamp output signal in Figure 1.

During testing, response of the analog processing module is scanned over a range of beam positions and intensities (Figure 4). This data is linearly fit against the input position. The deviation from this fit, converted to millimeters with detector scaling, is the position error (Figure 5).

CAMAC crates were chosen for these modules. The existing (but dedicated) crates saved money and are regularly maintained by the Fermilab Controls Department. The board's power requirements, low frequencies, and its I/O connector scheme also made the CAMAC format possible. Surface mount technology was used to increase density to four BPM channels per board. A single crate at a service building can have up to 20 modules and 80 BPMs, but each location has only 72–76 BPMs.

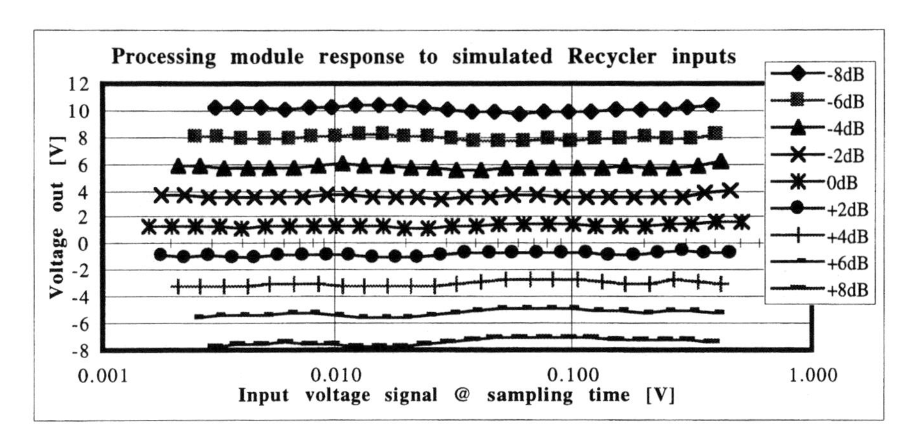

FIGURE 4. Response of an analog processing module to simulated Recycler signal, over a range of beam positions and intensities before offset nulling.

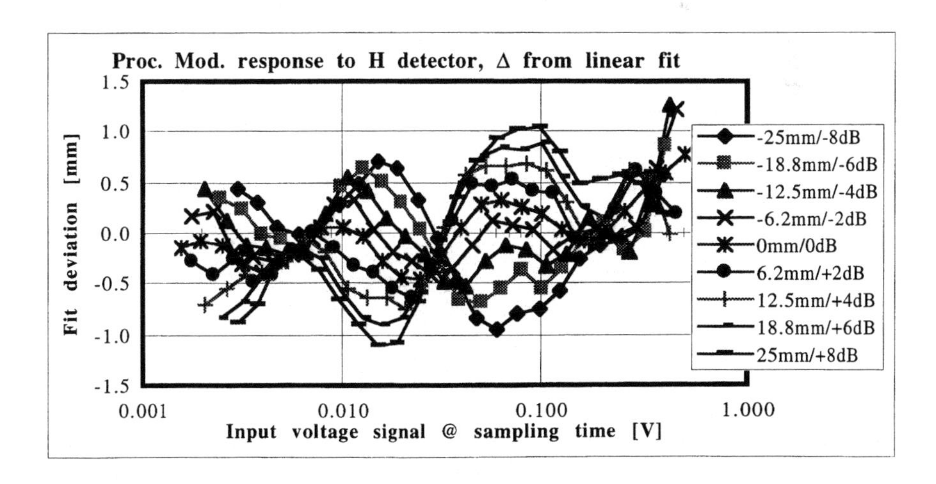

FIGURE 5. Deviation from a linear fit, or position error, for the analog processing module response in Figure 4. Scaling is to the 0.32 dB/mm horizontal detectors. Vertical detector scaling halves this error.

Bundled twisted-pair cables from four preamps are grouped at a racktop distribution PC board into a single, shielded, twist-and-flat ribbon cable, bringing signal inputs to the CAMAC modules. Preamp DC power supply modules connect to the racktop board to distribute power. Another ribbon cable, carrying position signals and external timing signals, connects to a separate transition module. The transition module groups these signals into two 50-pin cables destined for the VME digitizers, and four 50-pin cables from the VME timing signal generators. It also buffers the timing signals and provides test points.

Digital Electronics

A dedicated 3-slot VME crate is home to position signal digitization and processing, timing signal generation, and external communication. The VME processor board is a Motorola MVME162, which has four General Purpose Industry Pack (GPIP) sites. The digitizers are Industry Pack modules (IP320) purchased from Acromag Incorporated. These modules have a 12-bit successive approximation analog-to-digital converter chip (Burr Brown ADS774KE) that is multiplexed to accept 40 single-ended inputs and to digitize them at a rate of up to 100 kHz. Two of the Industry Pack sites are used for digitization, one for horizontal positions and one for vertical. The digitization is triggered either by software or an external TTL signal supplied by the in-house designed timing signal generators.

There are three data acquisition modes. The first mode provides first-turn orbit information for all channels. A beam transfer event will trigger the timing generator to send track-and-hold control signals to the channels, followed by a trigger to the digitizer. The second mode is similar but not tied to the first turn, taking orbits at approximately a 1 kHz rate. It is the default mode. The third mode provides turn-by-turn position of a single horizontal and vertical pair of channels. The mode and the turn-by-turn channels are selected by the user via an ACNET console application program.

The analog processing modules at each service building require TTL track-and-hold control signals. The control signals are created by multichannel timers implemented on Industry Pack modules. On each GPIP resides an ALTERA EPF10K40RC208-3 device, capable of implementing 20 timer channels. The 53 Mhz rf synchronous time base results in 19 ns resolution. Four GPIP modules located on a VME IP carrier module provide 80 track-and-hold signals per service building. All the timers are set high (Track mode) at once after a trigger and a delay. Each timer is set low (Hold mode) at a unique time, since beam signal occurrence varies according to detector location and cable length. For each timer, there is a separate "Hold" time for either antiprotons or counterrotating protons.

The trigger and delay to start the timing signals change according to operating conditions. The trigger can come from a beam transfer event, the Recycler beam synchronous clock, or an external trigger, used primarily to self-trigger during commissioning. The delay has a number of components. Each service building has an adjustable delay to normalize clock event and beam sync occurrence. A systemwide delay, broadcast over the Fermilab MDAT system, tracks the occurrence of the desired rf barrier bucket to be measured. Another common delay is a vernier adjustment of the sampling time within the valid position signal.

The ACNET console application program provides ring orbit and beamline readout for the different acquisition modes. The program also selects measured beam partition, sets timing delays, performs test signal scans, and controls ion clearing voltages.

Outside vendors were used to construct 460 detectors and assemble 460 preamp and 130 analog processing modules. Testing at Fermilab was done with three different Labview-based programs, checking operation within valid ranges and saving data to files.

Commissioning of the Recycler Ring and this system is expected to begin in September 1998.

ACKNOWLEDGMENTS

The following people are thanked for their contributions to the work described in this paper: Jim Crisp, Rupert Crouch, Bob Gorge, Dallas Heikkinen, Rich Janes, Rich Klecka, Dan Munger, Marvin Olson, Rick Pierce, Judy Sabo, Dan Schoo, Gianni Tassotto, and Lin Winterowd.

REFERENCES

[1] Jackson, G., "The Fermilab Recycler Ring Technical Design Report," *Fermilab TM-1991* (1996).
[2] Shafer, R., "Beam Position Monitoring," in *AIP Conf. Proc.*, **212**, pp. 26–58 (1989).
[3] Bhat, C., A. Leveling, private communications, 1996.
[4] Beynel, P., et al., "Compilation of Radiation Damage Test Data," CERN Report 82-10 (1982).
[5] Analog Devices, Inc., *Design-in Reference Manual* (1994).
[6] Aiello, G., M. Mills, "Log-ratio Technique for Beam Position Monitor Systems," *Nucl. Instrum. Methods A*, **346**, pp. 426–432 (1994).

Beam-Position Monitor System for the KEKB Injector Linac

T. Suwada, N. Kamikubota, K. Furukawa, and H. Kobayashi

KEK
High Energy Accelerator Research Organization
1-1 Oho, Tsukuba, Ibaraki 305–0801, Japan

Abstract. About 90 stripline-type beam position monitors (BPMs) have been newly installed in the KEKB injector linac. These monitors easily reinforce handling beam orbits and measuring the charge of single-bunch electrons and positrons which are injected to the KEKB rings. The design value of the beam position resolution is expected to be less than 0.1 mm. A new data-acquisition (DAQ) system has been developed in order to control these monitors in real time. The hardware and software of 18 front-end computers were tuned for the linac commissioning. This report describes the hardware and software system, the monitor calibration, and preliminary beam test results.

INTRODUCTION

The KEK *B* Factory (KEKB) project (1) is in progress in order to test CP violation in the decay of *B* mesons. KEKB is an asymmetric electron-positron collider comprised of 3.5 GeV positron and 8 GeV electron rings. The PF 2.5 GeV linac (2) is also being upgraded to the KEKB injector linac in order to inject single-bunch positron and electron beams directly into the KEKB rings. The injected beam charges are required to be 0.64 nC/bunch and 1.3 nC/bunch, with a repetition rate of 50 Hz, for the positron and electron beams, respectively. High-charge primary electron beams (~10 nC/bunch) are required in order to generate sufficient positrons. Therefore, it is important to easily control the orbits of the beams; most importantly, the beam positions and charges of the primary high-current electron beams have to be controlled so as to suppress any beam blowup generated by large transverse wakefields. A BPM system has been developed to perform this function since 1992. The goal of the beam position measurement is to detect the charge center of gravity with a resolution of 0.1 mm. The amount of the beam current needs to be precisely controlled in order to maximize the positron production and the beam injection rate to the KEKB rings. This facilitates a well-controlled operation of the injector, allowing us to reach an optimum operational condition with as a short a tuning time as possible and maintain such a condition through a long-term operation

CP451, *Beam Instrumentation Workshop*
edited by R. O. Hettel, S. R. Smith, and J. D. Masek

period. A new DAQ system based on VME/OS-9 computers, which are connected with the linac control system, has also been developed in order to control these monitors in real time. The performance of the new system has been tested using electron beams at an extended linac beam line during its commissioning.

HARDWARE SYSTEM

Design of a Beam Position Monitor

A conventional stripline-type BPM made of stainless steel (SUS304) was designed with a $\pi/2$ rotational symmetry. A drawing of the monitor geometry and a photograph are shown in Figures 1 (a) and (b), respectively. The total length (195 mm) was chosen so as to make the stripline length (132.5 mm) as long as possible, so that it can be installed into limited space in the new beam line of the linac. Each electrode is comprised of a 50 Ω transmission line. The angular width of the electrode is 60° in order to avoid strong electromagnetic coupling between the neighboring electrodes. A 50 Ω SMA vacuum-feedthrough is connected to the upstream side of each electrode, while the downstream ends are short-circuited to a pipe in order to simplify the mechanical manufacturing. Quick-release flange couplings (manufacturer's standard KF flange) are used at one end of the monitor for easy installation into the beam line. More detailed descriptions are given elsewhere (3–6).

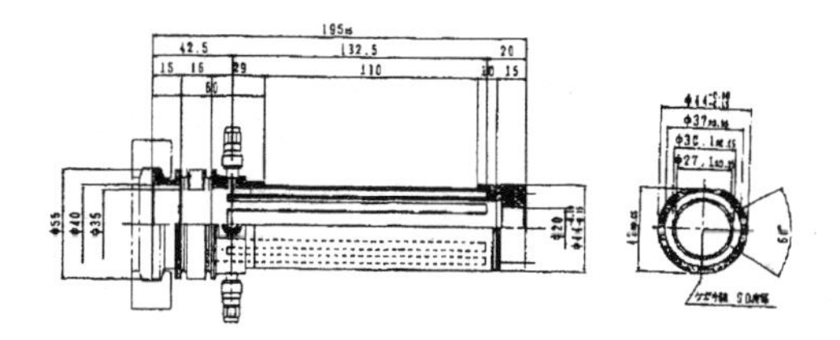

(a)

(b)

FIGURE 1. Geometrical drawing (a) and photograph (b) of the stripline-type beam position monitor installed in the new beam line.

Data-Taking System

All analog signals of the BPMs are connected with monitor stations in conjunction with those of wall current monitors (WCMs). Eighteen monitor stations, each of which is comprised of a front-end computer (VME/OS-9 with a MC68060 microprocessor at 50 MHz), a signal digitizing system (an oscilloscope), and a signal-combiner box, are located on the linac klystron gallery at nearly equal intervals along the beam line. Each monitor station can control a maximum of twelve BPMs. A schematic drawing of the monitor station is shown in Figure 2. Four signals of a BPM are sent directly to a signal-combiner box through 35 m-long coaxial cables. The frequency response and characteristics of the coaxial cable are given elsewhere in detail (5). Two signal combiners combine the horizontal and vertical signals from each BPM, respectively. In the combiner box, one of two horizontal (vertical) signals is delayed with a time of 7 ns in order to reject its signal-mixing. The two signals from the signal combiners are fed to two input channels of a digital sampling oscilloscope (Tektronics TDS680B) with a sampling rate of 5 GHz and a bandwidth of 1 GHz. The Unix workstations and the front-end computers communicate with each other through a network system. As shown in Figure 3, all of the front-end computers are linked to a switching-hub with a star topology. Fiberoptic cables are used for physical connections in order to avoid electromagnetic interference from high-power klystron modulators. The hub has a link to the linac control network, where Unix workstations (WSs) and man-machine interfaces (PCs and X-terminals) are connected.

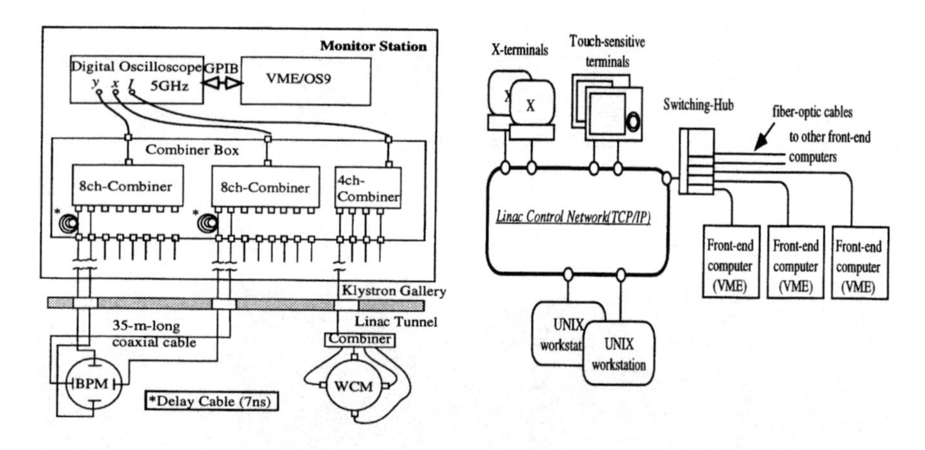

FIGURE 2. Schematic drawing of the data-taking system in a monitor station.

FIGURE 3. Block diagram of the linac control network and the new network (right side) system.

Calibration of the Monitors

All of the monitors were installed into the beam line after bench calibration. The bench calibration system for the BPMs is described in detail elsewhere (7). Here, only the bench calibration and the beam position calculation are briefly described. All of the BPMs have been calibrated by "mapping," which is performed on the test bench with a thin current-carrying wire (500 μmϕ), stretched through the center of the monitor, to simulate the beam. The calibration coefficients of the map function, which relates to the pulse-height information obtained from four pickups for various wire positions, are measured by the bench calibration. The horizontal (x) and vertical (y) beam positions are represented by map functions with a third-order polynomial, as follows:

$$x = \sum_{i,j=0}^{3} a_{ij}(\Delta_x / \Sigma_x)^i (\Delta_y / \Sigma_y)^j, \tag{1}$$

$$y = \sum_{i,j=0}^{3} b_{ij}(\Delta_x / \Sigma_x)^i (\Delta_y / \Sigma_y)^j, \tag{2}$$

$$Q = G \sum_{k=1}^{4} g_k V_k. \tag{3}$$

Here,

$$\Delta_x = g_1 V_1 - g_3 V_3, \ \Sigma_x = g_1 V_1 + g_3 V_3, \tag{4}$$

$$\Delta_y = g_2 V_2 - g_4 V_4, \ \Sigma_y = g_2 V_2 + g_4 V_4, \tag{5}$$

where a_{ij} and b_{ij} are the coefficients of the map functions (derived by fitting the map data to the map functions by using a least-squares fitting procedure); V_1 and V_3 (V_2 and V_4) are the horizontal (vertical) pickup voltages; and g_k (k=1–4) are the gain correction factors. Q is the beam charge, which is calculated by summing four-pickup voltages, and G is a conversion factor used to calculate the beam charge, which can be measured by wall-current monitors. The gain correction factors (g_k), which correct any signal-gain imbalance caused by attenuation losses of the cables and the difference of the coupling strength of the combiners, are measured by using fast test-pulses with a width of 500 ps. These parameters (a_{ij}, b_{ij}, g_k, and G) for each BPM are stored in the Unix workstations as a calibration table, and are loaded into each front-end computer at every startup.

SOFTWARE SYSTEM

Control Software

Several DAQ processes are running concurrently on the front-end computers and on Unix workstations. The processes, with data and control flow, are shown in Figure 4. DAQ processes for the WCM are almost the same as those for the BPM, except for the data format, as shown in the figure. The read-out process resides on each front-end computer. It reads waveforms of BPM signals from the digital oscilloscope, and then calculates the beam positions and currents while taking into account the calibration coefficients described above. Trigger pulses synchronized with the linac beam are provided to all of the oscilloscopes at the 0.67 Hz cycle. This rate is limited by the communication throughput between a front-end computer and a oscilloscope through a GPIB line. The calculated beam positions and currents are transferred to two Unix workstations over the linac control network with the UDP-protocol (8). In order to reduce the network traffic, data transfer is done only when the beam current data at the front-end is renewed. As a result, the total traffic between front-end computers and Unix workstations is always constant (24 packets per second: 0.67 Hz × 18 VMEs × 2 WSs). Unix workstations receive beam currents from eighteen front-end computers and store ten recent data for each front-end on shared-memory regions. The data servers for BPM use the data on the shared memory. It is worth noting that the data requests from applications do not increase the network traffic to the front-end computers.

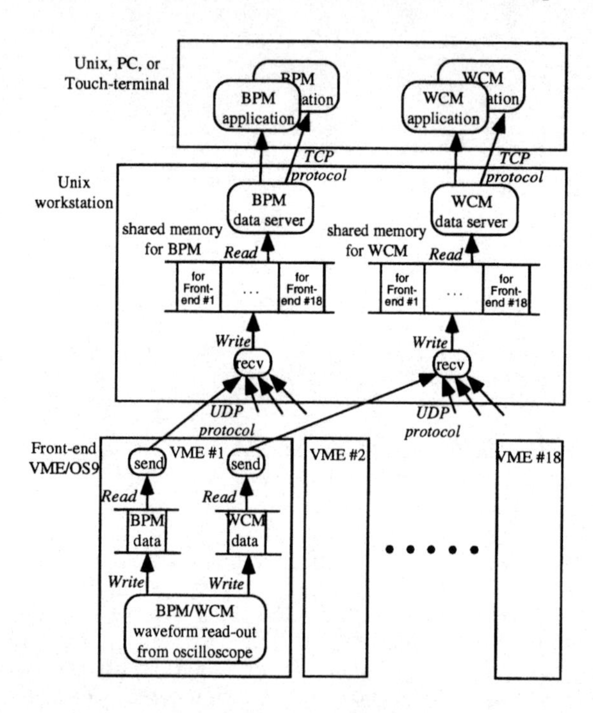

FIGURE 4. Block diagram of the control software and data flow for the data-taking system.

BEAM TESTS

Beam and DAQ System Tuning

DAQ system tuning and the beam tests have been performed (2) using single-bunch electron beams at the extended linac section (sectors A and B). Single-bunch electron beams can be generated by the new pre-injector (9), which is comprised of two subharmonic bunchers, a prebuncher and a buncher. The electron gun can generate a beam charge of about 18 nC/pulse with a repetition rate of 50 Hz. Single-bunch electron beams greater than 10 nC/bunch were stably accelerated from the outlet of the buncher until the end of sector B without any observed beam loss. The beam energies were about 500 MeV and 1.5 GeV, at the end of sectors A and B, respectively. The longitudinal width of the beams was measured by an optical transition radiation monitor to be about 16 ps with a full width at half maximum after tuning of the preinjector. The beam tests during the commissioning were performed at a 5 Hz repetition rate. Figure 5 shows an example of the beam tests at a beam current of 10 nC/bunch, in which the DAQ processes of four monitor stations (22 BPMs) concurrently worked well. It shows the horizontal displacements of the beam from the center (the top drawing), the vertical displacements (middle), and beam charge (bottom), simultaneously with a refresh rate at around 1 Hz after elaborate software tuning. The DAQ system operated stably during the beam tests. Figure 6 shows pickup signals obtained by a beam of 6.2 nC/bunch at the third monitor station.

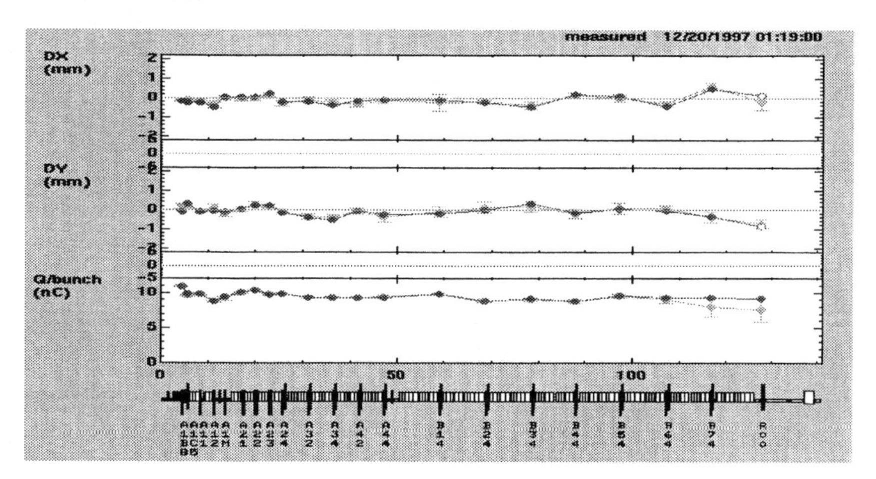

FIGURE 5. Variation of the beam orbits and charge intensity along the linac. The beam current of the single-bunch electrons is 10 nC/bunch.

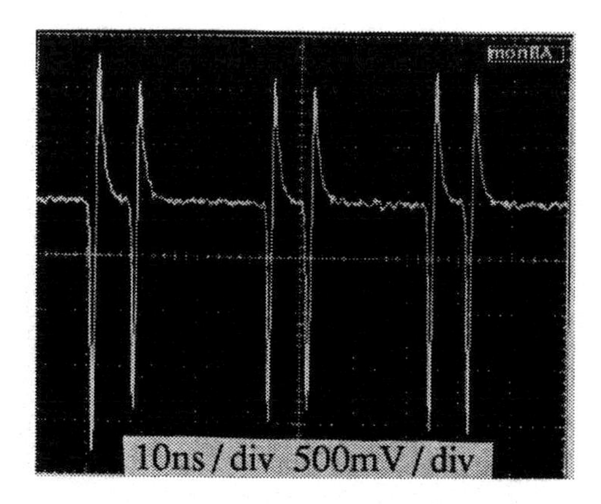

FIGURE 6. Pickup-pulse signals measured by the digital oscilloscope at a beam charge of 6.2 nC.

Position-Resolution Measurement

The position resolution of the BPM has been measured using single-bunch electron beams with a beam charge of 7 nC on the basis of the "three-BPM method". The principle of this method is simple. The beam orbit of a charged particle through a system with both magnetic lenses and acceleration can be generally represented by a multiplication process of transfer matrices (R) (10). If the beam orbit is represented by a coordinate vector (X), passage of the charged particle through the system can be represented by an equation neglecting the second-order transfer matices:

$$X_i(1) = \sum_{j=1}^{i} R_{ij} X_j(0), \tag{6}$$

where $X_i(0)$ and $X_j(1)$ are the components of the initial and final coordinate vectors through the system action, and R_{ij} are the components of the transfer matrix. Here, if any three beam positions (x_1, x_2 and x_3), which are the first components of the coordinate vector X, are measured in the system, the third beam position (x_3) is linearly correlated by the expected beam position (x^c_3) using the beam positions (x_1 and x_2) as follows:

$$x^c_3 = ax_1 + bx_2 + c, \tag{7}$$

$$x_3 = x^c_3. \tag{8}$$

Here a, b, and c are constants obtained by a least-squares fitting of the correlation plot in accordance with Equations (7) and (8). Thus, the position resolution can be obtained by calculating the standard deviation of the correlation plot and assuming that the resolutions of the three BPMs are the same. Beam tests were performed using single-bunch electron beams of 1 and 7 nC. The beam positions were changed within the range of ±2 mm by two correction dipole magnets placed just after the buncher. Figure 7 gives an example.

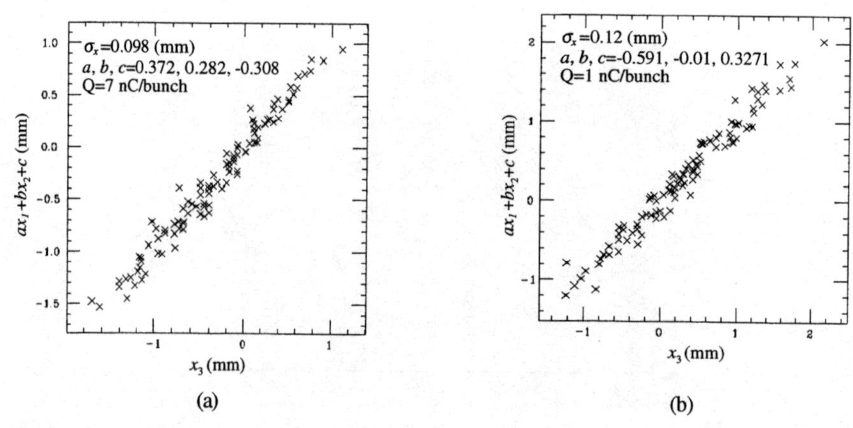

FIGURE 7. Correlation plots between the measured and calculated beam positions for the beam charge of (a) 7 nC/bunch and (b) 1 nC/bunch by using the three-BPM method.

CONCLUSIONS

The new beam position monitor system has been installed in the linac in order to reinforce the beam monitoring of the beam positions and currents. The present system was inspected with electron beams in an extended linac beam line. The data-taking rate became 0.67 Hz after elaborate system tuning of the new DAQ. The rate is mainly limited by the data-transfer rate from the oscilloscope through a GPIB line. The present DAQ system has been found to be sufficiently fast and stable. The position resolution has been measured by single-bunch electron beams for beam charges of 1 and 7 nC/bunch to be around 0.1 mm.

ACKNOWLEDGMENTS

The authors gratefully acknowledge Mr. T. Obata for his help in developing the software of the front-end computer. They also thank Drs. H. Fukuma and H. Koiso for their help in developing beam-orbit analysis programs, and the KEKB linac commissioning group for their help in the monitor calibrations and beam tests.

REFERENCES

[1] Kurokawa, S., et al., "KEKB B-Factory Design Report," KEK Report 90–24 (1991).
[2] Sato, I., et al., "Design Report on PF Injector Linac Upgrade for KEKB," KEK Report 95–18 (1996).
[3] Suwada, T., et al., *AIP Conference Proceedings* **319**, 1993, p. 334.
[4] Suwada, T., *Proceedings of the 10th Symposium on Accelerator Science and Technology*, (Hitachinaka, Japan) 1995, p. 269.
[5] Suwada, T., et al., *AIP Conference Proceedings* **390**, 1996, p. 324.
[6] Suwada, T., et al., presented at the First Asian Particle Accelerator Conference, KEK, Tsukuba, Japan, 1998.
[7] Suwada, T., in preparation.
[8] Kamikubota, N., et al., *Proceedings of the 1994 International Linac Conference* (KEK, Tsukuba, Japan), 1994, p. 822.
[9] Ohsawa, S., et al., *Proceedings of the XVIII International Linac Conference* (CERN, Geneva, Switzerland), 1996, p. 815.
[10] Brown, K. L., et al., "TRANSPORT: a computer program for designing charged particle beam transport systems," SLAC-91, Stanford Linear Accelerator Center, 1977.

The DELTA Beam-Based BPM Calibration System

A. Jankowiak, C. Stenger, T. Weis, K. Wille

Institute for Accelerator Physics
and Synchrotron Radiation
University of Dortmund, Germany

Abstract. A third-generation synchrotron light source like DELTA[1] (1) requires a measuring system to determine the beam position with high resolution and great accuracy with respect to the center of the quadrupole magnets. This paper presents the beam-based BPM calibration system developed for DELTA, providing a calibration accuracy of about 150 μm. We describe the basic idea of the measurements, the installed hardware, and present the results of an initial calibration of the closed-orbit measuring system (2).

INTRODUCTION

The conventional approach for calibrating a closed-orbit measuring system (CO system) requires several steps carried out one after the other. The first step is to measure each BPM on a test bench to obtain a calibration with respect to the center of the BPM. Possible offsets of the BPM electronics and the influence of varying damping factors of the measuring cables have then to be taken into account. This procedure is very time consuming and has to be performed with great accuracy. It is also very difficult to repeat this calibration after the final installation of the BPMs, cables, and electronics. Since most of the BPMs are normally mounted in or close to quadrupole magnets, the last step of calibration is to determine the position of the BPMs with respect to the axis of the quadrupoles.

The DELTA BPMs are assembled to the quadrupoles in two different ways. The heads for one half of the BPMs fit to the aperture of the quadrupole magnets with an accuracy of 70 μm ("fixed" BPMs). The other half of the BPMs have no mechanical connection to the quadrupole and can move to avoid any stress to the quadrupoles resulting from thermal movements of the vacuum chamber ("floating" BPMs). There is

[1] **D**ortmund **EL**ectron **T**est **A**ccelerator

CP451, *Beam Instrumentation Workshop*
edited by R. O. Hettel, S. R. Smith, and J. D. Masek

an uncertainty of the position of the floating BPMs with reference to the quadrupole axis of approximately 1 mm.

For these reasons, it was decided to install a system for beam-based calibration of the CO system.

BASIC IDEA OF BEAM-BASED CALIBRATION MEASUREMENTS

An electron beam passing through a quadrupole magnet (length l) displaced by Δx_Q with respect to the magnetic axis gains an orbit kick with angle $\vartheta_Q = l \cdot \Delta k \cdot \Delta x_Q$ if the focusing strength is changed by Δk (3). In linear approximation, this leads to a closed-orbit distortion of

$$\Delta x_i(\Delta k, \Delta x_Q) = \frac{\sqrt{\beta(s_Q)\beta(s_i)}}{2\sin(\pi Q_x)} \cos(\pi Q_x - |\psi(s_i) - \psi(s_Q)|) \cdot \vartheta_Q \tag{1}$$

where Δx_i is the orbit displacement in the i-th BPM and the usual nomenclature has been used. For a beam passing along the axis, this orbit deviation vanishes and therefore the magnetic center of the quadrupole can be determined as follows.The strength of the quadrupole magnet whose BPM is to be calibrated is varied by Δk and the resulting average quadratic orbit distortion

$$\overline{\Delta x^2(x_Q)} = \frac{1}{N} \sum_{i=1}^{N} \Delta x_i^2(\Delta k, \Delta x_Q) \quad \propto \quad \Delta x_Q^2 \tag{2}$$

$$N = \text{ number of BPMs}$$

is measured as a function of the measured beam position x_Q in the selected quadrupole. By steering the beam across the quadrupole and fitting a parabola

$$p(x_Q) = a \cdot (x_Q + b)^2 + c \tag{3}$$

to the measured data points the position of the center of the quadrupole can be determined from the position b of the vertex of the parabola (see Fig. 2). The parameter c takes into account the noise of the BPM system and should be the same for all calibration measurements.

HARDWARE OF THE BEAM-BASED CALIBRATION SYSTEM

The Closed-Orbit Measuring System

The DELTA closed-orbit measuring system consists of a total number of 44 BPMs, 40 of which are mounted in quadrupole magnets. They are made out of blocks of stainless steel with an inner geometry identical to the cross-section of the vacuum chamber. Each of them houses four capacitive pick-up electrodes (ESRF type) and is connected to the BPM electronics via double-shielded RG223U coaxial cables. The BPM-electronics have been fabricated by the French company BERGOZ (4) and allow for a resolution of less than 1 μm and a long-term stability (measured in the laboratory with a signal source and power divider over a period of 240 h) of better than 3 μm. Due to the control system connection, using CAN-bus modules with 12-bit accuracy, the present resolution of orbit measurements is limited to ≈10 μm. The maximum repetition rate for a complete orbit measurement is about 1 Hz.

Additional Cabling of Quadrupole Magnets

Since none of the DELTA quadrupole magnets has its own power supply, additional hardware had to be installed to change the *k*-value of individual quadrupoles. In order to obtain the most flexible solution and to avoid a distributed system which is difficult to maintain, it was decided to install an additional cabling for all quadrupoles. Therefore, we connected each quadrupole to one of a pair of 19" racks, each rack accommodating one half of the DELTA ring, with an extra cable ($2{\times}4mm^2$).

For calibrating the CO system, it would have been sufficient to install additional cables only for these quadrupoles where BPMs are mounted. Nevertheless we decided to connect all quadrupoles because the beam-based calibration system also provides the possibility of determining the local beta-functions by measuring the betatron tune shift as a function of the quadrupole strength. This allows us to compare the theoretical optics with those of the real machine and to detect deviations from the fourfold symmetry of the DELTA lattice.

Selection of the Quadrupole Magnets

To select a specific quadrupole, a relay cascade is used for both half rings (see Fig. 1). This setup allows using of one DC-power supply to add an additional current to the quadrupole to cause a change of strength. The main advantage of the design of the relays cascade is the inherent protection against short circuits between different quadrupole circuits. For a later extension, a second input port will be used to operate the system with two different power supplies. For a 1.5 GeV beam, the excitation currents of the strongest quadrupoles are in the order of 60 A. Together with a resistance of the quadrupole coils of 0.7 Ω this corresponds to a maximum voltage drop of 42V per quadrupole. Therefore, a voltage controlled DC current source with potential free output of 70V–10A was chosen as the additional power supply.

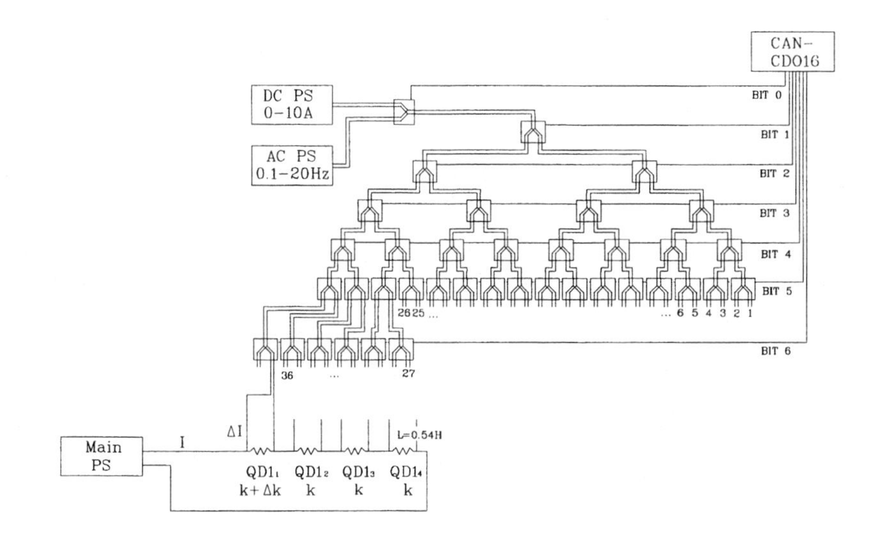

FIGURE 1. Sketch of the hardware installation for beam-based calibration measurements.

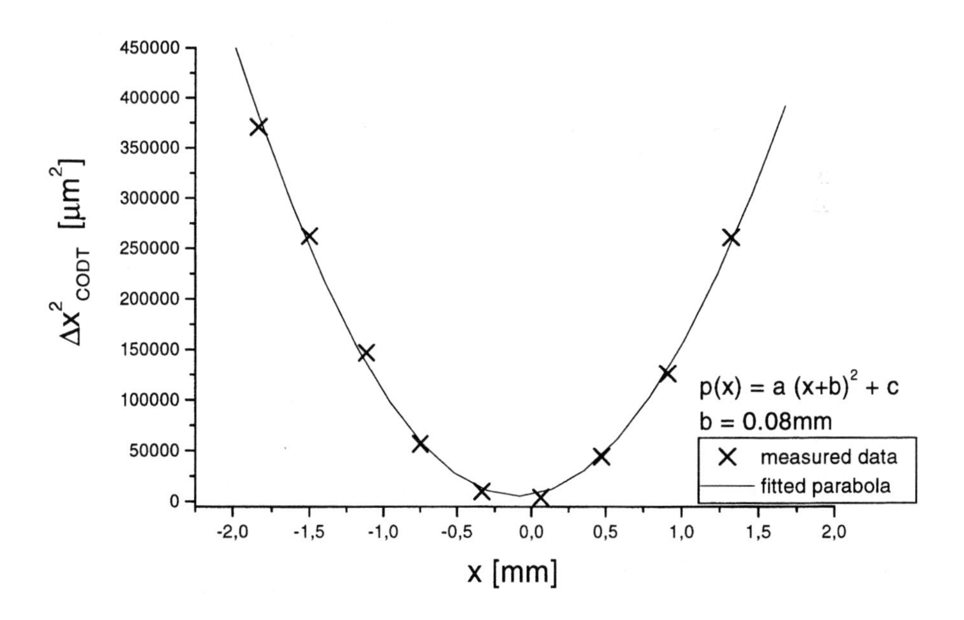

FIGURE 2. An example of the calibration measurement of BPM No. 35, Quadrupole QF5-4 (k=-2.62 m^{-2}, Δk=5%, β=18.7 m).

RESULTS OF THE INITIAL CALIBRATION OF THE CO SYSTEM

The first step of the calibration measurements is to correct the orbit using the uncalibrated CO system. This results in an orbit with position uncertainties with respect to the center of the quadrupole magnets of about 1 mm. Nevertheless this is a good starting point for further measurements. In the next step, an adequate k-variation of the selected quadrupole has to be calculated according to the present k-value and the local beta-function (see equation No. 1). This k-variation should be strong enough to measure a significant closed-orbit distortion, but should not disturb the operation of the storage ring. It turned out that for most of the quadrupole magnets $\Delta k \approx 5\%$ is a useful value.

In the last step, where the beam is steered in equidistant steps across the quadrupole, the variation of the k-value by Δk and the measurement of the average quadratic closed-orbit distortion $\overline{\Delta x^2(x_Q)}$ as a function of the beam position (see Fig. 2) is performed. The measurement is done automatically by a program written in Tcl/Tk (5). This program selects the corresponding quadrupole for the BPM which is to be calibrated via the relay cascade, calculates the necessary current the extra power supply must add to the quadrupole, and steers the beam with a local 3-magnet bump around the quadrupole. For each BPM we measure a total range of ±2 mm in steps of ~0.5 mm for the horizontal and vertical directions separately. At the moment, the calibration of one BPM in both directions requires four minutes; therefore a complete calibration of the CO system lasts four hours.

To determine the offsets automatically, it is necessary to have a quantitative measurement for the goodness-of-fit which allows one to determine bad data without plotting them. The incomplete gamma function $Q(\nu/2, x^2/2)$, where ν is the difference between the number of measured data points and the number of parameters to fit and x^2 is calculated as a result of the fitting routine, seems to be a promising candidate (6). For values of Q between 0.2 and 1, the fitted data is well represented by the parabola.

Figure 3 shows the results of the first complete calibration of the DELTA closed-orbit measuring system for both directions. The values of the average absolute offsets are:

$$\left|\overline{x_{offset}}\right| = 0.52\text{mm}$$

$$\left|\overline{z_{offset}}\right| = 0.63\text{mm}$$

These values show the necessity of the calibration measurements. It is astonishing that there is no significant difference between the offsets of the "fixed" and the "floating" BPMs. Naturally, we expected that the average absolute offsets of the fixed BPMs should be smaller than those of the floating BPMs. The calibration measurements, however, show that there is no difference. This is a hint that there are unknown uncertainties to be studied in the future.

By calibrating a BPM more than once, the accuracy of the calibration procedure is estimated to be on the order of 150 μm. This value can be decreased by repeating the complete calibration after an orbit correction which takes into account the first measured offsets. This will reduce the effect originated by a beam passing diagonally across the quadrupole. For a longitudinal distance of 120 mm between the pick-ups and the middle of the quadrupole, a beam passing under an angle of 1 mrad results in a deviation of the beam position of 120 μm. This can be reduced by a better orbit correction that brings the

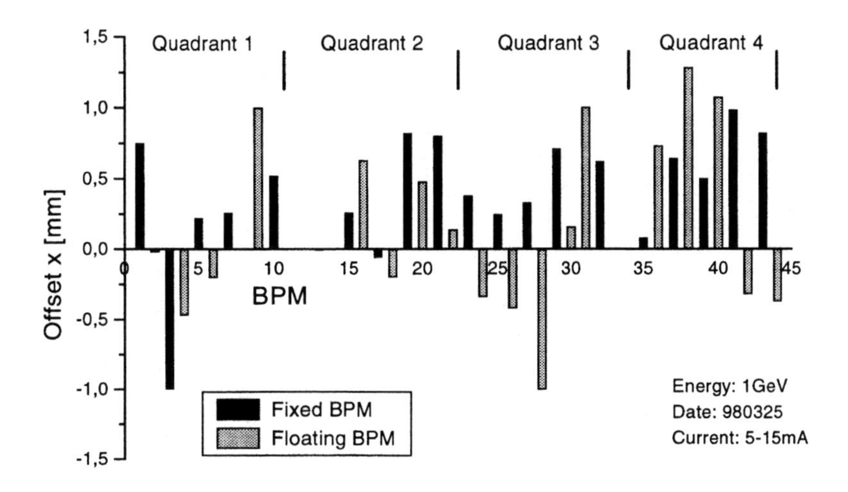

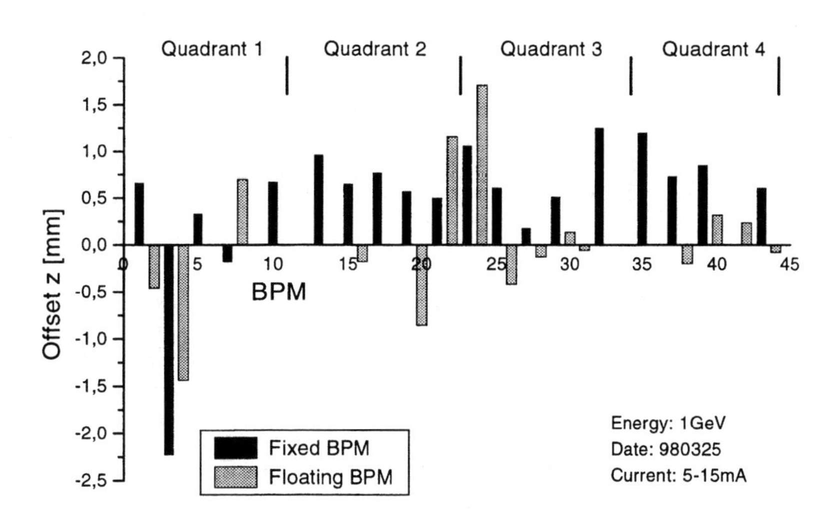

FIGURE 3. Results of the initial calibration of the closed-orbit measuring system. The diagram shows the offset of the CO system with respect to the quadrupole axis.

beam nearer to the center of the quadrupole magnets.

From the parameter c (Eq. (3)) of the fitted parabola, it is possible to estimate the noise figure of the BPM system. From all calibration measurements, we obtained an average value for the noise figure of 10 μm, which is in good agreement with the predicted value for the DELTA CO system.

FUTURE PLANS

Future work will deal with detailed studies of the performance of the calibration system. In particular, we will investigate the influence of the beam-based calibration on such beam parameters as emittance, lifetime, and orbit stability.

To study possible thermal movements of the BPMs when the vacuum chamber heated is by synchrotron radiation and to isolate them from real orbit drifts, we want to speed up the measuring time to less than 1 minute for both directions.

A better accuracy of the beam-based calibration should be possible by performing a harmonic modulation of the quadrupole strength (7) with the frequency f:

$$\Delta k(t) = \Delta k_0 \cdot \sin(2\pi \cdot f \cdot t) \tag{4}$$

This leads to an harmonic closed-orbit oscillation:

$$\Delta x_i(\Delta k_0, \Delta x_q, t) \;\; \propto \;\; \Delta k_0 \cdot \Delta x_q \cdot \sin(2\pi \cdot f \cdot t) \tag{5}$$

which can be detected in the frequency domain.

By measuring $\Delta x_i(\Delta k_0, \Delta x_Q, f)$ at a BPM with a suitable betatron phase advance as a function of the beam position in the selected quadrupole, the BPM offset can be determined. We will install a second AC power supply for the harmonic excitation of the quadrupole and connect an FFT analyser to a Bergoz BPM processor, which has a bandwidth of 2 kHz for such measurements. By using a lock-in amplifier, the sensitivity of this measurement can be drastically increased and a better accuracy or a lower value of quadrupole k-modulation is possible.

To increase the performance of the beam-based calibration and the closed-orbit measuring system, we have begun the development of a fast 16-Bit ADC interface board based on the CAN-protocol. This board will have a measuring speed of 6 kHz and, by the use of an integrated micro controller, averaging capabilities. Together with the higher resolution of the ADC, it is possible to reach a resolution of the orbit measurements of <1 μm. This will benefit the performance of the closed-orbit measurements and also the beam-based calibration system.

ACKNOWLEDGMENTS

Many thanks to the DELTA crew for spending some not-scheduled shifts in the DELTA control room and for making this initial calibration measurement possible before the workshop starts.

REFERENCES

[1] Friedl, J., DELTA group, *Recent Results of the Commissioning of the DELTA Facility*, PAC97, Vancouver, Canada, 1997.
[2] Jankowiak, A., "Status of the DELTA Beam-Based BPM Calibration System," DIPAC97, Frascati, Italy, 1997.
[3] Röjsel, P., "The QSBPM System Performance and a Comparison to Conventional BPM Systems," EPAC96, Barcelona, Spain, 1996.
[4] Bergoz Precision Beam Instrumentation, *Beam Position Monitor Users Manual, Rev. 1.4*, Crozet, France.
[5] Ousterhout, J. K., "*Tcl and the Tk Toolkit*," Addison-Wesley Publishing Company, 1994
[6] Press, W.H., et al, *Numerical Recipes in C,* Second Edition, Cambridge University Press, Reprint 1995
[7] Barrett, I., et al., "Dynamic Beam-Based Calibration of Orbit Monitors at LEP," *Proceedings of the 4th International Workshop on Accclerator Alignment,* Tsukuba, Japan, December 1995, KEK Proc. 95–12 and CERN-SL/95–97 (BI).

Beam Jitter and Quadrupole Motion in the Stanford Linear Collider*

Robert E. Stege Jr. and James L. Turner

Stanford Linear Accelerator Center
Stanford, California 94309

Abstract. Spectral analysis of beam jitter in the Stanford Linear Collider (SLC) has shown that some beam motion is confined to narrow frequency bands. Vibration analyses of linac quadrupoles using high-resolution accelerometers yield spectra having a similar footprint. It was found that motion at 59 Hz is driven by a pressure oscillation in the accelerator structure cooling water, while other frequencies were found to be vibrational modes of the structure itself (1). This paper presents motivating beam data, describes instrumentation used for vibration measurements, presents vibration-related data, and summarizes the solutions used to reduce quadrupole motion.

INTRODUCTION

Jitter measurements of the SLC beams during 1994/95 running show motion at around 59 Hz. This jitter was severe enough to have a significant detrimental effect on the luminosity. Subsequent measurements of the linac quadrupoles using sensitive accelerometers revealed a 59 Hz peak in vertical motion having sufficient magnitude to produce the observed beam jitter. The source of the 59 Hz vibration was traced to the pumps for the accelerator structure cooling water system, which rotate at 59 Hz (2). Spectral analysis using a water pressure transducer and a signal analyzer revealed a strong 59 Hz pressure modulation from some of the pumps.

Using a portable signal analyzer and a laptop computer, correlations were made between quadrupole motion, water pressure modulation, and specific pumps. A test stand was set up where pumps could be studied while they were out of service and possible solutions could be tried. Asymmetry in the pump impellers, which were crude sand castings, was suspected to be the main source of the pressure modulation. It was also found that a flexible rubber coupling or "boot" at the discharge of the pump would reduce the pressure modulations by a factor of about two. The experimental setup is

* This work supported by DOE contract DE-AC03-76SF00515.

CP451, *Beam Instrumentation Workshop*
edited by R. O. Hettel, S. R. Smith, and J. D. Masek

described, excerpts from the data are presented and solutions that have been tried or proposed are discussed.

SETUP

Beam jitter data was taken through the SLC control system, then processed and plotted offline in Matlab (3). To facilitate field measurements of quadrupole motion and water pressure variation, a portable Hewlett Packard 3560A Dynamic Signal Analyzer was used for data acquisition in conjunction with an Apple 5300c PowerBook for further data analysis. The entire set-up was packed in a rolling, foam-lined suitcase (Figure 1), which could be lowered down the linac access shafts to make quadrupole motion measurements. Quadrupole motion was measured with a PCB Piezotronics model 393B31 accelerometer, which has a sensitivity of 10 V/g and a resolution of 5 μg. It is a single-axis accelerometer and was placed directly on top of the quadrupole for the measurements. All data presented here represent vertical motion. Water pressure modulations were detected using an Ashcroft pressure transducer with a range of 150 psi and an output of 26 mV/psi with an upper frequency limit of at least 2 kHz. The water pressure measurements were taken directly at the discharge piping on the pumps.

Data was acquired with the signal analyzer operating in spectrometer mode, integrated twice to yield displacement, and displayed as a power spectral density (PSD) plot. The processed data was then transferred through the serial ports to the laptop where the data is further processed using Matlab.

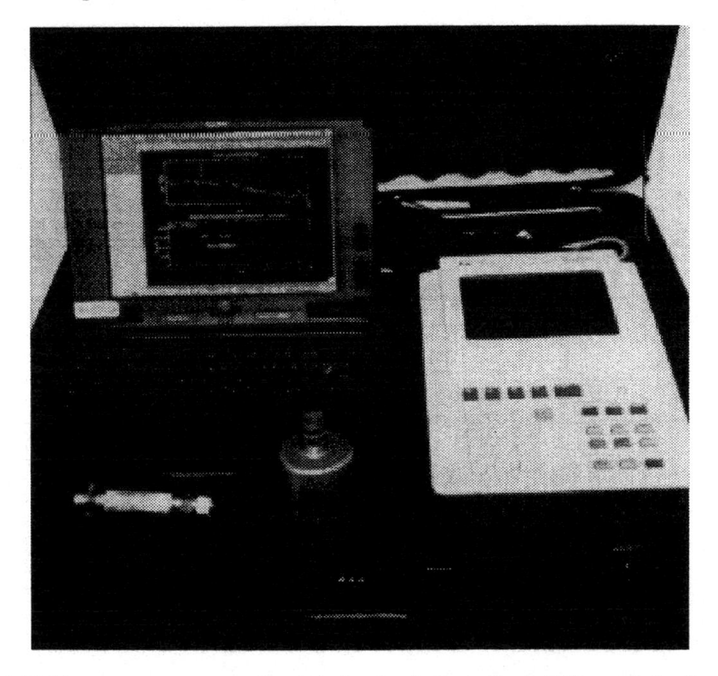

FIGURE 1. Field measurement setup showing the Apple PowerBook 5300c and the Hewlett Packard 3560A Dynamic Signal Analyzer packed in a rolling foam lined suitcase. In the foreground are the pressure transducer and the accelerometer that were used in the measurements.

MEASUREMENTS

Figure 2 shows an example of the beam data that prompted these studies. It shows the FFT of buffered vertical beam position monitor data taken pulse by pulse. The motion at 59 Hz is clearly seen. In the bottom graph of the figure, the right to left integral of the top plot is mapped. The advantage of this plot is that the vertical steps indicate the amount of power introduced at a particular frequency. The "Int" in this and subsequent figures indicates the integrated amplitude contributed between the "I" of "Int" and the next higher frequency "Int."

Investigation in the linac housing with the sensitive accelerometer revealed vertical motion on some of the quadrupoles. Data on one of the quads is plotted in Figure 3. The sharp step on the bottom plot indicates that most of the power is at 59 Hz resulting in 229 nm of motion at that frequency.

Measurements with the water on and off indicated that the source of the 59 Hz was in the accelerator structure cooling water system. This led to investigation using a water pressure transducer, which produced the data in Figure 4 where we see a very strong pressure modulation at 59 Hz. This data was taken on the pump for the same Linac sector as the quadrupole motion data presented in Figure 3. Further measurements at other Linac sectors show that where there was little 59 Hz pressure modulation there was no detectable 59 Hz motion on the quadrupoles.

The transfer function plotted in Figure 5 was obtained by comparing the magnitude of the quadrupole motion with the magnitude of the water pressure modulation for five different frequencies. The frequencies are the first five harmonics of 59 Hz.

During our measurements, a project was undertaken to install flexible rubber couplings or "boots" at the discharge piping of the pumps. These were intended to make removal and replacement of the pumps quicker and easier. Before and after measurements (see Figure 6) showed that they reduced the 59 Hz quadrupole motion by a factor of two.

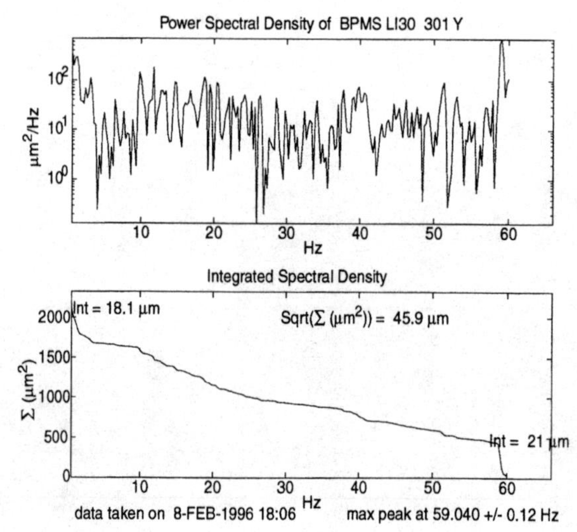

FIGURE 2. Jitter measurements of the SLC beams during 1994/95 running show a peak in the vertical motion at approximately 59 Hz. This jitter had a significant detrimental impact on the luminosity.

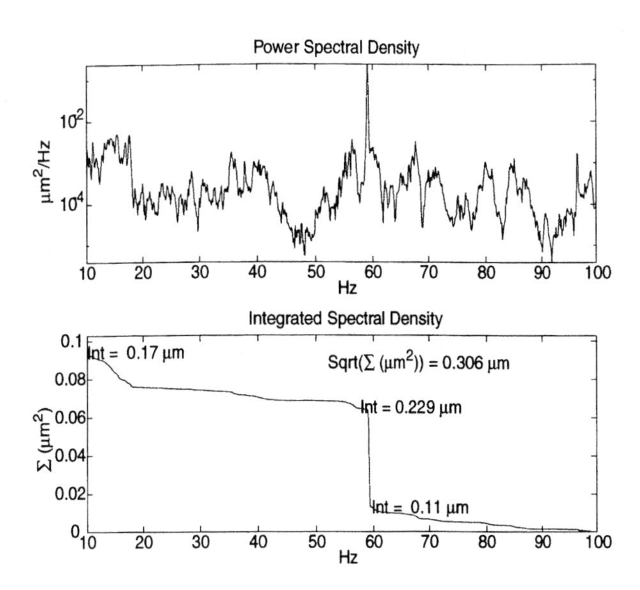

FIGURE 3. Accelerometer measurements on selected LINAC quadrupoles reveal a 59 Hz peak in the vertical motion of a magnitude sufficient to produce the observed beam jitter.

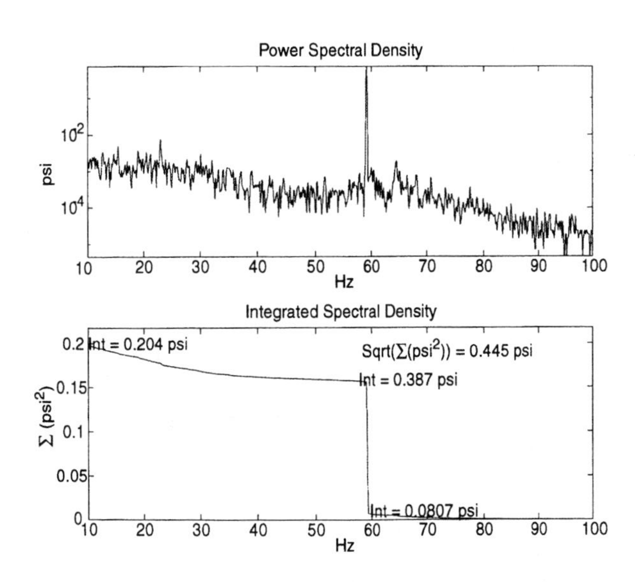

FIGURE 4. Spectral analysis of the accelerator structure LCW pressure modulation shows a strong 59 Hz component. This 59 Hz corresponds to the rotation rate of the pump motors.

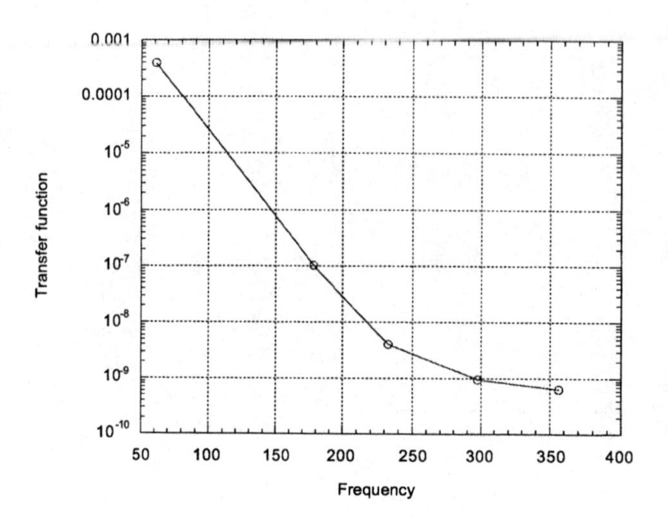

FIGURE 5. Transfer function from the accelerator cooling water to the vertical motion measured on the linac quadrupoles.

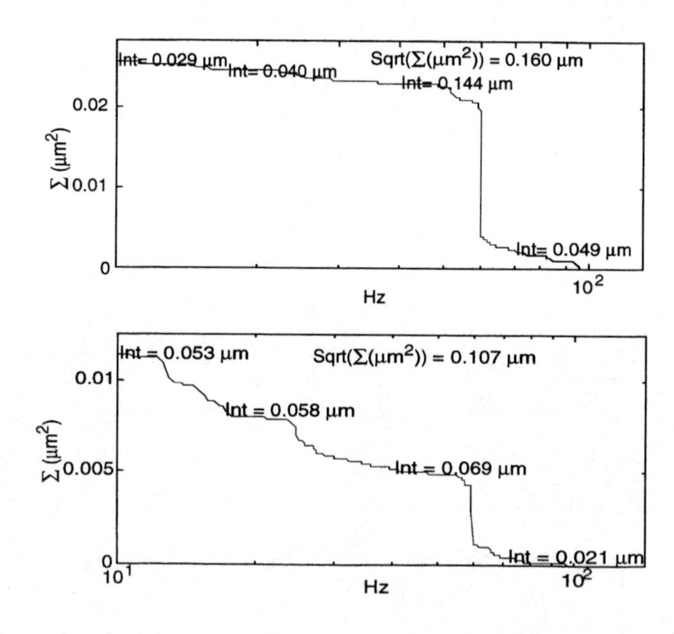

FIGURE 6. The 59 Hz motion of the quads can be reduced by a factor of two with the installation of a rubber "boot" at the output of pumps. The top plot shows quadrupole motion without the boot and the bottom plot shows the motion after installation of the boot.

SUMMARY AND CONCLUSIONS

Clearly the boots on the pumps help reduce motion by about a factor of two. An apparent major source of the 59 Hz is asymmetry in the impeller castings. A project to replace these with better impellers has not yet proved fruitful but measurements at the test stand have shown that other factors such as balancing and bearings must also be considered. Future accelerator installations should try to avoid this problem by specifying vibration criteria at the design stage.

REFERENCES

[1] Turner, J. L., et al, "Vibration Studies of the Stanford Linear Accelerator," presented at the Particle Accelerator Conference, 1995.

[2] Turner, J. L., Stege, R. E., "Vibration of Linac Quadrupoles at 59 Hz", Collider Note 399, SLAC internal publication, 1995.

[3] Matlab is a math software package by The MathWorks.

A Machine Protection Beam Position Monitor System*

E. Medvedko, S. Smith, A. Fisher

Stanford Linear Accelerator Center
Stanford, CA 94309, USA

Abstract. Loss of the stored beam in an uncontrolled manner can cause damage to the PEP-II *B* Factory. We describe here a device which detects large beam position excursions or unexpected beam loss and triggers the beam abort system to extract the stored beam safely. The bad-orbit abort trigger beam position monitor (BOAT BPM) generates a trigger when the beam orbit is far off the center (>20 mm), or rapid beam current loss (*dI/dT*) is detected. The BOAT BPM averages the input signal over one turn (136 kHz). AM demodulation is used to convert input signals at 476 MHz to baseband voltages. The detected signal goes to a filter section for suppression of the revolution frequency, then on to amplifiers, dividers, and comparators for position and current measurements and triggering. The derived current signal goes to a special filter, designed to perform *dI/dT* monitoring at fast, medium, and slow current loss rates. The BOAT BPM prototype test results confirm the design concepts.

INTRODUCTION

The maximum stored energy in the PEP-II rings, 200 kJ for the high-energy ring (HER) at 3 A current and 9 GeV energy, and 77 kJ for the low-energy ring (LER) at 3 A and 3.5 GeV, can melt through the vacuum chamber if the impact is localized. A beam abort trigger system (BATS) protects each ring by kicking the beam into a dump in one turn, spreading it across the exit window to avoid damage. The BATS has been installed around the rings to receive triggers from a variety of faults (such as loss of rf power, loss of dipole current, etc.) and abort the appropriate ring. A new addition to this trigger network is the bad-orbit abort trigger beam position monitor (BOAT BPM). This device acts like rescue boat, reacting quickly to a call for help. The BOAT sends a trigger to the abort system if the orbit is far off center, or if a rapid current loss is detected. One BOAT BPM will be placed on each ring. This paper discusses design of the BOAT BPM and prototype tests results.

---

* Work supported by U. S. Department of Energy, contract DE-AC03-76SF00515.

CP451, *Beam Instrumentation Workshop*
edited by R. O. Hettel, S. R. Smith, and J. D. Masek

DESIGN

To detect large beam position offsets for the beam-abort system trigger, two sets of capacitive pickup electrodes (buttons) are installed in HER and LER (Figure 1), separated by roughly 90° in betatron phase in each ring for the position measurements. A third set of buttons is used for additional measurements of the *X* position at a dispersive point, to trigger on a rapid change in position (*dX*/*dT*) due to energy loss in the first ten turns after an rf trip.

The sum of the signals from one of the sets of buttons is proportional to the total current in the ring. This signal is used to normalize position measurements and to measure the beam-current loss rate (*dI*/*dT*). To minimize the number of abort system inputs, the *X* position outputs are OR'ed together, separately from OR'ed *Y* position outputs. An AND gate disables the outputs if the current is below 100 mA. The BOAT BPM does not depend on external control; all parameters and thresholds are set in hardware in the chassis.

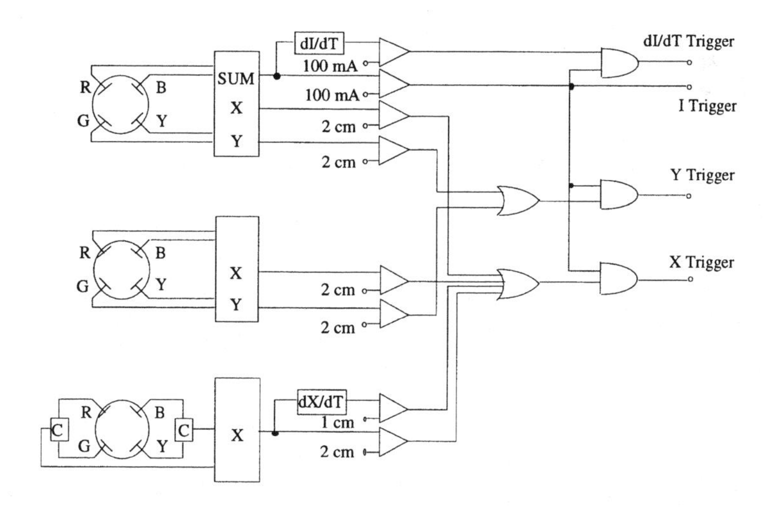

FIGURE 1. The BOAT bpm block diagram. R, G, B, and Y, refer to cable color codes. SUM is the sum of the signals from four buttons. C is a combiner. The comparator thresholds are labeled 100 mA, 2 cm, and 1 cm.

REQUIREMENTS

The BOAT BPM measures beam position, beam current, and beam current loss rate. The measurements must be averaged over one beam turn (the revolution frequency is 136 kHz) to respond to the total current in the ring and to avoid fill pattern sensitivity. There are three regimes of current loss (*dI*/*dT*): fast, where the current loss is greater than 20 mA in 20 μs; medium, where loss is more than 100 mA in a time between 50 μs and 1 s; and slow, where the loss is more than 100 mA in a time longer than 1 second.

The module must generate a trigger when the beam orbit is off center in *X* or *Y* by ±20 mm, $dX/dT \geq 10$ mm in a time less than 1 ms, or the fast or medium current loss threshold are exceeded. If the beam current is less than 100 mA, then no trigger is generated, since this current is below the threshold of damage.

The minimum required dynamic range, 30 dB, is the ratio of the maximum ring current, 3 A, to the lowest current which must be detected, 100 mA. Including a margin at each end of the range increases the requirement to at least 36 dB. Four triggers must be supplied to the beam abort system: *X* and *Y* positions, the current loss rate (dI/dT), and current status, to indicate whether the current exceeds 100 mA.

PROTOTYPE DESIGN

The BOAT BPM prototype was designed to accept signals from one set of buttons. The design and assembly were performed under time pressure in order to be ready to test the prototype and verify design ideas during the January HER run. The prototype was assembled in a chassis with connectorized components, evaluation boards, and a hand-wired circuit board.

The 476 MHz rf frequency is chosen as the input signal. Amplitude demodulation is used to produce a baseband voltage from the input signal.

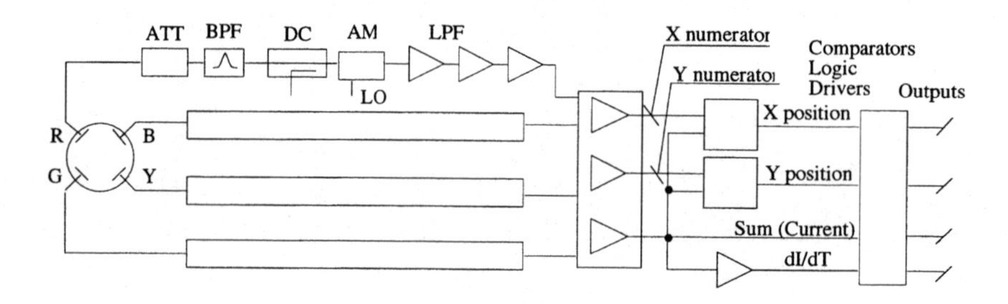

FIGURE 2. The BOAT BPM prototype block diagram.

The signal from a button (Figure 2) goes through a fixed attenuator to a 476 MHz rf filter with 14 MHz bandwidth. The signal from the filter is demodulated by an Analog Devices AD607 (1). This is an inexpensive rf receiver part that includes a UHF mixer, if subsystem, PLL, and an I&Q demodulator. While generally used for the communication industry, the specifications were suitable for our design. It first down-converts the rf frequency to a 10.7 MHz if. The if is then I&Q demodulated with a second LO generated by the on-board PLL. The I-component is taken as the amplitude of the button signal, since the PLL keeps the phase synchronized. A directional coupler in front of the AD607 is used to calibrate channel gains.

We intend the AD607 to operate with fixed gain, although it is normally used with internal automatic gain control. Useful linear dynamic range was an issue. Our first dynamic range measurements revealed an approximately 30 dB linear response, limited by a mismatch of the dynamic range of the I&Q converters to that of the PLL. After adjusting gains to the I&Q relative to the PLL, we achieved 38 dB of linear response.

After amplitude demodulation, the button amplitudes go through four-pole, low-pass filters to suppress the revolution frequency. Op-amps derive *X* and *Y* differences and the button amplitude sum. Analog divider chips are used to get *X* and *Y* positions. Position is calculated by:

$$X = K \times \frac{R+G-Y-B}{R+G+Y+B}$$

$$Y = K \times \frac{R+B-Y-G}{R+G+Y+B} \quad , \qquad (1)$$

where, for a round pipe,

$$K = \frac{d}{2\sqrt{2}} \; .$$

Here R, G, Y, and B are signals from the buttons in volts (Figure 2) and d is the beam pipe diameter. The position signals from the dividers go to comparators with adjustable threshold levels, which set the trip points for the beam-abort system.

The AC-coupled signal from summing amplifier feeds the dI/dT active filter. This filter has a response crafted to detect three regimes of current loss. The signal from the filter drives a comparator; the time-domain response of the filter enhances fast current losses (peaking) and suppresses very slow current responses (ac-coupling) so that a single threshold level provides all the logic for the three timescales of current loss.

TEST RESULTS

Bench tests showed that the BOAT BPM prototype is functional. X and Y have small offsets. After bench testing, the module was moved to the HER for tests with beam.

Revolution Frequency Filter Performance

Three active filters, one a 0.5 dB ripple, two-pole Chebyshev filter plus two one-pole filters, make a smooth frequency response. With a cutoff frequency of 50 kHz, the four-pole filter should give about 44 dB suppression at 136 kHz. The two following pictures (Figure 3), show a sine wave at 136 kHz frequency at the filter input (on the left) with 270 mV peak-to-peak amplitude, and at the filter output (on the right) at about 4 mV peak-to-peak. The filter gain is 1.9, so the revolution frequency suppression is 42 dB, close to the prediction of 44 dB.

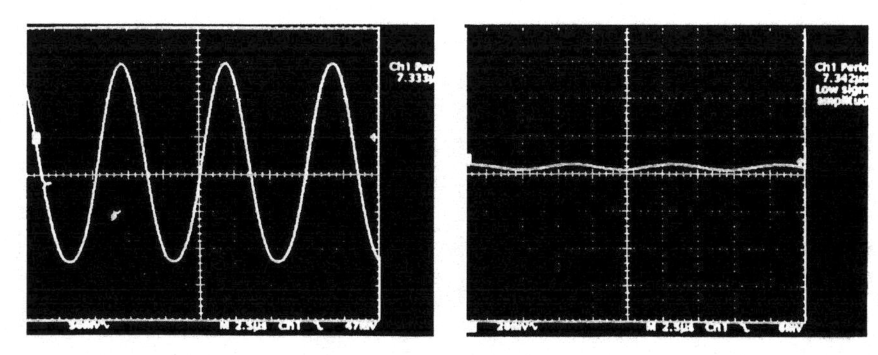

FIGURE 3. Signal at the filter input (left, 50 mV/div., 2.5 µs/div.) and at the output (right, 20 mV/div., 2.5 µs/div, AC coupled).

Beam Position Measurement

During the measurements, a few test points were observed: the output of the revolution frequency filter for each channel, the dividers' outputs, and the numerator and summing amplifiers' outputs. An example of the measurements taken with 360 mA of beam is shown in Table 1. The calculated results match the measured data well.

TABLE 1. The Example of the Beam Position Measurements

Test points*	R	B	Y	G	Xn	Yn	Xp	Yp	Sum
Measured data, V	−1.88	−1.47	−2.01	−1.81	0.21	−0.47	0.56	−1.07	4.07
Calculated data**, V					0.21	−0.47	0.53	−1.18	

* R, B, Y, G are the output of each channel's filter; Xn and Yn are the *X* and *Y* numerator voltages; Xp and Yp are the *X* and *Y* position voltages; Sum is the summing amplifier output (Figure 2).

** Data are calculated according to equation 1, based on the measured R, B, Y, G and Sum.

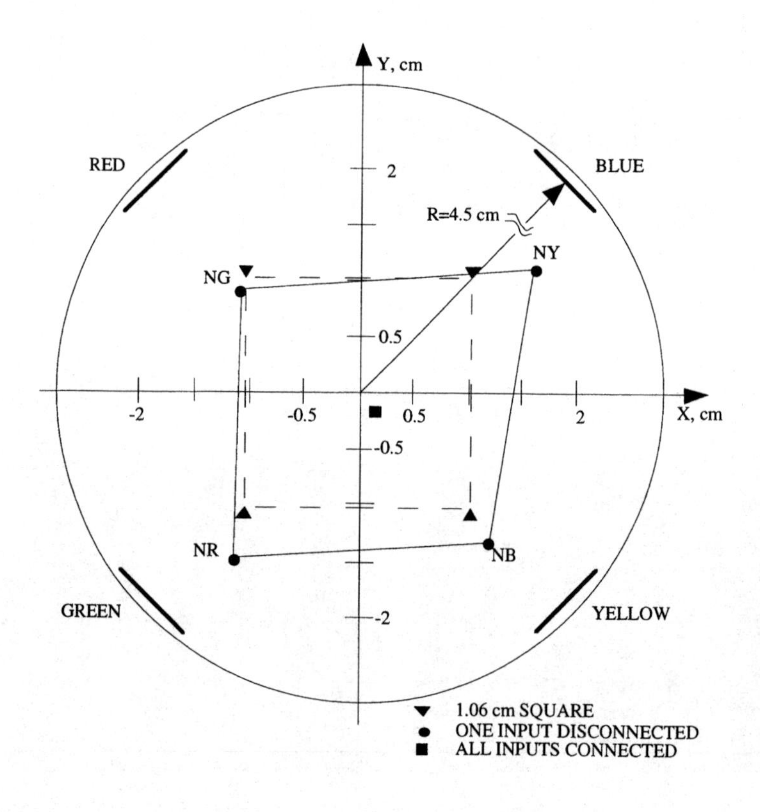

FIGURE 4. Measured position and beam offsets simulated with one input cable disconnected at a time. NR indicates no red cable, etc.

The position response was tested by disconnecting one button cable at a time. According to the position calculation equation (Eq. 1), if one of the signals is zero, the position values, both X and Y, should have the proper sign and a magnitude of $K/3$=1.06 cm for our 9 cm diameter beam pipe. The test results are shown on Figure 4, along with a square with corners at ±1.06 cm to guide the eye. The measured beam positions are in the correct quadrants, close to the ±1.06 cm coordinate square. A position offset can be seen.

dI/dT Filter Performance

The filter response to a step function input has peak at 12.5 μs (about two beam turns) in the time domain, followed by a "flat" part (Figure 5). The flat part is not really flat; it slowly decays to zero with a 1-second time constant, which roughly corresponds to the thermal conduction time for heat to spread away from a small hot spot. The peaking of the step function response makes the system more sensitive to rapid current loss. This defines the "fast" response sensitivity. The ratio between the peak and flat parts should be 5, according to the ratio of the requirements for the medium-speed to fast *dI/dT* trip thresholds, i.e., the ratio of 100 mA to 20 mA.

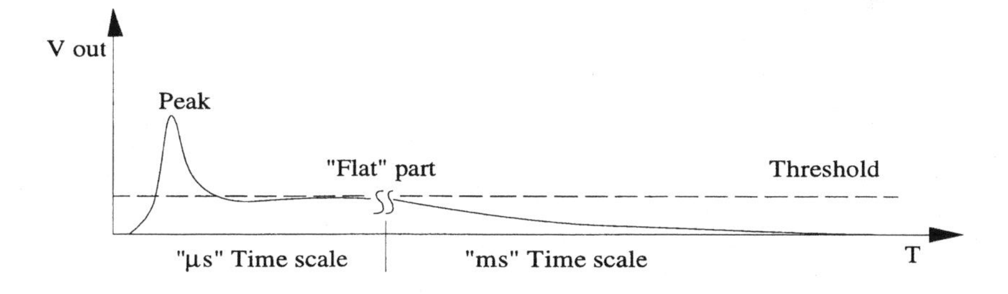

FIGURE 5. Sketch of the *dI/dT* filter pulse response in time domain.

During HER commissioning, fast beam losses were rare (and unpredictable, as we were not trying to lose the beam). Therefore, to test the response of the system, we simulated fast current losses by abruptly disconnecting one button input to the BOAT chassis. The transient response of the *dI/dT* signal is shown in Figure 6.

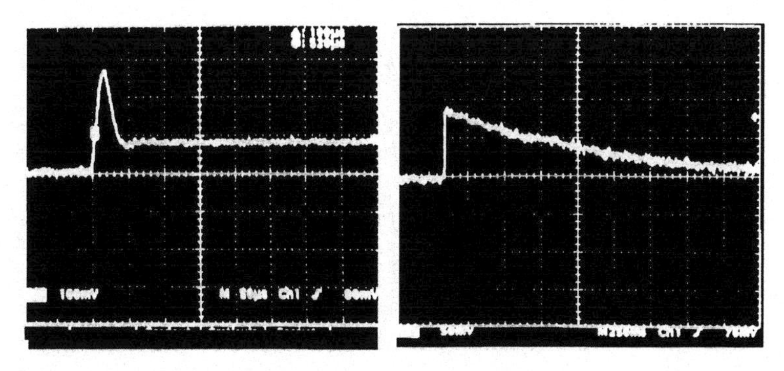

FIGURE 6. The *dI/dT* filter response seen on two different time scales. The left picture shows the peak and "flat" part at 50 ms/div. The right picture shows the "flat" part at twice the vertical sensitivity and 250 ms/div.

The ratio between peak and "flat" parts is approximately three (Figure 6, left). The bench test gave six for the peak to flat ratio. This disparity may be due to the finite fall time of an rf signal switched by a human hand disconnecting an SMA connector.

Dynamic Range

The BOAT BPM dynamic range is 37 dB. Figure 7 shows the dependence at the summing amplifier output signal in volts versus input signal, in volts rms. The non-linear parts correspond to the dynamic range limits of the AD607.

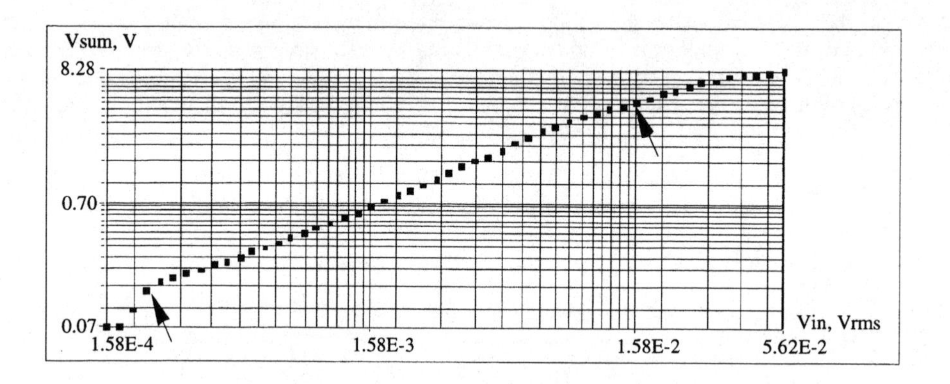

FIGURE 7. The Sum signal, Vsum versus input signal, Vrms.

FURTHER PLANS

Two BOAT BPM chassis will be built, one for each of the PEP-II rings. For the production devices, the rf circuitry for each opposing pair of buttons will be integrated on an rf daughterboard. Each rf board will produce one position signal and one sum signal, which are further combined and processed on a motherboard. The rf boards will be tuned and calibrated on a test bench. The mother board will carry five rf boards, four to measure position from two sets of buttons, and a fifth to measure the X position in the dispersive region. The comparators, logic, dI/dT and dX/dT filters, and Sum amplifier shall be located on the mother board.

SUMMARY

The BOAT BPM is intended to perform fast detection of large beam offsets or rapid current losses. The device's purpose is to trigger the beam-abort system to protect the PEP-II rings. AM demodulation allows us to use inexpensive commercial components. Prototype test results confirm design concepts.

ACKNOWLEDGEMENTS

For fruitful discussions we would like to thank Uli Wienands and Artem Kulikov. Thanks to Bob Noriega for nice module assembling. Thanks to our group for immediate help and good wishes. Thanks for those who fix fresh coffee in the morning.

REFERENCES

[1] Analog Devices, Inc., "Low Power Mixer/AGC/RSSI 3 V Receiver IF Subsystem," 1995, www.analog.com.

BPM Testing, Analysis, and Correction

James A. Fitzgerald, James Crisp, Elliott McCrory, Greg Vogel

Fermi National Accelerator Laboratory[1]
P.O. Box 500, Batavia, IL 60510

Abstract. A general purpose stretched-wire test station has been developed and used for mapping Beam Position Monitors (BPMs). A computer running LabVIEW software controlling a network analyzer and *x-y* positioning tables operates the station and generates data files. The data is analyzed in Excel and can be used to generate correction tables. Test results from a variety of BPMs used for the Fermilab Main Injector and elsewhere will be presented.

INTRODUCTION

The Main Injector (MI) Accelerator project at Fermilab required the design and construction of several new Beam Position Monitors (BPMs). Approximately 200 primary diagnostic MI BPMs were required. The MI design required these to be installed inside the quadrupole magnets, limiting design options. These BPMs use four striplines that are combined electronically to produce the horizontal and vertical positions. The Recycler Ring design uses BPMs of the split-plate design that are installed in the conventional manner. This BPM has two plates, and is used in either a horizontal or a vertical orientation. In addition, other varieties including *the 8 GeV line BPM*, a special *low-level rf BPM,* and a wide aperture BPM was also tested. The test stand is shown in Figure 1 and the BPMs are shown in Figure 2.

In order to test, characterize, align, and provide data for calibration, a general purpose test stand was designed and constructed. A similar stand was used for RHIC BPMs at Brookhaven (1). This stand was designed to accommodate a wide variety of devices, and is strong enough to support a small trim magnet. The LabVIEW (3) control system allows for easy development of custom routines for automatic testing.

[1] Operated by Universities Research Association under contract with the U. S. Department of Energy

CP451, *Beam Instrumentation Workshop*
edited by R. O. Hettel, S. R. Smith, and J. D. Masek

FIGURE 1. Photo of the test stand.

FIGURE 2. Recently tested BPMs, from left, Main Injector, Recycler Ring, Low Level rf, Wide Aperture, and 8 GeV line.

THE TEST STAND

The stand consists of two units, a rigid frame (3 ft × 3 ft × 6 ft) for the mechanical components and one standard six-foot relay rack. The rectangular frame contains the *x-y* positioning tables, a vertical stretched wire, and the BPM mounting fixtures. Both units are on wheels and can be easily relocated (Figure 1). A wire stretched through the BPM is driven at the appropriate rf frequency, to simulate beam, and produce output signals. A Macintosh Power PC, running LabVIEW virtual instrumentation software, controls *x-y* positioning tables, which moves the wire through a grid pattern. The number of points measured are approximately 50 to 160 producing reasonable file sizes. An HP network analyzer controlled through a GPIB bus collects the data.

The test stand was constructed from standardized aluminum "erector set" materials manufactured by 80/20 Inc. (5). The main frame rails are three inch square extrusions, with 1/2-in Al tooling plate for attaching components. The frame is aligned using gauge blocks and an optical survey. The BPM under test is located in a precision manner using hard drill bushings and matching alignment pins. The BPMs were constructed with precision-reamed alignment holes which mate with the test stand and are also used in the beam line for the alignment survey.

The stretched wire is driven from an HP network analyzer rf out port and is terminated to the R port. The wire has an impedance of approximately 200 ohms and required matching to the 50-ohm system. A small printed circuit board, with a 4:1 matching transformer, is used to couple to the wire at each end.

The Positioning tables used are "monolithic" *x-y* units (as opposed to single-axis units bolted together), which are more compact and have an orthogonality specification of 50 arc-sec (2). This eliminates construction and stand alignment steps. The tables have a three-axis position accuracy of .001 inch, and repeatability spec of .0005 inch (accuracy of the lead screws). The lead screws are metric to simplify calculations. The tables use stepping motors, controlled with nuLogic drivers and controllers. The motors can be used in full step, half step, or micro step modes, producing resolutions of 5 micron (.0002 in), 2.5 micron (.0001 in), and .5 micron (.00002 in) respectively (4). The tables also incorporate linear encoders that have a resolution of 2.0 microns (2). The Lab View measuring routine is set up with encoder feedback to position the tables to the desired coordinates. The ultimate limit on the accuracy is therefore set by the encoder resolution of 2.0 microns. Measurements have shown accuracy repeatable to within 20 microns (.0008 in). Shown in Figure 2 are some of the Fermilab BPMs recently tested.

Data Collection

The data is downloaded through a net connection for plotting and analyzing. An Excel spreadsheet is used to generate a set of plots, as desired, for each BPM. The files, charts, and serial numbers are recorded as permanent records for future reference. One sample BPM of each design is maintained as a reference. Plots generated include:

1. Electrically measured vs. mechanical position map.
2. Horizontal and/or vertical outputs represented as 20Log(A/B) vs. position.
3. Horizontal error from a least square fit vs. vertical position.
4. Horizontal error from a least square fit vs. horizontal position.

CORRECTION EFFORTS

We performed an initial off-line analysis on the BPM data for all detector types using Microsoft Excel. The data was imported into a spreadsheet, *A* and *B* values were calculated for each position, and a preliminary linear fit was performed using the built in LINEST function. The fit parameters were used with the *A/B* term to calculate a position that was compared to the measured position of the wire. A plot of the position error (calculated-measured) versus measured position became a primary tool in determining the validity of a measurement and the acceptability of the detector. Valid data sets are combined with BPM electronics data to generate the tables used by the control system database to display beam position. For the Fermilab Main Injector, a LabVIEW program is being used to calculate the coefficients of a fifth-order polynomial relating the output voltage for a given detector, combiner box, and rf module combination with the actual beam position.

The calibration data for the Fermilab Main Injector (FMI) Beam Position Monitors have been used in at least two ways. (1) It provides database tables of the mapping from raw (ix, iy) values to real (x, y) values. (2) It provides accurate raw readings from the beam physics model of the FMI to the controls system, when the control system is redirected to get device readings from this model. A pair of database tables have been created from the BPM calibration data so that one may obtain reasonable real (x, y) values, given the raw digitizations received from the electronics. This has been accomplished in the following manner. For each calibrated BPM, the mechanical and electrical offset is determined from the calibration data. Then, this offset is subtracted from the entire calibration data set for this BPM. Third, the calibration grid points are averaged for all BPMs at each point. Finally, these averaged gridpoint values are stored as a pair of database tables that, on average, represents the calibration of all the BPMs. The axes of the tables are raw (digital) x and y readings (0–255); one table produces the real x value (in mm) and the other produces the real y value.

A user can get an accurate reading of the actual (x, y) position of the beam within the BPM. He would use the raw readings obtained from the BPMs through the control system as keys to these two database tables. This gives the real x and the real y position. The offset of that BPM, determined from a separate control database, is then added. The second use of the FMI BPM calibration data is for producing reasonable raw BPM reading from a simulated FMI. Simulated positions are generated from a beam-physics model of the FMI at each BPM. Attached to each BPM object in the software is the calibration information for that BPM. Using a four-point interpolation, raw digitized values are reported back to the control system application that is running with its connection to the real world redirected to this model. Thus, a programmer, physicist, or engineer can test algorithms for removing the nonlinearities in the FMI BPMs long before the FMI is actually available.

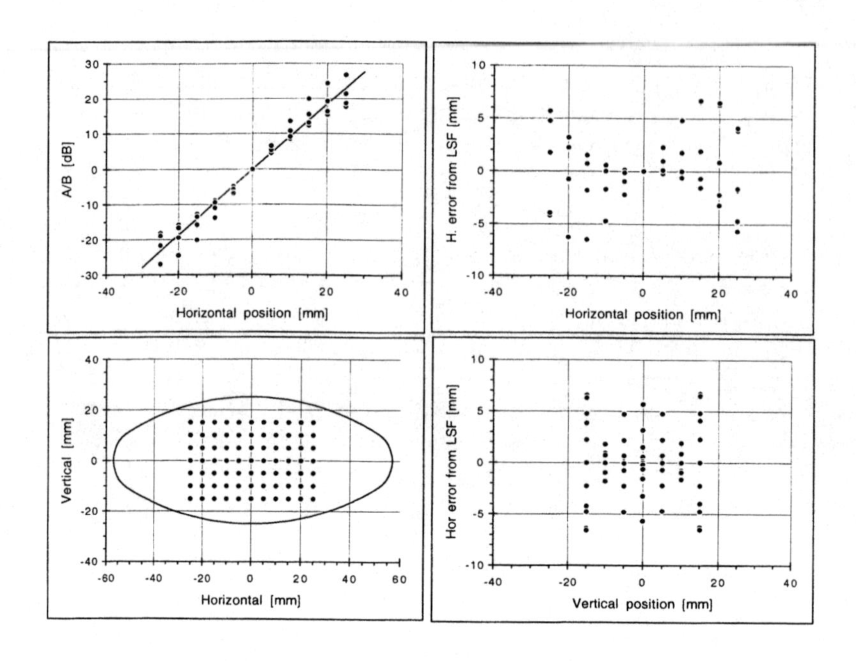

FIGURE 3. Main Injector BPM data.

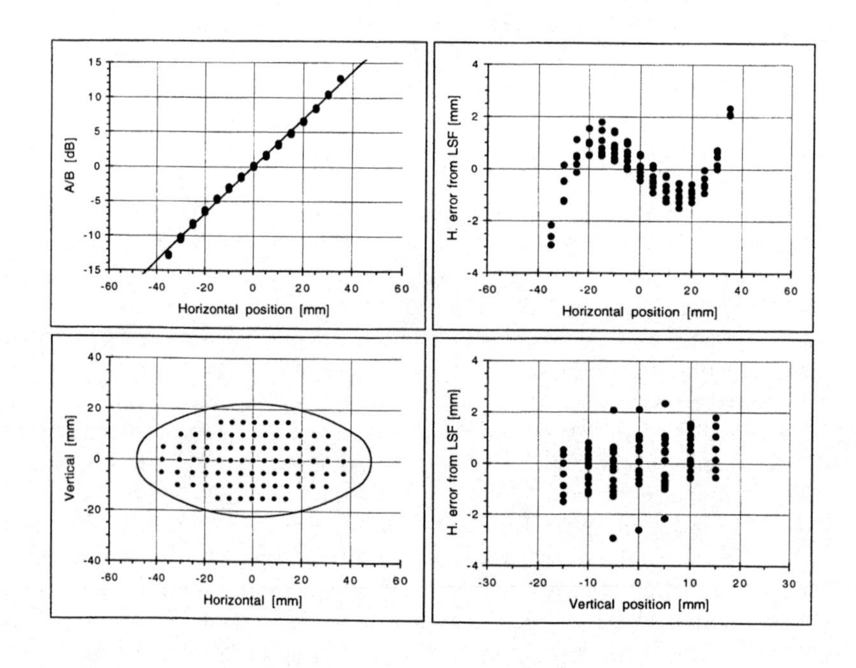

FIGURE 4. Recycler Ring BPM data.

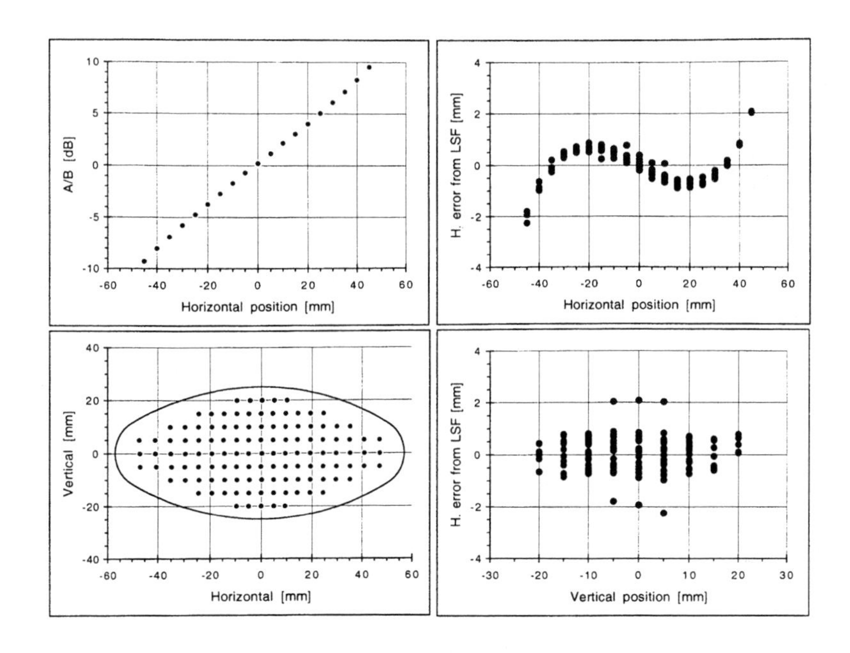

FIGURE 5. Low Level rf BPM data.

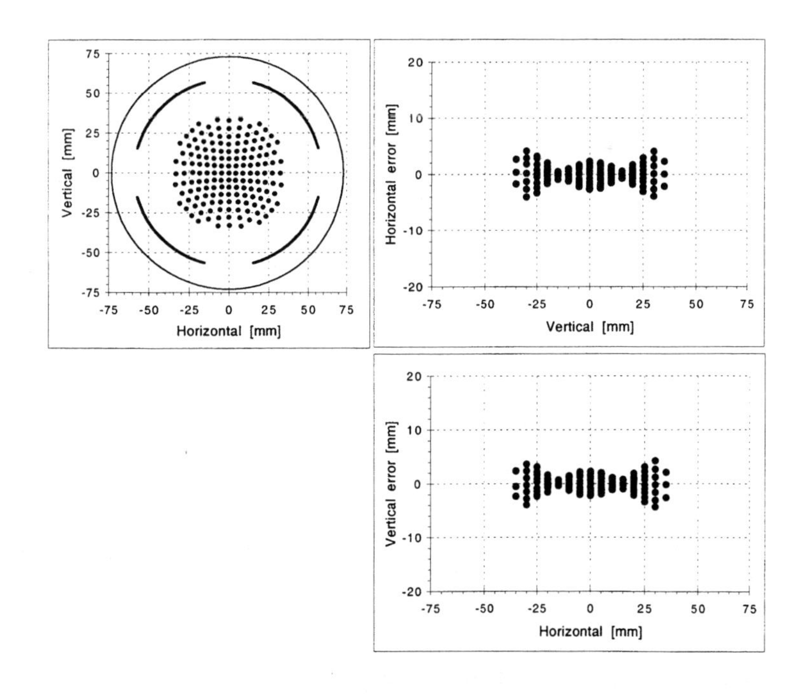

FIGURE 6. Wide Aperture BPM data.

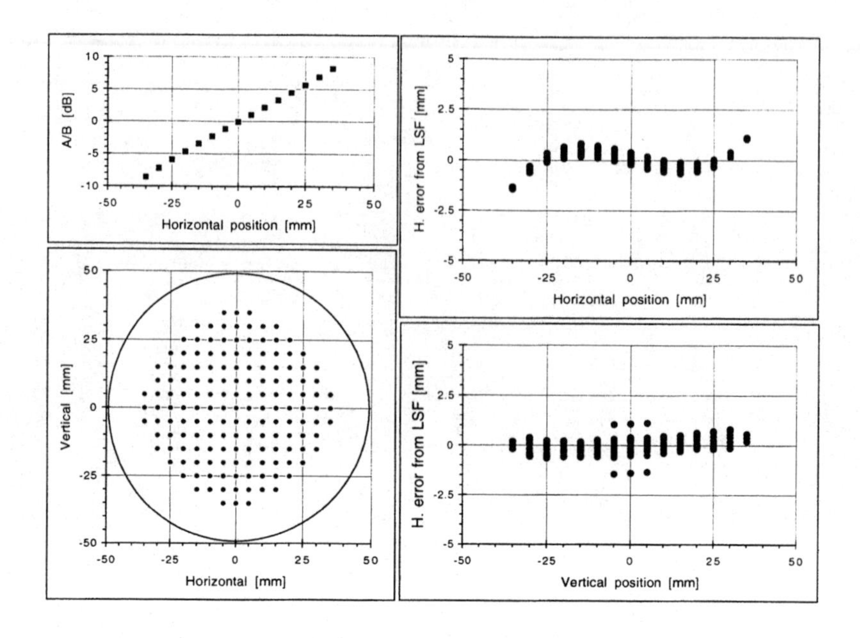

FIGURE 7. 8 GeV Line BPM data.

SUMMARY

The test stand has a theoretical resolution capability of 2 microns, as limited by the encoder. Although we have not verified to this accuracy, we have demonstrated its repeatability to at least 20-micron level with actual BPM measurements. We have shown that most BPMs are accurate near the center axis, but have increasing nonlinearities in the off-axis locations. Work is in progress to implement correction algorithms to compensate for these nonlinearities.

It was found that better results are obtained if the BPM ends have a 360-degree ground connection to the frame enclosing the wire. The proximity of a plate to an open end results in distorted position measurements in some cases. Sliding plates and bellows were used to make a ground connection to the tables. The split-plate BPMs are sensitive to standing waves on the stretched wire as each plate is effectively at a different location. The linearity mapping for some of the split plate BPMs was therefore done at a reduced frequency.

ACKNOWLEDGMENTS

The authors wish to thank Brian Fellenz and Shoua Moua, who operated the Test Station and managed the logistics of hundreds of BPMs. Thanks also to Tom Shea and Pete Cameron who let us inspect their test stand.

REFERENCES

[1] Shea, T.J.,et al., "RHIC Beam Position Monitor Assemblies" Proceedings of the 1993 IEEE Particle Accelerator Conference, pp 2328-2330.

[2] "Precision Positioning Components and Systems", New England Affiliated Technologies, Lawrence, MA, 01841

[3] "LabVIEW User Manual", National Instruments Corp., Austin, TX, 78730

[4] "nuLogic User Manuals", nuLogic Inc. (National Instruments Corp. 1997), Needham, MA,

[5] "80/20 Inc." Columbia, IN, 46725

Beam Position Monitor System of DAΦNE

A. Stella, A. Drago, A. Ghigo, F. Marcellini, C. Milardi,
F. Sannibale, M. Serio, C. Vaccarezza

INFN Laboratori Nazionali di Frascati
00044 Frascati (Roma), Italy

Abstract. The DAΦNE beam position monitor (BPM) system consists of 150 monitors installed all along the machine. Design issues, calibration procedures, experimental results and performance of the system are described. The closed orbit in the main rings is extracted from the BPM signals through narrowband receivers (realized by Bergoz Precision Beam Instrumentation for DAΦNE), then acquired and processed by a real-time task based on four independent processors dealing with different machine areas. The data acquisition system is integrated in the DAΦNE control system and measures five complete orbits in a second. Implementation criteria, measurements and results are reported.

OVERVIEW OF THE BPM SYSTEM

The Φ-Factory, DAΦNE, is a high-current multibunch e^+e^- double-ring collider, presently being commissioned at INFN-LNF. Electron and positron beams are generated and accelerated along a linac up to the nominal energy of 510 MeV, stored and phase-space damped in the accumulator ring before injection into the main rings through a ~100 m transfer line.

Table 1 summarizes some of the DAΦNE operating parameters relevant to the diagnostic system:

TABLE 1. DAΦNE Parameters

	Accumulator	Main Rings
Energy	510MeV	510MeV
rf frequency	73.65 MHz	368.25 MHz
Number of bunches	1	120
Single bunch current	150 mA	40 mA
Bunch length	100 ps	100 ps
Revolution frequency	9.2 MHz	3.06 MHz

CP451, *Beam Instrumentation Workshop*
edited by R. O. Hettel, S. R. Smith, and J. D. Masek

Because of the various requirements of each part of DAΦNE resulting from the different beam characteristics and the vacuum chamber geometry, several different pickup devices and monitor configurations have been designed and installed in the transfer-lines, accumulator and main rings. These include short-circuited strip-lines, matched striplines, button electrodes, and special monitors for use in the interaction regions.

The low intensity of the beam in the transfer lines (TL) and the accumulator requires a high-sensitivity BPM. For this reason, mainly stripline monitors are used. The four electrodes are 50 Ω, stainless-steel strips, short-circuited at the downstream end. This is mechanically convenient and has no relevant effect on the upstream signal. Single strips in the same monitor are matched within 0.05 Ω to each other.

The acquisition system in the accumulator and transfer lines is simpler than the main rings' system. BPM signals are multiplexed and acquired using an oscilloscope remotely controlled through GPIB and a LabView application, which also provides a versatile user interface.

TABLE 2. DAΦNE Beam Position Monitors Summary Table

	BPM	Total BPM
Damping Ring	short circuited strip line	4
	button electrodes	8
Transfer Lines	short circuited strip line	23
Main Ring e–	button electrodes	35
Main Ring e+	button electrodes	35
Interaction Region	button electrodes	9+9
	matched strip line	4+4

In order to correct the closed-orbit distortion and optimize the operation of the main rings, 35 BPMs have been installed all along each ring and a further 13 BPMs are situated in each interaction region. The beam position monitors of the main rings are detected and processed by a dedicated acquisition system, described in the next sections, which provides the closed orbit for both the positron and electron rings at a fast rate.

Matched stripline monitors and electrostatic monitors with six-button electrodes have been developed for interaction regions to allow simultaneous measurements of the transverse position of positron and electron beams (1). The stripline monitors are 50 Ω strips, matched at both ends and with a directivity of ~25dB.

MAIN RINGS SYSTEM

Pickup Characteristics and Calibration Procedures

The button electrodes, manufactured by Metaceram (FR), are 10 mm in diameter, have a matched impedance of 50 Ω and a typical capacitance of 4.2 pF. They were designed to produce acceptable signals and to keep the vacuum chamber impedance as low as possible. The capacitance of each electrode was measured before mounting. An example of reflected waveform from a capacitance measurement based on the TDR method (2) is shown in Figure 1.

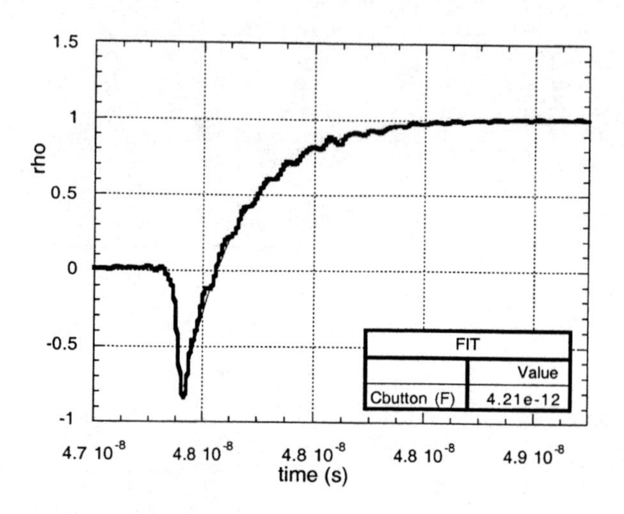

FIGURE 1. TDR measurements of the button pickup electrode.

Buttons with similar capacitance have been grouped together in each monitor in order to minimize the electric offset error. This error can be estimated deducing $\Delta V_b/\Delta C_b$ from the equation that represents, in the frequency domain, the voltage V_b induced by the beam at the external termination as the product of the transfer impedance Z_b times the beam current spectrum I :

$$V_b(\omega) = Z_b(\omega) \cdot I(\omega) = F\phi R_0 \left(\frac{\omega_1}{\omega_2} \right) \frac{j\omega / \omega_1}{1 + j\omega / \omega_1} \cdot I(\omega) \tag{1}$$

where $\omega_1 = 1/R_0C_b$ and $\omega_2 = c/2r$ with C_b the button capacitance to ground, r the button radius, c the speed of light, $\phi = r/4b$ the coverage factor, b the half height of the vacuum chamber and F a form factor which depends on the vacuum chamber geometry (3). In our monitors, the value of F ranges from ~ 0.6 in the rectangular types to 1 in the round ones.

The capacity values C_b in each monitor are matched within 0.01 pF, resulting in an electrical offset error for the BPM installed in the main rings within 50 μm, in the worst case.

Since the vacuum chamber cross section is variable along the ring circumference, several different configurations of BPM have been developed (Figure 2).

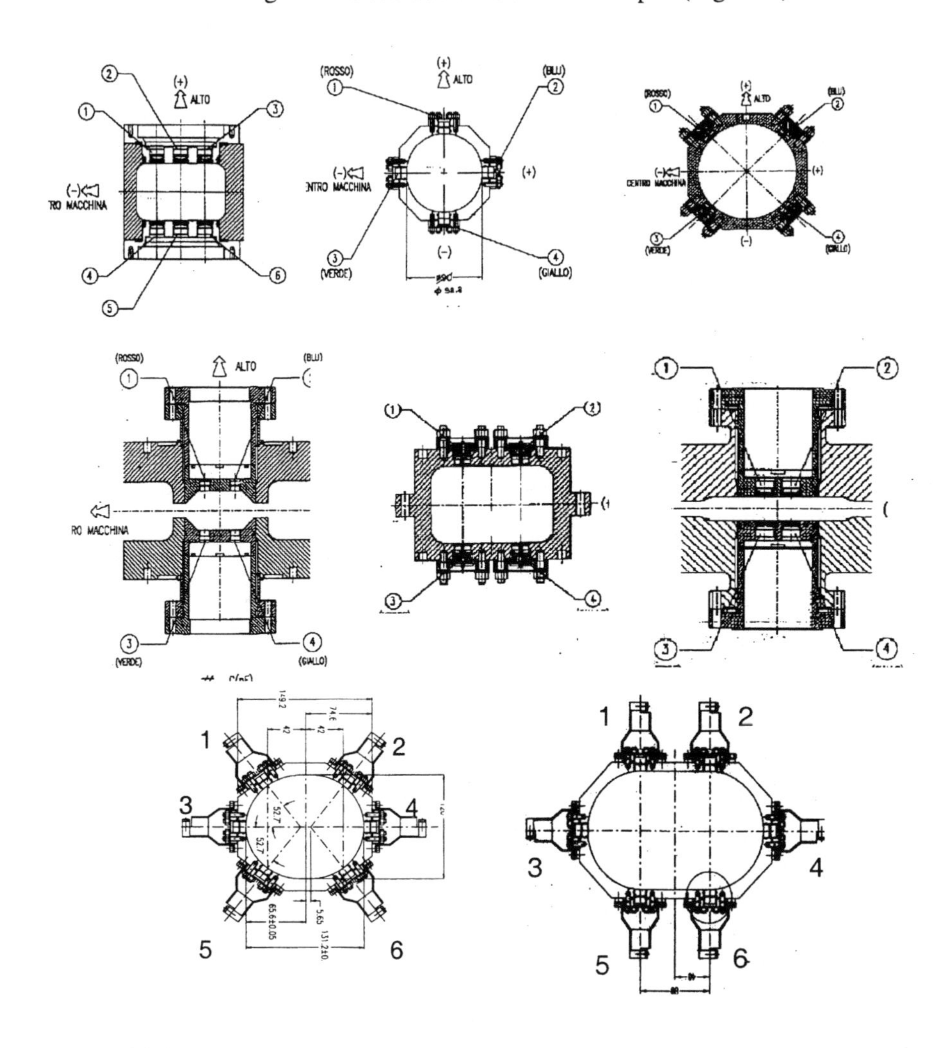

FIGURE 2. Schematic layout of the BPM installed in the DAΦNE Main Rings.

An accurate calibration of each type of BPM installed has been performed, with both numerical simulations and bench measurements (4), to recover the non-linearity of the transfer function. Starting from calibration data, a non-linear fit of two dimensionless quantities, derived from the signal induced on the electrodes, is used to reconstruct accurately the beam transverse position.

Table 3 summarizes, for various BPM configurations, the beam position reconstruction error (rms) applying a fourth-order polynomial fit in a 20 mm × 20 mm zone around the center of the monitor.

TABLE 3. Reconstructed Beam Position Error

BPM type	ΔX_{rms} fit error	ΔY_{rms} fit error
Round diagonal	21 μm	21 μm
Dipole	22 μm	39 μm
Wiggler	33 μm	93 μm
Rectangular	26 μm	29 μm
Round orthogonal	14 μm	14 μm

Detection and Data Acquisition

The beam signals from the pickup electrodes of each BPM are transmitted through independent good quality coaxial cable having an average length of 25 m up to the detection electronics.

The signal detection circuit has been developed by Bergoz Precision Beam Instrumentation with particular specifications for DAΦNE (Table 4).

TABLE 4. BPM Detection Electronics Parameters

rf detection frequency	736.515 MHz
if processing frequency	21.4 MHz
Minimum input signal detected	–73 dBm
Dynamic range	80 dB
Button sampling frequency	2.5 KHz

In order to measure beam position with any multibunch configuration (up to 120), the pickup frequency of 736.5 MHz has been chosen. It corresponds to twice the accelerating rf frequency and to a typical button transfer impedance of ~0.2 Ω.

The signal from the BPM electrodes are time-multiplexed into a superheterodyne receiver, which converts the beam-spectrum-selected harmonic to an intermediate frequency, if = 21.4 MHz, before amplitude detection.

The demodulated signal is demultiplexed into four values that are stored in analog memories. The four signals are summed and the sum is maintained constant by an automatic gain control, which makes it possible to obtain the pseudo-beam positions by simple sums and subtractions between the demodulated voltages (5, 6).

The Bergoz detectors and all of the hardware for the acquisition are distributed in four racks located in different areas of the Main Ring Hall. Each BPM board provides two analog voltages, in the range [–10 V, 10 V], which represent the linear combination of the button signals used to deduce the transverse beam position (x,y) with the fitting algorithm described above.

The signals coming from the detection electronic boards are grouped in one of the four independent racks, then multiplexed and acquired (Figure 3).

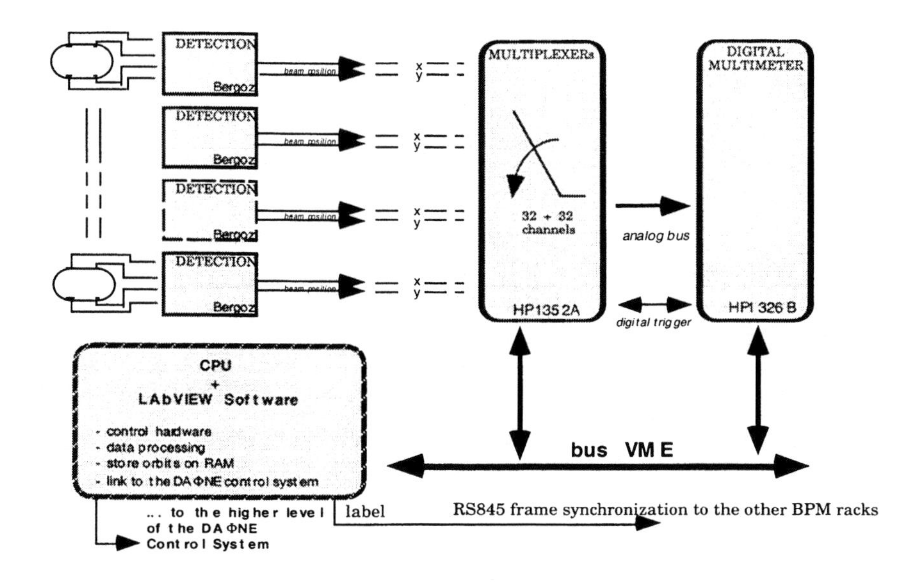

FIGURE 3. Schematic of the Main Rings BPM acquisition system.

The acquisition system which reads and processes the two analog outputs from each monitor has been developed following the VME standard, using mainly commercial hardware.

An independent processor, based on a Motorola 68000 CPU, running a purpose-built LabView application, controls two HP1352A FET Multiplexers and a HP1326B Digital Multimeter.

The two FET multiplexer modules provide high-speed switching up to 64 channels, and are directly connected to the digital multimeter. A scanning list of channels is downloaded into a RAM on the multiplexer modules at the startup and scrolled automatically during the acquisition (7).

The scanning operation does not require any intervention from the central CPU until the end of the whole acquisition of all the BPMs in the scan list. The trigger for channel advance comes from two handshake lines that directly link the HP modules, while an analog bus connector provides the link to the voltmeter for the signals to be measured.

This working mode allows a fast acquisition of the closed orbit since, during the measurements, each CPU is free to process the previously measured data in order to apply the linearization fit and to store the beam positions in a circular buffer memory. The acquisition rate is ~5 orbits/sec. Each processor deals with one quadrant of the closed orbit, both for the electron and positron rings. One of the four processors (the master) sends a "start" command along with the "actual time" information on an RS485 line to the other three processors (the slaves).

The four processors are fully integrated in the DAΦNE control system, both from the hardware and the software point of view. The orbit data are directly accessible from the control system user interface, which provides many tools to display the closed orbit throughout the whole machine, as well as in the interaction regions, where a local orbit analysis and correction is necessary in order to control the interaction point and then the luminosity.

Data from the BPM system are also accessed from automatic tasks recording the beam response to different localized kicks by the corrector magnets, the so-called response matrix, used for machine modeling.

System Performance

The response of the whole BPM system has been analyzed as a function of the stored beam current. In this way, all the different device parts—pickup, cable, detection and acquisition equipment—have been checked. Moreover the error affecting the position measurements at low current has been determined.

One set of measurements of one hundred consecutive orbits each, at several beam current values, has been recorded. The difference between each averaged orbit and the reference orbit, measured at ~18 mA, is reported for different BPMs located along the positron ring (Figures 4–5).

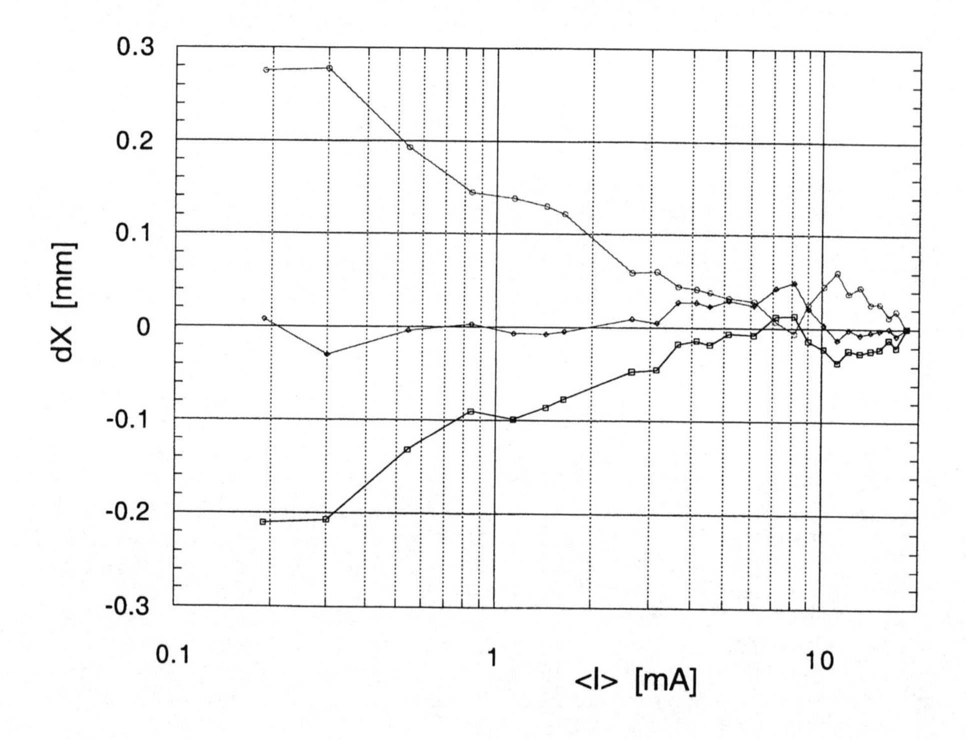

FIGURE 4. *X* mean position vs. beam current for different monitors.

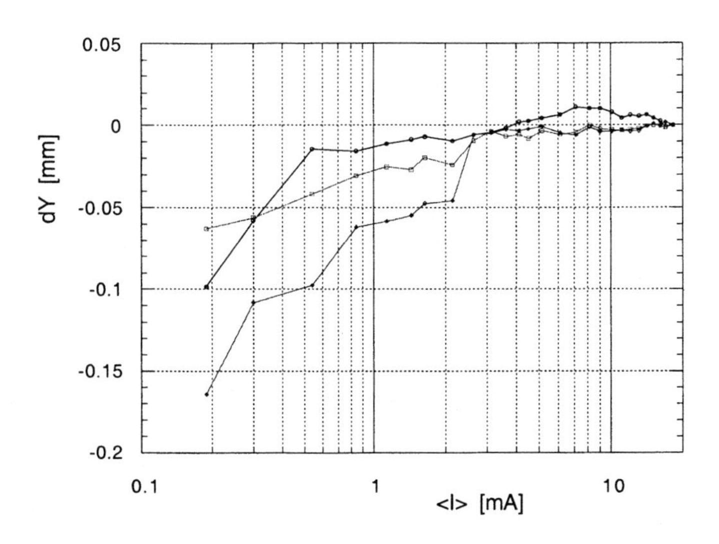

FIGURE 5. *Y* mean position vs. beam current for different monitors.

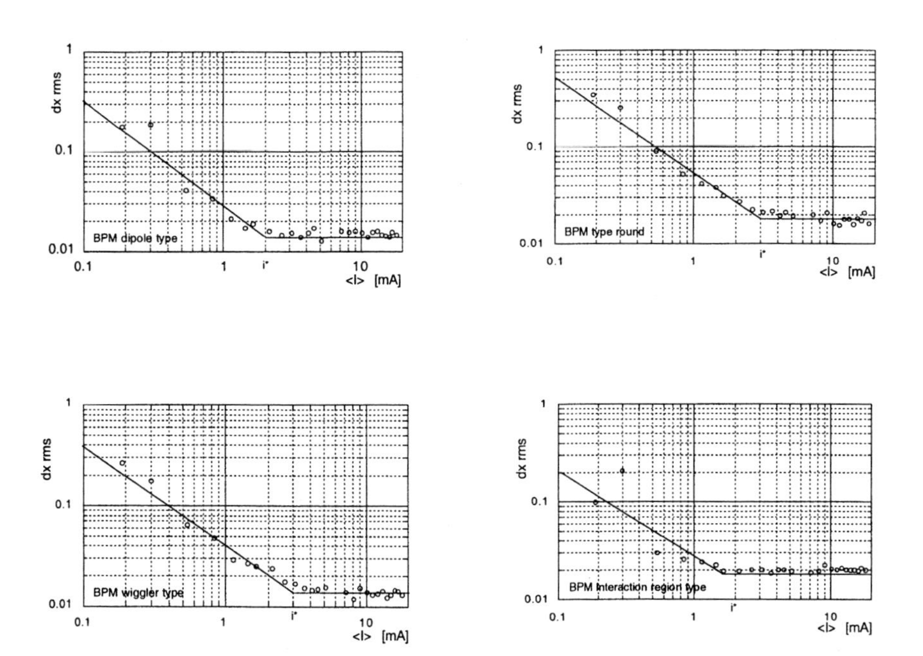

FIGURE 6. Error position rms vs. beam current for different monitors.

The rms beam position error δx (Figure 6) is initially inversely proportional to the beam current $< I >$, and asymptotically approaches about 0.02 mm for currents above the threshold I^*

$$\delta x<I>=k \text{ for } <I> \; <I^* \tag{2}$$

The k and I^* values for different type of monitors are reported in Table 5.

TABLE 5. k and I* Values for Different Types of Monitors

Type	k [mm mA]	I^* [mA]	δxrms[mm]
Round diagonal	0.05	3.1	0.02
Rect	0.04	3.1	0.018
Dipole	0.03	2.1	0.017
Wiggler	0.04	3.1	0.018
Interaction region	0.02	2.1	0.02

A stable measurement at a current value as low as 0.19 mA is the lowest limit of our analysis so far.

ACKNOWLEDGMENTS

The BPM system is the most extensive of all the DAΦNE diagnostics, numerous people cooperated in the various facets of the project.

The authors wish to thank Oscar Coiro, Donato Pellegrini, Olimpio Giacinti, whose support during commissioning has been of great importance in the system realization. The help of Carlo Marchetti for all the bench measurements is gratefully acknowledged.

We are grateful to Julien Bergoz and Klaus Unser for helpful collaboration.

REFERENCES

[1] Serio, M., A. Stella, C. Vaccarezza, "Beam Position Monitor for the DAΦNE Interaction Regions," Presented at the Diagnostic Particle Accelerator Workshop 1997, Frascati, Italy.
[2] Hewlett-Packard, "TDR fundamentals," App. note 62.
[3] Marcellini, F., M. Serio, A. Stella, M. Zobov "DAΦNE Broad Band Button Electrodes," *Nuclear Instr. Meth. in Phys. Res. A* **402**, pp. 27–35 (1998).
[4] Ghigo, A., F. Sannibale, M. Serio, C. Vaccarezza, "DAΦNE Beam Position Monitors," Presented at the Beam Instr. Workshop, Argonne, 1996.
[5] Unser, K. B., "New Generation Electronics Applied to Beam Position Monitors," Presented at the Beam Instr. Workshop, Argonne, 1996.
[6] Bergoz Precision Beam Instrumentation, "BPM Board User's Manual."
[7] Hewlett-Packard Company, HP E1326B and HP1352A Digital Multimeter and FET Multiplexer User Manuals.

A Two-Bunch Beam Position Monitor Performance Evaluation

Robert Traller, Evgeny Medvedko, Steve Smith, Roberto Aiello

Stanford Linear Accelerator Center
Stanford, CA 94309

Abstract. New beam position processing electronics for the Linear Accelerator allow faster feedback and processing of both positron and electron bunch positions in a single machine pulse. More than 30 electron-positron beam position monitors (epBPMs) have been installed at SLAC in various applications and have met all design requirements. The SLC production electron bunch follows the positron bunch down the linac separated by 58.8 nS. The epBPM measures the position of both bunches with an accuracy of better than 5 μm at nominal operating intensities. For SLC, the epBPMs have measured the position of bunches consisting of from 1 to 8×10^{10} particles per bunch.

For PEP-II (*B* Factory) injection, epBPMs have been used with larger electrodes and several BPMs have been combined on a single cable set. The signals are separated for measurement in the epBPM by timing. In PEP-II injection we have measured the position of bunches of as little as 2×10^9 particles per bunch. To meet the demands of SLC and PEP-II injection, the epBPM has been designed with three triggering modes:

1. As a self-triggering detector, it can trigger off the beam and hold the peak signal until read out by the control program.
2. The gated mode uses external timing signals to gate the beam trigger
3. The external trigger mode uses the external timing signals offset with internal vernier delays to precisely catch peak signals in noisy environments.

Finally, the epBPM also has built-in timing verniers capable of nulling errors in cable set fabrication and differences in channel-to-channel signal delay. Software has made all this functionality available through the SLC control system.

INTRODUCTION

The epBPM is a single-width standard CAMAC module functionally very similar to older processors used in the SLAC Linear Collider (SLC) but with many improvements and modern components. Beam signals are taken from four orthogonal striplines in a circular vacuum chamber, connected via double-shielded 50 Ω coax. The pulse doublet seen at the inputs goes through a 40 Mhz low-pass filter before being sent to both the

CP451, *Beam Instrumentation Workshop*
edited by R. O. Hettel, S. R. Smith, and J. D. Masek

internal trigger generator and the "sample and holds" (S/H). After filtering and amplification, each stripline signal goes to two S/Hs. Thus the four stripline inputs culminate in two sets of four S/Hs, with each set independently triggerable. In effect, the epBPM is like two processors connected to a single beam line device. The epBPM uses eight 14-bit ADCs (13 bits plus sign).

TABLE 1. Differences between epBPMs and Old Linac BPMs

Function	epBPM	Old linac processors
Measurements per pulse	2	1
Programmable Delay Triggers	2	1
Calibration	internal calibrator	beam
Bandwidth	40 Mhz	20 MHz
Ultimate resolution	5 μm	~30 μm
Trigger modes	3	2
Packaging	single width CAMAC	double width CAMAC

To form the trigger, each stripline signal is amplified and summed with the other three before the trigger is derived from the zero crossing of the summed wave form. The heart of this internal trigger generator is an Analog Devices AD891 Data Qualifier chip originally designed for recovering data from hard disk drives. The Data Qualifier produces an ECL-level pulse adjusted in our application to about 30 ns duration. In the self-triggered or gated modes this becomes the hold signal for the S/Hs. After generation of one pulse, the Data Qualifier is ready for another zero crossing event.

The ungated, self-triggered mode is not currently used by any epBPM installation. Gating the internal trigger with the distributed SLC timing system has enabled us to accurately measure any two of three bunches in the linac separated by approximately 60 ns. The need for a third trigger mode was anticipated for areas where signal might be weak and the pulses and/or reflections on cables might come in quick succession. Such is the case in PEP-II injection. The solution was to employ internal delay verniers which make it possible to use the distributed timing system to trigger the S/Hs at the precise peak of the signal.

Analog Devices AD 9500 digitally programmable delay generator has proven to be simple, effective and monotonic in providing a 10 ns vernier in 40 ps steps; six are used. One vernier is used for each of the two external timing signals brought in as a gate or trigger. In addition, in all trigger modes the hold signal for X+, X–, Y+ and Y– goes through additional AD9500s to compensate for any differences in internal signal path or mismatch of the cable set.

Unlike earlier SLC BPM processors, the epBPM has a built in calibrator. A CAMAC command causes the generation of an ECL pulse which is split. One of the pulses is then delayed 4 ns before both are applied logic gates to produce two narrow pulses of opposite polarity slightly staggered in time. An e^+/e^- select signal toggles a switch to determine which pulse is applied to the inverting and which to the non-inverting amplifier inputs. The result is a facsimile of an e^+ or e^- doublet which passes through a programmable attenuator before connecting to the regular signal input path via combiners. During calibration a pedestal measurement is made by internally generating a hold signal when no beam or calibration pulse is present, then the calibrator is triggered at three separate attenuator settings to accurately determine the difference in gain between channels.

The heart of the digital portion of the epBPM is the Xylinx field programmable gate array. All decoding of CAMAC commands, generation of calibrate and pedestal pulses, ADC control and handling of the serial bit stream from the eight ADCs takes place here.

B FACTORY INJECTION LINES

The completion of the first production epBPMs coincided with commissioning of the PEP-II (*B* Factory) injection. Before any beam reached these BPMs, the installation was used to test the epBPM displays and software and to insure proper calibration and response to all CAMAC functions. During this period we discovered a problem with low drop-out voltage regulators in the epBPMs which required minor modification. Subsequently it was discovered that the CAMAC power supplies purchased for PEP-II would oscillate when loaded with certain modules, including epBPMs. Vendor modifications were implemented (BiRa Systems).

The PEP-II injection lines require long cable runs to connect the BPMs to the closest CAMAC crate. Cable sets of from 200' to 500' of RG214 were used. The BPMs themselves are spaced 8 to 10 meters apart on the beam line. A series of 10 dB couplers were mounted on cable trays above the BPMs and linked with RG214. The BPMs were connected to the couplers with 10' sets of RG223. Up to six BPMs were connected in this way with an additional two BPMs added to the end of the line with a power combiner. This results in a train of eight beam pulses separated by approximately 75 ns.

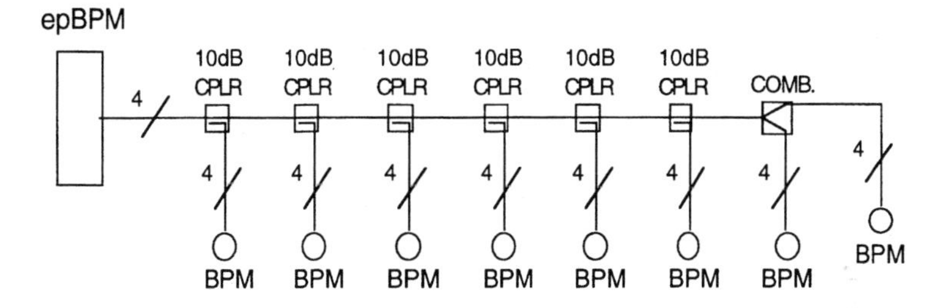

FIGURE 1. Typical section of *B* Factory injection BPM system.

Before the installation was complete, a test was devised using the epBPM as part of a TDR. We output a strong fast pulse (~35 volts peak, ~2ns) into a TEE. One side went to the BPM cable plant, the other side of the TEE went to a four-way power splitter and into the epBPM. Thus, the epBPM saw the pulser output and its reflections off the BPMs and couplers. Sweeping the timing in increments on 1 ns or less while reading the ADCs gave a picture very much like the view seen using a 400 Mhz oscilloscope.

Timing

The distributed timing signals for PEP are different from SLC. SLC distributes 119 MHz to its CAMAC via programmable delay units (PDU) and the control system can program triggers in delay steps of 8.4 ns. PEP's timing is based on 476 MHz and PEP's programmable delay units (PPDU) provide steps of 2.1 ns. The epBPM software and timing diagnostics must recognize which system it is using and increment the internal timing verniers to provide consistent timing steps with either delay unit. In fact, the entire handling of timing at SLAC had to be modified so that everything was referenced to nanoseconds instead of 8.4 ns "ticks" as was the previous practice.

Time-of-flight calculations were made to estimate initial delays to enter into database timing variables. We were able to estimate the timing of all the BPMs in a string quite accurately, relative to the first BPM in that string. The arrival time of beam at the first BPM was a bit harder to determine.

Although the external timing mode was designed largely with PEP injection in mind, we chose to leave the processors in "gated" mode for initial commissioning. We reasoned that even if some BPM readings were polluted by noise or reflections, timing in the gated mode would be less critical and it would be easier to find the first beam. We only had to get a beam signal within a 35 ns window rather than accurately find the correct peak out of a string of eight doublets. As it turned out, PEP injection worked so well that beam reached a profile monitor at the far end of the injection line before any other instrumentation was required to do its job. Nevertheless, several BPMs did read some signal and the timing diagnostics enabled us to insure that we were selecting the correct BPM signal.

For the most part, PEP injection continued working in the gated mode without problems. Only when we attempted to inject at intensities much below 2×10^9 particles per bunch did triggering become a bit erratic in gated mode. Eventually all PEP injection epBPMs were set to external trigger mode.

LINAC FEEDBACK SYSTEMS

Unlike *B* factory injection, commissioning of epBPMs for linac systems consisted of replacing units in working systems. Program managers are not anxious to shut down for wholesale replacement of critical feedback components. Even during scheduled downtimes, the fear is that good running configurations will be difficult to recover if critical systems are changed.

The linac BPM system consists typically of eight BPMs in each 100-meter section of linac. These BPMs are cabled to processing electronics in three CAMAC crates, thus two or three BPMs are processed in each crate. When the system was originally designed, an attempt was made to limit the need for programmable triggers. The original BPM processors were designed with a "hard-wired" connection to the same trigger in each crate. The cable sets were fabricated in lengths that insured all BPM signals arrived at a crate simultaneously. This has worked well enough but constrains modifications. Signal paths must remain the same for every processor in a crate unless independent triggering is available. Without independent triggering for the electronics, any timing adjustments must be made the same for all the units in a crate. If one BPM has noise or cross-talk from the other beam pulse, the trigger (gate) cannot be moved independently to find a time frame free of the unwanted signal.

Collimator Feedback System

Near the end of the linac is an area of collimators and a feedback system involving 12 BPMs with 24 processors to measure the position of both e^+ and e^- on each machine pulse. The 24 processors would be replaced by 12 epBPMs. We began the upgrade in stages, installing two or three epBPMs at a time and testing them before the next phase. In the collimator feedback area of the linac there are enough triggers for each epBPM to get two triggers: one for each set of S/Hs. However, these same two triggers apply to all epBPMs in a crate. For only about 1/3 of the units were there enough triggers to independently trigger each epBPM. Furthermore, where there are not enough triggers available to independently trigger the epBPMs, we cannot combine them with old linac processors in the same crate, as the internal delays are quite different for the two.

In the lab we used the built-in calibrator to find values for the internal timing verniers which would compensate for differences in signal delay from channel to channel. These are stored in a PROM and assume a perfectly matched cable set. On-line diagnostics were developed to scan a range of values for the verniers and select the optimum setting for the installation. We could not just scan the vernier delays for X^+, X^-, Y^+, Y^- and read the ADCs for a peak value as this could be polluted by position offsets. Since the BPM is symmetric, we assumed $X^+ + X^- = Y^+ + Y^-$. Therefore, since $X^+ = Y^+ + Y^- - X^-$, we calculated the ratios:

$$\frac{X^+}{Y^+ + Y^- - X^-}$$

$$\frac{X^-}{Y^+ + Y^- - X^+}$$

$$\frac{Y^+}{X^+ + X^- - Y^-}$$

$$\frac{Y^-}{X^+ + X^- - Y^+}$$

The vernier delays at which these ratios peak are the optima. The software allows one to write these values directly into the module and store them in the database. The database also stores the difference between the module's pre-programmed values and those calculated using the on-line utility. These then are the site-specific cable offsets will be added to values stored in module memory should the module be changed for maintenance reasons.

The collimator feedback area is one of the most important for good position resolution. After commissioning the utilities for adjusting vernier delays in the epBPMs we were able to use orbit-fitting algorithms to calculate resolution in this area of about 4 μm with beam intensities of 3×10^{10} to 4×10^{10} particles per bunch.

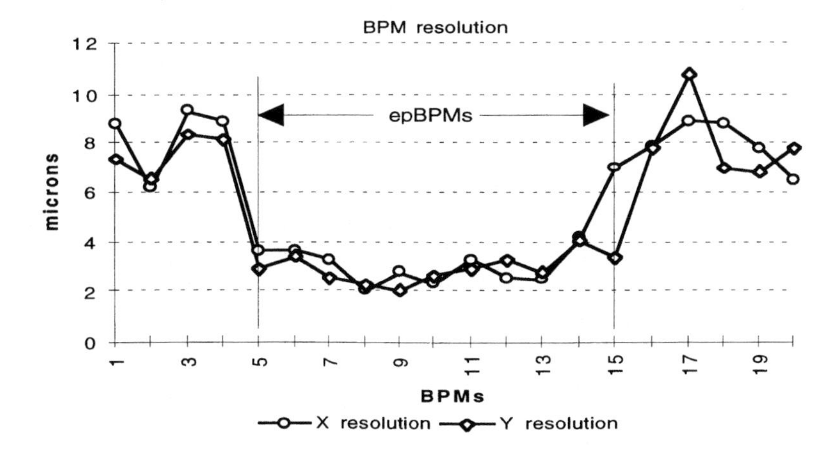

FIGURE 2. Resolution measured for 20 BPMs in the collimator section of the linac. Beam intensity is about 4×10^{10} electrons per bunch.

Ring-to-Linac Launch Feedback System

By the time the collimator feedback system was fully commissioned, the '97 SLC run was well under way and beam currents were rising as systems got tuned up. The next feedback system to be upgraded was the area where all three bunches enter the linac from the damping rings. In the collimator region, only positrons and electrons destined for collisions are present. Here, production electrons, positrons, and the scavenger bunch bound for the positron target all converge from North and South damping rings.

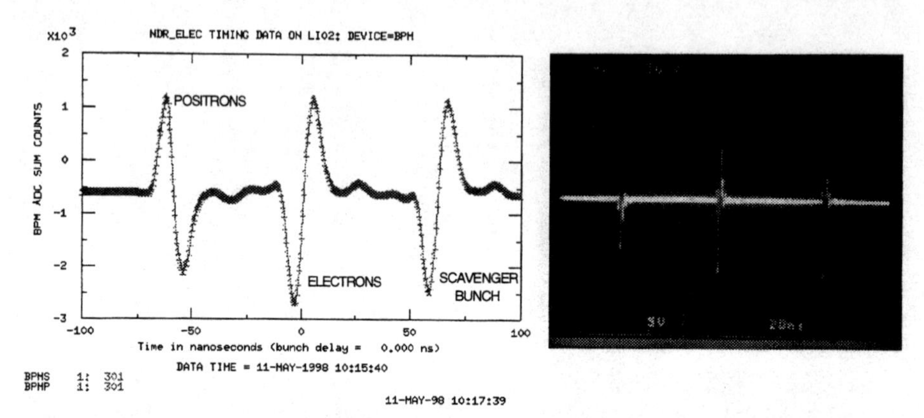

FIGURE 3. Left: plot made by reading ADCs while incrementing trigger in 500 ps steps. Right: 400 Mhz scope picture of input to epBPM

The entire linac is a noisy environment with klystron modulators firing at 120 Hz. Throughout the history of BPMs at SLAC, we have battled noise with ferrite toroids, timing, filtering...whatever.

Triggering and Noise

At this stage in the commissioning of the epBPMs, it became apparent that we had some vulnerability to noise. The epBPM was designed with a 46 dB dynamic range. It was specified to be able to measure beam currents from 5×10^8 to 1×10^{11}. This was achieved without the use of programmable attenuators, possibly a mistake. The ability of the internal trigger to fire down to 5×10^8 also makes it vulnerable to noise and reflections caused by the larger signals coming in quick succession. We began to see noise interfere with calibrations as well as the ringing of one beam signal interfering with our ability to read the following pulse in the gated mode. We can read the pulses accurately in the external trigger mode, but independent triggering is not always available. Certainly, we must be able to get good calibration data. We had to raise the trigger threshold about 6 to 8 dB to get good calibrations and reliable readings in the gated mode of narrowly separated beam pulses.

FINAL FOCUS

On either side of the SLC interaction point (IP) there is a feedback system using BPMs to measure the energy loss from collisions at the IP. It is beyond the scope of this paper to describe the functioning of this system except to say it was originally seen as an area for using epBPMs. The system currently employs electronics designed for the Final

Focus Test Beam (FFTB) which works well but requires complicated wiring and occupies much CAMAC crate space. Considerable effort was made to use epBPMs but we were not able to achieve the necessary resolution. BPM diameters in this region are about three times that of the linac and the resolution demands are at least as stringent. Two epBPMs remain in the Final Focus replacing conventional electronics in less critical applications.

INJECTOR

Not part of our original upgrade plan, a feedback system using a single epBPM has been implemented at the very first BPM in the machine. This unit is less than one meter from the polarized electron gun. Here, a laser fires two quick pulses at the cathode producing the two bunches of electrons that will become the bunch destined for collisions at the SLC IP and the bunch used to create positrons. Machine physicists use an epBPM in externally triggered mode to monitor the two pulses and and keep their relative amplitudes the constant.

Once, after a brief maintenance downtime, we were asked to look at a problem with this unit. In this area there are a lot of beam losses. The two bunches might each have 6×10^{10} electrons right out of the gun and lose 20% within a few meters. The epBPM was showing a significant degradation from the level before the maintenance access. Using the epBPM timing verniers, I noticed the separation between the two bunches had changed by some 800 ps. The problem turned out to be timing in the laser system which had been recently modified. This demonstrates the use of the BPM as a timing diagnostic.

CONCLUSIONS

The epBPM has met all design specifications and has proven to be an accurate and versatile processor of BPM stripline signals. It can easily measure the position of pulses separated by less than 60 ns. Resolution better than 5 μm at 5×10^{10} particles/bunch using 25 mm diameter BPMs is easily achievable. Although resolution is dependent on the amount of signal, bunches 2×10^{9} to 8×10^{10} particles have been measured in the internally triggered mode.

In the lab better than 48 dB dynamic range was measured for the internal trigger. To use the internal trigger in noisy environments this dynamic range may have to be limited by raising the internal triggering threshold. In the external trigger mode, the trigger threshold is of no concern. The problem is merely the availability of stable distributed timing signals.

The internal timing verniers and the SLC/PEP timing system have proven to be more than stable enough for externally timing S/Hs. The long term stability of the timing has thus far remained within a few hundred picoseconds of original settings

ACKNOWLEDGMENTS

Many people are necessary for the commissioning of a complete system.

From Controls Department Software Engineering, we are indebted to Tony Gromme the development of the core BPM software, Linda Hendrickson for her expertise in feedback systems, Mike Zelazny for implementing all our requests for diagnostics and utilities, and Nancy Spencer and Ken Underwood for database help.

From Controls Department we are indebted to Raymond Nitchke and Vern Smith for technical assistance with the physical plant and design of the PEP-II injection BPM cable system.

To Pantaleo Raimondi thanks are given for much help in evaluating BPM resolution.

Development Of Nanometer Resolution C-Band Radio Frequency Beam Position Monitors In The Final Focus Test Beam

G. Mazaheri, T. Slaton, T. Shintake*

Stanford Linear Accelerator Center, Stanford, California

**National Laboratory for High-Energy Physics, 1-1 Oho, Tsukuba-Shi*

Abstract. Using a 47 GeV electron beam, the Final Focus Test Beam (FFTB) produces vertical spot sizes around 70 nm. These small beam sizes introduce an excellent opportunity to develop and test high resolution Radio Frequency Beam Position Monitors (RF-BPMs). These BPMs are designed to measure pulse to pulse beam motion (jitter) at a theoretical resolution of approximately 1 nm. The beam induces a TM110 mode with an amplitude linearly proportional to its charge and displacement from the BPM's (cylindrical cavity) axis. The C-band (5712 MHz) TM110 signal is processed and converted into beam position for use by the Stanford Linear Collider (SLC) control system. Presented are the experimental procedures, acquisition, and analysis of data demonstrating resolution of jitter near 25 nm. With the design of future e+e- linear colliders requiring spot sizes close to 3 nm, understanding and developing RF-BPMs will be essential in resolving and controlling jitter.

Paper not available for publication.

CP451, *Beam Instrumentation Workshop*
edited by R. O. Hettel, S. R. Smith, and J. D. Masek

Performance of the Beam Position Monitor System for the SLAC PEP-II *B* Factory[1]

Ronald G. Johnson, Stephen R. Smith, and G. Roberto Aiello[2]

Stanford Linear Accelerator Center
P.O. Box 4349, Stanford, CA 94309

Abstract. the beam position monitor (BPM) system for the SLAC PEP-II *B* Factory was designed to measure the positions of single-bunch single-turn to multibunch multi-turn beams in both rings of the facility. Each BPM is based on four button-style pickups. At most locations the buttons are connected to provide single-axis information (x only or y only). Operating at a harmonic (952 MHz) of the bunch spacing, the BPM system combines broadband and narrowband capabilities and provides data at a high rate. The active electronics system is multiplexed for signals from the high-energy ring (HER) and low-energy ring (LER). The system will be briefly described; however, the main purpose of the present paper is to present operational results. The BPM system operated successfully during commissioning of the HER (primarily) and the LER over the past year. Results to be presented include on-line calibration, single-bunch single-turn resolution (<100 μm), and multibunch multi-turn resolution (<3 μm), multiplexing, and absolute calibration. Thus far, the system has met or exceeded all the requirements that have been tested. The remaining requirements will be tested when both rings are completed and commissioned this summer. In addition, typical results of beam physics studies relying on the BPM system will be presented.

INTRODUCTION

The PEP-II *B* Factory (1) at SLAC is nearing the end of the construction phase. At present the high-energy ring (HER) is complete and is being commissioned. The low-energy ring (LER) will be complete this summer and the commissioning phase will continue. The beam position monitor (BPM) system (2) for both rings is also nearly complete. All of the cables and active electronics are in place. As the last of the ring components are installed, it only remains to connect the BPMs. Since the active electronics are multiplexed between the HER and LER, much of the system for both rings has been commissioned along with the HER.

---

[1] Work supported by Department of Energy contract DE-AC03-76SF00515
[2] Present Address: Interval Research Corp., 1801 Page Mill Road, Bldg. C, Palo Alto, CA 94304

CP451, *Beam Instrumentation Workshop*
edited by R. O. Hettel, S. R. Smith, and J. D. Masek

The PEP-II storage rings are 2200 m in circumference and are located in a common tunnel. Both rings use rf of 476 MHz and fill 1658 buckets (every other bucket). The HER stores up to 1.0 A of electrons at 9.0 GeV and the LER stores up to 2.1 A of positrons at 3.1 GeV.

The BPM system is required to measure the positions of multibunch beam on a turn-by-turn basis (136 kHz) and single-bunch beam injected in a 200 ns ion-clearing gap. The electronics work in both a narrowband mode, for multibunch, and in a wideband mode, for single bunch. A simple difference over sum algorithm is used for position calculation. A 952 MHz bandpass filter selects the processing frequency in multibunch and generates an rf burst in single bunch. The system bandwidth is set at 10 MHz. Requirements for high accuracy and wide dynamic range led to development of an on-line calibration system which can operate in the presence of beam.

In the next two sections, the BPM system will be described, with particular attention to the active electronics. However, the main purpose of this paper is to present some of the operational results obtained in the commissioning. Finally, the status of requirements not fully tested will be presented.

BPM SYSTEM DESCRIPTION

Button-style pickups (3) were chosen as the basis of the BPM system. There are three basic types of vacuum chamber used in the rings, which necessitated somewhat different mechanical designs for the buttons. However, all buttons are 1.5 cm in diameter and have nearly identical electrical characteristics, 50 Ω impedance and 2.6 pF capacitance. The shapes of the vacuum chambers in the arcs of both HER and LER and the necessity to avoid the synchrotron radiation fan led to placing the buttons on the diagonal rather than the vertical and horizontal axes.

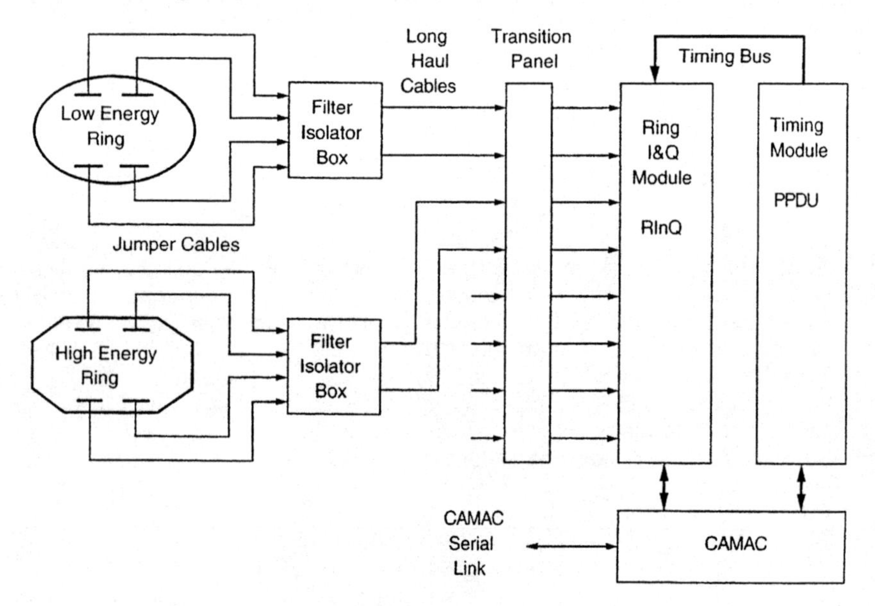

FIGURE 1. BPM system block diagram (4 of 8 channels fully shown).

In Figure 1, a block diagram of the BPM system is shown. Signals from the four electrodes are filtered (center frequency, 952 MHz and bandwidth 150 MHz) and in most cases combined by the Filter-Isolator Box (FIB) located close to the buttons. The signals from the buttons are combined to create *x*-only or *y*-only BPMs. In a few cases (25) the signals are not combined in the FIB, preserving both *x* and *y* information from the same set of buttons. The FIB absorbs out-of-band power and has provision to allow biasing the buttons (to 350 V). The signals from the FIB are processed in the Ring I&Q (RInQ) module and delivered to the control system through the CAMAC interface.

Timing for each BPM is supplied by the PEP-II Programmable Delay Unit (PPDU) which provides 16 independent channels of signals delayed relative to one of the two fiducials received from the PEP-II timing distribution system. The PPDU has a resolution of 2.1 ns. The RInQ module contains timing verniers with 40 ps resolution.

Measurement accuracy for single bunches is limited by reflection of the multibunch signal through the cable. The RInQ was designed to have a VSWR better than 1.2. An isolator was included in the FIB to reduce reflections.

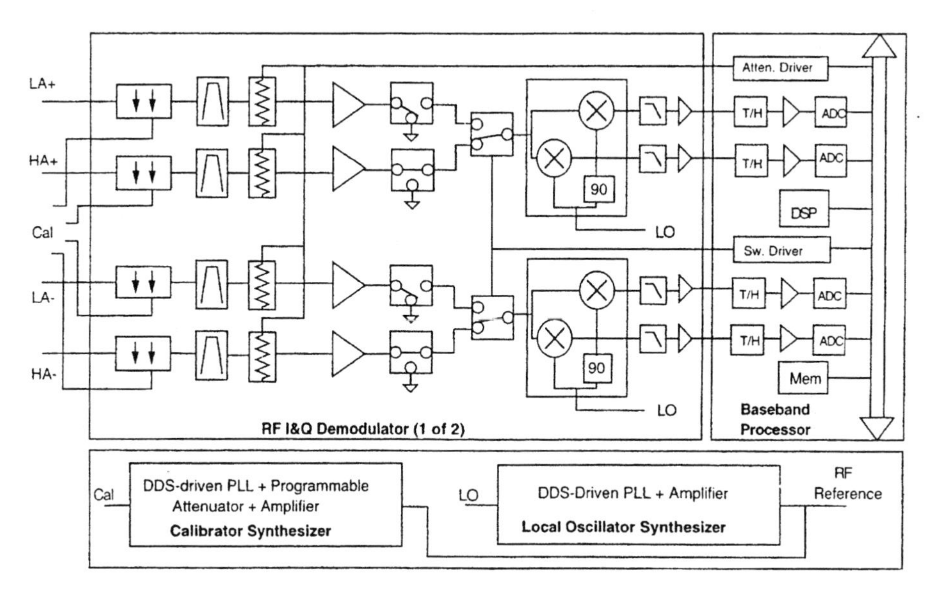

FIGURE 2. Ring I&Q (RInQ) processor block diagram.

ACTIVE ELECTRONICS

Processing of BPM signals by the RInQ module is based on baseband conversion using in-phase and quadrature (I&Q) demodulators. The RInQ module accepts two or four signals for HER and similarly for LER, where the processing is multiplexed between the two rings. A calibrator is also included on each RInQ. A diagram of the RInQ module is shown in Figure 2.

The 10 dB directional coupler receives signals from the FIB and the calibrator (through the coupling port). A 952 MHz bandpass filter selects the processing frequency. The programmable attenuator extends the dynamic range and an amplifier optimizes the signal to the I&Q demodulator. Switches multiplex HER and LER and provide the required 86 dB isolation. A low-pass filter sets the system bandwidth to 10 MHz. Track-and-hold and 14-bit analog-to-digital converters acquire the signals. A digital signal processor calculates the position and signal strength.

The RInQ module also contains a direct-digital synthesis (DDS) phase-locked loop (PLL) synthesizer, adjustable in amplitude and frequency, to provide calibration signals to each channel. Using the calibration signals, each channel can be calibrated for offset and gain mismatch. By operating the calibrator at a few kHz off the local oscillator frequency and using a fixed-frequency curve-fitting algorithm (4), the amplitude and quadrature phase unbalance can be added to the calibration parameters.

CALIBRATION RESULTS

The RInQ modules can be calibrated on-line in either narrowband or wideband mode. The wideband mode is calibrated by taking a series of measurements over the working bandwidth and taking a weighted average of the calibration parameters. Calibration parameters consist of offsets (4), gains (4), and phase errors (2) for both I and Q of the positive and negative channels, i.e., ten parameters. These calibration parameters have been shown to be stable over several hours; but they do depend on channel attenuation. Recalibration is performed whenever the beam current changes by 25% which would require a change in channel attenuation to achieve the optimum measurement.

Typically when a calibration of all BPMs is requested through the control system over 95% of the BPMs report a successful calibration for either HER or LER. Multiplexing works for calibration and for separate operation of HER or LER; but, both rings have not been run at the same time as yet.

Absolute calibration of the BPM system has been made. This calibration is described in Reference (5) and the on-line data base contains the offsets for calculating absolute positions. There have been no beam-based tests of the calibration to this time.

MEASUREMENT RESULTS

The BPM system was operational for each stage of PEP-II commissioning. Most of this commissioning has been with HER. After being timed, the BPMs produced position information which aided the tuning of the ring. There were of course some problems such as incorrect conversion constants for calculating position and a few incorrectly connected modules.

Each part of the system was tested and repaired if necessary before installation. Many of the FIBs and RInQ modules did require repair. However, after the system was installed problems have been minimal. About 15% of the RInQs installed have developed problems. All of these modules have been repaired.

The calculated minimum resolution of the BPM system is 0.2 μm which is dominated by the ADC quantization error. For measurements from the operating BPM system, the random error for multibunch multi-turn data closely approaches this limit. However, systematic errors have limited the resolution obtained thus far.

Noise which appears at the frequency difference between the measurement frequency (952 MHz) and the local oscillator, or twice that frequency, actually dominates the system resolution. Errors at these frequencies indicate that the calibration parameter set is not correct for the measurement. The source of this problem appears to be either the CAMAC power supply or the PPDU; but, the problem has not been isolated as yet.

In any case, single-turn measured resolutions for individual RInQ modules do not exceed 100 μm and the average resolution is about 50 μm. For multibunch multi-turn measurements the resolutions are 3 μm or better.

In addition to an aid in tuning, the BPM system was used as a fundamental measurement tool in a number of ring studies. In the paragraphs below two of those studies will be described.

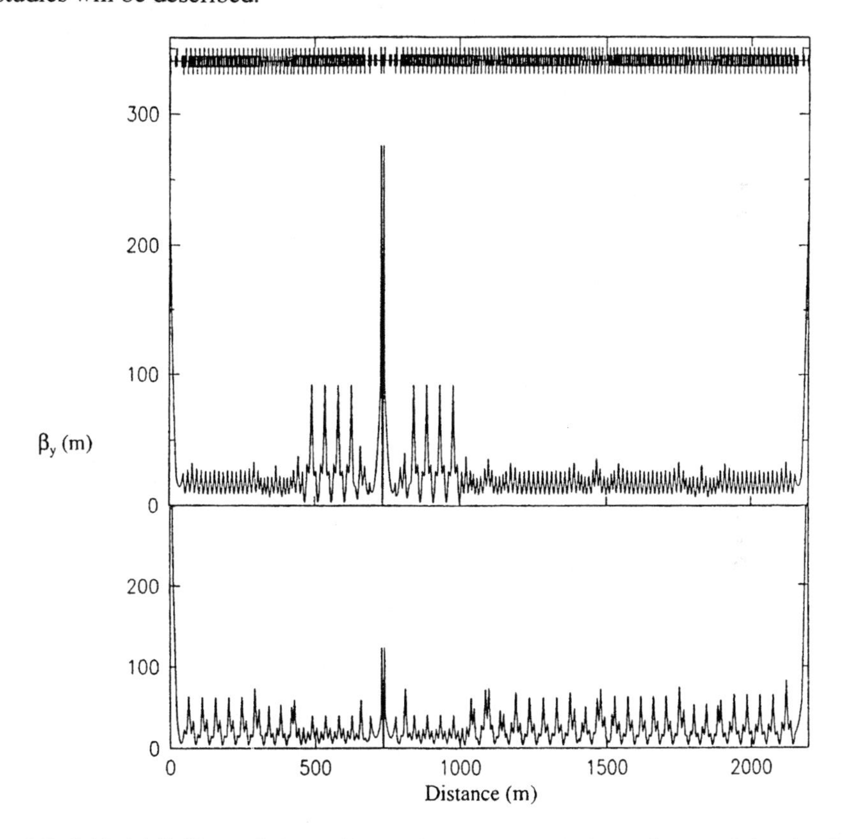

FIGURE 3. Model β_y: Upper, design and lower, fit. The representation at the top of the plots is the ring lattice.

Ring Study 1

The first study uses a very powerful method for determining the linear optics in a storage ring by comparing a model response matrix to a measured response matrix

(6). A computer code called LOCO (Linear Optics from Closed Orbit) varies the parameters in the model response matrix to minimize the χ^2 deviation between the model and the measured orbit response matrices. The measured response matrix is obtained by measuring the change in orbit at the BPMs with changes in steering magnet excitation. The method is quite powerful and can be used to determine quadrupole magnet gradients; the calibration of steering magnets and BPMs; the roll of quadrupoles, steering magnets, and BPMs; etc. The example cited here is the determination of quadrupole gradient errors. In Figure 3, β_y from the design model and the fit model are shown. In Figure 4, the quadrupole gradient error derived from the β-distortion evident in Figure 3 is shown.

The strong errors shown in Figure 4 are due to the four quadrupole magnets near the interaction point. The gradient error is only about 0.5%. Apart from the quadrupoles near the interaction point, the remaining gradient errors represent the resolution of the method. Subsequent adjustment of the interaction quadrupole magnets brought β_y around the ring close to the design value.

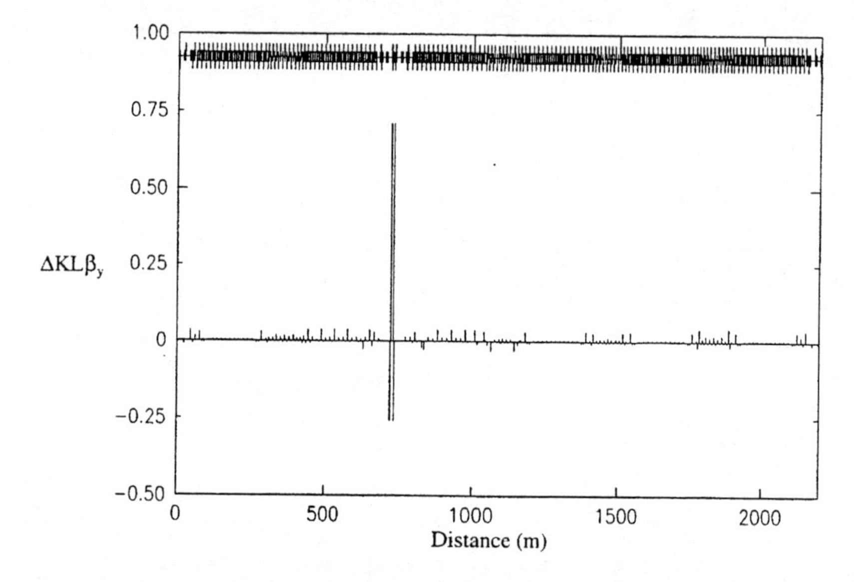

FIGURE 4. β- distortion due to quadrupole gradient errors.

Ring Study 2

In the second study, the BPM system was used to identify coherent bunch to bunch instabilities through the bunch train. The PEP-II rings have 3492 rf buckets of which every other one is filled. The ion-clearing gap comprises 5% of the ring circumference. This leaves 1658 filled buckets in a train, preceded by a gap. With feedback off a coherent bunch to bunch oscillation is seen to develop through the bunch train. This effect can be seen in the BPMs by measuring a particular bunch turn-by-turn at a point with some dispersion, and by observing the size of the transverse oscillation at the synchrotron frequency grow as a function of bunch number through the bunch train. The data in Figure 5 was taken in rather unstable conditions.

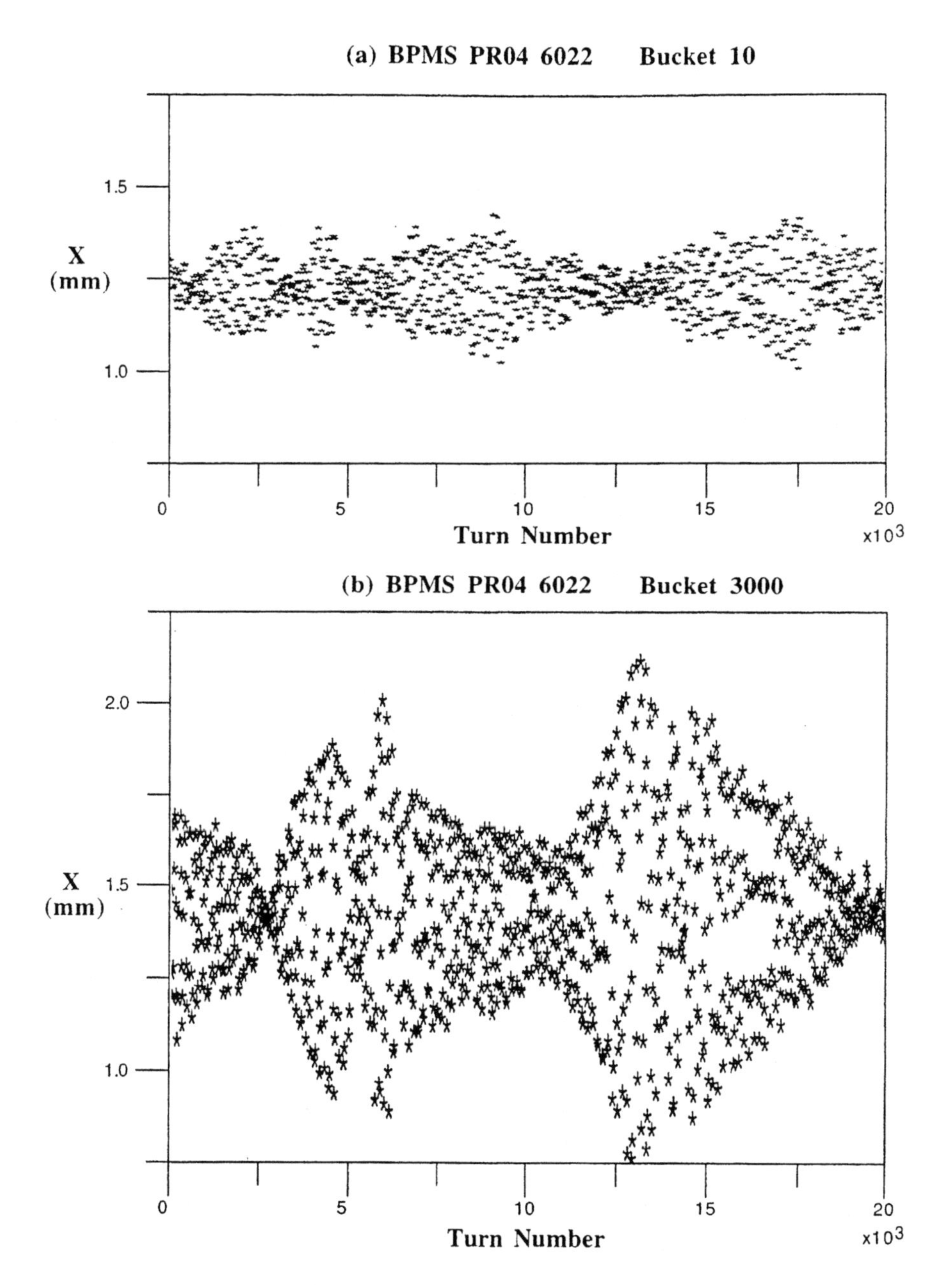

FIGURE 5. Coherent bunch to bunch instability: (a) x-position vs turn number for bucket 10; (b) x-position vs turn number for bucket 3000.

In Figure 5(a), 1000 measurements of the horizontal beam position of the 10th filled bucket, sampled every 20 ring turns, at a particular BPM are shown. In Figure 5(b), the oscillations of the 3000th bunch, at the same BPM, sampled at the same rate is shown. The amplitude in the second case is a factor of five larger. The amplitude is small following the gap and then builds up rapidly. The Fourier transform of these position sequences shows all of the signal is at the synchrotron frequency.

Even with the feedback on some coherent bunch-to-bunch instabilities were seen. However, it is expected that these instabilities can be controlled as the feedback systems are fine-tuned.

CONCLUSIONS

The PEP-II BPM system has performed very well thus far. It was operational from the start of commissioning of both HER and LER. The system has been an aid to the commissioning team in meeting all the goals set for PEP-II through this stage of the commissioning process. The BPMs have provided results which exceed all the requirements set for them. However, this is not to say that they have reached the limit of their performance. In addition, there are parts of the system which have not been tested completely or at all.

Isolating and fixing the problem which affects the position resolution is the most pressing issue. However there are a few individual BPMs that need to be repaired. About 15% of the RInQ modules have suffered infant mortality (of various components). All of these modules have been repaired. Problems with other parts of the BPM system (buttons, FIBs, and cables) have been negligible.

Further testing of the BPM system include: 1) full tests of multiplexing, 2) isolation between HER and LER channels, 3) measurement of a single low-intensity bunch in the ion clearing gap, and 4) absolute calibrations. Most of these tests require running both the HER and LER together.

The continued good performance of the PEP-II BPM system will be expected to contribute to the successful commissioning and operation of the PEP-II *B* Factory.

ACKNOWLEDGMENTS

The authors wish to thank Don Martin and Mark Mills for their vital developmental work on the PEP-II BPM system, Linda Hendrickson, Alan Cheilek, Mike Zelazny, and Tony Gromme for their help with the software, Bob Noreiga, Brooks Collins, and Vern Smith for their work in installation, James Safanek for providing some of the data presented, and Tom Himmel, Ray Larsen, and Alan Fisher for useful discussions.

REFERENCES

[1] PEP-II, An Asymmetric B Factory, Conceptual Design Report, SLAC-418 (1993).
[2] Aiello, G. R., R. G. Johnson, D. J. Martin, M. R. Mills, J. J. Olsen, and S. R. Smith, "Beam Position Monitor System for PEP-II, Beam Instrumentation," Proc. of the Seventh Workshop, 1996, eds. A. H. Lumpkin and C. E. Eyberger,. AIP Conf. Proc., **390** (AIP, Woodbury, NY, 1997) pp. 341–349.

[3] Shafer, R. E., "Beam Position Monitoring," Accelerator Instrumentation, 1989, eds. E. R. Beadle and V. J. Castillo, AIP Conf. Proc., **212** (AIP, New York, NY, 1990) pp. 26–58.
[4] Johnson, R., S. Smith, N. Kurita, K. Kishiyama, and J. Hinkson, "Calibration of the Beam position monitor System for the SLAC PEP-II B Factory," presented at the 1997 Particle Accelerator Conference, Vancouver, B.C., Canada, 12–16 May, 1997.
[5] IEEE Std. 1057, Digitizing Waveform Recorders (1989).
[6] Safranek, J., Nucl. Instrum. Methods A, **388,** 2736 (1997).

1 nA Beam Position Monitor

M. Piller, R. Flood, L. Hammer, M. Parks, E. Strong, L. Turlington, R. Ursic

Thomas Jefferson National Accelerator Facility

Abstract. A new BPM system, based on resonant cavities, has been developed for measuring the transverse position of very low-intensity electron beams delivered to Experimental Hall B at the Continuous Electron Beam Accelerator Facility (CEBAF) in Newport News, VA. The system requirements called for measuring down to 1 nA with a 100 m m resolution. The actual system is much better: it can measure down to 100 pA at the 100 m m required resolution. A 100 pA beam yields about 1 electron per bunch. Each 1 nA BPM utilizes three resonant RF cavities to determine the position of the beam: one cavity sensitive to X position, a second cavity sensitive to Y position, and a third cavity which measures intensity. The position cavities operate at room temperature in a dipole type mode at 1497 MHz and contain internal field perturbing rods in an arrangement similar to that of the CEBAF rf Separator cavities. The position cavities are electron beam welded assemblies made of copper plated stainless steel. The RF output signal from each cavity is processed using a down-converter and a DSP based commercial lock-in amplifier operating at 100 kHz. The lock-in amplifiers connect to the EPICS control system via an IEEE 488 bus. System features under development include intensity and position modulation measurement capabilities. This paper provides measured performance results and an updated overview of the installed and operational 1 nA BPM system.

Paper not available for publication.

CP451, *Beam Instrumentation Workshop*
edited by R. O. Hettel, S. R. Smith, and J. D. Masek

New Microwave Beam Position Monitors for the TESLA Test Facility — FEL

T. Kamps and R. Lorenz

DESY Zeuthen, Platanenallee 6, D-15738 Zeuthen

Abstract. Beam-based alignment is essential for the operation of the SASE-FEL at the TESLA Test Facility Linac. In order to ensure the overlap of the photon beam and the electron beam, the position of the electron beam has to be measured along the undulator beamline with a high resolution. Due to the severe space limitations, a new microwave concept is being considered. It is based on special ridged waveguides coupling by small slots to the magnetic field of the electron beam. The four waveguides and slots of each monitor were split into two symmetric pairs separated in beam direction. All waveguides are about 35 degrees apart in azimuth from the horizontal axis and will be fabricated using electro-discharge machining (EDM). Waveguide-to-coax adaptors were designed to couple the signal of each waveguide into a coaxial cable. The goal is to measure the averaged position of a bunch train in a narrowband receiver with a center frequency of 12 GHz. A prototype of this monitor was built and tested on a testbench, as well as at the CLIC Test Facility at CERN. The paper summarizes the concept, the design, and further improvements of the waveguide monitor.

INTRODUCTION

The construction of a free-electron laser (FEL) based on the self-amplified spontaneous emission mechanism (SASE) is under way at DESY [2]. The goal is to get a coherent, very bright beam of photons with wavelengths tunable between 6 and 20 nm. The high-intensity electron beam needed to drive the undulator will be delivered by the TESLA Test Facility Linac (TTFL, [1]). Some parameters for the FEL operation of the TTFL are listed in Table 1.

TABLE 1. Parameters for the TTFL-FEL.

Bunch length (FWHM) for Phase I / II	250 μm / 50 μm
Bunch charge and repetition rate	1 nC , 10 Hz
Bunch spacing	1 μs – 111 ns
Number of bunches (pulse length 800 μs)	1 – 7200

The position of the electron beam might vary inside the undulator, mainly because of field imperfections in the dipole and quadrupole magnets. Since a precise overlap of the electron and the photon beams is essential for the FEL operation, the

CP451, *Beam Instrumentation Workshop*
edited by R. O. Hettel, S. R. Smith, and J. D. Masek

position of the electron beam has to be measured and corrected along the undulator beamline. About 10 beam position monitors (BPMs) with a resolution of less than 5 μm averaged over the bunch train are required for the beam-based alignment of each undulator module. The same resolution has to be reached for single bunches during commissioning.

In Phase I the undulator consists of three modules, each containing a 4.5 m long vacuum chamber having a rectangular profile. These chambers, made of aluminum, will be equipped with an alternating arrangement of correction coils and monitors. All mechanical parts of the pickups have to fit inside the undulator gap of 12 mm; the magnets allow only horizontal access. The realization of two different monitor concepts is under way: an electrostatic pick-up and a microwave monitor (see also [3]). The scope of this paper is to discuss the idea and the concept of the latter. Since the signals are coupled into waveguides, we will call this structure a waveguide monitor. Emphasis will be on analytical and numerical aspects of the design, as well as on measurements and tests.

CONCEPT OF THE WAVEGUIDE MONITOR

Basic Idea

If a beam of charged particles is centered in a circular, conducting beam pipe, then there is a uniformly distributed electromagnetic field accompanying the beam. The closer the velocity of the beam is to the velocity of light, the more this field looks like a transverse electromagnetic (TEM) wave. An off-center beam produces a 'distortion', and the TEM-fields are no longer uniformly distributed. The wall current density of a beam at a position (r, ϕ, t) is given by [4]

$$i_w(r,\phi,t) = \frac{-I_b(t)}{2\pi R_0}\left[1 + 2\cdot\sum_{n=1}^{\infty}\left(\frac{r}{R_0}\right)^n \cos\left(n(\phi-\theta)\right).\right] \tag{1}$$

$I_b(t)$ is the beam current, R_0 the beam pipe radius, (r,θ) the beam position and ϕ the angular width on the inner surface of the beam pipe.

BPM systems mainly consist of four subsystems: the transducer close to the beam, transmission lines, the electronics and the software. Most of the transducers used in BPM systems detect the electric and/or the magnetic field around the beam pipe. Short (buttons) or long electrodes (striplines) are often used for this purpose. Their signals are detected and subtracted in the electronics or in a computer to measure the 'field distortion'. Because of the limited space inside the undulator gap and the extremely short bunches expected for Phase II, waveguides are used in the transducer for the microwave concept realized here (Fig. 1a). Since the wall current density is proportional to the magnetic field on the inner surface of the beam pipe, small slots can be used to couple this field into the waveguide. Therefore, Equation (1) can be used to describe the behavior of this new structure.

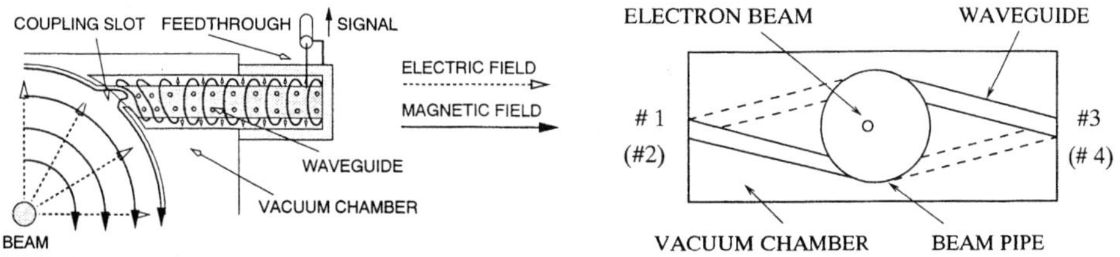

FIGURE 1. a) Principle sketch of the coupling mechanism; b) cross-section of the monitor. The waveguides #2 and #4 (dashed lines) are behind #1 and #3.

Fundamental Design Aspects

The design frequency should be as high as possible so that a waveguide fits into the vacuum chamber, and to get a reasonable coupling. Its upper limit is given by the cutoff frequency of the beam pipe (about 17 GHz). A compromise was $f_0 = 12$ GHz, the same frequency as used for the cavity monitors to be installed in the diagnostic sections [3]. This offers the possibility of developing both electronics in parallel. Another advantage is that the frequency band is used commercially (e.g., TV-sat, DBS).

The height of the vacuum chamber limits the flange size at each side and leads to a design in which the waveguide is in the middle of the flange. The result is that the four waveguides must be split into two symmetric pairs, separated in beam direction. Since the coupling slots are positioned at ± 34 degrees with respect to the horizontal plane, there is a slight angle for each waveguide. After the frequency f_0 was chosen, the size and the shape of the waveguide were studied (MAFIA 2D/3D, [5]). The goal was to have f_0 well above the fundamental waveguide cutoff frequency, but below the next higher cutoff. The special waveguide shape came out during the numerical 3D design: the ridge lowers the cutoff frequency and its shape enhances the magnetic field close to the slot, which results in a larger coupling.

The transmission to a waveguide port was calculated by using the S parameter macros of MAFIA. A 'beam' was simulated by a thin conductor in the beam pipe, thus forming a TEM line excited on one side. This method was used to estimate the coupling through a slot (depending on its size, its position/orientation and the wall thickness) and to study tolerances. In addition, the position of the inner conductor ('beam') was changed to estimate the sensitivity. Finally, the coupling from the waveguides into a 50-Ω system was designed and optimized.

Expected Signals

Let us assume a charge q in the center of the beam pipe. The signal V_s at the output port of each waveguide can be estimated by

$$V_S = k \cdot Z_0 \cdot q \cdot B = k \cdot 50\ \Omega \cdot q \cdot B. \qquad (2)$$

Z_0 is the impedance and B the bandwidth of the external circuit. The coupling factor k contains the coupling to the beam and from the waveguide into a 50-Ω system. For small displacements δx and δy from the center, the beam position in terms of voltages can be calculated for the structure in Figure 1 from

$$\delta x = \frac{1}{S_x} \cdot \frac{(V_{S_1} + V_{S_2}) - (V_{S_3} + V_{S_4})}{(V_{S_1} + V_{S_2} + V_{S_3} + V_{S_4})}, \qquad \delta y = \frac{1}{S_y} \cdot \frac{(V_{S_2} + V_{S_3}) - (V_{S_1} + V_{S_4})}{(V_{S_1} + V_{S_2} + V_{S_3} + V_{S_4})}. \quad (3)$$

S_y and S_y are signal functions depending on the monitor geometry. The reason for this complex form is that the angle between the slots is not 90^0. Neglecting this, the resolution δx can be estimated by the relations for electrode monitors [4]:

$$\delta x = \frac{V_N}{V_S} \cdot \frac{R_0}{2} \cdot \frac{1}{\sqrt{2}} = \frac{F \cdot \sqrt{k_b \cdot B_e \cdot Z_0 \cdot T}}{V_S} \cdot \frac{R_0}{2} \cdot \frac{1}{\sqrt{2}}. \quad (4)$$

F is the electronics noise figure, k_b the Boltzmann constant, and T the temperature. The linearity error estimated from Equation 1 is less than 1% for a beam position within ± 0.5 mm from the electrical center.

DESIGN AND TESTS OF PROTOTYPE I

Design

A first prototype [3] was built in 1997, its design is shown in Figure 2. The waveguide holes and the ridges of the waveguides were fabricated by electro-discharge machining (EDM). According to MAFIA-calculations, the coupling factor of a single slot is about 0.5% at 12 GHz. The right part of this figure shows the ridge, which has to be inserted into the hollow waveguide, together with a coaxial adapter containing a standard commercial vacuum feedthrough from KAMAN Corp. The whole piece is flange-mounted, and the SMA-connector is parallel to the beam.

The coupling factor k, measured using a Vector Network Analyzer, is slightly less than that expected from MAFIA simulations. This is probably due to problems in the fabrication of the coax-adapters: all feedthroughs were welded into the flange with an (unexpected) angle and the couplings into 50 Ω are not matched.

Test Results

Measurements in the Laboratory

For tests at DESY Zeuthen, the prototype was mounted on a system of two stepping motors and a 125 μm tungsten wire was stretched through the structure. By moving the BPM block instead of the wire, high frequency wire oscillations are minimized. With this assembly it was possible to move the structure under test with a precision of less than 1 μm. The whole setup is shown in Figure 3a.

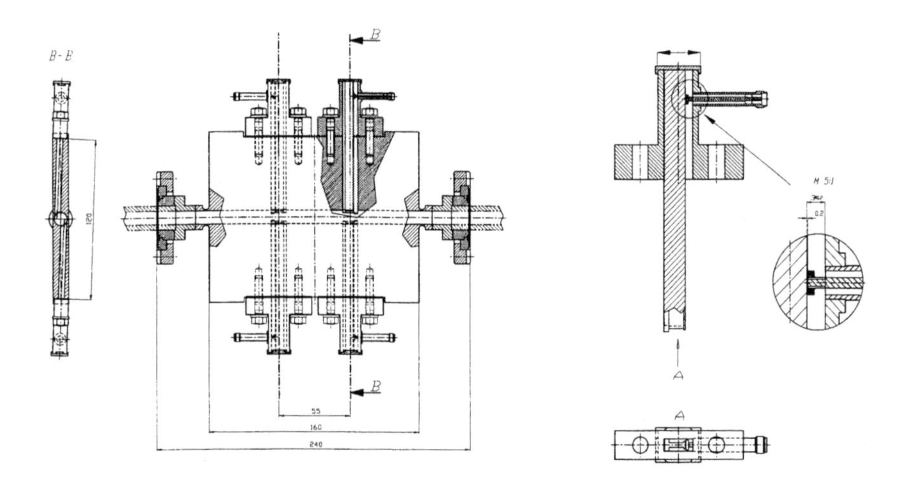

FIGURE 2. Design of prototype I. Both planes are separated in beam direction by 55 mm.

A cw signal of 12 GHz was induced by a signal generator into this coaxial system (wire and inner beam pipe surface). Impedance transformers were placed in front and behind the BPM to minimize rf reflections caused by impedance mismatches. The output signals of all four channels are amplified, filtered, and measured in a power meter. All components are controlled by a PC running a Labview application.

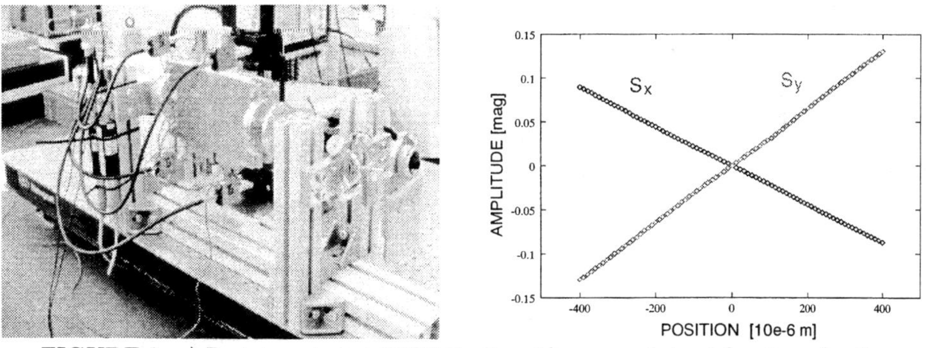

FIGURE 3. a) Prototype test at DESY Zeuthen; b) measured signal functions S_x, S_y.

First, the wire was moved on a 1 mm square around a point close to the electrical center with a stepwidth of 10 μm. With these data, three calibration algorithms were tested to obtain the signals functions S_x and S_y in Equation (3). The results for a range of ± 400 μm from the electrical center are shown in Figure 3b).

In the first method, the wire position is approximated on a sub-grid around the first guess of the position value by interpolation. The iteration stops when a reasonable value of convergence is reached. The other methods assumes a polynomial relation between the wire position and the signal function. With the knowledge of the signal functions the 'beam' position can be predicted better than 3 μm within a radius of 500 μm around the electrical center. The method following Equation (1) gives an estimation for the angular position of each of the coupling slots. The results obtained have to be further investigated by mechanical measurements, since this may offer a method to check 'real' structures after fabrication.

The slope of the curve in the linear part around the center is directly proportional to the sensitivity of the BPM. All calculated, simulated, and measured sensitivities are summarized in the last section (Table 2).

Tests at the CLIC Facility (CTF) at CERN

For measurements at the CLIC Test Facility (CTF) at CERN [6] the prototype BPM was installed in the beamline of the CTF structure's drive beam. The main purpose of this test was to study the rf behavior of the BPM and to measure the signals induced in the waveguides. The charge of a single bunch was about 2.6 nC, its energy 50 MeV and the repetition frequency 5 Hz. Some beam parameters were optimized for this test, especially the beam size at the BPM location.

The signals of the four channels were coupled into 15 m long cables, filtered at 12.0 GHz ($B = 730$ MHz; see Fig.4 b) and amplified. Additional attenuators were inserted at the electronics input for matching reasons and to avoid nonlinearities in the mixers and amplifiers. A major problem was to obtain a phase-stable reference signal related to the beam. Therefore, most of the measurements were done by bypassing the mixer stage and by displaying the amplified signals directly on a digital sampling oscilloscope (Tektronix 11801B).

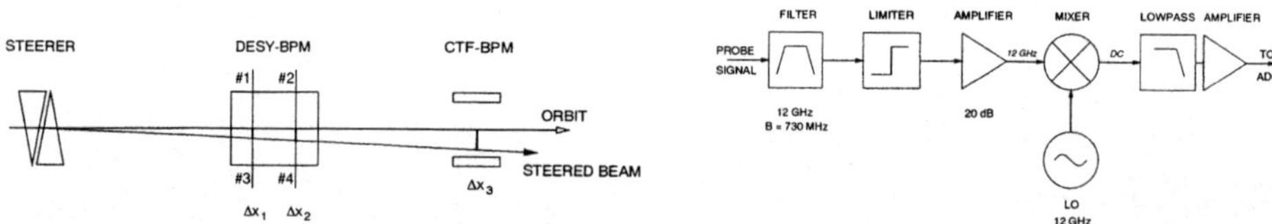

FIGURE 4. a) Installation/steering at the CTF; b) signal processing for the CTF tests.

The voltages of all four channels measured on the scope were 400 – 1100 mV (peak-to-peak), they differ roughly by a factor of 2. The largest voltage was measured on port #3, the lowest on port #1. This might be caused by two factors: the coupling factors k are not the same for all waveguides, and the beam was not centered in the monitor. Using Equation (2) and taking into account all attenuations (cables, attenuators, filters, limiters), the coupling factors are estimated to be 0.5 – 1.5%. For steering experiments, the electron beam was centered by quadrupole scans. Then it was steered in both transverse planes, and the output signals of every channel were detected on the oscilloscope. From these data the sensitivities in both planes were calculated.

SUMMARY

It has been demonstrated that this new waveguide monitor can be built. All the test results obtained are in reasonable agreement with analytical estimations and with MAFIA simulations. The signals coupled into a 50-Ω system and measured on a scope are large enough that the desired resolution can be reached.

The slope of the curve in the linear part around the center is directly proportional to the sensitivity of the BPM. All calculated, simulated, and measured sensitivities are summarized in Table 2. A reason for the lower horizontal sensitivity measured at the CTF might be the detection method and an offset from the real center position.

TABLE 2. Summary of all calculated and measured sensitivities.

Method, Measurement	Sensitivity [dB/mm]		Remarks
	horizontal	vertical	
Theory	5.67	3.89	wall current, Eqn. (1)
Simulation (Mafia)	5.41	4.25	slightly other structure
Laboratory (DESY)	5.59 / 5.6	3.83 / 3.89	before/after CTF tests
Tests at the CTF	4.3	3.96	

Further Developments

Recently, a new prototype of the waveguide monitor has been built, which will be tested soon. In Figure 5 one clearly sees that the waveguides are now completely fabricated by EDM. Further improvements include

- a higher coupling factor for each slot, now more than 0.9 %;
- a coupling into the 50-Ω system which is transverse to the beam direction, having a higher bandwidth and leading to a more compact design; and
- a separation of both planes by 41 mm (3/2 of the undulator wavelength).

In addition, the realization of a heterodyne receiver is under way. The signal of each waveguide will be filtered, amplified in a low-noise amplifier, and down-converted to less than 50 MHz in two stages. 12-bit ADCs will be used to detect the resulting signals.

According to MAFIA calculations one can design a BPM having a similar geometry and the same electronics frequency even for a reduced undulator gap width of 8 mm. Another interesting point is that the angle between two opposing waveguides can be further increased up to 45°. This would improve the linearity, reduce the position calculation algorithm in Equation (3) and simplify the electronics, too.

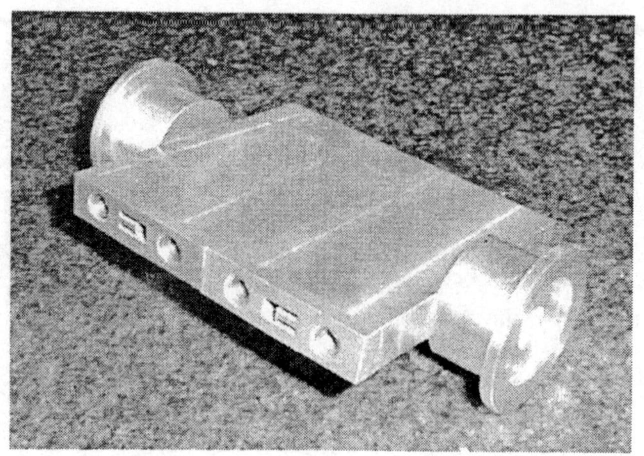

FIGURE 5. New Prototype fabricated by EDM.

ACKNOWLEDGMENTS

We would like to thank U. Hahn and H. Thöm for the mechanical design of the monitor. Furthermore, we would like to thank our colleagues from the TESLA group at DESY Zeuthen for their help in the prototype tests and for their work on the electronics. Special thanks are extended to Hans Braun and his CTF team at CERN for their help in the realization of all tests.

REFERENCES

1. *TESLA TEST FACILITY LINAC — Design Report*, edited by D. A. Edwards, DESY Hamburg, TESLA-Note 95-01, March 1995.
2. *A VUV Free Electron Laser at the TESLA Test Facility Linac — Conceptual Design Report*, DESY Hamburg, TESLA-FEL 95-03 (1995).
3. Lorenz, R., et al., "Beam Position Measurement Inside the FEL Undulator at the TESLA Test Facility Linac," presented at the DIPAC97, Frascati (1997).
4. Shafer, R., "Beam Position Monitoring," in *AIP Conf. Proc. 212*, p. 26–58 (1989).
5. Klatt, R., et al., "MAFIA — A Three-Dimensional Electromagnetic CAD System for Magnets, RF Structures and Transient Wake-Field Calculations," *IEEE Proceedings of the LINAC 86*, p. 276.
6. Braun, H. H., et al., "Results from the CLIC Test Facility," in *Conference Proc. of the EPAC94*, London, pp.42–46 (1994).

Investigation of Beam Alignment Monitor Technologies for the LCLS FEL Undulator†

R. Hettel, R. Carr, C. Field‡, D. Martin

Stanford Synchrotron Radiation Laboratory
‡Stanford Linear Accelerator Center
Stanford, CA 94309

Abstract. To maintain gain in the proposed 100 m long linac-driven Linac Coherent Light Source (LCLS) Free Electron Laser (FEL) undulator, the electron and photon beams must propagate colinearly to within ~5 μm rms over distances comparable to the 11.7 m FEL gain length in the 6 mm diameter undulator vacuum chamber. We have considered a variety of intercepting and non-intercepting position monitor technologies to establish and maintain this beam alignment. We present a summary discussion of the applicability and estimated performance of monitors detecting synchrotron radiation, transition and diffraction radiation, fluorescence, photoemission or bremsstrahlung from thin wires, Compton scattering from laser beams, and image currents from the electron beam. We conclude that: 1) non-intercepting rf cavity electron BPMs, together with a beam-based alignment system, are best suited for this application; and 2) insertable, intercepting wire monitors are valuable for rough alignment, for beam size measurements, and for simultaneous measurement of electron and photon beam position by detecting bremsstrahlung from electrons and diffracted x-rays from the photon beam.

INTRODUCTION

The Linac Coherent Light Source (LCLS) will produce intense pulses of coherent x-rays in the 15–1.5 Å range generated by self-amplified spontaneous emission (SASE) from a 4.5–14.4 GeV single-bunch electron beam passing through a 100 m long undulator (1). The pulse repetition rate is 10–120 Hz. LCLS parameters are given in Table 1.

† Work supported in part by Department of Energy Contract DE-AC03-76SF00515 and Office of Basic Energy Sciences, Division of Chemical Sciences.

CP451, *Beam Instrumentation Workshop*
edited by R. O. Hettel, S. R. Smith, and J. D. Masek

TABLE 1. LCLS Electron and Photon Beam Parameters

Electron Energy	4.5 GeV	14.4 GeV
Emittance (normal)	2 π mm-mrad	1.5 π mm-mrad
Charge/bunch	1 nC	1 nC
Peak current:	3400 A pk	3400 A pk
Bunches/pulse	1	1
Pulse rep rate	10-120 Hz	10-120 Hz
Bunch radius:	37 μm rms	31 μm rms
Bunch divergence	6.1 μrad	1.7 μrad
Bunch length	20 μm rms	20 μm rms
Photon 1st harmonic	15 Å (0.82 keV)	1.5 Å (8.2 keV)
FEL gain length	3.7 m	11.7 m
FEL peak pwr/pulse	11 GW	9 GW
FEL avg pwr	0.36 W	0.51 W
FEL beam radius	37 μm rms	31 μm rms
FEL divergence	3.2 μrad rms	0.38 μrad rms
FEL peak brightness	1.2×10^{32}	12×10^{32}
FEL avg brightness	0.42×10^{22}	4.2×10^{22}
Spontan.peak pwr/pulse	8.1 GW	81 GW
Spontan. avg pwr	0.27 W	2.7 W
Spontan. beam radius	52 μm rms	33 μm rms
Spontan. beam diverge	6.2 μrad rms	2 μrad rms
Spontan. critical energy	22 keV	200 keV

The undulator has 52 segments, each 1.92 m long, separated by 0.24 m gaps containing vacuum pumps, quadrupoles, and diagnostics (Figure 1). Quadrupoles are equipped with precision transverse movers that are used for beam steering. The undulator gap is 6 mm, and the vacuum chamber within has a 5 mm ID. This chamber dimension will be preserved as much as possible in the gaps between undulator segments to minimize impedance.

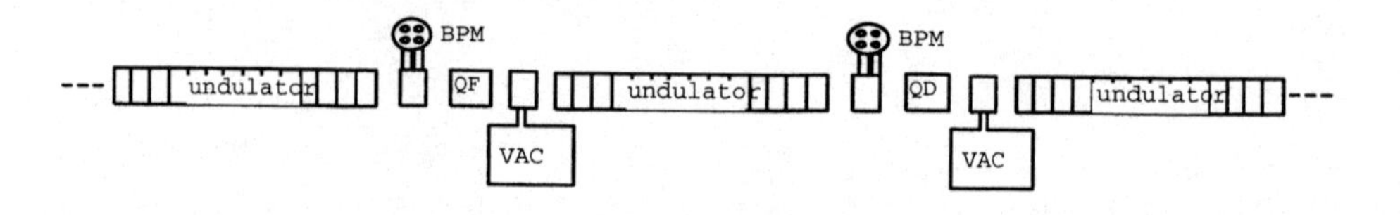

FIGURE 1. The 100 m LCLS undulator consists of 52 magnet sections (1.92 m) separated by 0.24 m gaps containing permanent magnet quadrupoles, vacuum pumping components, and BPMs.

To achieve FEL gain the electron beam must be continuously bathed in the photon beam it creates. For high gain, the two beams must overlap to within ~10% of the transverse beam size in the undulator. The absolute straight line trajectory of the electron beam must be maintained to this degree over distances comparable to an FEL gain length. For the 1.5 Å LCLS photon beam created by the 14.4 GeV electron beam, the overlap requirement and the 11.7 m gain length electron beam straightness tolerance is ~5 μm rms. For the 15 Å, 4.5 GeV case, the 10% overlap is only needed over a 3.7 m gain length.

Several position monitor technologies for aligning the LCLS undulator beams have been considered (2). The choice of beam alignment method determines which BPM types are the most appropriate as discussed in the following section.

BEAM ALIGNMENT METHODS

Techniques considered for achieving LCLS undulator beam alignment include:

1. Using a photon monitor located downstream of the undulator to align spontaneous radiation from individual undulator sections as they are steered in sequence;
2. Using absolutely aligned and stable non-intercepting monitors located in the gaps between undulator sections;
3. Using absolutely aligned insertable intercepting monitors to establish initial alignment and stable non-intercepting monitors to maintain it; and
4. Using non-intercepting monitors and beam-based alignment to establish and maintain absolute beam straightness.

The first method, using sequential steering of undulator radiation from individual sections, was used successfully for the 2m, 24–50 MeV CLIO infrared FEL at LURE (3), but the technique may not be practical for the higher energy and much longer LCLS system due to problems with detecting radiation from downstream undulator sections in the presence of the intense photon beam coming from aligned upstream sections. The method might prove useful as a secondary alignment technique, especially if a system of insertable filters can be used to absorb the upstream photons.

The second and third methods both rely on the ability to install monitors with 5 μm absolute measurement accuracy with respect to a straight line over 11.7 m gain length intervals and maintaining that accuracy over time for 1.5 Å FEL operation. These methods do not seem to be practical given the conclusion from SLAC alignment experts that they can only guarantee 25 μm accuracy over these distances. However they may suffice for 15 Å FEL operation where the gain length is only 3.7 m and the electron beam is not expected to deviate by more than a few microns from magnet errors over this distance. The third method may also work for 1.5 Å FEL operation if the intercepting monitor can simultaneously detect electron and photon beam positions and beam overlap with 5 μm or better relative accuracy; we discuss such a monitor below.

The fourth method employs a powerful beam-based alignment algorithm to achieve absolute beam straightness (1). By recording the readings of roughly aligned

BPMs as a function of beam energy (varied between 4.5 and 14.4 GeV) and by fitting a model of the undulator electron transport optics to those readings, offset errors for quadrupoles, BPMs, and incoming beam trajectory can be calculated and corrected. When this process is repeated 2–3 times (which may take a few hours), simulations indicate that BPM offsets and electron beam straightness in the 100 m long undulator can be established and maintained with better than 5 μm rms accuracy.

We conclude that we will use stable, high-resolution, non-intercepting beam position monitors in the gaps between LCLS undulator sections that can be absolutely aligned to the micron level using a beam-based alignment algorithm. In addition, we will install insertable intercepting monitors that provide an alternate means to measure position and to cross-check beam-based alignment results. As described below, the intercepting monitors will simultaneously measure electron and photon beam position to 5 μm. A spontaneous radiation monitor located downstream of the undulator after the electron beam dump will be available to check photon beam alignment.

MONITOR PERFORMANCE REQUIREMENTS

The LCLS undulator BPM system must be capable of establishing and maintaining electron and photon overlap in both transverse directions to 5 μm rms or better for 1.5 Å FEL operation. While beam-based calibration eliminates the need for micron level installation accuracy, an absolute BPM measurement accuracy of < 50 μm rms over 11.7 m gain length intervals after initial installation is desired to reduce the beam-based calibration time and to achieve FEL gain at low electron energies without that calibration. This specification includes nominal 25 μm absolute accuracy tolerances in alignment over 11.7 m and in knowledge of BPM electrical center location with respect to nearby external fiducials (a few cm away).

Micron resolution and stability is needed only in a bandwidth comparable to thermal drift frequencies (<< 1 Hz, over periods of days), implying that BPM readings from many beam pulses can be averaged for higher resolution. Single shot resolution of order 1 μm for a 1 nC bunch is desired to detect 120 Hz pulse-pulse trajectory instability. A dynamic range of 40 dB is needed for low and high intensity operation. Monitors must be mounted on precision translation stages so that their mechanical alignment can be adjusted and preserved to 1 μm rms with respect to a system of stretched wires running parallel to the undulator (1), similar to the system used for the SLAC FFTB.

The total longitudinal beam impedance of 52 BPMs (one per drift section between undulator segments) must be kept well below a loss factor of 1 kV/pC to keep the correlated energy spread of the electron bunch below 0.1%; otherwise the FEL saturation length would increase beyond 100 m. Insertable BPMs may have much larger impedance since they can be withdrawn for FEL operation. Insertable BPMs must be designed to have minimum impedance when withdrawn.

Intercepting monitors must be able to handle the power densities from both electron and photon beams. Some monitors may only be capable of low intensity operation, having to be withdrawn before operating with the high peak current needed for lasing.

BEAM POSITION MONITOR CANDIDATES

We have investigated several intercepting and non-intercepting beam position monitor technologies that might meet the performance needs for the LCLS undulator.

Intercepting Monitors

Precisely insertable fluorescent screens and crystal wafers, transition radiation monitors, and wire scanners were considered as intercepting electron beam position monitors for the LCLS undulator. The fluorescent and transition radiation monitors can measure horizontal and vertical beam position simultaneously, while the wire scanners require sequential measurements using one wire per plane.

Phosphor screens were eliminated as precision monitors because of the low resolution and dynamic range caused by finite grain size, deposition non-uniformity, and blooming of the phosphor. Fluorescing crystal wafers, such as CsI and YAG, overcome these limitations. YAG crystals in particular have recently been shown to have micron resolution and large dynamic range when visible fluorescence is viewed through a telescope with a CCD camera (4). The problem with using this type of monitor is that both the electron beam and undulator photon beams will excite the crystal, making it difficult to precisely measure position of just one of the beams. This problem might be reduced if the crystal wafer is mounted on the back of a photon-absorbing substrate that passes the electrons, or if absorbing filters (e.g., 100 μm tungsten) can be inserted upstream of the crystal. An alignment fiducial on the crystal holder, viewable by the monitor camera, may be needed for absolute accuracy.

Transition radiation (TR) from a precisely insertable thin foil provides a powerful way to measure beam size and position, especially at wavelengths comparable or longer to the electron bunch length (~30 μm rms) where the transition radiation is coherent. However, the performance of this type of monitor in the LCLS undulator is questionable since undulator radiation at TR wavelengths will be reflected from the foil and will obscure electron beam measurement. Again, this problem might be reduced using insertable tungsten filters upstream of the TR foil.

Wire scanners are used successfully at SLAC to measure micron or smaller rms beam sizes. Those in the FFTB (5) have been used with the same beam intensity as projected for LCLS. Overlap between the electron beam and a precisely positioned carbon wire is detected downstream of the undulator by measuring either bremsstrahlung gamma rays (having a 1/E spectrum extending up to the beam energy) or, in the event that excessive background radiation corrupts this measurement, degraded energy electrons produced by the bremsstrahlung process (in the range of 0.5 to 0.75 of the initial beam energy) that are magnetically deflected from the beam pipe. Radiation-hard Cherenkov detectors with thresholds above 15 MeV have been used to reject background synchrotron radiation having critical energy up to 1.5 MeV. For the LCLS undulator, both gammas and electrons will be detectable, and comparison of their results will give a good indication of systematic errors.

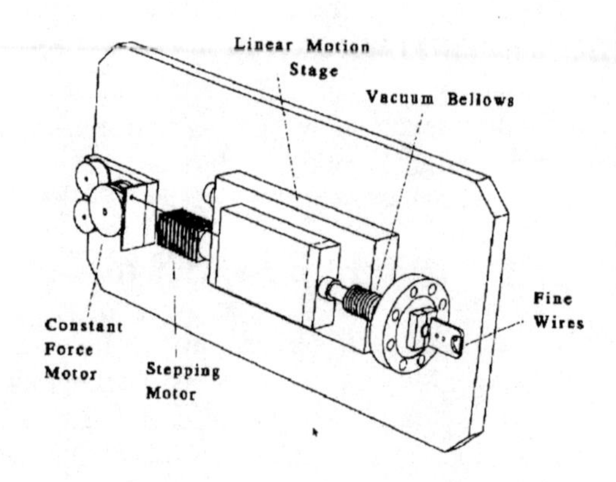

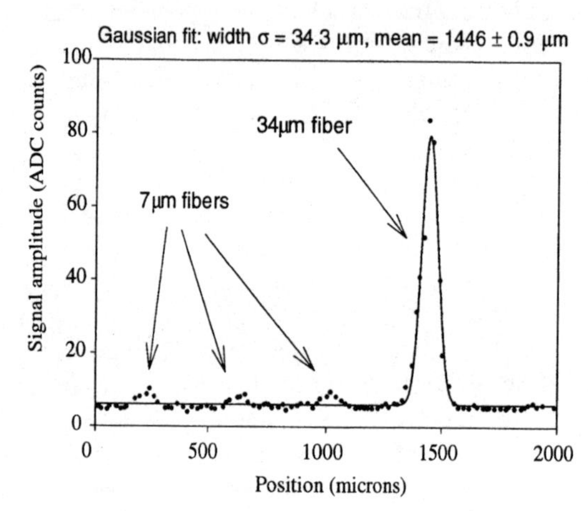

FIGURE 2. Wire scanner with micron resolution beam profile and centroid measurement (SLAC).

By stepping a wire across the beam, pulse by pulse for 10–20 pulses, using a linear motion stage (Figure 2), or by steering the beam across the wire, a profile of the beam can be measured. The beam shape is fitted on line, with a typical uncertainty of 2% of the width, and the center position obtained within 1–2 μm with respect to an external fiducial on the motion stage (Figure 2). Straight-line conventional alignment between stages can only be guaranteed to 25 μm over 11.7 m.

The LCLS beam intensity will be low enough that thinner wires of higher atomic number than carbon could be used without being destroyed. Their advantage is that the thinner the wire, the more accurately its can center be located relative to the fiducial marks outside the vacuum.

A distinct advantage of the carbon wire monitor is that it can be used for simultaneous measurement of electron and undulator photon beam position (Figure 3). While the impinging electron beam generates bremsstrahlung, the undulator photons will diffract from the wire in a powder diffraction pattern. An experiment at SSRL using 7 μm amorphous carbon wire filaments and 1.5 Å x-rays showed that an intensity maximum for Bragg scattering occurs at 25.8°. The energy range for practical Bragg angles is rather limited, though one could use third harmonic radiation when running the beam at lower energies. The energy dispersion caused by diffraction assures that a detector subtending a small angle will acquire x-rays with a narrow energy range.

We conclude that the preferred intercepting monitor for the LCLS is the wire scanner because of its ability to measure both electron and photon position at high operating intensities and because of its proven micron-level performance.

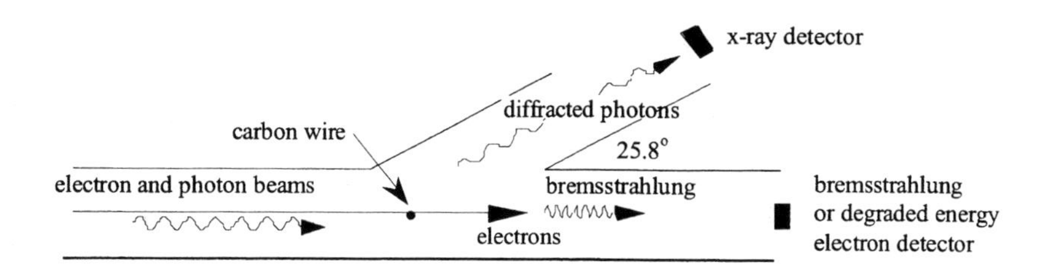

FIGURE 3. Combined electron/photon beam position monitor for one plane. Both beams strike the carbon wire; when they overlap, detectors record maximum signals simultaneously.

Non-Intercepting Monitors

Candidates for non-intercepting position monitors include diffraction radiation monitors, laser wire (or spot) monitor, and more commonly used rf BPMs.

A diffraction radiation (DR) monitor (6) having a 2 mm radius aperture within the 2.5 mm radius undulator vacuum chamber would produce micron wavelength DR (which, like TR, would be coherent at 30 μm or longer) that can be observed with a simple camera system to determine beam size and position. While the measured radiation pattern is sensitive to the transverse displacement of the electron beam from the center of the aperture, a derivation of position sensitivity in both planes has not been completed; it is premature to say this monitor would have the micron position resolution required. Furthermore, the monitor also has a high impedance (~75 V/pC loss factor), implying that only 10 monitors could be inserted during FEL operation.

The success of the laser wire monitor for measuring micron beams at the SLAC Linear Collider Final Focus (7) prompted us to investigate a method of measuring Compton scattering from a 1 μm × 10 μm laser "spot" (2). The spot would be created by focusing an intense pulse of 1.06 μm light from a high-powered laser (e.g., a 100 MW peak pulsed YAG laser). Because of the large background expected from bremsstrahlung and high-energy undulator photons, a measurement of degraded energy electrons at the end of the undulator might offer better performance. A principal problem with the laser spot monitor is that, due to possible changes in laser optical components over time caused by the high pulsed laser power and radiation environment, the absolute stability of the laser spot position is uncertain and there is no clear method for monitoring it. Another drawback is that if the electron beam is off the laser spot, there is no indication of which way to steer.

Uncertainties in performance of the DR and laser spot monitors led us to concentrate on specifying an appropriate non-intercepting rf BPM pickup and processing system for the LCLS undulator. Several high-frequency (rf) position monitor technologies were evaluated, operating either within the undulator gap or in the drift spaces between undulator sections. The devices and their calculated performance are identified and summarized in Table 2.

TABLE 2. rf BPM Design Parameters. BPM locations are either within the LCLS undulator pole gap (U) or in the drift spaces between undulator sections (D). Values for center accuracy are estimated.

Monitor Type	Parameters	Center Ac'cy	Resolution	Oper. Freq.	Issues
Wall Current (U)	z=6 mm R_B=2 Ω	100 μm	0.7 μm/nC	> 1 GHz	Ferrite saturation
Stripline (U)	z=9 mm Z_o=40 Ω	100 μm	0.2 μm/nC	2–5 GHz	Strips on ceramic cyl
Microwave Aperture (U)	3.0×1.5 mm slot to waveguide	100 μm	0.1 μm/nC	> 50 GHz	Op. freq > chamber cutoff; HOM errors
Cavity (U)	ϕ_{ID}=7 mm z=2.8 mm	50 μm	1 μm/nC	~ 32 GHz	f_o ~ cutoff; low Q
Stripline (D)	z=40 mm Z_o=50 Ω	50 μm	0.2 μm/nC	0.5–2 GHz	Technical maturity
Cavity (D)	ϕ_{ID}= 60 mm z=5 mm	5 μm	0.2 μm/nC	~ 6 GHz	Robust; TM_{010} mode

The region within the undulator gap considerably restricts the BPM mechanics that can be built. For example, the ferrite of the Wall Current Monitor cannot be allowed inside the undulator, nor will it fit. Feedthroughs for monitors within the undulator pole gap are difficult to accommodate. Monitors such as the Aperture Monitor, which operate on Bethe hole radiation, must have small apertures, and as such are strongly influenced by higher order modes. The Cavity BPM within the undulator gap, having beam pipe apertures nearly the size of the resonator end plates, would have a low Q. In addition, the relative compactness of any structure within the pole gap increases fabrication difficulty and raises the operating frequency, contributing to signal cable losses and higher component costs.

BPM structures in the drift regions offer superior performance with fewer design restrictions. Of those investigated, the Cavity BPM best meets the design requirements. Because of the natural symmetry of circular machining and the availability of ultra-precision diamond lathes, micron level absolute mechanical and electrical center accuracy can be achieved.

Excited by the passing beam, the cavity rings down in a set of characteristic frequencies, precisely determined by the cavity dimensions (8). Signal power may be extracted through four precisely machined apertures, each coupled to an external waveguide. The waveguide TM_{010} position-sensitive mode will exist, in two polarizations, only when beam traverses the cavity off axis. This position mode competes with the strong lower frequency (TM_{110}) dominant mode, which can be rejected using both frequency and symmetry discrimination. Presence of the dominant mode, not thermal noise, ultimately limits the achievable position accuracy. A cavity operating at 6 GHz was tentatively designed for the LCLS (1); its parameters are summarized in Table 3.

TABLE 3. LCLS Cavity BPM Parameters

Parameter	Value
Cavity radius	28.5 mm
Cavity length	5 mm
Beam pipe ID	5.0 mm
R/Q (TM_{110})	8.4 Ω at 6 GHz
V_{out} in 50 Ω (TM_{110})	15 μV/nm/nC
Peak E field at 1nC	7.7 MV/m
Long. Loss Factor	37.1 V/pC

CONCLUSION

We propose to install stable high resolution, non-intercepting cavity BPMs and intercepting carbon wire scanner units in the 52 drift sectons between LCLS undulator segments. The absolute position of the electrical centers of the cavity BPMs and of the intercepting wires will be known to < 50 μm rms with respect to a straight line over 11.7 m 1.5 Å gain length intervals after initial installation. This alignment accuracy in itself is likely to be sufficient to establish 15 Å FEL operation. It is also sufficiently accurate to launch a beam-based alignment algorithm which will straighten the electron beam and calibrate BPM offsets to < 5 μm rms with respect to a straight line over the 100 m undulator length, more than adequate for 1.5 Å lasing. The insertable wire monitors will provide an alternate means to measure position and to cross-check beam-based alignment results since they will capable of measuring electron and photon beam position overlap to within 5 μm. The wire monitors will also be used to measure beam profile and emittance. All monitors will be precisely movable and mechanical alignment stability will be maintained to 1 μm rms using a stretched wire positioning system along the undulator.

ACKNOWLEDGMENTS

The authors wish to express their appreciation for contributions to this article from J. Arthur, S. Brennan, H.-D. Nuhn, J. Sebek (SSRL); P. Emma, S. Smith, S. Wagner (SLAC); G. Lambertson, W. Barry (LBL); A. Lumpkin (APS); I. Ben-Zvi (BNL); and R. Fiorito (NSWC). They are also indebted to M. Cornacchia (SSRL) for his support.

REFERENCES

[1] LCLS Design Study Report, SLAC Report 521, April 1998.

[2] Hettel, R., D. Martin et al. "LCLS Undulator BPMs," internal SSRL report, Dec. 1996.

[3] Robinson, K., (STI, Seattle), private communication.

[4] Graves, W., E. Johnson, S. Ulc, "YAG Profile Monitor and its Applications," these proceedings.

[5] Field, C., *Nucl. Instr. & Meth. A* **360** (1995) p. 467.

[6] Rule, D., R. Fiorito, W. Kimura, "Non-Interceptive Beam Diagnostics Based on Diffraction Radiation," Proc. of the 7th Beam Instrumentation Workshop, *AIP* **390** (1996) p. 510.

[7] Ross, M. et al., "A Laser-Based Beam Profile Monitor for the SLC/SLD Interaction Region," Proc. of the 7th Beam Instrumentation Workshop, *AIP* **390** (1996) p. 281.

[8] Lorenz, R., "Cavity Beam Position Monitors," these proceedings.

Test Results of the LEDA Beam-Position/Intensity Measurement Module*

Chris R. Rose and Matthew W. Stettler

Los Alamos National Laboratory
P.O. Box 1663, M/S H805
Los Alamos, New Mexico 87545

Abstract. This paper describes progress in the design and testing of the log-ratio-based beam-position/intensity measurement module being built for the Low Energy Demonstration Accelerator (LEDA) and Accelerator Production of Tritium (APT) projects at Los Alamos National Laboratory. The VXI-based module uses four, 2 MHz if inputs to perform two-axis position measurements and one intensity measurement. To compensate for systematic errors, real-time error-correction is performed on the four input signals after they are digitized and before calculating beam position and intensity. Beam intensity is computed by using the average of the four log-amplifier outputs. This method provides a better off-axis intensity response than the traditional method of summing the rf power from the four lobes. Several types of test data are presented including results of the real-time error correction technique, a working dynamic range of over 80 dB, and achievable resolution and accuracy information.

INTRODUCTION

This paper reports progress on the design and testing of the LEDA beam-position/intensity module reported earlier (1). Key areas are the real-time error-correction technique, showing results with more complete data than was presented earlier, and an improved method of performing the intensity measurement. A block diagram of the module is shown in Figure 1. The position-measurement technique is based on the log-ratio transfer function which has been described by several authors (2,3,4). The log-ratio technique is defined in Equation 1:

$$V_{\log ratio} = \log(T) - \log(B) \quad (1)$$

* Work supported by the US Department of Energy.

CP451, *Beam Instrumentation Workshop*
edited by R. O. Hettel, S. R. Smith, and J. D. Masek
1998 The American Institute of Physics 1-56396-794-4/98/$15.00

where *T* and *B* represent the power in the intermediate-frequency (if) signals for opposite top and bottom lobes of a beamline probe. Subtraction is easier to perform than division and can be done either by digital or analog techniques.

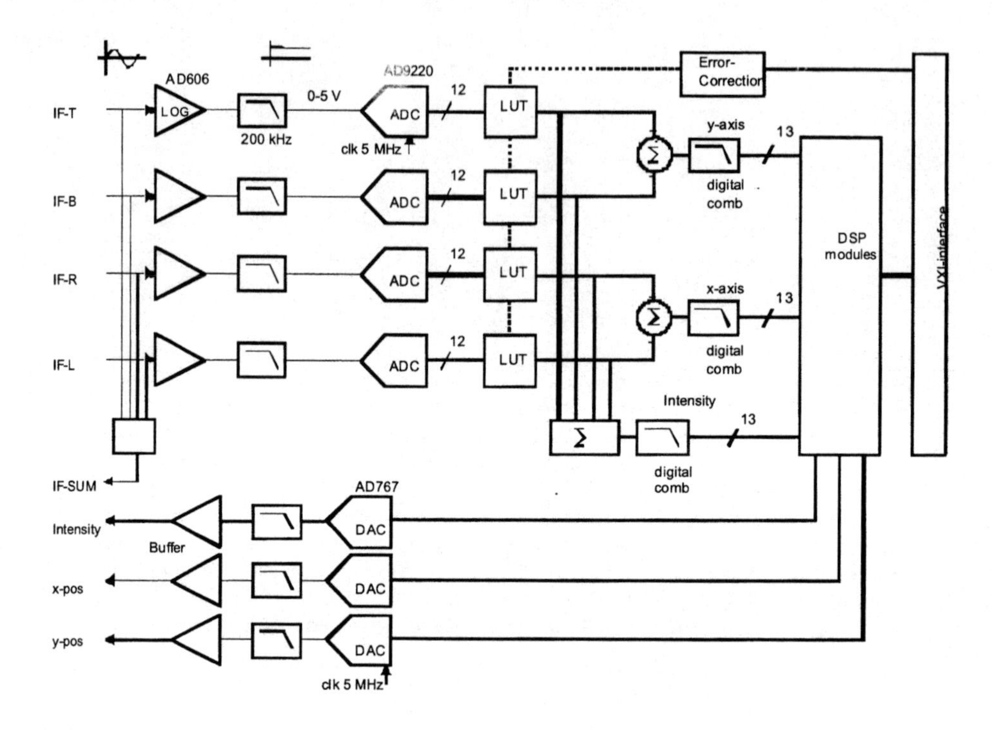

FIGURE 1. Block diagram of the LEDA Beam-Position/Intensity measurement module.

Referring to Figure 1, the 2 MHz if inputs are digitized at a 5 MHz sampling rate, digitally corrected, subtracted, filtered, and passed to the DSP modules for further processing. Three separate DAC channels provide front panel signals for monitoring. For a more detailed explanation of the module's operation, the reader is referred to Reference 1.

The module is divided into three sections: first, the analog front-end; second, the digital-to-analog section; and third, the DSP, error-correction, and VXI-interface section. Some of the important specifications of this module are listed in Table 1.

ERROR-CORRECTION PERFORMANCE

Each logarithmic amplifier exhibits some non-ideal behavior in the form of small perturbations in its transfer function, DC offsets, and distortion effects at the upper and lower ends of the dynamic range. By using digital error-correction, non-ideal performance in the log amps and other system non-linearities can be removed from the overall system transfer function. The signal level range at the input of the ADCs is 0–5 V. The ADC resolution in mm depends on the specific probe sensitivity in dB/mm or other suitable units.

TABLE 1. General Requirements for the LEDA Beam Position/Intensity Module

Item	Value	Units
Frequency input (if input)	2.00	MHz
Maximum input signal power	10	dBm
Input impedance	50	Ω
number of if inputs	4	each
number of axis measurements/module	2	each
Range, x-, y-axis outputs	±10	V
Range, intensity output	0 –10	V
Measurement bandwidth	≥ 180	kHz
Measurement resolution	0.03	dB
ADC resolution	12	bits
ADC sampling rate	5	MHz

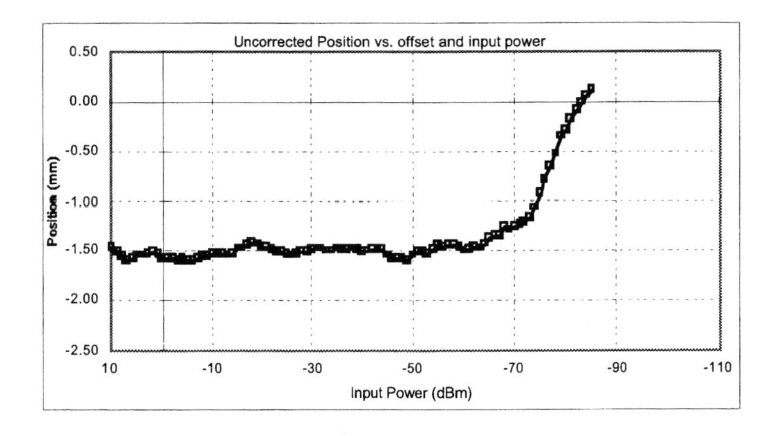

FIGURE 2. Plot of single-axis uncorrected position data for 5 dB offset inputs.

Digital data streaming from the four ADCs are the addresses to the look-up tables (LUTs). Corrected data are output from the LUTs and are used in the subtraction and intensity measurement processes. Figure 2 shows uncorrected position data for a 5 dB offset input ratio performed on the if-T and if-B input channels.

The probe sensitivity is assumed to be 3.484 dB/mm. Circuit offsets and other non-linearities cause the plot to be shifted slightly downward and exhibit significant ripple. Over the useable dynamic range of +10 to –70 dBm the mean is –1.493 mm, and the standard deviation is 0.069 mm. The same input offset, 5 dB, and power range using error-correction is shown below in Figure 3. Again, the same probe sensitivity value of 3.484 dB/mm is used to plot the data. The corrected position from +10 to –70 dBm has a mean of –1.437 mm, and standard deviation of 0.016 mm, a significant improvement over that of the uncorrected response. The data plotted in the above two figures is taken with input signals at the beam-position/intensity module if inputs and do not include other system components such as the beamline probes, down-converters, and cabling. These data serve as a useful baseline of the best possible performance from the system.

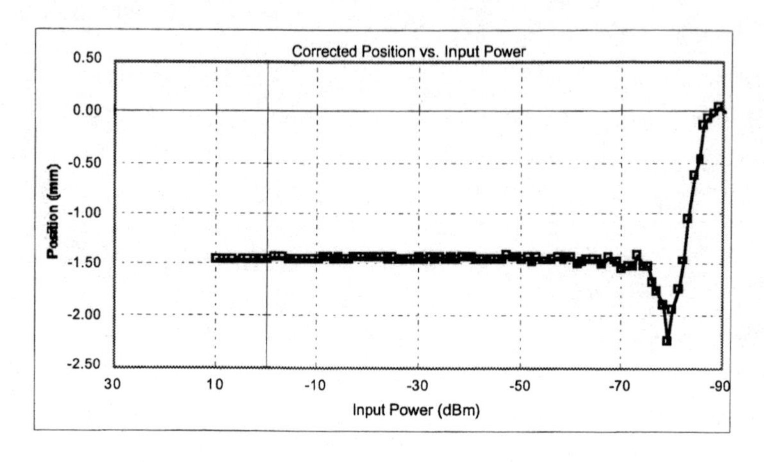

FIGURE 3. Plot of single-axis corrected position data for a 5 dB offset between the inputs.

INTENSITY MEASUREMENT

The original method chosen to perform the intensity measurement was to sum the four if inputs (IF-SUM), present the sum to a log amplifier, and then digitize the log amplifier output and perform digital error correction and calculations on that signal (1). This required a fifth log-amp/ADC channel. However, after examining constraints, such as limited board space, the need to have a fifth channel was eliminated by using a different intensity calculation method, an average of the log inputs (AVG-LOG). The IF-SUM signal is still available to other accelerator systems via the front panel. The IF-SUM and AVG-LOG techniques are defined in general terms as:

$$V_{IF-SUM} = \alpha_1 \log(T + B + R + L) \tag{2}$$

and

$$V_{AVG-LOG} = \alpha_2 \left(\log T + \log B + \log R + \log L\right) \tag{3}$$

where T, B, R, and L represent the input power of the respective lobes. Constants α_1 and α_2 are scaling factors used to normalize the respective responses.

The IF-SUM technique when used with the LANL circular probes yields a very non-linear response as a function of beam position (5). Ideally, the intensity measurement should be independent of beam position. For a centered beam, the IF-SUM technique provides an accurate measure of the beam intensity over a very wide dynamic range. However, as the beam moves off center within the probe, the probe's response distorts the accuracy of the measurement. A better intensity measurement method to use with these probes is the average log (AVG-LOG) technique.

Given the fact that each lobe signal is already digitized on the board, it is simple to sum and average them digitally as shown in Figure 1.

Using Gilpatrick's mathematical model of a LANL 10.25 mm radius probe response (6) and the above mentioned intensity techniques, the responses of the IF-SUM and AVG-LOG techniques are shown in Figure 4. Both curves are normalized to 100 mA beam current.

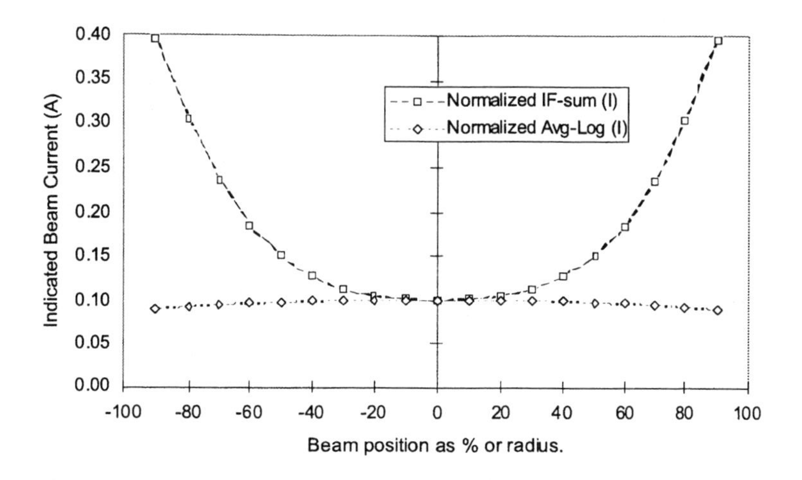

FIGURE 4. Comparison of the indicated current response of two intensity measurement methods: IF-SUM and AVG-LOG.

The AVG-LOG technique provides a response less correlated with position than the if-SUM technique. The x-axis in Figure 4 is beam position as a percentage of beam-probe radius out to 90%, and the y-axis is indicated beam current.

Even though the AVG-LOG technique still yields a beam-intensity measurement partially correlated to beam position, knowing beam position in x- and y-space, and indicated intensity, the real beam intensity can be calculated. This calculation is done in the main control system and not in the module.

REFERENCES

[1] Rose, C. R. and M. W. Stettler, "Description and Operation of the LEDA Beam-Position/Intensity Measurement Module," PAC, Vancouver, B.C., Canada, May 1997.

[2] Gilpatrick, J. D., "Comparison of Beam-Position-Transfer Functions Using Circular Beam-Position Monitors," PAC, Vancouver, B. C., Canada, May 1997.

[3] Wells, F. D., et al., "Log-Ratio Circuit for Beam Position Monitoring," American Institute of Physics, Accelerator Instrumentation, Second Annual Workshop, Batavia, Illinois, 1990.

[4] Aiello, G. R., and M. R. Mills, "Log-Ratio Technique for Beam Position Monitor Systems," *Nuclear Instruments and Methods in Physics Research*, 1994.

[5] Gilpatrick, J. D., private communication.

[6] *Ibid.*

Diagnostics Used in Commissioning the IUCF Cooler Injector Synchrotron

Mark S. Ball, Dennis L. Friesel, Brett J. Hamilton

Indiana University Cyclotron Facility
2401 Milo Sampson Ln., Bloomington, IN 47408

Abstract. Several new diagnostics systems were designed to aid in the commissioning of the IUCF Cooler Injection Synchrotron (CIS). Among them are a time of flight measurement system (ToF), a multi-wire profile monitor system (Harp) and a beam position monitor system (BPM). Pulsed beam from the 7 MeV linear accelerator is monitored using the ToF system. Several removable Harps are mounted in the injection beamline and ring which are instrumental for tuning ring injection and accumulation. BPMs are placed at the entrance and exit of the four ring dipole magnets to facilitate beam centering during injection and ramping. Fast and slow BPM displays are available to the operator for these functions. These diagnostics and their uses for CIS ring commissioning will be discussed.

ENERGY MEASUREMENT

A time of flight system (ToF) is being used in the 7 MeV injection beamline to detect changes in the beam energy from the RFQ/DTL. The system is similar to the one used at TRIUMF (1), where a change in energy is measured as a change in phase between two pickups of a fixed, known distance apart.

The 200 μs pulsed beam is accelerated through the RFQ/DTL using a 425 MHz, 300 kW rf amp. The non-interceptive, resonant, beam pickups are immediately downstream of the accelerator, 2.5 meters apart. A beam signal is detected, buffered and sent to the ToF electronics. An Analog Device AD607 (2) is used as an rf to if, 10.7 MHz, converter.

The AD607 is normally used in wireless communications as a down-converter amplifier. It has a mixer and log amp with AGC, as well as, an I & Q demodulator, all in a 20-pin surface-mount chip. It has been tested in the lab as an AGC beam position detector with a 75 dB range. The output of the AD607 is a constant 300 mV signal, which is fed into limiting amps and then a type II phase detector. The output is filtered to a DC level, amplified, and displayed on a scope.

In order to achieve an energy of 7 MeV, the DTL amplifier must be operated in an unregulated mode, relying on a large capacitance to hold the charge over the 200 μs pulse

CP451, *Beam Instrumentation Workshop*
edited by R. O. Hettel, S. R. Smith, and J. D. Masek

period. The resultant output energy can sag by as much as 500 keV over the span of the pulse period. Using the ToF monitor, the amplifier can be adjusted to minimize the sag (Fig. 1), flattening the output energy to the acceptance of the CIS injection aperture, 180 keV. The monitor also provides a good comparison between beam intensity on the stop in the CIS ring and the energy.

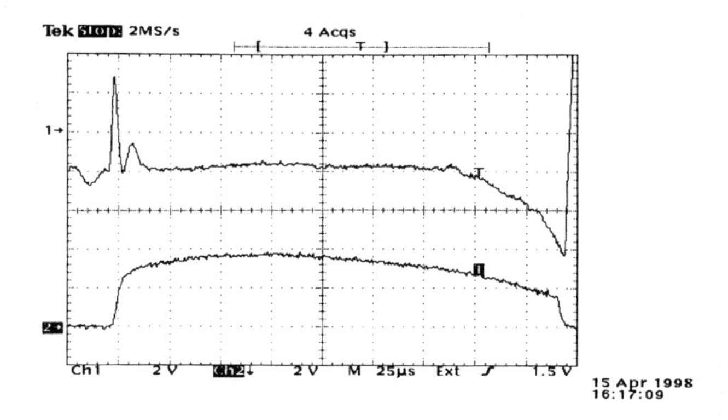

FIGURE 1. Top display is the ToF. The average energy is 7 MeV. The bottom display is the CIS stop located downstream of the first dipole. The energy sag at the end of the ToF pulse corresponds with the sag in intensity on the stop.

BEAM PROFILE MONITORS

Mechanically, the CIS Harps are very similar to those used in other labs. They are secondary emission multi-wire chambers using high-voltage cathode grids and multi-pin vacuum connectors. Each detector board uses 24 wire grids with 1 mm spacing. The circuit board is also a familiar design, using a large RC time constant at the input of the detector. The 48 signals are multiplexed through a single integrator. One interesting aspect of the Harp electronics design is the timing electronics, which are incorporated on the printed circuit board using a programmable array logic chip (PAL). A trigger from the devices on the board provides the ADC clock. The data acquisition parameters can be changed by reprogramming the PAL. This permits data acquisition and display of Harp signals to occur in two ways. For operations, the signal goes to a fast ADC with FIFO memory allowing the user to acquire data at a fast rate and display it whenever desired on the operations monitor.

During commissioning and troubleshooting, a stand-alone display system is used, incorporating a PC and graphical programming package from Hewlett-Packard, HP VEE, (3). The Harp has been the primary diagnostic during commissioning of the CIS ring. Two Harps have been placed in the ring injection beamline and one in the ring. The first Harp is mounted in front of a stop, allowing one to monitor intensity and shape of the beam over 200 µs pulse length while adjusting the profile. The Harps are moved in and out of the beam path using air actuated insertion devices. The flexibility of this setup, allows an operator to observe a Harp display of first-turn beam in the CIS ring along with multi-turns (Fig. 2) and also accumulate maximum intensity with the Harp out of the beam path. Plans have been made to remove the Harp presently used in CIS and replace it with a new low-profile design. There will also be five Harps in the beamline connecting CIS to the Cooler.

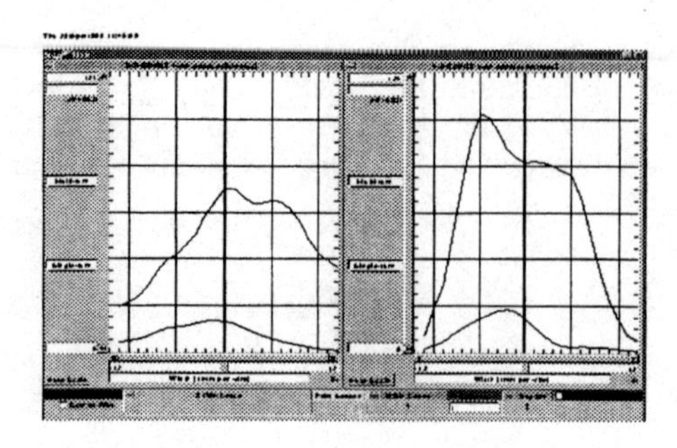

FIGURE 2. The CIS Harp as seen using HP-VEE. The small traces are first-turn beam in the ring. The large traces are multiple turns.

BEAM POSITION MONITORS

There are eight BPMs (4) in the CIS ring, positioned at the entrance and exit of each of four dipole magnets. All signal conversion is done at the pickup and the DC signals are multiplexed to different areas. The wide bandwidth of the position signals is useful for observing and changing ramp vectors, which are spaced every 11 ms (Fig 3). For the moment, vectors are changed by hand, but plans have been made to implement a feed forward loop to automatically adjust the vectors to keep the beam centered in the ring during ramping.

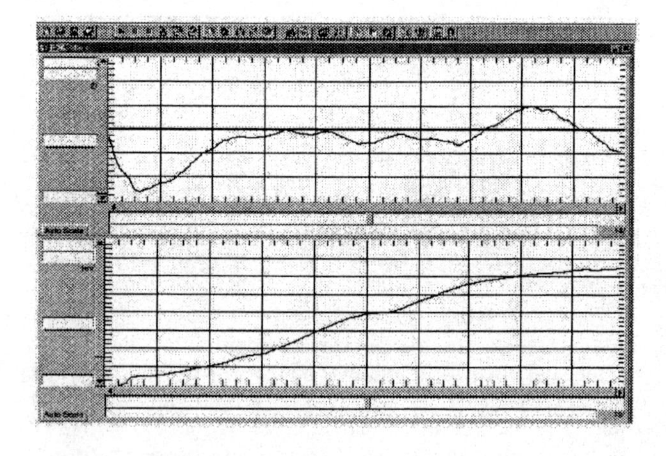

FIGURE 3. The upper display is the beam position (2 mm/Div.) in the CIS ring during a 7 MeV ramp. The bottom display is the intensity during the same ramp.

WALL GAP MONITOR

A wall gap monitor (WGM) is used to provide bunched beam information such as bunching factor measurements, and to observe beam intensity losses during the ramp (Fig. 4). A digital oscilloscope is used to capture the signal and display it for the operator. During ramping, losses are easily detected by comparing amplitude losses with BPM position distortions.

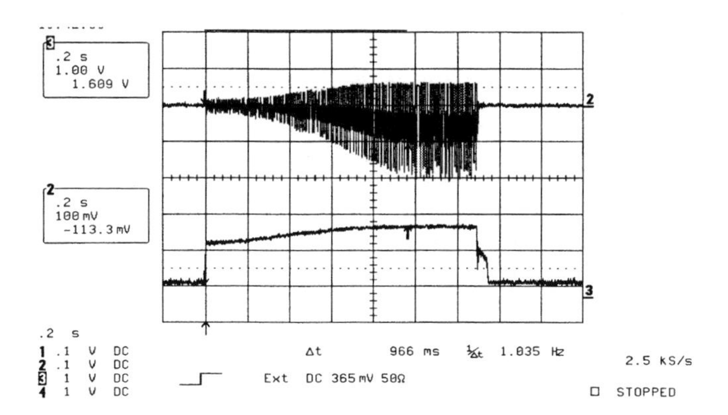

FIGURE 4. The top display is the WGM output during a 225 MeV ramp. The bottom display is the output from a logarithmic amplifier of the same ramp.

REFERENCES

[1] Yin, Y., R. E. Laxdal, A. Zelenski, P. Ostroumov, "A Very Sensitive Non-Intercepting Beam Average Velocity Monitoring System for the TRIUMF 300-keV Injection Line," Beam Instrumentation Workshop, Argonne, IL, May, 1996.
[2] Hewlett-Packard "Test and Measurement Catalog," pp. 47–49, 1996.
[3] Analog Devices "Linear Products Data Book," pp. 7.7–22, 1990/91.
[4] Ball, M., V. Derenchuk, B. J. Hamilton, "Beam Diagnostics in the Indiana University Cooler Injector Synchrotron" in *Beam Instrumentation Workshop* (Argonne, IL, May 1996), pp. 544–548.

Diagnostics Development in SRRC

K. T. Hsu, C. H. Kuo, Jenny Chen, C. S. Chen, K. K. Lin, C. C. Kuo,
Richard Sah

Synchrotron Radiation Research Center,
No. 1 R&D Road VI,
Hsinchu Science-Based Industrial Park,
Hsinchu, Taiwan, R.O.C.

Abstract. There are several new developments in diagnostics at the SRRC. These new developments include an orbit feedback system, tune monitor, filling pattern monitor, time-domain coupled-bunch oscillation monitor, and an improved synchrotron radiation monitor. A global orbit feedback system as well as a local orbit feedback system have been developed to eliminate excursions from the reference orbit that are caused by various perturbations. A digital receiver-based tune monitor provides a fast tune reading as a complementary tool to the commercial spectrum analyzer. Transient digitizers are used to acquire real-time filling patterns. Turn-by-turn and bunch-by-bunch beam signals acquired by the transient digitizer can extract information from coupled-bunch oscillations. Updated synchrotron radiation monitors provide a more convenient user interface.

INTRODUCTION

The accelerator complex of the SRRC is composed of a 50 MeV linac, a 1.3 GeV booster synchrotron, and a 1.3–1.5 GeV storage ring (Taiwan Light Source, TLS). The machine was dedicated in October, 1994. To meet stringent requirements, beam instrumentation system development is a continuous effort. The system includes all standard diagnostic devices that are used in modern synchrotron light sources. To accommodate the increased demands of machine operation and accelerator physics, various diagnostics and associated electronics are used to improve machine performance continually. Some of these new developments are highlighted in the following paragraphs.

ORBIT MEASUREMENT AND ORBIT FEEDBACK

The orbit is acquired by 52 sets of button-type beam position monitors (eBPMs). About 40 eBPMs are equipped with switched-electrode electronics developed in-house.

CP451, *Beam Instrumentation Workshop*
edited by R. O. Hettel, S. R. Smith, and J. D. Masek

Bergoz's BPM electronics are used for the other eBPMs. The orbit is acquired at a 1 kHz rate using high-performance data acquisition modules (Analogic DVX 2503) located in an VME crate. The CPU module in the VME crate is a PowerPC-based system running LynxOS 2.4. The acquired orbit is processed further and updated in the control database 10 times per second. The fast orbit information is sent to a corrector control crate through a 1.2 GB/s reflective memory network. Photon BPM (pBPM) information is also acquired by another VME crate and stored in the reflective memory. Global as well as local orbit feedback control algorithms execute in DSP modules in the corrector control VME crate. Figure 1 shows the configuration of the orbit acquisition and orbit control system.

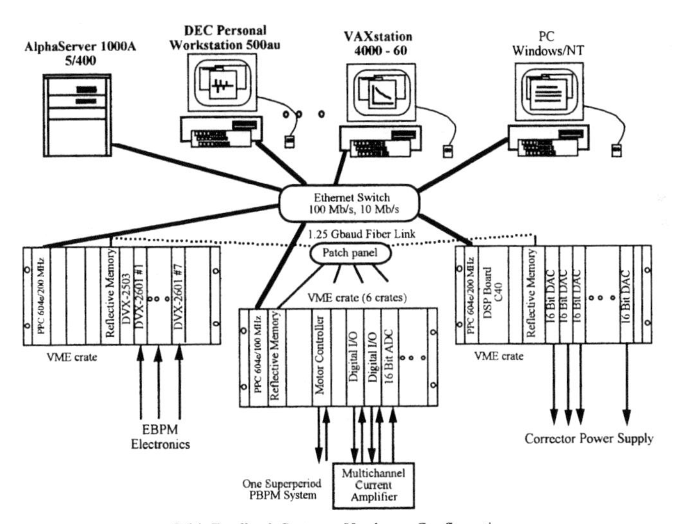

FIGURE 1. Functional block diagram of the beam position acquisition system and corrector control system.

The orbit feedback system has been tested to suppress various perturbations. Global orbit feedback maintains a reference orbit. Local feedback is used to keep beamline source point position and angle constant within stringent requirements. Feed-forward orbit control is widely used to reduce orbit excursion during the operation of the undulators. However, it has been shown that global orbit feedback is also effective in eliminating orbit excursion caused by the undulators' operation (1). For example, in the operation of exotic undulators, it is possible to change the phase and gap during the experiment scenario in an elliptical polarization undulator with the phase varying mechanically. It is not simple to do this two-dimensional feed-forward orbit control.

Figure 2 shows the global orbit feedback loop used to eliminate the orbit leakage of a dynamic local bump (2). This dynamic bump will be used to enhance the production of polarized light from the bending magnet (EPBM). A flip rate from 1 Hz to 10 Hz is tested. Trapezoid or sinusoid waveforms are available. The source of the bump leakage comes from a non-ideal local bump and small discrepancies in the characteristics of power supplies, correctors, and vacuum chambers at the corrector sites.

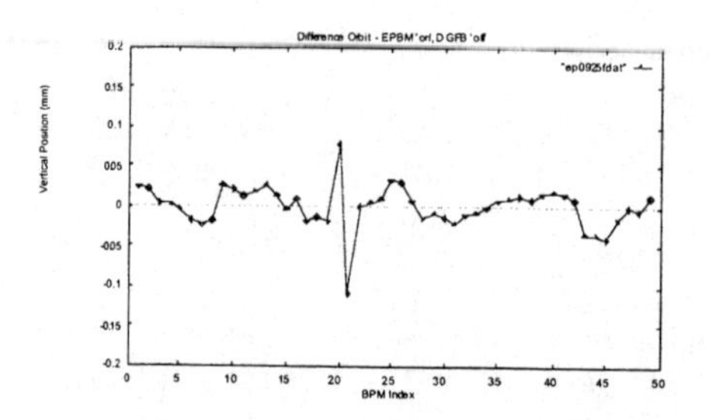

(a) Snapshot of the orbit with global feedback off, showing a small orbit leakage.

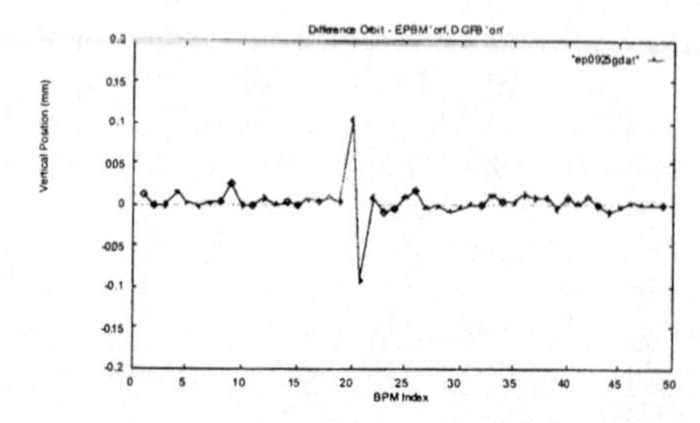

(b) Snapshot of the orbit with global feedback on. No orbit leakage is seen.

FIGURE 2. Difference orbit for EPBM operation with feedback off (a) and on (b).

The effectiveness of the orbit feedback system is dependent on the performance of the eBPMs/pBPMs, correctors, and control algorithms. Continuous upgrading of these crucial items, combined with third-harmonic cavity and coupled-bunch feedback projects will enable TLS to achieve high beam stability in the coming year.

DIGITAL RECEIVER-BASED TUNE MONITOR

A digital receiver-based tune monitor has been implemented as a dedicated system. The system includes rf front-end electronics and a digital-to-analog converter, multi-channel digital receivers, and a digital signal processor as shown in the upper part of Figure 3. A portion of the beam spectrum is processed by the rf front-end electronics

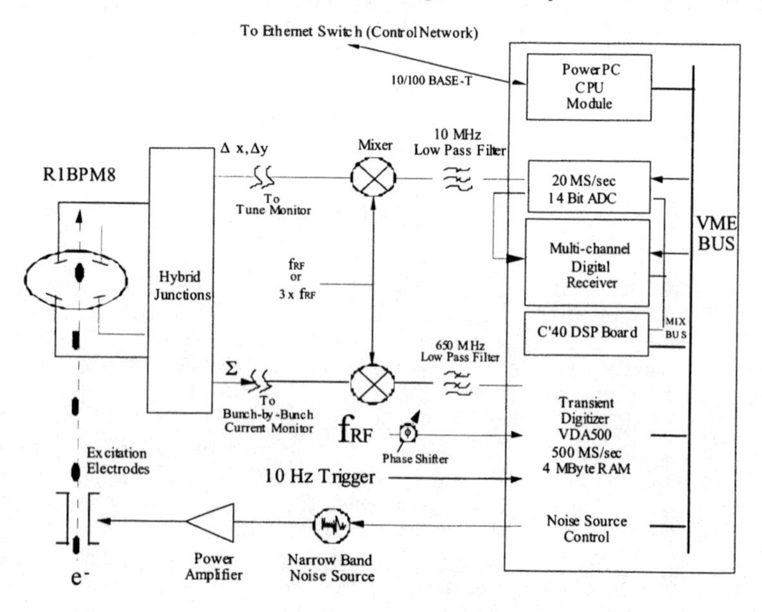

FIGURE 3. Functional block diagram of the digital receiver tune monitor and bunch-by-bunch current monitor.

and down-converted to a baseband signal. The ADC converts the baseband signal to digital data streams and sends them to the digital receiver. A digital tuner performs digital mixing, low-pass filtering, and data formatting. The output I/Q data stream is processed by the DSP to get the spectrum and perform peak identification. The ADC module, digital receiver module, and DSP module are contained in a VME crate. The 14 bits, 20 MS/sec ADC provides a more than 70 dB dynamic range and 8 MHz real-time bandwidth. Narrow bandwidth white-noise excitation is available if the stored beam is quiet.

The frequency and decimation factor of the low-pass filter are software programmable. The tune update rate is limited by the computation power of the DSP chips. Three channels of digital receiver are implemented on a single C40 DSP via a Fourier transform. The spectrum update time is less than 10 msec. To increase the speed, we will need to upgrade to multiple DSPs or a new generation DSP (eg., 320C6x). Processing gain for 1024 points FFT is about 30 dB for an input signal with 0 dB SNR. Appropriate data averaging algorithms are used for data smoothing. The measured tune is updated in the machine database every 100 msec (the data update rate of the control system). About ten spectra per second are displayed on the console computer. The user interface is based on the PV-WAVE package. Histogram and spectrogram displays are available to satisfy various needs. A fast tune update signal, with msec time resolution, is available at the local VME crate for accelerator physics studies or monitoring the power supply ripple. The booster will install a dedicated tune monitor to aid in tuning lattice parameters in the near future.

To examine the performance of the tune monitor, a sinusoidal current with 0.2 Amp peak-to-peak is applied to a quadrupole trim as shown in Figure 4(a). The tune variation due to this excitation is about 0.0009. It is estimated that the resolution of the tune monitor is about 0.0002. Results are consistent with the parameters of the tune monitor. These parameters include sampling rate, decimation factor, and data length.

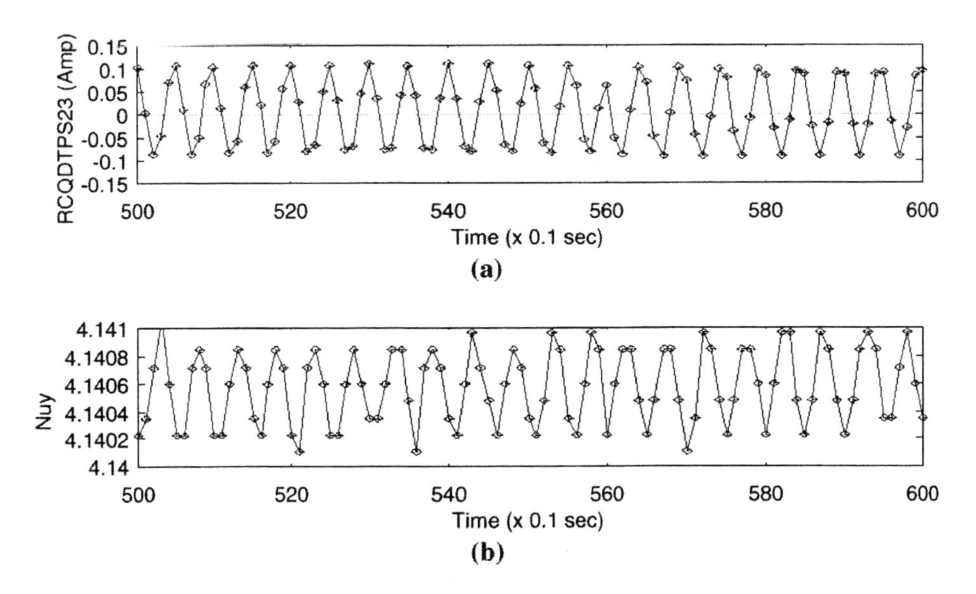

FIGURE 4. Tune variation due to a 200 mA peak-to-peak sinusoidal wave applied to the quadrupole trim (RCR1QDPS16).

The tune change due to the focusing error of the undulator U5 is shown in Figure 5. This operation first closed the gap to 18.5 mm, then opened the gap to 220 mm. The hysteresis loop near the minimum gap is due to an intentional 1 second delay in the tune value update. The gap opening was changed at 1 mm/sec for this test run.

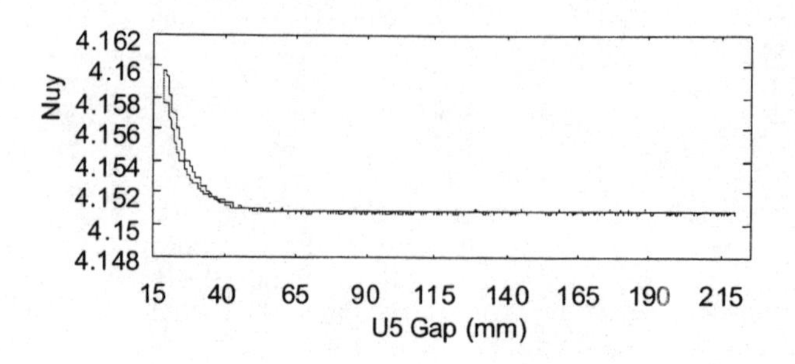

FIGURE 5. Vertical tune shift versus U5 gap.

Figure 6 shows the working point evolution on the tune diagram during energy ramping from 1.3 GeV to 1.5 GeV. In this test, power supply currents were increased linearly. Tune is spread over in a small region during this ramping test. The ramping route is not optimized for this test run.

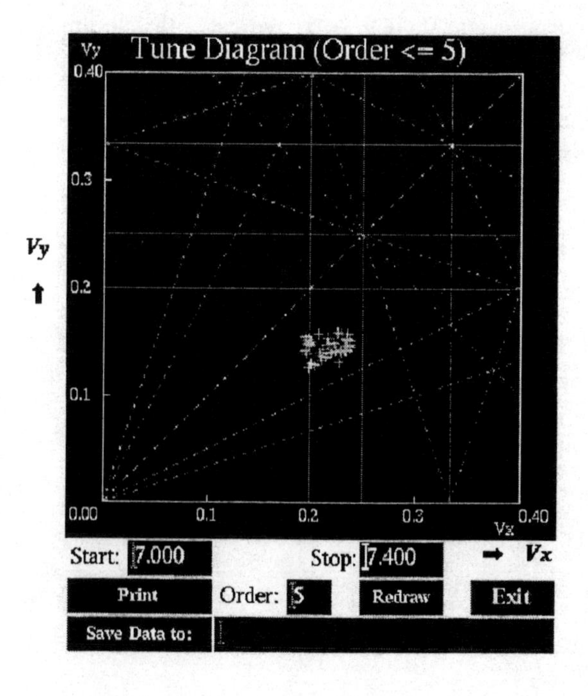

FIGURE 6. Tune evolution on tune diagram during energy ramping test run.

FILLING PATTERN MONITOR

The coupled-bunch oscillation is strongly dependent on the filling pattern of a stored beam. The population distribution in the stored beam can be easily observed by a wide bandwidth oscilloscope. However, a dedicated filling pattern measurement (3) is useful from an injection control, instabilities-study point of view. Controlling the filling pattern is possible if real-time filling pattern information is available. The injection parameters can be adjusted dynamically to achieve the desired filling pattern. The filling pattern acquisition system is shown in lower part of Figure 3. The digitizer uses 500 MHz rf as an external clock to synchronize with the bunch signal. One data point is acquired for each bunch. The next consecutive 200 data points represent the filling pattern. Post-processing the data of consecutive turns can be used to improve amplitude resolution.

COUPLED-BUNCH OSCILLATION OBSERVATION SYSTEM

Coupled-bunched oscillations in longitudinal as well as transverse directions are detected by a bunch-by-bunch, turn-by-turn phase and transverse oscillation detector. Data is digitized by a 500 MS/sec ADC and stored in 8 MB of onboard memory. The 8 MB memory is capable of storing signals for every bucket of TLS for up to 40,000 turns. The schematic diagram is shown in Figure 7. The control console accesses bunch oscillation data (stored on a local disk) via the NFS file system. Time-domain coupled-bunch oscillation data are an elegant tool for diagnostics (4). Such a system also plays an important role in the fast beam-ion instability (FBII) experiments in many facilities (5). This system will help the FBII study at TLS soon.

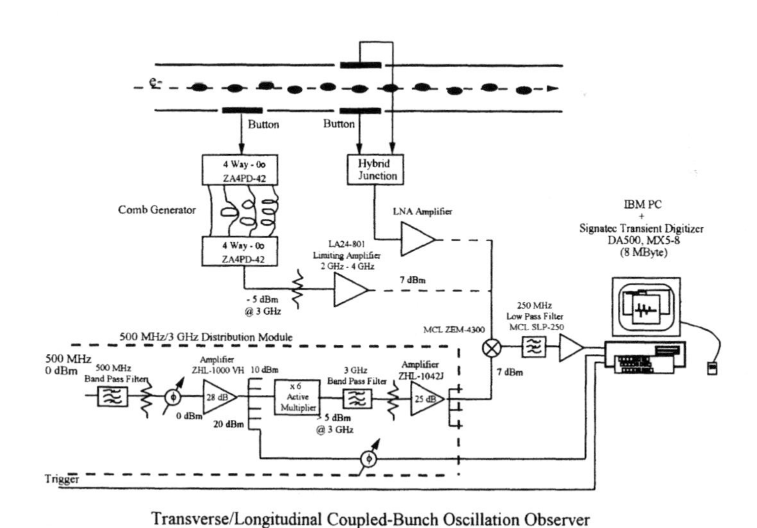

Transverse/Longitudinal Coupled-Bunch Oscillation Observer
(Turn-by-Turn, Bunch-by-Bunch)

FIGURE 7. Functional block diagram of the 500 MS/sec digitizer system.

To test the performance of this coupled-bunch oscillation observation system, the stable stored beam has been intentionally excited by a 50-cycle burst at the synchrotron oscillation frequency. Figure 8 shows the synchrotron oscillation of three consecutive bunches on 15,000 turns. Strong damping was observed for the test lattice.

The tune shift along the bunch train is also interesting in a study of multibunch instabilities. The tune of an individual bunch can be obtained by performing an FFT of acquired data. Since the frequency precision of the FFT is limited by the sampling rate, however, the theoretical resolution of a 32 K point FFT is about 0.00005 in tune.

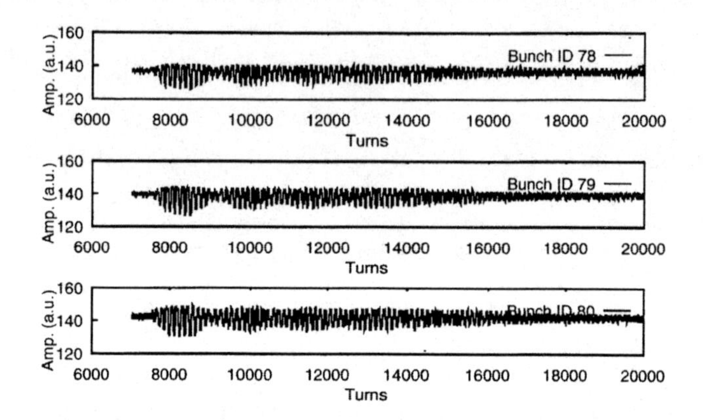

FIGURE 8. Observed synchrotron oscillation of three consecutive bunches. The beam is excited by a 50-cycle burst at the synchrotron oscillation frequency.

SYNCHROTRON RADIATION MONITOR SYSTEM

The synchrotron radiation monitor currently operates with visible light. The diagnostic station is equipped with a CCD camera imaging system. The working wavelength is 500 nm with a 10 nm passband. One-to-one conjugation ratio optics are used. The horizontal acceptance angle is 8 mrad, to minimize the depth of the field and diffraction effects. The beam profile is acquired by a PCI bus-based frame grabber. The captured real-time image (30 frames per second) is displayed on a local SVGA monitor. The image is sent to the console via the control network at about 10 frames per second. On the control console, a PV-WAVE package aided fast prototyping of the user interface. Time structure is observed by a streak camera with a low-jitter clock source.

The present diagnostic station shares the front-end with a VUV beamline. It is inconvenient for the operation of a shutter and valve during injection. To relieve this inconvenience, to eliminate heat load problems at the first mirror, and to improve the performance of the optical system, a dedicated diagnostics beamline is being planned. The new diagnostics station will be equipped with standard profile monitoring and time-structure measurement facilities. The thermal and optics designs will be improved.

SUMMARY

Diagnostics play a crucial role in pushing to ultimate machine performance and in doing advanced machine physics. To accommodate high orbit stability, the orbit acquisition and orbit feedback system will continue to improve its reliability and its performance. Orbit feedback has been tested in TLS and is expected to be useful in increasing orbit stability. A prototype digital receiver-based tune monitor provides a fast tune reading as a complementary tool to a commercial spectrum analyzer. Test results show that a rapidly measured tune is possible with high reliability. Transient digitizers are used to acquire a real-time filling pattern. Turn-by-turn and bunch-by-bunch beam signals acquired by the transient digitizer can be extracted from coupled-bunch oscillations. Time domain analysis will help provide diagnostics for the longitudinal and transverse feedback systems. Updated synchrotron radiation monitors provide a more convenient user interface. Planning of a dedicated diagnostics beamline is under way.

ACKNOWLEDGMENTS

The authors express their thanks to the staff of the operations group; their skillful operation of the machine during beam testing is most helpful. Thanks also go to K. H. Hu, K. T. Pan, J. S. Chen, and C. J. Wang for their technical assistance.

REFERENCES

[1] Kuo, C. H., K. T. Hsu, J. Chen, K. K. Lin, C. S. Chen, R. C. Sah, "Orbit Feedback Development in SRRC," *Proceedings of the International Conference on Accelerator and Large Experimental Physics Control System*, ICALEPCS'97, IHEP, Beijing, 3 to 7 November 1997.

[2] Lin, K. K., K. T. Hsu, J. S. Chen, C. H. Kuo, C. S. Chen, K. H. Hu, J. Chen, K. T. Pan, C. J. Wang, J. R. Chen, C. T. Chen, "A Dynamic Local Bump System for Producing Synchrotron Radiation with an Alternating Elliptical Polarization," *Journal of Synchrotron Radiation*, May 1, 1998.

[3] Chin, M. J., J. A. Hinkson, "PEP-II Bunch-by-Bunch Current Monitor," *Proceedings of the 1997 Particle Accelerator Conference*, 1998.

[4] Claus, R., J. Fox, H. Hindi, I. Linscott, S. Prabhakar, W. Ross, D. Teytelman, "Observation, Control and Model Analysis of Coupled-Bunch Longitudinal Instabilities," *Proceedings of the Fifth European Particle Accelerator Conference*, EPAC 96, Sitges (Barcelona), 10-14 June 1996, pp 346-348.

[5] Edited by Yong Ho Chin, *Proceedings of the International Workshop on Multibunch Instabilities in Future Electron and Position Accelerator (MBI97)*, Tsukuba, KEK, 15–18 July, 1997.

Wire Breakage in SLC Wire Profile Monitors*

C. Field, D. McCormick, P. Raimondi, M. Ross

Stanford Linear Accelerator Center
P.O. Box 4349, Stanford, CA 94309

Abstract. Wire-scanning beam profile monitors are used at the Stanford Linear Collider (SLC) for emittance preservation control and beam optics optimization. Twenty such scanners have proven most useful for this purpose and have performed a total of 1.5 million scans in the 4 to 6 years since their installation. Most of the essential scanners are equipped with 20 to 40 μm tungsten wires. SLC bunch intensities and sizes often exceed 2×10^7 particles/μm^2 ($3 C/m^2$). We believe that this has caused a number of tungsten wire failures that appear at the ends of the wire, near the wire support points, after a few hundred scans are accumulated. Carbon fibers, also widely used at SLAC (1), have been substituted in several scanners and have performed well. In this paper, we present theories for the wire failure mechanism and techniques learned in reducing the failures.

SCANNER OPERATION

Wire-scanning beam profile monitors (or wire scanners) are used throughout the SLC for beam size monitoring and optimization (2). A typical scan takes about 1 second, with the wire actually within the beam envelope for about 20% of the pulses that occur during that time. Two million scans have been done during the roughly 40 months of SLC operating time elapsed since most of the scanners were installed, averaging about 1 scan/minute. Approximately 0.3% of all SLC pulses have intercepted a wire, illustrating the utility of such phase space monitors in the linear collider.

The SLC is a prototype linear collider and our ability to control emittance propagation has developed as both the beam size monitors and the tools to effectively use them have developed. Four groups of wire scanners have proven most useful: 1) at the exit of the damping ring, 2) following the bunch length compressor at the entrance of the linac (RTL – S2), 3) near the end of the linac (S28), and 4) at the entrance to the final focus (FF), following the SLC arcs about 100 m from the IP. The 10 FF wires, installed as part of an FF upgrade (3) in 1994, are different from the others since they

* Supported by the U.S. Department of Energy under contract DE-AC03-76SF00515

CP451, *Beam Instrumentation Workshop*
edited by R. O. Hettel, S. R. Smith, and J. D. Masek

are located in a complex beamline close to the high-energy physics' detector, the SLD. Since the other scanners are located upstream of the collimation sections, the optimum wire size is determined by the expected beam size and limits on the scattered beam power. The FF scanner wire sizes are further constrained so that the scattered radiation produced during the scan does not harm sensitive SLD detector components, resulting in thinner, weaker wires. Several attempts were made to determine the optimum wire size and material that would allow both operation of the SLD and provide a signal strong enough for accurate emittance estimates.

TABLE 1. Parameters and locations for the critical SLC emittance scanners. The scanners are listed in roughly the order that the beam passes them during routine operation. Typical SLC beam intensities for 1994–1998 operation are 3.5×10^{10} particles/bunch with a bunch length of 1± 0.5 mm resulting in peak currents of 2 kAmp. The wire material is tungsten, except as noted. The units shown in *italics* are critical for emittance preservation control.

Location	Number	Wire diameter (μm)	Wires	expected beam size	Number of scans/ device	Purpose
RTL - ε	*2*	*40*	*x, y, u*	*200×50 μm*	*80000*	*Beam size*
Linac begin	*4*	*40(20y)*	*x, y, u*	*300×30 μm*	*140000*	*Emittance*
Linac end	*4*	*40(20y)*	*x, y, u*	*200×40 μm*	*150000*	*Emittance*
FF (e^+/e^-)	*10*	*15–40 & 34 C*	*x, y*	*10 – 200 μm*	*15000*	*Emittance*
Other	36	50–500		3mm – 3mm	430000 (total)	Emittance, energy spread and optical parameters
Total	56			Total no. scans	1930000	(age varies 3 to 7 yr.)

TUNGSTEN WIRES

Wire failures in the SLC scanners were reported in 1992 (2). The failure frequency was greatly reduced at that time with the introduction of a purely ceramic support mechanism. Since 1992, SLC peak beam intensities have increased 30% and beam sizes have dropped 10%. Wire failures, somewhat similar to those originally observed, have again become a concern. The rate is about 2% of that seen with the initial support mechanism but can be as low as 10% in the FF locations with smaller beam sizes and thinner wires. Figure 1 shows the number of scans before failure as a function of the beam charge density.

The thin tungsten wires always fail at the point of tangency to the cylindrical support stud. Figure 2 shows the mounting scheme. We have concluded that the failure must arise from a large number of high voltage discharges between the wire and the titanium-nitride (Ti-N) coated alumina support stud. The wires have always been eroded at both ends. Close examination of the wires prior to failure show weakening begins after the first few scans. Figure 3 shows a typical end of a failed wire. Each discharge displaces a small piece of tungsten. A small stripe is clearly visible on the stud under the wire tangent point. Since the capacitance of the wire must be quite small, we think that the high-voltage pulse must be quite high.

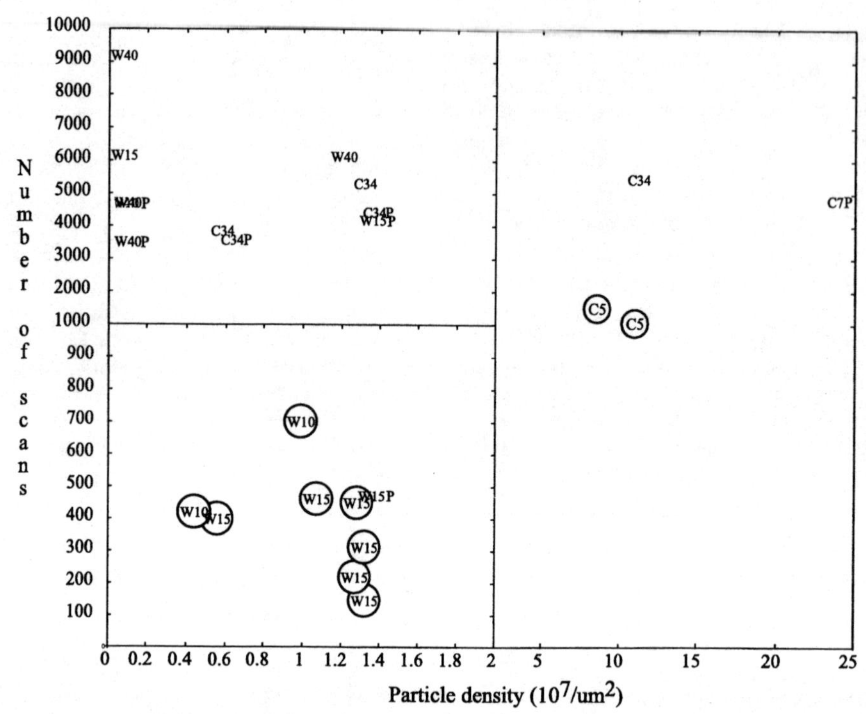

FIGURE 1. Number of scans performed with FF scanners vs. beam density. The symbol indicates the wire type and size in µm. Circles indicate failed wires. The plot shows roughly three regions: 1) failure prone wire sizes with low (1×10^7) particle density, 2) wires that survive indefinitely at lower charge density and 3) carbon wires that survive the highest charge density. The 'P' indicates scanners used only for positrons. These wires evidently fail much less often.

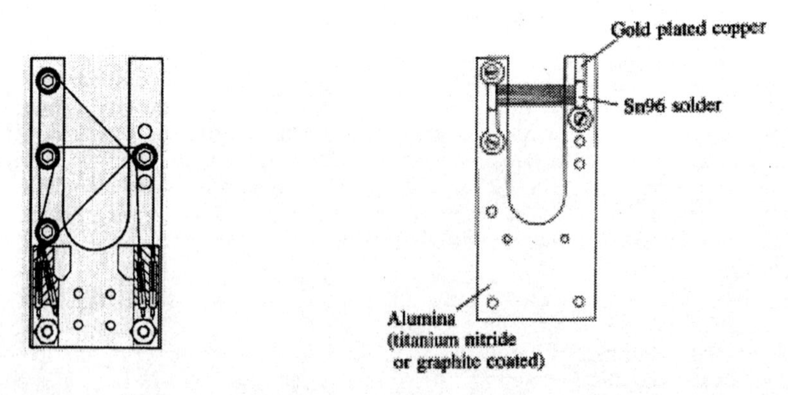

FIGURE 2. Wire mounting scheme for tungsten and carbon wires. The tungsten wire mounting scheme used on most scanners is shown at the left. In this figure, the beam passes into the page, roughly centered between the tines of the alumina fork. The wires are accurately positioned by the four ceramic alignment pins and tensioned by springs shown at the bottom of the figure. Ti-N coating provides a drain path for charges that may collect on the fork and alignment pins. The right side of the figure shows the carbon wire support scheme used when the tungsten wires were replaced. No tensioning springs or alignment pins are used. The separation between the fork tines is 1 inch.

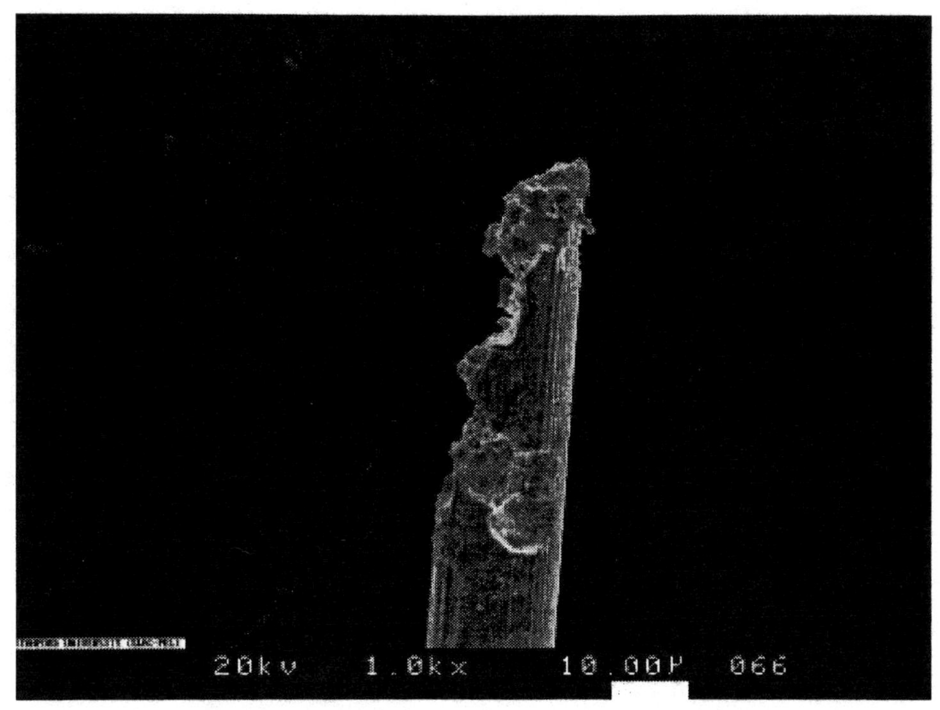

FIGURE 3. Failed 15 μm diameter tungsten wire showing the rough surface resulting from many discharges.

A reasonable estimate for charge depletion in a 10 μm tungsten wire, causcd by secondary emission, is 7.5×10^{8} charges (2% of the beam population). Assuming this occurs in a 10 μm length of the wire, its capacitance of about 10^{-15} farads means this has a potential energy of 10^{-6} Joules.

This depletion is made up by recharging through the wire. Since we are looking for a mechanism that causes damage at the ceramic roller pins which position the wires, it seems necessary that a fast recharge via the tensioning springs is inhibited. If the springs were not shorted, this would occur naturally because of their large inductance. Somehow we have to hypothesize that the back of the center pin does not make good enough contact with the tension block, or that some mutual inductance is enough to do the job.

For damage of the type observed to occur, it seems necessary also to assume that the wire does not make good contact with the alignment pin, at least in vacuum. Good contact in this case means electrical resistance less than that of the tungsten wire, a few ohms for a 10 μm example. In this case the surface of the alignment pin forms a relatively large capacitance which will drain the charge on the wire through the contact resistance.

The heat that is available from the flow of charge, through a constriction of a few tens of ohms, is enough to raise the temperature of a 10 μm length of the wire by 3000°C, or to melt perhaps a 7 μm length. Evaporation requires much more energy. There would be enough energy to remove about 1.5 μm length. However, it would indeed be hard to generate and transfer the heat efficiently, so the sites of melting/ evaporation, if any, would be expected to be much smaller.

The Ti-N coating provides a drain path to ground of about 100 kΩ which should prevent the accumulation of high charge between pulses.

As seen in Figure 1, the 10–15 μm tungsten wire is most prone to failure. In most instances, we have replaced it with 34 μm carbon wire. At this time, we have not observed any failures of the carbon wire. Some signal loss and resolution loss results from the lower Z, higher diameter wire.

CARBON FIBERS

Carbon fibers have been used at SLC since its inception. They are able to handle greater intensity than tungsten wire by roughly an order of magnitude. They are themselves limited, however, at bunch densities of about 3×10^9 for $\sigma_x\times\sigma_y = 1\ \mu m^2$. Their disadvantage lies in the weaker wire-scan signal, 1.8% of that of tungsten. Diameters of 34 μm, 7 μm and 4 μm have been used. Obviously the smallest diameter is used for the smallest beam spots, and encounters the most hostile conditions.

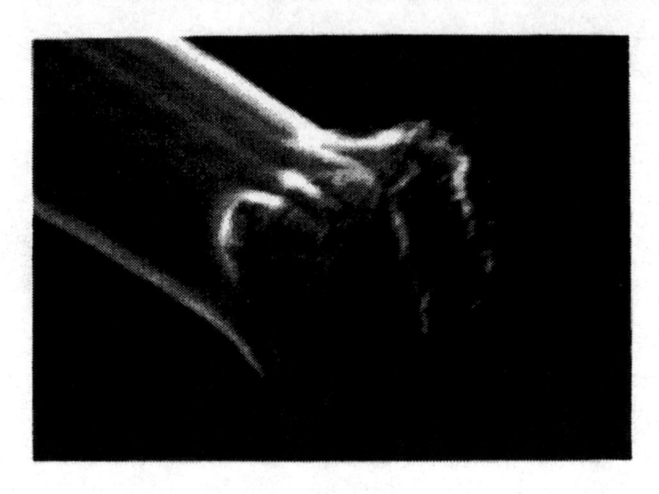

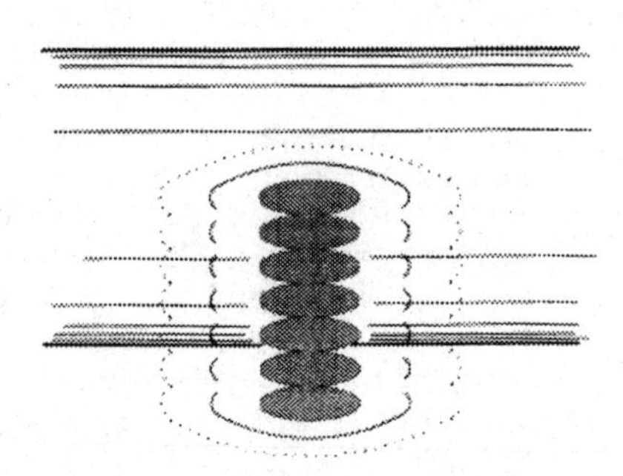

FIGURE 4. Failed 4 μm carbon wire with inset showing the progression of successive beam pulses scanning across the wire. This wire was broken at the point of intersection with a beam of 3×10^9 particles/μm^2.

Figure 4 shows a failed 4 μm carbon wire. All failures that can be associated with the beam have been directly at the beam impact point, and occurred after no more than a few pulses. From the remains that have been recovered, severe distortion of the carbon is observed. This is evidence for strong forces caused, perhaps, by shock thermal expansion somehow interacting with a softening of the material at temperature >2500 C.

The forks that hold the carbon fibers have, in most cases been made of MACOR ceramic. In a few cases aluminum was used. The fibers were positioned by laying them over a ledge or in a groove in the fork material and then holding them in tension by encapsulating the ends in SN96 flux-free solder. This, in turn, was connected to electrical ground. The thinnest fibers form a resistance of 10^4 ohm to ground for the (positive) secondary emission depletion charge at the beam collision point. No failures associated with beam scanning have been documented at the solder joint, or at the position where the fiber runs over the fork material. Tungsten wires mounted with this technique were observed to fail in a fashion similar to the pin-mounted wires.

CONCLUSION

Wire scanners have proven the most effective beam size monitor for linear collider operation. Recent SLC parameters and optimization challenges have forced us to develop scanners with thin wires or with low *Z* wires. This, in turn, has caused wire failure problems. While the hypotheses listed above may not prove to be the primary cause of the failures, they have helped us to develop useful solutions.

ACKNOWLEDGMENTS

We would like to acknowledge the efforts of the wire assemblers, G. Sausa, J. King and Y- Y. Sung; the wire electron microscopc photographers, E. Hoyt, L. Shere and R. Kirby; and the vacuum installation group under K. Ratcliffe.

REFERENCES

[1] Field, Clive, "The Wire Scanner System of the Final Focus Test Beam," *Nucl. Instrum. Meth. A,* **360**, pp. 467–475, (1995).

[2] Ross, M. C., E. Bong, L. Hendrickson, D. McCormick, M. Zolotorev, "Experience With Wire Scanners at SLC 1992," presented at the Accelerator Instrumentation Fourth Annual Workshop, Berkeley, CA, October, 1992 (in proceedings pp. 264–270).

[3] Zimmermann, F., T. Barklow, S. Ecklund, P. Emma, D. McCormick, N. Phinney, P. Raimondi, M. Ross, T. Slaton, F. Tian, J. Turner, M. Woodley (SLAC), M. Placidi, N. Toge, N. Walker, "Performance of the 1994/1995 SLC Final Focus System," *Proceedings of the IEEE Particle Accelerator Conference,* 1995, pp. 656–658.

An Improved Resistive Wall Monitor

Brian Fellenz and Jim Crisp

Fermi National Accelerator Laboratory
P.O. Box 500, Batavia, IL 60510

Abstract. Resistive wall monitors were designed and built for the Fermilab Main Injector project. These devices measure longitudinal beam current from 3 KHz to 4 GHz with a 1 ohm gap impedance. The new design provides a larger aperture and a calibration port to improve the accuracy of single-bunch intensity measurements. Microwave absorber material is used to reduce interference from spurious electromagnetic waves traveling inside the beam pipe. Several types of ferrite materials were evaluated for the absorber. Inexpensive ferrite rods were selected and assembled in an array forming the desired geometry without machining.

INTRODUCTION

Resistive wall monitors have been built and installed in each of the accelerators at Fermilab. They are used to measure longitudinal bunch shapes, calculate emittance, and diagnose instabilities. In the Tevatron, the signal is used by the Sampled Bunch Display as well as the Fast Bunch Integrator systems to measure and track the intensity of individual bunches. Recently, they have been used to monitor luminosity for colliding beams operation. Along with monitoring bunch manipulations such as cogging and coalescing, all of these functions will be useful in the new Main Injector.

An explanation of how a resistive wall monitor works along with a description of new features incorporated into the Main Injector design are described below.

HOW A RESISTIVE WALL MONITOR WORKS

A resistive wall monitor measures the image charge that flows along the vacuum chamber following the beam. The image charge has equal magnitude but opposite sign. Depending on the beam velocity, the image charge will lag behind and be spread out along its path. The ultimate bandwidth of such a detector is limited by this spreading of the electric field lines between the beam and the inside walls of the beam pipe. The spreading angle is approximately $1/\gamma$ for relativistic beams (γ is the ratio of total energy to rest energy). The estimated bandwidth limit from spreading is 47 GHz at injection to the Main Injector for a 3 cm radius pipe and 8 GeV proton energy. In practice, the detector response is difficult to maintain above the microwave cutoff frequency of the beam pipe,

CP451, *Beam Instrumentation Workshop*
edited by R. O. Hettel, S. R. Smith, and J. D. Masek

measured to be 1.5 GHz for the elliptical beam pipe used in the Main Injector. Above cutoff, the characteristic impedance of the beam pipe and the impedance of nearby structures such as bellows or changes in geometry can effect accuracy.

FIGURE 1. Resistive wall monitor showing circuit board and ferrite.

In order to measure the image current, the beam pipe is cut and a resistive gap is inserted (Figure 1). Various ferrite cores are used to force the image current through the resistive gap rather than allowing it to flow through other conducting paths. In addition to image current, other currents are often found flowing along the beam pipe. The gap and cores are placed inside a metal can to shunt these "noise" currents around rather than through the resistive gap. The inductance of the cores and the resistance of the gap forms a high pass filter with a corner frequency of $R/2\pi L$, typically a few kilohertz. Above this frequency, cores act to minimize the net current through their center by inducing a current through the resistive gap that just cancels the beam current.

The gap impedance is chosen to be well below the impedance of the cores inside the shielding can. Several types of ferrite and microwave absorbers are used to maximize the impedance and minimize resonances within the desired bandwidth. The Main Injector shielding can has an impedance greater than 30 ohms with the ferrite cores. In parallel with the 1 ohm gap impedance, 30 ohms can cause frequency dependent errors of $\pm 1.5\%$ or 0.15 db.

If the charge density around the circumference of the gap is not uniform, the voltage across the gap will vary around the circumference. The gap will act as an azimuthal transmission line transporting charge until the voltage equalizes. The time domain response of the detector would be distorted during this time. Position detectors have been made by exploiting this effect. The elliptical shape of the Main Injector pipe aggravates this problem (Figure 2). To overcome this problem, a round geometry is used for the gap and the signals from several monitor points equally spaced around the circumference are combined to form a single output.

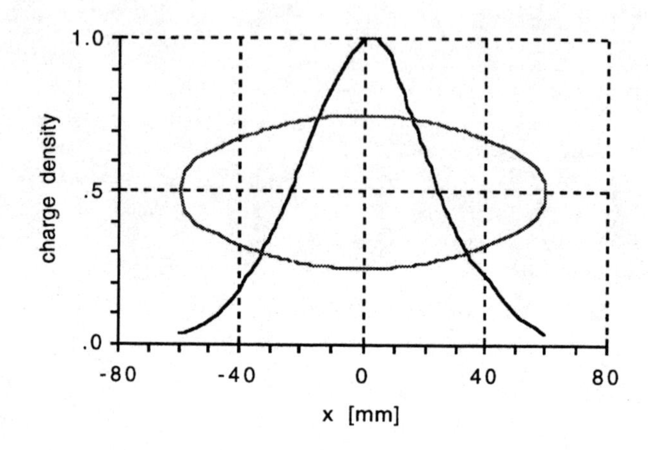

FIGURE 2. Cross section of Main Injector beam tube and charge density versus horizontal position induced by a line charge at the center.

DESIGN IMPROVEMENTS

Circuit Board

The resistive gap is formed with 112 equally spaced 122-ohm 1% ceramic resistors mounted on a flexible circuit board wrapped around the outside of a ceramic gap. The board material is Rogers RT/Duroid 5880 and is .020 inches thick with copper clad on both sides. The outside rings on the ceramic gap physically support the circuit board, and are plated with a conducting layer to make electrical contact with the pipe. The voltage across the gap is measured at four equally spaced positions and combined with a shunt resistive combiner using microstrip transmission lines on the board. Keeping the transmission lines as short as possible reduces the effect of small termination errors and insures a flat frequency response.

A calibration port was incorporated on the Main Injector resistive wall monitor. Similar to the combiner, microstrip transmission lines and series resistive splitters inject a calibration signal at four points evenly spaced between the monitor points. Resistive “L” pads are used to terminate the microstrip lines into the 1 ohm gap impedance. These resistors reduce the transmission from the calibration port to the output port by 40 db. The capacitance of the ceramic and the gap impedance limit the calibration bandwidth. This was extended slightly by shunting two of the series resistors with a small compensating capacitor.

On-line calibration is important to maintain accurate bunch intensity measurements. Small errors caused by impedance mismatch or changes in measurement electronics can cause significant errors. Desired accuracy of bunch intensity measurement is better than 1% or 0.09 db. This fidelity is required over large bandwidths to avoid errors caused by changes in bunch length. One hundred feet of 7/8 inch diameter foam dielectric solid jacketed cable is used to transport the signals from the beam enclosure. The cable attenuation at 1 GHz is 1.31 db/100ft and must be accounted for when calibrating the signal. The VSWR of this cable and connectors is specified to be less than 1.2, which can generate frequency dependent errors as large as 20%. The calibration port allows automatic calibration between beam pulses to correct these effects as well as any cable or component changes.

Ceramic Gap

A ceramic vacuum break was used at the gap to isolate the ferrite and other materials from the vacuum. The geometry was carefully designed to avoid resonances. The image current produces a voltage across the inside surface of the gap that propagates radially out to the resistors on the surface. The radial thickness of the ceramic is 1/4 wavelength at 5 GHz. The permitivity of ceramic is about 10, making the characteristic impedance 120 ohms per square. For a 1-ohm impedance, the optimum length of the ceramic gap is 0.14 inches (circumference*gap impedance/ohms per square). Three ceramic rings are used to form the gap to make them easier to manufacture. The center ring isolates the vacuum and the outside rings help balance the forces exerted by differential expansion between ceramic and metal as they are heated and cooled from brazing temperature. A round geometry was chosen to obtain more uniform image current around the circumference and is significantly cheaper to build.

Noise Reduction

Other currents not associated with the image current may flow along the beam pipe and interfere with its measurement. The current divider formed between the shielding can and the 1-ohm resistive gap reduce the amount allowed to flow through the gap by 100 db, the ratio of their impedances. This requires care in making the electrical connection between the shielding can and the beam pipe to insure a very low impedance. The cores inside the shielding can further reduce the noise current allowed to flow through the gap for frequencies above 0.03 hertz. This corner frequency is estimated from the resistance of the shielding can (about 10 μohms) and the inductance of the cores.

Microwave Absorber

When the beam passes a discontinuity, electromagnetic energy is launched into the beam pipe. This energy can travel in either direction but typically travels slower than the beam. The resistive wall monitor cannot differentiate between currents induced by beam and those induced by electromagnetic energy traveling along the beam pipe. In the Fermilab Main Ring, this spurious signal was as large as 10% of the beam signal. To reduce these signals, a microwave absorber is placed inside the beam pipe at both ends of the resistive gap. The transition from the elliptical beam pipe used in the Main Injector to the round gap is done with the microwave absorbers.

In previous resistive wall monitors, the absorber material was selected for its vacuum properties as well as its microwave absorbing characteristics. This application required a size that could not readily be made. Placing inexpensive absorber made from ferrite-loaded epoxy outside a ceramic pipe was tried, but, the required length became excessive because the material had to be placed out in a low-field region.

Several types of ferrite material that could be placed in the vacuum were obtained for testing. The material eventually selected was purchased in .375 inch diameter rods 7.5 inches long. An array of 74 rods held in position at each end with rexolite disks worked well (Figure 3). The inside clearance conformed to the elliptical Main Injector beam pipe. The assembly attenuated microwave signals traveling through a test set-up by 30 db. The amount of power deposited by the beam is estimated to be only 1.2 Watts assuming 10^6 protons in 2 nanosecond long bunches and a continuous 53 MHz bunch rate.

FIGURE 3. Microwave absorber made with an array of ferrite rods.

TESTING AND RESULTS

The most accurate method of measuring the fidelity of the resistive wall monitor was done by forming a 50-ohm transmission line through its center with the appropriate diameter conductor. The ends were gradually tapered to a standard type N connector.

Transmission through this line was flat to ±0.5 db below 4 GHz. Coupling from the transmission line to the output port is proportional to the ratio of the 1 ohm gap impedance to the 50 ohm characteristic impedance of the line, or –34 db. The signal at the output port was flat to ±0.5 db when normalized by the signal passing through it (Figure 4).

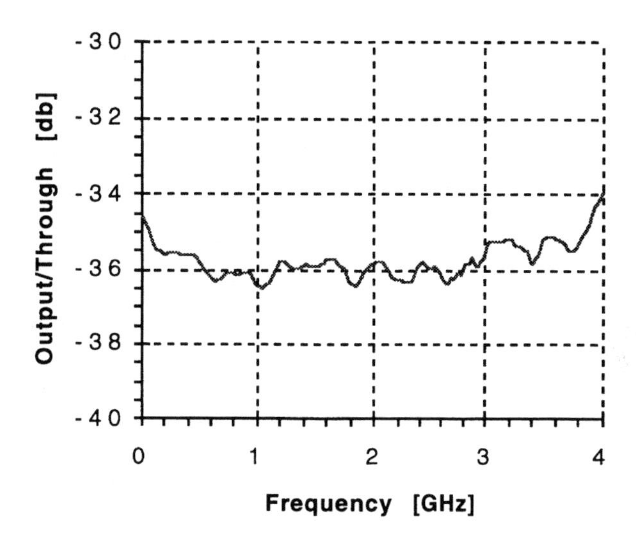

FIGURE 4. Output normalized to the signal passing through the test set-up. The nominal level is given by the ratio of the 1 ohm gap impedance to the 50-ohm test line, or –34 db.

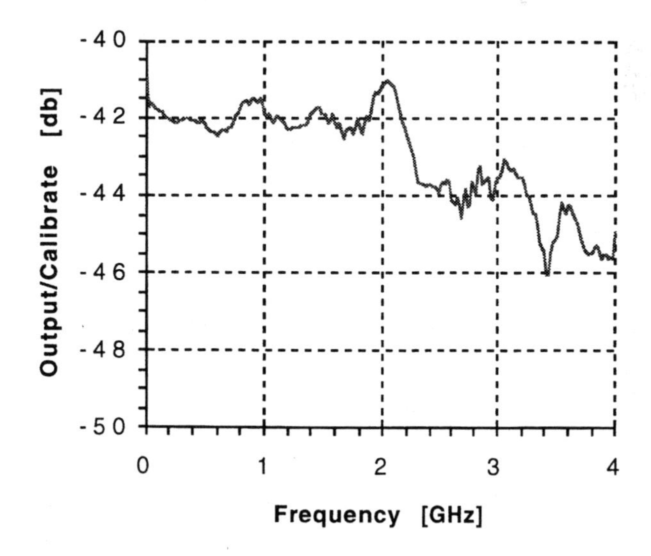

FIGURE 5. Coupling between the calibration port and the output port. Nominal coupling is given by the ratio of the 1 ohm gap impedance to the 100-ohm series resistor, or –40 db.

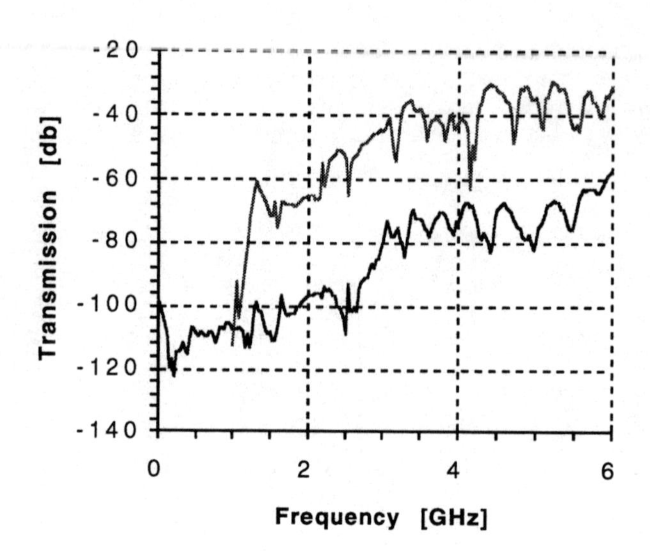

FIGURE 6. Transmission through an absorber with and without ferrite material installed. The cutoff frequency of the test set-up is 1.2 GHz.

The results are somewhat misleading in that the presence of the large center conductor significantly increases the cutoff frequency for microwave modes. The TE_{01} mode has the lowest cutoff frequency and was measured at 1.5 GHz in Main Injector beam pipe. The amount of coupling between the gap and microwave modes is not easily measured. TM modes can induce longitudinal currents that would flow through the resistive gap and thus be strongly coupled. However, the higher order TM modes have an odd symmetry around the circumference and their effect is reduced by combining the four equally spaced pick-off points.

The microwave absorber was measured by comparing the coupling between small loops at each end through the pipe with and without the ferrite absorbing material (Figure 6). Tests indicate that $600 of ferrite provides 30 db of attenuation, better than $7,500 of the previously used microwave absorber material.

CONCLUSION

It is virtually impossible to build a device with sufficient fidelity, accuracy, and stability to measure bunch intensity to 1%. Furthermore, frequency dependent errors would require the calibration to depend on bunch length. The calibration port will allow simple corrections to provide greater accuracy and reliability for measuring bunch intensity as well as easy diagnosis of system errors. The most cost-effective solution is to build a good device and correct measurements with calibration. This approach is commonly used to obtain optimum performance from test equipment.

Stacking up inexpensive ferrite rods allows flexibility in the shape of the absorber without expensive machining. Tests demonstrate $600 of ferrite worked better than $7,500 of microwave absorber used in previous detectors.

REFERENCES

[1] Webber, R., "Longitudinal Emittance: An Introduction to the Concept and Survey of Measurement Techniques Including Design of a Wall Current Monitor," *AIP Conference Proceedings on Accelerator Instrumentation*, No. 212, pp. 85–126, (1989).

[2] Chao, A. W., "Coherent Instabilities of a Relativistic Bunched beam," *AIP Conference Proceedings on Physics of High Energy Particle Accelerators,* No. 105, pp. 353–523, (1983).

Broadband FFT Method for Betatron Tune Measurements in the Acceleration Ramp at COSY-Jülich

J. Dietrich and I. Mohos

Forschungszentrum Jülich GmbH
Institut für Kernphysik
Postfach 1913, D-52425 Jülich, Germany

Abstract. A method for measurement of betatron tune without the need for feedback-driven excitation has been developed at COSY. A bandlimited broadband noise source was used for beam excitation. The transverse beam position oscillation was then bunch-synchronously sampled and digitized with a high resolution ADC. The Fourier transform of the acquired data represents immediately the betatron tune. A functional description of the measurement system and the evaluation of the measured data are presented.

INTRODUCTION

The cooler synchrotron and storage ring COSY, with a circumference of 184 m, delivers medium-energy protons. The corresponding revolution frequencies in the acceleration ramp are between 0.45 MHz (flat bottom) and 1.6 MHz (flat top). For beam diagnostic measurements, a magnetic impulse kicker or a broadband stripline exciter can be used. The 50W-stripline electrodes are mounted azimuthally in 45^0-positions; therefore, excitations in the diagonal direction are also possible and are useful, for example, to observe horizontal and vertical tune changes at the same time in one spectrum during beam optimization procedures. The mode of excitation and the strength can be automatically set. Beam position monitors (BPM) with low noise broadband amplifiers deliver signals proportional to the beam response on the excitation. The bunch-synchronous pulse, necessary for the sampling, is derived from the BPM-sum signal of the same BPM.

The betatron tune, Q, is the quotient of the betatron oscillation and the particle revolution frequencies. The betatron frequency ($f_\beta = Q * f_0$) is usually higher than the revolution frequency, but due to the bunch structure (undersampling), only the fractional part of the betatron tune (q) is measured:

$$f_\beta^n = n * f_0 \pm Q * f_0 = (n' \pm q) * f_0 \quad (1)$$

where n and n' are integers and q is the fractional part of the tune.

CP451, *Beam Instrumentation Workshop*
edited by R. O. Hettel, S. R. Smith, and J. D. Masek

MEASUREMENT CONFIGURATION

Via the stripline unit, resonant excitations of coherent betatron oscillations in the horizontal and vertical directions can be performed by means of a broadband white noise source with fixed cutoff frequencies. The betatron oscillation appears as an amplitude modulation of the beam position, causing bunched beam double sidebands around each harmonic of the revolution frequency and also around DC in the frequency spectrum. The frequency range of the noise source always covers at least one betatron sideband at the fundamental harmonic in the whole ramp without frequency feedback. A low-noise preamplifier and gain-controlled stage amplify the sum and difference signals of a beam position monitor. The gain-controlled amplifier for the difference signal ensures an optimal use of the 14-bit ADC. The sum signal is the reference for a clock generator. Figure 1 shows the block diagram of the FFT tune meter.

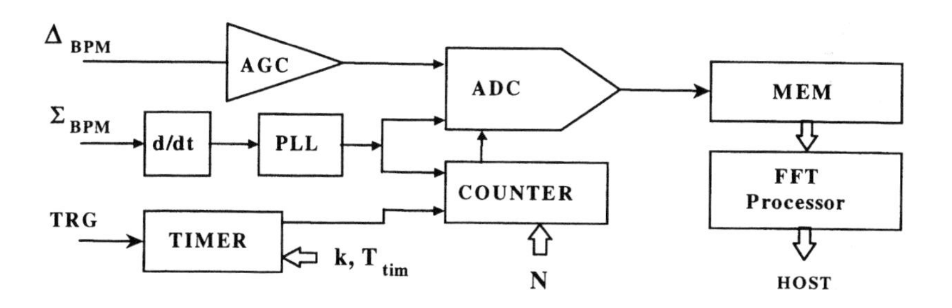

FIGURE 1. Block diagram of the FFT tune meter.

With proper signal processing, the clock generator tracks the bunch peaks and also the synchrotron oscillation. The bunch-synchronous clock pulse produces positive edges at the bunch peaks. For investigation of the synchrotron oscillations, a signal proportional to the synchrotron oscillation can be derived from the tracking circuitry of the clock generator. The peak value of the difference signal, proportional to the beam position, is sampled by means of a fast S&H circuit under control of the bunch-synchronous clock. A fast, high-resolution ADC digitizes the output in each clock period. This configuration combines the functions of a synchronous demodulator and a frequency normalizer. The magnitudes of the samples carry the position changes and the betatron oscillation; the sampling frequency is always equal to the revolution frequency. Due to the bunch-synchronous sampling, the frequency components of the synchrotron oscillation are suppressed. The sampled data, therefore, contain mainly the frequency component of the first sideband with

$$f_{\beta,n'=0} = q^*f_0 < 0.5^*f_0. \quad (2)$$

DATA PROCESSING

Performing the discrete Fourier transformation of N sequentially acquired samples gives:

$$S\left(\frac{m}{NT}\right)=\sum_{n=0}^{N-1} s(nT) * e^{-j(2\pi nm)/N} , \tag{3}$$

where T is the time interval of the samples, $s(nT)$the n-th sample of an array consisting of N samples, and $S\left(\frac{m}{NT}\right)$the m-th Fourier component at $f_m = \frac{m}{NT}$.

Due to the bunch-synchronous sampling, the frequencies of the resulting data are normalized to the revolution frequency. Oscillations appear as a peak in the normalized frequency domain. The value N must be properly chosen because it determines the frequency resolution of the FFT-spectra (equal to $1/NT$ with $1/T = f_{sample} = f_0$). As shown above, the bigger the samples in the array used for evaluation, the higher the frequency resolution and consequently the accuracy of the q-measurement. The lowest normalized frequency is zero (DC component); the highest according to Nyquist is $f_{sample}/2$, the corresponding tune range is between 0 and 0.5.

Because the revolution frequency is used as the sampling frequency, it follows that:

$$q * f_0 = f_q \Leftrightarrow f_{m'} = \frac{m'}{N} * f_0 \text{ therefore } q = \frac{m'}{N} \quad (0 < q < 0.5 \text{ or } 0 < 1\text{-}q < 0.5). \tag{4}$$

The acquired data blocks with N data words each are transformed by the FFT resulting in frequency spectra with $N/2$ datapoints.

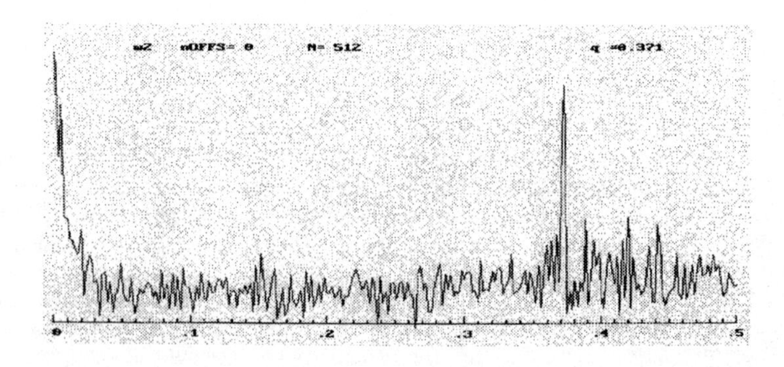

FIGURE 2. Betatron line in the normalized frequency domain.

Figure 2 shows a sideband line in the spectrum; the normalized frequency corresponds to the fractional betatron tune value. The spectra subsequently acquired with equidistant time intervals and displayed with frequency axis vertically, show the tune versus time in the acceleration ramp (Fig. 3).

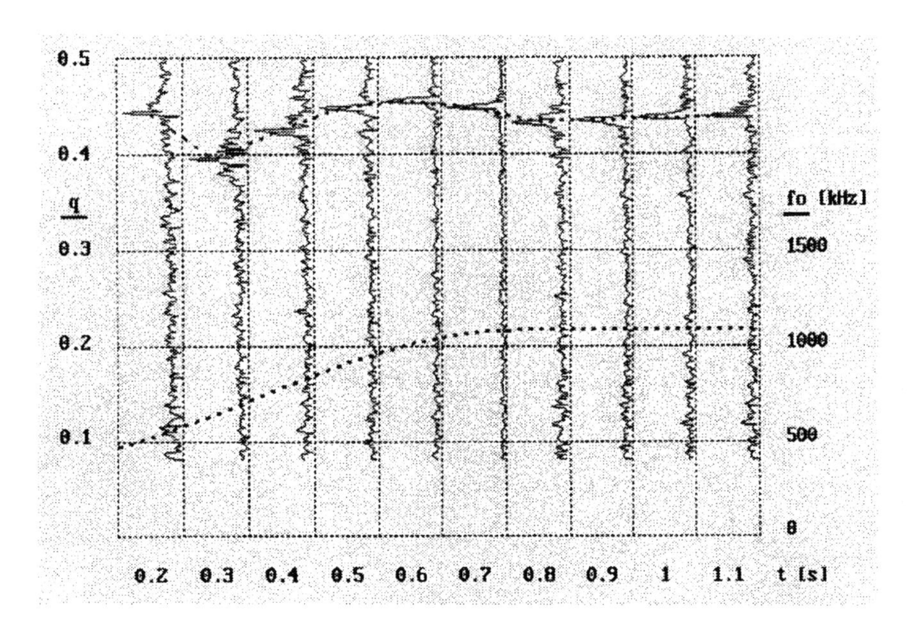

FIGURE 3. Display of a tune measurement in the acceleration ramp consisting of ten FFT- spectra. The sideband lines are clearly seen. The frequency ramp is shown in the lower part.

In the normalized frequency domain, the fractional tune value is directly shown. The frequency f_m of the m-th datapoint ($m = 1,..,N/2$) is $f_m = m/NT$, with $1/T = f_{sample} = f_0$. If f_m' is the frequency of the sideband, it follows $f_m' = m'/N*f_0$, and, since $f_m' = q*f_0$, then $q = m'/N$.

TIMING

The data are taken in blocks of N data words each and are stored sequentially in memory. To start the measurement, the COSY timing system triggers internal timing logic which, in turn, generates k timing pulses with constant time delay. The number k of timing pulses and their delay must be properly chosen in order to obtain a tune measurement time overlapping the total acceleration ramp time as desired. In the data acquisition cycle, $k*N$ samples corresponding to k tune value measurements are acquired. The acquisition time for a tune resolution of 1/1000 is less then 2 ms.

CONCLUSIONS

Several advantages of this method are remarkable. Spurious peaks with constant frequency can easily be recognized and separated because their time dependence shows an inverse normalized frequency behavior to the frequency ramp. The single measurements in the ramp are independent from each other. Therefore, an unsuccessful measurement does not disturb the results from the good data. The acquisition time is short so nonlinear changes of the tune have less effect on the accuracy. Because of the bunch-synchronous sampling, the sampling frequency corresponds to the fundamental frequency f_0 of the pulse spectrum and the FFT-spectra contain only the frequency range up to $0.5*f_0$. Consequently, all higher order sidebands are converted to the same (the first) sideband and all harmonics of the revolution frequency to DC; therefore, no disturbing aliasing components appear in the spectra. The sampling clock tracks the synchrotron phase oscillations of the beam; therefore, longitudinal and transverse spectra are separated .

REFERENCES

[1] Dietrich, J. et al., "Transverse Measurements with Kicker Excitation at COSY-Jülich," *Proc. of the 5th European Particle Accelerator Conference*, Barcelona, Spain, 1675–1677 (1996).

[2] Bojowald, J. et al., "Stripline Unit," Jül-2590, ISSN 0366-0885, 232 (1992) and Bojowald, J. et al., "Longitudinal and Transverse Beam Excitations and Tune Measurements," Jül-2879, ISSN 0944-2952, 174 (1993).

[3] Biri, J. et al., "Beam Position Monitor Electronics at the Cooler Synchrotron COSY-Jülich," *IEEE Trans. Nucl. Sci.* 41, 221–224 (1994).

The Design and Initial Testing of a Beam Phase and Energy Measurement for LEDA*

J. Power and M. Stettler

Los Alamos National Laboratory
Los Alamos, NM 87545

Abstract. A diagnostic system being designed to measure the beam phase and beam energy of the Low Energy Demonstration Accelerator (LEDA) is described and the characterization of the prototype presented. The accelerator, being built at LANL, is a 350 MHz proton linac with a 100 mA beam. In the first beam experiments, the 6.7 MeV RFQ will be characterized. Signals received from an rf cavity probe in the RFQ and capacitive pick-ups along the high-energy beam transport line will be compared in phase in order to calculate the beam phase and energy. The 350 MHz signals from four pick-ups will be converted to 2 MHz in a VXI-based down converter module. A second VXI phase processor module makes two, differential-phase measurements based on its four 2 MHz inputs. The heart of this system is the phase processor module. The phase processor consists of an analog front end (AFE), digital front end (DFE), digital signal processing (DSP) modules and the VXI bus interface. The AFE has an AGC circuit with a >60 dB dynamic range with a few degrees of phase shift. Following the AFE is the DFE which is uses an in-phase and quadrature-phase (I and Q) technique to make the phase measurement. The DSP is used to correct the real-time data for phase variations as a function of dynamic range and system offsets. The prototype phase module gives an absolute accuracy of ±0.5 degrees with a resolution of <0.1 degrees and a bandwidth of 200 kHz.

INTRODUCTION

A diagnostic system is being designed to measure the beam phase and beam energy for the LEDA accelerator. We have chosen to base the design of this new system on work previously done at Los Alamos on the Ground Test Accelerator (GTA) program (1). A simplified diagram of this approach is shown in Figure 1. The LEDA beam is bunched at 350 MHz by the RFQ. The energy of the beam is calculated by measuring the time-of-flight of the beam between a pair of pickups with a known separation. To calculate the time-of-flight we measure the difference in phase of 350 MHz beam signals from the two pickups.

* Work supported by the U.S. Department of Energy

CP451, *Beam Instrumentation Workshop*
edited by R. O. Hettel, S. R. Smith, and J. D. Masek

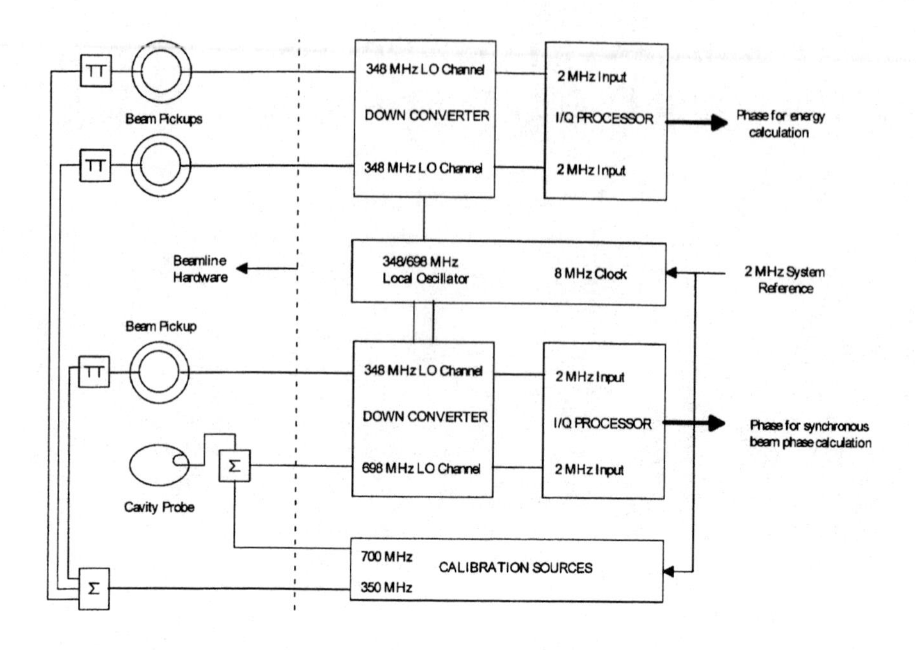

FIGURE 1. Block diagram of the beam phase and energy measurement system to be used on the LEDA accelerator.

To compare the performance of the RFQ with the theory, we also need to measure the phase of the beam, relative to the cavity field, as it exits the RFQ. This measurement is made by measuring the phase difference between signals from a cavity field probe and a nearby beamline pickup. We translate the phase measured at the pickup location back to the RFQ exit location, based on the measured beam energy and the separation distance.

The system is comprised of four major components. These are the capacitive beam pickups, rf down-converter module, I/Q phase processing module, and the calibration hardware. Though each of these components will be described, this paper will concentrate on the I/Q (in-phase/quadrature-phase) processor module.

Capacitive Beam Probes

The LEDA beamline of interest is a 1.87 in.-i.d. pipe assembled with 4.5 in.-o.d. Conflat® flanges. Each capacitive probe is assembled within a single Conflat® flange to conserve space. The capacitive probe consists of a ring, which is supported by two SMA feedthrough connectors. This ring is 0.2 inches long and slightly larger in i.d. than the i.d. of the beamline pipe. Figure 2 shows a probe assembled within a 4.5 in. diameter flange. The dimensions of the ring and the cavity into which it is mounted were selected such that the ring forms a 100 Ω transmission line. When the probe is connected between two 50 Ω cables it presents a very low impedance perturbation (two short lengths of near-100 Ω line in parallel). This design was chosen to allow a calibration signal to be injected in one connector, passing through to the normal output connector on the opposite side.

FIGURE 2. A capacitive beam probe within a 4.5 in. Conflat® vacuum flange. The impedance of the ring is 100 Ω to match the 50 Ω termination.

One potential problem with this type of probe is that the phase of the beam signal is dependent on the position of the beam to some degree. All of our probes are calibrated in a transmission line-test fixture using a centered center conductor. As the beam moves away from the center of the probe, an offset error is introduced. In our energy-measurement case, the beam is expected to be in essentially the same location in the two probes used for the measurement and hence the error is differentially subtracted. In the case of the output phase measurement, only one probe is used, but the beam is expected to be near the center of the probe where the error is minimal. For small variations in beam position we expect to see an error of approximately 0.5 deg/mm offset, at 350 MHz.

Down-converter Module

The 350 MHz and 700 MHz signals from the capacitive probes and the cavity sample probe are being down-converted to 2 MHz where more precise phase measurements can be made. We have built a prototype 4-channel unit that is packaged within a single wide VXI module. The I/Q phase processor is packaged similarly.

Each channel of down-conversion is housed in a small metal box for improved isolation. A standard design is used which includes input diplexers for broadband frequency matching, a double-balanced mixer and a low-gain output amplifier. The levels of the signals being down-converted are as large as 20 dBm and attenuation is actually required at the input of each down-converter.

One important aspect of the down-conversion is that the local oscillator frequencies must be phase-locked to the 2 MHz system reference frequency. These 348 MHz and 698 MHz sources have yet to be designed.

I/Q Processor Module

The phase measurements are all made at a frequency of 2 MHz in the I/Q processor module. This module has four input channels, configured as two differential-phase channels. A block diagram of a single phase-measurement channel is shown in Figure 3. This includes the rf input, down converter, 2 MHz phase measurement, DSP processing and the VXI interface to the control system.

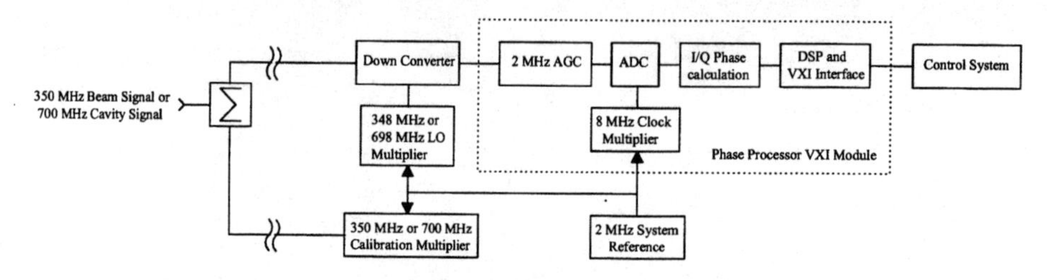

FIGURE 3. Block diagram of a single phase-measurement channel.

Phase Module Design

The phase-measurement technique employed is to sample the 2 MHz input signal at an 8 MHz clock rate which, is phase-locked to the 2 MHz system reference. By sampling at four times the input frequency, we produce a repeating pattern of I, Q, –I and –Q values (see Figure 4). This data stream phase would be arbitrary except for the fact that all phase measurement channels are sampled with the same 8 MHz clock. One of the important aspects of this technique is the high common-mode rejection of low-frequency noise. The digital process includes subtracting –I from I (and –Q from Q) removing the common mode error in both the signal and the ADC circuit. The phase is then calculated the as arctangent of the ratio I/Q.

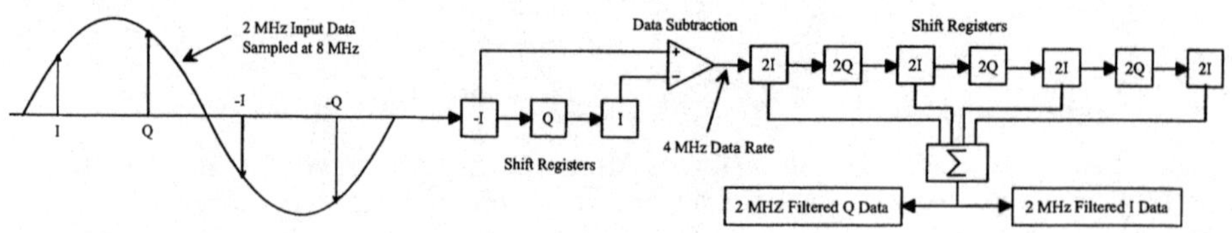

FIGURE 4. The I and Q data is created by sampling the 2 MHz input at 8 MHz. A simplified diagram of the digital process that follows the sampling is shown.

The I/Q processor contains two independent differential-phase channels. Figure 5 shows a block diagram of one differential phase channel. Two of these channels are in each processor module. The input circuits contain AGC stages to allow the input level to vary by over 60 dB while providing a constant level to the ADCs. Each AGC uses an Analog Devices AD600 dual, variable-gain amplifier as the heart of the circuit. This device is unique in that it has a fixed-gain amplifier proceeded by a 40 dB variable attenuator (2). This provides a more constant group-delay over the entire gain range. Two stages are used in series to achieve about 76 dB of constant-output range. This is considerably more than the 46 dB range in beam current that we expect.

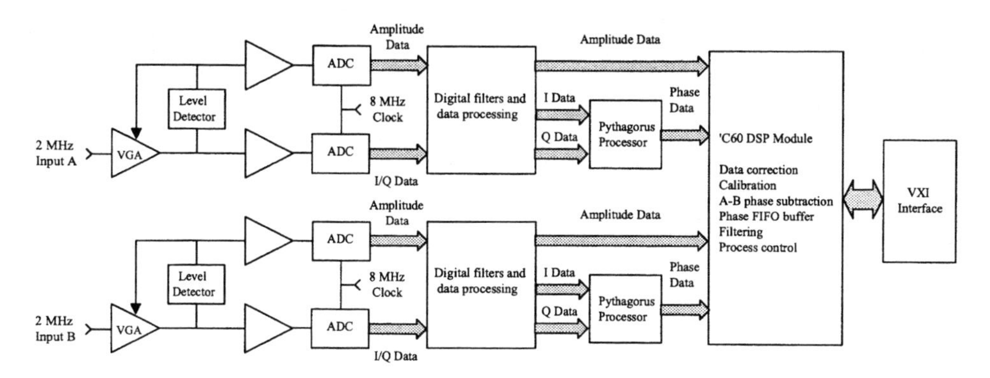

FIGURE 5. Two phase-measuring channels are processed and the data subtracted to produce a single differential-phase measurement block. The phase module contains two independent differential-phase measurements like the one shown.

The AGC circuit is followed by two Analog Devices ADS802, 12 bit ADC circuits. One samples the 2 MHz output to generate the I and Q data and another is used to digitize AGC control voltage. The gain control signal is linear at 32 dB/V. This amplitude data is used in the phase calibration algorithm.

The two data streams for each of the differential channels is processed in a custom programmable logic array, which contains the functions shown in Figure 4. These include the de-multiplexing of I and Q, data subtraction, and four-tap filtering to reduce the overall bandwidths to about 210 kHz. Though not shown in Figure 4, the amplitude data is filtered to 210 kHz as well. The most significant 16 bits of the I and Q data are passed to a Plessey Semiconductor PDSP16330 Pythagoras processor which calculates the arctangent of I/Q (as well as the magnitude, which is not used). This chip provides 12 bits of phase data for a resolution of $360/2^{12}$ or 0.0879 degrees (3).

Two DSP modules control the operation of the phase processor. These are Spectrum Signal Processors TIM-40 units with 40 MHz Texas Instruments TMS320C40 DSP chips. One processor will control each differential phase channel as well as pass data to and from the VXI interface logic. The DSP modules support several important functions. These functions include controlling the module calibration (as a function of signal amplitude), storing correction data into local RAM, and real-time correction of the phase data. The DSP modules also maintain a FIFO array of the corrected phase data and filter the data with various bandwidths down to below 10 Hz. Real-time, filtered data is available to the VXI interface as well as being output to the front panel via DAC circuits.

Performance of the AFE and DFE Circuits

At this time, the analog-front-end (AFE) and digital-front-end (DFE) circuits have been designed and tested and the complete phase processor VXI module is just now beginning initial testing. The overall performance of the phase measurement system is dominated by the characteristics of the analog- and digital-front-end circuits and these will now be presented.

The resolution and absolute accuracy of the phase measurement measure the performance of the AFE and DFE circuits. We expect the LEDA accelerator beam current to be tunable over a 46 dB range and we desire a absolute accuracy of ±0.5 degrees for the energy measurement. A resolution of 0.1 degrees should be adequate (defined as one standard deviation) at a bandwidth of 200 kHz. These requirements lead to the selection of the AGC front end and 12-bit ADC designs. The theoretical signal-to-noise ratio of the AFE is fixed at 79 dB for the first 40 dB of its gain range, and then decreases, dB for dB, as the second 40 dB of gain is utilized. At our lowest expected beam current of –46 dB, the S/N ratio should be 73 dB. For S/N ratios of 79 and 73 dB, the phase resolution should be 0.04 and 0.92 degrees respectively. The measured noise performance is shown in Figure 6. The measured resolution of about 0.05 degrees is fairly constant for the first 40 dB of signal range, increasing to about 0.12 degrees at –46 dB.

The absolute accuracy of the measurement system will depend on the linearity of the I/Q phase measurement, the accuracy of the calibration system, and the stability of the system between calibrations. At this point, we can only report on the linearity of the I/Q phase measurement. The linearity of the AFE/DFE circuits was measured by using two synthesized oscillators locked to a common reference frequency.

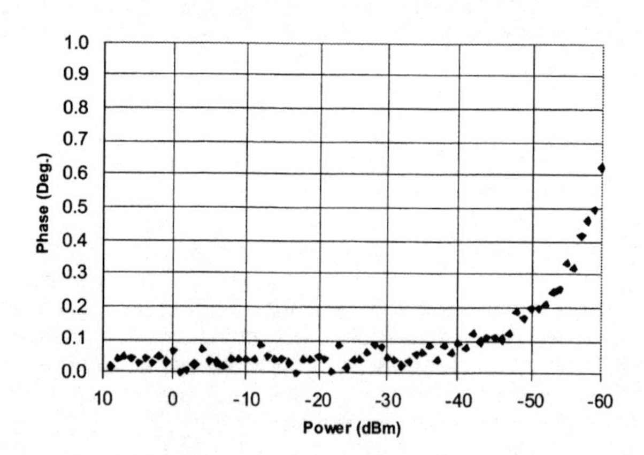

FIGURE 6. The resolution of the phase measurement as a function of input power is shown. Each point is the standard deviation of 1024 data points.

One is set to a frequency of 8 MHz while the other is set to 1.999 MHz. This provides a constant phase slew of 360 deg/ms resulting in 2000 phase measurements per cycle (2 MHz data rate). Both channels of the differential AFE/DFE circuit are simultaneously driven. The analysis includes subtracting the data from the two channels and comparing the result with a perfectly linear phase ramp. Analyzing the differential data removes a slight defect (actually a characteristic) of the programmable synthesized signal generators used. The results of this measurement are shown in Figure 7. The rms error is 0.06 degrees with a peak-to-peak error of 0.6 degrees.

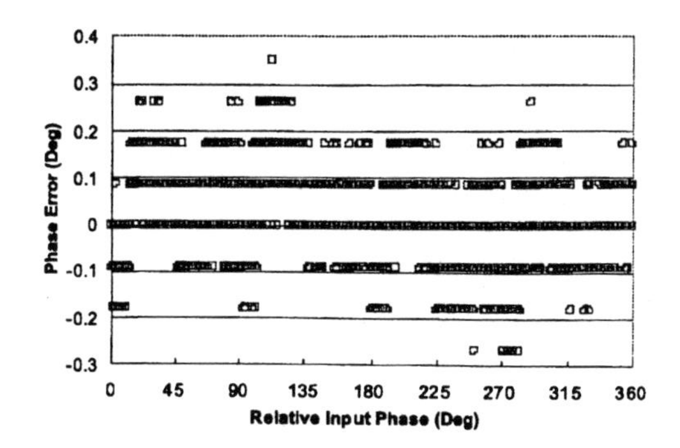

FIGURE 7. The resolution of the phase measurement as a function of input power is shown. Each point is the standard deviation of 1024 data points.

Calibration System

The calibration hardware and signal generation circuits will be located in a separate chassis that is controlled by the phase processor module over the VXI bus. The calibration system will generate 350 MHz and 700 MHz signal sources, which are phase-locked to the 2 MHz system reference and have programmable amplitudes over a 63.5 dB range. When enabled, these signals pass through the phase probes as well as being coupled into the cavity field probe. The calibration signals will pass through the entire signal chain to allow for a complete system calibration in situ. In most cases, the calibration signals for a pair of beam probes are derived from a passive rf splitter with short (<1 m) cable runs between the probes. In such cases, we need only be concerned with the phase stability of the splitter and two short cables and not that of the oscillator or the 150-foot-long cable runs between the electronics and the beamline. The single exception is the 700 MHz signal for the cavity probe and its associated 350 MHz beam signal. These calibration signals will include errors due to cable stability and oscillator phase drifts.

CONCLUSION

The design and testing of a new beam-phase and energy measurement system is in progress. Beam and cavity-field signals at 350 MHz and 700 MHz are down-converted to 2 MHz for phase measurement. Both the down-converter and phase processors are packaged in the VXI format. The I/Q phase processor has demonstrated a dynamic range well beyond the 46 dB required, with a resolution of near 0.1 degrees and an absolute accuracy of ±0.3 degrees. Testing of the DSP control hardware and software is underway.

REFERENCES

[1] Gilpatrick, J. D., K. F. Johnson, R. C. Connolly, J. F. Power, C. R. Rose, O. R. Sander, R. E. Shafer, D. P. Sandoval, V. W. Yuan, "Experience with the Ground Test Accelerator Beam-Measurement Instrumentation," *Proceedings of the Beam Instrumentation Workshop No.319*, Santa Fe, NM, 1993.

[2] Analog Devices, "Dual, Low Noise, Wideband Variable Gain Amplifiers," *AD600/AD602 Device Data sheet, Rev. A*, 1992.

[3] GEC Plessey Semiconductors, "PDSP16330/A/B Pythagoras Processor," *PDSP16330/A/B Device Data Sheet*, February, 1995.

On-line Phase Space Measurement with Kicker Excitation

J. Dietrich, R. Maier, I. Mohos

Forschungszentrum Jülich GmbH
Institut für Kernphysik
Postfach 1913, D-52425 Jülich, Germany

Abstract. A new method for on-line phase space measurements with kicker excitation at COSY was developed. The position data were measured using the analog output of two beam position monitors (BPMs) and directly monitored on a digital storage oscilloscope with an external clock (bunch-synchronous sampling). Nonlinear behavior of the proton beam was visible as well as were resonance islands. Typical measurements are presented.

INTRODUCTION

COSY is a cooler synchrotron and storage ring, delivering protons (unpolarized or polarized) with momenta between 300 MeV/c and 3300 GeV/c for experiments in medium-energy physics. It contains two cooling systems to shrink the beam phase space. An electron-cooling system reaches up to a momentum of 600 MeV/c and is complemented by a stochastic cooling system that covers the upper range from 1500 to 3300 MeV/c. Beam extraction is accomplished by the conventional resonant extraction mechanism as well as with the stochastic extraction method. Proton beams are routinely delivered to three internal and three external experimental areas (1). In this paper, special emphasis is given to the measuring technique of the transverse phase space. The knowledge of the phase space near the electrostatic septum is essential for optimization of the resonant extraction process and very useful for beam dynamics experiments.

EXPERIMENTAL SETUP

The experimental procedure starts with exciting the beam particles to collective transverse (in our case only horizontal) oscillations with betatron frequency by a fast diagnostic kicker magnet in the COSY ring. The beam bunch is short-time deflected (0.75 μs – 2 μs width, rise and fall time < 1μs) and the resulting bunch oscillations are measured using the beam position monitors (BPMs). The kicker excitation is

CP451, *Beam Instrumentation Workshop*
edited by R. O. Hettel, S. R. Smith, and J. D. Masek

synchronized with the COSY rf signal and can be adjusted in time by programmable delay, so that a single deflection of the total bunch can be performed (bunch synchronous excitation). The amplified and filtered sum and difference signals from the BPM electrodes are digitized by flash ADCs (20 MHz clock rate), stored in FIFO memories (4K or 64K width), and transferred to files. Depending on the FIFO width, data of about 200 or 3200 successive turns can be stored. Up to now, the phase space was calculated from the raw data of two BPMs and MAD calculations for TWISS parameters (α,β,γ) (2). Now a new method has been developed. The position data are measured using the analog output of two BPMs and directly monitored on a digital storage oscilloscope. The sum signal of a BPM is used to detect a passing bunch. The signal is differentiated and fed into a PLL-circuit. The differentiated sum signal has a zero crossing at each bunch peak, nearly independent of the bunch-shape and the bunch frequency. The phase loop tracks the zero crossing point and generates a clean, jitter-free clock pulse in phase with the bunch peak (Fig. 1). The output signal controls the sampling of the oscilloscope input stages (external sampling clock). Two BPM difference signals are displayed on line in the *xy*-display mode. The time of flight between the two BPMs is compensated by an electrical delay. The display represents the position of one BPM versus the other (except the calibration). To get the phase space diagram (angle x_1' versus position x_1), the transfer matrix between the two BPMs must be known. Another representation uses the normalized momentum $p_1 = \alpha \cdot x_1 + \beta \cdot x_1'$ versus position x_1 (canonical coordinates x_1, p_1). If the phase advance between the two BPMs is equal to $\pi/2$, the following expression for the normalized momentum p_1 is found:

$$p_1 = \sqrt{\frac{\beta_1}{\beta_2}} \cdot x_2; \qquad (1)$$

that means monitoring x_2 versus x_1 is similar to p_1 versus x_1 except the factor $(\beta_1/\beta_2)^{1/2}$.

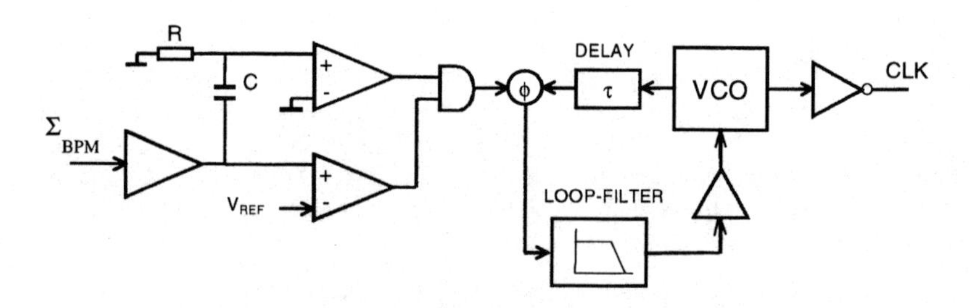

FIGURE 1. Bunch-synchronous tracking generator.

EXPERIMENTAL RESULTS

A problem for such measurements is the "damping" of the oscillations due to the finite betatron frequency spread of the particles. Typically, about 100 oscillations are seen in our case. To overcome this problem, the measurements were performed with a cooled beam. Figure 2 shows the on-line horizontal "phase space" plot (difference signal BPM_{x24} versus difference signal BPM_{x22}) near a third-order resonance for four different kick strengths (deflection angles) with an electron-cooled beam (approximately $5 \cdot 10^9$ circulating stored protons, momentum 1.675 GeV/c). Under these conditions, more than 40000 oscillations could be observed. The momentum deviation $\Delta p/p$ is about $2 \cdot 10^{-3}$ before and $1 \cdot 10^{-4}$ after cooling the beam. A sextupole (nonlinear) magnetic field is used to excite the third integer resonance (in this case the horizontal tune amounts $Q_x = 11/3$).

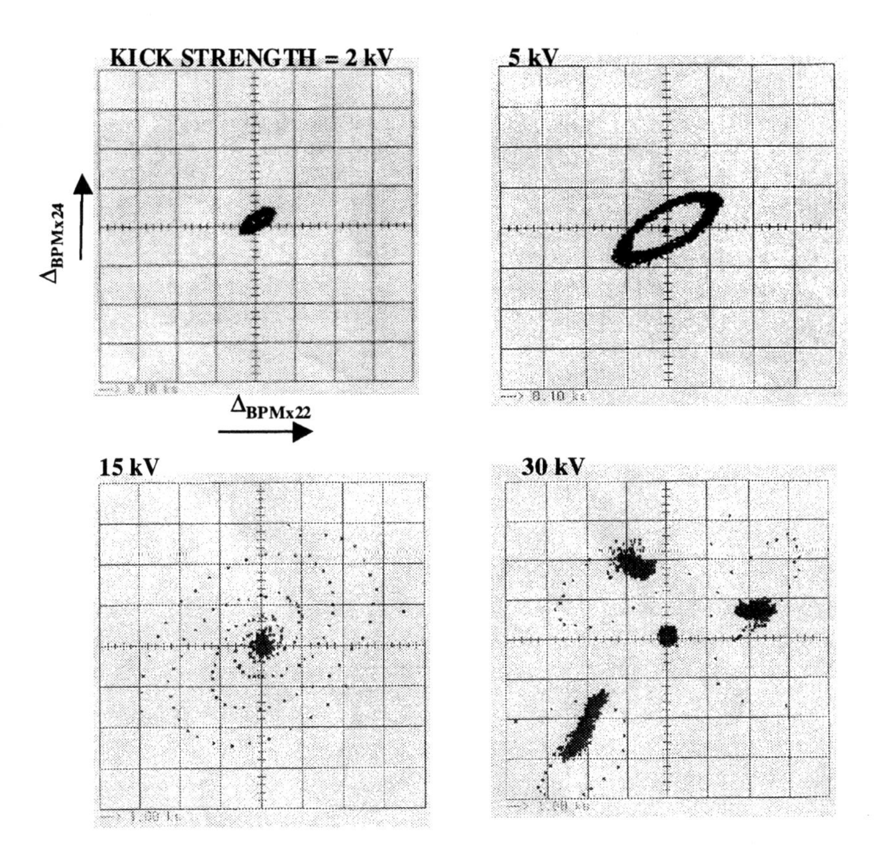

FIGURE 2. Horizontal "phase space" plots of an electron-cooled beam at four different kick strengths in kV (deflection angles) displayed on line with a digital storage oscilloscope in *xy*-display mode. Vertical direction: analog difference signal of BPM_{x24}, horizontal direction: analog difference signal of BPM_{x22}.

The effect of the nonlinearity makes the tune increase with increasing kick amplitude, so there is one amplitude for which the tune is exactly 11/3. Furthermore, there is a frequency entrainment effect causing all nearby amplitudes to lock onto exactly the same tune. This accounts for the existence of so-called resonance islands (3). When the beam is kicked with a small amplitude, the particles are not kicked upon the resonance. At a certain amplitude, the "lock-on" is visible and islands are formed (see Fig. 2). During the first 100 turns, the motion is damped before the particles are trapped in the island. The particles jump to another island at each turn and return to the starting island after three revolutions. The bunch within the island performs a circular motion around the center of the island, the so-called stable fixed point. After about 37 turns, the bunch returns to its original position in the island.

CONCLUSIONS

The studies of beam centroid motion after collectively perturbing the beam by a fast kicker yield important information about the lattice. This procedure is also useful in nonlinear beam dynamics studies. Due to the non-negligible beam size, the interpretation of the experimental results is difficult, especially if the beam center is displaced close to the separatrix. Some of the particles are stable here, some are instable. The degree to which the beam centroid motion accurately represents the motion of a single particle depends on the emittance of the beam; the smaller the emittance of the beam, the more accurate is its representation of single particle motion. Further limitations are the decoherence of the betatron motion and crossing of nonlinear resonances.

The shown method is extremely useful in determining the transverse phase space on line without analyzing the digitized FIFO memory of the beam position monitors.

REFERENCES

[1] Maier, R., "Cooler synchrotron COSY – performance and perspectives," *Nucl. Instrum. and Methods A* **390**, 1–8 (1997).

[2] Dietrich, J., et al., "Transverse Measurements with Kicker Excitation at COSY-Jülich," *Proc. of the 5th European Particle Accelerator Conference*, Barcelona, Spain, 1675 – 1677 (1996).

[3] Caussyn, D. D., et al., "Experimental studies of nonlinear beam dynamics," *Phys. Rev. A* **46**, 7942 –7952 (1992).

Streak-Camera Measurements of the PEP-II High-Energy Ring*

A. S. Fisher, R. W. Assmann†

Stanford Linear Accelerator Center, Stanford University
M.S. 17, P.O. Box 4349, Stanford, California 94309, U.S.A.

A. H. Lumpkin

Advanced Photon Source, Argonne National Laboratory
9700 South Cass Avenue, Argonne, Illinois 60439-4800, U.S.A.

B. Zotter

CERN, CH-1211 Geneva 23, Switzerland

J. Byrd, J. Hinkson

Advanced Light Source, Lawrence Berkeley National Laboratory
1 Cyclotron Drive, Berkeley, California 94720, U.S.A.

Abstract. The third commissioning run of the PEP-II High-Energy Ring (HER, the 9 GeV electron ring), in January 1998, included extensive measurements of single-bunch and multibunch fills using LBNL's dual-axis streak camera combined with Argonne's 119.0 MHz synchroscan plug-in. For single bunches, the dependence of bunch length on charge and rf voltage was studied from 0.5 to 2.5 mA and from 9.5 to 15 MV; the measured values ranged from 38 to 49 ps rms. The multibunch work focused on longitudinal instabilities as the current in the ring was raised to 500 mA, and the length of the bunch train was varied from 100 bunches (with 4.2 ns spacing) to a full ring. Large oscillations of up to 180 ps peak to peak were observed for bunches half a ring turn away from the start of the train, especially at higher currents and for trains filling roughly half the ring. These observations led to a new fill pattern with more gaps that allowed us to raise the current to 750 mA by the end of the run.

---

* Supported by the U.S. Department of Energy under contracts DE-AC03-76SF00515 for SLAC and DE-AC03-76SF00098 for LBNL.

† Present address: CERN, CH-1211 Geneva 23, Switzerland.

CP451, *Beam Instrumentation Workshop*
edited by R. O. Hettel, S. R. Smith, and J. D. Masek

INTRODUCTION

During the January 1998 commissioning run of the PEP-II *B* Factory at the Stanford Linear Accelerator Center (SLAC), the longitudinal characteristics of the electron beam in the High-Energy Ring (HER) were measured with synchrotron light and a dual-axis synchroscan streak camera, following techniques originally applied on LEP at CERN (1), on APS at Argonne (2), and more recently on ALS at LBNL (3). This paper describes measurements of both single-bunch and multibunch beams in PEP-II.

PEP-II and its diagnostics are discussed in another paper at this workshop (4). In particular, that paper describes the HER synchrotron-light monitor and the optical path added in December 1997 to transport some of this light to the streak camera, located in an optics room 11 meters above the tunnel.

STREAK-CAMERA EXPERIMENTAL CONSIDERATIONS

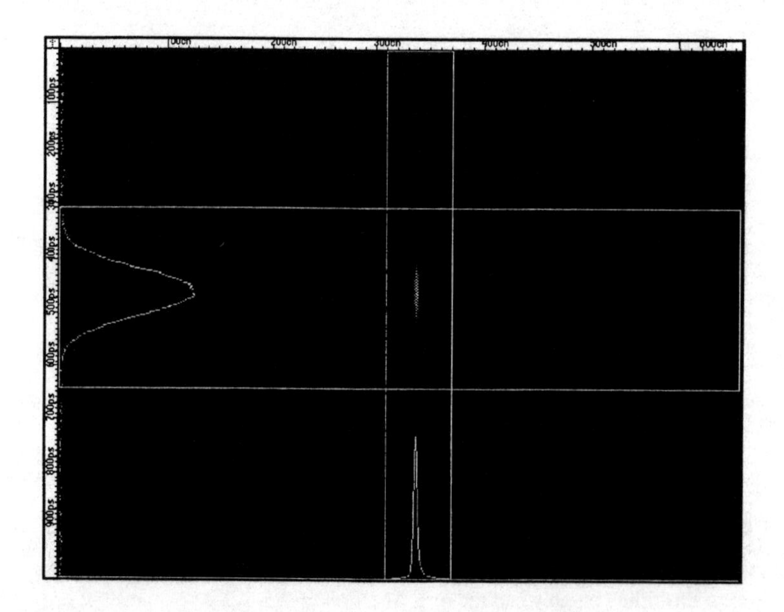

FIGURE 1. Typical single-axis streak image for bunch-length measurements, accumulated over many turns. Vertical: 1 ns full scale. Horizontal: position (channel number) along slit. The projections onto the axes, in the regions bounded by cursors, are shown along the sides.

A brief explanation of the dual-axis synchroscan streak camera (Hamamatsu model C5680) used for the PEP-II measurements is helpful for understanding the data. First, we consider a standard streak camera, with a single axis and a triggered sweep. The incoming light is imaged onto a slit that is narrow in the vertical direction, and is re-imaged onto a photocathode. While an axial voltage accelerates the photoelectrons toward a microchannel plate (MCP), a high-speed, high-voltage ramp is applied to a top-bottom pair of electrostatic plates to deflect the electrons as a function of arrival time across the MCP, converting the temporal distribution into a spatial one, in exchange for the loss of vertical spatial information. The MCP preserves the distribution while amplifying it. Another voltage accelerates exiting electrons onto a phosphor screen imaged by a video camera. The resulting image displays the temporal structure of the light pulse vertically, and its spatial distribution across the input slit horizontally. The time resolution can be as fast as 200 fs in the newest model.

Various effects limit these ideal characteristics, broadening the measured pulse duration. Chromatic dispersion in the lens between the slit and photocathode contributes an effect of typically a few percent for pulses of 40 ps rms. For broadband light, this problem is usually avoided by including an optical bandpass filter (≤10%) before the slit. A selection of filters allows measurement of the dispersion. For example, we found a 12 ps shift in the pulse centroid when changing the filter's center wavelength from 450 to 600 nm; if no filter were present, a measurement of a very narrow pulse extending over visible wavelengths would appear to be at least 12 ps wide.

If the input light is too bright, space charge between the photocathode and the MCP can spread the distribution. Consequently, the camera needs to be operated at a low light level, where shot noise is prominent. With repetitive signals (such as those from a stable storage ring), the statistics can be improved by summing several pulses, but any jitter in the trigger electronics for the ramp broadens the sum. Also, the retrigger rate is at best several kilohertz, preventing triggers on every ring turn.

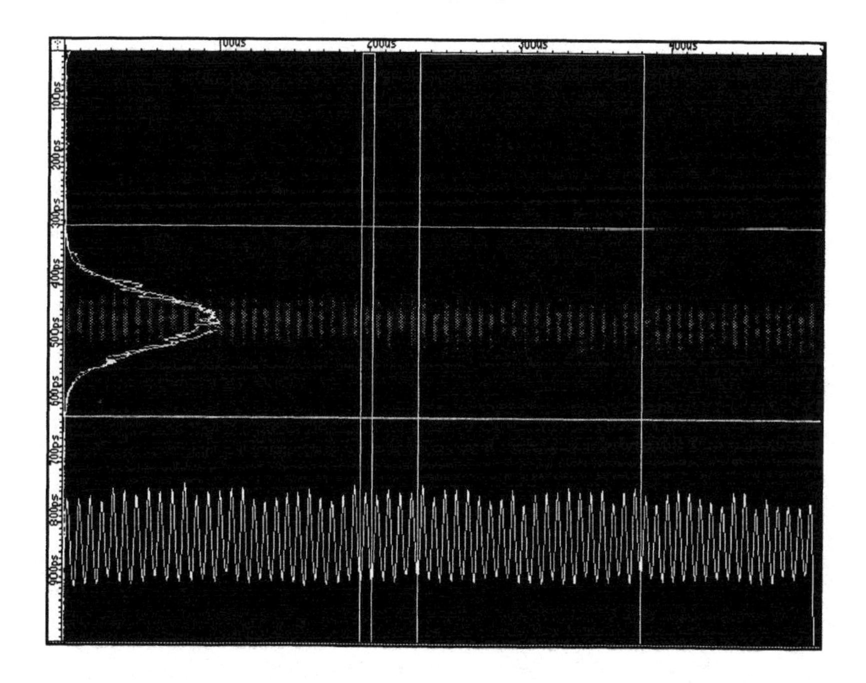

FIGURE 2. The light from a single bunch captured on consecutive turns. Full scale: 1 ns vertical, 500 µs horizontal. The projections on the left show a single bunch and the sum of the 20 bunches (determined by the two sets of cursors).

One common idea is to use a trigger derived from the light itself, picked up by a fast photodetector, in order to reduce jitter in the electronics external to the camera. However, some of the jitter is in the internal trigger circuit. In addition, the streak camera's long internal delay (50 to 100 ns) necessitates a substantial optical delay line after the trigger pick-off, and this delay varies with the time scale selected on the camera.

If the light is tightly locked to a stable rf signal (e.g., light from a mode-locked laser or from a storage ring without large synchrotron oscillations), the trigger jitter can be avoided by replacing the fast ramp with the sinusoidal output of a tuned rf circuit. This "synchroscan" option, available on some models (usually not the highest-speed units, especially in combination with dual sweep), provides acquisition at every zero crossing (that is, at twice the drive frequency), while the peaks of the sine wave are off scale. This approach allowed us to accumulate the low-noise sum of Figure 1 for measuring the length of a single HER bunch. The summing continues over a full 30 Hz (for the RS-170 American format) video frame interval. Our measurements at PEP demonstrated the value of this technique, and also one of the pitfalls: the camera displayed previously unknown, intermittent, 100 ps jumps, lasting for several milliseconds, in the 119 MHz reference output of the standard PEP timing module. We obtained an alternate source while the Controls Department investigates what appears to be a common malfunction (although few triggered devices are as sensitive to this jitter as a streak camera).

For accurate measurements, it is important to verify the factory calibration of the time axis, especially, as in our case, when the plug-in used was not the one that accompanied the camera from the factory. By putting an etalon—an flat optical plate with partly reflecting faces and with a known thickness and refractive index—in the light path, the camera sees the main pulse and a diminishing series of echoes, spaced by the round trip time inside the etalon. Since 1 mm of glass adds 10 ps to the round trip, a convenient thickness separates the pulse from its first echo for calibration of the time base.

For storage rings, a valuable option is the addition of a second time axis, by slowly (nanoseconds to milliseconds) deflecting the photoelectrons horizontally with a second pair of plates. A sweep rate corresponding to several ring turns lets us compare the same single bunch on consecutive turns (Figure 2). When both synchroscan and the slower horizontal axis sweeping are combined, synchrotron oscillations show up plainly as oscillations in bunch arrival time. With a multibunch fill, we were able to examine longitudinal oscillations along the bunch train.

Assembling the hardware for the PEP measurements required a three-laboratory collaboration. The Advanced Light Source at Lawrence Berkeley National Laboratory provided the dual-axis streak camera. Each synchroscan plug-in is tuned at the factory for narrow-band operation at a single frequency between 75 to 165 MHz; the ALS plug-in uses 1/4 of their 500 MHz ring rf frequency. This is too far from the corresponding PEP subharmonic, 119 MHz. Instead, a 119 MHz plug-in from the Advanced Photon Source at Argonne National Laboratory was brought to SLAC for the final week of the January run.

BUNCH-LENGTH MEASUREMENTS

We measured the length of a single bunch in the HER, and the dependence of length on the synchrotron tune (that is, the rf voltage) and the charge in the bunch. To reduce the noise, we used the summing procedure discussed above, which gave a clean Gaussian profiles like those of Figure 1.

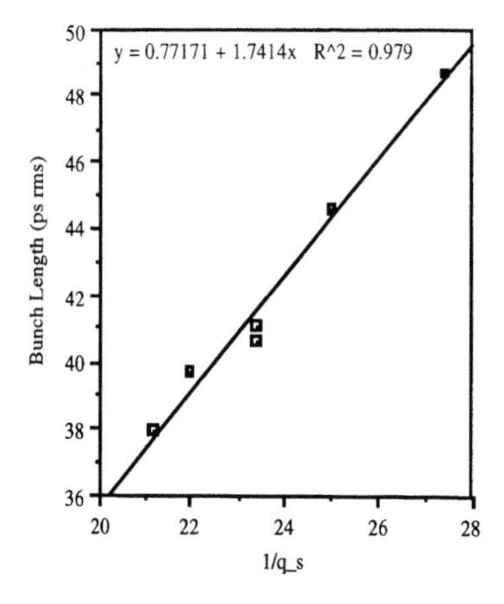

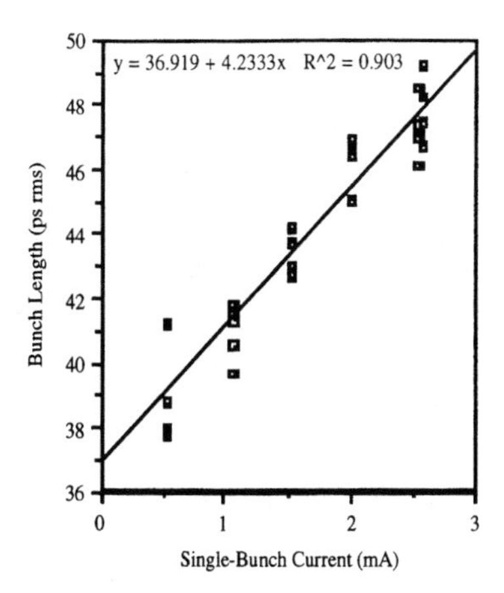

FIGURE 3. Variation of bunch length with the inverse of the synchrotron tune, for a single bunch at 1 mA.

FIGURE 4. Variation of bunch length with single-bunch current, at 14 MV of rf; 1 mA corresponds to 4.6×10^{10} electrons in the bunch.

Bunch length was measured with a current of 1 mA for rf voltages between 9.5 and 15 MV, corresponding to synchrotron tunes q_s from 0.0365 to 0.0472 and frequencies f_s from 5.0 to 6.4 kHz. The bunch length σ_t should be related to the synchrotron frequency, energy spread σ_E / E, and momentum-compaction a by

$$\sigma_t = \frac{\alpha}{2\pi f_s}\frac{\sigma_E}{E} \,. \tag{1}$$

The plot of Figure 3 shows this inverse variation with synchrotron tune. Using $\alpha = 0.00241$ for the HER, the measured bunch lengths at 1 mA correspond to an energy spread of $(6.31\pm0.08)\times10^{-4}$, in reasonable agreement with the design value of 6.14×10^{-4} and an independent measurement using the quantum lifetime.

Other scans studied bunch length versus single-bunch current in the ring, from 0.5 to 2.58 mA, with the rf held at 14 MV. We were restricted to this maximum to avoid peak-signal damage to the longitudinal-feedback system. The length appears to grow linearly (see Figure 4) over the range studied, consistent with potential-well distortion (5). There appears to be no sign of a knee due to the onset of the microwave instability; the calculated threshold lies between 1.8 and 6.4 mA (6).

In PEP's nominal parameter list, the standard HER fill—0.99 A in 1658 bunches (0.60 mA/bunch), with 14.0 MV of rf and a tune q_s of 0.0449—has a calculated bunch length of 38.4 ps, compared to 39.5 ps from the linear fits of these two plots.

MULTIBUNCH INSTABILITIES

Other measurements, made during the last two days of the January 1998 run, examined the onset of longitudinal instability in long bunch trains. For this work, we used horizontal sweep settings of 5 or 10 μs, close to the ring's 7.3 μs revolution time. Vertically, the fast axis was set to either 600 ps or 1 ns full scale, in order to resolve the lengths and especially the relative phases of the bunches.

The normal fill pattern puts charge in every second 476 MHz rf bucket of the ring, corresponding to a bunch spacing of 4.2 ns (238 MHz), except for a gap which we varied to study its effect on stability. The camera's synchroscan drive was at 119 MHz, half the bunch spacing; with a proper phase delay, consecutive bunches appeared in alternation on the rising and falling zero crossings of the sine-wave drive. Because the direction of the sinusoidal sweep (and hence the direction of the time axis) alternates, it is preferable to slightly offset the phase from the zero crossings, so that the image shows the even and odd bunches of the train slightly separated. The typical result (Figure 5) resembles a bunch train and its mirror image. Because this figure uses a 10 μs horizontal sweep, the gap in the train appears twice; the head of the train (Bunch 0) is just to the right of the first appearance of the gap, on the left; the tail is to the right. For our choice of phase delay, the time axis for the upper train points downward, while it points up for the lower train.

In the first multibunch observations, the ring was filled to currents between 400 and 525 mA, using a roughly uniform pattern with a gap of 5% of the ring's circumference. For these currents, the middle of the train exhibited significant phase instability, with 100 ps peak-to-peak oscillations, as Figure 5 shows for a 480 mA fill (and which were even more apparent in the rapid motion of the streak-camera video, captured on videotape). The head and tail of the train remained stable, with only a constant phase shift over the first 100 bunches due to the transient in loading of the rf cavities due to the gap. For part of that day, similar observations were made in tests of the longitudinal

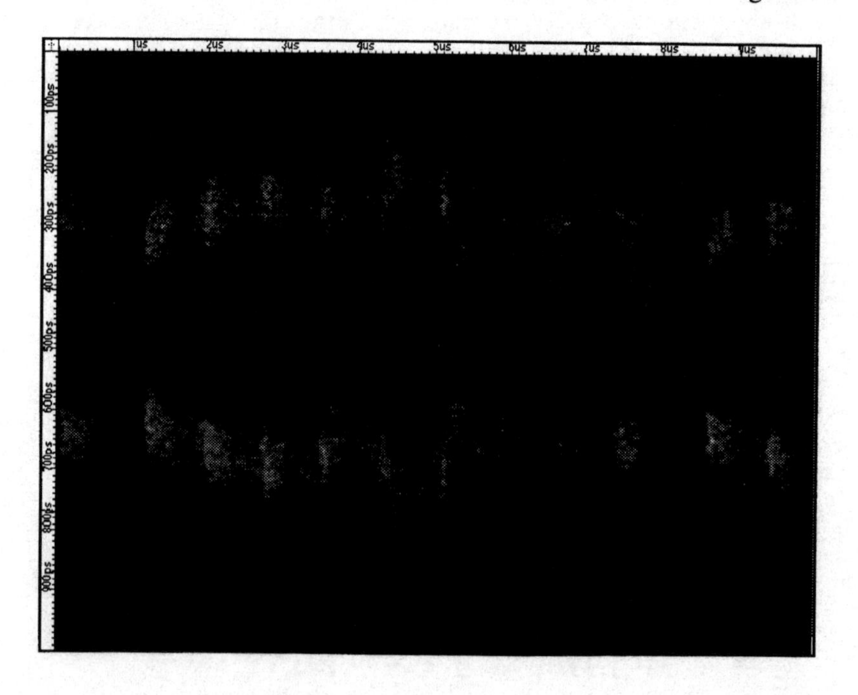

FIGURE 5. Longitudinal instability in the center of a 1658-bunch train with a 480 mA current. Filled in a 9-zone pattern with 4.2 ns spacing and a 5% gap. Full scale: 1 ns vertical, 10 μs horizontal.

feedback system (7), which provided turn-by-turn recordings of the phases of each bunch. (Because power amplifiers were out for repairs, the system could not be used to stabilize the train.)

Subsequently, we filled the ring with bunch trains of varying lengths, with the same charge per bunch and 4.2 ns spacing. The instabilities began after filling about one third of the ring. Figure 6 shows a 700-bunch train (about half of the ring, shown this time on a 5 μs scale) with 228 mA. The last 250 bunches oscillate even more than before, with up to 180 ps peak-to-peak motion, while the head remains still.

Later, we again filled the ring consecutively but at a lower current, so that it reached 330 mA when filled with the normal 5% gap. As before, the tail of the train was unstable for train lengths filling 1/3 to 2/3 of the ring, but once the train got longer it stabilized. The instability threshold depends on the length of the gap and the current.

FIGURE 6. Large longitudinal oscillations at the end of a 700-bunch train with a 228 mA current and 4.2 ns spacing. Full scale: 1 ns vertical, 5 μs horizontal.

The instability often led to rapid current loss. On several occasions, within a single video frame of the streak camera, most of the bunches in the ring were lost, with only those at the head of the train remaining. The fact that these instabilities were strongest half way around the ring from the gap eventually suggested that we could achieve a higher total current with more gaps than the nominal fill. In the final hour of the run, a new current record for the HER—750 mA—was set using six 5% gaps evenly spaced around the ring. These efforts will continue later this year, with the goal of reaching the intended operating current of 1 A. However, 750 mA is sufficient to achieve the design luminosity.

ACKNOWLEDGMENTS

We would like to thank Dan Alzofon, Bill Roster, and Vern Brown for their assistance in the rapid assembly of the optical transport line. We appreciate the support and advice given by Uli Wienands, the system manager for the HER. Both the ALS and the APS supported our efforts with their equipment and staff. We are also grateful to the entire PEP-II commissioning team.

REFERENCES

[1] Rossa, E., Bovet, C., Disdier, L., Madeline F., and Savioz, J.-J., "Real-Time Measurement of Bunch Instabilities in LEP in Three Dimensions Using a Streak Camera," *Proc. 3rd European Particle Accelerator Conf.*, Berlin, 1992, pp. 144–146.

[2] Lumpkin, A. H. "Advanced Time-Resolved Imaging Techniques for Electron-Beam Characterization," *Proc. Accelerator Instrum. Workshop*, AIP. Conf. Proc. **229**, 151 (AIP Press, Woodbury, NY, 1991); Lumpkin, A. H., "Time-Domain Diagnostics in the Picosecond Regime," *Proc. Microbunches Workshop*, Upton, NY, Sept. 1995, AIP. Conf. Proc. **367**, 327 (AIP Press, Woodbury, NY, 1996); Lumpkin, A. H., Yang, B. X., and Chae, Y. C., "Observations of Bunch-Lengthening Effects in the 7-GeV Storage Ring," *Nucl. Instrum. Methods*, **A393**, 50–54 (1997); Yang, B. X., *et al.*, "Characterization of Beam Dynamics in the APS Injector Rings Using Time-Resolved Imaging Techniques," *Proc. Particle Accelerator Conf.*, Vancouver, BC, May 1997 (IEEE, Piscataway, NJ, to be published). Yang, B., Lumpkin, A., Harkay, K., Emery, L., Borland, M. and Lenkszus, F., "Characterizing Transverse Beam Dynamics at the APS Storage Ring Using a Dual-Sweep Streak Camera," in this volume.

[3] Hinkson, J., Keller, R., Byrd. J. and Lumpkin, A. "Commissioning of the Advanced Light Source Dual-Axis Streak Camera," *Proc. Particle Accelerator Conf.*, Vancouver, BC, May 1997 (IEEE, Piscataway, NJ, to be published).

[4] Fisher, A. S. "Instrumentation and Diagnostics for PEP-II," in this volume.

[5] Heifets, S., private communication.

[6] Heifets, S., "Comments on the Results of the HER January Run," PEP-II AP Note 98.05, 1 Feb. 1998.

[7] Teytelman, D., Fox, Hindi, J. H., Limborg, C., Linscott, I., Prabhakar, S., Sebek, J., Young, A., Drago, A., Serio, M., Barry, W., and Stover, G., "Beam Diagnostics Based on Time-Domain Bunch-by-Bunch Data," in this volume; Prabhakar, S., Teytelman, D., Fox, J., Young, A., Corredoura, P., and Tighe, R., "Commissioning Experience from HER PEP-II Longitudinal Feedback," in this volume.

Periscope Pop-In Beam Monitor

E. D. Johnson[☆], W. S. Graves, and K. E. Robinson[⌘]

Brookhaven National Laboratory
Upton Long Island, New York, USA
[⌘]STI Optronics, Bellevue, Washington, USA

Abstract. We have built monitors for use as beam diagnostics in the narrow gap of an undulator for an FEL experiment. They utilize an intercepting screen of doped YAG scintillating crystal to make light that is imaged through a periscope by conventional video equipment. The absolute position can be ascertained by comparing the electron beam position with the position of a He:Ne laser that is observed by this pop-in monitor. The optical properties of the periscope and the mechanical arrangement of the system mean that beam can be spatially determined to the resolution of the camera, in this case approximately 10 micrometers. Our experience with these monitors suggests improvements for successor designs, which we also describe.

INTRODUCTION

At BNL our current FEL projects center on single pass designs, in particular the High Gain Harmonic Generation (HGHG) scheme (1). To obtain a high performance FEL, high peak current, low-emittance electron beams must be produced, transported, and controlled throughout the accelerator system. Clearly, one must possess metrology tools to measure these beam attributes in a precise and reliable manner. We have developed a number of intercepting diagnostics for these tasks, many of which utilize YAG scintillation crystals as described elsewhere in these proceedings (2). These screens are especially useful for our experiments since they provide the opportunity to measure both the beam position and profile for a single electron pulse.

The undulators we are using for our FEL experiments are fitted with integral correction stations (3) so beam monitors must be available to determine the proper settings of the steering-focusing magnets. To maintain overlap between the electron beam and the growing radiation field of the FEL, it was felt that the position resolution of the monitors should be at least 10 percent of the beam size and that they should image the beam with comparable resolution (4). For typical parameters of our experiments, spatial resolution and absolute precision on the order of 20 micrometers are required. Since there are no "breaks" in our undulators for diagnostic stations, the monitors must fit in the gap of the undulator vacuum chamber. For the design discussed in this paper the probe must fit inside an 11 mm chamber opening.

CP451, *Beam Instrumentation Workshop*
edited by R. O. Hettel, S. R. Smith, and J. D. Masek

DESIGN SYNTHESIS

It is fair to say that the design presented here is the product of a number of constraints imposed by the use of existing hardware, space limitations in the experimental hall, evolution of design ideas, and plain dumb luck. In the spirit of these workshops we freely admit this but, to preserve a modicum of dignity, will refrain from a complete discussion, pleading space limitations in the proceedings volume. Interested readers should feel free to contact the authors should they wish to be regaled with the lurid details. Whatever the process, we have arrived at a design that presents some very nice features which make it durable and precise. In Figure 1, the essential features of the monitor are shown schematically.

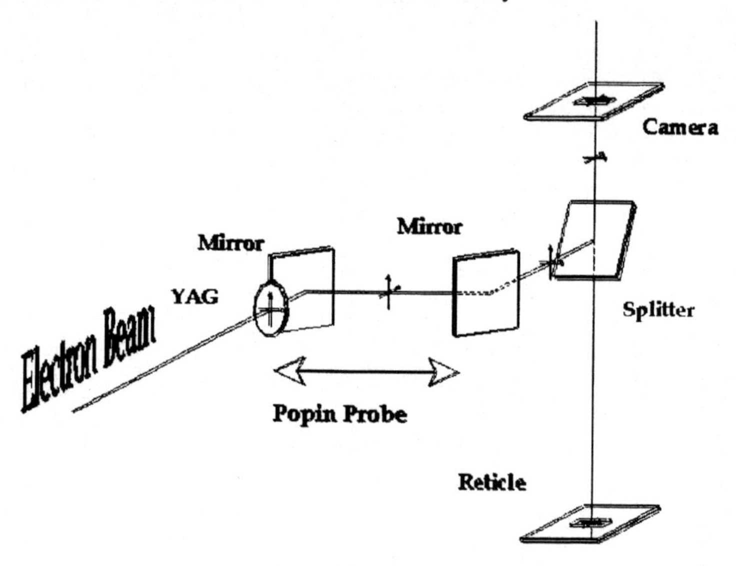

FIGURE 1. The YAG crystal and two mirrors constitute the pop-in probe. It is moved as an assembly in and out of the electron beam. Note the orientation of the arrow and paw figure. It traces the beam image through the optical system. The image presented to the camera is what one would view on the front surface of the scintillation screen.

The probe is configured as a periscope which has the effect of displacing the image from the axis of the electron beam by a distance determined by the mirror spacing. Reproducible insertion of the probe is not required to maintain the precision of the monitor since the camera is firmly mounted to the undulator structure. As long as the probe length is stable and the camera does not move, the precision of the monitor is preserved, so a "sloppy" actuator can be used. To provide fiducial reference marks, a fixed reticle is viewed through a beam splitter at a position for the YAG crystal. An outline drawing of the monitor is provided in Figure 2.

In this design the probe, actuator, and optics are all on one side of the undulator. The light from the screen is collected through the hollow shaft of the probe. The probe itself is made in two tubular sections, the inner of stainless steel and the outer of aluminum. Their relative lengths and materials were chosen to match the material path from the center of the undulator to the camera mount. In this way, temperature drifts in the experimental hall do not compromise the precision of the monitor. An increase in temperature makes the probe longer but it also shifts the camera from the undulator center by the same distance.

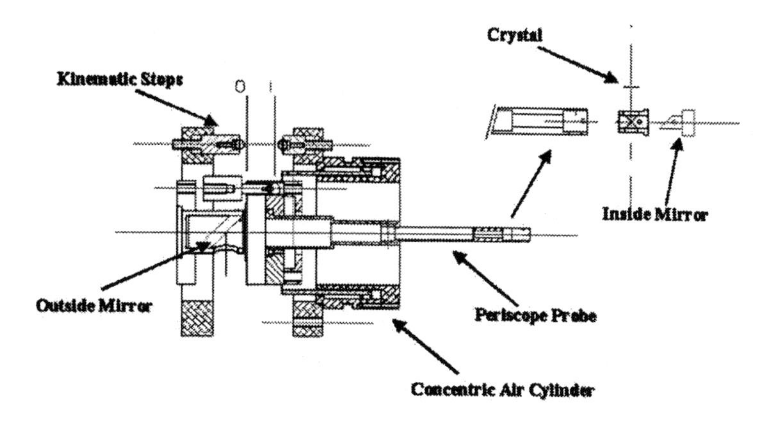

FIGURE 2. This cross-section drawing shows the probe assembly and actuator. In this plan view the electron beam would be traveling from the top to the bottom of the page. The vacuum bellows and chamber mounting have been omitted from this view but are shown in Figure 3. The inset to the figure shows how the crystal and inside mirror are fitted to the inner end of the probe.

The inner diameters of the probe sections become progressively larger as the distance from the screen increases to maintain a viewing angle of approximately 90 milliradians. The minimum outer diameter of the probe is 9.8 mm while the YAG crystal is 0.5 mm thick and 6 mm in diameter. To keep the crystal and inner mirror in place, a carrier was designed as shown in the inset of Figure 2. The crystal sits in a counterbore in the carrier that just relieves its outer edge below the inside diameter of the probe. The inner mirror slides inside the carrier with the assembly secured by a single stainless steel roll pin. The inner mirror was fabricated in its finished shape from aluminum and polished. In retrospect it might have been better to start with a large polished mirror and trepan out the mirror. Holding the small blanks stable to polish a flat surface with minimal edge roll off turned out to be rather time consuming.

A double-sided 2.75" conflat flange holds the probe. Three pairs of ground stainless steel rods are set into the edge of the flange, 120 degrees apart. They form three "vee" grooves that are used to provide kinematic stops for the probe in its inserted and retracted positions as well as providing the location for the outer mirror. A bellows on the inner side of the flange allows the probe to be moved. A non-magnetic window is mounted on the outer side of the flange. The window flange has three extra through-holes to allow the outer mirror adjusters to rest on the probe rods. In this way the entire probe is referenced from these rods. The outer mirror can be adjusted for tilt and rotation. During preliminary assembly, an alignment laser is used to ensure that the inner and outer mirrors are parallel.

The actuator for moving the probe is a concentric air cylinder. Inner and outer cylinders capture the tubular piston so it is a double-acting device. The piston carries the probe and overcomes the vacuum load provided by the bellows. The effective areas of the inner and outer drive have been adjusted so that at the cylinder operating pressure of 25 psig, the clamping forces against the tooling ball stops for the inserted and retracted positions are equal. The stops for the probe are small tooling balls mounted on brass shoulder screws. This allows adjustment for the stop position and tilt adjustment of the probe. The bellows flange is rotatable and the rings supporting the probe stops are dogged to the cylinder. This allows a rotational orientation of the assembly before it is bolted into place. The welded bellows were specified with extra segments so small adjustments of about 5 degrees or less can be made by rotating the stops after the assembly is under vacuum. This rotational adjustment is important to keep the light from the screen traveling down the center of the probe. A plan view of the monitor mated to the chamber is given in Figure 3.

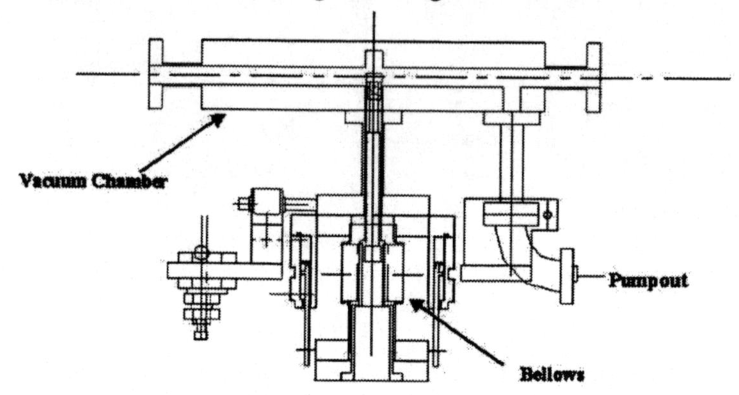

FIGURE 3. This cross-section drawing shows a plan view of a partially inserted probe assembly with its actuator attached to the vacuum chamber. The electron beam would move left to right on the page.

The optical arrangement is essentially as shown in Figure 1. The only additions are a 175 mm focal length achromat lens placed just above the beam splitter and the introduction of a neutral density filter wheel between the probe outer mirror and the beam splitter. The neutral density filters were selected as the method for attenuating the light so that the collection angles would remain constant; an iris would attenuate the light *and* change the aperture of the optical system. In practice an ND 1.0 filter seems to be used nearly all of the time.

A DC motor drives the filter change wheel and the position is encoded by microswitches. The motor drives the filter wheel via a belt which both conserves space close to the probe, and keeps the motor further away from the undulator and electron beam. Figure 4 provides a view of these components.

For imaging the beam, we use a COHU 4910 CCD camera (5). The reticle is also provided with an inexpensive back light (a 12-volt truck running-lamp; the white plastic mount works well as a diffuser!). If the light is turned on, the image of the reticle is superimposed on the beam image. If the light is turned off, only the beam is viewed. The beamsplitters we used were 70 percent reflective, 30 percent transmissive AR coated glass. The viewing scheme worked fine in the laboratory imaging the reticle and a surrogate beam image provided by a He:Ne laser. What we found in using the monitor on the experiment was a multiple image in the vertical of the beam spot. This image seems to be due to multiple internal reflections from the front and back surfaces of the beam splitter. Substitution of a first-surface mirror for the beam splitter provides a single undistorted image. After preliminary alignment of the monitors using a He:Ne laser, placed on the desired electron beam trajectory, we substituted mirrors for the beam splitters. We hope to find a modestly priced, durable beam splitter to use in this and future monitors that does not produce this artifact. The undulator with its five monitors is shown in Figure 5.

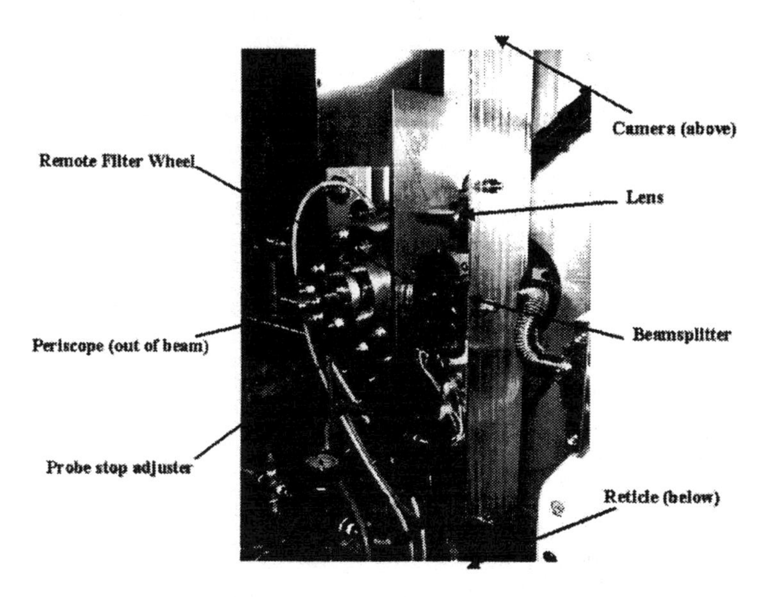

FIGURE 4. This photograph of one of our monitors shows the optical beam distribution components.

FIGURE 5. The five monitors mounted on the "Cornell undulator" at the BNL Accelerator Test Facility. This undulator, on loan from CHESS was rebuilt and measured by the ANL/Advance Photon Source. The FEL experiment is being undertaken in collaboration with the APS. Note that in this photograph, a Pulinex camera is being used for the first monitor.

EVALUATION AND CONCLUSIONS

On the whole, the monitors have performed quite satisfactorily. At the time of writing we have not had the opportunity to track down the beam splitter problem, although we are confident it can be resolved. Our operating experience seems to indicate that four separate filters are not required. Our next design will probably have two single filters on air actuators that could be put in the beam at the same time. This would give four possible filter settings (none, A, B, A+B) and simplify the design.

The inner mirror fabrication was more time-consuming than originally anticipated, and while convenient, precision machining might eliminate the adjustment of the external mirror. A probe is under development with a precision spacer with reference surfaces to locate two flat mirrors. While this simplifies the probe, it means both mirrors are in vacuum. The ability to remove and relocate the outer mirror turned out to be very useful for preliminary alignment of the monitors; a feature that is sacrificed by the new design. Whether this tradeoff is beneficial will have to await construction and testing of the new monitors.

The coaxial actuator turned out to be a pleasant surprise. We adjusted the design so it could be fabricated from standard tubing, so it was not nearly as expensive to make as we had first thought. One should also note that it does not have precision guides; only precision end stops. For this application the guiding provided by the piston seals was perfectly adequate. Compared with other types of through-vacuum actuators this seems to be a real advantage. We currently have plans to use these actuators on several other instruments at the National Synchrotron Light Source.

ACKNOWLEDGMENTS

We would like to thank Don Shea, Mike Lehecka, and Cheng Shu of the NSLS for their expert assistance in making and modifying the monitors and Jack Jagger of the APS for his help in integrating them into the design of the experiment. We would also like to thank Xijie Wang of the ATF and Li-Hua Yu of the NSLS for their comments regarding the operation of the monitors as they use them to perform the experiment. This work was performed under the auspices of the U.S. Department of Energy under contracts DE-AC02-76CH00016 and DE-AC02-98CH10886.

REFERENCES

[1] Yu, L. -H., "Generation of intense uv radiation by subharmonically seeded single pass free-electron lasers," *Physical Review A* , **44**, 5178 (1991).
[2] Graves, W. S., E. D. Johnson and S. Ulc, "A High Resolution Electron Beam Profile Monitor and its Applications," In these proceedings, 1998.
[3] Quimby, D. C., S. C. Gottschalk, F. E. James, K. E. Robinson, J. M. Slater and A. S. Valla. "Development of a 10-Meter Wedged-Pole Undulator," *Nucl. Instrum. and Methods A* , **281** (1989).
[4] Yu, L. –H., private communication.
[5] COHU 4910 series camera 1/2" CCD, specified sensitivity as used 0.65 lux. This information provided for reference only, since it was used in this instrument. Equipment by other manufacturers could, in principle, be used for this application.

The DAΦNE Luminosity Monitor

G. Di Pirro, A. Drago, A. Ghigo, G. Mazzitelli, M. Preger, F. Sannibale, M. Serio, G. Vignola

INFN Laboratori Nazionali di Frascati
00044 Frascati (Roma), Italy

F. Cervelli, T. Lomtadze

INFN Sezione di Pisa
56010 San Piero a Grado (PI), Italy

Abstract. DAΦNE, the Frascati Φ-factory, is an e^+/e^- collider with 2 interaction points (IPs). The center of mass energy is 1020 MeV and the design luminosity 4.2×10^{30} cm^{-2} s^{-1} in single bunch mode and 5×10^{32} cm^{-2} s^{-1} in multibunch mode. Between the possible electromagnetic reactions at the interaction point, single bremsstrahlung (SB) has been selected for the luminosity measurement. The SB high counting rate allows real-time monitoring, which is very useful during machine tune-up and moreover the narrow peak of the SB angular distribution makes the counting rate almost independent from the beam position at the IP. A description of the experimental set-up, calibration results and luminosity measurements is presented.

INTRODUCTION

DAΦNE, the Φ-factory under commissioning at Frascati (1), is an e^+/e^- collider with the center of mass energy tuned to 1020 MeV for the production of Φ mesons. Two interaction regions allow two simultaneous experiments. One of these in particular, the detector KLOE (2), will start data acquisition at the end of this summer and will investigate the CP violation in the K_0 meson decay. High luminosity is needed to measure this rare event with good accuracy. DAΦNE has been designed to obtain a maximum luminosity of 5×10^{32} cm^{-2} s^{-1}. The design strategy used to achieve this performance is to store up to 5 A in 120 bunches in each ring. This multibunch, high-current approach allows one to maintain the single bunch luminosity value to 4.2×10^{30} cm^{-2} s^{-1}, permitting a relaxation of the requirements on related machine parameters. At the present time, the single-bunch luminosity commissioning is nearing completion.

CP451, *Beam Instrumentation Workshop*
edited by R. O. Hettel, S. R. Smith, and J. D. Masek

Several luminosity measurement methods are possible in lepton colliders. Most of them use electromagnetic reactions at the interaction points (IPs) with well known theoretical cross sections. The most commonly used are:

Single bremsstrahlung (SB) (3):

$$e^{+} + e^{-} \rightarrow e^{+} + e^{-} + \gamma \tag{1}$$

Double bremsstrahlung (DB) (4):

$$e^{+} + e^{-} \rightarrow e^{+} + e^{-} + 2\gamma \tag{2}$$

Small-angle Bhabha scattering (SAS):

$$e^{+} + e^{-} \rightarrow e^{+} + e^{-} \tag{3}$$

Reaction (1) can be monitored by measuring the number of γ photons emitted in the direction of one of the colliding beams. The counting rate is typically very high, allowing one to perform 'on-line' luminosity measurements. Moreover, the very sharp SB angular distribution significantly simplifies the geometry of the monitor at the interaction region and makes the measurement weakly dependent on the position and angle of the two beams at the IP. On the other hand, the measurement must be carefully extracted from the background, which is mainly due to gas bremsstrahlung (GB), the interaction of the stored beam particles with the residual gas molecules. The GB counting rate depends linearly on the distance between the IP and the γ detector; for this reason the SB method is not used for large high-energy machines with very long interaction regions.

In reaction (2), two γ photons are emitted simultaneously, one in the direction of the e^{-} beam and the other in the direction of the e^{+}. The counting rate is much smaller than in the SB case and the angular distribution is broader. From the background point of view, GB contribution can be easily removed by measuring coincidence in the γ detectors. Accidental counts of simultaneous emission of two SB or GB photons must be removed on a statistical basis by the delayed-coincidence technique.

In reaction (3), the background subtraction can be performed in a very efficient way, but the counting rate is the lowest and the dependence of the cross section on scattering angle is strong.

For the DAΦNE luminosity monitor the SB technique has been selected but the system layout also permits the possibility of DB luminosity measurements (5). Complete information on the SB luminosity measurement technique can be found in Ref. (6); in the final part of this introduction only the relevant features of the method are described.

The SB γ photons are collected by a proportional counter situated near the IP. The integral of the signal coming from this detector is proportional to the energy of the γ photon. An energy analysis and counting system (Figure 1), counts and makes the energy spectrum of all the signals having an amplitude greater than a tunable threshold, fixed by a discriminator. The typical system also allows coincidence measurements with a selected rf bucket, very helpful for background subtraction, as explained later, and anticoincidence measurements of counts generated by charged particles on a scintillator plate placed in front of the proportional counter. This last feature allows one to filter undesired counts due to lost beam particles, pairs produced by the electromagnetic showers generated by γ, beam particles hitting the vacuum chamber, etc.

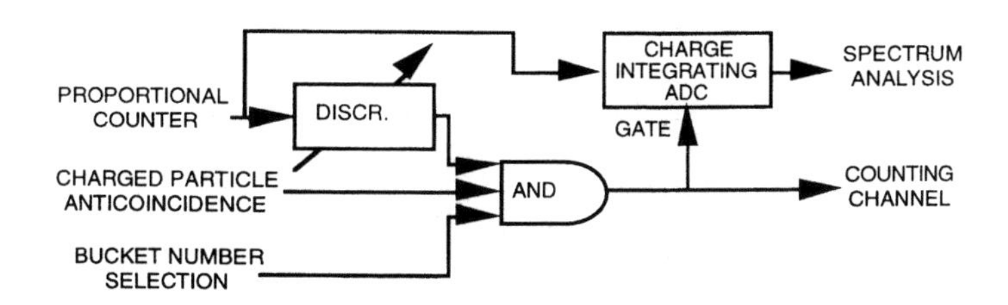

FIGURE 1. SB Measurement Typical Diagram.

The γ counting rate includes both SB and background photons. After background subtraction, as explained later, the SB counting rate N_{SB} can be obtained and used for calculating the luminosity value:

$$L = \dot{N}_{SB} \Big/ \int_{E_T}^{E_{MAX}} dE \int_{\Omega_T} d\Omega \frac{\partial^2 \sigma_{SB}}{\partial E \partial \Omega} \tag{4}$$

where E_{MAX} is the maximum energy a γ photon can have. (For relativistic particles it is practically half the center-of-mass energy of the beam.) σ_{SB} is the SB theoretical cross section, Ω_T is the solid angle viewed by the detector, as defined by a collimator placed in front of the proportional counter, and E_T is the minimum photon energy accepted by the system. The value of E_T is fixed by the discriminator threshold voltage level. In order to evaluate E_T, a calibration procedure is necessary. This is performed by using the GB spectra obtainable by the energy analysis system when only one beam is stored in the machine. The comparison between the well known GB theoretical spectrum shape and the experimental one allows E_T to be evaluated.

EXPERIMENTAL SETUP

Interaction Region Layout

DAΦNE is a ~100 m long collider with two independent vacuum chambers. Only at the two interaction regions (IRs) do the beams have a common chamber. Each IR, whose length is about 10 m, has splitter magnets at the extremes where the beams are horizontally separated into the different vacuum chambers. The beams collide at the IP with a horizontal angle tunable from 10 mrad to 15 mrad; standard operations are performed with 12.5 mrad. Figure 2 shows the splitter area and the position of the proportional counter. The layout allows the placement of two detectors in each of the IRs, but at the present time, only the one on the positron beam direction is installed.

The distance of the detector from the IP is ~6 m. At the splitter output, a 1.5 mm thick aluminum window allows the γ photons to come out of the vacuum chamber and to enter the detector. Between the thin window and the proportional counter, a 10-radiation-length thick lead collimator with a 10 mm radius circular aperture is used to fix the accepted solid angle within a cone of 1.6 mrad semi-aperture, centered on the crossing angle of 12.5 mrad. By integrating the SB cross section within this solid angle, it has been determined that 70% of the SB γ photons are accepted.

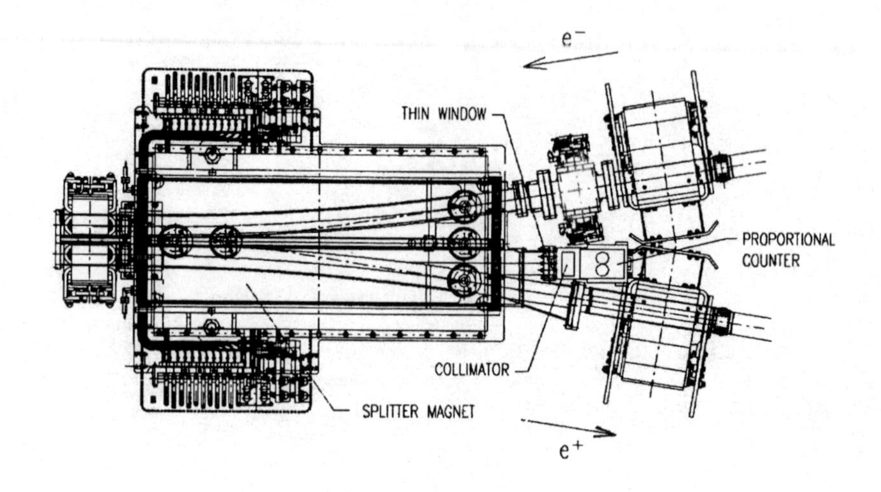

FIGURE 2. SB Luminosity Monitor layout at splitter area.

Special care has been used in shielding the monitor photomultipliers from the stray magnetic fields from nearby machine magnets.

Proportional Counter

The DAΦNE luminosity monitor proportional counter is a very fine-sampling lead-scintillating fiber calorimeter with photomultiplier read-out. The sampling structure is the same as the one used for the KLOE electromagnetic calorimeter (7).

The calorimetric module is built up by gluing 1 mm diameter blue scintillating fibers between thin-grooved lead plates, obtained by passing 0.5 mm thick lead foils through rollers of a proper shape. The grooves in the two sides of the lead are displaced by one half of the pitch so that fibers are located at the corners of adjacent, quasi-equilateral triangles, resulting in optimal uniformity of the final stack. The grooves are just big enough to insure that the lead does not apply direct pressure on the fibers. The blue-green scintillating fibers (Pol.Hi.Tech-46) provide high light yield, short scintillation decay time, and long attenuation length (8). The selected fiber pitch of 1.35 mm results in a structure which has a fiber:lead:glue volume ratio of 48:42:10 and a sampling fraction of ~15% for a minimum ionizing particle. The final composite has a density of ~5g/cm^3 and a radiation length of ~1.6 cm. It is self-supporting and can be easily machined. The resulting structure is quasi-homogeneous and has high efficiency for low-energy photons.

This kind of sampling calorimeter has been extensively tested (9) and has an excellent linearity and energy resolution for fully contained e.m. showers induced by photons, given by:

$$\frac{\sigma_E}{E} = \frac{4.4\%}{\sqrt{E_{(GeV)}}} \qquad (5)$$

Each of the luminosity monitor calorimeters has a squared face (122 × 122 mm^2) and is 184 mm long, corresponding to a 11.5 radiation length. Fibers are vertically positioned and are read on one side by two plastic light guides, each covering half the width (61 mm), but the whole length. Each light guide is equipped with a 2-inch photomulti-plier tube (EMI 9814B).

Energy Analysis and Counting System

Each of the IRs has a completely independent energy analysis and counting system (EACS). Figure 3 shows a diagram of this system.

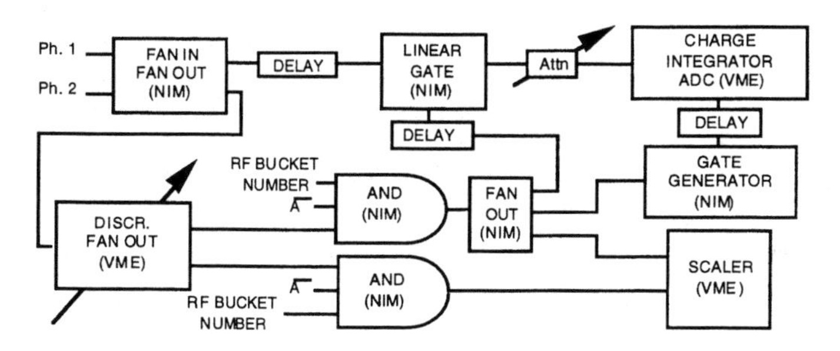

FIGURE 3. Energy Analysis and Counting System diagram.

The signals coming from the two proportional counter photomultipliers are summed at the EACS front end and then split into an analog channel for pulse integral analyses and into a logic channel for pulse counting and for shaping the system gates.

The charge integrating ADC needs a relatively long gate pulse, 120 ns FWHM. Most of the gate duration is needed for setting up the electronics. With such a gate, the rate of accidental counts gives a non-negligible contribution to the error in the energy spectrum acquisition. The 20 ns window of the linear gate upstream the ADC reduces this effect by a factor of six.

The EACS front end dynamic range accepts photomultiplier pulses with voltages down to −2.5 V, without saturating. The attenuator downstream of the linear gate permits matching the dynamic range of the ADC to this signal range. Figure 4 shows a linearity measurement of the EACS.

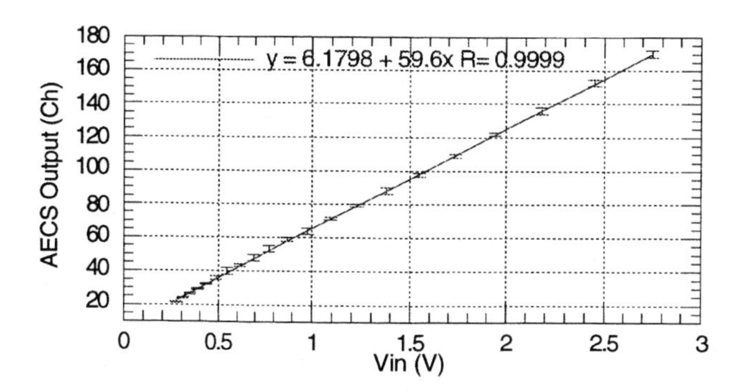

FIGURE 4. Energy Analysis System linearity.

The logic channel allows coincidence measurements with two different rf buckets, a feature used in one of the background subtraction schemes. The photomultiplier output pulse length at the pedestal is about 17 ns. In order to permit complete integration of the pulse, the linear gate ENABLE signal has been set with a duration of 20 ns FWHM. The distance between two contiguous buckets is 2.7 ns. This means that in measuring the coincidence with two different rf buckets, special attention must be paid to selecting and filling the two buckets at least 20 ns distant from each other.

The luminosity measurement in the multibunch mode must be performed without any coincidence with the rf. The 20 ns photomultiplier pulse length limits the counting system bandwidth to 50 MHz. At the maximum luminosity, 5×10^{32} cm^{-2} s^{-1}, the expected SB counting rate, with a threshold set to 0.7 times the beam energy, is 4 MHz, while the background rate is one order of magnitude lower. In this situation an underestimation of the counting rate of about 8% is expected. In order to improve this figure, photomultipliers with shorter output pulses must be used.

The two EACS are fully integrated in the DAΦNE control system. The components in Figure 3 labeled with "VME" are remotely controlled. The HV power supplies for the proportional counter photomultipliers are programmable VME boards. It is therefore possible to switch off one of the photomultipliers and perform the energy analysis of the signals coming from the other under computer control.

Background Subtraction Schemes

In the DAΦNE luminosity monitor, two background subtraction methods can be used: single counting channel pedestal subtraction (SCPS) and missing bunch background subtraction (MBBS).

The former uses a single counting channel of the EACS for getting the different counting rates when the two beams at the IP are in collision or separated. First the beams are separated and the background counting rate is recorded; then the rate with the beams again in collision is measured. The difference between these two values gives the SB counting rate that, when divided by the SB integrated cross section, gives the luminosity value. The DAΦNE IR has a low vertical beta configuration; the dimensions of the beam at the IP, at 1% coupling, are 20 μm rms and 2 mm rms in the vertical and horizontal planes respectively. This geometry makes the horizontal separation of the two beams impractical and longitudinal separation usable only with very low stored-current values. On the contrary, in the vertical plane, where bumps as high as ten sigmas can be obtained, the separation of the beams at the IP can be performed efficiently. Because of the finite lifetime of the beams, the background level must be periodically measured in order to maintain errors within an acceptable range. The SCPS method also allows luminosity measurements in the multibunch mode.

The MBBS method uses the following configuration. A single bunch is injected and stored into the electron beam. Two bunches are injected into the positron ring, one in the bucket colliding with the stored electron bunch and the other, usually with a smaller current, in a bucket not in collision. Assuming that the residual gas pressure does not change between the passage of the two positron beams and assuming that the background is due to the GB contribution only, it is possible to write:

$$\dot{N}_C = \dot{N}_{SB} + a_{GB} I_C^+ \tag{6}$$

$$\dot{N}_{NC} = a_{GB} I_{NC}^+ \tag{7}$$

where $\dot{N}_C$ and $\dot{N}_{NC}$ are the counting rates relative to the colliding and noncolliding bunches respectively, $\dot{N}_{SB}$ is the counting rate due to SB photons, a_{GB} is the GB coefficient and I_C^+ and I_{NC}^+ are the currents of the colliding and noncolliding positron bunches. It is straightforward to derive from expressions (6) and (7) the relation:

$$\dot{N}_{SB} = \dot{N}_C - \frac{I_C^+}{I_{NC}^+}\dot{N}_{NC} \tag{8}$$

By using formula (8) and knowing the current ratio of the positron bunches, it is possible to measure the SB counting rate independently of other machine parameters. Special attention must be paid in checking that the background contribution is being dominated by the GB term. For example, if the lifetime of one of the beams is short, then the e.m. showers generated by the beam particles hitting the vacuum chamber will give a significant contribution to the background. In this situation, formula (8) cannot be applied. The MBBS method does not allows luminosity measurements in the multibunch mode.

EXPERIMENTAL RESULTS

Gas Bremsstrahlung Threshold Calibration

The complete description of this threshold calibration method can be found in Reference (6). The principle is to analyze a measured GB spectrum by using results of the well known related theory. By multiplying the value of the spectrum at a channel by the channel number and repeating this operation for all the channels, the $1/x$ dependence of the spectrum can be removed (see an example in Fig. 5). In this modified spectrum, some properties are enhanced. First of all, for a relativistic beam, the maximum energy that a GB photon can have is practically the energy of the stored beam. The sharpness of this high energy limit of the GB spectrum is mitigated by the calorimeter resolution. In the modified version of the spectrum, the channel relative to this limit can be analytically derived. The knowledge of this channel and of the one related to zero energy, derivable from the linear fit of the energy analysis system response (see Figure 4), allows one to calibrate the value E_T of the low-energy threshold fixed by the discriminator. In the case of Figure 5, the high energy limit is 510 MeV at channel 142, the zero channel is 6, and E_T is 94 MeV at channel 25. The spectrum includes 1.5 Msamples.

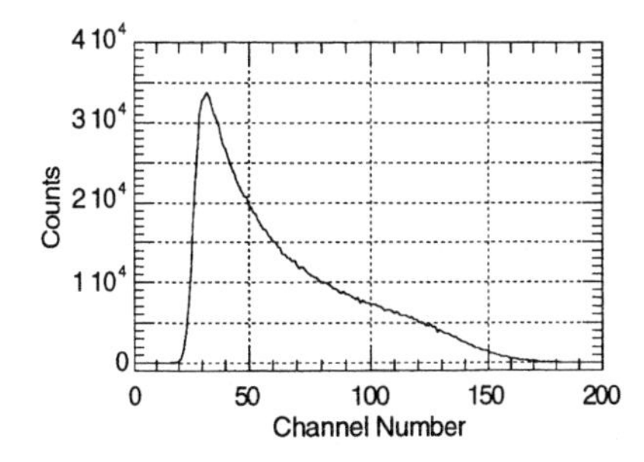

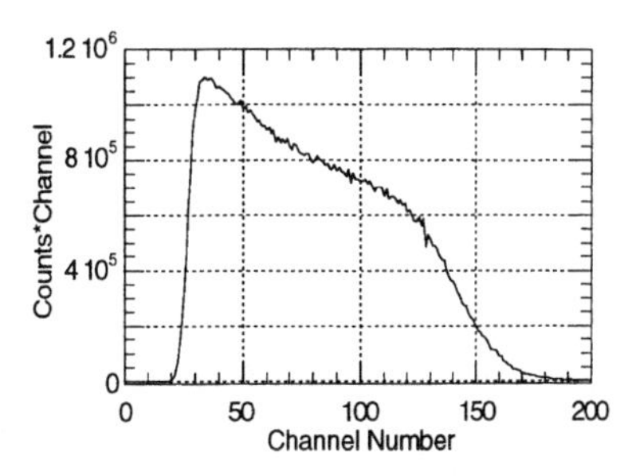

FIGURE 5. Gas bremsstrahlung measured spectra: raw and modified.

The rms resolution of the EACS is 1-2 channels out of 256. The resolution of the proportional counter at 510 MeV is about 6%; that corresponds, in the case of Figure 5, to about 8.2 channels. Thus in this case the global rms resolution of the monitor is about 8.4 channels and it is dominated by the calorimeter term. The total width of the spectrum is about 150 channels; assuming for a rough estimate a standard deviation of the GB modified spectrum of about 37.5 channels (150/4), then the contribution to the error in the threshold calibration can be estimated to be about 2.5%. In the measurement of the Figure 5, the energy 'width' of a channel is 3.75 MeV. This generates an additional indeterminacy on the threshold value of about 2%. Special care must be also taken in the determination of the zero energy channel. In the case of Figure 4 an error of 0.5 channels generates an indeterminacy of 2% on the threshold. The statistical error on the single channel counts must be kept lower than 1% so that the indeterminacy of the high-energy limit channel is negligible. This requires an average sampling per channel of at least 10^4 samples. The sampling rate of the EACS in the 256 channels scale is 1.1 kHz, permitting the acquisition of a 2 Msamples spectrum in half an hour. By summing all the contributions a total indeterminacy of ±6.5% of the threshold is obtained, generating an indeterminacy of ±5% on the SB integrated cross section and thus on the luminosity.

The above indeterminacy estimate has been done assuming that the gains of both the calorimeter photomultipliers are identical. The gain-equalizing procedure consists of measuring the GB spectrum of the individual tube and in regulating the gain in order to obtain the same high-energy threshold channel in both the photomultipliers. A bad equalization degrades the calorimeter resolution.

Luminosity Measurements

In Figure 6, a set of luminosity measurements performed with the SCPS method are shown. The typical accuracy is ±15%, with a 10% contribution coming from the threshold calibration due to the fact that the calorimeter photomultipliers were not well equalized. Other contributions are background evaluation (4%) and statistical fluctuations and accidental counts (1%). The measurements in Figure 6 were performed with different values of the positron beam coupling.

At the present time, luminosity measurements with the MBBS method are affected by larger errors. The main reason is that the necessary real-time measurement of the positron bunch current ratio has not yet been implemented.

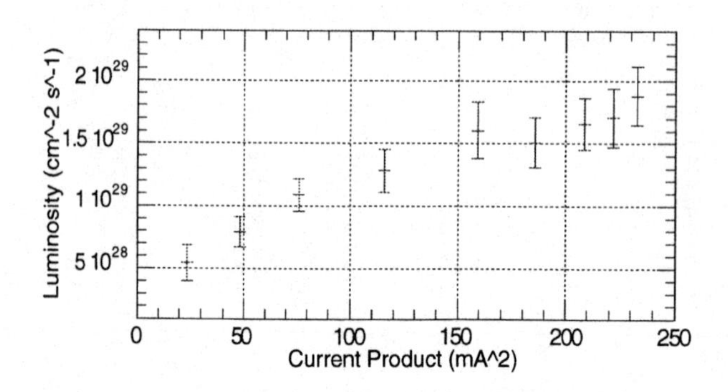

FIGURE 6. Example of luminosity measurements on DAΦNE.

ACKNOWLEDGMENTS

The authors want to thank O. Coiro, O. Giacinti and D. Pellegrini for their prompt and skillful assistance during the installation and the optimization of the monitor electronics, and M. Anelli for his precious advice on photomultiplier issues. Finally the authors are particularly indebted to P. Possanza for turning a chaotic manuscript into this paper.

REFERENCES

[1] Vignola G. and The DAΦNE Project Team, "DAΦNE the First Φ-Factory," *Proceedings of the V European Particle Accelerator Conference*, Sitges, Spain, 10–14 June 1994.
[2] The KLOE Collaboration, "KLOE, a General Purpose Detector for DAΦNE," Frascati Internal Note LNF-92/109, April 1992.
[3] Altarelli G., F. Buccella, "Single Photon Emission in High-Energy $e^+ e^-$ Collisions," *Il Nuovo Cimento XXXIV*, **5**, pp. 1337–1346, December 1, 1964.
[4] Di Vecchia, P., M. Greco, "Double Photon Emission in $e^+ e^-$ Collisions," *Il Nuovo Cimento L A*, **2**, pp. 319–332, July 21, 1967.
[5] Preger, M., "A Luminosity Monitor for DAΦNE," DAΦNE Technical Note IR-3, Dec. 17, 1993.
[6] Dehne, H. C., et al., "Luminosity Measurement at ADONE by Single and Double Bremsstrahlung," *NIM* **116,** pp. 345–359, 1974.
[7] Lee-Franzini, J., et al., "The KLOE e.m. Calorimeter," *NIM A* **360**, pp. 201, 1995.
[8] De Zorzi, G., *Proceedings 4th International Conference on Calorimetry in High Energy Physics*, La Biodola, Italy, 1993.
[9] Miscetti, S., *Proceedings 4th International Conference on Calorimetry in High Energy Physics*, La Biodola, Italy, 1993.

Diagnostics for a 1.2 kA, 1 MeV, Electron Induction Injector

T. L. Houck,* D. E. Anderson, S. Eylon, E. Henestroza, S. M. Lidia, D. L. Vanecek, G. A. Westenskow*, and S. S. Yu

*Lawrence Livermore National Laboratory (LLNL), Livermore, CA 94551
Lawrence Berkeley National Laboratory (LBNL), Berkeley, CA 94720

Abstract. We are constructing a 1.2 kA, 1 MeV, electron induction injector as part of the RTA program, a collaborative effort between LLNL and LBNL to develop relativistic klystrons for Two-Beam Accelerator applications. The RTA injector will also be used in the development of a high-gradient, low-emittance, electron source and beam diagnostics for the second axis of the Dual Axis Radiographic Hydrodynamic Test (DARHT) Facility. The electron source will be a 3.5"-diameter, thermionic, flat-surface, m-type cathode with a maximum shroud field stress of approximately 165 kV/cm. Additional design parameters for the injector include a pulse length of over 150 ns flat top (1% energy variation), and a normalized edge emittance of less than 200 π-mm-mr. Precise measurement of the beam parameters is required so that performance of the RTA injector can be confidently scaled to the 4 kA, 3 MeV, and 2-microsecond pulse parameters of the DARHT injector. Planned diagnostics include an isolated cathode with resistive divider for direct measurement of current emission, resistive wall and magnetic probe current monitors for measuring beam current and centroid position, capacitive probes for measuring A-K gap voltage, an energy spectrometer, and a pepperpot emittance diagnostic. Details of the injector, beam line, and diagnostics are presented.

INTRODUCTION

Induction accelerators are a unique source for high-current, high-brightness, electron beams. A collaboration between the Lawrence Livermore National Laboratory (LLNL) and Lawrence Berkeley National Laboratory (LBNL) has been studying the application of induction accelerators as drivers for relativistic klystrons. The relativistic klystron is a very high-power microwave source that could be used to power future linear colliders based on the Two-Beam Accelerator (RK-TBA) concept (1). As part of the collaboration, a prototype relativistic klystron is being constructed at LBNL to study issues concerning physics, engineering, efficiency, and cost (2). A major technical challenge to the successful operation of a full scale relativistic klystron is the transport of the electron beam through several hundred meters of narrow aperture microwave

CP451, *Beam Instrumentation Workshop*
edited by R. O. Hettel, S. R. Smith, and J. D. Masek
1998 The American Institute of Physics 1-56396-794-4/98/$15.00

extraction structures and induction accelerator cells. Demanding beam parameters are required of the electron source, an induction injector, to achieve the transport goals. The RTA injector is currently undergoing testing of its pulsed power system with beam tests scheduled in three months.

Induction accelerators are also used to produce intense x-ray sources for radiographic applications. The Dual Axis Radiographic Hydrodynamic Test (DARHT) Facility, under construction at the Los Alamos National laboratory (LANL), will use two induction accelerators. A LLNL and LBNL collaboration is designing the induction accelerator for the second axis. This accelerator, an upgrade over the first axis, will produce a 4 kA, 20 MeV, 2-microsecond electron pulse that can be "chopped" into shorter pulses to generate a series of radiographic images. The major challenge for this accelerator is achieving the required x-ray spot size. This spot size defines the maximum beam emittance at the bremsstrahlung converter. Once again, very demanding beam parameters are required at the electron source to achieve accelerator performance goals.

THE RTA INJECTOR AND DIAGNOSTIC LAYOUT

The RTA injector, depicted in Figure 1, is comprised of 24 three-core induction cells and a thermionic cathode. Beam focusing is accomplished by three large-bore solenoids installed on the central pumping spool and seven smaller solenoids located within the anode stalk. Cathode current and A-K gap voltage diagnostics are included.

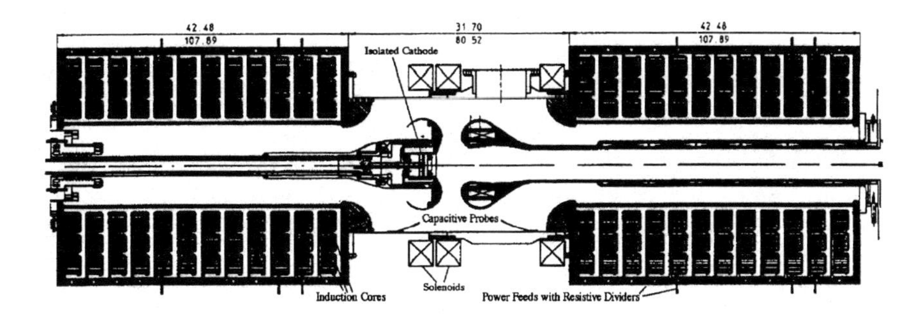

FIGURE 1. Depiction of the RTA injector indicating the locations of the isolated cathode, capacitive (dV/dt) probes, and resistive dividers for the power feeds.

The majority of the diagnostics will be installed after the injector as indicated in Figure 2. The first 1.4 m of beam line will include two beam position and current monitors to allow the offset and angle of the beam at the exit of the injector to be measured. A pop-in probe will be incorporated in a pumping port to allow the beam profile to be viewed. A pepperpot emittance diagnostic is shown in Figure 2. However, several different diagnostic packages will be installed depending on the beam parameters to be measured or diagnostics to be studied.

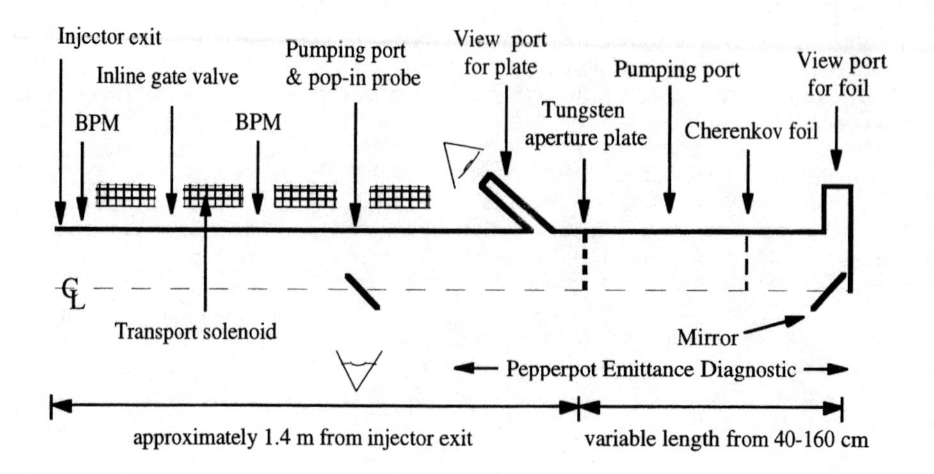

FIGURE 2. Beamline layout for emittance measurements. Other layouts will feature additional pop-in probes, an energy spectrometer, and/or an isolated beam dump.

DIAGNOSTICS

A variety of diagnostics will be used to determine the performance of the injector, both permanently installed monitors for general operational and temporary diagnostics specific to the injector commissioning and troubleshooting. A general issue for all the diagnostics is the pulse length. Recent induction accelerators at LLNL and LANL operate with pulse lengths of 50 to 80 ns. Many of the diagnostics and measurement techniques developed for those accelerators will need to be modified. For example, interceptive diagnostics such as Cherenkov foils are much more at risk of damage at the longer pulse lengths. The diagnostics to be used on the RTA injector will also serve as prototypes for the DARHT injector.

Cathode Current

An accurate measurement of the emitted current from the cathode is required both for determining the performance of the injector and benchmarking codes. Several techniques were considered and eventually rejected due to geometric, thermal, and/or reliability constraints. The final selection is to electrically isolate the cathode from the stalk and forcing the current to flow through several parallel, 0.25-inch-wide strips of 25 μm-thick nichrome foil that act as current-viewing resistors. The potential drop across the foil is measured and the current inferred. To improve the time response ($\approx L/R$), a parallel-strip shunt geometry is used where the foil is folded on itself to increase the resistance while lowering the series inductance. This well-known technique is particularly amenable for the restrictive geometry of the cathode housing.

A-K Voltage and Beam Energy

Three different methods will be used to determine the A-K voltage and beam energy. The first method involves measuring the applied voltage to the induction cores with resistive dividers at the connection of the power feeds to the induction cells. The resistive dividers are comprised of 15 similar resistors in series. This arrangement minimizes the problem of voltage dependent properties. The resistors should evenly divide the total potential drop, and the applied voltage is simply the number of resistors times the potential across any one resistor. Summing the applied voltage of all the power feeds gives the applied voltage to the A-K gap. Resistive dividers can be noisy and have a limited bandwidth. However, for our application, the bandwidth is not a limitation, as the voltage rise time is approximately 100 ns.

Capacitive *dV/dt* pickup probes are used for a more direct measurement of the A-K gap voltage and also to provide greater bandwidth with respect to the resistive dividers. The probe signal is approximately

$$V_s \cong (C_p Z_o) \, {}^{dV_o}\!/_{dt}, \tag{1}$$

where Z_o is the transmission line impedance and C_p is the capacitance between the electrodes and the probe. For our experiment Z_o is 50 ohms, C_p is 0.5 *pF*, the voltage rise time is 5×10^{12} *V*/s producing a signal voltage of 125 volts. The signal is then integrated to determine V_o. The minimum response time of the probe is $Z_o C_s$ where C_s is the stray capacitance between the probe and the wall. C_s is approximately 100 *pF* producing a response time of 5 ns. The maximum time is set by the integrator. We use printed circuits for our capacitive probes as illustrated in Figure 3. At least two probes, separated longitudinally, are required to resolve the contributions from the anode and cathode halves of the injector.

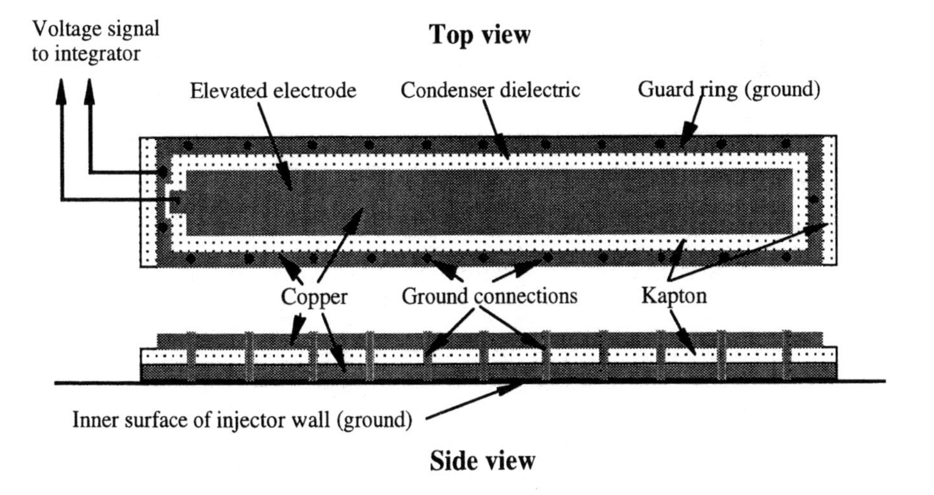

FIGURE 3. Schematic of the printed circuit capacitive dV/dt pickup probe.

A conventional energy spectrometer comprised of an on-axis collimator, dipole magnet, scintillator, and viewing port will be used to directly measure beam. By varying the beam radius at the entrance of the graphite collimator, the current density at the scintillator can be adjusted. The spectrometer is on loan from LANL where it was calibrated to 1% accuracy in absolute energy. Relative variations in beam energy can be determined to 0.1% and the total energy variation that can be imaged across the scintillator is about 5%. A streak camera will be used to look at energy variations across the flat-top portion of the pulse. A gated camera will be used to examine the beam energy during specific time slices of the pulse, e.g. during the rise time.

Beam Current and Centroid Position

Two different methods will be used to determine the beam current and centroid position. Magnetic pickup (B-dot) loops will determine the time derivative of the current pulse. The voltage induced on the loop can then be integrated to recover the current. Our B-dot loop diagnostics will consist of eight loops evenly spaced around the inner diameter of a one-inch-thick flange. The loops are recessed into the flange to avoid beam interception. Four evenly spaced loops are summed to minimize the error in total current due to offsets in beam centroid. Geometrically opposite pairs are differenced to generate signals that are proportional to the centroid offset. The B-dot loops have the advantage of simple construction and a broad frequency response. The principle disadvantage is the requirement for integrators.

Resistive wall current monitors measure the potential drop of the return wall current across a known resistance generating a signal proportional to the current. Our monitors will consist of 25 μm-thick nichrome foil that spans a short insulated break around the circumference of the beam tube wall and have a total resistance of a few milliohms. The diameter of the foil matches the inner diameter of the beam tube to avoid discontinuity in the beam pipe wall. An inductive material is placed outside of the foil to force the wall current to flow through the foil. The time response of the monitor, determined by the series inductance and resistance, is about 5 ns. The parallel inductance of the monitor will cause the signal to decay or "droop." This is expected to be small, but noticeable, for the 250 ns FWHM pulse length. The output signal can be corrected for known droop in the diagnostic through compensation circuits, either passive or active. Alternatively, for digitized signal data, the correction can be done with analysis software. Eight voltage pickoffs evenly spaced around the diameter of the foil are used to determine beam current and centroid position, similar to the B-dot loops. The sensitivity of the monitor is approximately a volt per kiloampere of current. The resistive wall current monitors have the advantage of producing a signal proportional to the current. However, the frequency response is less than that of B-dot loops.

The beam dump will be electrically isolated through nichrome film (current-viewing resistors) similar to the cathode to allow for the measurement of the total current deposited in the dump. However, the centroid position will not be determined.

The errors inherent in the above monitors for off-axis beams are easily calculated and not expected to be an issue. The major concern is in the calibration of the diagnostics. The nominal tolerance for commercial components used in the monitors would limit accuracy to about 5%. By calibrating complete systems, i.e. the individual diagnostic with cables, integrators, and compensation circuits, as applicable, we hope to achieve accuracies approaching 1%.

Current Density Profile

The current density profile will be measured using Cherenkov and/or optical transition radiation from intercepting foils. A primary concern with using foils is possible damage from beam energy deposition. Average heating of the foil can be controlled by adjusting the repetition rate of the injector. The difficulty lies in the single-shot heating where material can be melted and ejected before the heat is conducted away. If the foils are sufficiently thin that the radiation generated by bremsstrahlung escapes, assuming the beam pulses are short compared to the thermal conduction time, the temperature rise in the foil is approximately:

$$\Delta T = \frac{J\Delta t}{\rho C_{pe}}\left[\frac{dT_e}{dz}\right]_c, \tag{2}$$

where J is the current density, Δt is the pulse length, ρ and C_p are the density and specific heat of the foil, respectively, and $[dT_e/dz]_c$ is the collisional stopping power for the electrons. For a given foil material, the minimum beam radius at intercept can be determined from Equation 2. Since the collisional stopping power is reasonably constant with energy for relativistic electrons, Equation 2 may also be used to scale the measured performance of the foils to the higher energy of the DARHT injector. For example, the minimum beam diameter to avoid damage for a 1 kA, 300 ns, relativistic electron beam pulse on a thin quartz foil is about 2 cm. We anticipate using a number of foil materials, including kapton, quartz, graphite, tantalum, and tungsten.

The light generated at the beam/foil interaction will be recorded using both gated and streak cameras. The streak camera will be used principally to determine if the properties of the foil and/or beam change during the pulse. The significant levels of energy deposited in the foil could affect the dielectric constant or generate a surface plasma that could be confused as a variation in beam parameters.

Emittance

Measuring the beam emittance is expected to be very difficult as the beam is highly space charge dominated. A pepperpot emittance diagnostic is being constructed. The effect of space charge can be appreciated by considering the envelope equation for a round, uniform beam in a drift region:

$$\frac{d^2R}{dz^2} = \frac{\varepsilon^2}{R^3} + \frac{2I}{RI_A(\beta\gamma)^3}, \tag{3}$$

where R is beam radius (edge), ε is the unnormalized edge emittance, and I_A = 17 kA. The cathode is in a magnetic field free region. We desire for the emittance to dominate space charge effects for the beamlets in the region following the pepperpot, i.e.

$$\frac{\varepsilon_n^2}{R_b^2} >> \frac{2I}{\beta\gamma I_A}, \text{ or } \varepsilon_n^2 >> \frac{I_b h^2}{2\beta\gamma I_A}, \text{ where } I = \left(\frac{h}{2R_b}\right)^2 I_b. \tag{4}$$

R_b and I_b are the beam radius and current at the front of the pepperpot, and h is the aperture. The size of the aperture is the only variable for adjusting the relative contribution of emittance to space charge. For the designed RTA injector beam parameters, 1.2 kA, 1 MeV and 100 π-mm-mr, and using a 250 μm aperture, the emittance term is approximately an order of magnitude larger than the space charge. Our aperture plate will consist of a rectangular pattern of 121 (11×11) 250 μm apertures with 7 mm spacing on a 500 μm thick tungsten plate. The tungsten plate represents about two range thickness for 1 MeV electrons. A second 1 mm-thick tungsten plate with nominal 750 μm apertures will be placed in front of the thinner plate to improve thermal conduction. At four locations, apertures will not be drilled in the thicker plate to assist in determining background and orientation. The beamlets will strike a quartz foil located 80 cm (adjustable from 40 to 160 cm) after the aperture plate. The Cherenkov light generated at the foil will be imaged with a gated camera.

Small effects such as the aperture plate thickness (vignetting), fringe fields from the upstream solenoids, and beam waist location will be accounted for in the data analysis. A more serious problem concerns the effect of the conductive aperture plate on the beam. It has been demonstrated (3) that, while the local distribution in phase space can be determined, the global x–x' curve is dominated by the non-linear focusing of the aperture plate. E-Gun simulations performed for our beam parameters indicate that the beam emittance determined from the pepperpot data will be as much as six times the actual emittance in the absence of the aperture plate. This effect can be accounted for in the analysis. However, the accuracy in the final emittance value will suffer. Note that this is only an issue for space charge dominated beams where aperturing is required.

An alternative to the pepperpot diagnostic is to vary the focusing of the beam and look at changes in the radial profile of the beam. The most straight forward method will be to insert a nonconducting foil a short distance after a solenoid. The strength of the solenoid could then be increased causing the focus to move through the foil. Matching the observed radius variation with the applied solenoidal field to computer simulations will allow the emittance to be inferred. Once again, the issue will be the minimum beam spot size that the foil can tolerate. By operating in an over-focus regime, i.e. the beam waist always remains in front of the foil, it may be possible to limit the minimum beam radius at the foil, but still generate significant variations in radius. We are also studying the effect of using a quadrupole magnet to focus in a single plane. This will reduce the issue of current density, but limit the emittance measurement to a single plane.

SUMMARY

A variety of diagnostics will be used to characterize the beam produced by the RTA injector. These diagnostics have been extensively developed over the past decade for use on induction injectors/accelerators. However, the challenge for the RTA program is to adapt these diagnostics for pulse lengths a factor of four longer, and, in the development of diagnostics for the DARHT second axis, a factor of nearly 40 longer. Determining emittance of the highly space charge dominated beam will also require new levels of precision and/or technique.

ACKNOWLEDGMENTS

Many people have provided advice and insight regarding our diagnostic issues. In particular, we thank Tom Fesseden, Bill Fawley, and Mike Vella of LBL, and Dave

Moir and Mike Brubecker of LANL. The work was performed under the auspices of the U.S. Department of Energy by LLNL under contract W-7405-ENG-48, and by LBNL under contract AC03-76SF00098.

REFERENCES

[1] Sessler, A. M., S. S. Yu, *Phys. Rev. Lett.*, **54**, 889 (1987).
[2] Houck, T. L., et al., *IEEE Trans. Plasma Sci.*, **24**, 938–946 (1996).
[3] Hughes, T. P., R. L. Carlson, D. C. Moir, *J. Appl. Phys.*, **68**, 2562–2571 (1990).

Beam Current Monitors at the UNILAC

N. Schneider

Gesellschaft für Schwerionenforschung mbH
(GSI), D-64291 Darmstadt

Abstract. One of the most basic linac operation tools is a beam current transformer. Using outstanding materials, the latest low-noise amplifiers, and some good ideas, a universal current monitoring system has been developed and installed at the UNILAC at GSI. With a dynamic range of 112 dB, covering the low-current range down to 100 nA peak to peak at S/N=1, as well as 25 mA pulses, provided for high-current injection to the SIS synchrotron, a well-accepted diagnostic instrument could be placed at the disposal of the operaters.

INTRODUCTION

Due to the upgrading of the prestripper section of the UNILAC accelerator at GSI/Darmstadt, it was necessary to improve the performance of the current transformers in the beam lines.

To measure currents with good sensitivity (S/N>>1), it is mandatory to have a high current transfer ratio, i.e. a low number of secondary windings, while for low droop it is necessary to have high inductance, i.e. a high number of secondary windings.

This problem has been solved by using ultra-low-noise amplifiers, high-permeability cores with low magnetostriction, signal clamping and the application of a low-frequency feedback (see Fig. 2), which corrects the first 8ms of the signal to less than 0.5% droop.

HEAD AMPLIFIER

The "heart" of the head amplifier is an operational amplifier of outstanding performance: the input noise is less than that of a 50Ω resistor and the slew rate is 15V/μs. Integrated in the head amplifier are also a clamp and a differential signal driver, as well as the ability to perform range switching.

To accurately determine the mean value of the beam current, an integrating measurement must be performed. The integration time should correspond to the gate pulse duration.

CP451, *Beam Instrumentation Workshop*
edited by R. O. Hettel, S. R. Smith, and J. D. Masek

FIGURE 1. Picture showing core, housing, and head amplifier and digitizer, connected with a ten-wire twisted-pair cable of a maximum of 300 feet in length.

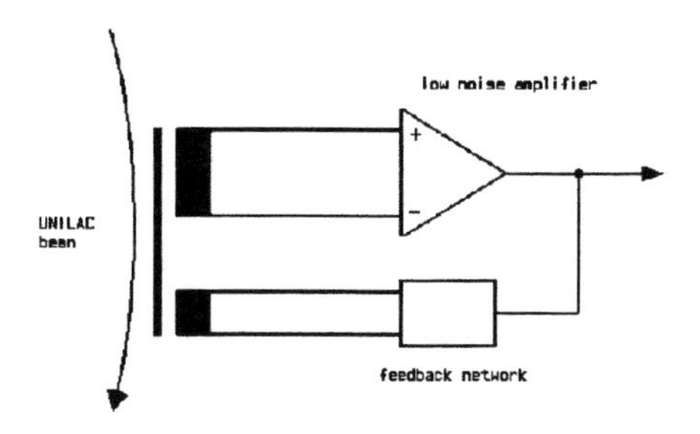

FIGURE 2. Schematic of head amplifier.

INTEGRATING DIGITIZER

A simple method of bringing the integration time into correspondence with the gate pulse duration is integration by quantified charge compensation. The signal to be digitized is integrated by an analog integrator, and compensated by pulses of constant amplitude and duration, but of opposite polarity (see Fig. 3).

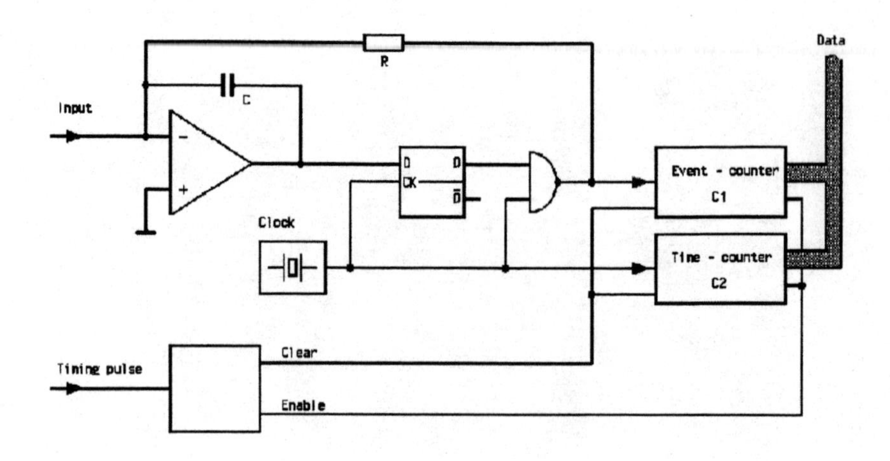

FIGURE 3. Schematic of integrating digitizer.

The digitizing sequence is started by a timing pulse, delivered by the UNILAC timing system, which is also used to disable the clamp in the head amplifier.

In the first stage a small capacitor (*C*) of an integrator is charged (see Fig. 3). When the output passes a threshold, a D-flipflop is set. With the following clock pulse, a well-defined charge amount is subtracted from the capacitor through the resistor (*R*) and, at the same time, a counter (*C1*) is incremented. A second counter (*C2*) is increased during the timing pulse by the clock pulses, representing the pulse duration.

The practical setup works with clock frequencies up to 24 MHz. Of utmost importance is a very stable power supply, because its value affects the result most directly.

After one measuring sequence (machine cycle) the two counters contain:

$$C1 = i \times t \times f_c \times C_f \tag{1}$$

$$C2 = t \times f_c \tag{2}$$

where i = current, t = pulse width, f_c = clock frequency and C_f = constant factor. The quotient $C1/C2 = i \times C_f$ gives the mean value of the current, independent of the pulse length.

As it is well known that laboratory values are something entirely different than the rough business at the beam line, we found that the noise floor due to (external) hum and microphony was measured to be more than 500 nA peak-to-peak, reducing the dynamic range.

Therefore, it was decided to further shift the range up to 50 mA, giving up the high sensitivity, but keeping the dynamic range. Nevertheless, the installed current monitors are very useful linac equipment and well accepted by the operating crew.

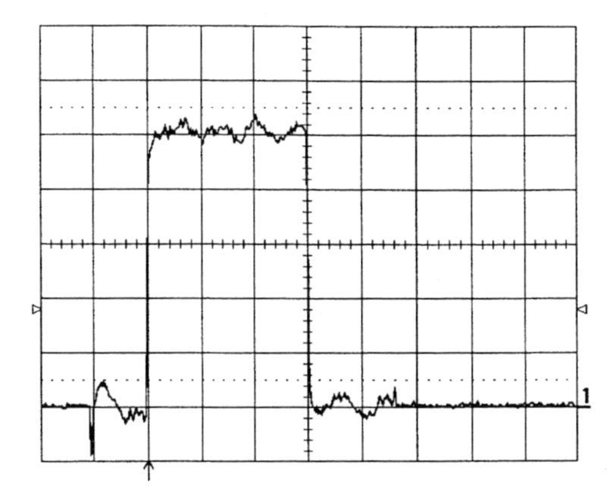

FIGURE 4. Scope picture showing a 5 μA pulse of 600 μs duration within the 1 ms clamp pulse, risetime < 8 μs.

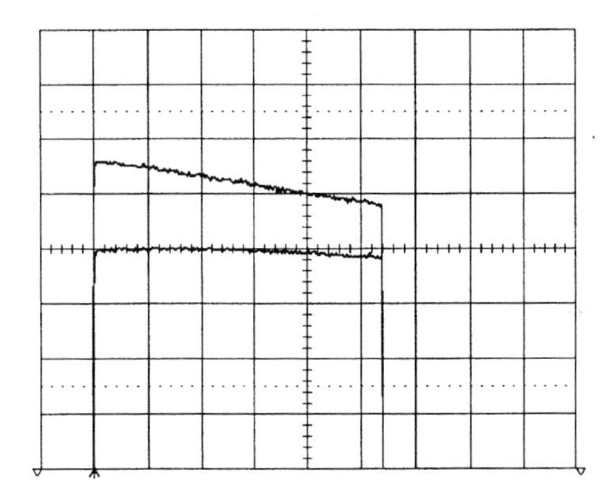

FIGURE 5. Scope picture showing a 11 ms pulse with and without using the feedback loop. Vertical scale is 5% of the signal magnitude.

In the future the equipment will be extended. The interlock system of the UNILAC has to be improved due to higher beam intensities. By monitoring the difference between subsequent beam current transformers, any beam losses between them can be detected and an interlock invoked.

CONCLUSION

It is possible to work near the physical limits of monitoring beam currents if some certain facts are attended to, such as:

- using high permeability cores

- good shielding
- differential signal transmission
- using low noise components
- stable power supplies

ACKNOWLEDGMENTS

Finally, I want to thank H. Walter, who managed the digital handshake to the control system and EPLD programming, as well as H. Reeg, who did all of the mechanical constructions, the boring exploring of core and shielding materials, and the very useful SPICE simulations.

Gated Beam Imager for Heavy Ion Beams

Larry Ahle and Harvey S. Hopkins

Lawrence Livermore National Laboratory
P.O. Box 808, Livermore, CA 94551

Abstract. As part of the work building a small heavy-ion induction accelerator ring, or recirculator, at Lawrence Livermore National Laboratory, a diagnostic device measuring the four-dimensional transverse phase space of the beam in just a single pulse has been developed. This device, the Gated Beam Imager (GBI), consists of a thin plate filled with an array of 100-micron diameter holes and uses a Micro Channel Plate (MCP), a phosphor screen, and a CCD camera to image the beam particles that pass through the holes after they have drifted for a short distance. By time gating the MCP, the time evolution of the beam can also be measured, with each time step requiring a new pulse.

INTRODUCTION

Lawrence Livermore National Laboratory has, for the past few years, been developing the world's first circular ion induction accelerator, or recirculator, as part of its heavy-ion fusion research program. A critical task of this development is measuring and understanding any change in beam quality as a pulse travels around the accelerator. Of specific interest is the measurement of the first and second moments and the emittance growth. The standard tool used to measure the emittance of heavy ion beams is the slit scanner. This device uses two parallel slits, one downstream of the other, to measure the two-dimensional, either x-x', in plane, or y-y', out of plane, phase space. Not only is this device not able to measure both simultaneously, it also requires many beam pulses for one emittance measurement, i.e. one beam pulse each time one of the slits is moved.

In an effort to improve on this measurement device, the Gated Beam Imager (GBI) was developed at LLNL. This device uses small holes instead of slits, a micro-channel plate (MCP) to convert ions to electrons, and a phosphor screen to produce images on a CCD camera. The result is a device which measures both x and y emittance for a given time interval with a single beam pulse. Changing the timing gate of the MCP allows one to measure the time evolution of the beam pulse. Thus, the Gated Beam Imager obtains the same information as the slit scanner with a factor of 20 reduction in the number of beam pulses required.

CP451, *Beam Instrumentation Workshop*
edited by R. O. Hettel, S. R. Smith, and J. D. Masek

THE DEVICE

The Gated Beam Imager is based on the pepperpot beam diagnostic method which has previously been used to diagnose electron beams (1). In the pepperpot method where a mask, or hole plate, with small holes (the pepperpot) is introduced into the beam, the beam ions are stopped by the hole plate except where they pass through the holes forming small beamlets. These beamlets pass through a drift region where they freely expand from space-charge and emittance forces. Making the holes small enough limits the space charge-induced expansion of the beamlets to a small percentage of the expansion due to emittance, the quantity of interest. The beamlets are intercepted by a detector which is excited and the spots are observed with a CCD camera, and digitized for analysis. Figure 1 shows a schematic of the GBI with several key dimensions listed.

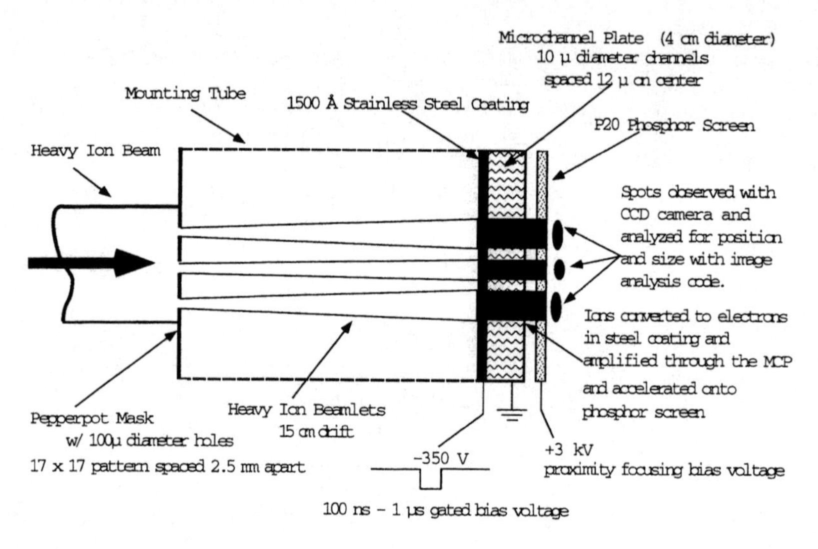

FIGURE 1. Schematic view of GBI.

Several dimensions of the GBI such as pepperpot hole size, hole spacing, and beamlet drift length must be sized to achieve the goal of maximizing the beamlet's growth due to emittance while minimizing the growth from remaining space-charge forces. Additionally, adequate signal strength must be present at the CCD camera and the beamlet image spot size must be much greater than the camera system resolution so that statistical averages of the CCD pixels are valid. These requirements place constraints on the pepperpot hole size, hole spacing, and beamlet drift length.

The application of a micro channel plate as the GBI detector allows consistent and repeatable signal output with increasing cumulative exposure to damaging heavy ions. The 1500-angstrom stainless steel coating on the input side of the MCP stops the heavy ions, resulting in a secondary electron signal proportional to the input heavy-ion signal (2). These electrons are then amplified through the MCP and proximity focused onto an output phosphor screen and viewed by a CCD camera through an optical lens. The optical focus allows the small CCD chip to see the entire phosphor screen at the cost of a large image pixel size. The proportional conversion of ions to electrons while maintaining the spatial relationship sidesteps the signal reduction with increasing dose observed when the phosphor output is directly exposed to heavy ions (3). The microchannel plate detector also allows amplifying weak signals making it possible to gate the beamlets rapidly and still have enough output signal at the phosphor screen.

THE ANALYSIS

After taking an image with the GBI, two corrections to the image must be made before it can analyzed to the find the emittance. First the dark charge contribution must be subtracted away. When the CCD chip has its bias voltages applied, a small amount of charge will build up in each pixel due to the thermal creation of electron-hole pairs. The longer the exposure, the more dark charge is built up. Since, the CCD chip reads out one pixel at a time, the amount of dark charge will be different for each pixel, less for the first pixel read out, more for the last. This effect is corrected for by taking an image with no incident light on the chip and subtracting that image from the real data image. Beyond this dark charge correction, the image must be corrected for any nonuniform response of the detector, or flat fielded. A uniform x-ray source is used as the baseline and an image of this source is taken. This image is used to calculate a correction factor for each pixel.

Once both of these corrections have been made to the data image, it can then be analyzed to determine the emittance. The first step is to search for beamlets on the CCD image. Contiguous groups of pixels above an intensity threshold are found. The exact value of the threshold is a little bit trial and error, but should be large enough so that pixels with only background do not contribute to these groups. If a single group consists of enough pixels, the group is said to be a beamlet. This cut on the number of pixels is to insure that holes in the hole plate which only have a fraction of its area overlapping with the cross section of the beam do not contribute to the emittance calculation. After finding all beamlets on the image, each one must be associated with the appropriate mask hole. This is facilitated by four extra holes around the center hole of the hole plate. This uniquely determines the center hole beamlet on the CCD image and the other beamlets are matched with holes relative to the center hole.

After this beamlet image to hole association, the moments are then calculated. Only a single row(column) of pixels for each beamlet is used in the $x(y)$ emittance calculation. This is done to more closely emulate a slit scanner analysis for comparison and to eliminate any correlations in the x and y moments which arise from the circular hole shape. The row and column used is determined by the pixel of peak intensity for each beamlet. The extent of pixels used in each row and column is determined by the first pixel on each side of the peak pixel, that has an intensity below 10% of the peak intensity. This cut is an exclusive one.

For each pixel that has passed this cut, there is an intensity, I_m, and position of the center of the pixel, x_m. Let x_h equal the center position of the hole, then

$$x'_m = \frac{x_m - x_h}{L} \tag{1}$$

where L is the drift length. Also, let

$$I_{tot} = \sum_m I_m \tag{2}$$

where this sum and all sum subsequent are over all pixels used for all beamlets. Given these equations, one uses Equations 3 through 7 below to calculate the moments.

$$\langle x \rangle = \frac{1}{I_{tot}} \sum_m x_m I_m \tag{3}$$

$$\langle x' \rangle = \frac{1}{I_{tot}} \sum_m x'_m I_m \tag{4}$$

$$\langle x^2 \rangle = \frac{1}{I_{tot}} \sum_m \left(x_m - \langle x \rangle \right)^2 I_m \tag{5}$$

$$\langle x'^2 \rangle = \frac{1}{I_{tot}} \sum_m \left(x'_m - \langle x' \rangle \right)^2 I_m \tag{6}$$

$$\langle xx' \rangle = \frac{1}{I_{tot}} \sum_m \left(x_m - \langle x \rangle \right)\left(x'_m - \langle x' \rangle \right) I_m \tag{7}$$

The definition of emittance used for this experiment is the rms emittance, defined below as,

$$\varepsilon_{rms}^2 = \langle x^2 \rangle \langle x'^2 \rangle - \langle xx' \rangle^2 \tag{8}$$

with the definition of normalized emittance as,

$$\varepsilon_{norm} = 4\gamma\beta\varepsilon_{rms} \tag{9}$$

where β and γ have the usual definitions of special relativity and whose product equals 0.00209 for this experiment.

VERIFICATION

The Gated Beam Imager was designed as an improvement to existing diagnostic devices, so it should give similar results. To verify this, measurements were taken on the Recirculator at similar positions for the GBI and the slit scanner. The configuration of the Recirculator was a 80 keV, 1.8mA, 4 μs long, K+ beam pulse. The measurement was done after the beam had traveled through a 90 degree bend without any acceleration. The slit scanner used had two 50 μm-wide slits with one 7.4 cm downstream from the other. The downstream slit had a Faraday cup attached behind it to measure the current of the beam that made it through both slits. Figures 2 and 3 show the results.

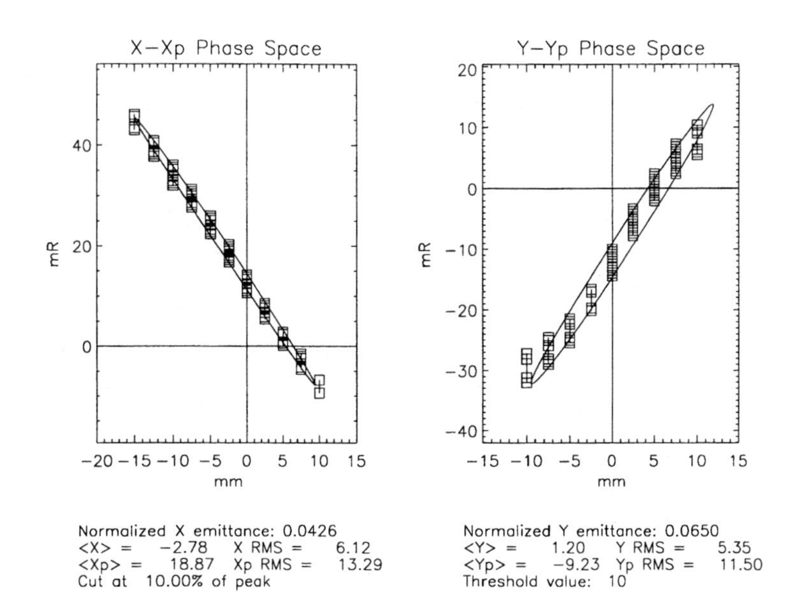

FIGURE 2. Gate Beam Imager data.

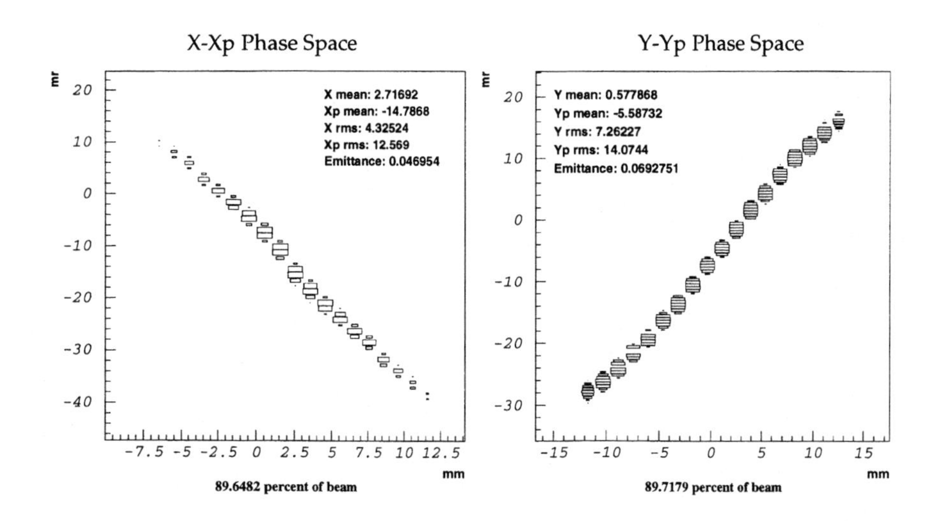

FIGURE 3. Slit Scanner results.

When comparing data from the two devices one should ignore the differences in the first moments, since this is a function of the inability to precisely align the beamline with the vacuum tank that housed both detectors. While the GBI is in this vacuum tank, slit scans cannot be done. The whole tank had to be rotated to do the slit scan in both directions. In addition, the second moments are also slightly different because the beamline position of the two devices differed by 18 cm. The Recirculator has a standard FODO design and so the beam has just exited a quadrupole before it is incident on the diagnostic devices. Thus, the downstream device, the slit scanner, will have a smaller x rms and a larger y rms than the upstream device, the GBI. This is verified by the data.

A real direct comparison can be made by looking at the normalized emittance values. The GBI values are 0.043 for the x emittance and 0.065 for the y emittance. This compares to 0.047 and 0.069 from the slit scanner. Given the systematic errors of the two devices and the different thresholding techniques used to ignore background, this is excellent agreement. Also, notice the qualitative agreement in the plots of the phase space measured by the two devices. It should be kept in mind that the GBI data was obtained from just one beam pulse, while the slit scanner required 800 pulses.

THE FUTURE

One drawback of the current GBI design is the fact the CCD camera is disjoint from the rest of the device. Thus, every time the GBI is inserted in the beamline, the camera has to be refocussed and the image pixel size must be determined. This not only costs time, but hinders flat fielding of the device, since the flat field image will have a different pixel size than the data image. To get around this, an improved model is currently being designed where the CCD chip sit in the vacuum and is directly coupled to the back of the phosphor screen, which implies the CCD chip will have to be as large as the phosphor screen. This will mean that the image pixel size will be the same as the physical pixel size of the chip for all images, including the flat field image. This design will also allow the device to be positioned at many different locations along the beam line, instead of just at the end of the machine.

Another improvement currently being planned is the use of a hole plate with square holes. It is currently possible to make 100-micron square holes where the corners have a radius of curvature of 12 microns. The use of this plate will allow a large increase in the number of pixels used without having to deconvolute the effect of the hole shape on the x and y moments.

CONCLUSION

The HIF group at LLNL has developed a new device, the Gated Beam Imager, to measure the first and second moments of an ion beam and its emittance. The device is an adaptation of the pepperpot design used for characterizing electron beams. It measures both the in-bend plane and out-of-bend plane transverse emittance simultaneously which the slit scanner cannot do. It also reduces the required beam pulse by a factor of 20 to fully map the transverse moments as a function of time. The emittance measurements from the GBI agree quite well with measurements from the slit scanner. Further improvement of the GBI are planned and the full usefulness of this device is just beginning to be exploited.

ACKNOWLEDGMENTS

Both authors would like to acknowledge the help of Art Paul in his help in the development of the software to analyze the data and Gene Lauer, Art Molvik, Perry Bell, Dean Lee, and Geoff Mant for the technical help in developing the GBI. This work has been performed under the auspices of the US DOE by LLNL under contract W-7405-ENG-48.

REFERENCES

[1] Paul, A. et al, "Probing the Electron Distribution Inside the ATA Beam Pulse," *Nucl. Instrum. Methods Phys. Rev. A*, **300**, 137–150 (1991).

[2] Kaminsky, M., *Atomic and Ionic Impact Phenomena on Metal Surfaces*, New York: Academic Press, Inc., 1965, pp. 325.

[3] Hopkins, H., "Experimental Measurements of the 4-D Phase Space Map of a Heavy Ion Beam," Ph.D. Dissertation, University of California, Berkeley, 1997.

Diamond Detectors with Subnanosecond Time Resolution for Heavy Ion Spill Diagnostics

P. Moritz, E. Berdermann, K. Blasche, H. Rödl, H. Stelzer, F. Zeytouni[†]

Gesellschaft für Schwerionenforschung mbH (GSI), D-64291 Darmstadt [*]
[†]*Institut für Kernphysik der Johann Wolfgang Goethe-Universität, Frankfurt am Main*

ABSTRACT

Abstract. The application of CVD diamonds as radiation-hard particle detectors with outstanding properties for heavy ion beamline diagnostics is presented. Synchrotron particle spills ranging from a single ion to well beyond 10^8 pps can be analyzed while maintaining single-particle time resolution below fractions of a nanosecond. With segmented electrode structures on the diamond surface, higher particle count rates and improved monitoring of x/y beam profiles can be achieved. Diamond detectors with areas up to 30×30 mm^2 for a precise measurement system for beam intensity, beam profiles, and spill time-structure are described.

INTRODUCTION

The GSI accelerators provide all kinds of ion species in a broad energy range up to 2000 MeV/u. In the high energy beam transport lines, the ion beams extracted from the SIS synchrotron have to be carefully monitored at very different intensities, from a few thousands up to 10^{11} ions per machine cycle. The beam extraction time varies between 10 ms and 10 s. Diamond beam detectors are under development as a unique monitor system especially for the following applications:

GSI is a member of the RD42 collaboration at CERN. The development of CVD diamond detectors for use in high-energy physics experiments, accelerator facilities, and investigations concerning material characterization and quality improvement are the main aims of the collaboration

CP451, *Beam Instrumentation Workshop*
edited by R. O. Hettel, S. R. Smith, and J. D. Masek

- Precise beam intensity measurement using particle-counting techniques ranging from single ions to more than 10^9 ions/s.
- Beam profile measurements by a suitable segmentation of the active detector area, with the ability for dynamic profile measurement during the spill.
- Analysis of the particle spill time-structure using the high time resolution of the new detectors, which is better than 1 ns for low and high beam intensities.

Ensemble of Particle Detectors and Synchrotron Beam Diagnostic Devices

An overview of the counting capabilities of established particle detectors, the cryogenic current comparator (CCC), and the diamond detector is given in Figure 1. The operation ranges of devices used for the observation and measurement of particle beams inside the synchrotrons SIS and ESR at GSI are also shown. One can see that the diamond detectors fit very well into the ensemble. A diamond detector, together with a CCC, can give complete intensity measurement capabilities for all particle beams in the SIS extraction lines. In the lower intensity range, the counting properties of diamond detectors can be verified using scintillators. In the medium range, they can be compared against beam current transformers (BCT) or calibrated ionisation chambers (IC) as well as secondary electron emission monitors (SEEM). In the highest intensity range, calibration can be achieved with the CCC.

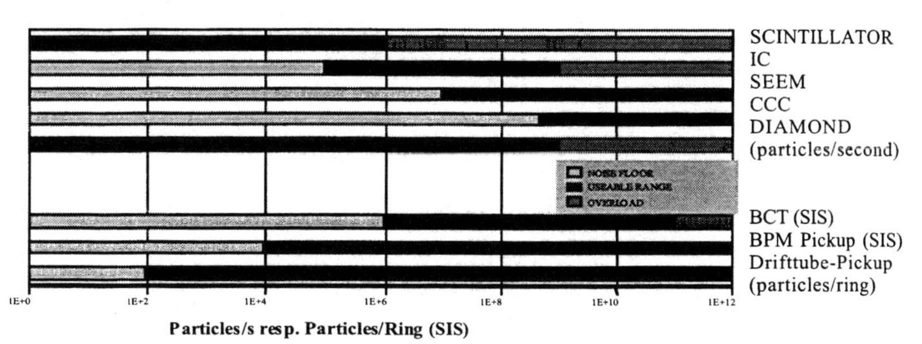

FIGURE 1. Intensity ranges of particle detectors for spill-measurements and of diagnostic devices in the SIS for measurements of the circulating beam

CVD Diamond Properties

Chemical vapor deposition (CVD) polycrystalline diamonds show very promising properties for use in electronics as well as particle detectors. The most important parameters are listed below, together with, for contrast, the corresponding values for silicon.

TABLE 1. Physical Properties of Diamond and Silicon

Physical Property at 300 K	Diamond	Silicon
Band gap [eV]	5.45	1.12
Electron mobility [cm^2/Vs]	2200	1500
Hole mobility [cm^2/Vs]	1600	600
Breakdown field [V/m]	10^7	3×10^5
Resistivity ρ [Ω cm]	$>10^{13}$	2.3×10^5
Dielectric constant ε_r	5.7	11.9
Thermal conductivity [W/cm K]	20	1.27
Lattice constant [Å]	3.57	5.43
Energy to remove an atom from the lattice [eV]	80	28
Energy to create an e-h pair [eV]	13	3.6

The highest thermal conductivity of all known materials (1), and the high energy of 80 eV needed to remove a carbon ion from the lattice lead to the diamond's excellent radiation hardness (2). Due to the large band gap, no p-n junction is required as in the case of silicon counters. After applying metallic electrodes, the diamond sample is ready for use as a detector. The high resistivity of the material allows the application of electric fields up to 6 V/μm. The low capacitance of diamond detectors in conjunction with high carrier mobility are basic characteristics for achieving narrow pulses.

Electronics for Diamond Detectors

A heavy ion passing a diamond detector produces a number of e-h pairs along its track, proportional to the energy loss in the material. The produced electrons and holes are separated by an applied electric field and move to the electrodes as long as they are not captured by chemical impurities or grain boundaries in the diamond bulk. This movement of the carriers within the detector's capacitance can be observed as a time-variant signal. The expected number of e-h pairs created by different ion species can be calculated according to the Bethe-Bloch formula. The estimate for the peak amplitude U_D is based on the relation:

$$U_D = \frac{dQ}{dt} \times R_e \qquad (1)$$

where Q is the generated charge, dt is the FWHM pulsewidth, and R_e is the amplifier input impedance.

The calculated peak amplitudes for a diamond detector of thickness 265 μm, an active area of 7×7 mm^2, a capacitance of 10 pF, and a corresponding pulsewidth of 1 ns (FWHM) are shown in Table 2.

TABLE 2. Response of a Diamond Detector for Various Source Particles

Particle Species	Generated e-h Pairs	Peak Amplitude
^{241}Am α-particles (5.45MeV) (laboratory source)	4×10^5	3.4 mV
^{12}C (1000 MeV/u)	5×10^5	4 mV
^{238}U (50 MeV/u)	5×10^8	4 V

The quality of today's CVD material makes the collection of about 30–40% of the electrical charge produced by heavy ions possible (3). Figure 2 shows a typical single particle signal obtained with a 10 GS/s single-shot digital storage oscilloscope (DSO) from a diamond detector with a thickness of 330 μm. As expected, the signal is a short pulse of 800 ps (FWHM) width, the risetime is less than 300 ps, and the (1/e-) falltime is 1 ns. The corresponding frequency domain spectrum occupies a bandwidth of more than 1 GHz. Depending on the nuclear charge of the projectile, its mass and the detector thickness, the signal has to be preamplified by a factor between 1 and 1000 to keep the signal level within a 0.1–1 V range for the electronics. In order to maintain the short pulse width, low-input impedance amplifiers have to be used. Careful matching of all transmission lines involved in the signal path is necessary to avoid signal reflections. The detector needs a bias voltage between 100 V and 3000 V DC decoupled from the preamplifier's input.

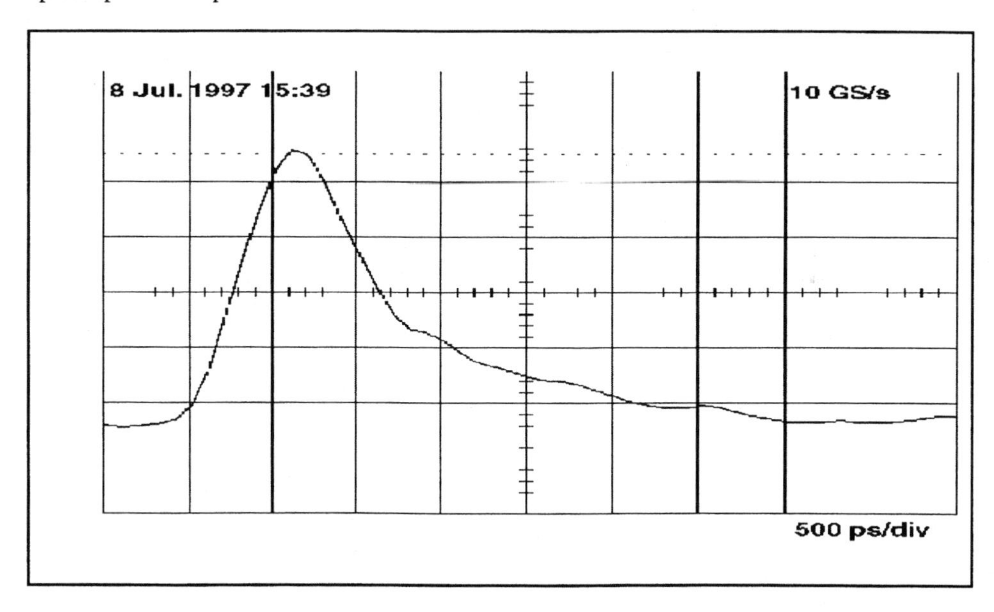

FIGURE 2. Diamond detector signal from a single ^{12}C ion at 200 MeV/u.

Detector Characterization

The short pulse width of the detector signal and the gap between electronic background noise and the lowest amplitudes found in the pulse-height distribution are the important parameters that characterize diamond detectors for beam diagnostic purposes. They define the counting rate capability and charge collecting efficiency.

While the beam intensity inside the synchrotron was observed by a beam current transformer, the extracted ^{20}Ne beam was measured by a diamond detector and simultaneously with an ionization chamber. In the test setup, the diamond detector and the ionization chamber were positioned at right angles to the beam's axis and both were located downstream. The pulses of the diamond detector have been counted using a 500 MHz time interval analyzer. The minimum time interval between two consecutive pulses was limited to 2 ns. The measured count rate was proportional to the beam intensity for rates up to 10^8 counts/s. Beyond this value, the random time distribution of the pulses reached the maximum time resolution of the time interval analyzer's electronics. It could also be observed that the idle count was zero. This zero noise particle counting capability for heavy ions has now been proved for more than 10 diamond detectors.

The pulse height distribution for the diamond detectors clearly shows a gap between background noise and lowest pulse amplitude. Figure 3 shows a typical pulse height distribution of a diamond detector with lead ions at 1 GeV/u used as projectiles.

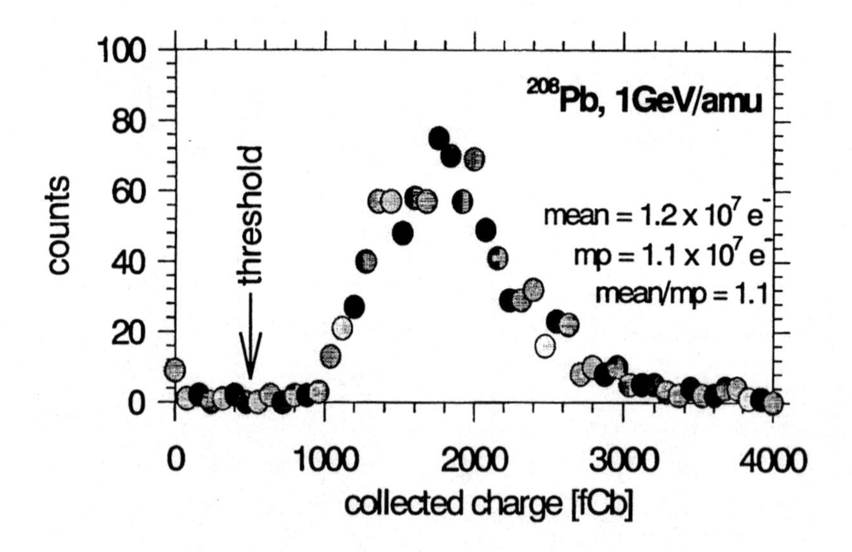

FIGURE 3. Pulse height distribution of ^{208}Pb ions in a 265 μm thick diamond

The improvement of the fast broadband amplifiers enabled us to characterize all diamond detectors with 5.45 MeV alpha particles. For more details on the characterization progress, see (4).

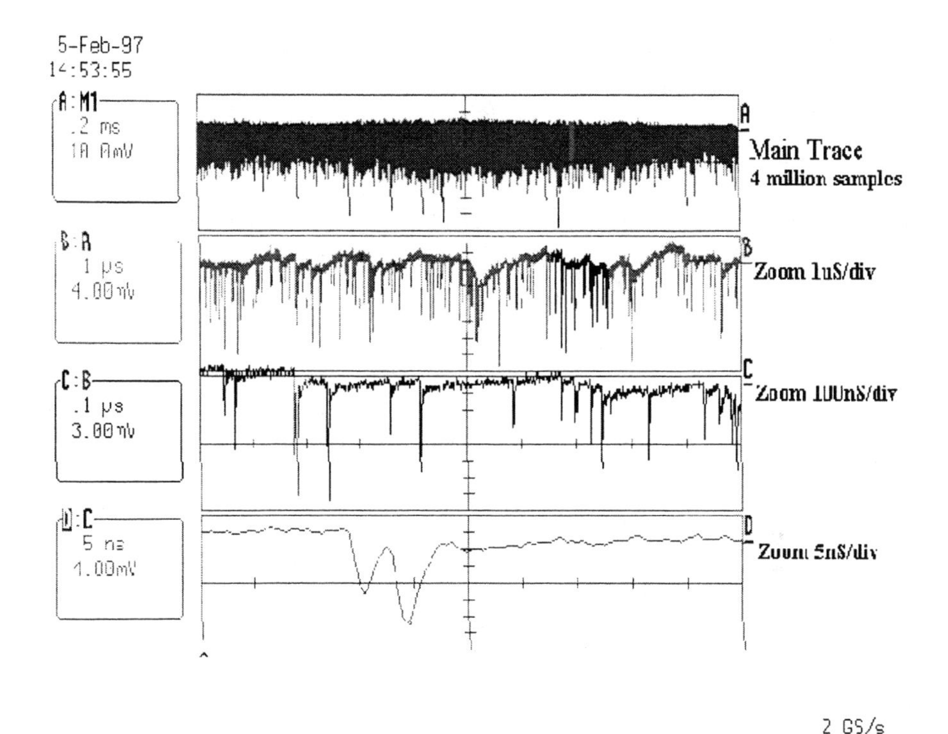

FIGURE 4. Carbon ions of 200 MeV/u from SIS passing a diamond detector. Traces B,C, and D are zooms of the 4 million data points of Trace A. Bottom trace (5 ns/div) indicates single particles.

Beamline Diagnostics

Heavy ions of more than 50 MeV/u passing a diamond detector undergo energy losses of approximately 1–2%. Thus, diamond detectors of 30–500 μm thickness permit transmission mode operation.

In a measurement shown in Figure 4, carbon ions of 200 MeV/u have been recorded with a diamond detector and a fast digital storage oscilloscope (DSO) (5,6). The registered particle count rate reached 10^8 particles/s. Even at the highest count rate, each particle could be registered with good time accuracy. Using mathematical signal processing, the spill structure in time and frequency domain could be analyzed.

Current Developments to Build a Complete Measurement System

The present work focuses on the development of segmented devices with a total active area of 30×30 mm^2. Figure 5 shows actual detector metalization layouts. The one being segmented into 8 parallel strips will be operated with fast 2 GHz electronics, and will increase the total count rate to beyond 10^9 particles/s. The 4×4 pixel detector with 20×20 mm^2 is also ready to use. Both detectors will be arranged together on a mechanical support that is moveable into the extraction line of the SIS. These diamond detectors are not only suitable to measure the intensity but also to obtain dynamic profiles of the particle beam.

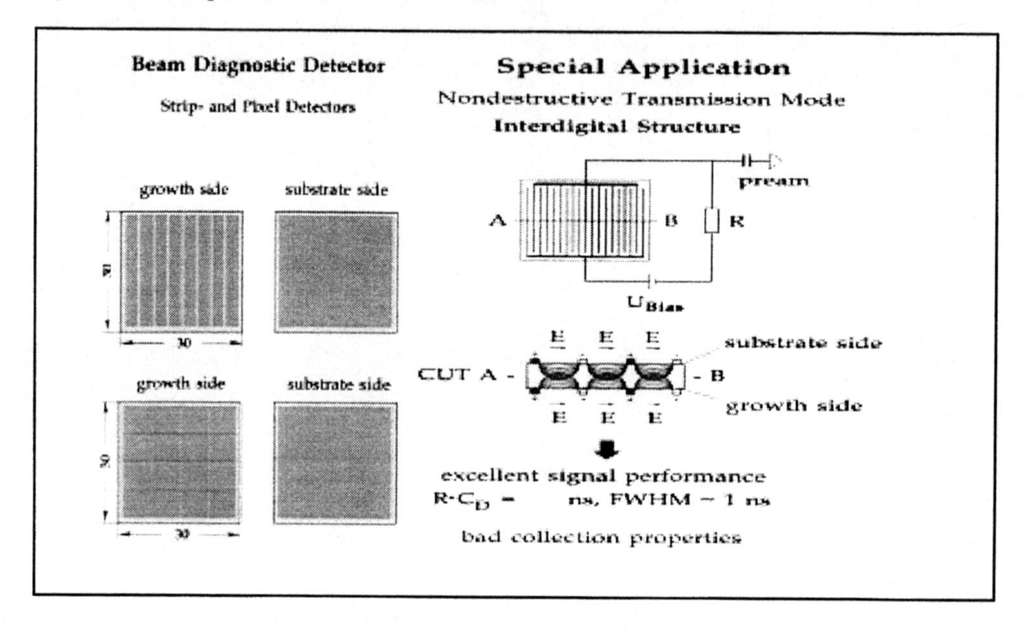

FIGURE 5. Strip , Pixel, and Interdigital Detector layouts.

CONCLUSIONS

CVD-diamond detectors

- are radiation hard devices
- allow big detector areas (presently up to a 4" wafer)
- can be metalized with any desired shape
- are easily assembled (just metalize and apply electrical contacts)
- can be wire-bonded
- are very fast detection devices

- can be operated as noise-free particle counters for heavy ions and alpha particles
- exhibit a wide counting range up to 5×10^8 particles/s per detector segment
- do not require electronic range switching when used together with scalers

REFERENCES

[1] Sauer, R., "Diamant als Elektronik Material," *Phys. Bl.* **51** (1995) Nr.5, p. 399ff.

[2] Meier, D. for the RD42 Collaboration, "5. Symposium on Diamond Materials," Electrochemical Society, Paris (1997).

[3] Berdermann, E., et al, "Diamond Detectors for Heavy Ion Measurements," GSI Scientific Report 1995, GSI 96-1.

[4] Berdermann, E., et al, "Diamond Detectors for Heavy Ion Beam Measurements," XXXVI International Meeting on Nuclear Physics, Bormio, (1998).

[5] Blasche, K. et al, "SIS Status Report," *GSI Scientific Report* 1996, GSI 97-1.

[6] Moritz, P., et al, "Diamond Detectors for Beam Diagnostics in Heavy Ion Accelerators," Presented at DIPAC III, Frascati (1997).

Design of a Tapered Stripline Fast Faraday Cup for Measurements on Heavy Ion Beams: Problems and Solutions

F. Marcellini* and M. Poggi**

** INFN, Laboratori Nazionali di Frascati,*
Frascati (Italy)
*and **INFN, Laboratori Nazionali di Legnaro,*
Legnaro (Italy)

Abstract. The design of a tapered stripline fast Faraday cup (TSFFC) to perform the impedance matching between the fast cup itself and the signal line (connector, cable, and amplifier) is reported here. The frequency response of the TSFFC as a high-pass filter is analyzed from a theoretical point of view and some solutions to achieve a broadband response are given.

INTRODUCTION

The design of a fast Faraday cup has to respect some rules regarding the required bandwidth and the characteristic impedance (1).

If only the TEM (or quasi-TEM) mode has to be transmitted along the line, the cut-off frequency is an important parameter. Using a stripline fast Faraday cup with a thickness between 0.5 and 1 mm, a bandwidth over 50 GHz can be reached.

The characteristic impedance, in planar geometry, given a dielectric medium, is proportional to the ratio between the dielectric thickness and the strip width. For instance, if the dielectric is teflon and its thickness is 0.5 mm, a width of 10 mm or more implies a characteristic impedance of 10 Ω or less, so a problem of impedance-matching with 50 Ω has to be worked out.

One solution is to use a resistor network to match the impedances (2). In our case, we sought to avoid resistor weldings on the strip by designing a tapered stripline fast cup. On the other hand, one can regard this as a high-pass filter, so other problems have to be solved. They will be shown in the next sections with the possible solutions.

CP451, *Beam Instrumentation Workshop*
edited by R. O. Hettel, S. R. Smith, and J. D. Masek

THEORY OF THE TAPERED STRIPLINE

An exhaustive study of the tapered transmission line theory is given in Reference 3. In the analysis of a tapered stripline, the starting point is to consider it as a transmission line with characteristic impedance changing continuously along the longitudinal direction as a result of the changes in the strip width.

These changes produce a change in the total reflection coefficient at the input as an addition of the same infinitesimal variations. The formula is the following:

$$\Gamma_i = \frac{1}{2}\int_0^L e^{-2jz\beta}\frac{d}{dz}(\ln Z)dz \tag{1}$$

where Γ_i is the total reflection coefficient at the input of the taper, L is the total taper length, z is the longitudinal coordinate, β is the inverse of the wavelength ($\beta=2\pi/\lambda$) and $\underline{Z}$ is the normalized impedance, as a function of the distance z along the taper. If the variation in $\underline{Z}$ with z is known, Γ_i may be evaluated from the above equation.

This was done for the fast cup. The strip width changes linearly with z, so the characteristic impedance is inversely proportional to the longitudinal coordinate (the dielectric thickness remains unchanged). The resulting integral does not have an analytical solution and was worked out numerically, taking β as parameter.

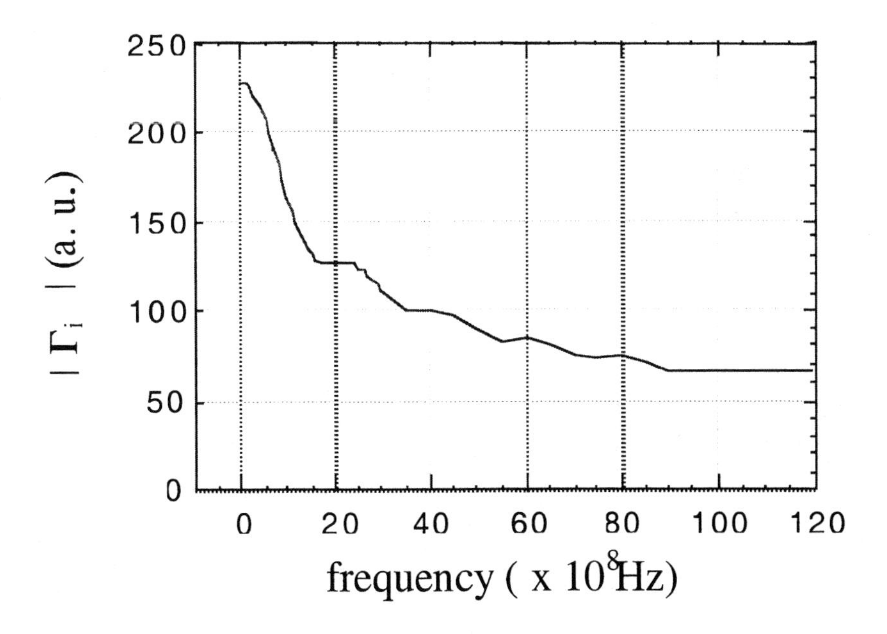

FIGURE 1. Calculated reflection coefficient vs. frequency

Figure 1 shows the calculated reflection coefficient magnitude vs. frequency. The high-pass effect is visible.

The tapered stripline was simulated using an HP program named "High Frequency Structure Simulator," and the plot of the transmission coefficient vs. frequency is reported in Figure 2, showing a similar high-pass filter effect.

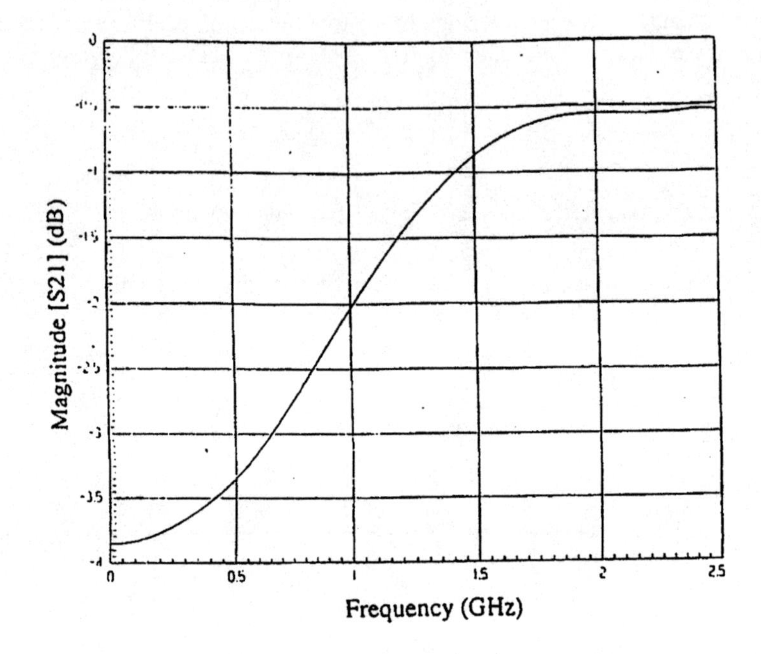

FIGURE 2. Simulated transmission coefficient vs. frequency

ANALYSIS OF THE ACTUAL TAPERED STRIPLINE

A drawing of the tapered stripline is shown in Figure 3. At the extremities, the width of the strip is matched with a 50 Ω characteristic impedance, but after a few millimeters it becomes larger to reach a width of 14 mm in the central region, which corresponds to an impedance of 7 Ω. This central region is the impact zone for the beam. The cup is 0.5 mm in thickness and the dielectric is teflon.

The frequency response between 0 and 6.5 GHz was analysed with a spectrum analyser, putting a signal into the 50 Ω matching section and looking at the other extremity, and the result is shown in Figure 4. In contrast with the simulation, where only half of this actual stripline was analyzed (from the 50 Ω matching strip to the center of the cup), the frequency response here shows an attenuation of 10 dB between 40 MHz and 1 GHz.

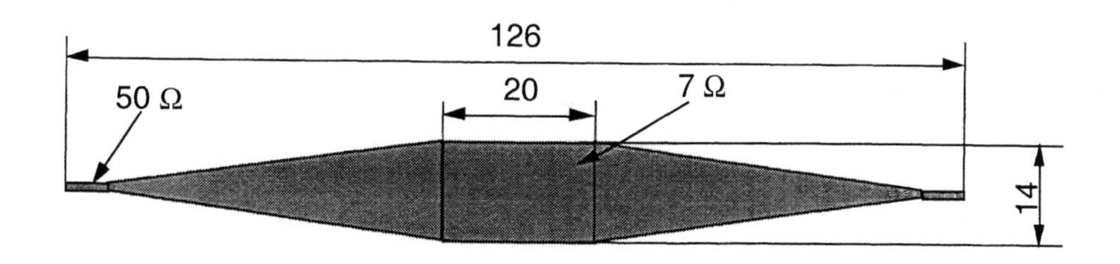

FIGURE 3. Fast Faraday Cup dimensions (in mm).

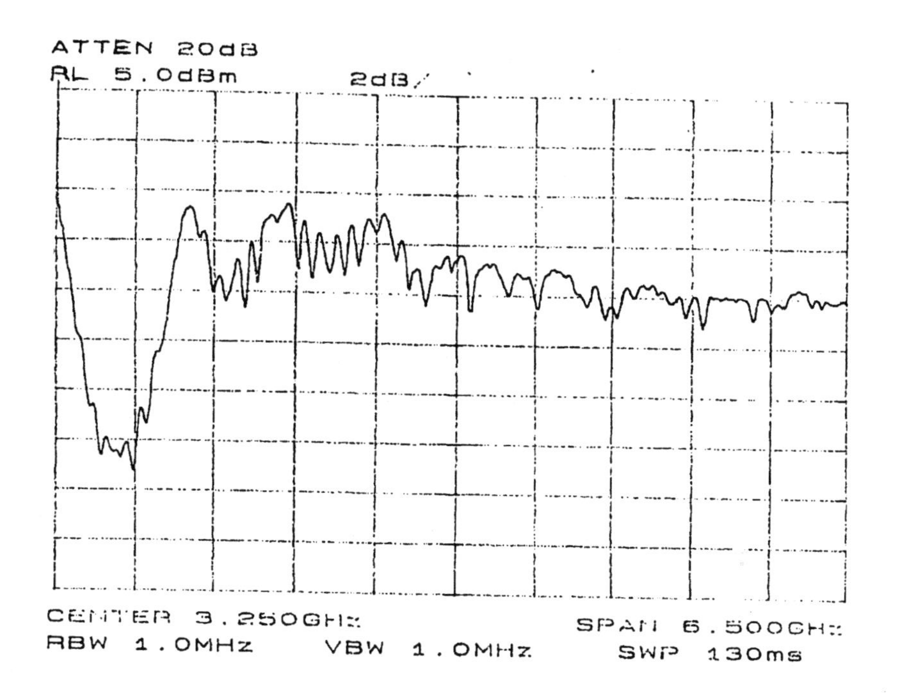

FIGURE 4. Frequency response of the TSFFC (range 0–6.5 GHz)

This kind of transfer function gives different responses at different input signal widths. However, for the beams involved, the cup must, be able to measure bunch widths between 10 ns down to 100 ps, so it is mandatory to have a broadband device. This has led to a more widespread problem of reconstructing the input signal, knowing the output signal and the transfer function of the intermediate device.

SIGNAL RECONSTRUCTION

The frequency response of the designed TSFFC can be approximated with a transfer function having two poles and two zeroes. These four points determine the 10 dB attenuation hollow mentioned before. By means of the "LAPLACE" function in the PSPICE code, the transfer function was successfully simulated. This is illustrated in Figure 5, where, using the same input signal, (actual in "A" and simulated in "B"), the output signal from the 20 GHz HP sampling oscilloscope is compared with the simulated output signal from PSPICE.

Once the transfer function is so characterized, the deconvolution of the output signal may solve the problem, but at present this is not useful, as the signal reconstruction cannot be done in real time.

Two different methods have been considered, neither yet realized, to achieve a result in real time: 1) the use of a Digital Signal Processor (DSP) to do a software reconstruction, 2) to carry out the transfer function inversion using a hardware filter with resistors and capacitors.

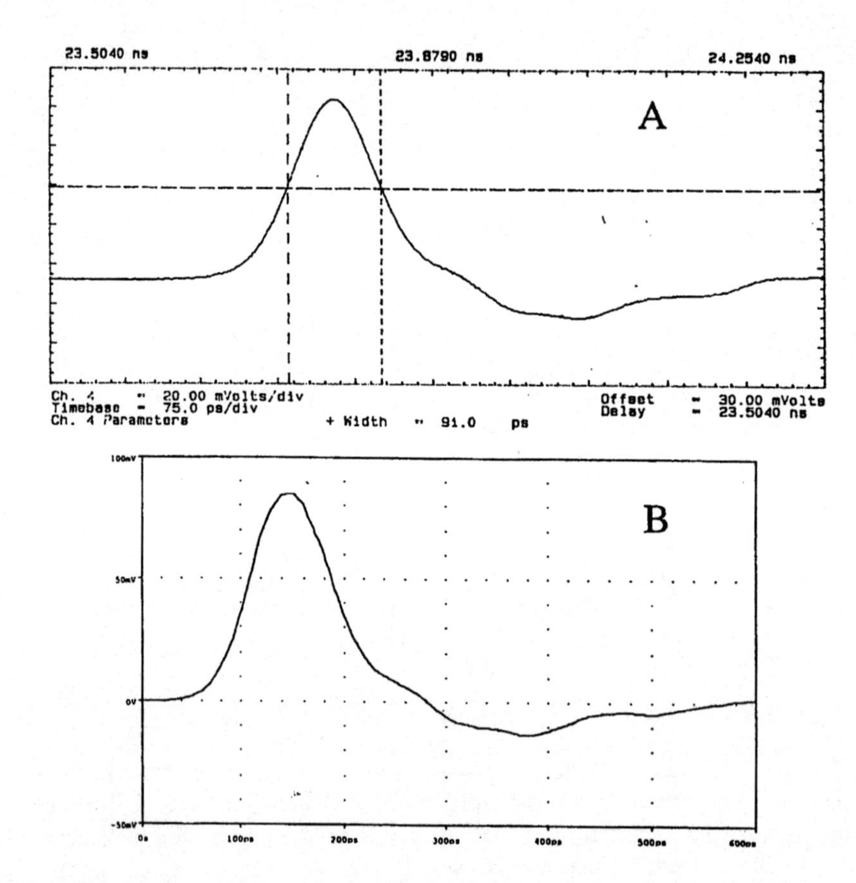

FIGURE 5. Signal reconstruction using PSPICE: "A" from the oscilloscope, "B" from PSPICE with simulated transfer function.

The first method implies taking the data from the HP oscilloscope with the HP-VEE program, working on them with a DSP, which performs the deconvolution with the inverse of the transfer function, and then displaying that processed data. The second method was simulated with PSPICE code. The inverse of the transfer function was approximated with only one zero and one pole at 20 dB/decade slope. The result is shown in Figure 6. With this second method there is a problem of signal attenuation due to the passive resistor and capacitor network.

CONCLUSIONS

Until now the problem of the input signal reconstruction has been successfully analyzed with simulations. The fast cup was implemented but has not yet been put in the beam. In a few months, a test with the hardware reconstruction will be performed. In the future, we hope to be able to do also the software reconstruction.

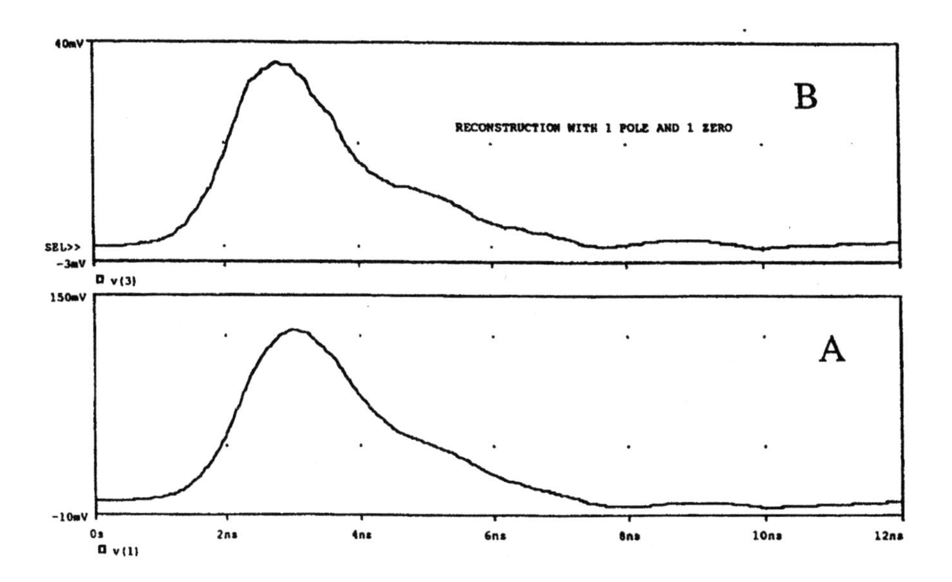

FIGURE 6. Signal reconstruction simulation using PSPICE with the "hardware" filter: "A" input signal, "B" output signal after simulated transfer function and filter.

ACKNOWLEDGMENTS

We would like to acknowledge Augusto Lombardi and Marco Bellato for their useful suggestions during the work and Roberto Ponchia for the design of the printed circuit board for the TSFFC.

REFERENCES

[1] Bellato, M. et al., *Nucl. Instrum. Methods A,* **382**, 118–120 (1996).

[2] Bollinger, L. M. et al., "Initial use of the Positive-Ion Injector of ATLAS," Presented at the 1989 SNEAP Conf., Oak Ridge, October 23–26, 1989, p. 477.

[3] Collin, R. E., *Foundations for Microwave Engineering*, McGraw-Hill International Editions, 1966, ch 5, pp. 237–238.

Commissioning Experience from PEP-II HER Longitudinal Feedback[1]

S. Prabhakar, D. Teytelman, J. Fox, A. Young,
P. Corredoura, and R. Tighe

Stanford Linear Accelerator Center, Stanford University, Stanford, CA 94309, USA

Abstract. The DSP-based bunch-by-bunch feedback system installed in the PEP-II high-energy ring (HER) has been used to damp instabilities induced by unwanted higher-order modes (HOMs) at beam currents up to 605 mA during commissioning. Beam pseudospectra calculated from feedback system data indicate the presence of coupled-bunch modes that coincide with a previously observed cavity mode (0-M-2).

Bunch current and synchronous phase measurements are also extracted from the data. These measurements reveal the impedance seen by the beam at revolution harmonics. The impedance peak at $3 \times f_{rev}$ indicates incorrect parking of the idle cavities, and explains the observed instability of mode 3. Bunch synchrotron tunes are calculated from lorentzian fits to the data. Bunch-to-bunch tune variation due to the cavity transient is shown to be large enough to result in Landau damping of coupled-bunch modes.

INTRODUCTION

The PEP-II high-energy ring (HER) electron beam has a design current of 1A. Beam currents up to 750 mA have been achieved so far. During commissioning, a variety of longitudinal and transverse beam dynamics experiments have been performed [1–3] with the help of the PEP-II longitudinal feedback (LFB) system [4]. The most useful diagnostic characteristic of this system is its ability to record the oscillations of each bunch over a 27 ms time window. The recorded data can then be archived and analyzed off line to extract detailed information about beam dynamics and beam conditions.

The two main sources of longitudinal motion identified in PEP-II are cavity impedance induced coupled-bunch instabilities and noise from the klystron. Coupled-bunch instabilities are usually caused by unwanted higher-order modes (HOMs) in the rf cavities, or by impedance sources elsewhere in the beam surroundings. At PEP-II however, the large beam current and small revolution frequency combine to produce low-mode instabilities within the bandwidth of the detuned rf

---

1. Work supported by DOE contract No. DE-AC03-76SF00515.

CP451, *Beam Instrumentation Workshop*
edited by R. O. Hettel, S. R. Smith, and J. D. Masek

cavity fundamental mode [3]. The HOM-induced instabilities have been successfully damped by the above-mentioned longitudinal feedback system. Low-mode motion is damped by a combination of rf feedback loops acting through the klystron.

In this paper we demonstrate the ability of the broadband LFB system to drive coupled-bunch motion with positive feedback, and damp it with negative feedback. We also present a novel beam-based method for measuring the longitudinal impedance spectrum [2]. This method involves calculation of the transfer function from fill shape (bunch current versus bunch number) to synchronous phase of a multibunch beam, which is shown to yield the longitudinal impedance seen by the beam at revolution harmonics. This technique has been used to measure the impedance of parked cavities at PEP-II, and explain the occasional instability of low-order coupled-bunch modes at unexpectedly low total currents. Multibunch synchronous phases and bunch currents are extracted from data stored by the LFB system.

In addition to providing impedance information, multibunch synchronous phase measurements are useful in themselves, since HER and LER phase transients need to be matched to achieve high luminosity. The synchronous phase transients are also an indicator of the amount of Landau damping afforded by bunch-to-bunch tune shifts.

Lorentzian fits to bunch motion spectra yield a bunch tune versus bunch number graph that correlates well with the synchronous phase transients. We examine the interesting features of one such graph in the final section, and demonstrate the decoupling of bunch oscillations at the ends of the bunch tune range.

BEAM PSEUDOSPECTRA

Coupled-bunch motion can be studied in the frequency domain by constructing beam pseudospectra from feedback system data [5]. A typical piece of HER data includes the sampled oscillation signals of each of the stored bunches over a 27 ms time window. These signals are interleaved and strung out into a single vector of successive samples detected by a stationary observer at the BPM location. The FFT of the resulting vector is nothing but the beam spectrum, with revolution harmonics suppressed. The pseudospectrum resulting from a single 27 ms transient covers the entire 119 MHz frequency range (DC to half the bunch crossing frequency) with a resolution of 37 Hz.

In the absence of feedback, HOM-induced coupled-bunch instabilities have been observed at beam currents above 550 mA in the HER. These instabilities are damped when negative feedback is switched on.

Figure 1 compares HER pseudospectra with positive and negative feedback, at beam currents of 317 and 330 mA respectively. It must be noted here that each of the lines in these spectra is a synchrotron sideband, since the revolution harmonics are suppressed. The beam is longitudinally stable in both cases, but positive feedback drives up the amplitudes of synchrotron sidebands in the 100–110 MHz frequency range. This frequency range coincides with the aliased impedance of the 0-M-2 mode, which is the largest measured cavity HOM [6].

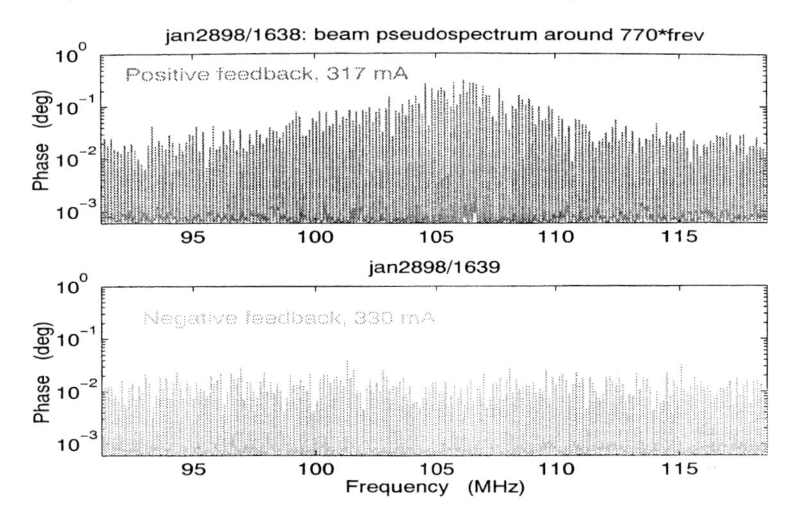

FIGURE 1. Beam pseudospectra with positive and negative feedback. Sidebands are driven up by positive feedback in the 100–110 MHz frequency range, which coincides with the aliased impedance of the 0-M-2 cavity HOM.

BUNCH CURRENTS AND SYNCHRONOUS PHASES

Line harmonics from the klystron impose the same low-frequency motion on all bunches. During commissioning, 360 Hz and 720 Hz lines from the klystron were large enough to be detected in the bunch data. These spectral lines afford a crude current monitor, since the bunch signals are proportional to charge times longitudinal phase. Bunch currents are estimated by projecting individual bunch signals onto a line harmonic spectrum calculated by averaging over all of the bunch signals.

Formal bunch current monitoring using the feedback system has been demonstrated at the ALS [7], and will be used at PEP-II during the next commissioning run.

Since the LFB system detects the product of charge and phase, multibunch synchronous phases are calculated by averaging the digitized signals for each bunch and dividing the averages by the corresponding bunch currents.

Figure 2 shows the averaged low-frequency bunch signal spectrum for a 291-bunch 96-mA fill. In this case, we calculate bunch currents by projecting individual bunch signals onto the 720 Hz line in the averaged spectrum and then scaling the result so that the calculated total beam current agrees with that measured by the DCCT (DC Current Transformer).

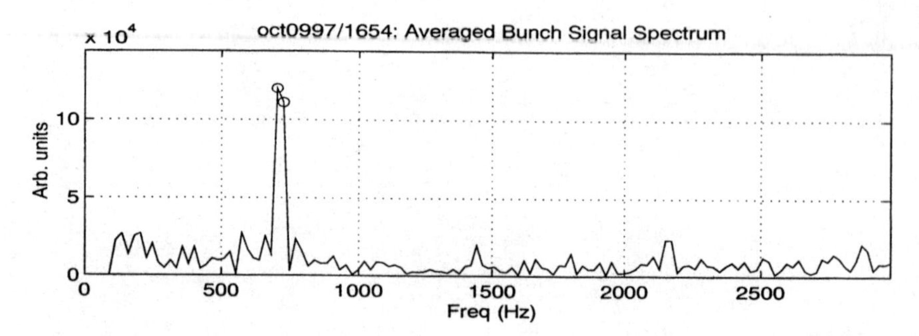

FIGURE 2. Low-frequency bunch signal spectrum: the 720 Hz line from klystron is demodulated to extract bunch currents from LFB system data.

The calculated bunch currents i_k are shown in Figure 3. There is an impulsive discontinuity near the beginning of the bunch train. The resultant synchronous phase ringing is shown in the same figure. We can see from the figure that the "impulse response" goes through about three oscillations and dies out in one revolution period. This indicates that the longitudinal impedance $Z(j\omega)$ has a strong resonance three revolution harmonics away from some multiple of the bunch frequency, which is a twelfth of the rf frequency in this case.

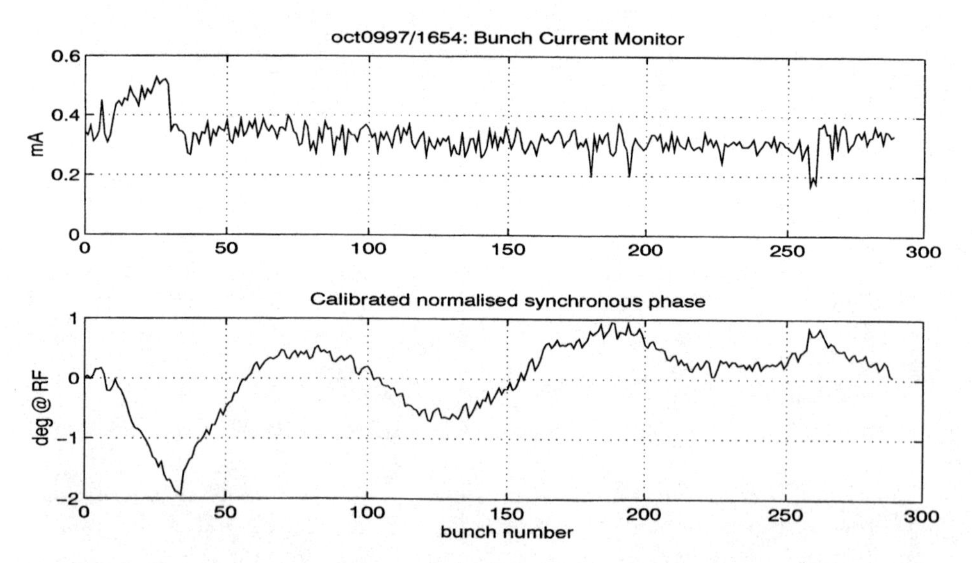

FIGURE 3. The upper figure shows the current variation in a 96-mA 291-bunch HER beam. We see a short impulsive discontinuity in the fill shape. This discontinuity produces a damped oscillation in the synchronous phase transient (lower figure). The transient completes three oscillations within the length of the bunch sequence.

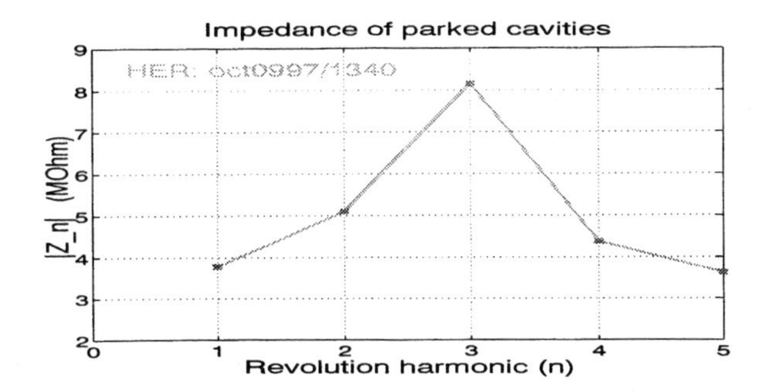

FIGURE 4. Impedance at the first few revolution harmonics, extracted from bunch currents and synchronous phases. The large value at n = 3 is due to the fundamental resonances of parked cavities.

EXTRACTION OF IMPEDANCE

If we increase the charge in a single bunch, its synchronous phase will ride up the rf voltage waveform to keep up with the increasing loss of energy to the wakefields. If we know the slope of the rf voltage, we can easily calculate the energy lost to wakefields per unit current from the synchronous phase increase. This gives us a measure of the integrated beam impedance. In itself, the integral reveals nothing about the shape of the impedance. However, some information about the frequency spectrum of the impedance can be gleaned from repeating this measurement at various bunch lengths.

In this note we demonstrate a new method of measuring the longitudinal impedance spectrum using synchronous phase data from multibunch fills. It can be shown that the discrete-time transfer function from bunch currents to synchronous phases is proportional to the aliased longitudinal impedance at revolution harmonics up to the bunch frequency [2].

Figure 4 shows a typical low-frequency impedance spectrum calculated from HER commissioning data using the method described above. We see a resonant impedance of $8M\Omega$ at $3\times f_{rev}$, calculated from the currents and phases shown in the previous figure. This indicates that the 12 parked cavities, which were supposed to be tuned 2.5 revolution harmonics away from the rf frequency, were actually parked much closer to the third revolution harmonic. If they were all parked exactly three revolution harmonics away from f_{rf}, their impedances would add up to $9.2M\Omega$ at the third revolution harmonic.

Coupled-Bunch Instability

Inaccurate parking of idle cavities is quite likely to have been the cause of low-mode coupled-bunch instabilities seen occasionally at currents below 100 mA during

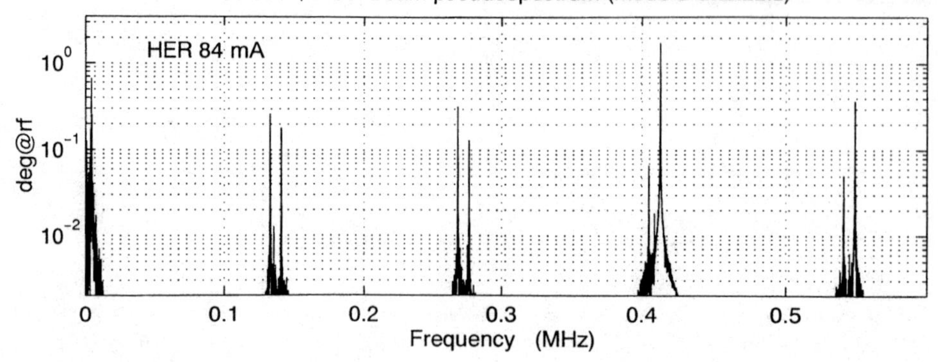

FIGURE 5. Beam pseudospectrum of a 84-mA 291-bunch beam, showing an unstable upper sideband at the third revolution harmonic (0.41 MHz).

commissioning. Figure 5 shows the low-mode beam pseudospectrum for a 291-bunch 84 mA fill, taken a few days before the data displayed in the previous figure. Each pair of lines in the figure is a pair of synchrotron sidebands. The pseudospectrum shows that mode 3 is unstable, with an average amplitude of 2° at the rf frequency. The shifting of the idle cavity tuners has been seen to attenuate this instability [8].

BUNCH TUNE VARIATION

The PEP-II beams are expected to have a 5% gap, to forestall conventional ion instabilities. This gap produces a transient in the rf cavity which results in synchronous phase variation across the bunch train. The beam-loading transient also causes the bunch tunes to vary across the train.

The upper graph in Figure 6 illustrates synchronous phase variation in a 368 mA HER beam with a 5% gap. The variation is fast in the initial portion of the bunch train, where the phase changes by 10° within the first 300 bunches. The corresponding variation in bunch tune is measured simultaneously using Lorentzian fits to bunch signal spectra (lower graph). As the bunch train loads the cavity fundamental, the bunch tune decreases from an initial value of 5900 Hz to a mid-train value of around 5800 Hz. Longitudinal oscillations at the beginning of the bunch train are decoupled from oscillations further down, i.e., we have some Landau damping of coupled-bunch instabilities. This contributes to the elevation of the instability threshold from the calculated value of 350 mA with no feedback to the measured value of 550 mA.

Although the bunch tune transient broadly matches the synchronous phase transient, we can clearly see local regions of flatness, within which all bunches seem

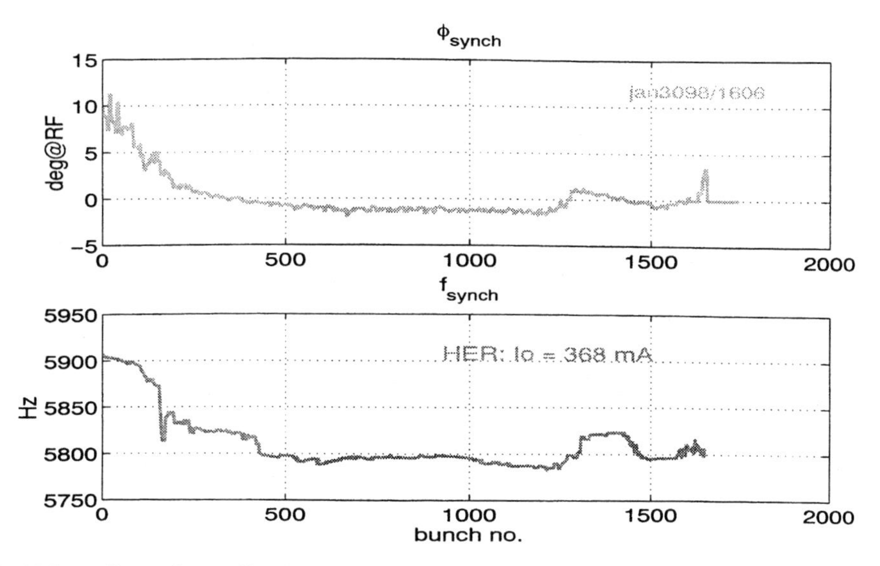

FIGURE 6. Upper figure: Synchronous phase variation across a 368-mA 1658-bunch HER beam due to a 5% gap in the bunch train. Lower figure: Corresponding bunch tune measurement.

to be oscillating at the same frequency. This is due to communication between bunches through wakefields. Bunches with approximately the same tune couple to each other and oscillate in the coherent mode most favored by the beam impedance. Each flat level in the tune transient thus represents at least one distinct coupled-bunch eigenmode.

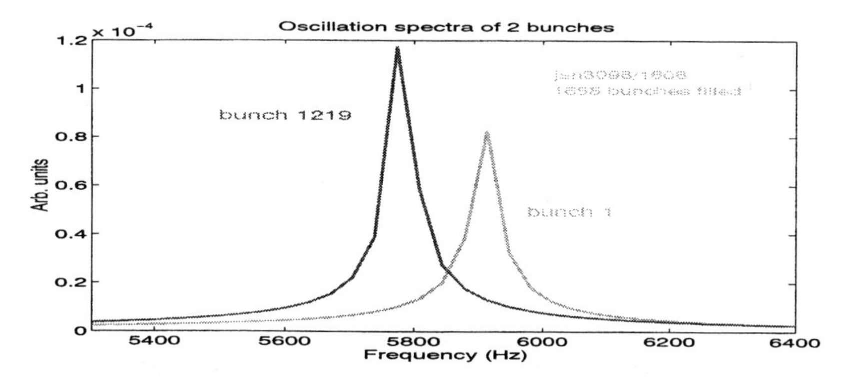

FIGURE 7. Decoupling of bunch spectra due to tune spread.

The decoupling of bunch oscillations due tune variation across a bunch train is illustrated by Figure 7. This figure shows Lorentzian fits to the oscillation spectra of two bunches at the two extremes of the tune spread. We can see that the spectra show almost no overlap.

SUMMARY

Coupled-bunch HOM-driven instabilities have been detected in the PEP-II HER at frequencies consistent with the largest measured cavity HOM. The modes are stable under the action of broadband longitudinal feedback.

A novel beam-based technique has been used to measure the longitudinal impedance spectrum. The technique involves calculation of the transfer function from fill shape to multibunch synchronous phase. Bunch currents and synchronous phases have been extracted from feedback system data by demodulating the line harmonics that are common to the signals of all filled bunches. The transfer function method has been used to identify inaccurately parked idle cavities as the cause of the mode 3 instability observed during commissioning. Multibunch synchronous phase measurements take on added importance at PEP-II because of the need to match gap transients in the two rings.

Bunch tune variation due to gap transients has been shown to be large enough to decouple the oscillations of bunches in the beginning and middle of a train with a 5% gap at total currents below 400 mA. This contributes to the elevation of the coupled-bunch instability threshold from the calculated value of 350 mA to the measured value of 550 mA (in the absence of longitudinal feedback).

ACKNOWLEDGMENTS

The authors would like to thank S. Heifets of SLAC and B. Zotter of CERN for helpful discussions and comments, and the PEP-II group for support.

REFERENCES

1. Teytelman, D., et al, these proceedings.
2. Prabhakar, S., et al, "Calculation of Impedance from Multibunch Synchronous Phases: Theory and Experimental Results," SLAC-PEP-II-AP-NOTE-98-04, (1998).
3. Prabhakar, S., et al, "Low-Mode Longitudinal Motion in the PEP-II HER," SLAC-PEP-II-AP-NOTE-98-06, (1998).
4. Fox, J. D., et al, "Observation, Control and Modal Analysis of Longitudinal Coupled-Bunch Instabilities in the ALS via a Digital Feedback System," in *Proc. 7th Beam Instrumentation Workshop (BIW96)*, Argonne, IL, (1996).
5. Prabhakar, S., et al, "Observation and Modal Analysis of Coupled-Bunch Longitudinal Instabilities via a Digital Feedback Control System," *Particle Accelerators*, **57/3**, (1997).
6. Rimmer, R., et al, "Updated Impedance Estimate of the PEP-II RF Cavity," in *Proc. 5th European Particle Accelerator Conference*, Sitges, Barcelona, Spain, (1996).
7. Prabhakar, S., et al, "Use of Digital Feedback System as a Bunch by Bunch Current Monitor: Results from ALS," SLAC-PEP-II-AP-NOTE-96-29, (1996).
8. Wienands, U., oral communication.

Longitudinal and Transverse Feedback Systems for BESSY-II

S. Khan and T. Knuth

BESSY-II, Rudower Chaussee 5, 12489 Berlin, Germany

Abstract. The commissioning of the high-brilliance synchrotron light source BESSY-II in Berlin started in April 1998. Within the commissioning period, bunch-by-bunch feedback systems to counteract longitudinal and transverse multibunch instabilities will be installed. This paper reviews their design and present status.

INTRODUCTION

BESSY-II [1] is a high-brilliance synchrotron radiation source currently being commissioned in Berlin. The first beam was stored on April 22, 1998.

The performance of a "third generation" light source can be seriously impaired by longitudinal and transverse multibunch instabilities leading to

- Limitations of the beam current,
- Increased beam spot and divergence due to transverse oscillations, and
- Broadening of the undulator line-width due to energy oscillations.

Multibunch instabilities are excited by long-range wakefields due to higher-order modes (HOMs) of the rf cavities or due to the finite wall conductivity. Presently, the BESSY-II storage ring will be operated with four DORIS-type pillbox cavities with a rich HOM spectrum. It is important to have control over the HOM position either by controlling the cavity temperature [2] or by adding a second plunger to the cavities. Both methods are currently under investigation. However, even in the most favorable case, radiation damping will not be sufficient to damp longitudinal instabilities at beam currents of several hundred mA. HOMs may also excite transverse instabilities under unfavorable conditions. However, the main source of vertical oscillations will be the resistive wall impedance, once insertion device chambers with small vertical apertures ($\pm$8 mm and $\pm$5.5 mm, made of aluminum) are installed.

This paper describes the design and status of a longitudinal and a transverse bunch-by-bunch feedback system presently under construction.

CP451, *Beam Instrumentation Workshop*
edited by R. O. Hettel, S. R. Smith, and J. D. Masek

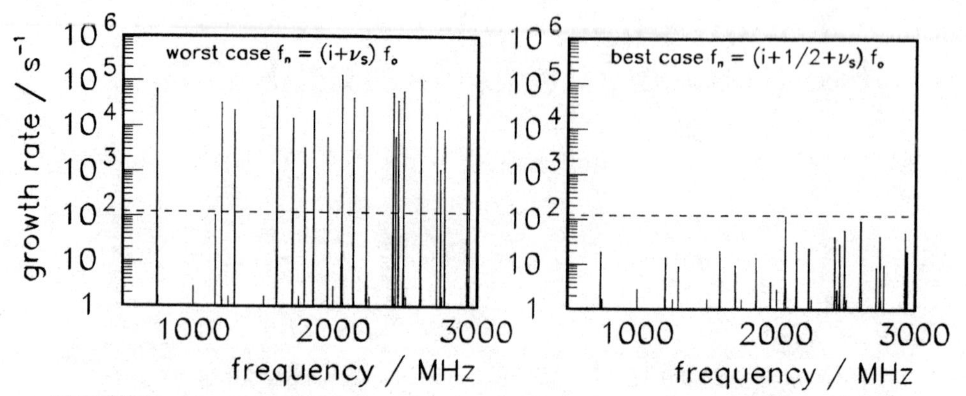

FIGURE 1. Longitudinal multibunch instability growth rates from HOMs at their most (left) and least (right) harmful position. The dashed line indicates the radiation damping rate.

LONGITUDINAL FEEDBACK SYSTEM

Longitudinal Multibunch Instabilities

Given the longitudinal shunt impedance $R_{\|n}$ of the nth HOM, the central frequency f_n, and quality factor Q_n, the impedance as function of frequency f is given by

$$Z_{\|n}(f) = \frac{R_{\|n}}{1 + iQ_n\left(\frac{f}{f_n} - \frac{f_n}{f}\right)} \tag{1}$$

and the growth rate of a longitudinal multibunch mode m is given by

$$\frac{1}{\tau} = \frac{I\alpha}{2\nu_s E/e} \cdot \mathrm{Re}\sum_p f_p e^{-(2\pi f_p \sigma_t)^2} Z_{\|n}, \quad \text{with} \quad f_p = f_\circ(p\,h + m + \nu_s). \tag{2}$$

Here, I is the beam current, α is the momentum compaction factor, ν_s is the synchrotron tune, E/e is the beam energy, σ_t is the bunch length in time, $f_\circ$ is the revolution frequency, h is the harmonic number, and the impedance is sampled at all integer values of p.

The growth rates from HOMs predicted by a 2D MAFIA simulation in the frequency domain [3] are shown in Figure 1 for a beam current of 400 mA, considering two limiting cases:

- The "worst case," where the HOMs coincide with multibunch modes, and
- The "best case," with the HOMs between two multibunch modes.

Measurements indicate that the simulation tends to overestimate the shunt impedance and Q-value of HOMs [4], [5]. Using a network analyzer, the HOM

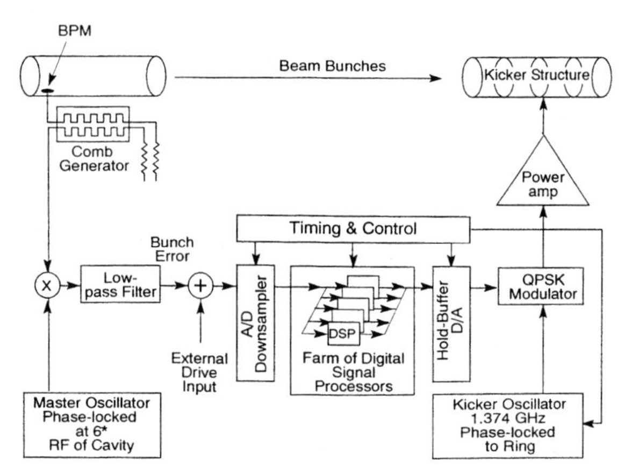

FIGURE 2. Block diagram of the longitudinal feedback system (courtesy SLAC LFB group).

frequencies and Q-values can be measured directly. Measuring the HOM frequency shift as function of the position of a small object in its field allows to identify the mode, while the integrated frequency shift measures the R/Q ratio. Measurements on a spare cavity are underway and first results indicate that the Q-values are indeed two to three times lower than predicted.

Longitudinal Feedback System Overview

The longitudinal system employs the digital electronics developed for ALS, PEP-II and DAΦNE [6]. Being developed and well tested over many years at the ALS, the advantages of this option for BESSY-II are minimum development effort, reliability, and well-maintained hardware and software. Furthermore, the digital system offers flexibility in the feedback algorithm and excellent diagnostic capabilities [7].

The block diagram in Figure 2 outlines the system. The bunch signal from a beam position monitor is fed into a comb generator to produce a 3-GHz signal for phase detection. The moment signal (phase·charge) is digitized at a rate of 500 MHz, downsampled and distributed to an array of DSPs, where a correction signal is computed for every bunch. The D/A-converted correction signal QPSK-modulates a carrier at 1374 MHz (11/4 times the rf frequency). More detailed descriptions of the electronics and the feedback algorithm can be found in [6], [8], [9], and [10].

Longitudinal Kicker Cavity

For the longitudinal kicker structure, a choice had to be made between a series of coaxial electrodes as used for the ALS and PEP-II [11] and a cavity-type kicker as developed for DAΦNE [12]. With only little space being available, the DAΦNE design, offering a larger shunt impedance in a single structure, was favored. For

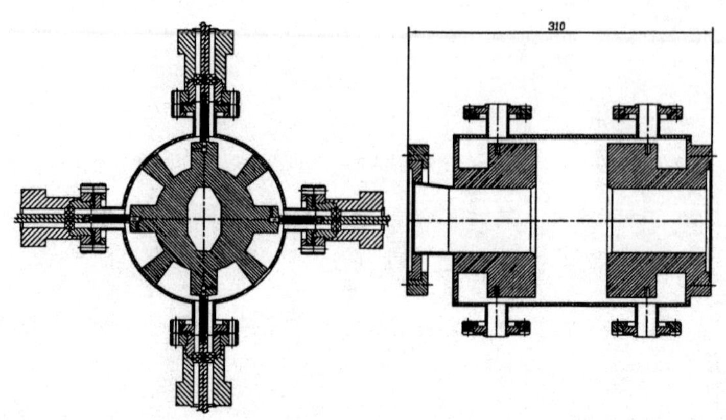

FIGURE 3. Strongly damped pillbox-type kicker cavity with eight waveguides.

BESSY-II, the kicker design required several modifications. The central frequency was moved from 1197 MHz to 1374 MHz, and the bandwidth was increased by using eight waveguides instead of six. Modifying the beam ports according to the BESSY-II chamber also increased the shunt impedance. Figure 3 shows the 31 cm long structure with a shunt impedance of $R_s \approx 1000\ \Omega$ and a Q-value of 5.6.

Longitudinal Damping Rates

The damping rate of the longitudinal feedback system can be expressed as

$$\frac{1}{\tau} = \frac{f_\circ\, h\, \alpha}{2\, \nu_s\, E/e} \cdot G, \tag{3}$$

with $G = \Delta U/\Delta\phi$ being the feedback gain, i.e., the kick voltage per unit phase deviation (radian), and all other symbols defined as before. The gain cannot be arbitrarily large, because the voltage is limited to $\Delta U \leq \sqrt{2NR_sP}$ for a total power P and N kickers. $\Delta\phi$ is limited by the phase resolution and beam noise.

Assuming a 250-W amplifier, one kicker, and 50% power loss, the maximum voltage is 500 V. As shown in Figure 4, saturation at $\Delta\phi = 7$ mrad leads to a maximum damping rate of 1000 s^{-1}. The maximum effective impedance $Z_{\|}^{\mathrm{eff}} = G/I$ is 180 kΩ. If necessary, performance can be improved by adding power, by using more kickers, or — subject to operational experience — by increasing the gain.

TRANSVERSE FEEDBACK SYSTEM

Transverse Multibunch Instabilities

The impedance of a given HOM and the resistive wall impedance is

$$Z_{\perp n}(f) = \frac{2\pi f_n^2}{c\, f} \cdot \frac{R_{\perp n}}{1 + iQ_n\left(\frac{f}{f_n} - \frac{f_n}{f}\right)} \quad \text{and} \quad Z_\perp^{\mathrm{rw}}(f) = \frac{c^2}{2\pi\, f_\circ\, f} \cdot \frac{1 - \mathrm{sgn}(f)\, i}{\pi\, b^3 \delta\, \sigma}, \tag{4}$$

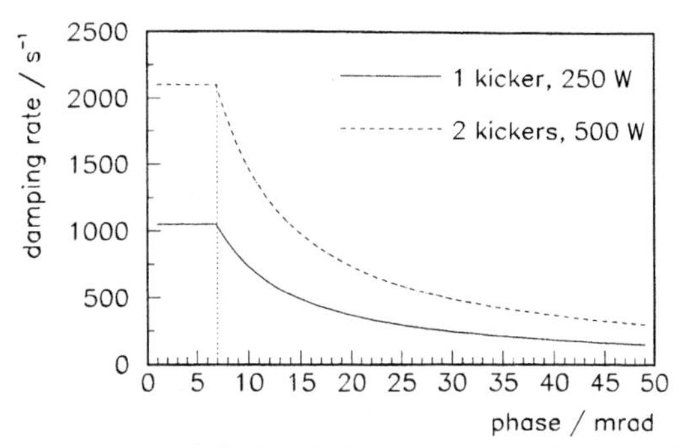

FIGURE 4. Longitudinal feedback damping rate as function of phase deviation.

where b is the half-aperture, $\delta \sim f^{-1/2}$ is the skin depth, and σ is the wall conductivity. The growth rate of rigid bunch transverse oscillations is

$$\frac{1}{\tau} = \frac{I\,c^2}{4\pi\,\nu f_\circ\,E/e} \mathrm{Re} \sum_p e^{-(2\pi\sigma_t)^2 (f_p - \frac{\xi}{\alpha} f_\circ)^2} Z_\perp(f_p) \quad \text{with} \quad f_p = f_\circ(p\,h + m + \nu), \quad (5)$$

where ν is the betatron tune, ξ is the chromaticity, and all other symbols are defined as before. Figure 5 shows the resistive wall growth rate as function of betatron tune and chromaticity.

Transverse HOMs can be obtained from a 3D MAFIA simulation, taking into account the three ports of the DORIS cavity that break the rotational symmetry. In a picture analogous to Figure 1, the "worst case" growth rates for some modes can reach 10^4 s^{-1}.

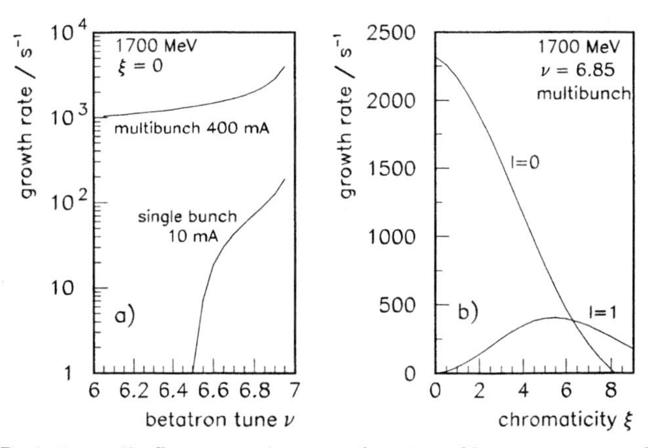

FIGURE 5. Resistive wall effect: growth rate as function of betatron tune and chromaticity.

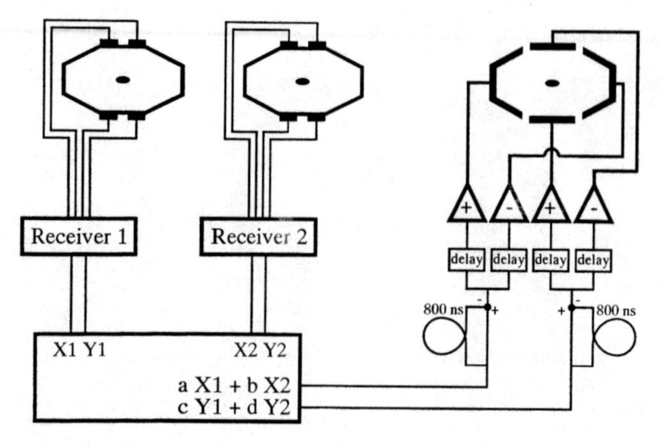

FIGURE 6. Block diagram of the transverse feedback system.

Transverse Feedback System Overview

For BESSY-II, the transverse feedback system will be modeled after the analog system developed for the ALS [13], [14]. Figure 6 shows a block diagram of the system, where signals from two sets of button-type pickups approximately 90° apart in betatron phase are used. The moment signals (displacement·charge) are detected at 3 GHz, differenced, mixed down to baseband, and combined in proper proportion. For offset rejection, the correction signals from subsequent revolutions are subtracted. The resulting kicks are provided by stripline kickers, where either one or — as shown in the figure — both electrodes are driven by a power amplifier.

Transverse Stripline Kicker

The transverse stripline kicker combines horizontal and vertical electrodes in one structure, which minimizes space requirements and leads to a low loss factor. The electrodes are shaped according to the octagonal vacuum chamber. Despite some coupling between horizontal and vertical electrodes, the small distance of the electrodes to the beam (32 mm and 17.5 mm) leads to a sufficiently large vertical kicker shunt impedance of 20 kΩ at low frequency, dropping to 10 kΩ at 250 MHz. In the horizontal coordinate, where no resistive wall effect is anticipated, the shunt impedance is lower by a factor of 2. The electrodes have a line impedance of 50 Ω and are 30 cm long, to minimize power picked up from the 500 MHz bunch sequence. A simplified MAFIA model of the kicker for wakefield and 3D electrostatic computation is shown in Figure 7.

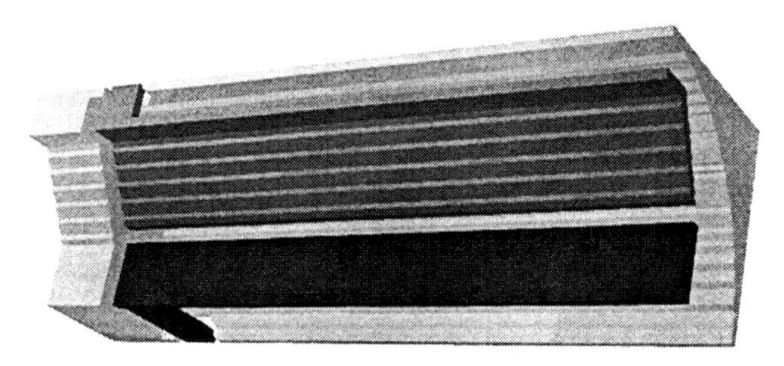

FIGURE 7. MAFIA model of the transverse feedback kicker (1/8 of the full structure).

Transverse Damping Rates

The damping rate of the transverse feedback system is given by

$$\frac{1}{\tau} = \frac{f_\circ \sqrt{\beta_1 \beta_2}}{2\,E/e} \cdot G, \tag{6}$$

where β_1 and β_2 are the beta functions at the pickup and at the kicker position, respectively, and $G = \Delta U/\Delta y$ is the gain, i.e., the kick voltage per unit displacement.

With a 100 W amplifier connected to each vertical electrode and assuming 50% power loss, the maximum voltage at low frequency is 2000 V. For saturation at $\Delta y = 1$ mm, the maximum damping rate is 4400 s^{-1}, dropping to 3100 s^{-1} at 250 MHz, as shown in Figure 8. The maximum impedance $Z_\perp = G/I$, that can be counteracted at a beam current of 400 mA, is 5 MΩ, whereas the resistive wall impedance is estimated to be 2 MΩ at the lowest frequency.

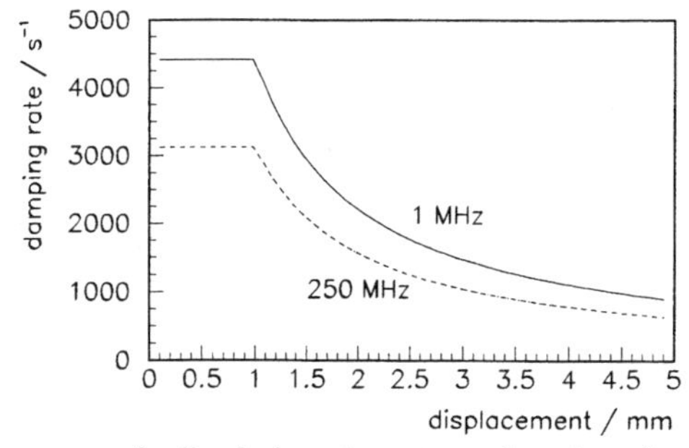

FIGURE 8. Transverse feedback damping rate as function of transverse displacement.

SUMMARY AND PRESENT STATUS

The design effort for feedback systems to counteract multibunch instabilities has been minimized by making use of the developments and experience that other facilities (SLAC, ALS, DAΦNE) have generously made available to BESSY.

The longitudinal feedback electronics have been fabricated and most of the commercially available components have been ordered. For the transverse feedback system, the components for the mixer and one receiver have been purchased and a prototype will be built during summer 1998. For both systems, existing button-type pickups can be used. The kickers for both systems will be installed in a straight section used for diagnostic purposes. First beam test opportunities will occur in several machine study periods toward the end of 1998.

ACKNOWLEDGMENTS

The help of W. Barry, J. Byrd, J. Corlett, and G. Stover (ALS, Berkeley); of J. Fox, H. Hindi, S. Prabhakar, and D. Teytelman (SLAC, Stanford); and of A. Gallo, F. Marcellini, M. Serio, and M. Zobov (INFN, Frascati) is gratefully acknowledged.

This work is funded by the Bundesministerium für Bildung, Wissenschaft, Forschung und Technologie and by the Land Berlin.

REFERENCES

1. Jaeschke, E., *Proc. of the 1997 Part. Acc. Conf., Vancouver*, in print (1997).
2. Svandrlik, M., et.al., *Proc. of the 1995 Part. Acc. Conf., Dallas*, 2762 (1995).
3. MAFIA Collaboration, *MAFIA Manual*, CST GmbH, Darmstadt (1996).
4. Corlett, J. N. and J. M. Byrd, *Proc. of the 1993 Part. Acc. Conf., Washington DC*, 3408 (1993).
5. Bartalucci, S., et al., *Part. Acc.* **48**, 213 (1995).
6. Teytelman, D., et al., *Proc. of the 1995 Part. Acc. Conf., Dallas*, 2420 (1995).
7. Prabhakar, S., et al., *Part. Acc.* **57**, 175 (1997).
8. Teytelman, D., et al., SLAC-PUB-7305 (1996).
9. Hindi, H., S. Prabhakar, J. Fox, and D. Teytelman, *Proc. of the 1997 Part. Acc. Conf., Vancouver*, in print (1997).
10. Young, A., J. Fox, and D. Teytelman, *Proc. of the 1997 Part. Acc. Conf., Vancouver*, in print (1997).
11. Corlett, J. N., et al., *Proc. of the 1994 Europ. Part. Acc. Conf., London*, 1625 (1994).
12. Boni, R., et al., *Part. Acc.* **52**, 95 (1996).
13. Barry, W., et al., *Proc. of the 1993 Part. Acc. Conf., Washington* 2109 (1993).
14. Barry, W., et al., *Proc. of the 1994 Europ. Part. Acc. Conf., London*, 122 (1994).

System for the Control and Stabilizing of OK-4/Duke FEL Optical Cavity[1]

I. V. Pinayev, M. Emamian, V. N. Litvinenko, S. H. Park, and Y. Wu

FEL Laboratory, Department of Physics, Duke University
P. O. Box 90319, Durham, NC 27708-0319

Abstract. The control system of an optical cavity is described. Usage of the piezoelectric actuators and position sensitive photodetectors in this system allows us to reach a resolution at a submicroradian level and to suppress mirror vibrations below 50 Hz.

INTRODUCTION

The OK-4/Duke free electron laser (FEL) has one of the longest optical cavities (53.76 m) among the existing FELs and the optical beam waist in the undulators is about 300 μm. This implies that the optical cavity control system should have a resolution better than a few μrad and be capable of keeping mirrors in the desired position with the same accuracy. Because the OK-4/Duke FEL has an almost concentric optical cavity with mirror radii of curvature of 27.27 m, it requires a resolution better than 1 μrad.

The previously used optical cavity control system, with an SL20A gimbal mirror mount (manufactured by Microcontrole) with 0.1 μm stepper motor actuators, had sufficient resolution. The existing tables supporting the mirror system, however, did not suppress vibrations down to the required levels. Also they had a long-term drift caused by the changing of the static load on the floor nearby (see Fig. 1).

It is possible to suppress high-frequency vibrations using passive damping, but the low-frequency vibrations and drift are still a problem for such systems. Therefore, we decided to replace the control system with a new one using active suppression of the mirror vibration and drift.

1. This work is supported by Office of Naval Research Contract #N00014-94-1-0818 and AFOSR Contract #F49620-93-1-0590.

CP451, *Beam Instrumentation Workshop*
edited by R. O. Hettel, S. R. Smith, and J. D. Masek
 1-56396-794-4/98/$15.00

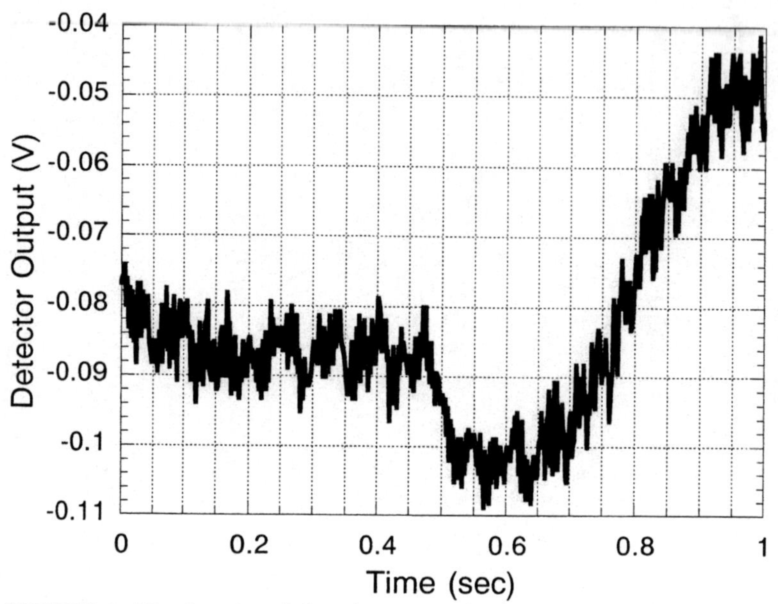

FIGURE 1. The changing of the mirror mount's tilt angle caused by the variation of static load on the table (10 mV=0.4 μrad).

GENERAL LAYOUT

The general layout of the optical cavity control system is shown in Figure 2. The mirror mount's tilt angle is measured with help of an auxiliary concave mirror (R=5.7 m) rigidly fixed on the mount. A small semiconductor laser installed inside the ring room is used as a reference. The three-foot concrete slab supporting the storage ring has pillars going down to the bedrock which provide a significantly more stable environment. Light emitted by the laser passes through the 0.6 mm diameter pinhole to the auxiliary mirror. Distance between the mirror and pinhole is equal to the radius of curvature of the auxiliary mirror. The image of the pinhole is focused on the two-dimensional position sensitive photodetector (PSD) S2044 manufactured by Hamamatsu. The pinhole diameter is chosen not only to provide a point-like light source but also to reduce the photon flux so that the PSD current does not exceed the maximal rating. This configuration has the advantage of being insensitive to the angular position of the laser beam. To prevent refraction caused by the airflow and to exclude the influence of ambient light, the entire optical path lies inside a sealed plastic tube.

The signal from PSD is processed by the analog unit shown in Figure 3 which provides output voltages proportional to the distance between the centers of the light spot and detector. The analog divider/multiplier AD734 is employed to normalize the signal and to make it insensitive to the variations of the laser power.

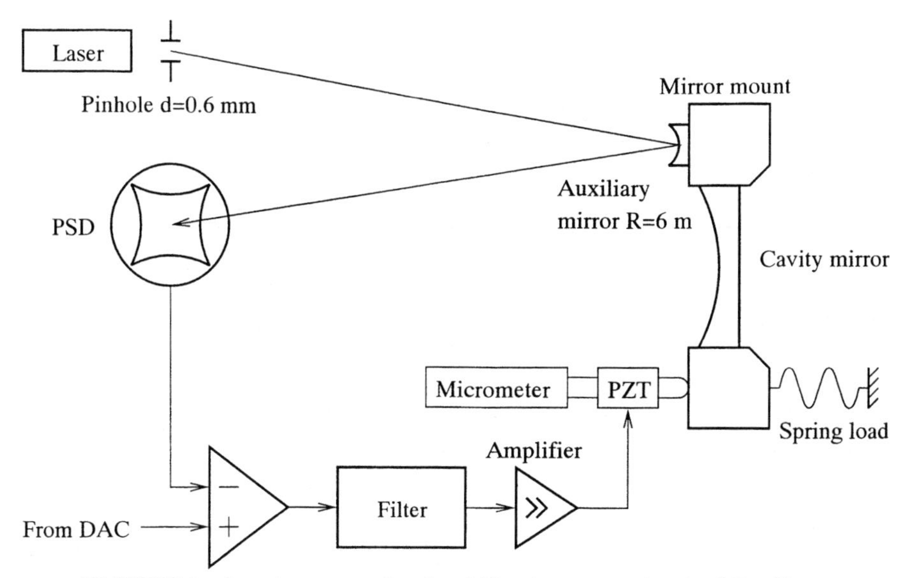

FIGURE 2. the mirror control and stabilization system for the OK-4/Duke FEL.

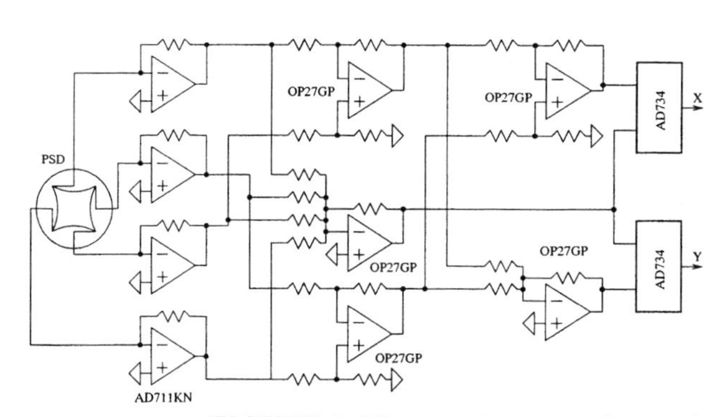

FIGURE 3. The operating circuit for two-dimensional PSD.

The precision of the mirror control system is defined by the PSD spatial resolution and the accuracy of the electronics. The main source of errors in the electronics is the bias current and offset voltage of the operational amplifiers being used. To minimize these errors, the precision operational amplifiers AD711 and OP27 are used. The electronics provide a geometrical accuracy better than 1 μm, assuming maximal values for the PSD dark current, bias currents, and offset voltages. According to the manufacturer's specifications, the PSD has a resolution better than 2.5 μm. The accuracy of the control system can be easily calculated to be better than 0.5 μrad. It should be mentioned that, as a result of the distortions, the PSD has a nonlinear response to the position of light spot. The typical deviation from linearity does not exceed 40 μm.

The S2044 PSD has a 0.3 μsec rise time and the analog chips in the electronics have a unity gain at a frequency of few MHz. The wide frequency band eliminates the influence of the position-sensitive detector on the gain-phase characteristics of the feedback loop.

We have chosen 30 μm P-830.20 piezotranslators manufactured by Polytec PI for our system. They are designed to withstand high loads and have a maximal operational voltage of 150 V. The small displacement range of the piezotranslators requires that the initial position of each mirrored be set by two manual micrometers. Fine adjustment of the mirrors is controlled by four 16-bit digital-to-analog converters (DACs). The signal from PSD is compared with the control voltage from the DAC and the error signal is amplified to change the voltage on the piezoelectric actuator.

Testing of the prototype feedback system with HP4194A Impedance/Gain-Phase Analyzer showed that the cutoff frequency is much lower than was estimated using the piezotranslator stiffness and the mirror mount inertia. We found a set of mechanical resonances of mirror mount itself. The strongest resonance which has the lowest frequency of about 100 Hz is caused by the ball bearing support (see Fig. 4). To provide a higher cutoff frequency with a substantial phase margin for stability we use a correction filter installed before the power amplifier. Its circuitry and characteristics are shown in Figure 5 and Figure 6, respectively. Each filter is tuned individually for better performance. This allows us to increase the cutoff frequency from 1 Hz to 50 Hz. The correction filter also integrates the error signal, which provides a much higher long-term stability. The gain-phase characteristics of the corrected feedback system are shown in Figure 7. The high-frequency vibrations are suppressed by mechanical absorbers incorporated into the table supports.

The optical cavity control system is brought under the EPICS storage ring control system. It provides individual mirror positioning as well as the control of the optical cavity axis. Observation of the output of the PSD electronics which has a sensitivity of 40 nanoradians per millivolt showed that the mirror vibration level is below 100 nanoradians with a negligible drift.

The system was used for two successful runs for lasing in the the wavelength ranges of 345–413 and 226–255 nm.

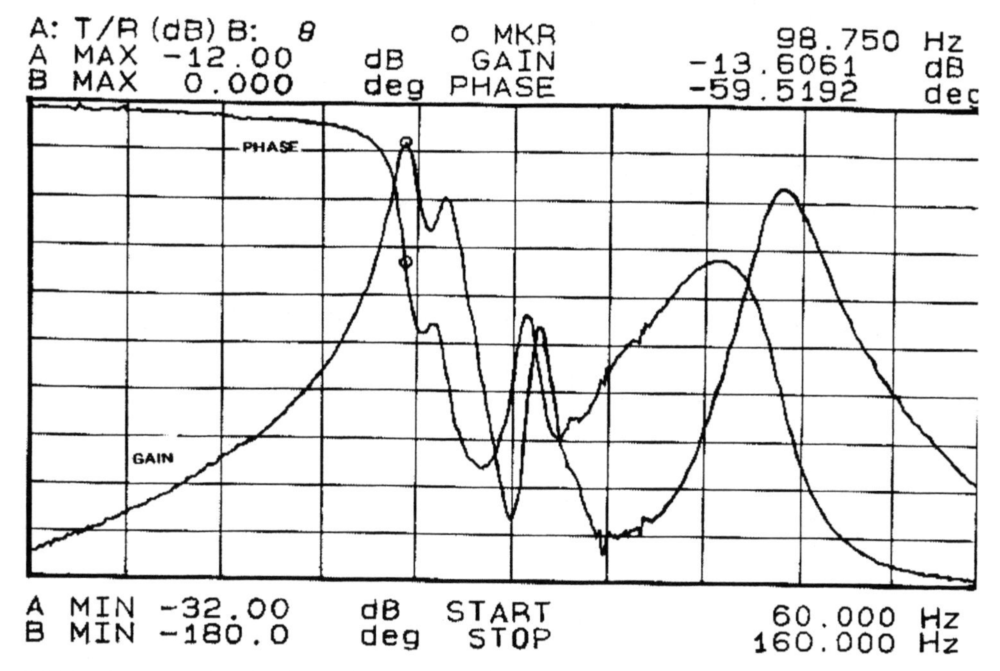

FIGURE 4. Amplitude-phase characteristic of the 100 Hz mirror mount mechanical resonance.

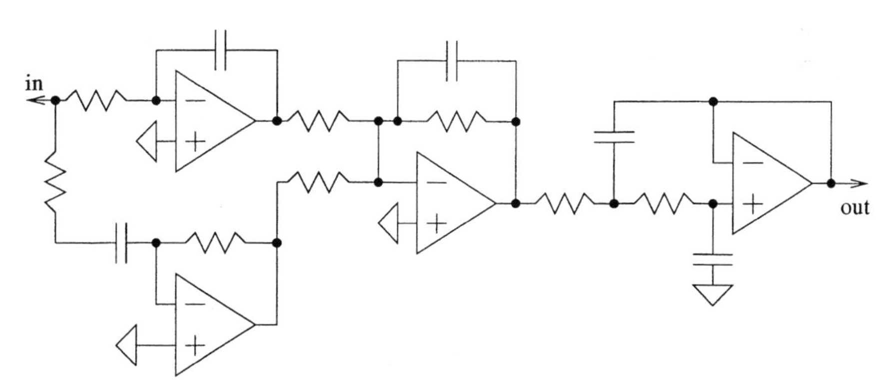

FIGURE 5. Circuitry of the correction filter.

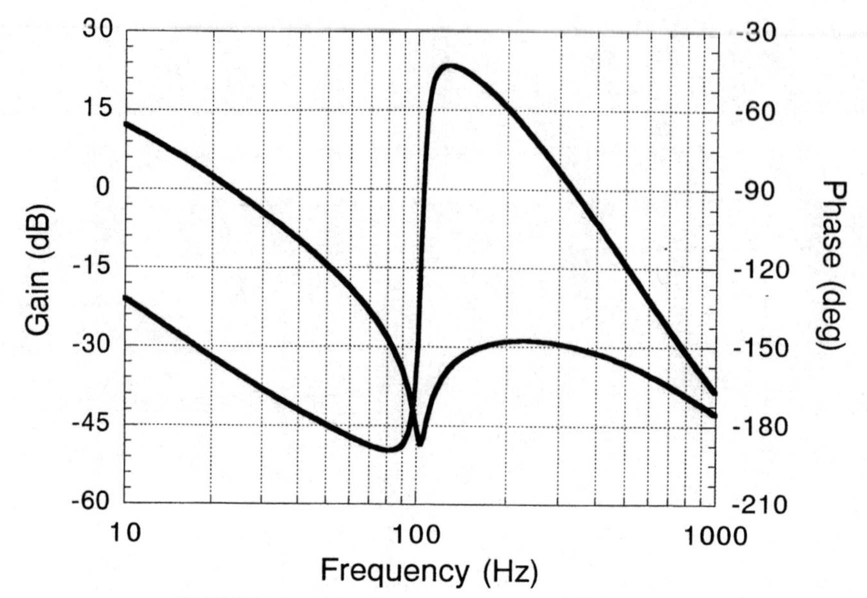

FIGURE 6. Gain-phase characteristic of the correction filter.

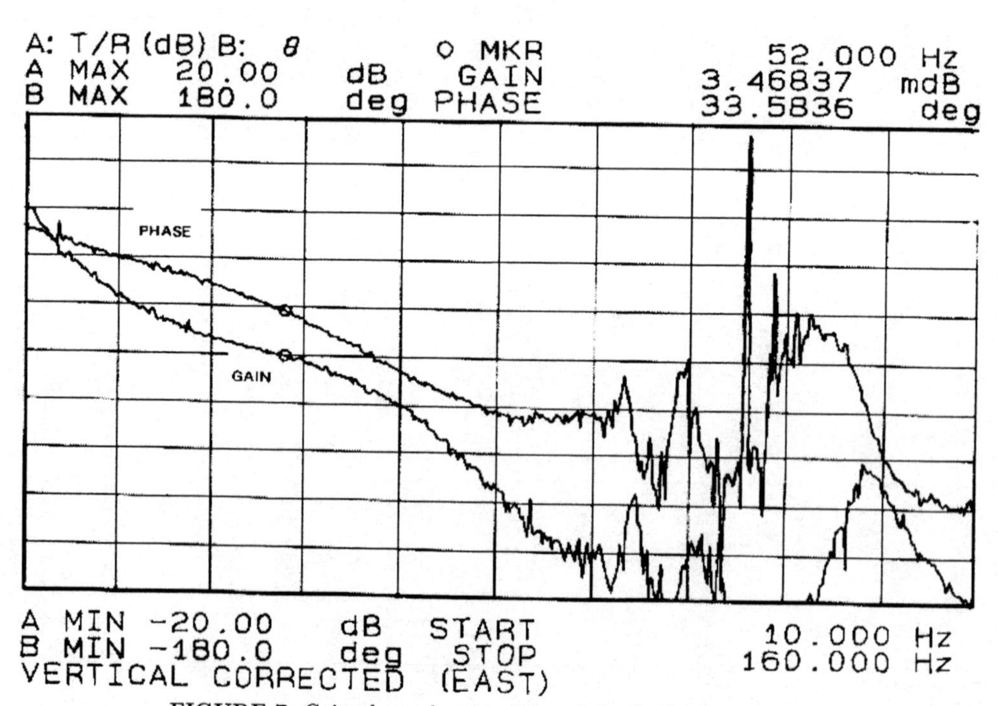

FIGURE 7. Gain-phase characteristics of the feedback system.

FUTURE IMPROVEMENTS

In the near future we plan to:

1. Replace the existing control system with a new system in which the PSD is moved with stepper motor driven translation stages. This will give us greater tuning range and will eliminate PSD non-linearities, which will be used as a zero position detector.

2. Design and manufacture a new gimbal mount for the optical cavity mirrors. This should allow us to eliminate low-frequency resonances and increase the frequency range of the feedback system.

We are also considering the possibility of replacing existing analog electronics with a digital system processing board for flexibility and higher accuracy.

CESR Feedback System Using a Constant Amplitude Pulser*

G. Codner, M. Billing, R. Meller, R. Patten,
J. Rogers, J. Sikora, M. Sloand, C. Strohman

Laboratory of Nuclear Studies,
Cornell University, Ithaca, NY 14853

Abstract. particle beam feedback system using constant-amplitude, 1000 V, 12 ns pulses has been built to provide longitudinal and horizontal feedback for stabilizing 14 ns spaced bunches for use in CESR (Cornell Electron Storage Ring). The pulse rate is modulated to obtain proportional amplitude control and the pulse arrival time is modulated to obtain both positive and negative kicks. The average repetition rate is limited by pulser power dissipation, but the instantaneous rate may be increased to full duty cycle for short periods of time to handle transients. The pulser drives a 50-ohm stripline kicker so the equivalent peak power at 1000 V is 10 kW. The characteristics of the pulser and its modulator will be described along with the system's operation.

BACKGROUND AND SYSTEM REQUIREMENTS

CESR (Cornell Electron Storage Ring) may store up to 45 bunches each of electrons and positrons in its present operational configuration. The electron and positron beams are arranged in trains of up to 5 bunches with a nominal spacing of 280 ns between lead bunches in a train. The bunches in a train have a minimum spacing of 14 ns. In high current operation, the unstable modes of oscillation of the beam must be suppressed using feedback.

The feedback system must sense bunch motion and deliver either deflection or acceleration independently to each bunch in order to damp all of the possible dipole multibunch instabilities. The desired peak amplitude for handling transients is approximately 1000 V. For a system delivering a sinewave into a non-resonant kicker, a 10 kW amplifier would be required. The power available in the present system with a linear amplifier is only 400 W (140 V peak) and is obtained by combining the outputs of a pair of 200 W commercial broadband power amplifiers.

---

* This work was supported by the National Science Foundation.

CP451, *Beam Instrumentation Workshop*
edited by R. O. Hettel, S. R. Smith, and J. D. Masek

In the linear feedback system, these 200 W amplifiers drive a 50-ohm stripline kicker which is shorted at one end (1). The plates of the kicker are powered differentially using a hybrid power divider driven by the combined amplifier output. Horizontal excitation is produced directly through deflection of the beam at the kicker. Longitudinal or energy excitation also uses deflection and requires dispersion at the kicker so that horizontal deflections translate into acceleration by changing the orbital path length (2). Since the kicker is driven in the same fashion for the two modes, they are distinguishable only by the modulation frequency. With the kicker shorted at one end, the incident power is completely reflected and must be absorbed by the amplifier.

A higher power driver is desirable, but purchasing additional high-power amplifiers and combiners is quite costly, especially since the high power combiners would be custom-built. Because an amplifier must be developed, a unique concept may be considered which is more tailored to accelerator feedback system parameters. The feedback system requires short periods of high output power due to transients followed by longer periods of lower output power as determined by the beam's damping rate and external excitation as well as the gain and noise level of the feedback system. An amplifier with the capability of short-term high power is suggested. It can be simpler and cheaper because its long-term power output is small. In CESR operating with 90 bunches, the average power may be only 10% of the peak power required for transients.

CONCEPT OF FEEDBACK WITH A PULSER

One type of amplifier suited to the feedback system uses a pair of rate-modulated, constant-amplitude pulsers, one for each kicker plate. Horizontal deflection is generated by pulsing one kicker plate or the other, and acceleration is produced with the plates driven in common-mode so the beam experiences the potential across the gap between the kicker plates and the beam pipe. Because the CESR feedback kicker is shorted, the applied pulse is completely reflected and the reflected pulse is the inverse of the incident pulse.

To minimize self-cancellation, the maximum allowed pulse length is 12 ns as measured at the 10% points of the waveform. This comes from the 3 ns transit time of the shorted kicker, from the bunch spacing of 14 ns and from the approximately triangular shape of the high-voltage pulse.

Since the pulse energy is small compared to the beam energy, the discrete nature of the excitation does not necessarily adversely affect the transverse beam size. (Computer simulations of feedback by Strohman calculated less than 100 μm RMS horizontal motion for 1500 V pulses.) Due to the balanced voltage excitation on both plates in the accelerating mode of operation, one would expect any horizontal deflection in this mode to arise only from imbalances in the pulse amplitude, shape, and timing.

For a given pulser amplitude, the maximum transient amplitude and the achieved damping rate of the beam determine the maximum pulse rate and duration over which the maximum rate is required. For the same fixed-pulse amplitude, the overall feedback loop gain will be limited by the system quiescent noise level, and these three parameters will largely determine an average power output for the pulser. In the presence of transient excitations of the beam, the repetition rate and peak amplitude of the transients will also determine the average reserve pulser power. CESR damping times may be as long as 3 ms so the feedback pulser must be able to provide full duty cycle for at least 3 ms. FETs now available can provide full duty cycle for 10 ms as determined by their transient thermal impedance. Full duty cycle is here defined as 90 pulses per CESR period of 2.56 μs. The main source of periodic transients is injection which can occur every 16.7 ms in CESR.

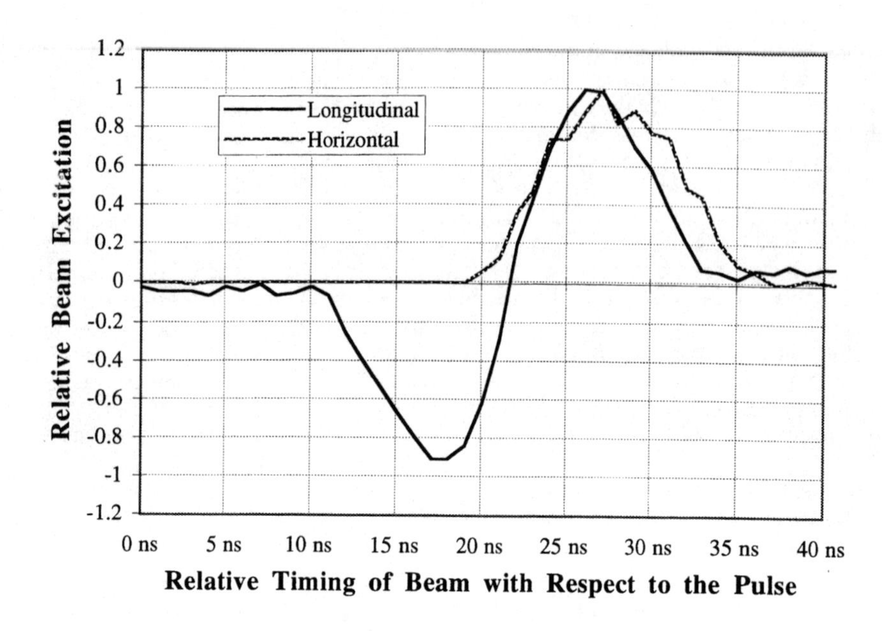

FIGURE 1. Beam excitation vs. relative timing of the beam and the pulse.

The horizontal deflection polarity is controlled by pulsing either one plate or the other. Polarity control for longitudinal excitation is obtained by timing the pulse so that the beam arrives either when the accelerating voltage across the gap in the kicker is negative due to the forward wave or positive due to the reflected wave. Figure 1 depicts beam acceleration or deflection versus timing (inferred from measurements of beam amplitude versus timing) for the two modes of beam excitation. For these measurements, the pulser rate is modulated at either the synchrotron or the betatron frequency, as appropriate. In the accelerating mode, the pulsers are driven simultaneously, whereas only one pulser is driven for horizontal mode. The pulse timing was not modulated, merely varied. A horizontal beam pickup was used in both cases to sense beam amplitude with dispersion at the beam detector necessary to sense longitudinal motion. A spectrum analyzer was used to measure the amplitude.

PULSER SYSTEM COMPONENTS

Overview

Figure 2 shows a block diagram of the pulser system. The beam-position monitoring electronics and digital signal processor (DSP) (3) provide 10-bit error words for e+ and e–, horizontal and longitudinal inputs to the rate modulator. The rate modulator sends synchronous data to the timing modulator located with the pulser. The timing modulators decode the data and send the appropriately timed triggers to the pulsers. The pulser outputs are transmitted to the kicker using coaxial cables. Lowpass filters (LPF) at the kicker inputs reduce the beam power transmitted back to the pulser.

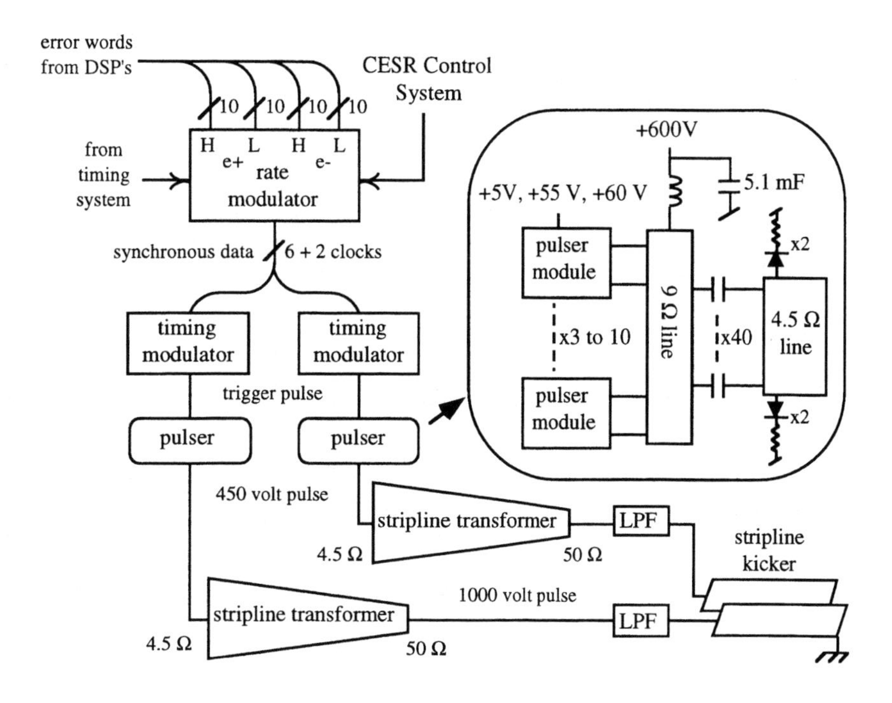

FIGURE 2. Pulser feedback system block diagram.

Rate Modulator

The rate modulator accepts two, 10-bit error words for each species of particle (e+ and e–) corresponding to horizontal and longitudinal bunch motion. Each word is compared to a 10-bit pseudo-random number generated by a 17-bit shift register with recursion. If the error word is greater than the random number, a trigger is generated. If both a horizontal and a longitudinal trigger are demanded, the horizontal trigger overrides, since horizontal deflections are expected to be required less frequently. The trigger information is encoded into a 3-bit word, one each for positrons and electrons, and is sent to the timing modulators along with a separate 36 MHz clock for each species. The 3 bits correspond to pulse enable, polarity, and mode (horizontal or longitudinal). The rate modulator also controls the long-term duty cycle of the pulsers, limiting it gradually to 10% of the 90 pulses per 2.56 μs. An analog input is provided for modulating the pulsers with an external source for diagnostics.

Timing Modulator

The timing modulators accept two sets of synchronous, three bit, parallel data and their corresponding 36 MHz clocks. The clocks and data are received by Hewlett-Packard HCPL-7101 optical isolators, and then the clocks are doubled to 72 MHz using S4402 phase-locked loop devices from AMCC. Transmission of the clock at half of the fundamental frequency allows the use of the TTL in, TTL out, 50-megabaud HCPL-7101 with the phase-locked loop providing narrowband filtering of the clock.

The data is decoded, registered, and used to select the appropriately timed trigger pulse to be sent to the pulser modules. For example, if the data corresponds to a positive horizontal pulse, one pulser would put out a pulse and the other would not. The timing modulator also includes fault detection circuitry which inhibits triggers if the pulser duty cycle limit is exceeded. The output trigger pulse width is adjustable from 4 to 6.5 ns, with a nominal value of 5.8 ns required to obtain a 12 ns high voltage pulse as measured at the 10% points of the waveform (7 ns FWHM).

Pulser Assembly

Each pulser assembly contains from three to ten pulser modules (printed circuit boards) connected to a 9-ohm strip transmission line whose center section is AC coupled to a 4.5-ohm stripline. The transmission lines are made of 3.2 mm thick G10 sandwiched between 6.4 mm thick copper of the appropriate width. Up to ten pulser modules may be mounted to the 51 mm wide, 406 mm long, 9-ohm line. A 600 V DC power supply and capacitor bank couple to the 9-ohm line through a 6 μH air-core inductor. The 5.1-mF capacitor bank provides energy storage for short-term high-power operation. A regulator board provides a common 60 V DC as well as regulation for the cascode nodes of each pulser module. The pulser is packaged in a welded aluminum enclosure with four separate EMI isolated compartments for the pulser modules, timing modulator, regulator, and capacitor bank. Control, interlock, and power supply connections between compartments and out of the pulser enclosure pass through rf filters.

Four, 20-ohm rf resistors (Florida RF Labs P/N 32-1143-20-5) are diode-coupled to the 4.5-ohm line to provide a termination for the reflected pulse from the shorted kicker. Each termination resistor is in series with a pair of UH840 ultra-fast recovery diodes. The power transistors, rf terminations, and all of the copper conductors are water-cooled.

The UH840 diode reverse recovery time is an important parameter since, in order to be properly terminated, the pulser must not be triggered when the reflected pulses return. In CESR, the cable length was chosen so that the reflected pulses arrive between trains. Nevertheless, the minimum unforced recovery time of currently available diodes mandates skipping every other train. The odd number of trains in CESR ensure that every train will be driven. (Note that the unforced recovery time of the diodes is approximately ten times the recovery time on the manufacturer's specification sheet because this specification applies to forced recovery.) One option for overcoming this limitation is to put out an early pulse which forces the diodes to recover. The "recovery pulse" would allow the pulser to excite successive trains in CESR but would lower its effective duty cycle capability.

Pulser Module

The pulser module consists of a pre-driver and a cascode driver and power stage packaged on a printed circuit board. The design, shown in Figure 3, was adapted from the CESR gun pulser (4) with significant changes to provide much higher duty cycle capability. The longer pulse width of the feedback pulser compared to the gun pulser allows the use of more readily available power devices since the relatively high inductance of the commonly available package can be tolerated by using high gate voltage together with the cascode arrangement.

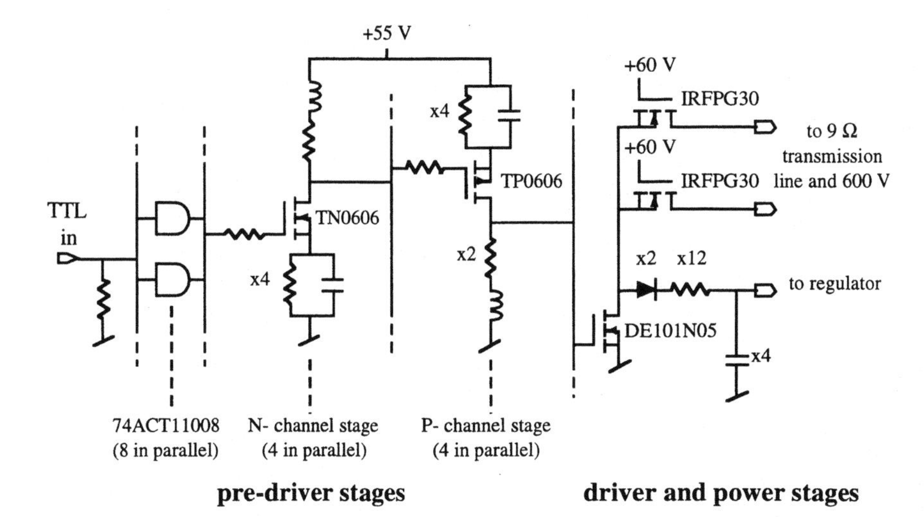

FIGURE 3. Pulser module schematic diagram.

Most of the board area is occupied by the pre-driver. This consists of eight 74ACT11008 AND gates in parallel driving an N-channel stage using four paralleled TN0606 MOSFETs (packaged in a single 16-pin ceramic DIP) which, in turn, drives a P-channel stage consisting of four TP0606 MOSFETs in parallel (also in a 16-pin ceramic DIP). Separate resistors in series with each transistor's gate prevent oscillations due to interactions between the paralleled devices. The pre-driver delivers a 20 V, roughly triangular pulse to the driver transistor.

The driver/power stage is in a cascode configuration. The driver is a DE101N05, an extremely fast, 100 V, RF MOSFET in a special low inductance package. The output stage contains a pair of IRFPG30, 1 kV MOSFETs in parallel. The gates of the IRFPG30s are biased at a nominal 60 V, although this voltage is adjustable to optimize pulse shape and power dissipation. The drains of the IRFPG30s are bolted to the 9-ohm transmission line, which also serves as a heat sink.

Transformer

The transformer is a tapered stripline device made of copper and G10. It provides an impedance transformation from 4.5 ohms to 50 ohms. The ground plane is 150 mm wide, 6.4 mm thick OFHC copper with water cooling. At the 4.5-ohm end, the G10 is 3.2 mm thick and the signal conductor is 102 mm wide. The G10 thickness tapers to 6.4 mm and the copper width to 12.7 mm at the 50-ohm end, 7.3 m away. The dielectric thickness is tapered because the signal conductor at the 50-ohm end of the line has to be wide enough for attaching cooling tubes. The entire assembly is packaged inside an aluminum enclosure for EMI shielding and safety.

The 4.5-ohm end of the transformer is intimately mated with the 4.5-ohm line in the pulser assembly. A type-N connector emerges from the 50-ohm end. The transformer provides a peak voltage transformation ratio of 2.4:1 for a 12 ns pulse, less than the ideal ratio due to dispersion and droop, so it delivers a 1080 V pulse out for a 450 V pulse in. Each transformer output is connected with coaxial cable to a lowpass filter and then the kicker.

PERFORMANCE IN CESR

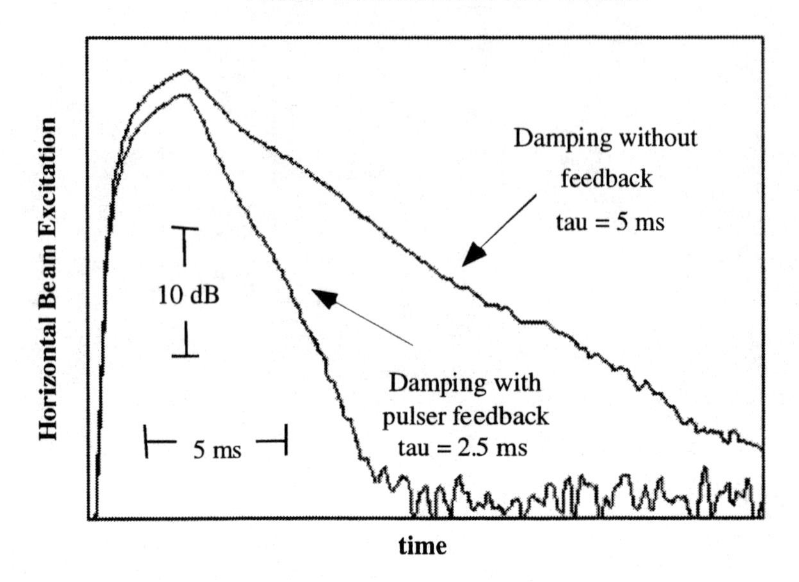

FIGURE 4. Measured horizontal damping in CESR.

The pulser feedback system has been installed and tested in CESR. Measurements of beam excitation versus timing show that the pulser's effect on the beam is as expected for both horizontal and longitudinal modes. Crosstalk between the two modes has been measured and found to be acceptable, although additional filtering of the sensed signals for each bunch is desirable. The feedback loops have been closed for both horizontal and longitudinal modes and they produce stable beams for both positrons and electrons. As an example, Figure 4 shows the horizontal beam response to transient excitation with and without feedback. The measurement was made using nine trains of positrons with one 6 mA bunch per train. A horizontal beam position monitor was observed with a spectrum analyzer set to zero span and centered on a betatron sideband. The beam is driven for about 4 milliseconds and then allowed to damp. The exponential decay appears linear with time on the log plot so the time constant is extracted by measuring the time it takes the amplitude to change by 8.7 dB. These preliminary measurements indicate that a duty cycle modulated pulser acting as a "digital amplifier" is a viable alternative to linear or CW phase-controlled amplifiers for providing particle beam feedback.

FUTURE DEVELOPMENT

A new signal processor (recently developed for CESR by Meller) incorporates digital filtering for each bunch and will facilitate pulser feedback by reducing both receiver noise and crosstalk between horizontal and longitudinal modes. Final installation of fault protection and interface circuits is complete, and the pulser feedback system will soon be commissioned for regular use in high-energy physics operation of CESR.

ACKNOWLEDGMENTS

Thanks to the operations and technical staff at Wilson Synchrotron Laboratory, especially to Buzz Metzler for his artistic plumbing. Thanks to John Stauffer for many long hours designing and assembling the stripline transformer. Thanks to Don Dawson, Bill Hnat, Mike Comfort, Mike Ray, Margee Carrier, and Bob Strohman for their applied expertise.

REFERENCES

[1] Rogers, J., "Coupling of a Shorted Stripline Kicker to an Ultrarelativistic Beam," Cornell CBN, 95–04, April 21, 1995.
[2] Sagan, D., M. Billing, "Using a Horizontal Kicker to Damp Longitudinal Oscillations," Cornell CBN, 96–06, May 23, 1996.
[3] Rogers, J., M. Billing, J. Dobbins, C. Dunnam, D. Hartill, T. Holmquist, B. McDaniel, T. Pelaia, M. Pisharody, J. Sikora, C. Strohman, "Operation of a fast digital transverse feedback system in CESR," presented at the International Particle Accelerator Conference, Dallas, TX, May 17, 1995.
[4] Dunnam, C., R. Meller, "Nanosecond MOSFET Gun Pulser for the CESR High Intensity Linac Injector," presented at the International Particle Accelerator Conference, May 1, 1993.

Main Injector Synchronous Timing System

Willem Blokland and James Steimel

Fermi National Accelerator Laboratory
Batavia, Illinois 60510

Abstract. The Synchronous Timing System is designed to provide sub-nanosecond timing to instrumentation during the acceleration of particles in the Main Injector. Increased energy of the beam particles leads to a small but significant increase in speed, reducing the time it takes to complete a full turn of the ring by 61 nanoseconds (or more than 3 rf buckets). In contrast, the reference signal, used to trigger instrumentation and transmitted over a cable, has a constant group delay. This difference leads to a phase slip during the ramp and prevents instrumentation such as dampers from properly operating without additional measures. The Synchronous Timing System corrects for this phase slip as well as signal propagation time changes due to temperature variations. A module at the LLRF system uses a 1.2 Gbit/s G-Link chip to transmit the rf clock and digital data (e.g. the current frequency) over a single mode fiber around the ring. Fiber optic couplers at service buildings split off part of this signal for a local module which reconstructs a synchronous beam reference signal. This paper describes the background, design, and expected performance of the Synchronous Timing System.

INTRODUCTION

The Main Injector ramps from 52.812 MHz to 53.104 MHz during the acceleration from 8 GeV to 150 GeV while increasing the speed of the particles. Given that there are 588 buckets in the Main Injector, a single turn goes from 11134 ns to 11073 ns, a difference of 61 ns. In contrast, the reference signal, transmitted over a cable, has a constant group delay. This difference leads to a phase slip that, at MI-30 or halfway around the Main Injector, is the group delay, 12 μs for a single-mode fiber, multiplied by the frequency sweep, 300 kHz, or more than 3 rf cycles. This is unacceptable for a damper system or a synchronously sampling beam position system.

The idea behind the Synchronous Timing System (STS) is to provide instrumentation around the Main Injector with a trigger synchronous to a particular bunch of the beam, even during acceleration. Previous work done at Fermilab on obtaining a beam synchronous trigger is described in (1). However, this system used a frequency counter on the rf signal to determine the frequency. The STS will transmit not only the rf but also a digitally encoded frequency value to simplify the design of the local receiver. A different approach to obtain a synchronous phase is to add a delay to the LLRF feedback

CP451, *Beam Instrumentation Workshop*
edited by R. O. Hettel, S. R. Smith, and J. D. Masek

loop that is equal to the cable delay from the LLRF to the local receiver. However, for the Main Injector LLRF such a delay would complicate the LLRF system and might even lead to instabilities.

An added complication is the temperature dependency of the signal carrying medium. For example, for a single-mode fiber this can be up to 30 ps/c/km. With a 10° C variance in duct temperature and a cable length of 5 km, the total time differs about 1.5 ns. That is enough of a variation that the STS needs temperature compensation. The STS design aims to provide a synchronous trigger with timing noise of around a 100 ps rms. This will satisfy the requirements for instrumentation such as dampers or synchronous beam position sampling systems and support beam instability studies.

The following sections in this paper describe the reconstruction of the phase, the signal distribution infrastructure, and the design of the receiver and transmitter boards.

SYNCHRONOUS TIMING

The STS must correct the phase to account for the difference in travel times between the reference signal and the particle speed as well as the variations in the group delay due to temperature changes. This section first analyzes the phase as a function of frequency (rf) followed by an analysis of the phase as a function of temperature.

Phase as a Function of Frequency

The LLRF is trying to keep the particles at the same phase for each turn no matter what the frequency, as the number of buckets remains the same. This is true for any location around the ring. As the frequency increases, the phase of the reference signal shifts as more cycles fit within the same cable length. Therefore the local receiver must "unwind" the phase of the rf signal to construct a synchronous phase. This is done by subtracting a delta phase that is the product of the change in frequency and the time-of-flight of the rf signal. In addition, a phase intercept is added to align the phase of the reconstructed rf with the phase of a detector:

$$\varphi_{sync} = \varphi_{RF} - \Delta\varphi + \varphi_{intercept} \quad (1)$$

$$\Delta\varphi = \Delta\omega \cdot \tau$$

with $\Delta\omega = \omega - \omega_{base}$, ω_{base} =radian frequency at beginning of cycle, τ = time-of-flight of rf, and $\varphi_{intercept}$ = phase intercept.

Phase as a Function of Temperature

The temperature dependency is modeled as an extra delay in the cable as a function of temperature:

$$\Delta\tau = \left(v(T_0) - v(T_1)\right) \cdot l \quad (2)$$

with l = length of cable, T_i= temperature at time i, and $v(T_i)$ = signal speed at T_i.

The $\Delta\tau$ will be measured indirectly by comparing the phase of the rf signal generated by the transmitter board and the phase of the rf signal received after it has gone around the ring. The receiver for this signal will be on the same board as the transmitter. The phase shift due to the $\Delta\tau$ is:

$$\Delta\varphi_{temp} = \Delta\omega \cdot \Delta\tau \tag{3}$$

Assuming that the distribution of the temperature dependent phase shift is linear along the signal path then the local phase shift is:

$$\Delta\varphi_{temp}^{local} = \Delta\varphi_{temp}^{total} \cdot \frac{\tau^{local}}{\tau^{total}} \tag{4}$$

with the value of a variable at an arbitrary location around the ring indicated by the superscript *local* and the location back at the LLRF after a full turn indicated by the superscript *total*.

The change due to temperature variation is very slow compared to the change due to frequency sweeping. The cable ducts are buried 4 to 7 feet and should change only about 10° C over the seasons. Fast changes between the generated and received phase will be regarded as errors that should be diagnosed. Even if the temperature varied by 1° C over a day this would only be a change of 0.1· ps per minute.

Reconstructed and Temperature Compensated Phase

The compensating phase shift to be locally added to the rf signal is the sum of the frequency phase slip and temperature compensation:

$$\Delta\varphi^{local} = \Delta\omega \cdot \tau^{local} + \Delta\varphi_{temp}^{total} \cdot \frac{\tau^{local}}{\tau^{total}} \tag{5}$$

This can be rewritten so that only one term, $\Delta\omega + \Delta\varphi_{temp}^{total} / \tau^{total}$, must be transmitted:

$$\Delta\varphi^{local} = \left(\Delta\omega + \frac{\Delta\varphi_{temp}^{total}}{\tau^{total}} \right) \cdot \tau^{local} \tag{6}$$

SIGNAL DISTRIBUTION

The distribution of the signal from the transmitter to the receivers is depicted in figure 1. The LLRF system provides the transmitter with a voltage controlled oscillator (VCO) synchronized strobe and the required digital information. We choose single mode fiber as the medium because of its low cost, low attenuation, and high bandwidth. The low cost, $1 per foot per 24 fibers, gives it a price advantage over copper cables especially considering the number of channels per cable. The low attenuation, 0.35 dB/km, makes it possible to split off passively a part of the signal at each of the 10 service buildings while avoiding the use of amplifiers and associated noise. Figure 2 depicts the loss calculation and includes losses due to connectors, splitters, and 5 km of cable.

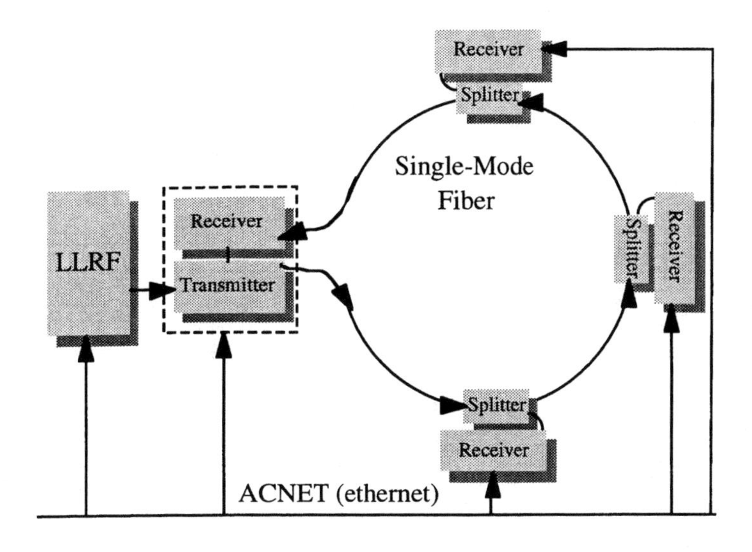

FIGURE 1. The signal distribution.

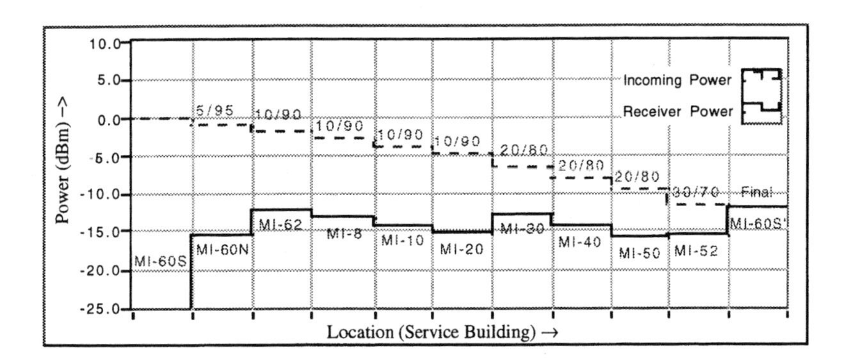

FIGURE 2. The passive splitting of a signal at each of the 10 service buildings. The top numbers indicate the splitter ratio while the top (dotted) line is the optical power in the incoming fiber to the service building. The lower (solid) line is the optical power to the receivers at each location with the name of the service building below the solid line. This calculation assumes a 1 dBm transmitter with a minimum receiver power requirement of -16 dBm.

The high bandwidth makes it possible to use the G-Link chip from Hewlett Packard (2) which supports a stream up to 1.2 Gbit/s. The STS will use the rf (53 MHz) as a clock and the 20 bits per clock cycle are used to carry the digitally encoded information. Finisar Optical modules (3) convert the G-Link signal to 1310 nm light waves and back. The STS will have, for redundancy reasons, two fibers with an active signal and a third fiber prepared for immediate use. The remaining 21 fibers are for future use.

The transmitters and receivers are VXI modules that, through the slot 0 processor, communicate with consoles using Fermilab's Accelerator Control Network (ACNET) protocol. This link is used for initialization, diagnostics, and calibration purposes.

RECEIVER AND TRANSMITTER DESIGN

Besides correcting the phase, the STS must be able to identify a particular bunch; otherwise a damper would not know which bunch to kick. A marker reset is generated by the LLRF on injection of beam into the Main Injector. The transmitter sends this marker to the receivers which then reset a local bucket counter. This counter counts each cycle (read bucket) of the synchronized rf and resets when all 588 buckets have passed by, thus always providing the same counter value for the same bucket.

Resolution Considerations

The maximum frequency change between updates, 10 μs, will be 20 Hz in the Main Injector, given the 2 MHz/s slew rate. This corresponds to about a 10 ps peak to peak jitter at the maximum cable delay of 25 μs.

The G-Link chip can transmit 20 bits per frame at a frame rate of 53 MHz. As the STS must transmit different messages, 4 bits are used to identify the type of message. This leaves 16 bits to describe the frequency sweep. Given a range of 500 kHz, (300 kHz is the current sweep range, 200 kHz extra range), the 16 bit word has a resolution of 500 kHz/65536 or 7.6 Hz. With the delay of 25 μs this results in a jitter of 4 ps peak to peak. These uniform white-noise quantization errors lead to an rms noise of $\sqrt{((10^2 + 4^2)/12)} \approx 3$ ps. This is small compared to the rms noise from the G-Link chip, about 42 ps. We do expect contributions from other parts of the electronics but not so high that we will come above a 100 ps rms noise total.

The rf signal from the LLRF, however, is not perfect. It is estimated that, depending on which cycle is running in the Main Injector, there is about a 100 ps of rms noise in this signal that is relevant to a damper system. Improvements on the quality of the rf signal will be for a future project.

The Receiver

The design of the receiver is shown in Figure 3. The STS signal comes in over the fiber and is converted into a parallel stream of 20 bits per data frame with a 53 MHz strobe by an optical receiver consisting of a Finisar FRM 1310 and a G-Link module. Diagnostic data from the Finisar and G-Link module are accessible over the VXI interface. The strobe is used as a clock to the phase shifter and phase calculation. The frame is decoded on an Altera chip into the correction term of equation (6) or a marker reset event. The phase calculation performs the multiplication in equation (6) and adds two more phase offsets. One is the intercept phase, $\varphi_{intercept}$, to align the synchronous phase with the phase of a beam pickup. The other is a phase from a local analog signal that could be part of a PLL on a beam signal to help improve the synchronous phase. ACNET can access the local intercept phase, $\varphi_{intercept}$, and time-of-flight ,τ_{local}, values over the VXI interface for calibration setup. The Phase Shifter is an I&Q modulator circuit. The Altera chip implements the sine and cosine functions of the corrected phase. These signals are converted to analog to be multiplied by the quadrature components of the strobe signal, summed together and bandpass filtered. The resulting signal is used to pulse the counter of the pattern generator and used as the synchronous rf output signal. The Pattern Generator contains a single-byte-wide memory bank with one byte for each bucket and one bit for each of the 8 output Triggers. The pattern generator's memory is set using ACNET. To allow easy calibration and diagnostics, the correction term, the calculated $\Delta\varphi$ shift, and the local signal are available through the Multiplexer (MUX) unit and the D/A converter as a analog signal or through the VXI interface as a digital value.

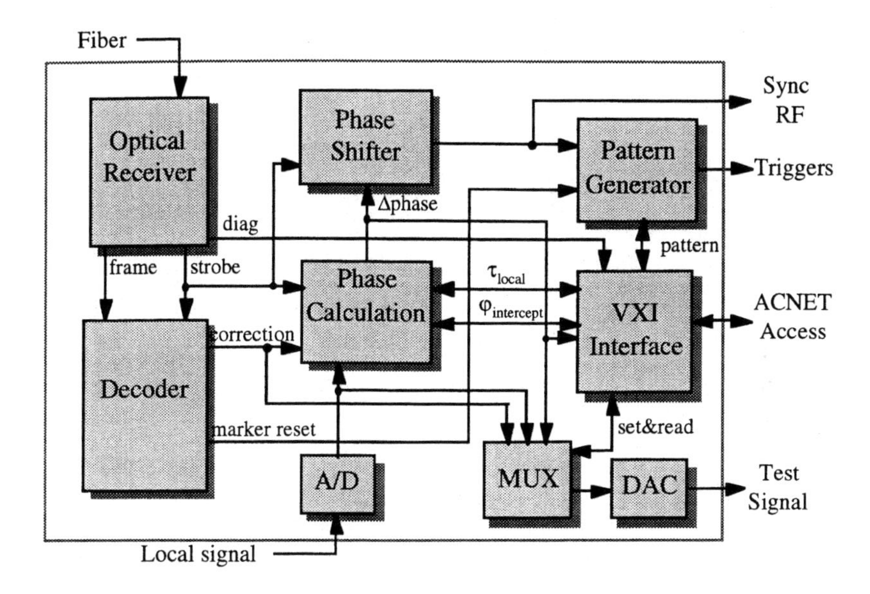

FIGURE 3. Diagram of the receiver.

The Transmitter

The transmitter module provides the optical signal used by the receiver modules. This signal is synchronized with the rf by the VCO of the LLRF and contains information on the frequency value, the change in fiber delay due to temperature, and the beam marker reset. The module is located close to the Main Injector LLRF system, and it contains the same DSP, an Analog Devices SHARC, as the LLRF system. (4) Using the same DSP simplifies communication between the two systems and also maximizes data transfer speed. The LLRF DSP transfers frequency information to the transmitter DSP through a fast link port.

Each transmitter module contains all the components of a receiver module except the pattern generator and the A/D converter. This receiver decodes the transmitter signal which has traveled the full circumference of the ring. The phase detector of the transmitter compares the synchronized output of the receiver module with the rf input of the transmitter module. Errors are accumulated and used to modify $temp_{adj}$. The phase calculation, implemented on the DSP, uses the digital frequency, $freq_{LLRF}$, the $temp_{adj}$, and the time-of-flight, to calculate the correction term. The marker reset is sent directly to the encoder to avoid processor delays, which might disrupt the single bucket precision of the signal. Access to the receiver and the τ^{total} value is handled by ACNET.

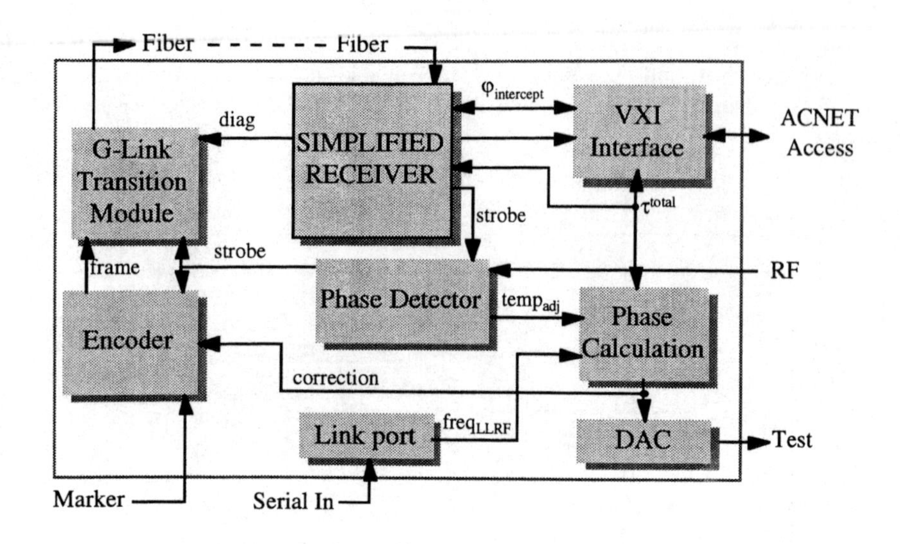

FIGURE 4. Block Diagram of transmitter.

CONCLUSIONS

We have designed a timing system that delivers a trigger synchronous to the beam during acceleration. The infrastructure has been installed and work to lay out the transmitter and receiver boards is in progress. We expect that the noise added by the STS to the synchronous trigger will be less than 100 ps rms. As the rf signal has about a 100 ps rms noise to it, the total noise is estimated around 140 ps. This can be improved locally by locking to a signal from a beam pickup for those applications that don't require first turn sampling. In the future, we hope to improve the rf signal from the LLRF to further reduce the noise in the synchronous trigger.

ACKNOWLEDGMENTS

Many thanks go to Brian Chase, Jim Crisp, and Keith Meisner for ideas and recommendations on the design issues.

REFERENCES

[1] Steimel, J., "Trigger Delay Compensation for Beam Synchronous Sampling," *Beam Instrumentation Proceedings of the Seventh Workshop*, 1996, pp. 476–482.

[2] Yen, C. S. et al., "G-Link: A Chipset for Gigabit-Rate Data Communication," *Hewlett-Packard Journal*, **43** (3), 103–116, (1992).

[3] "FTM/FRM-8510 Low Cost Gigabit Optical Transmitter/Receiver," Finisar Corporation, Mountain View, California.

[4] Chase, B. et al., "Current DSP Applications in Accelerator Instrumentation and RF," presented at the International Conference on Accelerator and Large Experimental Physics Control Systems (ICALEPCS '97), Beijing, China, Nov. 1997.

Diagnostic and Protection Systems for the Daresbury SRS Upgrade

J. A. Balmer, M. J. Dufau, D. M. Dykes, B. D. Fell, M. T. Heron, B. G. Martlew, M. J. Pugh, W. R. Rawlinson, R. J. Smith, S. L. Smith, B. Todd.

CLRC Daresbury Laboratory
Warrington, WA4 4AD, UK

Abstract. The UK light source, the SRS, is being upgraded by the addition of two multipole wiggler magnets. The reduced aperture of +/–7.5 mm within the titanium alloy tube has provided the opportunity to incorporate new sensitive electron beam position monitors. Due to investigations into the effects of synchrotron radiation striking uncooled surfaces, software and hardware vessel protection systems have also been incorporated for machine protection.

INTRODUCTION

The 2.0 GeV second generation Synchrotron Storage Ring at Daresbury, the UK's only synchrotron radiation light source, is now a mature machine. It provides facilities for over 2500 users from both academic and commercial backgrounds. The primary source of radiation is the 16 dipole magnets. In order to provide significant improvements, insertion devices have been added. Currently two superconducting wavelength shifters and one undulator have been completed and are now fully exploited. As a continuing policy to develop and improve the existing facility, further upgrades are scheduled to take place later this year. These involve the addition of two identical 2.0 Tesla permanent magnet multipole wigglers (MPWs) to the lattice. Due to the small space available (1.1 m) in the straight section locations, it was necessary to design the devices with as short a period as possible and with a small gap. To accommodate this requirement, a long, thin-walled titanium vessel has been designed, prototyped, and installed in advance of the MPW magnets, in order to demonstrate successful operation of the SRS with the 15 mm inertial gap. This small gap has presented problems not previously encountered with the current SRS machine vessels. Small-angle, mis-steered incident photon radiation from the upstream dipole presents a serious danger to this vessel both along its inner flank and along the vessel input flange aperture. Since cooled surfaces were impractical due to space limitations, an

CP451, *Beam Instrumentation Workshop*
edited by R. O. Hettel, S. R. Smith, and J. D. Masek

active machine protection system was required. This has involved the implementation of several new diagnostic, control, and interlock systems to protect this vessel and machine integrity.

MPW STRAIGHT EBPM DIAGNOSTICS

Some years ago, the SRS EBPM detector system was upgraded from two multiplexed single-plane homodyning detectors to a distributed system utilizing a two-plane detector at each location. These down-converting detectors (1) are connected directly to local 16-bit ADCs, which communicate with the control system via a G64 plant highway. This detector system is now used to routinely servo the orbit position horizontally to maintain beam position to ±10 μm for the duration of the fill.

EBPMs Within Existing Vessels

Due to the age and design of the SRS, the EBPM pickups themselves were installed when beam position servoing was unnecessary. For this reason they are not ideal in that they are located in separate, single plane measuring vessels (i.e. vertical measurement is made in the de-focusing quadrupole vessel and horizontal measurement in the focusing quadrupole vessel). This set-up is unsuitable for use around the new MPW vessel since they do not provide for the critical vertical measurement at each end of the tube, which is necessary to provide angular control across the straight. In order to overcome this problem, a revised pickup arrangement has been installed in the existing vessels. Figure 1 shows a cross section of each revised EBPM within the focusing and de-focusing quadrupole vessels.

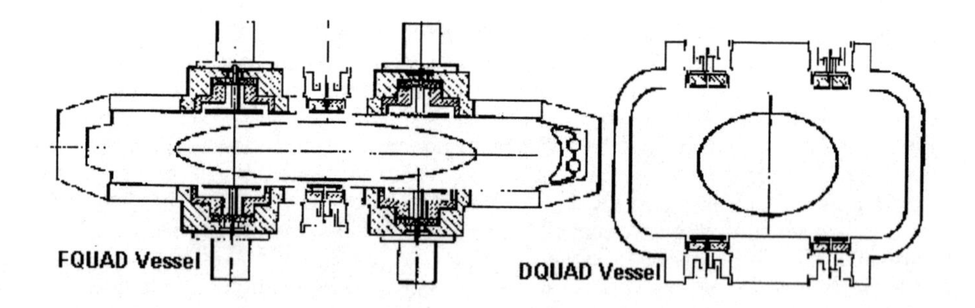

FIGURE 1. Revised two-plane EBPM button scheme, the focusing (FQUAD) and de-focusing quadrupole (DQUAD) vessels.

With this arrangement, the FQUAD vessel still measures the horizontal position using the existing EBPM buttons. Vertical position is now provided using a pair of on-axis, north/south buttons (modified ESRF design). For the DQUAD vessel, the original on-axis, north/south EBPMs were removed and plates carrying four modified ESRF buttons were installed, giving good horizontal and vertical response.

EBPMs Within The MPW Vessel

Since the MPW vessel is to be mounted within the magnet support system, located and isolated by vacuum bellows such that its position is fixed, it is advantageous to measure beam position within the limited aperture of the MPW vessel itself. To perform this function, ESRF-type buttons with titanium bodies were procured and fitted within a pocket at the rear of each flange, such that they and their cabling are prevented from fouling the MPW magnets. This allows the magnet poles to move past the rear of the flange as the gap is closed. Figure 2 shows the location of the buttons within the vessel and also shows an end view taken during electrostatic analysis using the QuickField™ 4 finite-element analysis (FEA) package.

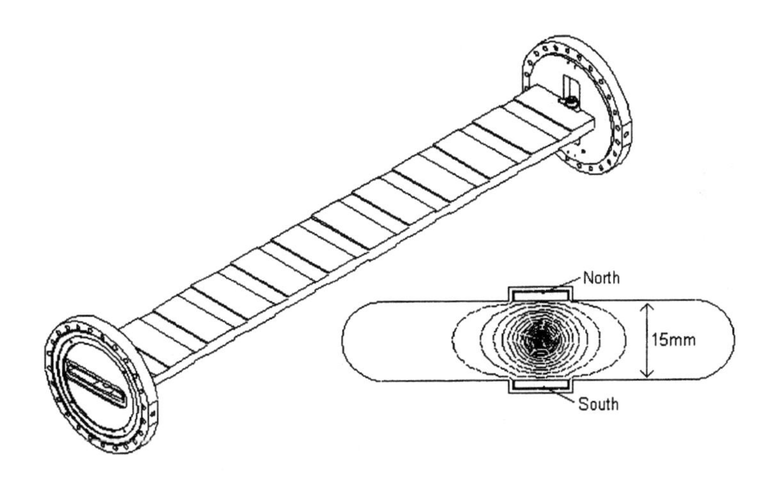

FIGURE 2. MPW vessel showing vertical EBPM layout and the equipotential distortion during simulations.

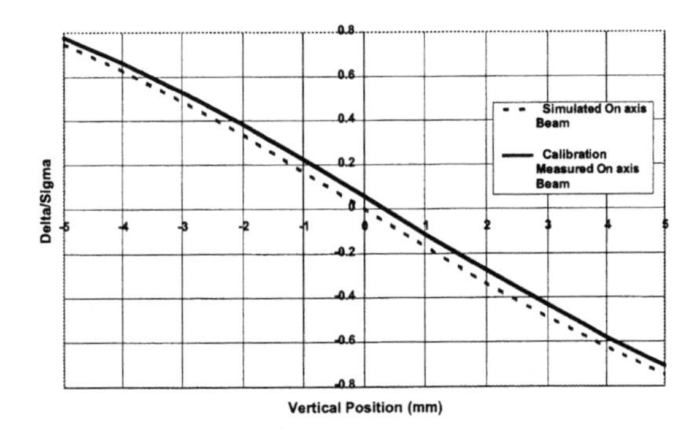

FIGURE 3. Comparison of simulated and measured MPW EBPM response.

The FEA simulation showed excellent correlation with measured calibration results using a swept wire technique on the prototype (Figure 3). The combination of such a small aperture with a relatively large button area (pickup plate diameter = 10 mm), gives a highly linear response with a low electrical calibration factor. Table 1, which compares EBPM calibration factors, shows the MPW EBPM to be less than half the value of other vessels. This means the control of beam position through the MPW device will be significantly improved, giving a higher stability source for users. The EPBMs will be incorporated in a local vertical position control system after the magnet's installation.

TABLE 1. Summary Showing MPW Straight EBPM Electrical Calibration Values

Vessel	Vertical Cal. Factor	Horizontal Cal. Factor
DQUAD EBPM	14	14
FQUAD EBPM	12	13
MPW EBPM	6	

Beam Position Processing and Machine Protection

Processing for the modified two-plane EBPMs is done with the existing Daresbury down-converting detector. For the new EBPMs within the MPW vessel, a commercial system manufactured by Bergoz Instrumentation using the switched-button amplitude measurement technique has been installed. This system offers advantages over the Daresbury design in that it is designed onto a single Eurocard, has a fast position update rate (1 kHz as standard), produces smoothed DC level outputs representing beam position, and has sufficient side-band rejection to allow single-bunch beam mode operation. Furthermore, this system lends itself to the simple introduction of a position-based hardware beam trip in the event of a mis-steering of the photon beam from the upstream dipole.

Facilities on the card allow the output response to be directly related to the beam position. Since the EBPM response is very linear, the system has been set up to provide a DC output representing 1 V/mm. A pair of simple windowing comparator systems is used on both the upstream and downstream MPW EBPMs to provide active interlocks along with a lower-level audible alarm in the event of a slow position drift towards the trip level. Furthermore, the cards produce three other signals which are used to guarantee their operation. Figure 4 shows a schematic block diagram of the MPW beam position hardware interlock.

The active interlock interrupts the low level rf drive to the klystron amplifier. A trip initiated by the beam position or by one of the other interlocked card test signals aborts the beam in 20 mS, thus protecting the MPW vessel. To allow filling of the storage ring, a simple override relay, actuated from the storage ring dipole DC current, bypasses the interlock chain, becoming automatically active at beam energies above 0.65 GeV (450 A).

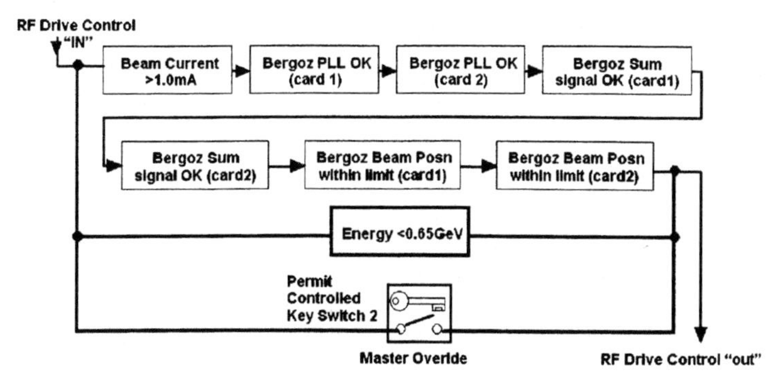

FIGURE 4. Block diagram of MPW beam position trip system.

SECONDARY INTERLOCK PROTECTION SYSTEM

Temperature Interlock

A temperature interlock system has been installed in addition to the BPM fast interlock in an attempt to give protection against slower temperature rises due to the photon beam striking the MPW vessel surface at a glancing angle in the horizontal plane. Thermocouples have been installed on the prototype ID vessel and surrounding equipment along the top and bottom of the vessel.

The system is comprised of 24 K-type thermocouples feeding into individual linearizing amplifiers that provide a 0–10V output signal proportional to temperatures from 0 °C to 1000 °C. A DL3000 ΔT Data Logger is used to compare the temperatures against the trip temperature level, which has been set at 50 °C. If any part of the vessel reaches the trip limit, the rf will immediately be tripped off.

Figure 5 shows how the thermocouples and data logger are integrated into the control system. Aside from the independent hardware temperature trip system, diagnostics and monitoring is provided through a 3U VME system, accessed over 10 Mbit Ethernet.

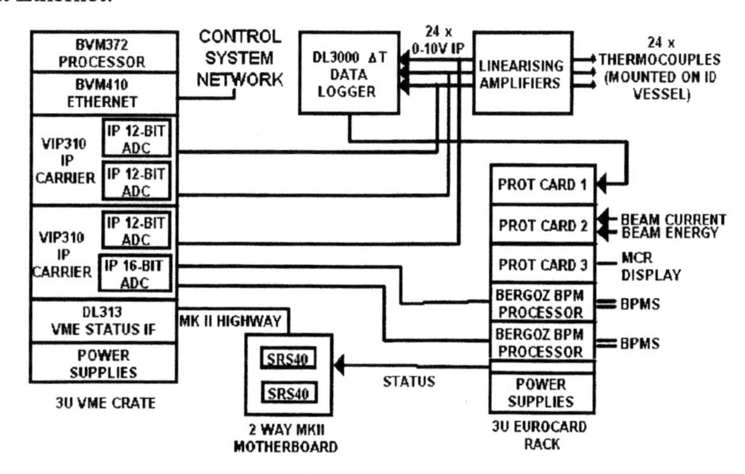

FIGURE 5. Schematic of control system interface to MPW protection systems.

The VME system is treated as a standard SRS control system Front End Computer (FEC) (2) with Industry Pack (IP) modules, using a Daresbury Status Interface module to access the analog signals and system status. Standard applications on the operator consoles can be used to monitor and log the additional BPM position and temperature information.

Software Protection Systems

A number of software-oriented steps have been taken at the SRS to further assist protection of the hardware systems. These include developing a software interlock system and making modifications to the existing software to enforce tighter orbit control and prevent dangerous electron beam mis-steering.

The software interlock and diagnostics provide a secondary safety system for the prevention of damage to the MPW vessel. The interlock continuously monitors orbit position from the existing 16 horizontal and vertical electron BPMs and, in the case of an excessive drift in position, will trip off the rf system.

The electron and photon beam position monitors, along with the storage ring power supplies, are interfaced to the control system via a network of VME systems via an Ethernet LAN as shown in Figure 6. The BPM software resides on the Gateway Processor (GP) along with the steering system front end computer (FEC) interface.

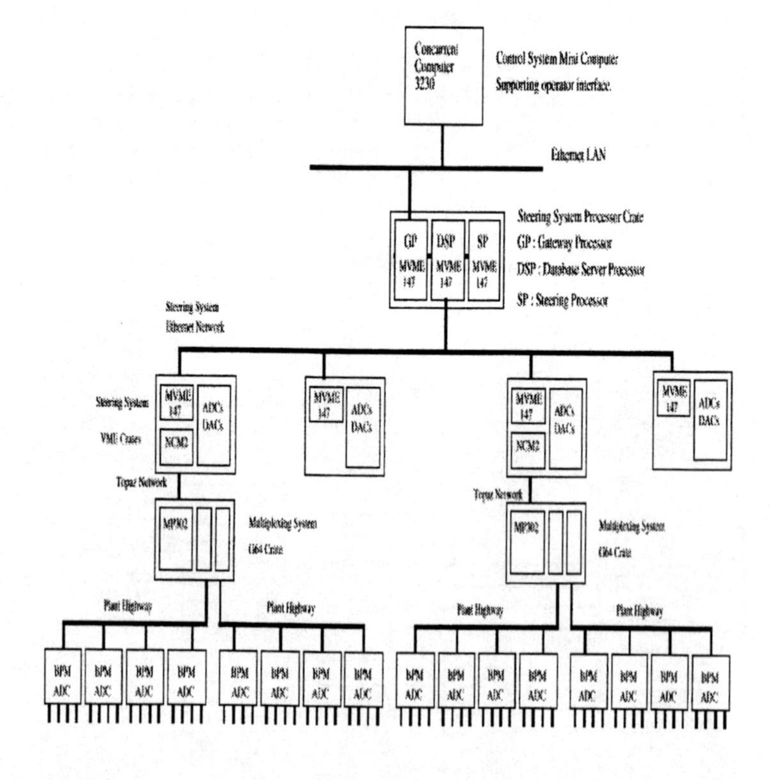

FIGURE 6. Schematic showing beam position monitoring data paths and hardware.

The system can be operated in two modes:

1. Ramp mode: all BPMs are monitored every 2 seconds during energy ramping and, if any one BPM is outside its own predefined limit, the interlock trips off the rf.
2. User beam mode: similar to ramp mode except that the orbit is captured on entering this mode and limits are taken relative to the start orbit.

Warning and trip limits are flexible and can be set at any time for each individual BPM. Warning limits are set just inside the trip limits and, when reached, will light an LED on the diagnostic control panel. Update rates faster than 2 seconds can cause deleterious competition for processor time with the ramp servo process.

The software trip diagnostics, written to log files on the VME steering system, include details of orbits when the trip occurred, along with times and mode of operation. Software has been written for the Windows NT to access this data and integrate it with the rest of the PC control system.

Both the ramping and user beam modes have been commissioned. The software interlock system is routinely used in Ramp Mode for both ramping and user beam conditions, although the user mode is expected to be used in the future. The system has also proven to be useful when a beam is dumped due to the hardware interlock tripping; the software system is then triggered, recording additional orbit diagnostic information.

ADDITIONAL SOFTWARE PROTECTION

Controlled Application of Bumps

The existing bump application software has been modified to prevent applying bumps that could produce harmful effects. All bumps applied now are absolute and have pre-defined limits in both the horizontal and vertical planes. A bump reset is performed only by cycling the magnets or by applying fixed steering file settings.

Ramp Servo Control

Improvement of the energy ramp software has been necessary to provide servo control of the orbit from EBPM readings whilst ramping the stored beam.

Jaw Settings

Ramping of the stored beam is permitted only when the horizontal and vertical jaw positions are within defined limits. This restriction gives limited protection, but does trap some orbits which would irradiate the MPW vessel if the jaws were left fully open.

Refill Sequence Manager

A change in the philosophy of SRS operation has required introducing sequencing software to control the precise order of events during refills. It utilizes the universal scripting language Tcl/Tk 8.0™ with a custom in-house library to interface the software to the control system. Each step of the refill process is defined by a Tcl scriptlet, written in consultation with accelerator physicists. This software has allowed an efficient refill procedure to be established and provides additional safety by ensuring that correct settings are applied in each operating mode.

REFERENCES

[1] Smith, R. J., P. A. McIntosh, T. Ring, "The Implementation of a Down Convertion Orbit Measurement Technique on The Daresbury SRS," presented at the 1994 European Particle Accelerator Conference, London, UK, June 1994.

[2] Martlew, B. G., M. J. Pugh, and W. R. Rawlinson, "Present Status of the SRS Control System Upgrade Project," presented at the 1996 European Particle Accelerator Conference, Sitges, Spain, June 1996.

Design of the Digitizing Beam Position Limit Detector*

Robert Merl and Glenn Decker

Advanced Photon Source
Argonne National Laboratory
9700 South Cass Avenue
Argonne, Illinois 60439 USA

Abstract. The Digitizing Beam Position Limit Detector (DBPLD) is designed to identify and react to beam missteering conditions in the Advanced Photon Source (APS) storage ring. The high power of the insertion devices requires these missteering conditions to result in a beam abort in less than 2 milliseconds. Commercially available beam position monitors provide a voltage proportional to beam position immediately upstream and downstream of insertion devices. The DBPLD is a custom VME board that digitizes these voltages and interrupts the heartbeat of the APS machine protection system when the beam position exceeds its trip limits.

INTRODUCTION

Insertion devices in the APS storage ring are powerful enough to damage the vacuum chamber in the event of a missteered beam. Beam position monitors that are located immediately upstream and downstream of insertion devices produce an analog signal proportional to beam position. Position signals are available from each BPM for both horizontal and vertical planes. These BPMs also generate signals that can be used to determine if stored beam is present and whether or not the BPM is functioning. The Digitizing Beam Position Limit Detector (DBPLD) samples the analog output of these BPMs and notifies the machine protection system (MPS) (1) if the beam is missteered. Once notified, the MPS will dump the stored beam by temporarily interrupting the rf system for 100 milliseconds. This paper describes the design of the DBPLD with specific attention to design methods and features that support reliable operation. The DBPLD was designed specifically to work with the commercially available Bergoz BPM

* Work supported by U.S. Department of Energy, Office of Basic Energy Sciences, under contract No. W-31-109-ENG-38.

CP451, *Beam Instrumentation Workshop*
edited by R. O. Hettel, S. R. Smith, and J. D. Masek
1998 The American Institute of Physics 1-56396-794-4/98/$15.00

(2), but is compatible with any device that can supply a voltage that is proportional to beam position.

ARCHITECTURE

The DBPLD is a 6U VME board that supports two separate insertion devices (IDs) and provides an interface to the Machine Protection System and EPICS (3), the APS control system. Figure 1 shows a diagram with DBPLD inputs and outputs.

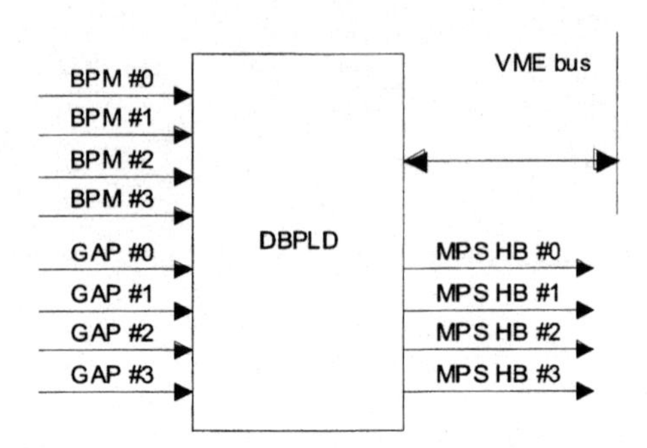

FIGURE 1. DBPLD inputs and outputs. One DBPLD board supports four BPMs that cover two insertion devices. Insertion device gap open/closed status is sensed through the GAP inputs. An optical heartbeat (MPS HB) is sent to the machine protection system when no missteering condition exists. The DBPLD is interfaced to the APS control system through an onboard VME interface.

Four BPM channels provide coverage for two IDs, with one BPM immediately upstream of an ID and one BPM immediately downstream. At the APS, IDs have redundant gap interlock switches that detect when the gap is fully open. The DBPLD is disarmed in this case. The four GAP interlock inputs on the DBPLD cover the two insertion devices. Four heartbeat (HB) connections allow the MPS to detect which of the four channels generated a beam abort.

The DBPLD has analog-to-digital converters (ADCs) at its front end, but is essentially a digital device. Internally, the DBPLD can be divided into three conceptual blocks: the datapath, the controller, and the VME interface/working RAM. These three sections communicate with each other over internal address, data, and control buses. A 5 MHz, board-wide clock drives every register on the board with the exception of the working RAM in the VME interface. The working RAM is a cycle-shared, dual-ported memory that requires a clock that runs at twice the frequency of the board-wide clock. Both of these clocks are derived from the same crystal and are in phase for fully synchronous operation. Throughout the system, data may only change at the rising edge of the board-wide clock. Asynchronous inputs run through registered synchronizers before they are sampled by the DBPLD. These and other techniques eliminate the hazards associated with metastability and provide robust, glitch-free operation.

A block diagram of the DBPLD is shown in Figure 2. Most of the detail shown is in the datapath. Data flows through the datapath under the direction of the controller. The VME interface doubles as a board-wide working memory where data can be stored and retrieved. This memory is dual ported, with the VME interface on one side and the datapath on the other. This architecture allows the APS control system full monitoring capability of DBPLD operation. With a mechanical key inserted in the DBPLD front panel, the control system may also write to the working memory through the VME interface.

DATAPATH

At the front end of the datapath are eight analog-to-digital converter sections, one for each channel and plane. These sections consist of a low-cost, eight-bit analog-to-digital converter coupled with a support circuit. The support circuit triggers the ADC and synchronizes its data to the DBPLD board-wide clock. A block diagram showing the support circuit and its relationship to the ADC and the DBPLD internal buses is shown in Figure 3.

The ADC requires 30 μs to make an analog-to-digital conversion, but it is only triggered once every 51.2 μs. Compare these times to the 200 ns period of the board-wide clock. The clock runs many times faster than the ADC conversion cycle. The support circuit simplifies the architecture by hiding the slower ADC from the rest of the design. A register in the support circuit stores the last valid output of the ADC for immediate transfer on the data bus any time it is addressed. This means that other parts of the circuit do not need to worry about accessing the ADC in the middle of a conversion, since data is always valid at the output of the support circuit. This method insures that the data bus is never driven with partially latched data.

Each of the eight analog to digital converter sections shares a common tri state data bus, address bus, and control bus. These shared buses run throughout the DBPLD and allow a simple time division multiplexed datapath to be used. Data flows from inputs at the top edge of Figure 2 to outputs on the bottom edge. The controller directs data on their way through the datapath with signals on the control and address buses. Notice that the data bus does not connect to the controller.

Analog signals that are delivered to the front panel of the DBPLD originate from BPMs that can be several feet away. These signals are transmitted over cables that could possibly become disconnected. A disconnected cable must generate a fault and cause a beam abort. The clock signal from the commercial BPM is shipped out over the same cable as the position data. Watchdog timers on board the DBPLD monitor clock signals from each BPM. These watchdogs look for changes in state of the BPM clock and assert an alarm if the clock gets stuck in any particular state. If a cable is disconnected, the watchdog senses a non-changing state and raises an alarm. This mechanism also detects failures at the BPM, such as loss of power. There are four watchdogs, one associated with each channel. Each is addressable by the controller over the address bus.

Two other external signals are brought in through the front panel. They are insertion device gap open/closed status and beam present/not present. The DBPLD is armed only when the gap status is closed and when beam is present in the APS storage ring. ID gap status and beam status bits exist for each channel and are address selectable by the controller.

The DBPLD supports four BPMs or eight planes, four horizontal and four vertical. Each plane may have unique upper and lower trip limits. This means that there are two limits associated with each plane. The DBPLD retrieves the upper and lower limits from its working RAM one at a time and places them in the "high" and "low" datapath registers. It places the most recent beam position data associated with these limits in the "live" datapath register and at the same time writes this position data to a unique location

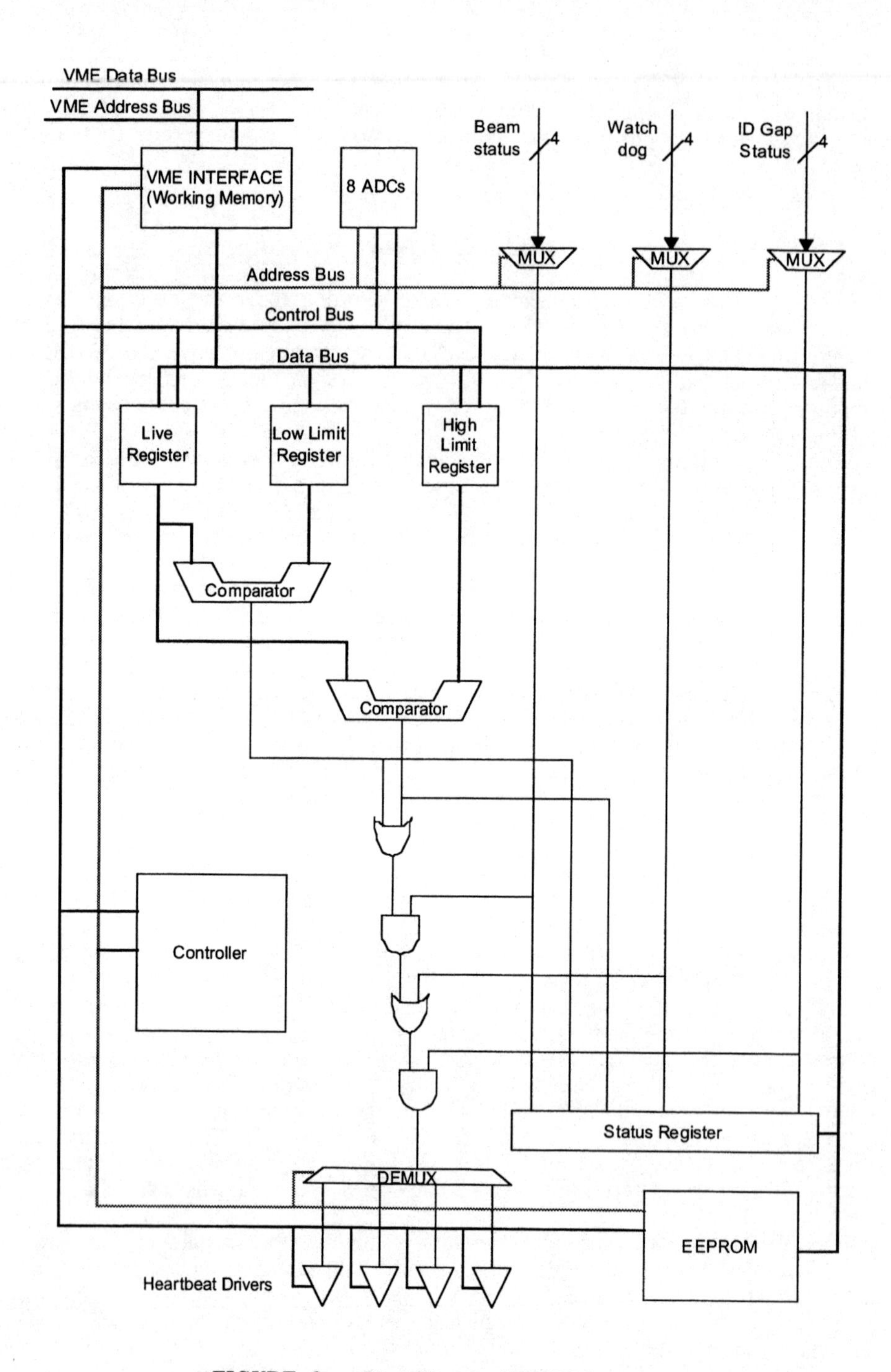

FIGURE 2. Block diagram of DBPLD internals.

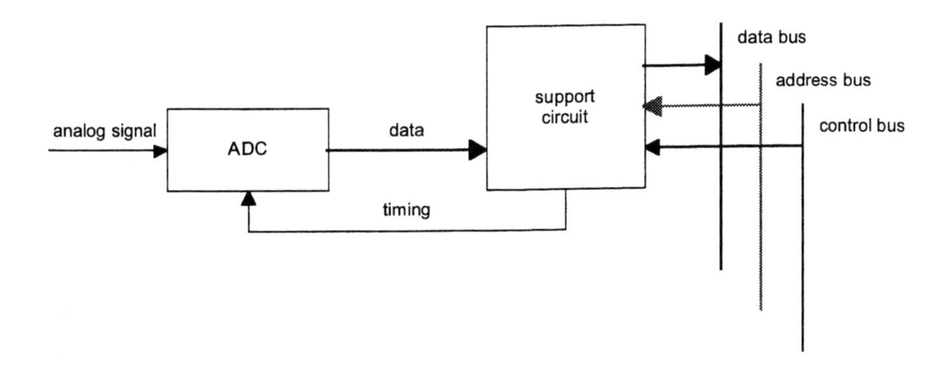

FIGURE 3. ADC section, analog-to-digital converter with support circuit. The support circuit simplifies timing between the ADC and the rest of the DBPLD.

in the working memory. The position data are compared with the high and low limits simultaneously by the two comparators. Combinational logic examines ID gap, beam present, and watchdog status and generates a heartbeat stop pulse, if necessary. The status register latches these signals. The contents of the status register are then stored to the appropriate working memory location. Since the working memory is dual ported with the VME interface, this mechanism allows the APS control system to access status, position, and limit information for all eight planes at any time.

A heartbeat generator is provided for each of the four channels. The heartbeat generators each supply a 1 MHz optical signal to the machine protection system. Absence of the 1 MHz signal indicates that a beam position limit has been exceeded or that there has been a fault on that channel. The machine protection system causes a beam abort when the heartbeat is absent.

During maintenance periods at the APS, VME crates are often powered down for service. To prevent limit settings from being lost under these circumstances, an electrically erasable-programmable read only memory (EEPROM) is used. When new trip limits are written to the VME interface from the EPICS-based control system, a dirty bit is set. The controller watches for this dirty bit and commits newly written limits to the EEPROM and then clears the dirty bit. At power up, limits are read out of the EEPROM.

CONTROLLER

Information on the internal address and control buses originates from the controller. The controller is a finite-state machine that is implemented in programmable logic using a hardware description language. Data flows through the datapath, but it is the controller that directs this flow.

At power up, the DBPLD enters an initialization mode where it loads the system RAM with limits from the EEPROM. When completed, it begins limit-checking operation. In the limit-checking mode, the DBPLD controller cycles through ten repeating states. These ten states constitute a minor cycle in which one of eight positions is compared against its trip limits. When the minor cycle completes eight times, a major cycle has been completed. Both horizontal and vertical positions for all four channels are checked during a major cycle. In Figure 4, the ten states are abstracted as five steps, with a distinct task completed in each step. The variable, m, denotes the minor cycle number and can range from 0 to 7. The table on the right side of the figure indicates the focus for any particular minor cycle. The address counter rolls over to 0 automatically after reaching a value of 31 so that major cycles repeat without any special action by the controller.

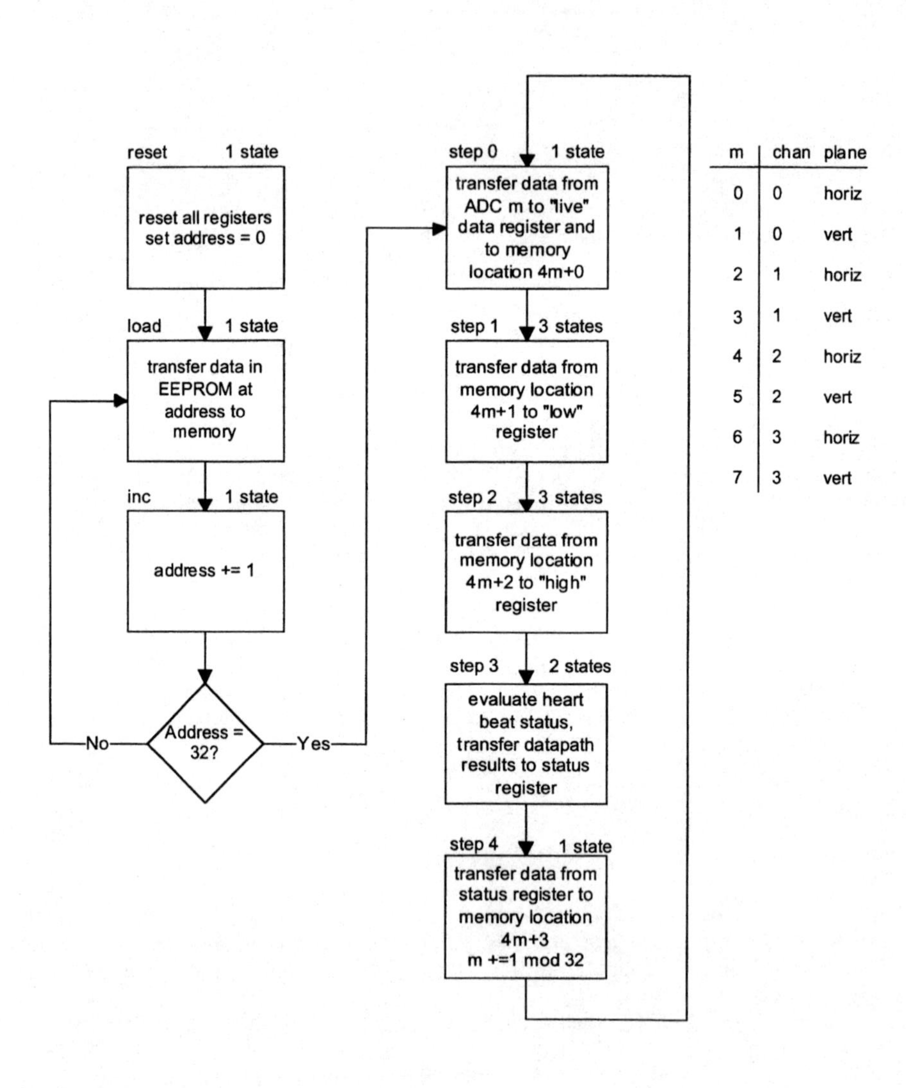

FIGURE 4. Controller state diagram.

RESPONSE TIME

The board-wide clock runs at 5 MHz, corresponding to a clock period, t_{clock}, of 200 ns. The time for one minor cycle to complete, t_{minor}, is computed as

$$t_{minor} = t_{clock} \times N = 200ns \times 10 = 2\mu s, \quad (1)$$

where N is the number of clock periods required for a minor cycle. The time for one major cycle to complete, t_{major}, is computed as

$$t_{major} = t_{minor} \times M = 2\mu s \times 8 = 16\mu s, \quad (2)$$

where M is the number of minor cycles in every major cycle. According to Equation (2), the DBPLD completes a major cycle every 16 μs.

The response time of the DBPLD is limited by the 51.2 μs conversion time of the ADC section, t_{ADC}. Two conditions must be met for the DBPLD to respond in this minimum amount of time. First, the change at the input must occur immediately before the ADC receives a convert pulse. Second, the ADC support circuit has to latch this sample on the clock cycle immediately before the controller transfers that same data to the "live" register. The maximum or worst-case response time results when the step in the input occurs during an ADC conversion cycle and the controller has just finished examining data from that ADC. In this case, two full conversion cycles and a major cycle of the state machine must take place. The worst-case response time, $t_{response}$ is computed in Equation (3):

$$t_{response} = t_{major} + 2 \times t_{ADC} = 16\mu s + 2 \times 51.2\mu s = 118.4\mu s. \quad (3)$$

Measurements of best-case and worst-case response times are shown in Figure 5. The worst case measured time is slightly better than the calculated worst-case time. It is possible that the ADC can tolerate some change at the input immediately after the conversion cycle begins.

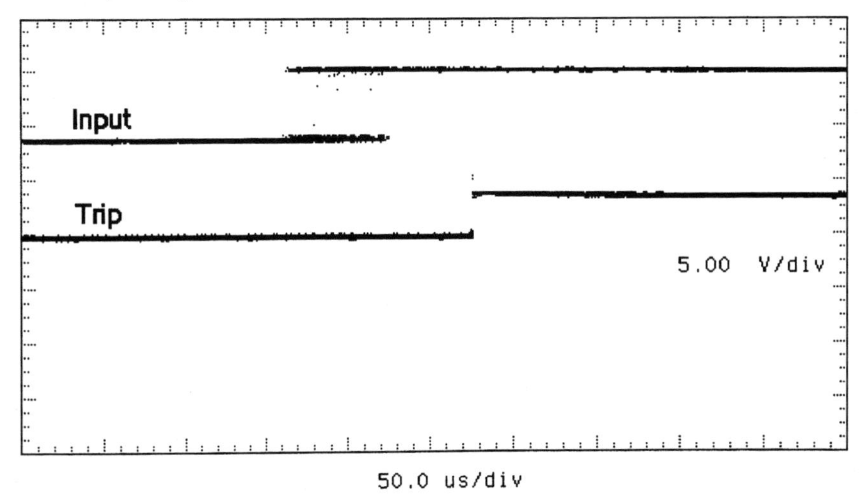

FIGURE 5. Response time measurement with a step input. The measurement shows a best case of 52 μs and a worst case of 113 μs. The figure shows 500 traces overlaid.

There is also a step response time associated with the BPM that supplies the position signal to the DBPLD. In the case of the Bergoz BPM, it is 580 μs. The end-to-end response time is then the sum of the BPM and DBPLD response times, which is less than 700 μs.

REFERENCES

[1] Lumpkin, A. et al., "Overall Design Concepts for the APS Storage Ring Machine Protection System," *Proceedings of the 1995 Particle Accelerator Conference*, Dallas, TX, May 1–5, 1995, 2467–2469 (1996).

[2] Unser, K. B., "New Generation Electronics Applied to Beam Position Monitors," Proceedings of the 7th Beam Instrumentation Workshop, Argonne, IL, May 1996, *AIP Conference Proceedings* **390**, 527–535 (1997).

[3] McDowell, W. P. et al., "Standards and the Design of the Advanced Photon Source Control System," *Proceedings of the International Conference on Accelerator and Large Experimental Physics Control Systems*, Tsukuba, Japan, November 1991, 116–120 (December 1992).

Radiation Safety System (RSS) Backbones: Design, Engineering, Fabrication, and Installation[1]

J. E. Wilmarth, J. C. Sturrock, F. R. Gallegos

Los Alamos Neutron Scattering Center, LANSCE Division, Los Alamos National Laboratory, Los Alamos, NM 87545 USA

Abstract: The Radiation Safety System (RSS) backbones are part of an electrical/electronic/mechanical system ensuring safe access and exclusion of personnel to areas at the Los Alamos Neutron Science Center (LANSCE) accelerator. The RSS backbones control the safety-fusible beam plugs which terminate transmission of accelerated ion beams in response to predefined conditions. Any beam or access fault of the backbone inputs will cause insertion of the beam plugs in the low-energy beam transport. The backbones serve the function of tying the beam plugs to the access control systems, beam spill monitoring systems and current-level limiting systems. In some ways the backbones may be thought of as a spinal column with beam plugs at the head and nerve centers along the spinal column. The two linac backbone segments and the experimental area segments form a continuous cable plant over 3500 feet from the beam plugs to the tip on the longest tail. The backbones were installed in compliance with current safety standards, such as installation of the two segments in separate conduits or tray. Monitoring for ground-faults and input wiring verification was an added enhancement to the system. The system has the capability to be tested remotely.

OVERALL SYSTEM DESCRIPTION

The Radiation Safety System (RSS) backbones are part of an electrical/electronic/mechanical system insuring the safe access and exclusion of personnel to the Los Alamos Neutron Scattering Center (LANSCE) Accelerator and the accelerator's numerous experimental areas (1,2).

The backbone system of cabling and control nodes consists of two heads (dual beam plugs), two backbones (redundant, identical cable runs), four major nerve clusters (node box input points), three tails (three major experimental areas' backbone segments), with

[1] Work performed under the auspices of the U. S. Department of Energy.

CP451, *Beam Instrumentation Workshop*
edited by R. O. Hettel, S. R. Smith, and J. D. Masek

a nerve cluster at the tip of each tail (the node box input point) forming a skeletal network of over 3500 feet from the head to the tip of the longest tail. (See Figure 1.)

The two beam plugs (the two heads) are located at the beam entry point into the 201 MHz rf accelerating tanks downstream of the three-energy-transport beam lines. The linac backbone segments are connected to the two RSS safety beam-plugs, 01BL02 and 01BL03 (3). Beam plug 01BL02 is connected to linac "A" backbone segment. Beam plug 01BL03 is connected to the linac "B" backbone segment. The two backbone segments and beam plugs are identical but are physically separated. The beam plugs are held out of the beam by signals from backbone segments when selectable conditions are satisfied.

Several safety design features were included in the system. These are methods for maintaining configuration control of the dual safety beam-plugs, (use of armored cable, redundancy of circuits, physical separation and sensing of faults) and are described in the following paper.

There are two methods, mechanical and electrical, for inhibiting the withdrawal of the RSS beam plugs, when configuration of the facility is to be maintained in an operating mode with no beam delivery beyond the low-energy-beam-transport areas. The plug mechanism has a metal plate that can be installed and held in place by a padlock. This inhibit requires a beam tunnel entry and is generally used for prolonged periods of time. The second method of inhibiting withdrawal is by a push-button switch at the junction boxes at the upstream end of the backbone. These switches and associated junction boxes have padlock attachments that allow the installation of the padlocks outside the beam tunnel. These are generally used for short periods when control of the safety beam plugs is needed. Configuration control is an important consideration in the design of the system.

The color "school bus yellow" was selected for all RSS backbone junction boxes and cable so that they could be easily distinguished. The cable used for the primary backbone is armored one-pair, shielded, #16 AWG with unique color codes for the four different runs of cable comprising the linear accelerator (linac) segment. (See Figure 2.)

The tails or experimental area segments are armored one-pair, shielded, #16 AWG cable. The three experimental area segments all use the same type cable and are called "Line D (LD)" backbone, "Line X (LX)" backbone, and "Line A (LA)" backbone. These cables connect the node junction boxes together with special termination of the armored jacket (yellow), which is similar to .75"-diameter oil-tight flexible conduit. The insulation of the one-pair within the armored jacket runs inside a 44" wire-way, from the armor cable termination to the junction box terminals, where connection is made at the junction box back-plate input terminals.

Physical separation of the linac segment of the armored-cable runs is accomplished by using the three vertically stacked horizontal cable trays that run the length of the accelerator-equipment building. The accelerator equipment building is a continuous structure over 2500 feet long. The cable runs were separated into three trays. The "A" backbone (hot) was installed in the top tray. The "B" backbone (hot) was installed in the middle tray, and the "A" and "B"returns were installed in the bottom tray. The physical separation of the cable in the three trays was to prevent any common mode damage to the cable on the long horizontal runs.

The input cabling to the backbone node boxes is constantly verified by a "sense" circuit board, designed locally. Each input is monitored for shorts and ground faults. If a fault is detected, it is displayed locally and remotely from each node box of the system.

The linac-backbone segments and experimental-area segments are powered by a plus/minus, 24 V power supply attached in the node boxes at the end of the linac and the experimental areas. The voltage is fed over the cable running toward the front of the accelerator. Dual isolated power supplies allow ground-fault detection to be used over the entire backbone cable segments (Figure 2).

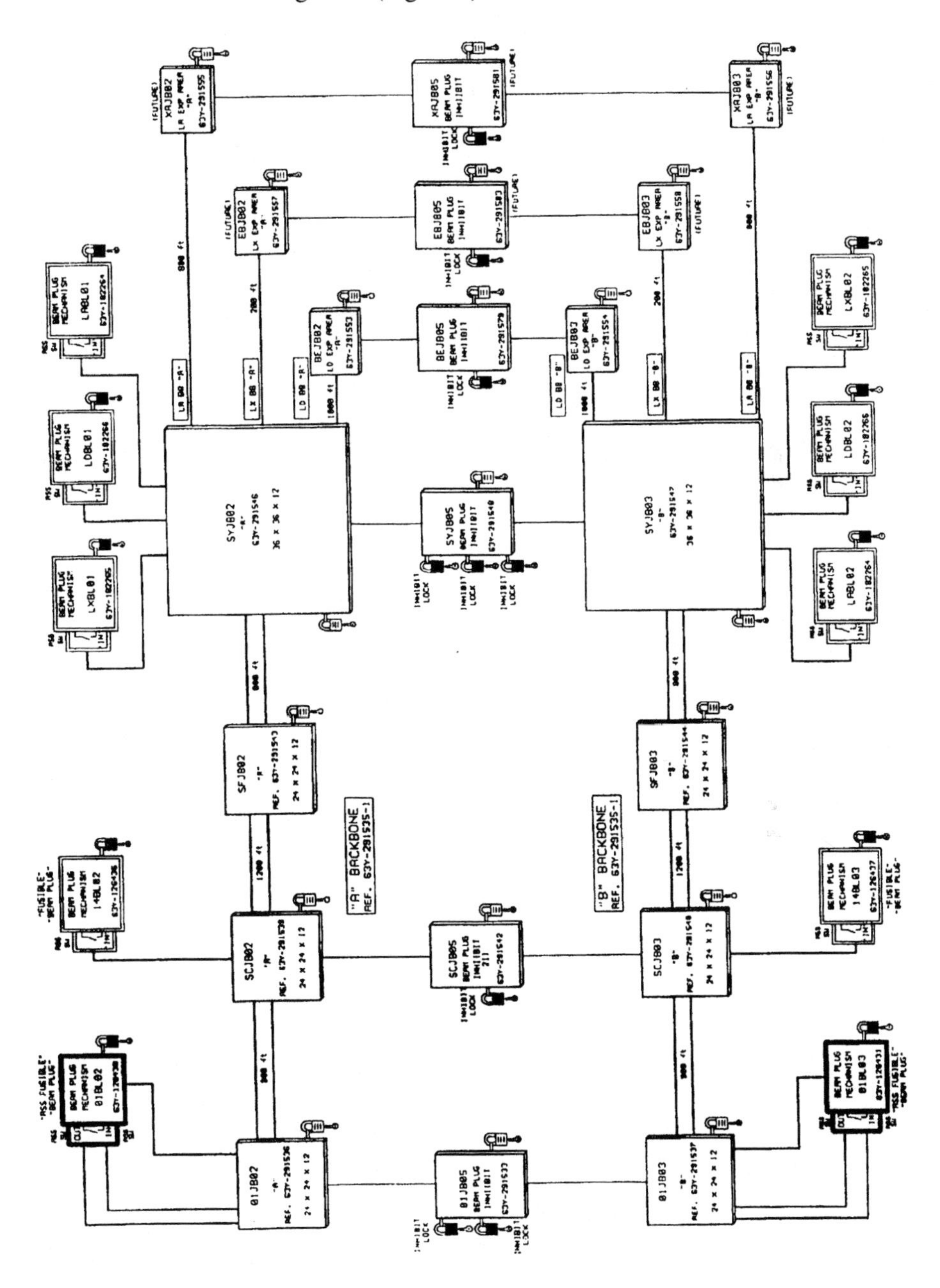

FIGURE 1. Overall RSS block diagram.

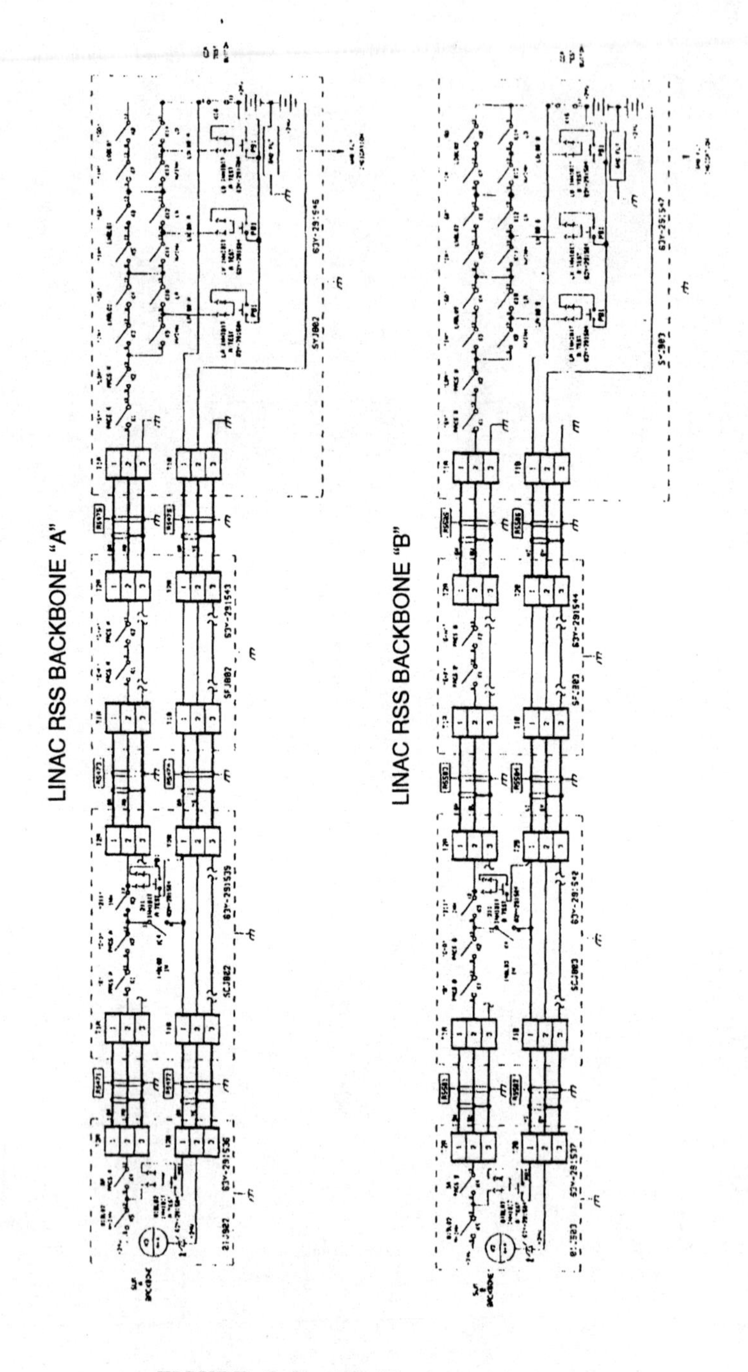

FIGURE 2. Simplified linac segment schematic

System Junction Box Description

Ease of maintenance, separation of components, and configuration control issues were considered in the system layout within the node junction boxes. LEDs are used to convey information through a window in the box lid, thus permitting information as needed without compromising access to components.

The interior of the box has a swing-out bent panel with a window and power supply mounting hardware. The side of the enclosure has openings for mounting two surge-protected outlets outside the enclosure. The node boxes are two basic sizes. The smaller version is 24 inches by 24 inches by 12 inches deep and the larger is 36" × 36" × 12". All boxes are Hoffman-Enclosure-Concept style, specially ordered with a window in the hinged lid of the box. The smaller version is used where space may be a limiting factor and few inputs are required. The window on the lid is for information display that is developed within the node. Decals are fabricated and applied to the inside of the windows to provide appropriate display information. The larger enclosure is used where space is not a limiting factor and as many as 12 inputs can be accommodated. The bottom plates of both boxes are used for the RSS wiring and RSS component-mounting surface. The boxes have a Hoffman-keyed lock in the upper latch to secure the lid of the enclosure. The keys are issued to maintenance personnel. The swing-out panel is secured by an RSS padlock, which is issued and controlled by the RSS Engineer.

The wiring located on the bottom plate is physically isolated from the rest of the junction box by the swing-out plate. The swing-out plate contains power supplies and their associated sense- and ground-fault-detection printed wiring boards (PWB). The power supply PWB boards are for distribution of voltages and detection of ground faults. There is direct wiring of the relay outputs on the bottom plate to the display PWBs located on the inside of the window found in the box lid. (See Figure 3.)

The bottom plate is the mounting panel for the quad-sense-evaluation PWBs which are wired between the terminal strip input and the coil of the relay associated with that particular input. There are four circuits contained on each PWB. There is a single output contact from each board, wired in series with the other sense PWBs, if installed, that is the sum signal, used to trip the fault relay and set the corresponding fault indication. The operation of the quad-sense PWB requires each input to the backbone to have a 160-ohm, .25 W carbon-film resistor installed at the end of cabling, preferably on the switch contacts of the input device. The output of the quad-sense circuit must be loaded with a resistance or coil of a relay to operate the sense circuit. (See Figure 4.)

The display of information at each junction box or node location is accomplished by a universal-display printed circuit board attached to the rear of the junction box window with cabling run to one pole of each backbone relay. A decal with the information for a particular box is installed between the window and the printed-circuit board with small, clear openings for the display LEDs.

The concept boxes were ordered from Hoffman Enclosures with several special features. The door of the enclosure also serves as the window for information display. The swing-out panel is used for the mounting of components and provides a barrier to access of the bottom plate. The bottom plates were standard, with the hardware mounting done after delivery. All incoming cable to the boxes is protected by wire-way, conduit, or tray until it is terminated inside the enclosure. The layered concept of the boxes allowed the fabrication of bottom plate wiring and component mounting in parallel with the swing-out plate by outside fabrication vendors. The mounting of boxes, connecting of the wire-way, conduit and tray, and installation of 110VAC power was completed simultaneously by electrician crews. The armored cable installation was accomplished before the mounting of boxes due to the long lead-time of fabrication by the Hoffman Enclosures and the early ordering and timely arrival of the cable.

The wiring from the terminal strips to the relay sockets is the least modular feature of the backbone enclosures. The wire is terminated in wiring ferrules at the terminal strips and relay sockets. The appropriate-sized ferrule is crimped on the wire to match the gauge. The ferrules provide the best mechanical attachment for electrical connections, prevent strands from fraying, and secure the capture of exposed wire in terminal wells and under screw clamps.

The physical-separation design criteria was used on the component layout throughout the junction boxes. Terminals are snap-on style with three colors (blue, orange, grey), used with mounting rails specified by European standards (DIN-EN 50 022) separated by barriers and nomenclature plugs for each input or output set (DIN-rail).

A single Allen-Bradley programmable controller (PLC) is installed at each node-box location to allow remote read-out of the status of both backbones at that location. The Allen-Bradley PLCs are connected together by a local area network exclusive to the RSS backbones.

System Installation

The accelerator facility's cable plant location allowed for the installation of cable and boxes during operating cycles of the accelerator. The linac equipment-aisle is a single, continuous building, 2500 feet in length and 10 feet wide at the narrowest doorway. The pulling of cable was accomplished by the use of a radio-controlled 4-wheel-drive, battery-powered vehicle that traveled in the twelve-inch-wide cable tray. This vehicle pulled a string, which was used to attach a pull-rope. The pull-rope was pulled by a battery-powered golf cart which is small enough to travel the hallway yet strong enough to pull bundles of cable from spools over the 2500-foot installation. The use of these vehicles allowed for safe and efficient cable installation. The vehicle remote-control capabilities reduced the need for climbing with ladders to access the cable-tray system (located in the ceiling area of the equipment building) to a minimum.

The current cable plant was installed in 1997/1998. There were several earlier prototype installations that were partially reused in the final version. Single, armored-cable runs with one cable for "A" and one cable for "B" were installed in 1995. They connect the experimental area "LD" to the switchyard (SY) equipment aisle node boxes. Two other experimental areas have temporary cables installed, awaiting expansion of the experiments in those areas for the node boxes and armored cable.

The design philosophy for the linac backbone was based on the backbone installation of 1995. The linac backbones were to be as physically separated as the facility and existing equipment would allow. The design criteria for simple, rugged, and expandable installations were used where possible. Relay logic was selected for ruggedness. The design of the safety-beam-plug control elements was incorporated into the linac Backbone concept. The 1995 backbone voltage was fed from the junction boxes that controlled the beam plugs.

The backbone components were selected or designed so that they are easily removable and replaceable. The relays are plug-in style with DIN-rail mountable sockets. The PWB's are tray-contained for DIN-rail mounting and have connectors for all input/output wiring to allow quick disconnection and replacement. The power supplies are screw-mounted to the swing-out plate by use of a mounting frame. Connectors are installed on the end of the cabling from the power supply to sense and distribution PWBs.

The wiring of the backbone circuit within each junction box is a larger AWG wire size and is color coded to match the color code of the armored-cable landing on the terminals from outside the box.

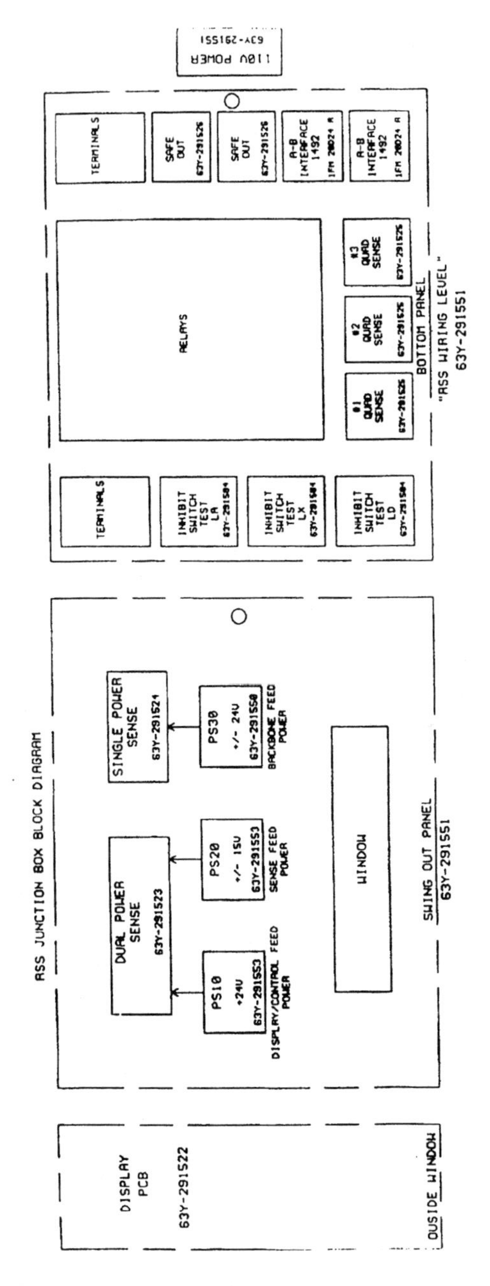

FIGURE 3. Typical junction box block diagram.

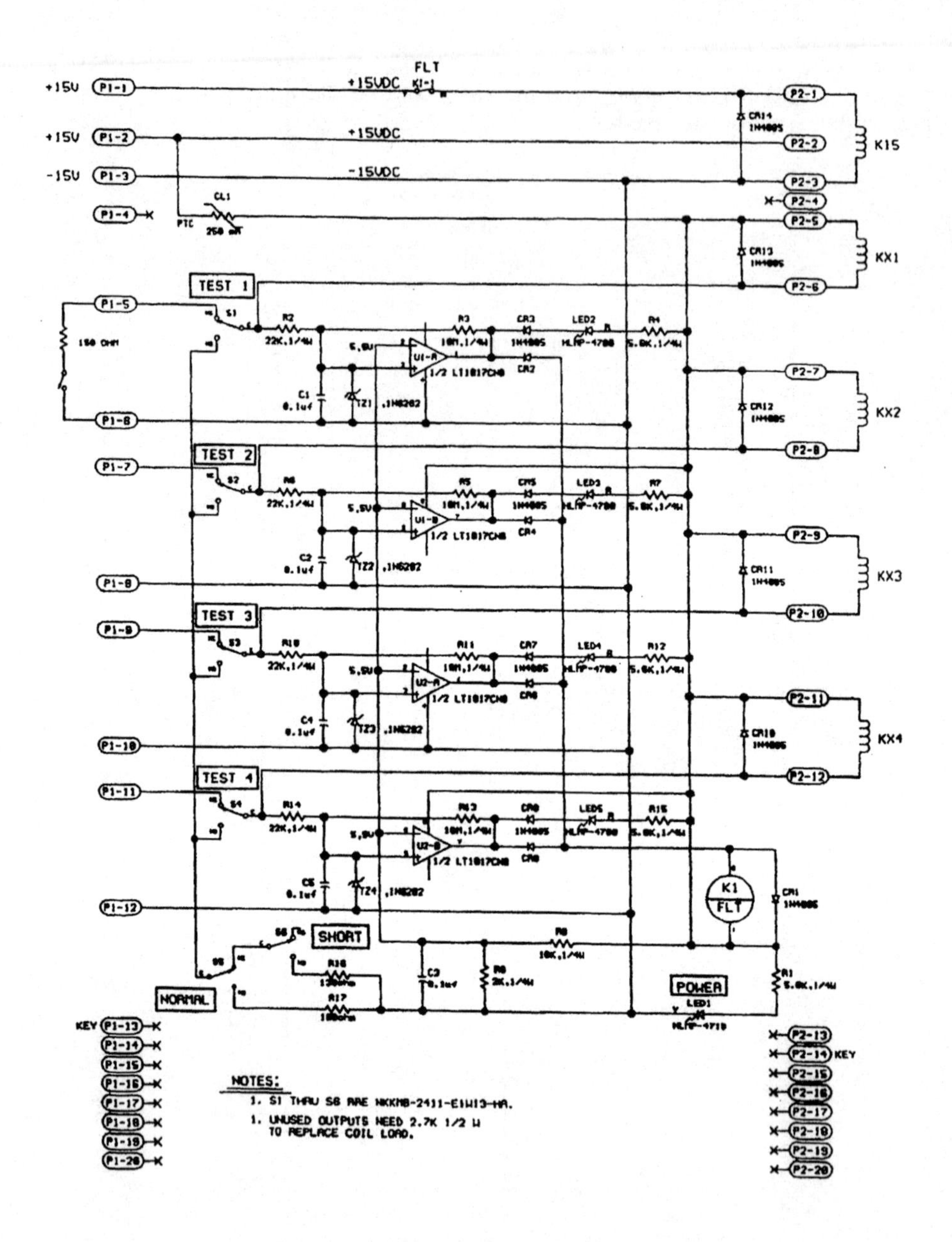

FIGURE 4. Quad sense PNB schematic.

Summary of Design Features

1. Redundant. Two separate cable plants with two separately controlled safety devices, i.e., RSS-rated beam plugs in the low-energy-transport section of the accelerator.

2. Robust and strong. Junction boxes, armored cable, large AWG cable, tray and wire way.
3. Expandable. The capability to add future backbones to the existing system. The capability to add inputs to existing node junction boxes.
4. Simple. The backbones consist of wire and relay contacts.
5. Easily identified components. The color "school-house yellow" is used for all cabling and junction boxes. All internal wiring is color coded and documented.
6. Physically separated. The separation of cabling in tray and wire way. The separation of terminations and components within junction boxes.
7. Testable. Accelerator operational testing and verification ability. Input-by-input or component-by-component.
8. Capable of remote configuration. The ability to select different operational modes from the Central Control Room by operators, i.e. computer control of equipment affecting operational modes.
9. Interruptible. Operations has the ability to interrupt the backbones with a single crash button. The automatic insertion of beam plugs with loss of any input to the backbone.
10. Verifiable backbones wiring. Ground fault indication at both local/CCR locations.
11. Verifiable input wiring. Testing available for short, open, and ground fault with indication at both local/CCR locations.
12. Displayable logic. The remote display of information about the backbones is accomplished with Allen-Bradley PLCs to VME via local area networking.
13. Access controllable. All junction boxes have a padlockable RSS wiring level. Access granted by RSS Engineer and documented controlled key.
14. Visually accessible. The design of windows in the door and on the swing out panel allows visual inspections of the quad sense boards LED's for problem identification. Logical display and information is on the outside window.
15. Modular. Design consideration for all components was plug-in replaceable. PWB's are in trays that are DIN-rail mountable. Relays are mounted in sockets. Terminals are DIN-rail mountable with screw connections. DIN-rail-mountable identification plugs and test equipment.
16. Label. Everything is labeled. Cables, PWB/ trays, terminals, power supplies, relays, connectors, junction boxes, switches, areas, and nodes.
17. Relay pole assignments. All relays used in the backbones have the same assignments. Pole 1, BB logic wiring. Pole 2, Allen-Bradley wiring. Pole 3, safe-out wiring. Pole 4, local display wiring.
18. Event counting. With the Allen-Bradley PLC it is possible to count the times that beam plugs are inserted. To measure the time between the two safety plugs reaching their in or out limits. The timing of any operational parameter associated with the backbones. The Allen-Bradley PLC's are read only, this means the PLC could be removed from the systems with no loss of the backbones operation other than remote display.
19. Organized documentation. The schematics for the junction box point-to-point wiring and the decal for the door window are under one drawing number. The number is displayed on the window decal. There is a different drawing number assigned for each junction box.
20. Dual outputs. Each system feeding inputs to the backbones requires dual outputs. A outputs are fed to the A backbone. B outputs are fed to the B backbone. PACS systems have redundant individual testable outputs.

21. Ferrule terminated wiring. All backbone wiring is terminated with ferrules which require a crimping tool and ferrules for different size wire. The ferrules offer a more solid contact with screw terminals. Stranded-wire fraying and loose ends are eliminated.
22. Protected 110VAC. The use of surge-protected receptacles with individually assigned and padlock-compatible circuit breakers for each node box.
23. Regulatory Guidance. DOE Order 5480.25, Safety of Accelerator Facilities. LANL Laboratory Standard LS107-01, Accelerator Access-Control Systems. LANSCE 6 AOT-6-95-QA-4, Radiation Security System Quality Assurance Management Plan.

Summary of Fabrication and Installation Features

1. Johnson Controls Northern New Mexico (JCNNM) electrician teams were specially trained in the requirements of backbone installation with the development of armored cable termination. JCNNM and Los Alamos Neutron Science Center (LANSCE) Protective System Team members developed cable pulling techniques that allowed for faster and safer cable installations over lengths of 2500 feet.
2. Dawn Electronics, vendor for fabrication of backbone RSS level wiring bottom plates, was instructed in specialized wiring requirements. The construction of the components was accomplished off-site.
3. Northern Design, vendor for fabrication of backbone printed circuit assemblies with DIN rail-mounting of tray components with assembled printed circuit boards.
4. Protective Systems Team, LANSCE-6, contributed with fabrication, installation, oversight and the specialized fabrication of Central Control Room display equipment. The installation of the Allen-Bradley equipment, with development of the logic and interfacing to VME equipment, was accomplished by team members.
5. Fabrication, installation, training and checkout was a three-year project. Associated with the installation of the RSS backbones were seven personnel access control systems (PACS) area installations which were connected to the linac backbones.

REFERENCES

[1] Gallegos, F. R., "LANSCE Radiation Security System (RSS)," presented at the Conference on Health Physics of Radiation Generating Machines—30th Midyear Topical Meeting San Jose, CA, January 5–8, 1997.

[2] Sturrock, J. C., F. R. Gallegos, M. J. Hall, "LANSCE Personnel Access Control System (PACS)," presented at the Conference on Health Physics of Radiation Generating Machines—30th Midyear Topical Meeting San Jose, CA, January 5–8, 1997.

[3] Jones, K. W., W. Boedeker, A. Browman. "Use of Fusible Beam Plugs for Accident Mitigation at the LANSCE Complex," presented at the Conference on Health Physics of Radiation Generating Machines—30th Midyear Topical Meeting San Jose, CA, January 5–8, 1997.

DISCUSSION GROUP AND CLOSEOUT SUMMARIES

Summaries of Discussion Groups and Closeout

Discussion Groups

BIW registrants were asked to select their top six choices for group discussions from the following topics:

Commercial rf Technology and Beam Instrumentation
4th Generation Light Source Instrumentation
Feedback Systems
Challenges in Beam Profiling
Beam Loss Monitors
Calibration Methods
High Resolution and Highly Stable BPM Methods
Closed Orbit Monitoring
Polarimeters and Applications
Low Current Monitors
Linear Collider Instrumentation
Colliding Beam Instrumentation

The first seven topics were the most popular and discussion groups were held accordingly. Summaries of these discussion sessions and the closeout session, as submitted by the chairs, are given below.

Commercial Applications of rf Technology

Ralph Pasquinelli, FNAL

In general, discussion sessions that do not have a "seed" problem to solve are very difficult to conduct. This session was certainly one of those. I started out with a few anecdotal stories about how we have been building a 16×16 channel multiplexer for connecting various diagnostic signals to a myriad of spectrum and network analyzers, counters, and power meters. Our requirement was for 1 kHz to 200 MHz bandwidth with 40 dB minimum isolation. After a year of prototyping we came up with a unit that utilized 8 layer circuit cards and lots of components. That same year, Analog Devices came out with a chip that did exactly this function, only better than our discrete unit. We bought the chip and are using it now.

Most of the rest of the discussion centered on requests for peoples' experience with industry providing solutions to our accelerator needs. There was some discussion that industry would indeed get involved when the number of channels required was high enough to justify NRE costs and still make a profit. Industry is not willing to

CP451, *Beam Instrumentation Workshop*
edited by R. O. Hettel, S. R. Smith, and J. D. Masek

create components for sale, but only complete systems or stand-alone equipment. Many of the instrumentation needs are such that the number of channels is low and very customized to the particular machine. For instance, even at laboratories where there are multiple accelerators, the diagnostics for the accelerators are typically customized for each, not allowing for the economy of producing in quantity.

Beam-Loss Monitors

Alan Fisher, SLAC

I began the discussion by reviewing the different applications for beam-loss monitors (BLMs). There are at least four tasks that a BLM may be required to do; some machines may need only one, while others need all four. The type of BLM of course depends on the requirements.

1. Detection of steady, low-level losses in a ring. For tuning, loss histories and correlation studies may be required. A counter works well in this case.
2. Detection of sudden, high losses in a ring for rapid tuning (or aborts; see item 4 below). Depending on the time scale involved, a counter may not count fast enough (due to pulse pile-up), may not get adequate statistics in the short interval involved, or may not be read often enough. An integration of the analog loss signal may be preferable.
3. Detection of one-pass losses, for measurement of injection loss in a ring or for a linac. A counter is certainly not adequate for this job: it would count either 0 or 1. Some sort of integration is again needed.
4. Machine protection, by aborting a stored beam or stopping a linac when losses are excessive. This requirement could accompany the other three cases. It is also helpful to record the loss distribution at the time of the event to diagnose the source.

PEP-II (SLAC). I then described the Cherenkov BLMs built for PEP-II (1). Each consists of a small, fast photomultiplier and a fused-silica Cherenkov radiator, all surrounded by 1 cm of lead and enclosed in a steel can for magnetic shielding (although this PMT can work in a 100 G field). The Cherenkov threshold (about 1 MeV) prevents response to synchrotron radiation, and the lead provides additional filtering. Despite the lead and steel, the detector weighs less than 2 kg and can be moved around the ring.

The PEP BLMs perform all four tasks above, using a processor that sends the signal through two input circuits, which together provide a wide dynamic range. The PMT pulses are counted and also RC-integrated over one ring turn. A peak detector then finds the worst turn in each 8 ms interval, and the result is digitized.

If the integrated signal exceeds a programmable threshold, we can choose to abort one or both rings. We need less than 30 μs to detect the loss and abort the beam, to avoid having the large stored energy in the beam burn through the vacuum chamber.

(The beam spirals inward rapidly after an rf trip, and the BLMs provide a second level of protection for such an event.) The BLM processor then records the triggering channel and, through a daisy chain linking all the processors, causes all BLMs to freeze their most recent readings. We have not made use of the abort feature yet, but we expect to use it in the next run, as we scrape the beam tail against some new collimators.

Another daisy-chain signal provides a 100 μs gate during injection, when the BLM network is inhibited from aborting the stored beam, since faulty injection is a more likely source of a large loss (and stored-beam losses will persist after the gate). To measure the loss on the worst turn around the injection time, the peak detector is digitized and read within 3 ms after the inhibit interval.

HERA (DESY). Next, Karl Hubert Mess spoke of the BLMs they built for HERA's proton ring. A commercial version of these devices is now available from Bergoz, under license from DESY. Their original function was to prevent a quench of the superconducting magnets in the proton ring by aborting the beam when loss increases, and they have been very successful in that role. Since that time, they have also been used on the electron ring, to locate the source of sudden decreases in lifetime. This problem has been traced to dust from the distributed ion pumps getting into the beam.

Quench protection requires an 8 ms response time, and since the bunch spacing is 96 ns, a counter can offer good statistics. To generate the loss signal, DESY chose a PIN diode, which emits a pulse when a particle from the beam loss shower passes through. To avoid responding to the synchrotron radiation from the electron ring in the HERA tunnel, two diodes are placed back-to-back. A coincidence circuit ensures that only harder particles (from beam loss) can pass through both diodes and cause an output pulse.

Loss histories are recorded and can be studied after an event to look for correlations with other machine parameters, such as lifetime. To strike a balance between machine protection and false aborts, four detectors must exceed their thresholds for an abort.

APS (Argonne National Laboratory). The Advanced Photon Source uses a long coaxial cable to detect losses. By filling air-dielectric RF cable with a suitable gas mixture, the cable becomes an ionization chamber. The pulse traveling with the beam integrates the total loss, while the pulses heading upstream arrive spread out in time, allowing a determination of the loss site. This device, known at SLAC as a PLIC (Panofsky's Long Ionization Chamber), thus has two advantages: it can be run along the machine, to get good coverage, and provides good localization (in a one-pass device like a linac or transport line). SLAC uses PLICs running along the 3 km linac, along the various transport lines, and through each PEP injection tunnel and around the first arc.

However, a single PLIC cannot localize losses from stored beam. At the Advanced Photon Source, several PLICs run around the ring in segments, to provide some position information. Robert Merl presented some transparencies on behalf of Glenn Decker to show one of their PLIC's responding to a scraper in the booster-to-storage-ring transport line. However, the PLICs have been less successful in the APS ring, due to background from synchrotron radiation (It's hard to shield a PLIC with lead!). APS is now testing a PEP Cherenkov monitor to look for injection losses at insertion devices.

ALS (Lawrence Berkeley National Laboratory). Jim Hinkson has been using the Bergoz version of the HERA BLM at the Advanced Light Source. He showed plots of typical loss signals near scrapers and insertion devices for different lifetimes. Recently, when it was difficult to store a beam at ALS, Jim used the BLMs to quickly determine that the scraper had not retracted out of the beam. A problem that might have taken a day was resolved in half an hour.

We also discussed the behavior of the Bergoz BLM when saturated. When Jim injected beam into a fully inserted scraper, the first turn produced an unusually high and wide BLM pulse. The BLM showed nothing for the second turn, 660 ns later, and recovered to show a signal on the third turn. Signals were observed on the fourth, fifth and sixth turns. I then noted that in my own tests of this device, I observed a similar saturation for about 1 μs after a heavy blast.

References

[1] Fisher, A.S., "Instrumentation and Diagnostics for PEP-II," in these proceedings.

Calibration Methods

Robert Webber, FNAL

The discussion session opened with about a dozen people in attendance. We found ourselves in a room with tables and chairs set up as in a lecture hall with everyone facing the front. The discussion chairman immediately involved participants by re-arranging the room set-up to form a "round table" type arrangement with participants facing each other. This activity was a good ice-breaker to facilitate discussion.

Discussion began with a suggestion by the chairman that DC beam current monitors (DCCT or PCTs) are ideal devices to calibrate because simple instruments are able to yield a very accurate calibration of the monitor. The discussion then turned to questions of the frequency response of such devices and how to assure calibration at frequencies other than DC.

Two potential problems with DC transformers were presented. This device operates by modulating the magnet bias of toroidal transformers at some frequency, typically a few hundred to a few thousand Hertz. It effectively operates as a chopper amplifier to translate the DC signal to be measured up to a frequency twice the modulating frequency. That second harmonic amplitude is then detected and interpreted as the DC component of the beam current. The resulting frequency response is like that of a sampled data system with the accompanying ability to alias "out of band" signals into an "in band" frequency. If the transformer is excited with beam or test current signals at frequencies that are harmonics of the modulating frequency, an output can be created which is indistinguishable from that caused by a DC beam excitation. This problem is generally avoided by packaging an AC transformer with the DC monitor to provide the desired high frequency response while nulling any currents in the DC section that could be aliased. A problem noted during construction and testing of a DC monitor at Fermilab revealed amplifiers in a high-gain feedback section of the electronics that saturated in response to signals at certain frequencies. This non-linearity also resulted in an output indistinguishable from that of a true DC beam current. Frequency response adjustments in the electronics were required to assure that saturation was avoided. A straightforward method was described to test for such problems. In the lab, a test current is injected into the DC monitor using a simple wire or, for very high frequencies, an impedance matched coaxial center conductor. A blocking capacitor of suitable value is included in series with the test current conductor to assure that no DC current is applied to the monitor. Test currents over a wide frequency range are injected into the monitor using an amplifier of suitable bandwidth and power capability. The monitor output is low-pass filtered and if any DC signal component is observed at the monitor's output in this condition, a problem with saturation or aliasing effects is most likely present. Signals within the monitor electronics can then be investigated to discover the source of the problems.

George Coutrakon of Loma Linda mentioned discussion at Loma Linda about a current monitor for accurately measuring the slow spill beam extracted from that machine. The difficulty is that the current is in the low nanoampere range. There was little discussion and no good idea for a solution to this problem.

Coutrakon then brought up the question of BPM calibration, describing the observation of beam motion in the Loma Linda transport lines that seems to correlate with the beam intensity in the ring. It was noted that the Loma Linda machine uses a BPM signal for feedback in the low-level rf system to control beam energy and position. It was hypothesized that, if this BPM signal were intensity sensitive, the orbit in the machine and therefore the extracted beam position would be intensity-dependent. It was suggested that any intensity sensitivity of the BPM electronics could be determined under controlled conditions on a test bench. Another suggestion was made to use the actual beam signals in a controlled manner as test inputs. The signal from one pick-up electrode, or the combined signals from several electrodes, could be amplitude-controlled by an attenuator and then equally split into the electronics inputs to simulate an "on center" beam. A suitable amplifier might be included upstream of the signal splitter, if necessary, and attenuators downstream of the splitter could be added to simulate off-axis beam positions. This method assures that testing is done with a signal spectrum identical to that present in actual operation. A momentum-selective extraction process at Loma Linda could cause the bunch length and therefore signal spectrum to change during the extraction interval at the same time as the ring beam intensity diminishes. It was noted that the original CERN LEP BPM system manifested a systematic measurement error due to the presence of signal frequencies from the beam which had not been accounted for during the design and testing of the electronics.

Steve Smith showed transparencies of the PEP-II BPM circuitry including couplers on the beam signal inputs to facilitate "on-line" calibration. There was discussion of the operation of that calibration system and questions regarding its accuracy. Steve's presentation of one or two transparencies greatly facilitated discussion on the topic, and attendees of future discussion groups should be encouraged to contribute such visual enhancements.

Gianni Tassotto of Fermilab mentioned the potential need for a one-meter diameter beam current toroid for the Fermilab NUMI experiment. The transformer would measure secondary beam current downstream of the production target in the vicinity of the focusing horn. There was a short discussion about the possibility of such a large device and how to deal with its potential sensitivity to electrical noise from the high-current pulsed horn and to stray particles or beam halo that might intercept the toroid itself.

Active discussion continued right up to the end of the allotted time. About two dozen people had ultimately shown up to participate. No earthshaking conclusions were forthcoming but there was a good sharing of experiences, problems, solutions, problems to avoid, and diagnostic methods. The session closed with participants restoring the table and chair arrangement in the room to the state we found it.

Challenges in Beam Profiling

Marc Ross, SLAC

Three types of beam profile monitor were picked for discussion of design challenges:

1. Ion- and residual gas-based proton synchrotron monitors.
2. Phosphor screen monitors.
3. High resolution wire monitors.

J. Zagel presented a summary of the FNAL program to improve the performance of the micro-channel plate (MCP) based ion monitor. The calibration/aging of the MCP is a challenge. They are considering the collection of electrons instead of ions. This device, of which there are several dozen in existence, is clearly the best way to check the beam profile in a medium energy proton synchrotron. Use of electrons instead of ions, pioneered at BNL, removes the requirement for very high voltage. Resolution is limited due to broadening during the drift to the collection electrode. Typical resolution for this device is a fraction of a millimeter, appropriate for such machines.

On the phosphor screen topic, W. Graves of BNL showed results from doped YAG disks. We discussed the optics of the monitor and tried to estimate depth of field and other resolution degrading effects. The material may have an intrinsic resolution of close to 1 μm, but its thickness and associated imaging problems will make it hard to realize performance below 8 μm. The BNL material is 0.2 mm thick.

On our final topic, M. Ross showed the performance of the SLC laser-based profile monitor. J. Frisch provided presentation material concerning the development work for the future laser-based profile monitors with very good resolution. The purpose of the laser-based profile monitor is to allow monitoring of very small, high charge density beams. Conventional wire scanners, using carbon, tungsten, or silicon carbide wires will not work adequately for beams smaller than 1 μm, or beams with particle density greater than $1\times10^{10}/\mu m^2$. These limits have been tested at SLAC. Other wire scanner challenges include motion control, wire supports and scattered radiation detectors.

Feedback Systems

Mario Serio, LNF-INFN

The discussion session was held in the main Auditorium and attended by about 30 people with a good representation of accelerator laboratories from the U.S. and from the rest of the world.

No request of advice on specific topics was asked by the audience and nobody but the Chairman had transparencies or other material to present. In spite of this and of the usual shy start-up of this kind of session, especially in a large auditorium, the discussion became very lively with many interesting contributions which enabled the mutual clarification of different terminologies and points of view.

At the beginning of the session, a short time was spent commenting on the accelerator-oriented definition of feedback, which differs from control theory's definition of returning a fraction of the system output to its input in order to maintain prescribed relationships between selected system variables. In fact, there was general agreement on an alternate definition of feedback as the technique (or the art) of exploiting the natural tendency of accelerator beams toward misbehavior in order to produce beneficial corrective actions. The most spectacular example of this is the stochastic cooling of (anti)proton beams, which leads to an apparent violation of the Liouville theorem.

The discussion was organized in matrix fashion, ordered by different families of feedback systems: longitudinal, transverse, position/orbit control and closed loop control of beam parameters and, for each family, by:

- Identification of suitable observable signals, detector noise rejection, common mode rejection, signal/data processing, speed of response (or, "How fast is fast?").
- Identification of suitable corrective actions, choice of power amplifier, actuator noise, kicker issues, magnets issues.
- Requirements, system configuration and necessary gain.
- Merits and demerits of time domain (bunch by bunch) vs. frequency domain (mode feedback), and, for orbit control, local vs. global correction.

It was pointed out that if an offending cause is known a-priori, it may be possible to remove it. In any case, the realization of a suitable specialized feedback system may be simplified to a large extent by a detailed knowledge of the offending mechanism.

On the other hand, if a feedback system is conceived during the design stage of an accelerator, a tendency to design in greater flexibility may result in excessive system complexity. However, such a system generally pays off in terms of redundancy and diagnostic capabilities.

Interesting examples of the two extremes were, among others, the implementation of a specialized loop (CESR) for damping transients with a low duty cycle, constant amplitude pulser and of a DSP based longitudinal bunch by bunch system (PEP-II, ALS, DAFNE).

Given the high feedback gain generally specified, feedback builders often face the problem of how to get rid of any useless common-mode signal present in the front-end before amplification and data treatment in the subsequent stages. For example, walk of synchronous phase along the bunches of an uneven fill in a longitudinal system, or stationary orbit in a transverse one, may easily lead to saturation. In similar cases, as pointed out by J. D. Fox (SLAC), speaking of DC rejection or "notching" at revolution harmonics is only a matter of taste, or of terminology, but is truly equivalent.

In practically all of the synchrotron light sources it is necessary to rely on a closed stabilization loop of some kind. While at Elettra (Trieste) the beam lines are equipped with local loops, at APS (Argonne) a global system has been implemented. In the Elettra system, additional loops tuned to harmonics of the main frequency are able to reduce the residual beam motion up to high frequencies. An overview of the APS system was presented earlier during an oral session at this Workshop.

Eddy currents in the vacuum chamber shield the magnet corrector field and ultimately limit the frequency response of the orbit loops. Modeling and compensating the eddy-currents effect in the control loop transfer function is not trivial. The obvious cure is to reduce the vacuum chamber thickness at corrector locations as much as practical.

Moreover, it was pointed out that in pulse-to-pulse systems, such as those in linear colliders, the effectiveness of feedback systems is impaired by the lack of a suitable model.

In conclusion, this session gave the opportunity to several experts to confront their experiences and to some newcomers to approach the fascinating subject of accelerator feedback. All participants contributed to a lively and interesting discussion.

4th Generation Light Source Instrumentation

Alex Lumpkin, APS

This working group was a follow-up to the opening discussion, Challenges in Beam Profiling. It was run in parallel with the Feedback Systems session. We filled the SSRL Conference Room with about 25 participants.

The session opened with an introduction by Lumpkin. The target beam parameter values for a few-angstrom, self-amplified spontaneous emissions (SASE) experiment and for a diffraction-limited soft x-ray storage ring source were addressed. Instrument resolution would need to be 2–3 times better than the value measured if possible. The nominal targeted performance parameters are: emittance (1–2π mm-mrad), bunch length (100 fs), peak-current (1–5 kA), beam size (10 mm), beam divergence (1 mrad), energy spread (2×10^{-4})= , and beam energy (tens of GeV). These are mostly the SASE values; the possible parameters for a diffraction-limited soft x-ray source would be relaxed somewhat. Beam stability and alignment specifications in the sub-micron domain for either device are anticipated.

Vinod Bharadwaj (SLAC) then presented the specific design parameters for the LCLS SASE project at 15 angstroms (0.8 keV) and 1.5 angstroms (8.2 keV) using beam energies of 4.5 and 14.4 GeV, respectively. Roger Carr (SSRL) presented comments on the undulator and beam-based alignment techniques for the LCLS. Since the BIW'96 meeting, the strategy of steering and making position measurements every few meters between undulator sections has been accepted. Vinokurov pointed this out in the APS SASE design, and it was confirmed by adjusted calculations by Kim at LBNL for the LCLS. This revised strategy reduced the instrumentation challenges over the 100 m length undulator for keeping the photon beam and particle beam adequately in line. The overlap of the beams should be held to about 5 mm over 10 m.

Transverse beam size measurements were then discussed. Alex Lumpkin (APS) presented the results of using a 3.5 m long diagnostics undulator with the 7 GeV beam to measure a particle-beam divergence of 3.3 μrad. The fundamental radiation was at 0.5 angstroms (26 keV), close to the 4GLS wavelength. The technique should scale to the 1 mrad regime and possibly to a single, few nC micropulse charge. Lumpkin proposed such a device as one line of the array of undulators at an eventual 4GLS user facility. Additionally, the x-ray pinhole imaging technique with an x-ray streak camera was shown to measure tens of microns with projected 1 ps (sigma) resolution in an earlier presentation. The issues related to signal strength for slices of the micropulse at sub-ps regimes remains an area for development, but the transverse size averaged over the micropulse seems to be solved.

The discussion moved to the measurement of sub-picosecond microbunches. In particular, the temporal profiles at the sub-100 fs regime are an issue. Most of the correlation methods using coherent transition radiation (CTR), coherent diffraction radiation (CDR), coherent Smith-Purcell Radiation (CSPR), or coherent synchrotron radiation (CSR) will provide a measure of pulse duration. The temporal profile is much more ambiguously determined and often relies on an assumed shape. William Graves (BNL) commented on laser gating of a material's transmission or reflection property to provide sampling of converted visible radiation from the particle beam, e.g., an OTR signal. This could work at the 100 fs level with a ultra-fast laser probe. Other laser-based techniques have been suggested. As a side note, the differential optical gating (DOG) technique has been demonstrated recently at the Stanford FEL by Schwettman, Smith, et. al. Temporal profiles on the sub-ps domain were obtained although not on a single pulse. Still, the technique avoids phase-jitter averaging effects, so it can be used over many pulses. Development still proceeds for the x-ray streak camera at the 100–200 fs (sigma) regime. This has the potential for longitudinal profiles with a spatial profile (submicropulse) as a complementary approach.

The discussion moved to the task of maintaining the photon beam and electron beam overlap to a few microns over a gain length of a few meters. (In the LCLS case this is 5 mm over 10m.) Suk Kim (APS) presented the rf BPM button configuration that is planned for the APS visible/UV SASE project. By rotating the axis of a pair of 4 mm diameter buttons, very high sensitivity to motion in one plane is calculated which should scale to sub-micron resolution. Tests are expected in the coming year. An LCLS person brought up the combined electron and photon beam diagnostic based on the interactions with a single, 10 to 30 μm thick carbon wire. Photons are diffracted to an x-ray detector positioned off-axis at 25.8° to the particle beam direction, and the bremsstrahlung radiation is detected by another detector at a more forward angle.

As the session's allotted time was over, we adjourned to allow further discussion among small groups. The challenges in the spatial, temporal, charge and position domain plus the preservation of beam quality were duly noted.

High-Performance Beam Position Monitor

Steve Smith, SLAC

This discussion group met in the auditorium as a plenary session of the Beam Instrumentation Workshop. The first topic was to identify what we mean by "high performance" in the context of beam position monitors (BPMs). Areas of interest included precision, accuracy, sensitivity, bandwidth, data rate, data reduction capability, maintainability, performance per unit cost, stability with respect to beam intensity, and mechanical stability.

Don Martin from SSRL presented a list of required parameters for the LCLS (Linac Coherent Light Source) beam position monitors to stand as an example of BPMs that require high performance in several aspects. The system must be consistent with the specifications in the following table:

Parameter	Value or Requirement	Comments
Repetition rate	120 Hz	
Bunch length	$\sigma_z = 0.15$ ps	
Peak current	I = 4 kA	
Resolution	< 1μm rms	@ 1 nC/bunch
Repeatability	< 5 μm	over hours
Accuracy	< 50 μm	

PThis started discussions on several topics, principally concerning the need for beam-based alignment (BBA). Jim Hinkson (LBNL) commented, "We must build in beam-based alignment from the beginning." Absolute accuracy requirements become untenable if one tries to build accuracy into (or calibrate it into) the instrumentation. Herman Schmickler (CERN) warned, however, that it takes months to get a machine ready for beam-based alignment; one can't count on beam-based alignment to get a machine working. Steve Smith (SLAC) added that there are capture tolerances for accuracy before getting BBA. Glen Decker (Argonne) seconded Hinkson; one must build in BBA. Douglas Gilpatrick (Los Alamos) noted "If you have stability and resolution, you can figure out absolute alignment." He added that when specifying stability, it is always with respect to a time period and a length scale. For example, one person's resolution may be another's accuracy due to differences in time scales. Bob Hettel commented that, for FELs, there is an additional requirement for the photons to be aligned with the electrons.

The discussion turned to the benefits of single-channel switched receiver vs. parallel receiver designs. Julian Bergoz stated "Resolution is easy. Stability with respect to beam current is hard. Accuracy is intermediate in difficulty." Hinkson added "Stability with respect to fill pattern is difficult, too." Schmickler asked that, for comparison purposes, one should "specify resolution as a fraction of aperture, so we can translate between very different geometries." Hinkson added that users care about stability in terms of fraction of beam size, saying "they don't know how big the pipe is."

Manfred Wendt (DESY) then presented requirements for BPM's to be used in an FEL undulator in the Tesla Test Facility. The system must meet the requirements outlined in the following table:

Parameter	Value or Requirement	Comments
Bunch charge	0.5–1.0 nC	per bunch
Bunch spacing	111 ns	9 MHz
Pipe size	10 mm	diameter
Pickup elements	tiny fingers	
Signal spectrum	rising to 4 GHz	
Resolution	< 5μm rms	average over 7000 bunches

Wendt asked the group how might they read this out with limited time and manpower. He proposed using an AM/PM monopulse receiver in the style of Vismara and Cocq with a 200 MHz low-pass filter. Glen Decker suggested a log-ratio system consisting of bandpass filters, downconverters, log-amps, and difference amplifier. Bob Shafer (Los Alamos) commented "It's not well known that AM/PM and log ratio techniques have poor noise figures."

A lively discussion envolved many participants on the question "Is it better to have striplines terminated or shorted?" The cost and reliability of extra feedthroughs were discussed as reasons to not terminate striplines. Warnings were issued about filter response surprises and unexpected amplifier noise when working with unmatched sources. Several stories of open vs. shorted striplines ensued.

The possibility of another type of BPM, based on acoustic waves propagating around a conducting beam chamber wall was brought up and briefly discussed. We had no difficulty in consuming the time allotted for discussion.

Closeout Discussion and BIW Y2K

Steve Smith, SLAC

The 1998 Beam Instrumentation Workshop ended with the traditional closeout session, where, after acknowledging those who made the Workshop successful (see preface), we attempt to identify our strengths and weaknesses, and poll the attendees for suggestions to improve future Workshops.

The discussion groups generated substantial comments. The discussion group format is popular, but a common complaint was voiced about the difficulties in getting full participation in a discussion group held in an auditorium; a smaller room with participants facing each other is more conducive to discussion. A suggestion was made that each participant be asked to bring one slide, presenting a problem they want to have solved, in order to get discussion going.

Publication on CD-ROM was suggested, as was a request to keep publication on paper. The consensus was that the Proceedings should be available at least on paper, CD-ROM publication could only supplement paper. A participant suggested that the publisher of the Proceedings, the American Institute of Physics, be invited to send a salesperson to the next workshop along with a stack of the Proceedings of earlier Workshops, as the back issues are always in demand.

Several topics for future tutorials were proposed. Tom Shea suggested one tutorial on "Grounding, Shielding, and Isolation" for instrumentation and another on magnetics, magnetic shielding, and magnetic sensors (such as Hall effect and NMR probes). Keith Jobe also suggested two topics, one on "Lasers, and how to use them as a diagnostic tool", and another on "Radiation Practices in the Tunnel", i.e. how to build things that must live in the machine. Julian Bergoz suggested "Beam vacuum interaction mechanisms" as a topic. Robert Merl suggested, since beam instruments are becoming increasingly complicated devices, often containing dedicated processors, a "DSP Tutorial" would be in order.

Bob Averill, BIW Organizing Committee member from MIT-Bates Laboratory, announced that the next Beam Instrumentation Workshop, BIW 2000, will take place in Cambridge, Massachusetts. The Massachusetts Institute of Technology (MIT), the Laboratory of Nuclear Science (LNS), and the Bates Linear Accelerator Center (BLAC), will host the Workshop on the MIT campus. Bob presented a brief overview of the MIT facilities and the Boston area, inviting participants to attend BIW 2000 and to experience this fascinating and historic region.

APPENDICES

List of Participants

Chris Adolphsen
SLAC
PO Box 4349
MS 65
Stanford CA 94309
Phone: (650)926-3560
Fax: (650)926-4892
E-mail: adolphsen@slac.stanford.edu

Larry Ahle
Lawrence Livermore National Lab.
Physics & Space Tech.
7000 East Ave.
Livermore CA 94551
Phone: (510)422-1621
Fax: (510)423-2664
E-mail: ahlel@llnl.gov

Jim Andrews
Picosecond Pulse Labs
PO Box 44
Boulder CO 80306
Phone: (303)443-1249
Fax: (303)447-2236

Robert Averill
MIT Bates Linear Accl. Ctr.
PO Box 846, 21 Manning Rd.
Middleton MA 01949
Phone: (617)253-9254
Fax: (617)252-9599
E-mail: averill@Bates.Mit.edu

Mark Ball
Indiana University Cyclotron
Dept. of Beam Diagnostics
2401 Milo B. Sampson Ln.
Bloomington IN 47401
Phone: (812)855-5162
Fax: (812)855-6645
E-mail: ball@iucf.indiana.edu

Edward Barsotti
Fermilab
PO Box 500
MS 308
Batavia IL 60510
Phone: (630)840-8104
Fax: (630)840-3754
E-mail: ebarsotti@fnal.gov

Julien Bergoz
Bergoz Instrumentation
Rue du Jura
Crozet, 01170 FRANCE
Phone: 33-450-410089
Fax: 33-450-410199
E-mail: bergoz@bergoz.com

Erika Bisgard
Pearson Electronics, Inc.
1860 Embarcadero Rd.
Suite 200
Palo Alto CA 94303
Phone: (650)494-6444
Fax: (650)494-6716

Willem Blokland
Fermilab
BD/RF & Instrumentation
PO Box 500
Batavia IL 60510
Phone: (630)840-2681
Fax: (630)840-3754
E-mail: Blokland@fnal.gov

Daniele Bulfone
Sincrotron Trieste
Accelerator Division
SS 14 KM 163.5 Basovizza
Trieste 34012 ITALY
Phone: 39-40-375-8579
Fax: 39-40-375-8565
E-mail: bulfone@elettra.trieste.it

Karel Capek
Jefferson Laboratory
Accelerator Development
12000 Jefferson Ave., MS 59
Newport News VA 23606
Phone: (757)269-7197
Fax: (757)269-6266
E-mail: capek@jlab.org

John Carwardine
Argonne National Laboratory
Dept. of Power Supplies
9700 S. Cass Ave., MS 401
Argonne IL 60439-4800
Phone: (630)252-6041
Fax: (630)252-5291
E-mail: carwar@aps.anl.gov

Daniel Cheever
MIT Bates Linear Accl. Ctr.
Dept. of Engineering
21 Manning Rd.
Middleton MA 01949
Phone: (617)253-9533
Fax: (617)253-9799
E-mail: cheever@mit.edu

Vinod Chohan
CERN
PS Division
1211 Geneva 23 SUISSE
Phone: (41)22-767-2719
E-mail: V.chohan@cern.ch

Gerald Codner
Cornell University
CESR
Dryden Rd.
Ithaca NY 14853
Phone: (607)255-5749
Fax: (607)255-8062
E-mail: gcodner@lns62.lns.cornell.edu

Roger Connolly
Brookhaven National Laboratory
RHIC
Bldg. 1005
Upton NY 11973
Phone: (516)344-4698
Fax: (516)344-2588
E-mail: connolly@bnl.gov

George Coutrakon
Loma Linda Univ.
Medical Center
Loma Linda CA 92354
Phone: (909) 558-6024
E-mail: coutrak@dominion.llumc.edu

Richard Davies
Polaron CVT LTD.
4-6 Cartery Lane
Kiln Farm
Milton Keynes, MK11 3ER UK
Phone: (44)1908-563267
Fax: (44)1908-568354

Ted Debiak
Northrop Grumman Corporation
MS K03-14
Stewart Ave.
Bethpage NY 11714
Phone: (516)346-3406
Fax: (516)575-2140
E-mail: ted.debiak@atdc.northgrum.com

Franz-Josef Decker
SLAC
PO Box 4349
Accelerator Dept.
Stanford CA 94309
Phone: (650)926-3606
Fax: (650)926-2407
E-mail: decker@slac.stanford.edu

Glenn Decker
Argonne National Laboratory
APS/Diagnostics
9700 S. Cass Ave., MS 401
Argonne IL 60439-4800
Phone: (630)252-6635
Fax: (630)252-4732
E-mail: decker@aps.anl.gov

Winfried Decking
Lawrence Berkeley National Laboratory
ALS, MS 80-101
1 Cyclotron Rd.
Berkeley CA 94720
Phone: (510)486-4588
Fax: (510)486-4960
E-mail: wdecking@lbl.gov

Jean-Claude Denard
CEBAF
MS-87
12000 Jefferson Ave.
Newport News VA 23606
Phone: (757)249-7555
Fax: (757)249-7658
E-mail: denard@cebaf.gov

Jurgen Dietrich
Forschungszentrum Julich GmbH
Institut fur Kernphysik
Leo-Brandt Str. 1, PO Box 1913
Julich, NRW D-52425 GERMANY
Phone: 49-2461-61-2678
Fax: 49-2461-61-2670
E-mail: j.dietrich@fz-juelich.de

Danny Dotson
Jefferson Laboratory
Ops/Rad. Con., MS 52A
12000 Jefferson Ave.
Newport VA 23606
Phone: (757)269-7296
Fax: (757)269-5048
E-mail: dotson@jlab.org

Kirsten Drees
Brookhaven National Laboratory
RHIC
Bldg. 1005-3, PO Box 5000
Upton NY 11973
Phone: (516)344-2348
Fax: (516)344-5729
E-mail: drees@bnl.gov

Roger Erickson
SLAC
Accelerator Dept.
PO Box 4349, MS 55
Stanford CA 94309
Phone: (650)926-2830
E-mail: roger@slac.stanford.edu

Brian Fellenz
Fermilab
BD/RF Instrumentation
Pine St., MS 308
Batavia IL 60510
Phone: (630)840-2512
Fax: (630)840-3754
E-mail: fellenz@fnal.gov

M. Ferianis
Sincrotrone Trieste
Accelerator Division
SS 14, Km 163.5 Basovizza
Trieste 34012 ITALY
Phone: 39-40-3758545
Fax: 39-40-3226338
E-mail: mario.ferianis@elettra.trieste.it

Clive Field
SLAC
PO Box 4349
MS 65
Stanford CA 94309
Phone: (650)926-2694
Fax: (650)926-4892
E-mail: sargon@slac.stanford.edu

Alan Fisher
SLAC
PEP-II
PO Box 4349, MS 17
Stanford CA 94309
Phone: (650)926-2436
Fax: (650)926-3882
E-mail: afisher@slac.stanford.edu

James Fitzgerald
Fermilab
BD/RFI
PO Box 500, MS 308
Batavia IL 60510
Phone: (630)840-4978
Fax: (630)840-2677
E-mail: jfitz@fnal.gov

John Fox
SLAC
PO Box 4349
MS 18
Stanford CA 94309
Phone: (650)926-2789
Fax: (650)926-8533
E-mail: jdfox@slac.stanford.edu

Josef Frisch
SLAC
Accelerator Dept.
PO Box 4349
Stanford CA 94309
Phone: (650)926-4005
Fax: (650)926-2407
E-mail: frisch@slac.stanford.edu

David Gassner
Brookhaven National Laboratory
Dept. AGS
Bldg. 911B
Upton NY 11973
Phone: (516)344-7870
Fax: (516)344-5954
E-mail: gassner@bnl.gov

Hank Gerwers
Princeton Scientific Corporation
PO Box 143
Princeton NJ 08542
Phone: (609)924-3011
Fax: (609)924-3018

Abiel Ghebremedhin
Loma Linda University Med. Ctr.
Loma Linda CA 92354
Phone: (909) 558-6024
E-mail: ghbreme@dominion.llumc.edu

Andrea Ghigo
INFN-LNF
Accelerator Division
V.E. Fermi, PO Box 13
Frascati, 00044 ITALY
Phone: (39)6-940-32213
Fax: (39)6-940-32256
E-mail: Ghigo@lnf.infn.it

Douglas Gilpatrick
Los Alamos National Laboratory
LANSCE-1
PO Box 1663, MS H817
Los Alamos NM 87545
Phone: (505)667-3159
Fax: (505)665-2904
E-mail: gilpatrick@lanl.gov

William Graves
Brookhaven National Laboratory
NSLS
Bldg. 725D
Upton NY 11973
Phone: (516)344-2095
Fax: (516)344-3238
E-mail: wsgraves@bnl.gov

Jim Hardin
Hewlett Packard
351 E. Evelyn Ave.
Mt. View CA 94041
Phone: (650)694-2439
Fax: (650)694-3594
E-mail: jim_hardin@hp.com

Leigh Harwood
Jefferson Laboratory
Accelerator Development
12000 Jefferson Ave.
Newport News VA 23606
Phone: (757)269-7686
Fax: (757-269-7658
E-mail: harwood@jlab.org

Robert Hettel
SSRL
PO Box 4349
MS 69
Stanford CA 94309-0210
Phone: (650)926-3489
Fax: (650)926-4100
E-mail: hettel@ssrl.slac.stanford.edu

Jim Hinkson
Lawrence Berkeley National Lab.
MS-46-125
1 Cyclotron Rd.
Berkeley CA 94720
Phone: (510)486-4194
Fax: (510)486-5775
E-mail: jahinkson@lbl.gov

Albert Hofmann
Chemin de l'Erse 20
GrandSaconnex CH1218 SUISSE
Phone: (41)22-767-6972
Fax: (41)22-767-5460
E-mail: Albert.Hofmann@cern.ch

Timothy Houck
Lawrence Livermore National Lab.
Defense Sciences Engineering
7000 East Ave., PO Box 808
Livermore CA 94550
Phone: (925)423-7905
Fax: (925)422-1767
E-mail: houck1@llnl.gov

Curt Hovater
Jefferson Laboratory
Accelerator Division
12000 Jefferson Ave., MS 58
Newport News VA 23606
Phone: (757)269-7685
Fax: (757)269-7658
E-mail: hovater@jlab.org

Kuotung Hsu
SRRC
Light Source
No. 1 R&D Rd. VI
Hsinchu Science-Based TAIWAN
Phone: 886-3-5780281 ext: 7210
Fax: 886-3-5789816
E-mail: kuotung@srrc09.srrc.gov.tw

Andreas Jankowiak
University of Dortmund
Dept. of Physics
Maria-Goeppert-Mayer 2
Dortmund, 44221 GERMANY
Phone: (49)231-755-5374
Fax: (49)231-755-5383
E-mail: janko@prian.physik.uni-dortmund.de

Keith Jobe
SLAC
PO Box 4349
Accelerator Dept.
Stanford CA 94309
Phone: (650)926-2084
Fax: (650)926-2407
E-mail: keith.jobe@slac.stanford.edu

Erik Johnson
Brookhaven National Laboratory
NSLS
Bldg. 725D
Upton NY 11973
Phone: (516)344-4603
Fax: (516)344-3238
E-mail: erik@bnl.gov

Ronald Johnson
SLAC
PO Box 4349
MS 50
Stanford CA 94309
Phone: (650)926-8520
Fax: (650)926-3800
E-mail: ron_johnson@slac.stanford.edu

Kevin Jordon
Jefferson Laboratory
FEL
12000 Jefferson Ave.
Newport News VA 23606
Phone: (757)269-7644
Fax: (757)269-6355
E-mail: jordan@jlab.org

Roland Jung
CERN
SL Division
1211 Geneva 23 SUISSE
Phone: (41)22-767-3298
Fax: (41)22-767-7740
E-mail: Roland.Jung@cern.ch

James Kamperschroer
General Atomics
Los Alamos National Laboratory
H838
Los Alamos NM 87545
Phone: (505)665-3970
Fax: (505)665-2509
E-mail: kamper@gat.com

Tony Kershaw
Rutherford Appleton Laboratory
Dept. of ISIS
Chilton St.
Didcot, Oxfordshire 0X11 0QX UK
Phone: (44)1235-446646
Fax: (44)1235-446615
E-mail: a.h.kershaw@rl.ac.uk

Shaukat Khan
BESSY
Rudower Chaussee 5
Berlin, 12169 GERMANY
Phone: (30)6392-4638
Fax: (30)6392-4632
E-mail: khan@bii.bessy.de

Suk Kim
Argonne National Laboratory
APS/Diagnostics
9700 S. Cass Ave., MS 401
Argonne IL 60439-4800
Phone: (630)252-6567
Fax: (630)252-4732
E-mail: shkim@aps.anl.gov

Theo Kotseroglou
SLAC
PO Box 4349
Accelerator Dept.
Stanford CA 94309
Phone: (650)926-5334
Fax: (650)926-2407
E-mail: theo@slac.stanford.edu

Patrick Krejcik
SLAC
PO Box 4349
Accelerator Dept.
Stanford CA 94309
Phone: (650)926-2790
Fax: (650)926-2407
E-mail: pkr@slac.stanford.edu

Erich Kugler
CERN
EP Division
ISOLDE, 1211 Geneva 23
CH-1211 SUISSE
Phone: 41-22-7673183
Fax: 41-22-7678990
E-mail: erich.kugler@cern.ch

Frank Lenkszus
Argonne National Laboratory
Dept. of Controls & Computing
9700 S. Cass Ave., MS 401
Argonne IL 60439-4800
Phone: (630)252-6972
Fax: (630)252-4732
E-mail: frl@aps.anl.gov

David Lesyna
Optivus Technology Inc.
1475 S. Victoria Ct.
San Bernardino CA 92408
Phone: (909)799-8312
Fax: (909)799-8348
E-mail: lesyna@optivus.com

Steve Lidia
LBNL
Center for Beam Physics
1 Cyclotron Rd.
Berkeley CA 94720
Phone: (510)486-6101
Fax: (510)486-5392
E-mail: lidia@rta.lbl.gov

Robert Lill
Argonne National Laboratory
APS/Diagnostics
9700 S. Cass Ave., MS 401
Argonne IL 60439-4800
Phone: (630)252-0265
Fax: (630)252-4732
E-mail: blill@aps.anl.gov

Ivan Linscott
Stanford University
Dept. of Elec. Eng.
215 Durand St.
Stanford CA 94305
Phone: (650)723-3676
E-mail: linscott@nova.stanford.edu

Marco Lonza
Sincrotrone Trieste
Accelerator Division
SS 14, Km 163.5 Basovizza
Trieste 34012 ITALY
Phone: 39-40-3758565
Fax: 39-40-3758565
E-mail: lonzam@elettra.trieste.it

Ronald Lorenz
DESY Zeuthen
Platanenallee 6
15738 Zeuthen GERMANY
Phone: 49-33762-77350
Fax: 49-33762-77330
E-mail: rlorenz@ifh.de

Alex Lumpkin
Argonne National Laboratory
Bldg. 362. MS 401
9700 S. Cass Ave.
Argonne IL 60439
Phone: (650)252-4879
Fax: (630)252-7187
E-mail: lumpkin@aps.anl.gov

Yousef Makdisi
Brookhaven National Laboratory
RHIC Project
Bldg. 1005-4
Upton NY 11973-5000
Phone: (516)344-4932
Fax: (516)344-2588
E-mail: Makdisi@bnldag.ags.bnl.gov

Donald Martin
SSRL
Accelerator Development & Controls
PO Box 4349, MS 99
Stanford CA 94309
Phone: (650)926-3006
Fax: (650)926-4100
E-mail: martin@ssrl.slac.stanford.edu

Dexter Massoletti
Lawrence Berkeley National Laboratory
ALS
1 Cyclotron Rd.
Berkeley CA 94720
Phone: (510)486-6857
Fax: (510)486-4960
E-mail: djmassoletti@lbl.gov

Gholam Mazaheri
SLAC
PO Box 4349
MS 65
Stanford CA 94309-0210
Phone: (650)926-2716
Fax: (650)926-4892
E-mail: gholam@slac.stanford.edu

Doug McCormick
SLAC
PO Box 4349
Accelerator Dept.
Stanford CA 94309
Phone: (650)926-2470
Fax: (650)926-2407
E-mail: djm@slac.stanford.edu

Evgeny Medvedko
SLAC
Controls Dept.
PO Box 4349, MS 50
Stanford CA 94309
Phone: (650)926-2685
Fax: (650)926-3800
E-mail: medvedko@slac.stanford.edu

Robert Merl
Argonne National Laboratory
APS/Diagnostics
9700 S. Cass Ave., MS 401
Argonne IL 60439-4800
Phone: (630)252-9356
Fax: (630)252-4732
E-mail: rmerl@aps.anl.gov

Karl Mess
DESY
Dept. of FEB
Notkestr. 85
Hamburg, D22607 GERMANY
Phone: (49)40-8998-3055
Fax: (49)40-8994-3055
E-mail: mess@desy.de

Michiko Minty
SLAC
PO Box 4349
MS-66
Stanford CA 94309
Phone: (650)926-3650
Fax: (650)926-2407
E-mail: minty@slac.stanford.edu

Istvan Mohos
Forschungszentrum Julich GmbH
Institut fur Kernphysik
Leo-Brandt Str. 1, PO Box 1913
Julich, NRW D-52425 GERMANY
Phone: 49-2461-61-2629
Fax: 49-2461-61-3930
E-mail: i.mohos@fz-juelich.de

Hiroshi Morimoto
Hitachi America LTD.
Procurement & Tech. Div.
50 Prospect Ave.
Tarrytown NY 10591-4698
Phone: (914)333-2972
Fax: (914)333-2782
E-mail: hiroshi.morimoto@hal.hitachi.com

Peter Moritz
GSI Darmstadt
Beam Diagnostics
Planckstr. 1
Darmstadt, 64291 GERMANY
Phone: 6159-71-2228
Fax: 6159-71-2104
E-mail: p.moritz@gsi.de

Raymond Muller
Hamamatsu Corporation
2875 Moorpark Ave.
San Jose CA 95128
Phone: (408)261-2022
Fax: (408)261-2522
E-mail: rmuller@hamamatsu.com

Christopher Nantista
SLAC
ARD-A
PO Box 4349, MS 26
Stanford CA 94309
Phone: (650)926-4906
Fax: (650)926-5368
E-mail: nantista@slac.stanford.edu

James O'Hara
Allied Signal Fed. Manu. & Tech.
MPF-6/LANL
Los Alamos NM 87545
Phone: (505)667-1950
Fax: (505)665-2904
E-mail: ohara@lanl.gov

Thomas O'Malley
Nuclear Research Corporation
125 Titus Ave.
Warrington PA 18976
Phone: (215)343-5900
Fax: (215)343-4670

Al Odian
SLAC
PO Box 4349
MS 65
Stanford CA 94309-0210
Phone: (650)926-2938
Fax: (650)926-4892
E-mail: odian@slac.stanford.edu

Ralph Pasquinelli
Fermilab
MS-341
PO Box 500
Batavia IL 60510
Phone: (630)840-4724
Fax: (630)840-4552
E-mail: pasquin@fnal.gov

Roslind Pennacchi
SLAC
PO Box 4349
MS 55
Stanford CA 94309
Phone: (650)926-3665
Fax: (650)926-5138
E-mail: shelter@slac.stanford.edu

Andreas Peters
GSIAbt
Abt. Strahldiagnose
Planckstrasse 1
D-64291, Darmstadt GERMANY
Phone: (49)6159-712313
Fax: (49)6159-712104
E-mail: A.peters@gsi.de

Chip Piller
Jefferson Laboratory
Accelerator Division
12000 Jefferson Ave., MS 58
Newport News VA 23606
Phone: (757)269-7534
Fax: (757)269-7658
E-mail: piller@jlab.org

Igor Pinayev
Duke University
FEL Laboratory
La Salle st ext, PO Box 90319
Durham NC 27708-0319
Phone: (919)660-2657
Fax: (919)660-2671
E-mail: pinayev@fel.duke.edu

Marco Poggi
INFN-LNL
Via Romea 4
Legnaro ITALY
Phone: (0039) (49) 8068480
Fax: (0039) (49) 641925
E-mail: poggi@lnl.infn.it

John Power
Los Alamos National Laboratory
LANSCE-1
PO Box 1663, MS H808
Los Alamos NM 87545
Phone: (505)667-7045
Fax: (505)665-2904
E-mail: jpower@lanl.gov

Thomas Powers
Jefferson Laboratory
Accelerator Division
12000 Jefferson Ave.
Newport News VA 23606
Phone: (757)269-7660
Fax: (757)269-7658
E-mail: powers@jlab.org

Shyam Prabhakar
SLAC
ARD-A
PO Box 4349, MS 18
Stanford CA 94309
Phone: (650)926-8529
Fax: (650)926-8533
E-mail: shyam@slac.stanford.edu

Martin Pugh
CLRC Daresbury Lab.
Keckwick Lane, MS B82
Daresbury, Warrington
Cheshiregton, WA44A UK
Phone: (44)1925-603120
Fax: (44)1925-603124
E-mail: m.j.pugh@dl.ac.uk

Willi Radloff
DESY
MKI
Notkestr. 85
Hamburg, D-22607 GERMANY
Phone: (49)40-8998-3355
Fax: (49)40-8998-4303
E-mail: radloff@desy.de

Jeff Reed
Pearson Electronics, Inc.
1860 Embarcadero Rd.
Suite 200
Palo Alto CA 94303
Phone: (650)494-6444
Fax: (650)494-6716

Randy Reiger
Princeton Instruments, Inc.
3660 Quakerbridge Rd.
Trenton NJ 08619
Phone: (609)587-9797 x170
Fax: (609)587-8970

Luigi Rezzonico
Paul Scherrer Institute
Accel. Rsch. & Dev.
Wurenlingen & Villigen
CH-5232 Villigen PSI SUISSE
Phone: 56-310-33-77
Fax: 56-310-32-94
E-mail: Rezzonico@psi.ch

Peter Röjsel
Lund University
Max Lab.
PO Box 118
SE-221 00 LUND SWEDEN
Phone: 46-46-2229717
Fax: 46-46-2224710
E-mail: peter@maxlab.lu.se

Chris Rose
Los Alamos National Laboratory
PO Box 1663
MS H805
Los Alamos NM 87545
Phone: 505-665-0950
Fax: 505-665-2676
E-mail: crose@lanl.gov

Marc Ross
SLAC
PO Box 4349
MS 66
Stanford CA 94309
Phone: (650)926-3526
Fax: (650)926-2407
E-mail: mcrec@slac.stanford.edu

Fernando Sannibale
INFN-NF
Accelerator Department
Via E. Fermi 40
CP 13, Frascati, 00044 ITALY
Phone: (39)6-94032213
Fax: (39)6-94032256
E-mail: sannibale@lnf.infn.it

Kees Scheidt
ESRF
Avenue des Martyrs
PO Box BP2200
Grenoble 38043 FRANCE
Phone: (33)4-76-882091
Fax: (33)4-76-882054
E-mail: scheidt@esrf.fr

Hermann Schmickler
CERN
SL Division
1211 Geneva 23
Geneva 1211 SUISSE
Phone: (41)22-76-74619
Fax: (41)22-76-77740
E-mail: Hermann.Schmickler@cern.ch

Norbert Schneider
GSI Darmstadt
Beam Diagnostics
Planckstr. 1
Darmstadt 64291 GERMANY
Phone: (61)59-71-2228
Fax: (61)59-71-2104
E-mail: norbert@alice.gsi.de

Nicholas Sereno
Argonne National Laboratory
Operations Analysis
9700 S. Cass Ave., MS 401
Argonne IL 60439-4800
Phone: (630)252-6867
Fax: (630)252-4732
E-mail: sereno@aps.anl.gov

Mario Serio
INFN-LNF
Accelerator Division
PO Box 13
Frascati (RM) 00044 ITALY
Phone: (39)6-94032276
Fax: (39)6-94032256
E-mail: mserio@inf.infn.it

Robert Shafer
Los Alamos National Laboratory
LANSCE-1
MS 808
Los Alamos NM 87545
Phone: (505)667-5877
Fax: (505)665-2904
E-mail: rshafer@lanl.gov

Thomas Shea
Brookhaven National Laboratory
RHIC
Bldg. 1005
Upton NY 11973
Phone: (516)344-2435
Fax: (516)344-2588
E-mail: shea@bnl.gov

Albert Sheng
Argonne National Laboratory
ASD Department
9700 S. Cass Ave.
Argonne IL 60439
Phone: (630)252-5837
Fax: (630)252-5948
E-mail: shengic@aps.anl.gov

Bradford Shurter
Los Alamos National Laboratory
LANSCE-1
Bikini Rd.
Los Alamos NM 87025
Phone: (505)665-1122
Fax: (505)667-0449
E-mail: rshurter@lanl.gov

Charles Sinclair
Jefferson Laboratory
12000 Jefferson Ave.
Newport News VA 23606
Phone: (757)269-7679
Fax: (757)269-5024
E-mail: sinclair@jlab.org

Tim Slaton
SLAC
PO Box 4349
MS 87
Stanford CA 94309
Phone: (650)926-3717
Fax: (650)926-4055
E-mail: slaton@slac.stanford.edu

Gary Smith
Brookhaven National Laboratory
Bldg. 911C
35 Lawrence Ave.
Upton NY 11973
Phone: (516)344-3473
Fax: (516)344-5954
E-mail: smith8@bnl.gov

Robert Smith
CLRC Daresbury Lab.
Keckwick Lane, MS B74
Daresebury, Warrington
Cheshiregton, WA44AD UK
Phone: (44)1925-603341
Fax: (44)1925-603124
E-mail: r.j.smith@dl.ac.uk

Steve Smith
SLAC
PO Box 4349
MS-50
Stanford CA 94309-0210
Phone: (650)926-3916
Fax: (650)926-3800
E-mail: ssmith@slac.stanford.edu

Todd Smith
Stanford University
Hansen Labs
Stanford CA 94309
Phone: (650)723-1906
Fax: (650)725-8311
E-mail: todd.smith@stanford.edu

Vernon Smith
SLAC
Controls Dept.
PO Box 4349, MS-64
Stanford CA 94309
Phone: (650)926-3519
Fax: (650)926-3800
E-mail: vrs@slac.stanford.edu

Robert Stege
SLAC
Accelerator Dept.
PO Box 4349, MS 39
Stanford CA 94309
Phone: (650)926-3915
Fax: (650)926-5389
E-mail: stege@slac.stanford.edu

Greg Stover
Lawrence Berkeley National Laboratory
MS 46-125
1 Cyclotron Rd.
Berkeley CA 94720
Phone: (510)486-7706
Fax: (510)486-5775
E-mail: gdstover@lbl.gov

Tsuyoshi Suwada
KEK
Accelerator Laboratory
1-1 Oho, Tsukuba
Ibaraki 305-0801 JAPAN
Phone: (81)298-64-5684
Fax: (81)298-64-7438
E-mail: tsuyoshi.suwada@kek.jp

Thomas Tallerico
Brookhaven National Laboratory
AGS/EPS
Bldg. 911B
Upton NY 11973
Phone: (516)344-4642
Fax: (516)344-5954
E-mail: tallerico@bnl.gov

Gianni Tassotto
Fermilab
Beams Division
PO Box 500, MS 308
Batavia IL 60510
Phone: (630)840-4325
Fax: (630)840-2677
E-mail: tassotto@fnal.gov

Peter Tenenbaum
SLAC
Accel. Resch. Dept. A
PO Box 4349, MS 26
Stanford CA 94309
Phone: (650)926-4429
Fax: (650)926-5368
E-mail: quarkpt@slac.stanford.edu

Dmitry Teytelman
SLAC
Dept. of ARDA
PO Box 4349, MS 18
Stanford CA 94309
Phone: (650)926-8532
Fax: (650)926-8533
E-mail: dim@slac.stanford.edu

Robert Traller
SLAC
Controls Dept.
PO Box 4349, MS 50
Stanford CA 94309
Phone: (650)926-4063
Fax: (650)926-3800
E-mail: bltsys@slac.stanford.edu

Teresa Troxel
SSRL
PO Box 4349
MS 99
Stanford CA 94309
Phone: (650)926-3135
Fax: (650)926-3600
E-mail: ttroxel@ssrl.slac.stanford.edu

Klaus Unser
Bergoz Instrumentation
Rue du Jura
Crozet, 01170 FRANCE
Phone: 33-450-410089
Fax: 33-450-410199
E-mail: unser@bergoz.com

Rok Ursic
Instrumentation Technologies
Srebrnicev Trg 4a
Solkan 5250 Slovenia
Phone: 386-65-132300
Fax: 386-65-132305
E-mail: rok@i-tech.si

Willem VanAsselt
Brookhaven National Laboratory
AGS
Bldg. 911B
Upton NY 11973
Phone: (516)344-7778
Fax: (516)344-5954
E-mail: vanasselt@bnl.gov

Giuseppe Vismara
CERN
SL Division
1211 Geneva 23
Geneva 1211 SUISSE
Phone: (41)22-76-75347
Fax: (41)22-76-77740
E-mail: giuseppe.vismara@cern.ch

Wolfgang Vodel
F. Schiller University Jena
Inst. of Solid State Physics
Helmholtzweg 5
Jena, D-07743 GERMANY
Phone: (49)3641-947421
Fax: (49)3641-947422
E-mail: p7vowo@rz.uni-jena.de

Ian Walker
GMW Associates
PO Box 2578
Redwood City CA 94062
Phone: (650)802-8292
Fax: (650)802-8298

Defa Wang
MIT-BATES
21 Manning Rd.
Middleton MA 01949
Phone: (617)253-9574
Fax: (617)253-9599
E-mail: dwang@aesir.mit.edu

Chris Waters
Pearson Electronics, Inc.
1860 Embarcadero Rd.
Suite 200
Palo Alto CA 94303
Phone: (650)494-6444
Fax: (650)494-6716

Robert Webber
Fermilab
Proton Source Dept.,
PO Box 500, MS 341
Batavia IL 60510
Phone: (630)840-5415
Fax: (630)840-8738
E-mail: webber@fnal.gov

Manfred Wendt
DESY
Dept. MKI
Notkestr. 85
Hamburg, D-22607 GERMANY
Phone: (49)40-8998-2517
Fax: (49)40-8998-4303
E-mail: wendt@ux-bello.desy.de

H-Ulrich Wienands
SLAC
PEP-II
PO Box 4349
Stanford CA 94309
Phone: (650)926-3817
Fax: (650)926-3882
E-mail: uli@slac.stanford.edu

David Williams
CERN
PS/BD
CH-1211 Geneve 23 SUISSE
Phone: (41)22-767-3580
Fax: (41)22-767-8200
E-mail: David.John.Williams@mail.cern.ch

James Wilmarth
Los Alamos National Laboratory
LANSCE 6
PO Box 1663, MS H812
Los Alamos NM 87545
Phone: (505)667-6959
Fax: (505)665-0046
E-mail: wilmarth@lanl.gov

Richard Witkover
Brookhaven National Laboratory
Bldg. 911B
35 Lawrence Ave.
Upton NY 11973
Phone: (516)282-4607
Fax: (516)282-5954
E-mail: witkover@bnl.gov

Bingxin Yang
Argonne National Laboratory
APS/Diagnostics
9700 S. Cass Ave., MS 401
Argonne IL 60439-4800
Phone: (630)252-9821
Fax: (630)252-4732
E-mail: bxyang@aps.anl.gov

Yan Yin
Y.Y. Labs, Inc.
PO Box 60176
Palo Alto CA 94306
Phone: (650)857-1245
Fax: (650)857-1249
E-mail: yanyin@triumf.ca

Reuben Yotam
SSRL
PO Box 4349
MS 69
Stanford CA 94309
Phone: (650)926-3167
Fax: (650)926-4100
E-mail: yotam@ssrl.slac.stanford.edu

Andrew Young
SLAC
MS 18
Stanford CA 94309
Phone: (650)926-4448
Fax: (650)926-8533
E-mail: ayoung@slac.stanford.edu

Alexander Zaltsman
Brookhaven National Laboratory
Bldg. 911-B
Upton NY 11973
Phone: (516)344-2967
Fax: (516)344-5954
E-mail: zaltsman@bnl.gov

Jim Zagel
Fermilab PO Box 500, MS 308
Batavia IL 60510
Phone: (708)840-4076
Fax: (708)840-2677
E-mail: zagel@fnal.gov

List of Vendors

GMW Associates
P.O. Box 2578
Redwood City CA 94062
Ian Walker
phone: (650) 802-8292
fax: (650) 802-82998

Hamamatsu Corporation
2875 Moorpark Avenue
San Jose CA 95128
Raymond Muller
phone: (408) 261-2022
fax: (408) 261-2522
email: rmuller@hamamatsu.com

Hewlett Packard Company
351 E. Evelyn Avenue
Mountain View CA 94041
Jim Hardin
phone: (650) 694-2439
fax: (650) 694-3594
email: jim_hardin@hp.com

Nuclear Research Corporation
125 Titus Ave.
Warrington PA 18976
Thomas E. O'Malley
phone: (215) 343-5900
fax: (215) 343-4670

Pearson Electronics, Inc.
1860 Embarcadero Road, Ste 200
Palo Alto CA 94303
Jeff Reed
phone: (650) 494-6444
fax: (650) 494-6716

Picosecond Pulse Laboratory
P.O. Box 44
Boulder CO 80306
Jim Andrews
phone: (303) 443-1249
fax: (303) 447-2236
email: jandrews@picosecond.com

Polaron CVT Limited
4-6 Carters Lane
Kiln Farm
Milton Keynes
MK11 3ER
United Kingdom
Richard Davies
phone: +44 (0) 1908 563267
fax: +44 (0) 1908 568354
email: cvtsales@polaron-group.co.uk

Princeton Instruments, Inc.
3660 Quakerbridge Road
Trenton NJ 08619
Jay Moskovic
phone: (609) 587-9797 Ext 170
fax: (609) 587-8970
email: jay.moskovic@prinst.com

Princeton Scientific Corporation
P.O. Box 143
Princeton NJ 08542
Han Gerwers
phone: (609) 924-3011
fax: (609) 924-3018

AUTHOR INDEX

J

K

L

M

N

O

P

R

S

T

U

V

W

Y

Z